DATE DUE

DE 1 8 98		
AE 1 4 03		
06 17 05		
DE 1 6 06		
8/12/05		

A History of Scientific Thought

Contributors

Michel Authier teaches mathematics in Paris.

Paul Benoît is a lecturer in medieval history at the University of Paris I.

Bernadette Bensaude-Vincent is a philosopher and lecturer in the history of science at the University of Paris X.

Geof Bowker teaches at the Graduate School of Library and Information Science at the University of Illinois, Urbana.

Jean-Marc Drouin is a lecturer at the Muséum National d'Histoire Naturelle in Paris

Catherine Goldstein is continuing her research in the history of mathematics at the Max Planck Institute for the History of Science in Berlin.

Bruno Latour is a professor of sociology at the School of Mines in Paris and the University of California at San Diego.

Pierre Lévy is a lecturer at the University of Paris X.

Françoise Micheau is a lecturer in medieval history at the University of Paris I.

James Ritter is working on the history of mathematics at the Max Planck Institute for the History of Science in Berlin.

Michel Serres is a professor of philosophy at the University of Paris I and Stanford University, California, and a member of the Académie Française.

Isabelle Stengers is a chemist and a historian of science who teaches at the Free University of Brussels.

A History of Scientific Thought

Elements of a History of Science

Edited by
Michel Serres

Translated from the French

BLACKWELL
Reference

ench text copyright © Bordas, Paris, 1989
on copyright © Blackwell Publishers Ltd, 1995

First published 1995

Original French edition first published by Bordas under the title
Éléments d'Histoire des Science

Blackwell Publishers Ltd.
108 Cowley Road, Oxford OX4 1JF, UK

Blackwell Publishers Inc.
238 Main Street
Cambridge, Massachusetts 02142, USA

British Library Cataloguing in Publication Data

A CIP catalogue record for this book is available from the British Library.

Library of Congress Cataloging-in-Publication Data

Eléments d'histoire des science. English
A History of scientific thought : elements of a history of science
/ edited by Michel Serres : [contributors] Bernadette Bensaude-
Vincent . . . [et al.].
p.   cm.
Translation of : Eléments d'histoire des science.
Includes bibliographical references and index.
ISBN 0–631–17739–6 (alk. paper) : $39.95
1. Science—History.   2. Science—Philosophy—History.
I. Bensaude-Vincent, Bernadette.
Q125.E4313   1995
509—dc20   94–7571
CIP

Typeset in 11/13 pt Garamond by Graphicraft Typesetters Ltd., Hong Kong
Printed in Great Britain by The Bath Press

This book is printed on acid-free paper

. . . doubtless we shall never know
how Knowledge came to us.
There are so many possible sources:
seeing, hearing, observing;
speaking, pleading, contradicting;
counterfeiting, imitating, desiring, hating, loving;
fearing and defending ourselves,
risking, taking chances, wagering,
living and working together or apart;
wanting to dominate through possession or control;
relieving pain, curing disease
or killing through murder and war;
astonishment in the face of death,
praying to the point of ecstasy;
creating with our own hands,
cultivating the Earth or destroying it . . .
. . . and we are troubled that we do not know
towards which of these acts, these words, these conditions
or towards what other unknown ends
it now unknowingly speeds . . .

Contents

Introduction   MICHEL SERRES   1

1   Babylon −1800   JAMES RITTER   17

2   Measure for Measure: Mathematics in Egypt and Mesopotamia
    JAMES RITTER   44

3   Gnomon: The Beginnings of Geometry in Greece
    MICHEL SERRES   73

4   Archimedes: The Scientist's Canon   MICHEL AUTHIER   124

5   Stories of the Circle   CATHERINE GOLDSTEIN   160

6   The Arab Intermediary   PAUL BENOÎT and FRANÇOISE MICHEAU   191

7   Theology in the Thirteenth Century: A Science Unlike the
    Others   PAUL BENOÎT   222

8   Algebra, Commerce and Calculation   PAUL BENOÎT   246

9   The Galileo Affair   ISABELLE STENGERS   280

10 Refraction and Cartesian 'Forgetfulness'   MICHEL AUTHIER   315

11 Working with Numbers in the Seventeenth and Nineteenth
    Centuries   CATHERINE GOLDSTEIN   344

12 Ambiguous Affinity: The Newtonian Dream of Chemistry in
    the Eighteenth Century   ISABELLE STENGERS   372

13 From Linnaeus to Darwin: Naturalists and Travellers
    JEAN-MARC DROUIN   401

14 Paris 1800   MICHEL SERRES   422

CONTENTS

15 Lavoisier: A Scientific Revolution   BERNADETTE BENSAUDE-VINCENT   455

16 In Defence of Geology: The Origins of Lyell's
    Uniformitarianism   GEOF BOWKER   483

17 Mendel in the Garden   JEAN-MARC DROUIN   506

18 Pasteur and Pouchet: The Heterogenesis of the History of
    Science   BRUNO LATOUR   526

19 Mendeleyev: The Story of a Discovery
    BERNADETTE BENSAUDE-VINCENT   556

20 Manufacturing Truth: The Development of Industrial
    Research   GEOF BOWKER   583

21 Joliot: History and Physics mixed together   BRUNO LATOUR   611

22 The Invention of the Computer   PIERRE LÉVY   636

Chronology   MICHEL AUTHIER   664

Notes   724

Bibliography   734

Index   753

Introduction

Michel Serres

Which the reader is invited not to neglect as it sets out the authors' aims and explains the structure of this book.

The history of science is making considerable progress, arousing a growing interest in France and elsewhere. This is undoubtedly because, living in a world dominated by science and technology, we increasingly question the whys and wherefores of its recent advent and sometimes even its legitimacy. Fluctuating political and military situations and the economy are not sufficient explanations of how our contemporary ways of life came about: we need a *history of science and technology.*

It is a great paradox that in schools and universities the history of science is not taught in the same way as the usual disciplines: it is only haphazard, depending on the inclination of the individual teacher. We generally learn about our history in isolation from that of the sciences. We study philosophy devoid of any scientific reasoning and great literature in splendid detachment from its scientific context; we study various disciplines uprooted from the soil of their history, as if they had happened by chance. In short, our entire learning process is inappropriate to the real world in which we live, a world which is a confused mixture of technology and society, of insane or wise traditions and useful or disturbing innovations. We are only just beginning to think about laws to encompass the advances of chemistry and biology.

1

This book aims to help resolve the cultural crisis caused in part by this estrangement; we are struggling to find a solution to it. This divorce between two worlds is sometimes expressed as hostility and sometimes as adoration, both of which are excessive. It would also like to encourage the introduction of teaching of the discipline as a normal part of the curriculum, both in secondary schools and at university.

The book is therefore aimed at a very wide readership:

— first of all, at honest people who question their environment and who have never been told to what extent the predominance of science and technology, whose power is omnipresent, was an active component of their past. Things which are considered great innovations today are sometimes two thousand years old, and what appeared to be irrational often paved the way for the triumph of reason. As this perspective is corrected, the modern world acquires depth; it becomes familiar;

— at teachers, at students of all disciplines who would like to give their work or their specialized studies a broader framework and a relevant context;

— at historians, philosophers, specialists in the arts, the legal profession, human and social scientists, who sometimes lack a grounding in science;

— at scientists themselves who are interested in the history of their discipline, which sometimes advances so fast that developments of twenty years ago are already considered out of date. Beneath this history, divided into such brief segments and so quickly forgotten, is there a continuous movement or are there broader perspectives? Where and how can scientists' predecessors be read and understood in their own language?

Sometimes we are astonished to find that people who live hundreds of miles away make the same gestures or have similar customs to our own, or, on the other hand, that there are vast differences between ourselves and our neighbours. In the same way, readers will delight in discovering practices which are very close to our own, although alien, in the Egypt of two thousand years ago, or conversely, the infinite gulf that separates them from the previous generation. That is how the history of science expands the limited ideas of our disciplines and our times: it most certainly forms the basis of a culture. The divorce between these two worlds and two cultures is probably very recent, but we think it has always been so and always will be.

Ideally, the history of science brings together several groups of people: scientists, authentic specialists in different disciplines; historians, of course, authorities on certain periods; sociologists, ethnologists and anthropologists, even psychologists from the field of social and human sciences, fascinated by the question of invention; and lastly philosophers accustomed to encounters of this kind through the nature of their work. This open-ended list goes on and happily includes engineers, masters of technology, doctors dealing with remarkable organisms, economists, teachers, lawyers and journalists, the task of the latter being to publicize or divulge knowledge. Together they work at a discipline seeking unity. Whether these encounters take place or not, the participants, virtual if not actual, retrace the history of science in splendid yet regrettable isolation. We sometimes seek in vain the connection that would bring together an algebraist keen to reconstruct Plato's teachings in Greek or a natural scientist and botanist on the trail of the great adventurers of the Age of Enlightenment and a political scientist. One of the aims of this book is to encourage these isolated individuals to collaborate with each other.

How was the book written? And by whom? Its authors, a very disparate group of people, include a theoretical physicist converted to hieroglyphics and cuneiform tablets, a mathematician who has come to Greek and Italian, a chemist philosopher and a philosopher chemist, a medieval historian versed in technology and conversely a natural scientist versed in medieval theology, a female researcher into the theory of numbers with a keen interest in the history of mathematics outside the Western world, a geologist engineer and a physicist of the Earth, a positivist sociologist who is also a theorist, a specialist in communications and computers and so on. French, American, Australian – each one assisted in the construction of this edifice, hoping to help cement it. In what way?

First of all, and thanks to the generous hospitality of Madame Annette Gruner-Schlumberger, the group lived together for several weeks to draw up a programme of work and appraise its execution. Once written, each contribution was read by everybody, gone over with a fine-tooth comb and discussed. Put in the hot seat, all of the contributors willingly submitted to the tough ordeal of having questions fired at them by the others. And they rewrote their chapters, taking into account these requests for clarification. In other words, such-and-such an article on mathematics or geology was judged by about ten other people, non-specialists from very different fields, and vice versa.

Second, and again in front of all the others, each contributor presented his or her article as a lecturer to an audience of students whose overall level was about that of a second-year undergraduate. Towards the end of the class the students offered their comments, which were sometimes very harsh and often pertinent. The piece was then rewritten, taking these criticisms into account, and delivered again the following year to the subsequent class of students, again to test its clarity. And so the group was rigorous, both in its life and in its written and oral work, in putting to the test the inevitable diversification implicit in the discipline itself, faithfully trying to reduce it. Likewise the group wanted to assess the clarity of its writing to facilitate communication. Naturally, the group was not without tensions between those who believe in science, those who believe in history, those who do not trust either and those who partially trust both. And if competence, goodwill and even friendship are not enough to give coherence to a body of learning shared by fellow thinkers, in our case we are helped by the common conviction that the history of science is becoming the cornerstone of contemporary culture because it plunges positive knowledge, the framework and driving force of our world, into the living and collective fabric of human experience.

But the main tension within the group and the book comes, we suspect, less from the disparity of the specialities but from the global idea that each of us had of the discipline. And that is one of the truly original aspects of this book: each chapter is not content just to relate what happened for such-and-such a subject at such-and-such a time, for example, the development of chemistry in the nineteenth century or of geometry in ancient Greece, but actually supports one theory among all those defended by historians of science. The reader is not subjected to one school of thought alone through buying and consulting a whole work devoted to a single, implicit tendency, but on the contrary is able to choose from the wide spectrum of opinions faithfully put forward.

Before planning and writing this book, and especially on the eve of its publication, we constantly asked ourselves the fundamental question: how do we conceive and record the history of science? Should we acknowledge one or several histories? How should they be presented?

The first solution and the more usual option was to write about all the sciences and their development throughout the general course of history. Such an account would begin with the Chinese or the Babylonians or even Stonehenge, a construction with no written language, and finish with the latest Nobel prize, relating the development through the centuries of the complete encyclopaedia of known exact, experimental and social sciences. Together with their techniques, from numeration or primitive astrology to the latest refinements of even the human sciences of today, the sciences would be immersed in the prevalent historical conditions and circumstances. Supposing we had added such a manual to the many guides and introductions already on the market which claim to give a clear picture of a lucid body of knowledge in a given period, we would not even have begun to answer the question. In a book of this kind of course you find some scientific statements, sometimes even a little history, but not quite the history of science.

It is appropriate to speak of it as an independent discipline, with its divisions and its specialisms, vacillating between several styles, often overwhelmed by its specific problems, with its divergent methods and schools at variance with each other. It does not act as a transparent window. Because the subjects with which it is concerned abound in information, it sometimes questions or challenges ordinary history and the generally accepted ideas of the sciences themselves.

There is a spontaneous history of the sciences, as Auguste Comte[1] would have said: it would be practised by a history ill-informed of the sciences and sciences with very little knowledge of history. In fact, this unhampered progress of total knowledge through global, homogenous and isotropic time is the essence of that unthinking spontaneity. Looking at it more closely, hundreds of complications arise: from the large-scale map of a coastline indented by the erosion of the rocks and the path of a hiker walking over the stones on the ground, we go from a continuous curve to haphazard, blind jumps which explore and meander, as scientists often do. Similarly, no science is unique, recognizable and coherent throughout a period of time, even in the medium term, which itself bifurcates and fluctuates. Thus it would seem naïve to speak of the march of reason in science history.

This spontaneity assumes that hundreds of other things are true: that it is sufficient to relate the succession of solutions to problems and experiments culminating in inventions; to paint the portraits of the geniuses who made the discoveries; to recognize in the past the embryos, dreams, seeds or foundations of contemporary achievements; to indicate clearly the watersheds or revolutions which mark the birth of a science or the milestones of its development; to describe the quarrels, arguments and polemics whose fire fuelled the engine of intelligent advance. And, on the other hand, it is deemed sufficient to hitch the science chapter to the current history book, to define the social, institutional, economic, cultural and political framework of the science content. Above all, it assumes this retrograde movement of the truth which projects the knowledge of today onto the past in such a way that history irresistibly becomes an almost programmed preparation for the knowledge of today. In truth, there is nothing more difficult than imagining a time that was free and fluctuating and not completely determined, when scientists looking for something did not know exactly what they were looking for while at the same time knowing it instinctively. This spontaneity actually has a dual root: the mute, literally religious admiration, which is sometimes justified, of everything that claims to be scientific and which, therefore, remains untouchable, and an equal admiration for history. Even if they profess to be atheists or liberated, our contemporaries willingly offer up sacrifices at these twin altars or bow down before this dual hierarchy. Anybody who questions the reliability, the reason, the knowledge or the work of science or history is immediately accused of departing from the rational. These two subjects are present-day taboos. Consequently, the spontaneous history of science is often reduced to a holy or rather a sacred history. Geniuses take on the role of prophets, breakthroughs are revelations, while the controversies and debates oust the heretics and symposia emulate councils. Science is gradually becoming incarnate in time as the spirit once did.

Now the authors of this book belong to a generation which was trained in science without being made inflexible by scientism, a generation which, having witnessed the problems of science at the same time as its growing power, holds it in quiet esteem combined with a certain agnosticism without bitterness. For them it represents neither absolute good nor absolute evil, neither God nor devil; it remains a form of knowledge and power, a means, no more and no less. Likewise history remains a discipline among others, neither a dogma nor a hell. Consequently, the history of science begins, as did the reading of sacred literature in the past, with a critique. It requires a brave effort to bring down the two authorities, the two statues, from the epistemological pedestals on which they have been placed by their worshippers. There will not be many generally accepted ideas left in the reader's mind after reading this book.

This is a true history of science, envisaged as an autonomous discipline with its choices, its intentions, its divisions and its own style and methods. It aims to be more of an entity in itself than the deceptively clear account given in a complete encyclopaedia of science covering the whole of history.

Far from tracing a linear development of continuous and cumulative knowledge or a sequence of sudden turning-points, discoveries, inventions and revolutions plunging a suddenly outmoded past instantly into oblivion, the history of science runs backwards and forwards over a complex network of paths which overlap and cross, forming nodes, peaks and crossroads, interchanges which bifurcate into two or several routes. A multiplicity of different times, diverse disciplines, conceptions of science, groups, institutions, capitals, people in agreement or in conflict, machines and objects, predictions and unforeseen dangers, form together a shifting fabric which represents faithfully the complex history of science.

There is nothing simpler or easier than this apparent complexity. Let us imagine a kind of road map showing the various routes which cross a country – minor footpaths and major roads connect the villages and cities in hundreds of different ways and lead to unknown parts. Careful, let us not forget to change maps from time to time to bring ourselves up to date, for new roads are being built all the time to improve, transform and disrupt the network and the country, and they can make yesterday's route map obsolete. Tomorrow we will not travel from A to B by the same means as today. We can even superimpose several maps, on different scales, of roads, railways, rivers, sea routes, air routes, telephone networks, electronics networks, maps charting satellite paths and so on, to have a choice of kinds of transport and the time we spend travelling, according to our means, our aims and what we want to carry.

When we compare the maps of this game which we keep in a drawer and which we all have in our minds, it is clear that what is most important, what remains stable in nearly all of them and makes them resemble each other, are the poles or summits, the nodes of these networks, the interchanges or almost mandatory routes, often the locations of cities founded a very long time ago. Capital cities are built around a cathedral or a major road inter-section, the convergence of a number of roads which soon branch off into a number of other roads. They comprise among other things seven stations and four airports, several miles of quays along a waterway, emitting waves in a thousand directions. In other words, they are the focus of a whole series of routes. No map emphasizes the details of the road, sea or air routes: they are unimportant; they are rarely described. But this book draws such a map with precision.

Translation: while the sciences, separately or together, accumulate and fragment into hundreds of disciplines, while they are constantly changing and shifting, producing different times often unpredictable in their progress, what does remain relatively invariant in their dramatic and turbid history are the points of convergence and bifurcation where problems are posed and decisions are or are not taken. What problems? What decisions? These are the nodes or summits of the various networks, relatively stable intersections, which are the chapters of this book.

Where does science come from? Where and when was it born? In Greece, in Egypt, dating back to farthest Antiquity? The first open question, that of its source, therefore the first chapter of this book, dating back to Babylon in the year minus 1800. Should we think in terms of one or several origins? Here is the first bifurcation: the discussion is of major importance, as for several centuries a controversy has raged between historians and scientists, most of whom favour the Greeks. We have changed that particular decision by showing many nuances and reading the actual source material itself. Yes, the reader may well be surprised: at the very beginning, why indicate a crossroads rather than the source itself? For the above reason, no doubt – was that source the Mediterranean or the Fertile Crescent: the Greeks or the Egyptians? But beginning with an intersection also shows, as honestly as possible, that in turning left towards the West we have chosen to neglect the history of the Orient on the right, in other words, ignoring the sciences as they were developing there, especially in China. That is in no way a value judgement, but it would take a completely different work to include these. Besides, have you ever seen the source of any stream that does not already form a confluence?

But we must specify what did appear: alongside astronomy and medicine, mathematics certainly emerged. But why this plural? Is there one or are there several? This is the second open question, the second bifurcation, the sec-ond chapter of the book, where we will read about the comparative history of algorithms written in hieroglyphics on papyrus and in cuneiform on clay tablets. From this we conclude that there are several origins and several sciences related to the cultures from which they sprang. But again we must specify what type of abstract ideas. Greece rightly boasts of inventing pure science and proof by demonstration. How? Are there one or two Greek mathematics? In the third bifurcation and third third chapter of the book we see how geometry, on Greek soil, was drawn from algorithms, older, but also, in our eyes, more recent.

Apart from a few anonymous scribes, the legendary Thales and the studious Euclid, there are quite a few portraits missing from a book about a discipline which traditionally likes to include them. Here is Archimedes, the prince of geniuses, as drawn by Plutarch and Polybius. What science did he practise? Pure, undoubtedly, yet applied according to the judgement of an exacting Platonist, but above all marked by the shadow of the tyrant ruling his city and the war machines defending it. The fourth bifurcation was already of tragic importance, as it still is today: since when, how and why has the purest science gone hand in hand with death and destruction or, conversely, with the protection of populations under attack? A crucial choice is offered to the scientist and his historian: knowledge or power, detached contemplation or unbridled violence? The ethical problem, as we can see, is not new: war or peace in the world of sciences? Just as maps do not give details of the sea route from Bordeaux to Montevideo or Boston, but indicate the ports where ships converge and diverge, so this book deals with the issues and allows readers to choose their own routes through the chronology to be found at the back of the book. Fast, slow, short or lengthy routes connect the crossroads, in other words the doubts, the hesitations, the major questions of the history of science, its true elements.

Similarly, once mathematics have been put on the path that Edmund Husserl said was impossible to miss, can we say that in the very long term a single concept retains the same meaning? Is it always a question of the same form or the same definition? Does geometry connect us directly with the remotest past? We are not very familiar with the Greek gods and goddesses, but what about Pythagoras' theorem? Has it been shining, unchanged, for more than two thousand years, the sole example of durability? In the fifth bifurcation, a new chapter, let it be the circle, for example, the great figure of the circle. Is it possible to decide whether we have been talking about the same idea for thousands of years? Yes or no?

Babylonians, Greeks and Egyptians – their legacies have been reference points for so long that we have come to see them as one and the same. We knew that they had been lost to us for a long time and that we rediscovered them through the traditions and cultures of the Arab-speaking world, so should these people be considered as intermediaries? Is there such a thing as Arab science? We lost one of our parents, but when we found them, we found two. How is it possible to describe accurately the originality of the second parent who is as close to us and to our ways of thinking as the first?

A sixth bifurcation is a new confluence of a river already joined by another tributary. A junction of major importance through the history of the Mediterranean countries, dare I say the cradle of the sciences, has constantly swung between Semitic and Indo-European influences: our body of knowledge draws on both. Thus the history of science is swelled by some contributions and sometimes discards others like forgotten branches. To gain a greater understanding, it is enough to stand at the confluences where decisions are made.

All maps have an index – always more or less the same, whatever the map and whenever it was drawn – of towns and sites, crossroads of the network. This book establishes a precise map index. It pinpoints the summits and interchanges, it describes the recurrent problems, the hesitations, the major issues of the history of science, its bifurcations, once again its elements. As for the routes, the reader, we repeat, is invited to dip into the chronology at leisure. And we will see the natural scientist adventurers of the Age of Enlightenment seeking and finding, widely scattered, the species which Darwin later listed chronologically. We have distributed or classified the major issues in a representational space which likewise will reconstitute some day the temporality appropriate to the history of science.

Are we so certain that our overall idea of science never changes? And what is that idea essentially? In a most unforeseen bifurcation our predecessors, in our very own universities, responsible for teaching similar to our own, called by that name a discipline which they held sovereign but which the Age of Enlightenment, on the other hand, saw as ignorance and obscurity. All the routes followed up to this point turn back, depending on whether we accept or refuse this kind of decision. Once again then, what is a science? While in Paris, Oxford and elsewhere the medieval university taught theology under this excellent title, arithmetic and algebra, ignored by all and scorned by the learned, were practised in the street and at fairs, under the name of algorism, for accounts, transactions and purchases. To the question of what is science, history sometimes replies with another question: where is it found? In the market square or in the classroom? And in what language is it spoken? By those who spout jargon or by those who speak coarsely? What is being said that is new in these different idioms? Here are two further bifurcations, the seventh and the eighth.

These decisions plotted on the map are accepted or rejected by an authority which often acts as a tribunal. Only such a court decides. The Greek schools, the Church councils and university symposia function in a similar way and claim a monopoly of the definition of words, of knowledge and the truth. They are tribunals and they hold the crucial power of cardinal importance in the history of science. It is here that orientations are defined and set down.

When a tribunal or authority acquits or sentences a defendant, two fundamental things change: time and truth. An alleged offence which possibly or probably occurred is subject to deliberation then judged, suddenly becoming true or false. There are clearly defined moments when the jury gives its verdict and the sentence is delivered, thus there is a before and an after. The clerk records it. What is the history of science? The time, human or social, of a class of truths duly recorded. Some will say the time of the truth, universally speaking. But how can the truth appear, change, vanish and give way to another?

The universal reply is through tribunals. Human societies have not invented many other means than these courts for constructing their history, their own environment and their remarkable culture. From the moment they appeared the sciences were subject to them, resisted them, adopted them, changed them, used them and controlled them. They themselves have become courts where judgements are made.

The history of science in turn describes these courts for itself, is subject to them, resists them, adapts to them and adopts them, changes them, uses them and will end up controlling them. A tribunal itself, it constantly revises the judgements pronounced by science by prescribing them.

Once again, there is our question: how do we write or present a history of science? Answer: by organizing a critique, by setting up tribunals. The bifurcations will remain fixed, without the flexibility of railway points.

In his second preface to the *Critique of Pure Reason*, Immanuel Kant presents Thales, Stahl, Torricelli and Galileo, inventors or founding heroes in the fields of geometry, chemistry and mechanics of the first truths, such as those making time or history: a chain of successive revolutions in which the different bodies of uncertain knowledge each in turn take the path of the sciences. Mathematics begins, and then physics and so on. Thus Copernicus, in changing what was fixed and what was mobile, founded scientific astronomy. The entire history of science suddenly constructs a magnificent vision induced by the Age of Enlightenment and leading to all our generally accepted ideas, but especially deduced from the fact that the same Kant establishes a critique and founds a tribunal of reason. As soon as the court is set up a time and a truth appear, a time of the true, a before and an after, a history of science. The philosopher emulates wonderfully the inner movement of each discipline and believes us naïve enough to make us discover as fact what he produces as a judge. The history of science is not like this, nor is it told like this, other than from the point of view of this universal Reason which makes up the tribunal that we have subconsciously set up over the centuries and from which we pronounce our rulings.

This book, which draws on original texts and is made up of bifurcations, is ever alert to the continuous function of certain tribunals in determining orientations. It relentlessly reopens the files on past cases and makes hundreds of revisions. Decisions in science are, fortunately, never irreversible, which is why the network fluctuates. Thales or Stahl and others are sometimes placed on one side of the scales, sometimes on the other, depending on the times and the influence of a mode of thought.

Let us take, for example, 'The Galileo Affair'. This chapter symbolizes and acts as a model for our enterprise. Yet again, it neither describes nor re-counts the famous trial, but relates it several times and in several voices, so to speak. We hear not only the cause of freedom of thought, that of stubborn facts versus writing, even the cause of the Church, for the evidence of the experiments does not shine as clearly as people think, but also, more par-ticularly, the debates of a new science versus traditional knowledge, those of mathematical physics, of rational mechanics . . . as well as the anxiety of the author, who even confesses coming to the Galileo affair via the gulf which separates the principle of thermodynamics and the equality between cause and effect as demanded by classical mechanics. How many numerous and complex trials battle within an affair like that, how many protagonists, stakes and historians have been involved from the start and still are today? That is the true interest of this chapter, not taking sides with one or other of the protagonists, which would amount to repeating the trial indefinitely, whereas it is better to understand it. The author of this chapter has written the histories of history cutting across the external time of causes and con-frontations and the internal time of things and equations. From this ninth bifurcation there are a hundred possible paths branching off.

Another crossroads which it is appropriate to mention here is the chapter on Lavoisier, in the fifteenth position. It is again a symbol and a perfect model of our enterprise: the French Revolution, which was political, is mixed up in the chemical revolution, just as in the Galileo affair religious, ideologi-cal and strictly scientific trials were all interwoven. A tax farmer under the *ancien régime*, Lavoisier displayed the same talents in this post as he did in scientific research; sentenced to death, he became a victim of his political enemies as much as of his jealous colleagues. In short, through assiduous and meticulous control, he managed his weights and measures, both in science and in society, but also in history whose future, present and past he controlled by appropriating them; like an objective tribunal, the scales he used settled the truths of chemistry and the time of chemistry. And so we see how the different trials this book is concerned with deal with both causes and things, in other words, the balance of power between various conflicting parties and with nature itself. From this one can even reach a satisfactory definition of science as the human court where causes and things are so close that they sometimes overlap and hence trials determine the boundary between the collective and the objective. The French word *chose*, meaning 'thing', is derived from the word 'cause': the history which causes the former to stem from the latter is no doubt bound up with the history of science. Scientific language, miraculously, results in an almost perfect superimposi-tion of the performative of the trials with the objective of the experiments.

The tenth aspect of the bifurcation annihilates one of its branches: the age-old Greek, Arab, Latin and modern history of refraction is challenged by Descartes. He claims to have invented it all when in fact he simply made no reference to anybody else: an operation which became standard in the sciences and in philosophy, which was practised by a number of brilliant thieves. All before me was ignorance and neglect of the fundamental questions; how lucky that I came along to reawaken reflection, discovery and invention; and then there was science – or philosophy. As a result of this publicity coup Descartes' ego certainly emerges inflated but ethically it is diminished. A set of railway points guarantees the signalman a place in posterity with the endorsement of history while his predecessors are consigned to oblivion. It is a legal decision in a way but a fairly complex one. A political tribunal killed Lavoisier but his scientific tribunal decided on the historical death of other chemists, who did not oppose his sentence. A religious court sentenced Galileo but the victim sentenced his judges as well as his predecessors. So who decides and what do they decide? Who makes rulings on time and on the truth? That is how the history of science is constantly revising trials by setting up a sort of mobile tribunal, leaving verdicts open and bifurcations undecided.

What is science? Where is it found? And now, who practises it? Who makes the decisions, of course, but also, who invents it? The eleventh bifurcation is a new branch: in the classical age, rich, enlightened enthusiasts with endless leisure played with numbers, not far from the salons, as others played roulette. A century later, learned professors in German universities revived the same discipline and transformed it into a profound, almost metaphysical theory. Academia systematized what idle nobles did for relaxation. How are the sciences transformed when those who practise them change? Theorems go from being challenges in letters to the school textbook or treatise which becomes a reference work. Everything changes at this intersection, even the idea of what is serious or fundamental. Town or gown, the salon or the university: each to his own truth, even for numbers.

Is it true that beyond a certain threshold ideas change? Yes or no? Yes and no. Newton discovered universal attraction, evidenced in the magnitude of the stars, but became bogged down by the details of chemical reactions and lost his way. Affinity made people laugh but it fuelled the search for the force which repels or attracts certain bodies in relation to others. How could an idea which seems outdated lead to the discovery of the greatest explanation of the world ever to appear in history? What today seems of little consequence yesterday created enormous shifts; but careful, it may create even greater shifts again tomorrow. Yesterday chemistry altered physics, today it appears to be part of it. But what about tomorrow? Who assures us that knowledge that can only be learned will never understand science which, however, understands itself? The twelfth bifurcation remains open and the possibilities keep shifting. The tribunal sometimes judges one way, sometimes another, depending on constraints which are now forgotten: causes and things exchange places and change.

In another example, for the last hundred years nobody has been able to think of time without referring to Darwin. Everything evolves according to the model of the species: the stars, things, the world, even history and the history of science. Is this a gigantic breakthrough and the thirteenth bifurcation? Yes and no, and more no than yes. In the century before Darwin's voyage to the Galapagos in the *Beagle*, hundreds of explorers had carried out the *devisement*[2] of the world and collected species so as to identify them, to name them: all that remained was to list them, to rearrange the classification table. There is no better preparation for time than space. There is no better preparation for evolution than a series of bifurcations of the species: there is no better preparation for the history of science than a series of points crucial for the problems and the decisions. Through the very examples of the problems, we gradually touch on the problem of the history of science itself.

But for two centuries, nobody has been able to think of the era of the sciences without referring to what happened, albeit unseen, during the French Revolution, when the scientists quite plainly took power. An astronomer was Mayor of Paris, the inventor of topology was at the head of the Committee for Public Health, the scholars occupied the institutions before the people did and in their place, and a geometrician, although a minor, gained the title of Emperor. The nobility and the clergy collapsed, society no longer lived according to the same divisions or the same offices, scientists at last formed a class or a genus, replacing the clerics and forming a new Church. After this fourteenth intersection, the history of science often becomes the new history of a new but old clergy. But the one will forget the other while going through the same motions.

Oblivion or memory, the same again or another bifurcation, the seventeenth in this book. Often, avant-garde science makes a dazzling rediscovery of a predecessor which it states emphatically had remained or become lost in obscurity. True or false? Usually both. The neglected are not those we believe to be so and those we rediscover were rarely lost. Mendel cannot be described as that unknown scientist. A tribunal can also redress false injustices.

Who forgets? Who remembers? Not only people and institutions, but also things, especially theories. The offspring of peas mark their parents with or on their organs; Euclid's system organizes the memory of ancient geometric works; likewise with Mendeleyev, whose table charts the past history of chemistry. Does it plot the future or does it summarize what went before? Both, but rather the second branch of our nineteenth bifurcation. We neither learn it nor admire it for the same reasons that constructed it, almost naturally. Various sciences are full of such tables where memory is recorded, providing rich sources for the history of science. As the book progresses, each chapter serves better as a model for the whole discipline as if, in imitation of science itself, it capitalized on its own experience.

The earth and even rocks remember, like science and the history of science. Mythical texts said they were 4004 years old, dating back to 9 o'clock on that Monday morning when God created the earth with one word. A sudden bifurcation, the sixteenth, is there: Lyell judges the earth eternal, as old as the hills. Then begins the secular quarrel about the synchronization of the different times: that of the sky, of the stars, of things, of humanity, of the world and of its history. Now we are talking about geology as well as the history of science, for, since the various sciences were born, we have not learned how to synchronize the times of various sciences or the different rhythms of inventions, divisions and lapses, in other words, of our many bifurcations and halts before an equal number of tribunals.

Why? As a result of the conflicts, the causes again. Scrutinize the causes and see how things follow or are transformed with them, and vice versa. Pasteur fights against Pouchet with germs, which themselves begin to fight against spontaneous generation at Pasteur's side. This is perhaps the greatest bifurcation within the history of science, going far beyond the wrangle between internal contents which are exclusively scientific and external conditions which are exclusively social, since, far from setting them apart, it links and combines them. This canonical chapter again symbolizes and on this point embodies our enterprise in that it shows how science too sets up tribunals, but in such a rational clarity that the burning issues are concealed. Human struggles and those of institutions are resolved in a glass jar with a distorted neck. Here, in the assembly convened by Pasteur at the Sorbonne to disprove once and for all his opponent's theory of spontaneous generation, causes became things and things became causes, as the Romance languages have always wanted them to be since their beginning. This eminently modern bifurcation has been stable since our languages have made sense. Pasteur's germs behave, all things being equal by the way, like Joliot's atoms. On the eve of the Second World War the atomic scientist behaves, all causes being equal by the way, like the biologist. A Commissariat à l'Energie Atomique was founded just as an Institut Pasteur was built and microphysics exists just as microbiology does. The author of the two chapters in the eighteenth and twenty-first positions links together so closely the conflicts and the interests – more generally known as the circumstances – that this link itself, which is becoming stronger and stronger, this crossroads or intersection begins to resemble, in a strangely convincing way, the things themselves that experimental science challenges and considers. The route map, through bifurcations and movable sets of points, emphasizes the works of men and of groups – paths traced, roads constructed by complex building works – but these roads also follow the thalwegs of the natural relief, and gradually these atlases become the atlases of the world. Suddenly the intrinsic nature of things and our constructions converge.

But once again what science, what scientists, who, where, when? First of all: how many? For a long time now there have been more scientists in industry than anywhere else. The history of science would be unrealistic if it stopped at the universities and the institutions officially devoted to research. The bifurcation of the sciences and of society begins with many highly sophisticated techniques of great added value developed by private companies. Here is another confluence, the twentieth, along which flow not only people and capital, but also demand, the market, the whole of the modern economy and its fluctuations, which are even more unpredictable and unstable than those of the network outlined since the beginning. Here knowledge adapts in real time, as does expertise, to the unforeseen hiccups of circumstances and demand. The relationship between the map and the real world becomes more refined.

No science without technology, without machines, especially those which Jacques Louis Lions called universal tools, in that they have the efficacy of tools and the universality of science: computers. Everybody believes and has countless reasons to believe that those who invented them, from Leibnitz and Pascal down to Turing and von Neumann, started out with an idea of the finished product before building their elements, be it hardware or software. No. Those who are searching do not know, they grope, dart about, waver and keep their options open. No, they do not build the calculator of tomorrow thirty years before its time, because they cannot foresee it, as we who know and use it might infer. The truth is, they mime, like all the protagonists, both individual and collective, material and intellectual, of this book, with its bifurcations and its shifting network. They obtain, yes, almost miraculously, a result which they did not foresee at all, and yet which they sought, thus foreseeing it in some obscure way.

If readers accept this openness, this quest, basically this ignorance of authentic scientists who seek, who know without knowing, they will understand the unexpected arrival on the market of these machines which were planned but unforeseeable. Likewise, they will suddenly grasp the whole of this book map, this book network, this book index, stable and fluctuating, structured but open, constructed in an exact imitation of what history and the history of the sciences actually were, mixing up their products and their inventors, their legacies, tributaries and confluences, the bitter quarrels between their pressure groups, many and splendid landscapes, things of the world and human causes mixed up together to the point of indecision, their temporarily final conclusions, what they have forgotten and their changing or misleading memories, their institutional organization, their strict rational determination and their exhilarating improbability: our fascinating wanderings over these maps.

To identify the fluctuation of movements at the intersections, open or closed, forcing a right turn or a left or beckoning indifferently in both directions, I followed, first of all out of convenience, the usual thread of time, but I have sometimes also bent it a little, on purpose. As we have perhaps just discovered to our surprise, ordinal numbers do not always appear in this preface in the order they do in the book. That is because at the moment of the departure from that order the subject led to the development of another thought, that of memory and oblivion or, as Bergson would have said, the retrospective movement of truth: it is only after the French Revolution, he says, that the storming of the Bastille, which sparked it off in a way, takes on any significance or even exists in history. At the precise moment when it took place that history was not there. We believe that the truth flows with the tide of time: it also travels up it, hence the loop in the numerical order.

More generally, that means that this book could have followed a different order. Thematic classification, for example, would probably have extended the open range of options and theories and distributed their tensions better. The history of science meets here its historian and his medieval parchments, its archaeologist, with his tablets and papyrus, its sociologists and their institutions, its philosophers and their concepts, its anthropologists . . . but, all in all, the order finally chosen emerged as the most faithful to the things themselves: it is as if the map comes slowly down to Earth and merges with what it is trying to represent. The tail of a shifting river basin with many confluences and beds, which as if by chance meets obstacles and dams, stopping altogether or gathering speed along corridors and wide stretches, on through disasters, without counting the frequent but fairly stable turbulences and counter-currents which beset its course, the losses and the forgotten branches . . . could there be a better model of mingling and of *percolation*[3] to understand the flow of history and the work of time?

Les Treilles, 1985–8.

1

Babylon −1800

James Ritter

Prologue: Nineveh, April 18,672 BC

Esarhaddon, Great King, Mighty King, King of the Universe, King of the Land of Assyria, is a worried man. His health has been failing for some time and, at all costs, he wishes to avoid the bitter dynastic struggle that has marked the Assyrian royal house for generations. He has therefore decided to divide the kingdom between two of his sons. One, Assurbanipal, will succeed to the overlordship of the Assyrian Empire; the other, Šamaš-šumu-ukīn, will inherit the recently acquired throne of Babylonia. To this end it is crucial that all the important officials and dignitaries of the Empire be informed and commanded to swear a loyalty oath to these two crown princes as soon as possible. The chief royal scribe, Ištar-šumu-ēreš, is placed in charge of the project.

On 18 April this official writes to the king concerning the arrangements for the swearing of fealty by one class of officials, the *ummânū*, the 'experts':[1]

> To the king, my lord, (from) your [servant] Ištar-šumu-ēreš.
> [Good] health to the king, my lord! May the gods Nabû and Marduk bless the king, my lord!
> The *ṭupšarrū*, the *bārû*, the *āšipū*, the *asû*, and the *dāgil-iṣṣūrē* staying

17

in the palace (and) living in the city (of Nineveh) will enter into the treaty on the sixteenth of Nisannu.

They should take the oath tomorrow.

It is not with kings and crown princes that we shall be concerned in this chapter but rather with this little list of 'experts'. Who were these people? What was their role in Mesopotamian society? Clearly they were important to Esarhaddon and clearly they were also considered to form a group apart, needing a special day for the oath, so we shall seek them out in the evidence provided by the clay tablets left to us by the civilization that developed and flourished between the Tigris and the Euphrates.

In particular we shall want to examine their origin: when and where did they arise as a professional group? To this end we shall have to leave Nineveh at the height of the Assyrian Empire, and turn to the first moment where they enter into history together – more than a thousand years earlier and to the south, in the land of Babylonia. It is during the time that Assyriologists call the Old Babylonian period, the first 400 years of the second millennium, that we can begin to trace the development of a field of study and practice that the Mesopotamians themselves seemed to see as somehow unified and privileged. Our evidence for this is often indirect and scattered, but there is enough to piece together the outlines of this special intellectual domain, perhaps the first of a kind we might be tempted to call 'rational'.

And Assurbanipal and his brother did become kings, for which we must be at least partially grateful, since the first spent much of his time collecting one of the largest and most complete libraries of the ancient world. Indeed, it was the discovery of the remains of just this library in the ruins of Nineveh by Austin Layard in the middle of the nineteenth century that led to the birth of modern Assyriology. And among those thousands of pieces, aside from the letter of Ištar-šumu-ēreš, there is a laudatory hymn about Assurbanipal himself and his own claims to be to an intellectual, on the same level as the *ummânū* at his and his father's court:

> Marduk, the wise one of the gods, bestowed on me wide understanding and wise comprehension;
> Nabû, the *ṭupšarru* of the Universe, gave me as a gift the precepts of his wisdom;
> Ninurta and Nergal endowed my body with heroic strength and un-equalled power.
> The deeds of the wise Adapa I learned, hidden wisdom, the whole of the art of the *ṭupšarru*;
> I can interpret the omens of heaven and earth, I participate in the Council of the *ummânū*;
> I can discuss (the treatise) 'If the liver is a mirror of the sky' with skilled *apkallū-šamni* (= *bārû*);

I can find the difficult reciprocals and products which are not of easy access²;

I can read the elaborate texts, in which the Sumerian is obscure, the Akkadian difficult to interpret;

I can decipher the stone inscriptions which date from before the Flood.

As we shall see, in the Old Babylonian period the accomplishments of these 'experts' were already esentially the same.

Babylon −1800

When we turn back to the first half of the second millennium, to the Old Babylonian period, we are in quite a different world from the one that we have just left. The Sumerian supremacy of the third millennium is still present, at least as a memory. The great empire founded by the Sumerian-speaking (or at least writing) kings of the Third Dynasty of Ur maintained a hold on the imagination of Mesopotamians for more than two centuries after its disappearance, swept away by the waves of West Semitic peoples moving into the area.

But this immigration brought a new vitality to the banks of the Tigris and Euphrates and, when our written sources again become frequent, a new symbiosis had been established. Numerous city-states, like those prominent during the largest part of the third millennium, are again in place. But this time it is the Akkadian language that is dominant; Sumerian has been relegated to the rank of a liturgical and erudite language of a small group, somewhat like Latin in the medieval period in the West. The various city-states vie for power, with sometimes Larsa sometimes Isin on top. In fact in what was heretofore a small village, Babylon, will be the most successful of these warring states – and that under a West Semitic dynasty founded in the great migrations of the twentieth century.

The accession to the throne of Ḥammurāpī around 1800 BC corresponds to the constitution of a new centralized state on a scale and for a time not known before, even during the fabled third millennium empires. Babylonia represents a vast, pacified area from Anatolia in the north to the Gulf in the south and from Palestine in the west to Elam in the east.

But the task of unification on the political, cultural and linguistic levels demands the establishment of a new commercial and military network. It is now that we find the first systematic references to various groups of experts and scholars in the texts at our disposal. Although there is some reason to believe that these were already in the process of constitution in the Ur III period, it is in the rich documentation left to us by Ḥammurāpī and his contemporaries that we can first really trace the nature and role of our protagonists. To find out more about them we shall see what we can discover about their professional activities from the letters that have come down to us from the Old Babylonian royal archives.

Professionals

Let us take the experts mentioned in Ištar-šumu-ēreš's letter in their reverse order and seek their equivalents, where they exist, under the Empire of Ḫammurāpī and his contemporaries.

The last, the *dāgil-iṣṣūrē* or 'bird observers', are the easiest to dispose of. These were specialists who foretold the future by the behaviour of birds and there is no evidence at all for their existence prior to the Neo-Assyrian period. This was a new field of expertise, like astrology, that was founded in the first millennium, showing by the way that new specialities continued to form during the whole of Mesopotamian history. But their time, from our point of view, lies in the future, and we must leave them aside.

The *asû* and the *wāšipū* (the Old Babylonian form of *āšipū*) were called in, either separately or together, in cases of sickness or injury. They divided what we call the field of 'medicine' between them. If modern Assyriologists have a tendency to translate the first by 'doctor' and the second by 'exorcist', the way the Babylonians themselves viewed their exact distinction is none too clear for us today. For example, from the period not long after the fall of Babylon, when the invading Kassites had renamed the country Kar-Dunyaš, we have this letter written by the king of the Hittites in Anatolia to his ally, the Kassite king of Babylonia:

> A message from Ḫattušili, the Great King, the King of the Land of Ḫatti, your brother: to Kadašman-Enlil, the Great King, the King of Kar-Dunyaš, my brother.
> I am well, my palace, my wife, my children, my soldiers, my horses, my chariots, also everything in my land, are well indeed! May all be well with you, may your palace, your wives, your children, your soldiers, your horses, your chariots, and everything in your land be well! (. . . . . .)
> And I have more to say to [my brother]. Concerning the *asû* whom my brother has dispatched here: people accepted him and he performed [. . .] cures on them but a disease befell him. I took great pains with him and I performed extispicies for him, but when [his time] came he died. Now one of my messengers will take (the *asû*'s) servants along (to Babylon) and my brother [should] question them, and they will tell about the cures which the *asû* often performed. (. . . . . .) I would never have thought to detain the *asû*, in view of the fact that when they received, during the reign of my brother Muwatalli, an *āšipu* and an *asû* (from Babylon) and detained them [in the Land of] Ḫatti, I was the first to argue with him, saying: 'Why do you want to detain them?', telling him that it

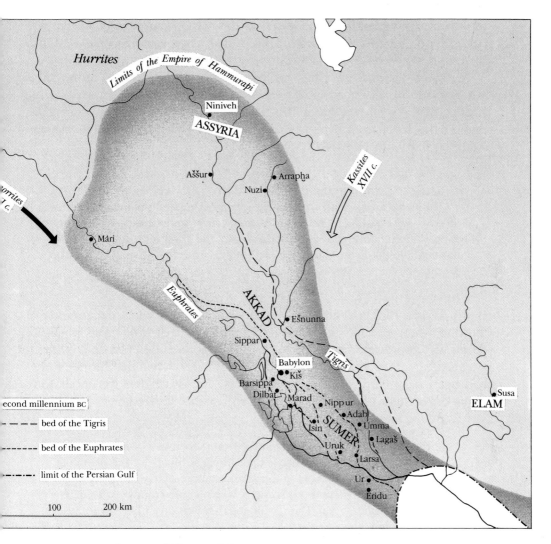

Mesopotamia in the time of Ḫammurāpī.

is not according to our custom to detain (such persons); and now should I have detained your *asû*? Of the former [*ummânū*] whom they received here, only the *āšipu* died [. . .]. The woman he married here was of my own family and he had a lovely house, [but if he had said]: 'I want to leave for my native country,' he could have gone right back [home]. [Would I] have held back a famous *asû* of Marduk?

We are much less well informed about the activities of the *wāšipum* in early Babylonia than those of his confrère, the *asûm*. And even for this latter, it is principally from the kingdom of Mari that we have the largest amount of

21

information, gleaned from the letters of the royal archives, whose clay was baked hard during the burning of the palace when Ḫammurāpī added it to his empire towards 1760 BC.

The *asûm* was master of his craft by virtue of his knowledge of the drugs which made up his pharmacopoeia. It was thus important, at least to the patient, that any knowledge in this field be shared as widely as possible, as witness this letter from Išme-Dagan, king of Assyria, to his brother Jasmaḫ-Addu, at that time viceroy of Mari:

> Speak to Jasmaḫ-Addu: thus Išme-Dagan, your brother.
> The medications with which your *asûm* has treated me are excellent.
> if any *simmum*-illness occurs, this medication immediately masters it.
> Now, I am sending you Šamšī-Addu-tukultī, the apprentice *asûm*, so that he may rapidly learn about this medication. Afterwards, send him back to me.

But there is a further point to be made. For us today each amelioration of the pharmacopoeia has to be judged empirically; that new drugs might be tested was appreciated at the time but the trials remained personal, without systematic control. A little while after our last letter, when the throne of Mari had been recovered from the Assyrians by its native dynasty under King Zimri-Lim, a high official, Dāriš-lībūr wrote to his sovereign:

> Speak [to my lord: thus] Dāriš-lībūr, [your servant]:
> [As to the] medications to use against 'dryness fever', [of the *asûm* of] Mardamān [and of the *asûm*] of the administration office, [. . .] wrote to me in these terms: 'These medications which come from [. . .], I have sealed them with my seal. Then, with their medications, I have sent these *asû* and Lagamal-abum to my lord.'
> (Now) my lord has already seen what the medication against 'dryness fever' of the *asûm* of the administration office is worth. I myself have seen what the medication against 'dryness fever' of the *asûm* of Mardamān is worth, and it is excellent. With Ḫammi-šagiš [. . .], and it is excellent. [. . .] drank it, and it is excellent.
> Now, perhaps the [. . .] will want to mix these medications in order to drink them. They must try these medications separately so that my lord may drink the one which he should drink.

The king guarded the power to attach the *asûm* to the palace or to another residence; in Mari this could be a certain kind of border city called a *pattum*, as witness this letter from a local official, Jawi-Ilā, to King Zimri-Lim:

Speak to my lord: thus Jawi-Ilā, your servant. (. . .)
Another thing: it is in a *pattum* of my lord that I reside and the *asûm* died a little while ago. May my lord not retain any *asûm* who will come to establish himself in the 'house' here, in your city, so that he may cure your servants, the citizens of Talḫajûm.

We have already seen, in the letter addressed by Ḫattušili to Kadašman-Enlil, that, faced with the illness of his Babylonian doctor, the Hittite king had an 'extispicy' performed. This was exactly the domain of the *bārûm*, often translated as 'diviner'. An extispicy, the examination of the organs of sacrificial animals, was one of the recognized techniques of *divination*, foretelling the future, along with pouring oil in water and observing the forms it took, or examining the direction in which smoke blew when an incense burner was lit.

The Old Babylonian king had the same power over the *bārûm* as he did over the *asûm*. Though principally a royal affair, other citizens, and particularly those who inhabited the *pattum* cities of Mari, could use his services. Here, during the Assyrian domination, Išme-Dagan writes once again to his brother at Mari:

Speak to Jasmaḫ-Addu: thus Išme-Dagan, your brother.
On the subject of Narām-Sîn, the *bārûm* of whom you wrote me, the king has assigned him to the district of Šitullim. Now you know yourself that this district is a *pattum*. Now I have written to the king as follows:
'Narām-Sîn suits Jasmaḫ-Addu. He wrote me to tell me: "Ibāl-pī-EI [and his son] are staying with my lord! [If not] Narām-Sîn, who will ever stay in my service?" Now [my father] must send [another] *bārûm* to Šitullim [so that my brother will have nothing] to say.'
[This] is what I have written to the king. (But) he will [not let this man] go. [The district] of Šitullim is a *pattum* [and] without a *bārûm*. [. . .] the king will assign a *bārûm* [to] this district. (. . . . . .)

The main task of the *bārûm* was to predict the future for military affairs. A diviner was attached to each army and no battle would be joined before the omens were taken. From Babylon this time, here is a memo from a functionary to a subordinate in times of trouble:

Speak to Bēlšunu: thus Qurduša.
May the god Šamaš keep you in good health!
As you have certainly heard, the open country is in confusion and the enemy is prowling around in it. I have dispatched letters to Ibni-Marduk, to Warad-[. . .], and to yourself. Take a lamb from the flock for the

bārûm and obtain a divination concerning the cattle and the flocks, whether they should move into my neighborhood; if there will be no attack of the enemy and no attack by robbers the cattle should come to where I am − or else bring them into the town of Kiš so that the enemy cannot touch them.

Furthermore bring whatever barley is available in Kiš and write me a full report.

The world of divination, even military divination, knew its specializations too, as witness this reproach from Zimri-Lim to his wife, Queen Šibtu, whom we often see consulting the *bārû*:

Speak to Šibtu: thus your lord. (. . . . . .)
On the subject of your writing me as follows: 'I have obtained some results in divination relative to weapons.' That is what you wrote me. Why is it in divination relative to weapons that you have obtained results and not in divination relative to the taking of cities? And (why) did you not (rather) write me as follows: 'This city will be taken' or 'This city will not be taken'?

That the people of Mesopotamia drew a distinction between the (serious) activities of the experts, and the (more doubtful) enthusiasms of priests and women can be seen in many letters from Mari written by a member of the second category who claims to have had a dream, inspiration or divine possession with a message for the king. This information is dutifully passed on, but with the expectation that the matter will be reliably looked into by a *bārûm*. Here Šimatum, possibly a daughter of Zimri-Lim, writes to the latter with the news of a dream that has been reported to her:

Speak to my lord: thus Šimatum, your servant. (. . . . . .)
And on the subject of the daughter of Tepāḫum, in my dream, a man rose to declare this: 'May the little girl, the daughter of Tepāḫum [be called] Tagīd-Nawū.' This is what he said. Now, may my lord have a *bārûm* look into the affair and if [this] dream was (really) seen, my lord, may little Tagīd-[Nawū] receive that name. (. . . . . .)

Dreams are all very well, but if you want to know if such information is trustworthy, you must turn to a real expert.

The first category mentioned in Ištar-šumu-ēreš's letter, that of *ṭupšarrum*, is, paradoxically, the most difficult. The word simply means 'scribe' but exactly what it was meant to cover here is not clear. For example, when the letter was written, the term *ṭupšarru* was often used as an abbreviation for *ṭupšar Anu Enlil* – 'scribe (of the book) "When Anu and Enlil"', that is 'astrologer'. But we know that astrology, though flourishing under the Neo-Assyrian Empire, was a late comer to the Mesopotamian world, coming into its own only in the first millennium. Indeed, when the chief *bārûm* of Old Babylonian Mari, Asqudum, observed an eclipse, his first thought was to perform an extispicy, to see just what that ominous event might signify. By this reckoning, the *ṭupšarru*-astrologer, like the *dāgil-iṣṣurê*, would have been absent from the Old Babylonian world.

On the other hand, *ṭupšarrum*, at all epochs, also means simply a professionally literate person, capable of reading, writing and, perhaps here especially, performing calculations. There is no word in Akkadian for 'mathematician', though, as we shall see, all scribes received some training in arithmetic. It is thus in the domains of health care, of prediction of the future and of calculations that, – at least in the Old Babylonian period – our 'experts' worked.

Now that we have caught a glimpse of the daily practice of these experts in the Old Babylonian period, it is appropriate to turn to that part of their lives in which they first became *ṭupšarrū*, their school days. Hopefully this will help to fill out the somewhat meagre information we have been able to glean from the official correspondence among the powerful that has come down to us.

Education

> A house with a foundation like heaven,
> A house which, like a **pisan** vessel, has been covered with linen,
> A house which, like a goose, stands on a (firm) base.
> One with his eyes not opened has entered it,
> One with open eyes has come out of it.
> > Its solution: the school.

This riddle is itself a school exercise, part of the obligatory Sumerian language course which formed the basis of the Old Babylonian curriculum. And if the description is a little self-serving, it remains none the less true that education was the cornerstone in the construction of an intelligentsia in the service of power, then as now.

But to what exactly were the apprentice scribe's eyes opened? The question of the course content of the 'House of Tablets' (**eduba** in Sumerian or *bīt ṭuppim* in Akkadian) is not an easy one to answer. We have effectively three sources of information:

- The exercise tablets of the students themselves. These are most easily identifiable when they are the lenticular-shaped tablets used by the beginning student. Consisting primarily of exercises in the writing of cuneiform signs and simple arithmetic tables (multiplication, inverses, etc.), as well as some elementary Sumerian literary compositions, they present literally the ABC (or more properly, the *tu-ta-ti*) of the school curriculum.
- The Royal Hymns. We have already seen a late example in the hymn of Assurbanipal (p. 18). Those in use in Old Babylonian times were composed during the preceding Ur III period in honour of the reigning kings of that first durable Mesopotamian Empire. They were copied and recopied as school exercises in the Old Babylonian school and give some information on the subjects studied by the king (and presumably others) when at school. (An example is cited on p. 47).
- Finally, and most revealing, are the so-called '***eduba*** texts'. In reality these are Sumerian literary exercises of an advanced nature cast in the form of a dialogue between two students or a student and his teacher or father:

> The time that I spend at school has been established (thus):
> my days off are three per month,
> the different holidays are three per month,
> altogether there are twenty-four days per month
> that I spend at school; the time is not long.
> In a single day the teacher has given my section four times.
> The count (of days made), my knowledge of the art of writing will not disappear,
> henceforward I can apply myself to (my) tablets, multiplications and accounting, the art of writing, the placement of lines, avoiding the splitting of words . . .
> my teacher has corrected the beautiful words.
> We should enjoy the company (of fellow students)!
> I know perfectly my art of writing,
> everything is easy for me.
> My teacher shows me a sign,
> I add others from memory.
> After having been at school for the time required,
> I can deal with Sumerian, the art of writing, the contents of tablets, the calculation of accounts.
> I can speak Sumerian!
> (. . . . . .)
> I want to write tablets:
> the tablet (of measures) from 1 ***gur*** of barley to 600 ***gur***,
> the tablet (of weights) from 1 shekel to 20 minas of silver,
> with any marriage contracts that may be brought me,
> business contracts, I can choose the verified weights of one talent,

the sale of houses, fields, slaves,
securities in silver, contracts for renting fields,
contracts for raising palm trees, [. . .],
even adoption tablets; I know how to write all that.

Note here, aside from the emphasis on a good Sumerian style, the con-
stellation of subjects mentioned at the end: metrology, legal contracts and
economic calculations.

Putting together all the evidence from the practice of professionals and
from the subjects studied in the Old Babylonian school, we may draw up a
tentative list of three subjects which seem to have had a special role to play
in Babylonian intellectual life: divination, medicine and mathematics.

But the information contained in the documents we have looked at does
not suffice to draw a really clear picture of the relations among these fields.
To go further we must look at the school texts and the handbooks of the
practitioners. We will see that they share many formal similarities in grammatical and organizational structure which mark them off from other texts,
literary or religious ones, for example, and provide all the evidence we can
now recapture as to the ways in which the Babylonians themselves viewed
these domains.

This is not to say that other fields could not be plausibly included here.
Texts in the domain of jurisprudence, for example, seem to have shared
many formal similarities with the others and, from a later period, astrology
must certainly be added to the list.

In examining this material, we must guard against the natural tendency to
read into these texts our own judgement of their contents. Whatever the
Babylonian appreciation of them was, we may be almost sure that it did not
coincide with out own. There are no Mesopotamian 'philosophical' texts to
render explicit for us the thought of the Babylonians; the plausibility of our
hypothesis that these subjects formed a special field for the scribes and
officials of the time of Ḥammurāpī must rely principally on the evidence of
the internal, formal relations among the texts themselves.

Divination

For the Babylonians the gods could and did write intimations of the future
on all kinds of materials and by means of all kinds of signs; birthmarks on
the skin, conformations of the organs of sacrificial animals, smoke forms
rising from an incense burner, etc. This was the domain of the *bārûm* − literally
'seer' − who was trained to interpret these signs as a function of the problems presented to him by his clients. The 'art of the *bārûm*', which modern
Assyriologists call 'divination', is a striking component of Babylonian thought
throughout its entire history. Indeed the Old Babylonian period alone has

left us over a hundred such texts. Here is the first part of one that treats of the forms that oil takes when mixed with water by the *bārûm* and what they signify for the person, private or public, who has consulted him:

1 If the oil, I have poured it on water, and the oil has descended and (then) risen and has surrounded its water –
for the (military) campaign: the arrival of calamity
for the sick man: the hand of the god, the hand is heavy.

2 If the oil has divided into two parts –
for the (military) campaign: the two camps will march against each other
I do (it) for the sick man: he will die.

3 If from the middle of the oil a drop has come our towards the east and has stopped –
I do (it) for the (military) campaign: I will carry away spoils
for the sick man: he will recover.

4 If, from the middle of the oil, two drops have come out and one was large and the other small –
the man's wife will give birth to a boy
for the sick man: he will recover.

5 If the oil has dispersed and has filled the bowl –
the sick man will die
for the (military) campaign: the army will be defeated.

6 If the oil has dispersed towards the east and 3 drops have come out –
Šamaš will demand of the man a solar disc for his life.

7 If the oil has dispersed towards the east and 4 drops have come out –
an old (debt of) silver towards Šamaš weighs on the man.

8 If the oil has dispersed towards the east and 5 drops have come out –
an old (debt of) silver towards Sîn weighs on the man.

9 If the oil has dispersed towards the east and 6 drops have come out –
situation of the councilor of Dingir-maḫ for . . . . . .

10 If the oil has descended and (then) has risen in the direction of my thigh –
seizure by Dingir-maḫ

11 If the oil was green –
seizure by Išḫara.

12 If the oil has formed a bubble in the direction of my thigh – situation of the personnel god of the man.

13 If the oil has come out to the right and to the left – situation of Sîn and Šamaš.

14 If the oil has dispersed towards the east and 7 drops have come out – situation of Kubu for . . .

15 If the oil has dispersed towards the east and 2 drops towards the right and 2 drops towards the left have come out – situation of the Twins.

16 If the oil, its two ramifications were broken to the right and to the left – the wife of the man will leave.

17 If the oil, its bubble was torn towards the east – the sick man will die.

18 If the oil has divided into 2 parts – the sick man will die for the (military) campaign: the army will not return.

19 If the oil has split towards the east – the sick man will die.

20 If the oil has attached itself to the right side of the bowl – the sick man will recover.

21 If the oil has attached itself to the left side of the bowl – the sick man will die.

22 If the oil was red – it will rain.

23 If the oil has formed a *nēkemtum* towards the right – for the (military) campaign: my army will conquer the enemy.

24 If the oil has formed a *nēkemtum* towards the left – for weapons: the god of the enemy will conquer the army.

25 If the oil has produced foam – the sick man will die.

26 If the oil has divided into 4 parts – my enemy will dismantle the fortresses of my army.

27 If the oil has let a drop escape towards the east which then attached itself to the edge of the oil – for the sick man: he will recover for the (military) campaign: I will conquer the enemy.

28 If the oil has become a film −
the sick man will die;
(but if the oil) has contracted at the moment of my second pouring −
even if he is very ill, he will remain alive.

29 If the oil has carried along its water −
the curse of the next world will seize the man.

30 If the oil, at the moment of his second pouring, has freed its water −
the curse of the next world will <not> seize the man.

31 If the oil has spread out and has formed a star at the centre of
its water −
seizure by Šamaš.

There are several points to be made here. First of all, the text is formally very rigid; almost every entry begins with a standard phrase − 'If the oil . . .' − followed by a description of a possible configuration of the oil in water. The formula which begins the entry is so fixed that it must be there even where (as for example in **16** and **17**) the grammatical structure of the rest of the phrase must be twisted a bit to accommodate it. Note that this part is always presented in the past tense; it describes a statement of fact, it presents a situation. This, in turn, is followed by a description of the future, introduced normally by 'for the (military) campaign' or 'for the sick man'. Here the significance of the conformation is announced for two possible sorts of inquiry, the king or palace official wishing to know how one of the perpetual military campaigns was going to turn out or any individual concerned about the prognosis for his illness.

Second, the text is systematic on several different levels, for example, in the order followed by the presentation of the omens: e.g. **3–4**, **6–9**, **14**, where it is a question of the oil dispersing to the east while an increasing number of drops, from one to seven, separate from it. It is possible that the few intervening entries which interrupt this order are the traces left by a composition modified over time. We also glimpse some structure in the linkage of the configuration of the oil and the future events: **20** 'if the oil has attached itself to the right side of the bowl − the sick man will recover'; **21**: 'if . . . the left side − the sick man will die'; the right side is positive, the left side negative (see also **23** and **24**). Or consider **29** and **30**; if the oil 'has carried along its water − the curse of the next world will seize the man' but if, on the contrary, the oil 'has freed its water − the curse of the next world will not seize the man'; here, evidently, playing on the dichotomy of capturing or freeing water and thus, by extension, maledictions touching the patient.

The *bārûm* was supposed to be equally skilful in interpreting the ominous significance of the general appearance or actions of his client. Though, unlike the oil texts, this is not a 'provoked' form of omen – there is no process necessary other than observation of a 'natural' phenomenon – the nature and arrangement of the omens show the same influences as the text we have just studied. An extract from one of these texts is given here:

1   If a man does not recognize another man when looking at him –
    this man is given into the hands of death.

2   If a man recognizes another man from (a distance of) 1 cubit to
    30 **nindan** –
    his deity will always be with this man.

3   If a man cannot focus(?) his eyes . . . when looking –
    the mind of this man is deranged.

4   If a man, his eyes stare all the time when he looks –
    confusion of mind will be inflicted upon him.

5   If a man, his way of looking is oblique –
    he will die of constriction.

6   If a man, his way of looking is straight –
    his god will always be with him for his luck; he will live in truth.

7   If a man, he winks with his right eye when looking –
    he will live in hardship.

8   If a man, he winks with his left eye when looking –
    he will live in righteousness.

9   If a man, he winks with both his eyes –
    his skull has been hit; exactly as his skull so will his mind be.

10 If a man, his eyebrows cover his eyes –
    food is given to him from the gods.

11 If a man, his eyebrows are non existent –
    this man is not (even) given *lubkum*.

12 If a man, his hair is as red as dyed wool –
    this man will live without peace of mind.

13 If a man, his hair is as black as ashes –
    his god will give food to eat to this man.

14 If a man, his hair has a white spot and it assumes the form of a
    flame –
    this man will live in good health.

15 If a man, his flesh shows white spots and is dotted with *nuqdum* –
    this man is rejected by his god and rejected by mankind.

31

We notice here the same concern to cover all the cases, to exhaust the realm of the possible. Entries **7–9**, for example, show the different consequences attendant upon a man who has a tic in the right eye (bad), left eye (good) and both eyes (literally split). Similarly for the colours of a man's hair (**12–15**) – red (bad), black (good), white (dependent on the specific details).

The other omen texts are always of the same nature, whatever may be the subject or the kind of prognostication indicated.

We are now in a position to draw a first conclusion. The purpose of the systematic nature of these texts – and indeed this will be true for all the texts in all the fields we shall look at here – is to cover the domain of the possible by a network of typical examples: a process which permitted the student (and later the practicing *bārûm*) to place any new problem, any new augural sign, in this framework. The Babylonian approach to the question of generalization is not, as with us, in the discovery and application of a 'general rule' in which each case may be enveloped, but rather through interpolation in a pattern of known results. And we shall see that this method was as applicable in medicine and mathematics as in divination.

Medicine

Let us turn to our second field of interest: medicine. Although we do not have an enormous wealth of documents in this field for the Old Babylonian period – there are only a handful of Old Babylonian medical texts – these are sufficient to show the formal structure in this domain, particularly when they are compared with the same sort of documents from later periods.

One of the Old Babylonian texts, though in a rather bad state of preservation, does present a text which is of considerable interest. The two most legible entries are given here:

1 If a sick man, his regard is fearful more than in the time when he was
   healthy, and his face appears fresh,
   this sick man will not recover.

2 If a sick man, his hands and his feet hurt (lit. 'eat') him, and he cannot
   stop crying, and his body is not at all warm,
   (it is) 'the hand of sorcery'.

The resemblance to the divination texts that we have just seen is striking. Each entry has again two parts; the first, beginning with an invariable 'If a sick man . . .' continues with a description of a state of affairs – here a series of medical symptoms; the second a prognosis, either direct (**1**) or an identification of the illness (**2**). In this particular case these amount to the same thing, since from other sources we know that the illness mentioned was

The imprint of a doctor's cylinder seal. These seals, rolled on tablets of fresh clay, served as signatures for their owners. This one, ornamented with a sphinx, is engraved with the following inscription: 'In the name of Sîn and Marduk, the gods of his lords, let whoever prints this seal be satisfied with life. Seal of Makkur-Marduk, asû, son of Sin-asared, asû. (AO 4 485, Kassite or Neobabylonian epoch (?). Louvre, Paris. Photograph, Réunion des Musées nationaux.)

considered mortal. But the classification of this text as 'medical' rather than 'divinatory' is modern and the difficulty of such a distinction can be appreciated by comparing this text with the last text studied in the preceding section. It is our idea of what is acceptable as a causal chain, rather than any formal difference, which underlies this division.

This single fragmentary example does not permit us to judge the degree of systematization achieved in the Old Babylonian period. But we may judge its development by the following brief extract from the final (Neo-Assyrian) 'canonical' version (running for forty consecutive tablets!) which was one of the jewels of Assurbanipal's library:

> If his right buttock is red − [. . . . . .]
> If his left buttock is red − he [will drag out] his illness.
> If his buttocks are red − [there is no] 'stroke'.
> If his right buttock is yellow − his illness will change.
> If his left buttock is yellow − his illness will be painful.
> If his buttocks are yellow − he will be anxious.

If his right buttock is black − his illness will be painful.
If his left buttock is black − he will be anxious.
If his buttocks are black − [. . . . . .]
If his right buttock is bruised − he will drag out his illness, then he will
    die.
[If his left buttock is bruised − . . . . . .]
If his buttocks are bruised − he will die.
[If his right buttock is dark − . . . .]
If his left buttock is dark − his illness will change.
[If his buttocks are dark − . . . .]
If his right buttock is inflamed − his illness will change.
[If his left buttock is inflamed − . . . .]
If his buttocks are inflamed − his illness will change.
[If his right buttock is sunken − . . . . . .]
If his left buttock is sunken − his illness will be long.
If his buttocks are sunken − he will be anxious.
If his right buttock is raised − his illness will change.
If his left buttock is raised − his illness will be painful.
If his buttocks are raised − he will be anxious.
If his right buttock is slack − his illness will change.
If his left buttock is slack − his illness will be a source of anxiety[2]
If his buttocks are slack − he will die.
If his buttocks are in good shape − he will recover.
If his buttocks are bruised to the point that he cannot leave the place
    where he is, nor pass water − he has been 'struck' from behind;
    he will die.

The kind of structure apparent here is repeated, in the full text, for each part
of the body, from the skull to the ankle. Clearly the desire to place any
possible symptom for any possible part of the body has led to a huge and
all-encompassing network in which the symptoms are sometimes multiplied
to unrealistic or even impossible lengths. Ringing the changes on the colour
of the buttocks: red, yellow, black; on whether they protrude or recede, etc.;
and integrating this with a different prognosis depending on whether it is
the right, the left, or both buttocks which are affected, the aim of the text is
twofold. Not only does it provide the practitioner with a mesh so fine that
any real symptom can be easily interpolated, but also it generates an implicit
level of generality by the systematic way in which it works through these
permutations. All other things being equal, the right buttock is less serious
than the left, yellow is a less worrying colour than red while black is quite
dangerous, and so on. As in divination, the general and abstract is not said
but shown − and in exactly the same way.

The medical field does, however, present us with a new kind of text, one
that divination apparently does not provide − at least not among the extant

34

Old Babylonian texts. We have three Old Babylonian texts of this kind −
here is a translation of the obverse of the best preserved among them:

1   If a man is bewitched −
    (. . . . . .) the kidney of a lamb which has not yet eaten grass (and)
      the plant *ernīnum* you will dry
    (together), he will eat and he will recover.

2   If a man is sick with jaundice −
    you will soak the root of licorice in milk, you will leave out overnight
      under the stars, you will mix in oil, you will give it to him to drink
      and he will recover.

3   If a man, his tooth hurts (lit. 'has a worm') −
    you will grind (the plant) 'sailor's excrement';
    if his right tooth is sick you will pour (it) on his left tooth and he will
      recover;
    if his left tooth is sick you will pour (it) on his right tooth and he will
      recover.

4   If a man is covered by an eruption −
    you will mix flour of malt little by little in oil, you will apply (it) and
      he will recover;
    if he is still not cured, you will apply hot *šimtum* and he will recover;
    if he is still not cured, you will apply the warm residue and he will
      recover.

5   If a man, a scorpion has bitten him −
    you will will apply the excrement of an ox and he will recover.

6   If a man, his eyes are sick −
    you will crush anemone (?), you will apply (it) and he will recover.

7   If a man has 'dryness fever' −
    [. . .] ash, *isqūqum* flour, the plant *ammaštakal* [. . .] (and) an old
      brick, you will mix in sesame oil, he will drink and he will recover.

This is really a new category of texts. Of course, it is still about diseases, their
treatment and their prognosis. Like the other texts, each entry begins with
a very rigid grammatical form − 'If a man . . .' − followed by a description of
symptoms or, where the cause is known ('scorpion bite'), by the malady
itself. The succeeding part, however, is new and is addressed directly to the
practitioner. For this reason, it is formulated in the present/future tense and
the second person singular; it tells the expert how to treat the disease, the
nature of the solution of the problem posed. This solution in these medical
texts is generally a series of directions involving the preparation and application of drugs or simples, or as we might say, a 'prescription'. Finally there
is an invariable ending, the prognosis 'and he will recover'.

As before, the text is systematic. The exhaustive nature of the presentation as a representation of generality is illustrated by e.g. entry **3** − 'if his right tooth is sick, you will pour onto his left tooth . . . and if his left tooth is sick, you will pour onto his right tooth' to express what we might say by 'apply to the opposite side to the sick tooth.'

But we shall seek in vain in this text (or in the others of this period) for a systematization by illness. Is this an accurate reflection of the state of the development of the domain of Old Babylonian medicine or a simple affair of the 'luck of the spade'? I personally think it is the latter and that we probably have to do here with selections, extracts, compilations. For in the periods succeeding this, side by side with explicitly titled 'extract' texts, we do indeed find texts with a rigorous, systematic ordering of symptoms and diseases, as witness this (typical) extract from a Neo-Assyrian text (I have suppressed the last prescriptions in the text the better to show the structure of the arrangement):

1 If a man suffers from a 'stroke' on the cheek −
   you will knead (8 medications) in . . . and strong beer, you will boil in a little copper casserole, you will spread on a bandage and you will bandage his mouth.

2 If a man is ill from a 'stroke' in the middle, so that he can no longer walk (. . .) −
   you will dry, crush, filter (4 medications), you will mix (them) in a little copper casserole, you will spread on a bandage and you will bandage and leave in place for three days (. . .)

3 If a man suffers from a 'stroke of death' − (. . . . . .)

4 If a man is ill from a 'stroke' on the side − (. . . . . .)

5 If a man is ill from a 'stroke' on the foot − (. . . . . .)

The introduction of this second kind of text, typified by the last two documents and distinct from those we have discussed earlier, necessitates the introduction of different names to distinguish them. In analogy with what we shall find in mathematics, we shall call the first type of text *table texts*, while the second kind will be known as *procedure texts*. The existence of these two categories and the relations between them is, in a sense, the heart of the argument that, for the Babylonians, there did exist an approach unique to certain privileged domains.

Mathematics

We have several hundred mathematical texts dating from the Old Babylonian period and these divide rather neatly into our two major categories. The first, and by far the most numerous, are the tables; multiplications, inverses, squares, square roots, etc. The procedure texts (and there are about a hundred of them) consist of texts in which a mathematical problem is given and, in many cases, the solution as well. It is from among the procedure texts that we choose our first example.

The tablet we shall be examining is an Old Babylonian text which consisted, when it was complete, of about 24 problems, separated by ruled lines. Here is the first problem:

I added the surface and my side of the square: 45.

You will put 1, the *wāṣitum*.[2] You will split in half 1: (30). You will multiply 30 and 30: (15). You will add 15 to 45: 1. 1 is the square root. You will subtract the 30 that you have multiplied from 1: (30). 30 is the side of the square.

The nature of the problem is clear; find the side of a square for which the sum of the surface and one side is given. (All the problems on the tablet are variations on this theme.) The problem divides rather neatly into two parts, separated as much by their grammar as by their content. The presentation of the problem is expressed in the first person singular and in the past tense ('I added . . .') while the solution is given in the second person singular and the future tense ('You will put . . ., you will split . . .'). This is why we can identify these problem texts as procedure texts.

To see more clearly the structure of this procedure, let us rewrite it in the following schematic form (keep in mind that this is an entirely modern representation):

1 1
2 $\frac{1}{2}(1)$ = 0;30
3 0;30 × 0;30 = 0;15
4 0;15 + 0;45 = 1
5 $\sqrt{1}$ = 1
6 1 − 0;30   = 0;30

There is thus a clear linear order in which each step consists of one operation, using as raw material the results of the preceding operations and the data of the problem; what we today call an algorithm. Thus, starting from 1 (step **1**), we must divide it in half (step **2**), multiply the result by itself (step **3**), add this result to the original datum (step **4**), extract the square root of the result (step **5**), and finally subtract from this result the result obtained in step **2** (step **6**).

This is not the place to enter into a long digression on the technical details of Babylonian mathematics, such as the fact that all numbers were expressed in base 60, explaining the apparently bizarre (but totally correct) values for the results of each operation (the interested reader is referred to p. 61 for a short explanation). I will only say here that the operations themselves were effected by reference to the series of tables. For example, the multiplication of **3** was performed by using a table of multiplication (or squares) and the square root in **4**, though trivial in this particular case, generally made use of a table of square roots. (We will return to these tables a bit later.)

The specificity of mathematics enabled the Babylonians to push the development of mathematical procedures beyond what could be done for medical procedures. The question arises of just how the Babylonian student was expected to bring the knowledge acquired in learning this particular problem to bear when faced with a new problem. It is all very well if he were given a problem of exactly the same nature with only the numbers changed; a sort of substitution is quite simple, but suppose he was to be given the difference rather than the sum of a square and its side. In fact just this variant provides the second problem of our tablet (from now on we divide the text into numbered procedure steps):

> I subtracted my side of the square from the surface: 14 30
>
> **1** You will put 1, the *wāṣitum*
>
> **2** You will split in half 1: (30)
>
> **3** You will multiply 30 and 30: (15)
>
> **4** You will add 15 to 14 30: 14 30 15
>
> **5** 29 30 is the square root
>
> **6′** You will add the 30 that you have multiplied to 29 30: (30)
> 30 is the side of the square

Notice that from step **1** to step **5** the two texts are identical in form (but not, of course, in the numerical values used). Step **6** is replaced by **6′**, a sum rather than a difference. In modern terms we might say that **1–5** form a sub-algorithm, common to both problems, with the different versions of step **6** distinguishing between the problems posed (sum or difference of area and side).

We now turn to the third problem on the tablet:

> I have subtracted a third of the surface and then I have added a third of the side of the square to the surface: 20

1 You will put 1, the *wāṣitum*

2 You will subtract (from it) a third of 1, the *wāṣitum* (i.e.) 20: (40)

3 You will multiply 40 by 20: you will write down 13 20

4 You will split in half 20 . . . : (10)

5 You will multiply 10 and 10: (1 40)

6 You will add 1 40 to 13 20: 15

7 30 is the square root

8 You will subtract the 10 that you have multiplied from 30: 20

9 Its fortieth: (1 30)

10 You will multiply 1 30 by 20: (30)

   30 is the side of the square

The problem here is equivalent to giving the sum of two-thirds of the surface and one-third of a side. The solution procedure consists of a first part (**1–3**) which, in effect, changes the scale of the problem, transforming it into one of the same kind as the first problem. Indeed, steps **4–8** are recognizable as the algorithm for solving a simple sum of the surface and side of a square, identical to the procedure given in that first problem. Having found the solution for this transformed square, the remaining part (**9–10**) is dedicated to a return to the original scale, which gives the answer to the original problem. In short, the aim of this part of the text is to show how the basic procedure (with its two variants), given in the first two problems, is the base from which to solve problems that, on the surface, appear much more complicated. The student has been taught to interpolate by 'embedding' his sub-algorithms with the appropriate changes of scale. This approach is repeated in all the rest of the text, with the problems given ringing the changes on first one, then two, then three squares with appropriate relations among them.

   An indication of the richness of the text can be seen by a simple list of the opening lines of each of the (surviving) problems on the tablet:

   I I added the surface and my side: 45

   II I subtracted my side from the surface: 14 30

   III I subtracted a third of the surface, then I added a third of the side to the surface: 20

   (. . . . . .)

   IX I added the surface of my two squares: 21 40
         One side was greater than the other side by 10

X   I added the surface of my two squares: 21 15
One side was smaller than the other side by a seventh

XI   I added the surface of my two squares: 28 15
One side was greater than the other side by a seventh

XII   I added the surface of my two squares: 21 40
I multiplied my sides: 10

XIII   I added the surface of my two squares: 28 20
One side was a fourth of the other side

XIV   I added the surface of my two squares: 25 25
One side was two-thirds of the other side and 5 ***nindan***[3]

(. . . . . .)

XVI   I subtracted [a third of the side] from the surface: 5

XVII   I added the surface of my three squares: 10 12 45
One side was a seventh of the other side

XVIII   I added the surface of my three squares: 23 20
One side was greater than the other side by 10

XIX   I multiplied my sides, then I added the surface
As much as one side exceeded the other side, I squared it and
I added it [to the surface]: 23 20
I [added] the sides: [50]

(. . . . . .)

XXIII   I added the four sides and the surface: 41 40

XXIV   I added the surface of my three squares: 29 10
One side was two-thirds of a side and 5 ***nindan***
(The other side) was a half of a side and 2 30 ***nindan***

Apart from a few apparent strays, such as **XVI** and **XXIII**, the systematic arrangement of this tablet is apparent at first glance. After a section treating of a single square (**I–VIII**?), there is a second part concerning two squares (**IX–XIV**), then three (**XVII**–).

Of course, not all the mathematical procedure texts are so systematically arranged. Many of them contain but a single problem or, if a multi-problem tablet, the examples range over different subjects. But even in the latter case, the subgroups of problems often show the same patterning as the text we have looked at.

We see here, once again, the difference between the Babylonian and the modern approach. Where we resolve questions – particularly in mathematics – by first creating a general rule and then specializing down to particular cases, the Mesopotamians could do the same thing by constructing their network of typical examples and then interpolating the new problem; either numerically − in the case of new values − or structurally by embedding previously learned algorithms in new procedures.

As was mentioned earlier, the calculations called for by the procedure texts were generally looked up on table texts. These tables were also arranged systematically as the following example of a multiplication table shows:

30	times	1	30
	times	2	1
	times	3	1 30
	times	4	2
	times	5	2 30
	times	6	3
	times	7	3 30
	times	8	4
	times	9	4' 30
	times	10	5
	times	11	5 30
	times	12	6
	times	13	6 30
	times	14	7
	times	15	7 30
	times	16	8
	times	17	8 30
	times	18	9
	times	19	9 30
	times	20	10
	times	30	15
	times	40	20
	times	50	25

Even where the systematization would have been less obvious, for example, in those tables in which are listed the *igigubbû*, the constants associated with different metrological problems, the arrangement of the tablet is by the subject-matter under discussion − surfaces, volumes of piles of clay, bricks, earth, walls, silos, etc. (for an example of this sort, the reader is referred to p. 64).

Thus we see that the world of mathematics in the Old Babylonian period consisted of procedure texts and tables. The first directly interpellated the student (or scribe) and showed him how to follow a given algorithm. Each step of this algorithm made an implicit reference to a table on which he could find the actual numerical value needed for that step of the calculation.

A table of square roots. This tablet carries, in two perfectly aligned columns typical of tables, first the number N then its root: '1: 1 is the root; 4: 2 is the root; 9: 3 is the root; etc.' It is clearly a school tablet. The reverse contains a literary exercise. (CBS 14 233, Old Babylonian epoch. University Museum, Philadelphia.)

The procedure texts themselves were often systematically ordered, this serving in lieu of our general formulæ and rules. In fact, the systematization of both procedure and table texts served as a means to the same end: that of providing a network or grille through which the mathematical world could be seized and understood, at least in an operational sense. Clearly, for such a method, the more complete and exhaustive the network, the more general and efficient the use.

Conclusion

Most societies have a cognitive domain which is privileged by them, by which they order and classify the world about them; this choice is a function of time and of space. Unfortunately for us, the Babylonians left no explicit introduction to their own views of the contours of this domain; self-conscious, reflexive thought did not constitute part of their written culture. Instead we have been forced to reconstruct this by looking first at the social practice and, in more detail, at the formal structure of texts in some of the candidate areas of this domain. In particular, we have seen that in the case of Old Babylonian Mesopotamia this area was structured by two series of texts, identifiable by their sylistic and grammatical particularities and, sometimes, by their spatial organization: procedure texts instructing the user in the means of solving a given problem; tables, to which the first referred, as an index or means of carrying out calculations.

Both types of texts found their efficacy in the creation of a network covering the corresponding part of the world in an 'exhaustive' manner. Any given problem could be situated, in theory at least, on this grille, either directly or by interpolation. The constant expansion and detailing that these series underwent in the course of time are an indication of the understanding and degree of mastery of the power of systematic compilations attained by the Babylonians.

Armed with this analysis, it would be possible to look at other texts. Astrology, as we have already pointed out, is an obvious candidate. But even where the subject is one not usually connected by most Assyriologists with those we have looked at, such as analysis may prove fruitful. It is no coincidence that the 'law codes' that we have from Mesopotamia and, in particular, the very long 'Code' of Ḥammurāpī, have all the markings of a table. The laws have a standard opening; 'If a man . . .', written in the past tense, with a 'prediction' – here a punishment – in the future tense following. The whole 'Code' is divided up into sections by subject: false witness, theft, bodily injury, etc., and each subject is analysed by means of a co-ordinate system whose dimensions are social class, sex, age and gravity of the offence. The direct filiation with the table texts of divination and medicine strikes the eye.

But this category of documents is not infinitely extendible and the vast bulk of cuneiform texts do not offer themselves to this kind of analysis. Neither mythological poetry on the one hand nor economic texts on the other show structuring of this kind.

Ordering and classifying; these, more than the actual content, have been the touchstone in our selection of fields to investigate. In doing so we have tried to avoid words like 'scientific' in order to avoid the temptation to rank these qualities with reference to our own understanding of them. This is not to admit a total relativism in which any culture's structuring of their world is equated to any other. We have not sought to understand a developmental stage of one or another science, nor to weigh its efficiency, limits, validity or achievements. Rather we have attempted to describe and characterize the practices of Babylonian reason, to show the coherence and underline the conclusion that the very exercise of knowledge does not escape the questioning of history. The Babylonians of the second millennium may not have chosen as we choose now – their criteria were not ours – but their choices, like ours, are part of history, theirs *and* ours.

2

Measure for Measure: Mathematics in Egypt and Mesopotamia

James Ritter

Mathematics has a paradoxical reputation among historians of science.

For some it is a river: a flowing stream of gradual, cumulative progress, each individual, generation, civilization adding its rivulet to the larger body of water. There are, it is true, the constraints of the countryside through which the river passes and boulders of ignorance or upheaval can momentarily divert or even block its flow. But soon the inner strength and impulse of the domain has reasserted itself and the great river has rolled on . . .

For others, however, mathematics is the study of great men, brilliant geniuses who, by their unfathomable originality and profundity of thought, have created in their image that which lesser men have been left simply to develop. It is the choices made by these fortunate few which have created the topography of the domain, its peaks of achievement, its neglected lowlands . . .

In either case the origins of the field have little allure as a field of study; for the first group, a tiny rivulet can give little advance information as to the nature of the wide flood with a strong current into which it will develop; for the second, the unknown, unrecorded geniuses who created the field lived too early to leave satisfying evidence of their work. Thus it is that the few historians who work on the earliest traces of mathematics are generally considered by their colleagues to be exotic specimens, content with childish babblings long since surpassed and quite rightly forgotten by both working mathematicians and those who study them.

I should like to argue, of course, that such a view is misleading, that there is something to be learned by a careful study of the beginnings of things; that even mathematics can be illuminated by a view from the source. And such a view, I maintain, will show that the two positions introduced above have both missed an essential point. For there is an internal dynamic to the development of mathematics that, at least partly, generates the kinds of problems and solutions that are the very body of the field (though this is not the only nor, at least for certain periods, even the principal source of such development). And there is a sense in which the choices made by those who do mathematics help to determine the direction the field will take (though again, it is not 'free, individual choice' but historically determined communal techniques which operate here).

All this can best be seen by looking at the earliest direct evidence that we possess in the field of mathematics – that bequeathed by the first literate civilizations to have left us records: Egypt and Mesopotamia.

Reading, Writing and Arithmetic

It has become clear in recent years that mathematics and writing have a close, symbiotic relationship. Born at the same time, their destinies have always been closely linked, even if the latter has to a large extent liberated itself from the constraints of the former. We are so used to seeing modern writing systems as reflections of spoken language that it is salutary to recognize that in the beginning this was not at all the case.

For a society to develop a mathematics that goes beyond simple counting, a material support of some kind is essential. Without writing, the limitations of human memory are such that only a limited degree of numerical sophistication can be achieved. This is a point which has been known for some time. What is new, however, is that we now know that the inverse is equally true. That is, for a society to develop writing, material needs and, in particular, the need of record-keeping are central. This point has only been understood since the archeological discoveries of the last few decades have allowed us to follow the development, virtually from scratch, of two writing systems, one used in southern Mesopotamia towards the middle of the fourth millennium, the other in the area around Susa in Iran, only slightly later.

In both societies the material support is clay, virtually indestructible, and the first documents are accounts. It was thus the need to measure, divide and distribute their societies' material wealth that gave birth to the first writing systems. In particular, the first, called by us 'cuneiform', was to know

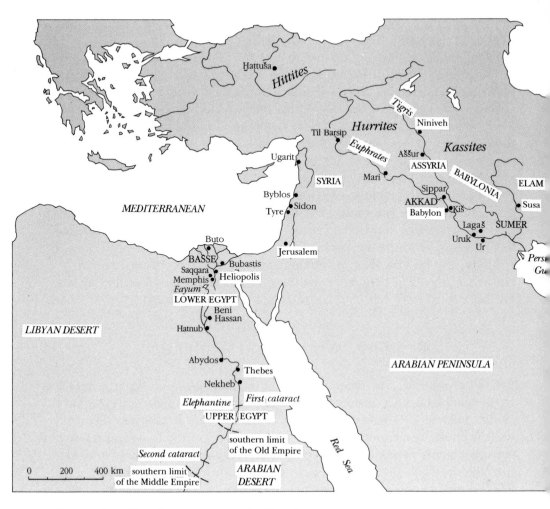

Egypt and the Near East in the second millennium BC

a great success over the succeeding three thousand years. Used to write not only the original Sumerian, but also Akkadian, Hittite, Elamite, Hurrian and many other languages of the Ancient Near East, it died out only at the beginning of our era.

In Egypt, where an independent civilization was in rapid growth towards the end of the fourth millennium, the situation concerning writing is less clear. First of all, the material support – for other than monumental inscriptions – was principally papyrus and, to a lesser extent, other perishable materials. Furthermore, the history of archeological excavation is quite different. The fact that the Nile Valley is narrow and heavily populated has meant that the vast majority of digs have been situated in desert areas, far from the mud-brick settlements, and have concentrated almost exclusively on the stone-built cemeteries and temples. Egypt has thus yielded fewer documents than Mesopotamia by a factor of several orders of magnitude. But what evidence remains for the nature of the everyday use of writing shows that here too the early existence and importance of economic and accounting texts is clear. The early pictographs, conserved in form when carved on stone and called, since the Greeks, 'hieroglyphic', became the cursive 'hieratic' script used almost exclusively on all current documents.

It is, perhaps, not unnecessary to point out that both the Egyptian and Akkadian cuneiform scripts, in which the documents we shall study here are written, are basically phonetic writing systems. In the former each sign represents one or several consonants (the vowels were not written in Ancient Egyptian, as they are not in modern Arabic or Hebrew), in the latter each sign is a syllable.

By the end of the third millennium these scripts had achieved something like a stable form. The signs and their combinations, the formation of words, numbers, etc., were taught in schools to a population of children strongly focused by class and sex, reserved almost exclusively to ruling-class boys. We possess, from both civilizations, examples of school exercises and pedagogic texts of various kinds (see p. 26). Mathematical texts and exercises are among them. We know that the learning of arithmetic began early in the schoolchild's career, along with reading and writing, and that mathematics, then as now, was considered one of the 'hardest' subjects.

Around 2000 BC, Šulgi, one of the kings of the Ur III Empire in Mesopotamia, was the subject of a literary hymn which became a model text used as a school exercise during the first half of the second millennium. In this document he boasts of his academic achievements and in particular:

> When I was a child (at) school,
> On the tablets of Sumer and Akkad I learned the scribal art.
> Among the young, no one could write a tablet like me.
> . . . . . .
> I am perfectly able to subtract and add, (clever in) counting and
>   accounting.
> The fair Nanibgal, (the goddess) Nisaba
> Has lavishly provided me with wisdom and intelligence.

Over a thousand years later the Assyrian king Assurbanipal was to repeat much the same thing in one of his hymns (see p. 18). We find the same

47

sentiments expressed in Egyptian school texts. The following, a 'scribal dispute' – a literary and pedagogic genre as popular in Mesopotamia as in Egypt – has one scribe taunting another with:

> You come here and fill me up with your office. I will show up your boastful behaviour when a mission is given to you. I will show up your arrogance when you say: 'I am the scribe, commander of the work gang.'
>
> You are ordered to dig a lake and you come to me to inquire about the troop rations. You say to me: 'Calculate them', abandoning your job; instructing you has fallen on my shoulders. (. . . . . .)
>
> I will explain to you the command of Pharaoh, though you are his royal scribe. You will be conducted under the palace balcony because of your brilliant achievements, when the mountains have disgorged great monuments for the king, lord of the Two Lands. For you are an experienced scribe, at the head of the work gang.
>
> A ramp, 730 cubits (long) and 55 cubits wide, must be built, with 120 compartments filled with reeds and beams; with a height of 60 cubits at its peak, 30 cubits in the middle; with a slope of 15 cubits; with a base of 5 cubits. The quantity of bricks is demanded of the troop commander.
>
> The scribes are all assembled but no one knows how to do it. They put their faith in you and say: 'You are a clever scribe, my friend! Decide quickly for us, for your name is famous.' (. . . . . .) Let it not be said: 'There is something he does not know.'

By the early second millennium the two cultures each had a sophisticated and efficient mathematics in place, which could be and was applied to the problems of society. Since both cultures at this time were highly centralized bureaucratic states, with an intensive, irrigation-based agriculture and a developed inland and foreign trade, one might expect that the problems treated would be similar. Indeed, the mathematics of both cultures are generally treated as 'practical' or 'empirical' by almost all historians of mathematics, and to a certain extent this is true.

However, the assumption behind this remark calls for a more detailed analysis. The historians of mathematics who make this claim, and they include members of both the schools mentioned at the start, also assume that there is just one mathematics, everywhere the same and only more or less developed. Thus, comparisons between Egyptian and Babylonian mathematics generally restrict themselves to the question 'Who was better?' But a closer look at the mathematics during, say, the first half of the second millennium in the two civilizations will show that there were in place two mathematics, so that even where the 'same' problem is treated the methods used are quite different. And most important, this difference in method has a determining effect on the development of the two mathematics. This, in turn, says something quite important about the very nature of mathematics.

The Two Granaries

We now turn to our first two texts; both date from the second millennium BC.

TEXT 1

Egyptian Text	**Babylonian Text**

Egyptian Text

Example of making a round silo of 9 (and of) 10.
You will subtract $\frac{1}{9}$ of 9: 1. Remainder: 8. Multiply 8 by 8: it becomes 64. You will multiply 64 by 10: it becomes 640. Add it to its half: it becomes 960. Its quantity in *ḥar*. You will take $\frac{1}{20}$ of 960: 48. The amount in 100-quadruple-*ḥeqat*: wheat: 48 *ḥeqat*.

Form of its procedure:

1	8	1	64	1/10	96
2	16	\10	640	\1/20	48
4	32				
\8	64	\1/2	320		
		Total 960			

Babylonian Text

The procedure for a 'log'.
5, a cubit, was its diameter. How much is it in grain measure?
In your proceeding: as much as the diameter put the depth. Convert 5: that amounts to 1. Triple 5, the diameter: that amounts to 15. 15 is the diameter of the 'log'. Square 15: that amounts to 3 45. Multiply 3 45 by 5, the *igigubbûm* of the circle: that amounts to '18 45 as the surface'. Multiply 18 45 by 1, the depth: that amounts to '18 45 as the volume'. Multiply 18 45 by 6, (the *igigubbûm* of) the grain measure: that amounts to 1 52 30. The 'log' contains 1 *pānum*, 5 *sūtum*, 2 1/2 *qûm* of grain.
That is the procedure.

Before any detailed analysis let us make a few remarks with respect to the form rather than the contents of the texts.[1]

One thing strikes the eye immediately: how unfamiliar these texts appear to us. There is none of that mathematical symbolism – equations, formulae, etc. – that characterizes modern mathematics. Instead there are only words and numbers – and, in the Egyptian case, marks '\'.

But in the course of reading we find ourselves in a more familiar universe: there is a clear problem presented – the calculation of the capacity of a cylindrical silo of given dimensions – though the metrological units may be difficult to identify. This presentation of the problem is followed by a solution, given step by step with the answer at the end.

The formal structure of the two texts, Egyptian and Babylonian, presents a certain number of traits in common, traits that are to be found in mathematical texts from ancient China and India as well. This structure is characterized by a presentation which is:

- *rhetorical* – the problems are expressed in (often everyday) words and not in symbols;
- *numeric* – data and results are concrete numbers and not abstract quantities;
- *algorithmic* – resolved by a specific series of steps without general proofs.

We can go further in our study of these formal similarities: the solution or algorithm of each problem is written in the second person (the student scribe is personally interpellated) and is in either the future tense or the imperative (it proposes a sequence of steps which, if carried out, will then furnish a correct result).

The context in which these texts are found permits us to see other points of resemblance. Each is extracted from a collection of problems written on papyrus or clay. These collections were surely used as school textbooks – almost all our mathematical sources, both Egyptian and Babylonian, come from such a school context. If there were ever anything in the Ancient Near East analogous to our research results or scientific communications they have disappeared without trace, not even a casual mention in other kinds of texts.

Having now noted the formal and contextual similarities between our two texts, we turn to the (substantial) differences which distinguish them.

The Egyptian Problem

The problem is introduced as an 'example' – a typical case. We are presented with a 'round' (i.e. circular in cross-section) granary whose dimensions are characterized as being '9 and 10'. As is typical in ancient mathematics, the metrological units used in a problem are only mentioned explicitly if these differ from the base unit. For linear measures in Egypt this is the *cubit* (about 52 cm) and this is the measure used here. We have therefore a cylinder of 9 cubits in cross-section and 10 cubits in height. The problem (though not stated explicitly, it is clear from the working) is to find the volume, not in 'cubic cubits', a measure that did not exist save as an intermediate step in the solution of a problem, but in 'practical' volume units of grain capacity. The algorithm used will be clearer for us if we represent the successive steps in the form of a lightly symbolized sequence as follows:[2]

	CALCULATION	OPERATION	
		CALCULATION OF THE VOLUME	
1	$\frac{1}{9} \times 9 = 1$	Multiplication	
2	$9 - 1 = 8$	Subtraction	
3	$8 \times 8 = 64$	Multiplication	area of the base
4	$64 \times 10 = 640$	Multiplication	volume in cubits cubed
		CONVERSION TO *HAR*	
[5	$\frac{1}{2} \times 640 = 320$	Multiplication]	
6	$640 + 320 = 960$	Addition	volume in *har*
		CONVERSION TO *ḤEQAT*	
7	$\frac{1}{20} \times 960 = 48$	Multiplication	volume in 100-quadruple-*ḥeqat*

What we have transcribed above is of course only the first part of the problem. The six lines of principally numerical calculation which follow,

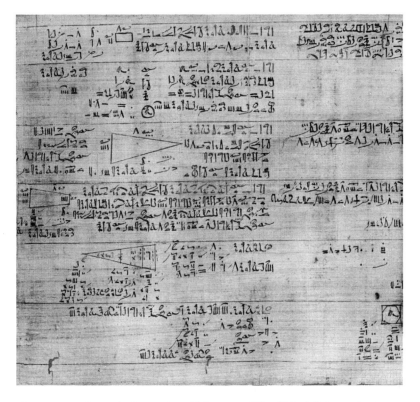

Egyptian mathematical exercises. This page of the Rhind Papyrus proposes a series of problems, each one with an algorithm for its solution. The exercises here concern calculations of the surfaces of different shapes. (For other problems from this Papyrus see pp. 49 and 55.) BM 10 057 and 10 058: copy dating from the second intermediate period, after an original from the Middle Kingdom, Thebes. (British Museum, London. Photograph Eileen Tweedy.)

headed by 'Form of its procedure', constitute the working of the problem and is something to which we shall return shortly.

The resolution of the problem consists then of seven well-defined steps, of which the fifth, though not explicitly stated in the rhetorical part of the algorithm, is understood; to 'add its half' it is clearly necessary to have calculated it first. Indeed this step is present in the numerical working that follows.

But where do all these numbers come from? There are exactly three sources for them.

First of all, the *data* of the problem: 9, the diameter of the cylinder, is used twice (**1** and **2**); 10, the height, once (**4**). The vast bulk of the other numbers are the *results* of the steps of the algorithm; for example, the 1 which is the result of **1** is used in **2**; the 8 which is the result of **2** is used in **3**, and so forth; in short, the result of each step is (normally) used in the step which immediately follows. Finally there remain a small number of numbers which come from neither of the above sources. In our example these are the $\frac{1}{9}$ of **1**, the $\frac{1}{2}$ of [**5**] and the $\frac{1}{20}$ of **7**. They are the *constants* of Egyptian mathematical theory, fixed numbers that must be learned (or looked up) and used each time a particular problem is to be solved. For example, $\frac{1}{9}$ is the fraction by which the diameter of a circle is to be multiplied as the first step in the algorithm for the determination of its area (our **1–3**). Similarly, $\frac{1}{2}$ and $\frac{1}{20}$ are constants used in the conversion of 'cubic cubits' to *ḫar* units (**5–6**) and *ḫar* to 100-quadruple-*ḥeqat* units (**7**) respectively.

We will sum up what we have just said by rewriting once again our algorithm in a yet more abstract form. In this version, the two initial data will be represented by $\mathbf{D_1}$ (9, the diameter) and $\mathbf{D_2}$ (10, the height, the result of step number **N** by **N**, and the constants by their numerical value.

STEP NUMBER	CALCULATION
1	$\frac{1}{9} \times \mathbf{D_1}$
2	$\mathbf{D_1} - \mathbf{1}$
3	$\mathbf{2} \times \mathbf{2}$
4	$\mathbf{3} \times \mathbf{D_2}$
5	$\frac{1}{2} \times \mathbf{4}$
6	$\mathbf{4} + \mathbf{5}$
7	$\frac{1}{20} \times \mathbf{6}$

When we turn to the series of numerical operations headed by the expression 'Form of its procedure' we are, in fact, looking at the calculational methods used by the Egyptians to actually perform the operations demanded by the foregoing algorithm. All the steps are detailed there save **2** and **6** which call for a subtraction and an addition respectively. In general additions and subtractions are seldom detailed; their results are simply given. However the other three major operations used in the Rhind Papyrus – multiplication, division (not used in our problem) and 'inversion' – are generally written out. Let us see how these work.

Taking as our first example the multiplication (8×8) indicated in step **3**, we find the calculation in the leftmost column under 'Form of its procedure'. Like all Egyptian calculations, the work is disposed in a double column, that on the left starting invariably at 1, that on the right with the value with which one is going to operate, here 8.

3

CALCULATION		TECHNIQUE
1	8	initialization
2	16	doubling
4	32	doubling
\8	64	doubling

The operation being the multiplication of 8 by 8, and having initialized with 8, the object is to arrive at 8 in the column on the left. This is done simply by successive doubling of both columns, three such doublings being sufficient to arrive at the desired number – 8 – on the left. A mark is placed against this entry, the answer – 64 – being read off the column on the right.

Of course this case is particularly simple. Were the problem to multiply 8 by 12, say, the scribe would have marked off, in the left-hand column, the entry containing 4 and the entry containing 8, the sum of their corresponding right-hand column values being 32 + 64 = 96. Indeed it is not hard to show that if N and M are any two integers whatsoever, this technique will give their product; simply initialize with respect to either of them and then double a sufficient number of times.

We are so accustomed to conceive of an operation as a single step that it is necessary to mark clearly the distinction between operations and what I shall call techniques. The Egyptian operations, which include our familiar four and others besides, have special formulations in Egyptian; e.g. multiplication of N by M is termed 'Calculate, starting from N, M times' (N and M being each time, of course, concrete numbers). On the other hand, the way in which any given multiplication is actually carried out depends absolutely on the choice of numbers to multiply together. For integers, as in the previous case, the method is always by doubling. But if the multiplicand is not an integer other methods must be employed. Take, for example, step **5**. Here one must multiply 640 by $\frac{1}{2}$. The working is explicitly shown in the text and looks like this:

5

CALCULATION		TECHNIQUE
[1	640	initialization]
$\frac{1}{2}$	320	halving

The technique introduced here is that of halving, the reciprocal of the doubling we have seen before. This technique can, of course, be continued to give $\frac{1}{4}$, $\frac{1}{8}$, $\frac{1}{16}$, etc. of the initial number. However, unlike the case of multiplication of integers by doubling, it is *not* possible to obtain any fraction by such a method; $\frac{1}{7}$, for example, cannot be decomposed into fractions of this kind.

Other techniques are represented in this problem. The multiplication of step **4** (64 times 10), for example, is carried out in the following fashion:

4

CALCULATION		TECHNIQUE
1	64	initialization
10	640	decupling

Of course, this result could have been obtained by a series of doublings (8 + 2 = 10), but the short cut produced by multiplying by 10 (decupling) is clearly useful and simple in a decimal number system such as the Egyptian – and our own. The inverse operation, dedecupling, is also used in our problem:

7

CALCULATION		TECHNIQUE
[1	960	initialization]
$\frac{1}{10}$	96	dedecupling
$\frac{1}{20}$	48	halving

Here we see how a combination of techniques, dedecupling and halving for example, allows the scribe to reach fractions such as $\frac{1}{20}$ which are out of reach of a simple halving system. And of course other techniques are used for the other steps in the algorithm, even though, for this problem, they are not detailed explicitly in the 'Form of the procedure'.

We may sum up by introducing a certain vocabulary. Each step of an algorithm corresponds to what we have called an operation. Thus, in each of the examples above, a single operation (multiplication) corresponds to a single line of the algorithm. However when detailed in the 'procedure' part of the text, this operation is seen to require different techniques. Thus we have seen initialization, doubling, halving, decupling, dedecupling, etc. We shall see yet others in the course of this chapter.

These techniques form the heart of the Egyptian mathematical system. They permit the scribe to carry out all the basic arithmetic operations necessary for the solution of numerical problems. He must choose, of course, in each case, the set of techniques appropriate to his particular numerical values. The numbers used in the problem, and especially the 9-cubit diameter makes the calculations straightforward; the pedagogic point, the learning of the algorithm, is achieved with a minimum of calculational difficulty.

The Second Silo

But the picture changes when we look at the very next problem in this same Rhind Papyrus. (Once again we have restored the initialization lines in the working left out for economy by the scribe.)

A round silo of 10 (and of) 10.

You will subtract $\frac{1}{9}$ of 10: $1\frac{1}{9}$. Remainder: $8\ \frac{2}{3}\ \frac{1}{6}\ \frac{1}{18}$. You will multiply $8\ \frac{2}{3}\ \frac{1}{6}\ \frac{1}{18}$ by $8\ \frac{2}{3}\ \frac{1}{6}\ \frac{1}{18}$; it becomes $79\ \frac{1}{108}\ \frac{1}{324}$. You will multiply $79\ \frac{1}{108}\ \frac{1}{324}$ by 10; it becomes $790\ \frac{1}{18}\ \frac{1}{27}\ \frac{1}{54}$. Add its half; it becomes 1185. Multiply 1185 by $\frac{1}{20}$: $59\ \frac{1}{4}$. This is what goes into it in 100-quadruple-$\underline{h}$eqat; wheat: $59\ \frac{1}{4}$ 100-quadruple-$\underline{h}$eqat.

Form of its procedure:

1	8	$\frac{2}{3}$	$\frac{1}{6}$	$\frac{1}{18}$		
2	17	$\frac{2}{3}$	$\frac{1}{9}$			
4	35	$\frac{1}{2}$	$\frac{1}{18}$			
\8	71	$\frac{1}{9}$				
\$\frac{2}{3}$	5	$\frac{2}{3}$	$\frac{1}{6}$	$\frac{1}{18}$	$\frac{1}{27}$	
$\frac{1}{3}$	2	$\frac{2}{3}$	$\frac{1}{6}$	$\frac{1}{12}$	$\frac{1}{36}$	$\frac{1}{54}$

\ $\frac{1}{6}$	1	$\frac{1}{3}$	$\frac{1}{12}$	$\frac{1}{24}$	$\frac{1}{72}$	$\frac{1}{108}$
\ $\frac{1}{18}$		$\frac{1}{3}$	$\frac{1}{9}$	$\frac{1}{27}$	$\frac{1}{108}$	$\frac{1}{324}$

Total	79	$\frac{1}{108}$	$\frac{1}{324}$	
1	79	$\frac{1}{108}$	$\frac{1}{324}$	
\10	790	$\frac{1}{18}$	$\frac{1}{27}$	$\frac{1}{54}$
[1	790	$\frac{1}{18}$	$\frac{1}{27}$	$\frac{1}{54}$]
$\frac{1}{2}$	395	$\frac{1}{36}$	$\frac{1}{54}$	$\frac{1}{108}$

Total	1185	
[1	1185]	
$\frac{1}{10}$	118	$\frac{1}{2}$
\ $\frac{1}{20}$	59	$\frac{1}{4}$

The problem is virtually the same as before; to calculate the volume (in grain units) of a cylindrical granary. Only the diameter has been changed; from 9 (cubits) to 10 (cubits). But what a difference on the calculational level! The procedure is the same as before: subtract $\frac{1}{9}$ of the diameter from itself, square the result, multiply this by the height and convert to *ẖar* and then to *ḥeqat* measure. Given in the schematic form we have introduced before, the algorithm looks as follows:

CALCULATION	*OPERATION* CALCULATION OF VOLUME	
1 $\frac{1}{9} \times 10 = 1\ \frac{1}{9}$	Multiplication	
2 $10 - 1\ \frac{1}{9} = 8\ \frac{2}{3}\ \frac{1}{6}\ \frac{1}{18}$	Subtraction	
3 $8\ \frac{2}{3}\ \frac{1}{6}\ \frac{1}{18} \times 8\ \frac{2}{3}\ \frac{1}{6}\ \frac{1}{18} = 79\ \frac{1}{108}\ \frac{1}{324}$	Multiplication	area of base in cubits squared

4 $79 \frac{1}{108} \frac{1}{324} \times 10$ Multiplication   volume in cubits cubed

= $790 \frac{1}{18} \frac{1}{27} \frac{1}{54}$ (*sic!*)

CONVERSION TO *ḤAR*

5 $\frac{1}{2} \times 790 \frac{1}{18} \frac{1}{27} \frac{1}{54}$ Multiplication

= $395 \frac{1}{36} \frac{1}{54} \frac{1}{108}$

6 $790 \frac{1}{18} \frac{1}{27} \frac{1}{54}$ Addition   volume in *ḫar*

+ $395 \frac{1}{36} \frac{1}{54} \frac{1}{108}$ = 1185

CONVERSION TO *ḤEQAT*

7 $\frac{1}{20} \times 1185 = 59 \frac{1}{4}$ Multiplication   volume in 100-quadruple *ḫeqat*

The second step of the algorithm, subtracting $1 \frac{1}{9}$ from 10, introduces us to the heart of Egyptian mathematical techniques: the writing and manipulation of fractions. For the Egyptians – and with the exception of $\frac{2}{3}$ – there are essentially no fractions other than 'unit fractions'. Where we would say $10 - 1 \frac{1}{9} = 8 \frac{8}{9}$, the scribe must write $8 + \frac{2}{3} + \frac{1}{6} + \frac{1}{18}$, written not as a sum but in simple juxtaposition. Of course the decomposition of $\frac{8}{9}$ into these 3 fractions is not unique but certain choices of these sums were apparently

---

Egyptian Fractions

There were not, in Egypt – nor indeed in Mesopotamia – general fractions in our sense of the word. The only forms of fraction recognized were what we today call *unit fractions*, that is fractions that we write $1/N$ plus a special symbol used to represent our $\frac{2}{3}$. The results of any calculation, intermediate and final, had to be expressed using these basic building blocks.

Egyptian notation for a fraction therefore centres on that number N which we call 'denominator'; this means that relations which may be 'obvious' for us, such as that $\frac{2}{3} \times 3 = 2$ were less so for the Egyptians.

---

made quite early in Egyptian history and then adhered to quite strictly. (One choice that was never made by the Egyptians was to write $\frac{8}{9}$ as a string of eight $\frac{1}{9}$ths. Unit fractions were never repeated within a number.) No working is shown and the technique by which this subtraction was accomplished must be reconstructed by us by means of knowledge gained elsewhere in the mathematical papyri.

To see what a difference is made by the need to deal with fractions, we turn to step **3**, the squaring of $8 \ \frac{2}{3} \ \frac{1}{6} \ \frac{1}{18}$, for which the working is given in some detail in the first block of calculations after the algorithm. This working (with a missing step restored) shows the following structure:

CALCULATION							TECHNIQUE
1	8	$\frac{2}{3}$	$\frac{1}{6}$	$\frac{1}{18}$			initialization
2	17	$\frac{2}{3}$	$\frac{1}{9}$				doubling
4	35	$\frac{1}{2}$	$\frac{1}{18}$				doubling
8	71	$\frac{1}{9}$					doubling
$\frac{2}{3}$	5	$\frac{2}{3}$	$\frac{1}{6}$	$\frac{1}{18}$	$\frac{1}{27}$		two-thirds
$\frac{1}{3}$	2	$\frac{2}{3}$	$\frac{1}{6}$	$\frac{1}{12}$	$\frac{1}{36}$	$\frac{1}{54}$	halving
$\frac{1}{6}$	1	$\frac{1}{3}$	$\frac{1}{12}$	$\frac{1}{24}$	$\frac{1}{72}$	$\frac{1}{108}$	halving
$[\frac{1}{9}$		$\frac{2}{3}$	$\frac{1}{6}$	$\frac{1}{9}$	$\frac{1}{27}$	$\frac{1}{162}$	two-thirds]
$\frac{1}{18}$		$\frac{1}{3}$	$\frac{1}{9}$	$\frac{1}{27}$	$\frac{1}{108}$	$\frac{1}{324}$	halving
Total	79	$\frac{1}{108}$	$\frac{1}{324}$				sum

The operation is, of course, multiplication. We recognize many of the techniques we have already seen. Halving, for example, of a 'unit fraction' is simple since it suffices to multiply the 'denominator' by 2. However the doubling of a fraction can cause problems, since the technique is most directly effected by a division of the 'denominator' by 2 and this is only possible when that number is *even*. How then did the scribe effect line 3 for example? That twice $\frac{2}{3}$ is $1 \ \frac{1}{3}$ can be easily expressed. Similarly, twice $\frac{1}{6}$

58

is simply $\frac{1}{3}$. But what to do with twice $\frac{1}{9}$? How to write this in the only way acceptable to an Egyptian, as the sum of 'unit fractions'? The solution adopted in Egypt (and as we shall see later also in Babylonia) is to construct tables for the difficult bits of their mathematics; that is a tabular array of results organized for easy reference. Now this same Rhind Papyrus possesses just such a table, occupying practically all the recto of the papyrus, giving the double of 'odd fractions' from $\frac{1}{5}$ to $\frac{1}{101}$. And we find a fragment of the same table among the Middle Kingdom papyri found at el-Lahun. A (simplified) extract is given here of this second document:

2 (/)			
3	2/3		
5	1/3	1/15	
7	1/4	1/28	
9	1/6	1/18	
11	1/6	1/66	
13	1/8	1/52	1/104

We shall not enter here into the long-debated problem of how the Egyptians developed this table, only that it was copied and recopied during all the Middle Kingdom at least. The scribe would simply have referred to the entry in the table for twice $\frac{1}{9}$ which is given as $\frac{1}{6}$ $\frac{1}{18}$. So that the double of line 2 would now read 35 $\frac{1}{3}$ $\frac{1}{6}$ $\frac{1}{18}$. But not content with this, the scribe knows (from, of course, yet another table) that $\frac{1}{3}$ $\frac{1}{6}$ is the same as $\frac{1}{2}$. Thus simplified, we now have line 3.

In line 5 we have a new technique, the two-thirds of a number. This is a single-step technique and is never shown broken down into simpler steps. The calculation of $\frac{2}{3}$ of an integer does not appear to have caused any difficulty for the Egyptian scribe. But the problem of taking $\frac{2}{3}$ of a fraction is more delicate. As an example, let us look at the determination of line 5. We must take $\frac{2}{3}$ of all the numbers in line 1: 8 $\frac{2}{3}$ $\frac{1}{6}$ $\frac{1}{18}$ (with of course the answers to be expressed as sums of 'unit fractions'). Two-thirds of 8 is not too difficult to calculate: 5 $\frac{1}{3}$. But what to do with the fractions? The answer, of course, is yet another table:

2/3 of 2/3:	1/3	1/9
1/3 of 2/3:	1/6	1/18

2/3 of 1/3:		1/6	1/18		
2/3 of 1/6:		1/12	1/36		
2/3 of its 1/2:		1/3			
1/3 of its 1/2:		1/6			
1/6 of its 1/2:		1/12			
1/12 of its 1/2:		1/24			
1/9	its 2/3:	1/18	[1/54]		
[....................]					
1/7	its 2/3:	1/14	1/42		
1/9	its 1/2:	1/14			
1/11	(its) 2/3:	1/22	1/66	its 1/3:	1/33
1/11	its 1/2:	1/22		its 1/4:	1/44

The same procedure was followed in the other lines to find $\frac{1}{6} = \frac{1}{2}$ of $\frac{1}{3}$ (line 7), [$\frac{1}{9} = \frac{2}{3}$ of $\frac{1}{9}$ (line 8)] and $\frac{1}{18} = \frac{1}{2}$ of $\frac{1}{9}$ (line 9) of the original number before assembling all the results marked with a \ for the final calculation.

In short, whenever the techniques used by the Egyptians necessitated arduous calculations, the scribes had recourse to already established tables. But it must not be thought that thereby all problems were resolved. Looking at the next working shown, the decupling of the square of $8 \frac{1}{6} \frac{1}{18}$, shows the difficulty in manipulating complex fractions, even with tables; the scribe has simply left out the final fraction $\frac{1}{81}$!

We leave the Nile valley now and turn to the question of how contemporaries in Mesopotamia dealt with their version of the problem of the granaries.

The Babylonian Problem

We will follow the same method of analysis for the Babylonian text as we did for the Egyptian. But before we look at the algorithm, a word about the metrology. Note that the diameter of the Babylonian cylinder is given as '5, (that is) one cubit'. We expect then, correctly, that since the '5' has no unit attached, it must be being expressed in the basic linear unit of the Mesopotamian system. This is, in fact, the **nindan**, which equals 12 cubits. But then why does the text seem to state that 5 **nindan** make up 1 cubit? This introduces us to the first of many differences between the mathematics of our two cultures. Whereas the Egyptians, like ourselves, used a base 10 arithmetic, the Babylonians, in their mathematical texts (and less commonly in their economic ones), employed a base of 60. Here the '5' is to be read $0;5 = \frac{5}{60} = \frac{1}{12}$ of a **nindan**, i.e. one cubit.

The Mesopotamian sexagesimal system

The choice of 'sixty' rather than 'ten' as the base of the Mesopotamian number system goes back to the Sumerian metrological systems of the fourth and third millennia. It is not difficult to see how to read these numbers.

Let us look first at our own decimal place system. We use the digits 1, 2, 3, ... , 9 plus 0 and the value of a digit depends on where in the number it is to be found. Each place in a number represents a particular power of 10. For example the number 642 can be analysed as:

$$6 \times 100 + 4 \times 10 + 2 \times 1 = 6 \times 10^2 + 4 \times 10^1 + 2 \times 10^0.$$

Now in the Mesopotamian sexagesimal place system there are 59 digits (but no zero before the third century BC) and each place in a number represents a power of 60. The number 6 4 2 in this system is to be understood as:

$$6 \times 60^2 + 4 \times 60^1 + 2 \times 60^0 = 6 \times 3600 + 4 \times 60 + 2 \times 1$$
$$= 21600 + 240 + 2 = 21842 \text{ in our system}$$

Of course there are 59 digits for the Babylonians and one commonly sees numbers such as 32 26 (a two-place number!) which represents $32 \times 60 + 26 \times 1 = 1946$ in our system. Modern Assyriologists write these numbers with the places separated by a blank or a point to avoid confusion with our decimal numbers.

Note that this sexagesimal representation is exactly that which we use for our time measures:

1 hour 4 minutes 23 seconds $= 1 \times 3600 + 4 \times 60 + 23 \times 1$
$= 6023$ seconds.

Our decimal fractions are written with the 'decimal point' preceding negative powers of ten:

$$3.54 = 3 \times 10^0 + 5 \times 10^{-1} + 4 \times 10^{-2} = 3 + \frac{5}{10} + \frac{4}{100}.$$

In the same way, sexagesimal fractions can be written as negative powers of 60; in the modern Assyriological convention, the 'decimal point' is replaced by the 'sexagesimal semicolon':

$$0;30 = 0 \times 10^0 + 30 \times 60^{-1} = \frac{30}{60} = \frac{1}{2}$$

or another example

$$0;5 = 0 \times 10^0 + 5 \times 60^{-1} = \frac{5}{60} = \frac{1}{12}.$$

There also existed a Babylonian unit fraction representation rather analogous to the Egyptian case (see p. 58), though it was less frequently used in mathematical texts.

We can now rewrite the algorithm in modern form:

CALCULATION		OPERATION	
	CALCULATION OF VOLUME		
0	height = diameter	Convention	
1	0;5 × 12 = 1	Multiplication	**nindan** → cubits
2	3 × 0;5 = 0;15	Multiplication	circumference
3	0;15² = 0;3.45	Square	
4	0;3.45 × 0;5 = 0;0.18.45	Multiplication	area of base in **nindan**²
5	0;0.18.45 × 1 = 0;0.18.45	Multiplication	volume in *mūšarum*
	CONVERSION TO GRAIN UNITS		
6	0;0.18.45 × 6.0.0 = 1.52;30	Multiplication	volume in *pānum, sūtum, qûm*

Steps **0** and **1** of the algorithm determine the height of the cylinder. **0** says that the height is to be taken equal to the diameter; **1** that this height, given in **nindan** since the diameter is so given, must be converted into cubits. Here we have a second point of difference with regard to Egypt. In the Nile Valley, volumes are first calculated (as in our modern systems) in homogeneous units (1 cubit × 1 cubit × 1 cubit) and then, if necessary, converted into grain measure. However, the Babylonians have a larger base unit – the **nindan**; though length and breadth are conveniently measured in this measure, the height or depth of a structure seldom needs such a large unit. Instead the basic unit of volume – the *mūšarum* – has the dimensions 1 **nindan** × 1 **nindan** × 1 cubit, hence the need for the conversion of the height of the cylinder into cubits.

The actual calculation of the volume is the subject of steps **2–5**. As in the Egyptian case, first the area of the base is calculated (in **nindan**²) and then multiplied by the height (in cubits) to give the volume (in *mūšarum*). The Babylonian calculation of the area of the base is not done directly but passes first by a calculation of the circumference (**2**), followed by a squaring of this circumference (**4**) and its multiplication by the constant 0;5 (**5**). Finally the conversion to grain units is effected in **6**, with the multiplication by 6.0.0, yielding a number (1.52;30) from which the standard grain units can be read off directly; the number of *pānum* is given by the figure in the sixties place (= 1), the number of *sūtum* by the number of tens (= 5) and the *qûm* by the number of units (= 2;30 = $2 \frac{30}{60}$ = $2 \frac{1}{2}$).[3]

During this process we have seen once again the three classes of numbers that are already familiar to us from the Egyptian calculation: data, results (of the calculation in each step) and constants. Unlike the Egyptians, the Babylonians had a special name for this last group; they are the *igigubbû* (the plural of *igigubbûm*) which are mentioned as being 'of the circle' for the 0;5 of **4** and 'of grain' for the 6.0.0 of **6**. Note, however, that the 12 of **1** and the 3 of **2**, which, though constants for us, were never called *igigubbû* by the Babylonians. Indeed they are not given as numbers in the text; instead the student is invited to 'convert' **nindan** into cubits and to 'triple' the diameter of a circle to find its circumference. That is, they are verbs rather than numbers! Evidently they were seen by the Babylonians as operations in themselves, like squaring, and not as objects having the same status as the *igigubbû*. (Incidentally, this is yet one more reason why it is anachronistic to speak of the Babylonian – or Egyptian – 'value of pi'; see p. 168.)

Using the same notation as in the Egyptian case with the addition of the operations (3) for tripling and → for conversion of **nindan** to cubits, we may rewrite the Babylonian algorithm in the following 'abstract' form:

0	D_1	=	D_2
1	→	D_2	
2	(3)	1	
3		2^2	
4	3	×	0;5
5	4	×	D_2
6	5	×	6.0.0

The only operations used in this algorithm, aside from the tripling and conversion mentioned above, are multiplication (**4–6**) and squaring (**3**). This is quite similar to the Egyptian example (though an Egyptian operation of squaring exists, it is often replaced – as in the texts we have looked at – by an explicit command to multiply N by N). But another important difference between the two civilizations reveals itself when we come to ask how the Babylonians arrived at their numerical results. For there is nothing in the Babylonian text which corresponds to the Egyptian calculations at the end of the algorithm, no 'form of its procedure'. What then were the techniques used in Mesopotamia to perform any given operation?

The answer is that the realm of technique in Babylonia was coextensive with the realm of tables.[4] We have already seen that the Egyptian scribe had recourse to tables for the techniques associated with fractions. But between the Tigris and the Euphrates the majority of techniques were referred, at least in principle, to such tables. Multiplication tables (from 1×1 to 59×59 . . . and beyond), tables of squares, square roots, cubes, cube roots, etc. . . . A typical table of squares, such as might have been used for step **3**, is given here:

1	1	1
2	2	4
3	3	9
4	4	16
5	5	25
6	6	36

7	7	49
8	8	1 04
9	9	1 21
10	10	1 40
11	11	2 01
12	12	2 24
13	13	2 49
14	14	3 16
15	15	3 45
16	16	4 16
17	17	4 49
18	18	5 24
19	19	6 01
20	20	6 40
30	30	15
40	40	26 40
50	50	41 40

Even the constants, the *igigubbû*, were collected in tables. Here is one example:

5	*igigubbûm*	of a circle
7 30	*igigubbûm*	of a ...
2 13 20	*igigubbûm*	of a basket
1 40	*igigubbûm*	of a load of earth
4 30	*igigubbûm*	of a load of bricks
7 12	*igigubbûm*	of a pile of bricks
6	*igigubbûm*	of a wall
5	*igigubbûm*	of bricks
6	*igigubbûm*	of the grain measure
[...]	*igigubbûm*	of a conduit
[...]	*igigubbûm*	......
4 [...] 8	*igigubbûm*	of a boat
3 45	*igigubbûm*	of an earthen wall
30	*igigubbûm*	of a load of square bricks
3	*igigubbûm*	of a load of half-bricks
45	*igigubbûm*	of a route

Note that the number 5 is given here with the entry '*igigubbûm* of a circle' while 6 (for 6.0.0) is the '*igigubbûm* of the grain measure' just as indicated in our text.

As in the Egyptian case, addition and subtraction were considered too elementary to need tables or special techniques.

The Other Way Around

Let us now look at the last part of another problem taken from the same tablet. We shall give the text and the analysis of the algorithm side by side:

Text 2

If one ***sìla*** was the (content in) grain and 1 6 40 was my depth, how much were my diameter and my circumference?

Convert 1 6 40, that amounts to 13 20.

Find the inverse of 13 20, that amounts to 4 30. Return.

Find the inverse of 5, (the *igigubbûm* of) the circle, that amounts to 12.

Find the inverse of 6, (the *igigubbûm* of) the grain measure, that amounts to 10.

Multiply 10 by 12, that amounts to 2.

Multiply 2 by 1 ***sìla***, the (content in) grain, that amounts to 2.

Multiply 2 by 4 30, that amounts to 9.

Calculate its square root, that amounts to 3.

3 is the circumference of the *qûm*-measure.

Take one third of 3, that amounts to '1 is the diameter'. This is the procedure.

1	→ 1,6,40 = 13,20	***nindan*** → cubit
2	13,20 – 1 = 4,30	subtraction
3	$0;5^{-1} = 12$	inverse
4	$6,0,0^{-1} = 0;0,0,10$	inverse
5	$0;0,0,10 \times 12 = 0;2$	multiplication
6	$0;2 \times 1 = 0;2$	multiplication
7	$0;2 \times 4,30 = 9$	multiplication
8	$\sqrt{9} = 3$	square root
9	$\frac{1}{3}\,3 = 1$	detripling

In this problem we recognize the (partial) inverse of the previous problem: given the capacity and depth of a cylinder, to calculate its diameter and the circumference of its circular section. Note that where a calculated result is to be kept aside for later use (for example, the result of step **2** is not to be used until step **6**) the text signals this fact by the command 'Return'.

The new operations are *detripling* – the inverse of tripling – (step **6**), the *square root* (step **8**) and the *inverse* (steps **2–4**). We shall only detail the last of these. To see more clearly how things work we state here the symbolic algorithm for this problem:

1	$\rightarrow \mathbf{D_1}$		
2	$\mathbf{1}^{-1}$		
3	$0;5^{-1}$		
4	$6.0.0^{-1}$		
5	**4**	×	**3**
6	**5**	×	$\mathbf{D_2}$
7	**6**	×	**2**
8		$\sqrt{\mathbf{7}}$	
9	$\frac{1}{3}$	**8**	

We see that in every case the result of an inverse is used in a multiplication (**2** is used in Step **7**, **3** and **4** in Step **5**). That is to say, the inverse serves jointly with a multiplication to form what we call division ($N/M = 1/N \times M$). In Mesopotamia the functional role of division is always played by these two coupled operations. The calculation of these inverses, as we might guess by now, was effected by means of a table like the following (just the obverse is shown here):

1	its $\frac{2}{3}$	40
	its $\frac{1}{2}$	30
	its $\frac{1}{3}$	20
	its $\frac{1}{4}$	15
	its $\frac{1}{5}$	12
	its $\frac{1}{6}$	10
	its $\frac{1}{8}$	7 30
	its $\frac{1}{9}$	6 40
	its $\frac{1}{10}$	6
	its $\frac{1}{12}$	5
	its $\frac{1}{15}$	4
	its $\frac{1}{16}$	3 55
	its $\frac{1}{18}$	3 20
	its $\frac{1}{20}$	3

There are, of course, 'holes' in the table, values of N for which no inverse exists in the form of a finite sexagesimal (just as such fractions as $\frac{1}{3}$ = 0.33333 . . . have no finite decimal expansion). Such values as 7, 11, 13, 14, etc. have no inverse in this sense and are simply skipped in Old Babylonian tables of inverses. The mathematical texts either avoid using such numbers or develop alternate algorithms to deal with these cases, such as reading multiplication tables 'backwards':

> The inverse of 7 cannot be found (in the table).
> What must I put which will give me 1.10?
> Put 10.

That is, 7 has no sexagesimal inverse but 70 (= 1.10) divided by 7 can be found by seeking 1.10 in the multiplication table for 7 and writing down the multiplicand of 7, i.e. 10. A similar situation occurs in the case of square and cube roots. With the question of inverses, essential and frequently used in mathematical texts, we arrive at one of the principal 'hard' bits of Babylonian mathematics. Though elaborate tables were available, and though in later Seleucid times, the number of places that a table would give for inverses or roots was inordinately long, the question of manipulating these inverses was always a point of difficulty for the scribe, both apprentice and professional.

A First Summing Up

We have seen enough of Egyptian and Babylonian mathematical techniques to draw some preliminary conclusions. First of all, we should underline the differences in the techniques developed in the two civilizations.

Treating of the same problem, for example the volume of a cylinder (p. 49), the operations used are distinct. The core of the problem lies in the determination of the area of the base: the Egyptians used an algorithm equivalent to the determination of the area of a square whose side is $\frac{1}{9}$ less than that of the diameter of the original circle; the Babylonians, one equivalent to multiplying the square of the circumference by the fixed constant $\frac{1}{12}$. Thus, already on the level of operations there is a considerable difference.

However, it is around the question of techniques that a real distinction comes into play. Where the Egyptians used the fundamental processes of doubling, decupling and their inverses, as well as the techniques I have called inversion and two-thirds, the Babylonians appealed to tables of products, roots and inverses. Thus, even where the operation is identical, say multiplication, the ways in which that operation is carried out can be quite distinct. Most important for our purposes, this difference in technique had far-reaching consequences on both the pedagogic and conceptual levels, as we shall now see.

Since, for any technique, there are easy and difficult calculations, any choice of technique is also an implicit choice of what kind of calculations will be difficult to carry out. That is, calculations are not intrinsically easy or difficult but are so only with respect to a choice of techniques.

In the case of Egypt, we have seen that working with their techniques leads quickly to fractional numbers and for the Egyptians – and Babylonians – that meant 'unit fractions'. Now, as we have also seen, the halving of any 'unit fraction' leads simply to another 'unit fraction' $\frac{1}{N} \rightarrow \frac{1}{2N}$, as does the doubling of an even fraction $\frac{1}{N} \rightarrow \frac{1}{N/2}$, but the doubling of an odd fraction is not at all easy. Similarly the calculation of two-thirds of a number is not trivial in this context. Finally the summing of fractions is centrally difficult. We recall that the manner of dealing with each of these problems was to construct tables in which the difficult calculations involved could be done once and for all and the results simply copied and looked up when the need arose.

For the Babylonians, on the other hand, the conversion of fractions into sexagesimals from the outset avoided this obstacle. But, in return, other problems arose. The use of sexagesimal numbers is useful only when the number can be written in a finite form. This means that the techniques of inverse and square and cube roots pose problems, because these are the major techniques which risk transforming finite sexagesimals (such as integers) into non-finite ones. Here is where the use of the tabulation of results really came into its own. Though multiplication tables – where such difficult problems do not arise – also existed for ease in manoeuvring in the world of base 60, it was the tables of inverses and of roots that were more than a convenience; they were an absolute necessity if these techniques were to be at all effective in the resolution of practical problems.

In short, the inevitable existence of areas in which a given choice of techniques becomes problematic gave rise, independently in the two civilizations, to the collection of results in the form of tables. But the mere existence of of tables for certain domains provides a privileged space for reflection on the nature of the results thus tabulated. Regularities, patterns, relationships become clearer; they seem to impose themselves on the eye of the user. Techniques cease to be merely tools for treating problems coming from the 'outside', those posed by the productive needs of the society. Instead, new problems, arising from a study of the tables underlying the implementation of these techniques, start to become the origin of further problems. That is, new problems, arising from the 'inside' of mathematical practice, mark a new level of autonomy and abstraction in mathematics. Of course, this was not the only source of such autonomization. Pedagogical needs, drills in the use of the various techniques which became more and more divorced from 'practical' problems, are probably behind many of the algorithms of the mathematical texts. But even here, it is the techniques developed to deal with difficulties that generated the dynamic of ancient mathematics.

The Next Step

On the Egyptian Side

We have seen that certain calculations, those related to the techniques of two-thirds and doubling of odd fractions, caused problems for the Egyptians and that recourse was had in just these cases to tables. Side by side with the $\frac{2}{3}$ table on the Rhind Papyrus (p. 55) we find the following text:

To make $\frac{2}{3}$ of a fraction.

If someone says to you: 'What is $\frac{2}{3}$ of $\frac{1}{5}$?':

You shall make its 2 times and its 6 times. This is its $\frac{2}{3}$.

Behold, one will do likewise for any fraction which occurs.

Now this little text is unique among all known Middle Kingdom mathematical problems. Like many others it starts off announcing the type of problem: the calculation of $\frac{2}{3}$ of a fraction. Then follows, as usual, the announcement of the particular case which will be treated: $\frac{2}{3}$ of $\frac{1}{5}$. But what follows is not usual. The text reads: 'You will make its 2 times and its 6 times'; that is, the student is told to multiply the 'denominator' of his original fraction by 2 and then separately by 6. The two fractions thus formed, added together, will be 'its $\frac{2}{3}$'. As we would express it today: $\frac{2}{3} \times \frac{1}{N} \rightarrow \frac{1}{2N}$

69

+ $\frac{1}{6N}$. Note that the original concrete case of $\frac{1}{5}$ has completely disappeared! What we have instead is a general rule. The real difficulty for an Egyptian in writing such a rule can be judged by the maladroitness of the formulation: the introduction of a numerical example which, in fact, is never used; the ending 'Behold, one will do likewise for any fraction which occurs,' a standard ending of many procedures in mathematics (and medicine) to indicate that the preceding numerical (!) example can be extended to all cases, superfluous here since the algorithm is already in a general form. All this tends to show that, though it was clearly possible for the Egyptians to express this level of generality, it was probably new and consequently difficult to manipulate. The futile attempt to force this new idea into the old mould of numerical algorithms indicates just how hard it really was. In fact, this example remains isolated among the texts which have come down to us; even much later, in the demotic texts of Graeco-Roman times, the experiment was only rarely repeated.

But why then was the attempt made, and why at this spot? I have already mentioned that this problem is to be found just after the small table concerning fractions of fractions, among which are prominent a number of cases of $\frac{2}{3}$ of a fraction. Aside from the easy case of $\frac{1}{2}$, we have $\frac{2}{3}$ of $\frac{1}{3}$, of $\frac{1}{6}$, of $\frac{1}{9}$ and of $\frac{1}{7}$. Whether this table was copied or calculated is irrelevant to our concerns here. The point is that this table was necessary because Egyptian calculational techniques had both need of and difficulty with just such manipulation of fractions, specifically $\frac{2}{3}$ (and $2/N$ for N odd) – precisely the core of the Egyptian area of difficulty. And the Egyptians had recourse then, as we have seen, to the construction of tables. Construction and/or perusal of tables might well have given rise to the observation of regularities and, at least in this case, to an attempt to express these regularities in the language of algorithmic mathematics. In short, the 'area of difficulty' of Egyptian mathematics, with its consequent tabulations, provides both the possibility and the motivation for conceptual advance such as that evidenced by this text.

On the Babylonian Side

Turning now to Mesopotamia, we recall that the problem of determining inverses was one of the principal elements of the Babylonian 'area of difficulty'. As in the Egyptian case, we shall see that the construction of tables of inverses could have inspired reflection and work on this subject. Here, for example, is a complete Old Babylonian tablet together with the first level of symbolic rewriting:

Text 3

The *igibûm* was greater than the *igûm* by 7. What were the *igûm* and the *igibûm?*

You: halve the 7, by which the *igibûm* was greater than the *igûm*: 3 30.

1 $\left(\dfrac{1}{2}\right) 7 = 3,30$ halving

Multiply 3 30 times 3 30: 12 15.

2 $3,30 \times 3,30 = 12;15$ multiplication

To 12 15, to which it amounted for you, add 1 [. . .]: 1 12 15.

3 $12;15 + 1,0 = 1,12;15$ addition

What is the root of 1 12 15? 8 30.

4 $\sqrt{1,12;15} = 8;30$ square root

Write down 8 30 and 8 30, its equal.

5 $8;30 \qquad 8;30$ branching

Subtract 3 30, the *tākiltum*, from one, add (it) to the other. The first is 12, the second is 5.

6 $8;30 - 3;30 = 5$ subtraction

7 $8;30 + 3;30 = 12$ addition

The *igibûm* is 12, the *igûm* is 5.

This algorithm introduces us to some operations that we have not seen before; the halving in **1**; the square root in **4**; and especially the bifurcation of the algorithm in steps **5–6**. The first two are familiar to us from the Egyptian sources. The last is something new; operating on the structural level of the algorithm, it permits one to operate differently on the same value and is of frequent occurrence in Babylonian algorithms. But the most important point for our present purposes lies elsewhere.

Note that the data appear to be incomplete; we are given two numbers, called *igûm* and *igibûm*, to be determined but only their difference, 7, is given. In fact, however, the missing datum is given implicitly by the very nature of the two numbers, for these are precisely the names of the two columns in a table of inverses; hence the student knows immediately that their product is 1. The entire problem is seen to stem directly from such a table; the vocabulary, the implicit nature of the product, all clearly indicate the source. In short, the problem is based on the inherent relations existent in an inverse table.

This type of problem, though in a different garb, had a brilliant future in Mesopotamia. Problems presenting a situation where the product and sum or difference of two numbers is given and the student asked to find the numbers is a classic of Babylonian mathematical education. Often this is posed in terms of the area of a square being added or subtracted from its side (see the examples on pp. 39–40). The standard explanation among historians of mathematics is that such problems represent a nascent algebra and it is true that if I add areas to lengths I am already on the way to another

71

level of abstraction about number, involving a certain liberation of numbers from their dimensional straitjacket. But what is particularly clear is that such an approach is indicative of the importance of techniques, developed to deal with one kind of problem, in inventing new avenues of investigation, here specifically the suggestive power of tabular arrangements.

Conclusions

After this discussion of ancient mathematics a number of partial conclusions present themselves which may have some bearing on mathematics in general. The advantage of looking at the early stages of any science is that often the combination of historical and cultural distance rids us of some of our own prejudices about how that science must be based on how it is now. The problems we have seen may appear elementary to our eyes, their study less gratifying than, say, the 'rational reconstruction' of the mode of fabrication of the tables used. But, as in mathematics, there is no royal road in the history of science and if, for example, an important part of mathematical activity was centred around the reading and consultation of tables, it is primordial to study in detail the instances of this type of work on the development of the domain. I would simply suggest the following ideas as first object lessons.

There is no 'inner necessity' to the way in which any given mathematical problem is solved. Techniques of solution are bound to their cultures and different cultures will solve the 'same' problem in different ways, even though the answers of course will turn out the same, which is not to say that there do not exist shared kinds of problems.

Nor does there appear to be a 'straight line' leading from 'practical' problems to 'abstract' ones. Different techniques may suggest different directions of exploration and these, in turn, may well present different levels and kinds of problems and approaches further removed from the immediate productive needs of society. Our examples drawn from Ancient Egypt and Babylonia point also to the importance of the pedagogic need for drill exercises as playing an important role.

Finally, the development of the earliest stages of mathematics point to the need for a more sophisticated analysis of the interplay between the practical needs of society and the 'freely generated' nature of mathematical investigation. If ancient mathematics was never 'simply' practical and empirical, it may be equally true that even the mathematics of today is not 'purely' abstract and speculative. The fact that techniques mediate advances suggest a way in which mathematical problems that arise in society are ultimately in some relationship with the techniques which that society has forged. This, in turn, suggests that mathematicians, like societies, can only pose those questions to which a potentiality of a response exists.

3

Gnomon: The Beginnings of Geometry in Greece

Michel Serres

Greek geometry emerges, perhaps from astronomy and the algorithms which were common in the Fertile Crescent.

As ports grew up in ever more far-flung places, from Apollonia on the Black Sea to Cyrene in Africa, or from Perga in Asia Minor to Sicily and Italy, the originators of knowledge were concentrated in rival schools. From the very beginning, the society of teachers and scholars was a microcosm of society in general. City-states broke up and confronted each other along the sea-shores. For example, the little Athenian city of the Academy, led by Plato, waged fierce battles against the ten sophists, Hippias, Protagoras and others, and concluded temporary alliances with foreigners from Croton, Cnidus, Locri and Elea: Pythagoras, Eudoxus, Timaeus, Parmenides and Theodorus of Cyrene.

The Greek Empire

The Greek world never achieved unity, neither when Athens, Thebes and Sparta were at the height of their power nor even when they were threatened with destruction by the great powers of the four cardinal points, Medes and Persians, Macedonians, Carthaginians and Romans. No league lasted long because the Greeks on the coast, relentless rivals, confined themselves,

73

like Alcibiades, to dreaming of a united empire. The cities and petty rulers hated each other as intensely as the philosophers. Meanwhile, the shores became hellenized. Coast-dwellers of three continents – Asia, Africa and Europe – spoke Greek. But the common language of sea trading died, as did the brief hegemonies, the schools, the little gods, and what we call the economy. Nothing was left of anything. This decline is known as Antiquity.

But in less than four centuries, Greek thinkers from Thales of Miletus to Euclid of Alexandria, from rival cities and schools, from rival economies and religions, intent on disputing with each other, sons of the land against lovers of forms, exponents of change versus those who believed in eternity, built between them, albeit unwittingly, in a stunning and unexpected way, a unique and invisible empire whose greatness has survived intact until our times. It is an edifice with no parallel in history, where over a span of more than two thousand years, their influence still causes us to work in the same way as themselves. Nor have we abandoned it on the pretext that we speak a different language, even if our enmity has increased. Is there any equivalent consensus in the annals of human history? The name of this edifice is mathematics.

The eastern Mediterranean

Space

Concentrated in a very limited area are Samos, the birthplace of Pythagoras, Thales' Miletus and Ephesus, the home of Heraclitus, not to mention Patmos, the island to which St John the Baptist withdrew towards the end of his life. This region is the source of arithmetic, of geometry and of physics – and thus the definition of logos, number, ratio or invariance. All this without even counting the origin of the word.

If we extend the area a little further, we include yet more homes of mathematics or mathematicians, the island of Chios and the entire shore of Asia Minor, from Cnidus to Cyzicus. It is said that those same places were the cradles of alphabetic writing, money and coins, iron metallurgy and, a little further south, monotheism was born.

If we extend the radius a little more, we find the eastern Mediterranean shown on the map opposite includes Ionia, Egypt, Greece, Italy, not to mention Palestine. It is the intersection of Africa, Asia and Europe – the seat of maritime cities feeling the thrust of the great empires of the Egyptians and the Medes and Persians and later the Romans, and meeting each other in sea battles. From this physical and human chasm, active since the beginning of time, emerged science, our religions, history and most of the traditions which have inspired us until now. Violent struggles are still going on there today.

Time

The most active period runs from the end of the seventh century to the end of the third century BC and a little later, in other words more than 300 years, equivalent to the length of time between Descartes and ourselves. During the centuries following this period, Hipparchus, Ptolemy and Diophantus respectively invented trigonometry, a classic model of the world and the first algebra, but the momentum slowed down considerably towards the end of the millennium which separated Thales and Proclus.

We have no direct sources for the achievements which preceded and informed Euclid. We have reconstructed our information based on writings by Plato and Aristotle, on the *Elements*, and even later authors, commentators or others, except for very rare fragments. Our only witnesses speak from a distance sometimes as great as that between ourselves and the Renaissance, hence the tenuousness of our reconstructions.

With a few exceptions, the originators of knowledge congregated in Schools. Were these like centres for research and teaching, or philosophical sects, religious communities, orders, brotherhoods, pressure groups, political parties, clubs or gangs? We do not know. But any collective resembles to a greater or a lesser degree all of these, even today.

The Schools

End of seventh century BC
1 Physicists of Miletus: Thales, Anaximander, Anaximenes. *Nature as the subject-matter of science.*
2 Pythagoreans of Croton: Pythagoras of Samos. *Numbers; duplication of the square; arithmetic, basic science.*
End of sixth–fifth century BC
3 Eleatic School: Xenophanes of Colophon, Parmenides, Zeno, Melissos. *Unity.*
Mid-fifth century BC
4 Chios School: Oenipides, Hippocrates. *Squaring the circle; duplication of the cube; trisection of the angle; first elements.*
Fifth century BC
5 Sophists and Megarians: Hippias of Elis, Euclid of Megara. *Quadratrics.*
Fifth and sixth centuries BC
6 The Atomists of Abdera: Leucippus, Democrites. *First infinitesimal alogorithm.*
Fourth century BC
7 Athenian School: Plato, Speusippus. *Polyhedrons.* Related: Theodorus of Cyrene, Theaetetus. *Irrational numbers.*
8 Cyzicus School: Eudoxus of Cnidus (Egypt, Tarentum). *Arithmetic; conic sections.*
9 Peripatetics: Aristotle, Autolycos of Pitane, Eudemus. *Encyclopaedia; history.*
End of the fourth century BC
10 Alexandrian School: Euclid. *Elements.*

Third century BC
Archimedes of Syracuse (287–212): *spiral; large numbers.* Eratosthenes of Cyrene (276–195): *geodesy; prime numbers.*
Apollonius of Perga (262–180): *conic sections.*
Second century BC
Hipparchus of Alexandria: *trigonometry.*
1st and 2nd centuries AD
Ptolemy of Alexandria (90–168): *solar system.*
End of 3rd century AD
Pappus of Alexandria: *geometry.*
4th century AD
Diophantus of Alexandria: *arithmetic and 'algebra'.*
5th century AD
Athenian School closed by Proclus (412–485): *commentator.*

GNOMON

Tradition

History as it is written today prohibits referring to the origin of geometry on Greek soil as a miracle, as the French philosopher Ernest Renan did. The scientists of today accept the existence of very rare occurrences in certain disciplines, though historians no longer encounter such things, finding only laws. It is as if monotonic time had changed sides. And yet, the birth of abstract space constitutes a most unexpected event even for those who know what was happening in the arithmetic of Egypt or of Mesopotamia. The building of that Greek empire which continues to rule over us could be considered even more improbable: despite its tangible, living reality, it does not feature in any history book.

As children we all travelled that road from Samos to Miletus, from calculating integers to demonstrating the equality of triangles. We journeyed from Miletus to Chios or Abdera, towards measuring the circle or the cone or the cylinder, and, if we went further, our odyssey took us to all the ports on the map, going back to the beginning of the era of the contruction of these transparent ideal objects. Is there any school anywhere in the world that does not teach its children the same elements in the same language? The word mathematics in ancient Greek means: that which is taught or learned. Where and when has it not been taught? Iranians, Spaniards, French, English and Tamils – we all speak Greek when we say parallelogram, logarithm or topology. This language within the system lives on and unites us. Nothing remains of the cities of Cyrene or Perga, or of the Elean School or that of Croton, not a temple or a weapon, no trade or workshop, but the list which runs from integers to conic sections has not aged one bit, even though sometimes we do not understand the terms number or diagonal in exactly the same way as the ancient Greeks. Who better triumphed over history and its fluctuations than that little collective which so quickly established this signal subject which has resisted erosion by time? Who better scorned the battles than this little group of irreconcilable enemies forging a common language – the only one capable of halting conflicts and which never needs translating? All the cultural hegemonies of the world are impotent against this community and against the universality of this teaching. We are cut off from Antiquity in every possible way; yet through mathematics, it remains contemporaneous, without alienation, since there cannot be any misinterpretation.

Hard and Soft

Did Thales come to the foot of the Pyramids to assess the conditions for durability? What does it take to endure? War, the fatal game in which the strongest survives, tyranny, trade, slavery, tools, production – everything comes to an end and is wiped out at some point. The strongest is never

77

strong enough to withstand time's flow for ever. The huge mass of stones crumbles or is covered in sand by the winds, yet the tomb of Cheops had the maximum benefit of all the elements: strategy, power and capital, religion, armaments and wealth. This mass, whose rubble Bonaparte calculated could surround the whole of France with a high, continuous wall, does not however attain the dimension of time. What empire will succeed in doing so? In the times of Thales, the old Pharoah was doubly dead, almost forgotten. Even the strongest does not last.

Just as in order to survive other cultures played the victim rather than the victor, so Thales inverted the game where only the hardest survive: only the softest lasts. All materials and powers wear out, but what will become of pure form, of the most faded image, the least tangible, the lightest, the least definable? Pure form, whose written expession has no importance, the very trace of which can be lost without damaging the meaning, whose memory itself can be lost or die without affecting history? Draw it wrongly, it does not matter. Do not draw it, do not even write it; it still does not matter. What is more, destroy sources and testimonies, obliterate the monuments, burn partial manuscripts or whole libraries, wipe out almost entirely the period when this form was created and, despite all attempts at annihilation, it remains unchanged from the moment it was first formulated, ever present in our oblivion. Even its concept can falter without much harm being done: we no longer understand reason or resemblance the same way, however no significant change has occurred. All that remains of the pyramid may be a shifting in the space of homothetic transformations, a theory as fleeting and soft as a ray of sunlight with its shadows – yet at last it fills the dimension of time.

By comparing the shadow of the pyramid with that of a reference post and his own shadow, Thales expressed the invariance of similar forms over changes of scale. His theorem therefore consists of the infinite progression or reduction of size while preserving the same ratio. From the colossal, the pyramid, to the small, a post or body, decreasing in size *ad infinitum*, the theorem states a logos or identical relation, the invariance of the same form, be it on a giant or small scale, and vice versa. Height and strength are suddenly scorned, smallness demands respect, all scales and hierarchies are demolished, now derisory since each step repeats the same logos or relation without any changes![1]

Thales demonstrates the extraordinary weakness of the heaviest material ever worked, as well as the omnipotence, in relation to the passing of time, of a certain logical structure: of the logos itself as long as we redefine it, no longer as a word or statement, but, by lightening it, as an equal relation; even softer because the terms balance each other, obliterate each other so that all that remains is their pure and simple relation. From the maximum remains of the maximal power of history optimally preserved, Thales extracts the minimal softness or lightness. Even measurement is forgotten in the new logos of similarity where a relation between small things equals another between large ones. A miracle: from almost negligible means the most durable empire possible was born, mocking history and never going into decline. We are only just beginning to value such economy, a cornucopia which yields infinitely from almost nothing.

Sun and Earth

Did the whole adventure begin with astronomy? How did the astronomers of Antiquity observe the stars in ancient times?

The shaft of the sundial or gnomon casts shadows on the ground or on the face of the sundial according to the positions of the stars and the Sun throughout the year. From Anaximander on, apparently, Greek physicists knew that these readings indicated certain occurrences in the sky. The light from above describes on the earth or on the page a pattern which imitates or represents the forms and real positions of the universe, through the intermediary of the stylus.

As nobody in those days really needed a clock, and as the hours varied enormously since summer and winter days, whatever their length or brevity, were always divided into twelve, the sundial was rarely used for telling the time. Thus it was not replaced by the timepiece but was used as an instrument of scientific research in its own right, demonstrating a model of the world, giving the length of shadows at midday on the longest and shortest days, and indicating the equinoxes, solstices and latitude of place, for example. It was more of an observatory than a clock. We do not really know why the shaft or pin is called a gnomon, but we do know that this word designates that which understands, decides, judges, interprets or distinguishes, the rule which makes knowledge possible. The construction of the sundial brings natural light and shadow into play, intercepted by this ruler, a tool of knowledge.

According to an oft-cited platitude in Herodotus, it seems that the Greeks inherited the gnomon and the division of the day into twelve from the Babylonians: who can say whether the division of these twelve parts into sixty is a result of the year being divided into 360 days or vice versa? In short, every angle or thirty-degree segment divides the sky into zones which the Greek language calls ζῴδιον (zodion), from ζῷον (zoon), animal, and ὁδός (odos), way, in other words the figure of a beast or of any other living being; the corresponding adjective designates the orbit, the route, the zodiac path. On the other hand, the noun expresses the signs of the zodiac. The sky fills with living forms, point by point.

Work back from the shadows to the light which cast them, and from that light to its unique source; this is one of Plato's lessons, when he speaks of knowledge. It is not a question of poetic imagery, but of the everyday actions of astronomers or, to be precise, their method which infers masses of information from the length or position of the line or the dark shadow. To this end they were able to construct a ruler as precise as the stylus which writes. The black ink on the white page reflects the ancient shadow cast by the sun via the pointer on a sundial. This point writes unaided on the marble or the sand as if the world knew itself.

Who knows, who understands? Never did Antiquity ask these two questions. Where should the head, the eye be placed in this observatory? In the band of shadow, at the source of light, or at the tip of the sundial pointer? These are modern problems. For example, the use of the telescope assumes the invention of the subject, which will place itself on the right side of the viewfinder, contemplating, observing, calculating, arranging the planets: it does not exist in the ancient Greek language. In those days the world as such was filled with knowledge in the same way as it is said the skies sang to the glory of God. For this culture the gnomon knew, discerned, distinguished, intercepted the light from the Sun, left lines on the sand as if it were writing on the blank page and, yes, understood. In external space and its bright or black events lie knowledge and the whole body; life, destiny and the group are plunged into the vastness or into the world and they are indistinguishable from it. The world represents itself, is reflected in the face of the sundial and we take part in this event no more and no less than a post, for, standing upright, we also cast shadows, or as seated scribes, stylus in hand, we too leave lines. Modernity begins when this real world space is taken as a scene and this scene, controlled by a director, turns inside out – like the finger of a glove or a simple optical diagram – and plunges into the utopia of a knowing, inner, intimate subject. This black hole absorbs the world. But before this absorption, the world as such remains the seat of knowing. We can no longer understand this phrase, we who, furthermore, destroy what we know.

Working back from the shadows to the light and from the images which have been reproduced or projected to their model, are lessons common to both Greek astronomy and to the Platonic theory of knowledge. The fact that the instrument which makes this former operation possible is called a gnomon in Greek astronomy helps us place the active centre of knowledge solidly outside ourselves.

What is more, the firmament is peopled with living forms, the signs of the zodiac. If light comes from the sun, even when it vanishes at night, who then carries on their back the wooden or stone statues of the animals, on the elevated path of the zodiac, so that their huge images are projected on the dark wall of the sky? Plato's cave describes the world itself. We will never know whether Plato first saw the Bear or the Dog on the starry vault above his head, before conceiving in his philosophy of the intelligible sky of forms

Gnomon: the stationary shaft which projects the shadow on a sundial. (First century AD, Egypt. Deutsches Museum, Munich.)

preceding or conditioning the intelligence of the things of this world, but we certainly see that in appearance the constellations are reduced to sets of points. Nobody has ever really seen Libra or Aries here or there but simply a simplex: never a continuous flowing image, but juxtaposed studs, as if the celestial models remained faithful to the theory of the Pythagoreans for whom all things are numbers. But where do these statues which cast their twinkling shadows on the black sky come from?

The profile of the Universe

The gnomon or sundial served less as an instrument for telling the time, which nobody wanted to know, from classical Antiquity to the days of our grandparents, but rather as a tool for constructing a geometric model of the Universe: it was both an observatory and a cosmographical map of the world.

AB shows the shaft of the gnomon, BC measures the shadow of the midday sun at the summer solstice, BE that of the winter solstice, BD the shadow at the equinox. The straight lines and the circle are then traced on the meridian and define it. FG represents the horizon and the point A the earth floating at the centre of the sphere of the world. Then, the two lines MJ and KH follow the tropics and LI the equator, as NO, which is perpendicular to it, represents the earth's axis. The angle ENO, equal to BAD, gives the precise latitude of the place and the angle DAE, equal to DAC, the inclination of the ecliptic, estimated at 24°, i.e. of the circular segment intercepted by the side of the regular pentadecagon.

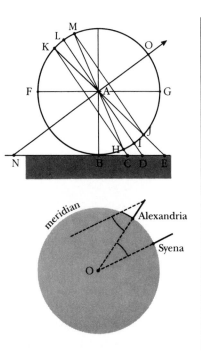

This body of information, discovered successively from Anaximander by Vitruvius (Roman architect of the first century BC) and Pytheas of Marseilles (Greek navigator and geographer of the fourth century BC), Ptolemy as well as Hipparchus, dates back for the most part to remotest Antiquity. Thales wrote two books on the equinoxes and the solstices; it was probably Oenopides who estimated the inclination of the ecliptic at 24°. The diagram should be read as a profile of the world as the Greek scholars saw it, but also as a summary of the history of their knowledge: every generation since the fifth century has drawn at least one thread from it.

To give a more precise idea of the results the Greeks achieved from the gnomon, here is an example of how Eratosthenes (276–195 BC) calculated. He placed one at Syene in Egypt, not far from the first cataract of the Nile, and situated on the Tropic of Cancer. Here there is no shadow at midday on the day of the summer solstice. On the same day, at the same time, Eratosthenes measured the angle made by the Sun and a second gnomon in the city of Alexandria, which he thought was situated on the same meridian. The two alternate-internal angles on the face are the same; now the one that he measured is equal to one fiftieth of a circle; thus all you have to do is multiply the distance from Alexandria to Syene by fifty to obtain the length of the earth's meridian. An impressive result was obtained with minimal means. To improve the measurement, Eratosthenes estimated the shadow of the gnomon not projected on to a plane but on to a sphere or perhaps the πόλος (polos) which Herodotus mentions in the place already cited.[2]

Machine and Memory

We find it difficult to translate the word gnomon because it resonates harmonically with the thing it designates and because knowledge sparkles at the tip of its shaft. Literally it means, apparently in an active form: that which discerns, which determines, but the word always designates an object. In his commentary on the second definition in Euclid's second volume, Thomas L. Heath describes it as 'a thing enabling something to be known, observed or verified'. The closeness of these two things or the repetition of them makes sense: there is a relationship between them, on their own. In this thing or through it, in the place it occupies, the world shows knowledge.

As the arm of the sundial rose perpendicular to its surface, the expression 'like the gnomon' meant to the Greeks of long ago the right angle or the plumb line. Suddenly, we could almost translate it as ruler or set square, especially as Euclid, in the work already mentioned, uses the word gnomon to describe the areas of complementary parallelograms of a given parallelogram, such that their addition or subtraction leaves them all similar to each other. Thus a set square shows two rectangles or two squares complementary to a given square or rectangle: the French word for set square, *équerre* itself seems to mean 'the extraction of the square' (*carré*) or sundial face (*cadran*).

Once again, how should we describe the gnomon? As an object, an arm which, placed in an appropriate position, will give astonishing results, latitude, solstice, equinox, which it provides automatically. That means it works all by itself, without any human intervention, like an automaton, without anyone to operate it: mechanical knowledge, since it intercepts a movement, that of the Sun. Let us call it a machine rather than an instrument, as tool strongly implies the person who uses it, or the conscious and finalized action for which that person designed and made the tool. On the contrary, the mental activity indicated by the word gnomon, in Greek, refers here to the machine, to an object. The gnomon instantiates one of the first pieces of automatic knowledge production in history, it is the first machinery to combine hardware and software. The role of the subject, and his or her knowing or thinking function, have nothing in common here with what they were to become in what we have called until now scientific knowledge.

The calculation of latitudes based on the shadow of the Sun at the solstices and the equinoxes, the

Gnomon

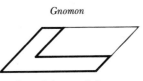

Case of the parallelogram

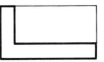

Case of the rectangle

Case of the square

first mathematical link between astronomy and geography, also led to the establishment, by Ptolemy or earlier by Hipparchus, of what Antiquity called tables of chords: long lists of ratios between the measurement of the sides of right-angled triangles and that of their angles, where we can see the beginnings of trigonometry. Here is memory, there is the gnomon: the table corresponds to the machine, mnemotechnics is associated with automatic knowledge. Likewise, in the science of the Babylonians, automatic calculation procedures exist alongside tables of measurement. In other words, and more generally, an algorithmic way of thinking always has two components, one which could be called mechanical and the other which must be called mnemonic. These can be described as the accumulation or recapitulation of the results of mechanical procedures or conditions of their repetition; the automaton and tables or dictionaries; hardware and software.

Another Reason?

All the knowledge signalled by the word gnomon and accumulated around its shaft, all this objective and tabular knowledge, is very different from the types of knowledge we traditionally associate with mathematical demonstration or deduction or with physical experiment, based as these are on criteria of rigour and precision, as well as being knowledge associated with the individual or collective subject. There is another logos here, a different episteme, in short, another reason that could be called algorithmic. Algorithmic thought, efficient and present among the Egyptians and Babylonians, coexisted in Ancient Greece with the new geometry, although concealed beneath its transparency; thus obscured by official mathematics, hellenic by tradition, fertile, it was to persist for many centuries before acquiring, in modern times, a status on a par with that of geometry.

Astronomy without Eyes

A person educated in modern science is astonished that astronomy with no means of looking or seeing, unlike our contemporary astronomy, could have existed so long ago. If the sundial was hardly ever used as a clock, if we ought to consider it more as an observatory, that word itself, anachronistic and ill chosen, is misleading. The gnomon is not the precursor of the theodolite any more than the sundial is the forerunner of the watch, for Greek astronomy did not observe in the way of the classical or modern ages in which telescopes are housed under domes. The act of seeing does not and did not fulfil the same need or hold the same importance for the act of discovering.

We are used to interpreting knowledge as a doublet of sensation and abstract formalities and philosophers repeat like parrots the assertion that

Anthypheresy or Euclid's algorithm (procedure)

PGCD. Take two numbers, say 20 and 12. If you divide the first by the second you have 8 left over. If you divide 12 by 8, you have 4 left over and if you divide 8 by 4, which goes exactly, you have nothing left over. So we say that 4 goes into both 20 and 12 on the grounds of their highest common factor.

To find it, we divided the two numbers into each other and the second by the remainder from that division, and that remainder by the second remainder, the third by the second and so on until there is no remainder. The last number of the series is called the greatest common denominator.

Euclid. *Elements. Anthypheresy* consists of a subtraction which takes the smaller of two magnitudes away from the larger and compares the smaller with the remainder, and so on.

'VII, 1: If, when the less of two unequal magnitudes is continually subtracted in turn from the greater, that which is left never measures the one before it, the magnitudes will be incommensurable.'

'X, 2: Two unequal numbers being set out, and the less being continually subtracted in turn from the greater, if the number which is left never measures the one before it until a unit is left, the original numbers will be relatively prime.'

Music (table or machine)

In *The Beginnings of Greek Mathematics*, Arpad Szabo describes the *Sectio Canonis* (the theory of the intervals) attributed to Euclid.

The whole chord is divided to produce the fourth or the fifth. The smaller segment is then subtracted from the larger. The remainder is subtracted from the smaller segment. This operation could be carried out out twice for the fifth and three times for the fourth ($\frac{2}{3}$ and $\frac{3}{4}$). Thus, after subtracting the smaller segment from the larger, the remainder was substracted from the smaller until the final disappearance of any remainder.

That is, in his view, the origin of Euclid's algorithm.

knowledge can only be based on the senses. That presupposes a subject, then a body and a whole training which sharpens the senses by means of sophisticated hardware. In those days, the gnomon and the projection surface received the information unaided, independently of the human eye. The objective receiver, shaft and marks, was later to give way to the sentient body, but it was there first. When Greek historians and enthusiasts relate the story of Thales who stood at the foot of the pyramids and tried to measure their height, significantly they confuse the shadow of any old stick with that of a body. Whether it is a question of the colossal building, of the post or of what we thought he was observing, what does it matter, each in its own way, stone, wood or flesh, assures the canonical role of the gnomon, the function of being discerning, objective. This is science without a subject, science which dispenses with the senses or does not operate through them. Put a stick in its place and nothing will change, build a stone tomb on the spot where the corpse decomposes, and the knowledge persists, unchanged.

It cannot be disputed that we can see here the division into light and shadows, a whole sensory scene, but nothing of it is conveyed through a subject, the possessor of faculties. Nor is it filtered through a theory or ending up in the construction of one. In the diagram of the Sun, at the source of light, rays, of the shaft and the message on the ground, there is no place for the eye, no position which we can call a viewpoint. And yet, theory is present. Exact or approximate, sometimes rigorous measurement, abstract reduction, the theoretical passage from the volume to the midday meridian plane and from that to the line and from that to the point, the geometric model of the world emerges here without the intervention of organs, functions or faculties. The world lends itself to be seen by the world that sees it: that is the meaning of the word theory. Even better: a thing – the gnomon – intervenes in the world so that it might read on itself the writing it traces on itself. A pocket or fold of knowledge.

In the literal sense, the gnomon is intelligent because it puts together situations chosen among thousands of others, therefore it discerns and understands. A passive receiver, it sees the light and then it actively traces the fringe of shadow on the page while theoretically showing the model of the sky. In order that we moderns might gain access, once more informed by the gnomon, to this automatic science, to this artificial intelligence, we should forget the philosophical prejudices of the modern interlude: man at the centre of the world, in the place of the gnomon, the subject in the middle of knowledge, its universal receiver and driving force, together with the imaginary reconstruction in all its dark intimacy which nobody ever penetrated, other than a few transcendental philosophers equipped with a mythical golden bough, of that same scene of light and shadow that they sketched from a real eye towards the filter of a legendary understanding. In fact, nothing is easier than abandoning that complicated faculty to read simply what the Sun writes on the ground.

The gnomon is not a tool in the sense of a stick held by a monkey which thus extends its grip, nor in the sense of a magnifying glass which enlarges the lens and enhances the performance of the eye. The artifice does not refer to the subject who directs it, but remains an object among objects, between the Sun and the ground, a thing made intelligent by its position in a specific place in the world which passes through it to be reflected itself. Through the gnomon the Universe thinks αὐτὸ καθ᾽ αὐτὸ (auto kath'auto), and knows itself by means of itself.

The nascent mathematical ideal thought never referred in Greece to a thinking subject nor was it thought itself. On the contrary, the most pregnant thought remained realism. Now the reality of idealities, knowing the form thing or the thing form, is revealed at the foot of the gnomon in the scene where things see things. The point, the line, the angle, the surface, the circle, the triangle, the square . . . are born there as ideal forms in the shadow and the light, in the middle of things themselves, in the world as it is, as real as the rays of light, the fringes of shadow, but above all their common sides.

Tables or Canonical Lists

A historian of science should not be surprised that tables of numbers and an instrument of observation from which they are drawn or in which they reappear correspond, as he is accustomed in a way to the fact that a science begins in that state: for example, the telescope shows thousands of positions of thousands of stars and a register records them. Welcome, but late, a comprehensive theory cancels that state: thus the laws of Kepler and of Newton obliterate in a single phrase this jumble, because this phrase enables anybody to find such and such a local detail immediately, through a numeric calculation.

An identical hope drove the nineteenth-century chemists who were led by their own material to draw up experimentally tables of bodies, and they caught themselves dreaming, like astronomers, that a general law would erase them by including them all at once. This coexistence of lists, tables or columns and of simple or complicated equipment seems to us to character-ize a pre-theoretical era, where observation prevailed over laws, in anticipation of the induction to come.

When we see coexisting in Antiquity the tables of chords which give the values of an arc or an angle based on the lengths of the sides of a triangle, and that instrument of observation which the Greeks called a gnomon, we bear in mind the historic pattern caused by the arrival of Newton or Kepler among the Alphonsine or Toledan tables recording the positions of the stars.[3] We then perceive the face of an experimental knowledge which associates an instrument and tables of numbers in anticipation of a theory whose unitary power makes the former redundant as well as the latter. Through this pattern, we understand the situation of the ancients and in this way we can make sense of it, of course. Here is the gnomon: it precedes the telescope; here are the tables of chords: they resemble Toledan tables. The whole constitutes a pre-modern pre-astronomy in anticipation of the theory of trigonometry.

Now we have just contracted a new way of looking since we see around us the coexistence of a machine and its memory, an automatic instrument and programmes. This is the same pattern, in a way, but also an entirely different one since we are not waiting for a theoretical law, the global understanding of which would erase in a single stroke of the pen our software and its relation to the hardware.

We are dealing with an authentic and original way of knowing, and not pre-knowledge or a state preceding knowledge. We are dealing with knowledge, and not its incomplete working-out. Greek astronomy provides an example of the second model rather than a paradigm of the first.

Geometry

Having reached the foot of the pyramids, Thales – but what does it matter who he is – demonstrates the similarity of the triangles formed, the first by Cheops and its shadow – but what does it matter which one is chosen from the three similar tombs or the name of the pharoah who lies here – and the second by a post driven in here and its dark half. A legend cites this post while another refers to the shadow cast by the standing geometer. Which should one choose, the body or the post? The angles are equal and the sides proportional. The same ratio makes the pyramid and the two erected elements match each other, an identical ratio which can be expressed in three statements.

First, or rather last, it defines homothesis, to the letter, the same way of being there, of being placed, or better, an area of movement, displacements with or without rotation. That is the statement of rigorous science, which can be read from now on in this story which tells of Thales' measurements during the course of his journey. Second, or rather mediately, it expresses this patent fact that each of these normal upright pegs on the horizon can act as a gnomon: the moment of midday recorded by one of the legends marks the principal function of the sundial as being that of fixing the meridian, and on it, the solstices and the equinoxes, solemn moments when the shadow lengthens towards its extremities. Thales, they say, wrote two books about them. To fulfil this function, the pyramid here is equivalent to the axle-tree or to the stick driven into the ground which in turn is equivalent to this motionless passer-by, frozen in his contemplation of the apical light: all gnomons.[4] And the tomb includes a funereal shaft which is directed at the absence of the star which, in the sky, indicates the north. This middle law expressing the resemblance or the similarity or even better the homothesis in the literal sense between everything that can act as a shaft or axis for an observatory of this kind, must be called historic, because it recounts the astronomy of the Ionians and their first models of the world, as well as what happened afterwards in geometry. Undoubtedly, the

equivalence of gnomons of varying heights leads to the homothesis of associated triangles for a same stable world, according to the statement of canonical geometry, and the right angles of the one probably come from the
Sun's rays of the other or from the blind edges of shadow, and the circles
of the orbits and the points without dimensions from the imperceptible
marks at the solstices or at the equinoxes. The Greek miracle falls and
descends from the sky, the old question of the origins of geometry is resolved in this luminous and dark passage of the stars to this axis whose
name says that it knows.

But third, or rather first and archaic, the anthropological meditation slowly
conducted in *Statues* renders coherent and thinkable, without the firmament
and before geometry, a fundamental similarity[5] between the tomb and its
mummified pharoah, the erect living body, half light and half dark, and the
post driven into this precise spot. Markings by death and by what ensues
from it, from this singular place, of the being there, markers represented by
the post and Hermes standing at the edges, here are three statues, in the
sense of the word as defined in this book, three markers which are precisely
homothetic (in other words placed there in the same way) mummies, living
bodies, cairn, obelisk or menhir, staff or stock, assuming the same function
of designating a deposit, burial place, habitat or frontier – oh! wonders! – to
trace soon, thanks to the Sun, the precise latitude of this place. This statement surpasses history and founds the law of science, for it says the same
thing in another language. The mean law in astronomy says the same thing
in the same language, metric, precise, exact, quasi-formal, and geometry is
found in it already born, embryonic. But the third or the first, the most
deeply buried and original, revealing three statues in these three apparently
dissimilar bodies, displays the rigorous homothesis in the literal sense of
these three local and deceased witnesses, of these three burial markers, and
say it in a primordial language so full of shadows that all our efforts at
thinking since the beginnings of geometry have not sufficed to rediscover,
retranslate or decipher it, to find it in the light of theorems.

Now this blinding light comes out of this darkness as statues rise from the
earth, from this primary, fundamental earth which has been repeating, without
knowing it, for more than two millennia, the word geometry. The earth
devastated by the Nile floods likewise returns to chaos and to the original
shadows from which measurement brings it into the light. The former do not
prevent the latter from emerging, but the light always prevents us from ever
seeing the darkness. Geometry shines forth and dazzles us, thus concealing
its dark origins. Yes, it falls and descends from the sky, through the flowing
history of astronomy, a simple and easy fall and cathode;[6] but it rises from
the earth, anabasis[7] and procession, emerges from the tomb, from the cave
where the shadows of statues dance, rises from the dead. Always ready to
laugh and burst into mocking ridicule, the Thracian peasant women in the

89

fable knows that the observer of the stars falls into the well: we learn from them that Thales' place gives way under his feet, undermined. Yes, geometry bears the name of its mother, the earth on which what falls from the sky is measured. Marked out with the help of the gnomon, it remains in the shade like a foundation, like a foundation dug beneath science; here lies the mummy, in a black womb where the stake is driven from which rises knowledge. Ἐπιστήμη (ἐπιστημα cf. note 5, p. 725)

The laws of geometry develop in new, modern times of scientific knowledge; the laws of astronomy are recorded in the time of the history of science which was born before the beginnings of geometry; the laws of statuary are formulated in the time of anthropology or that of the foundations which support the two others.

Artifices

Euclid defines a gnomon as that angled complement of a square which carpenters commonly call a set square, a word of the trade which wonderfully describes the extraction of a square right in the middle of its right angle (cf. p. 84). Whether it veers from the normal and bends towards the acute or the obtuse, the inner parallelogram remains similar to the outer, obtained by adding the gnomon to the first: a band or crown around a form which is thus reproduced as often as is required.

We will understand the geometric arithmetic of the Pythagoreans when we know that they called a gnomon the complement of successive square numbers expressed in odd numbers. Far from writing this situation as we do:

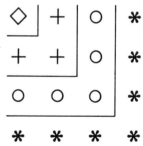

$$1^2 + 3 = 2^2$$
$$2^2 + 5 = 3^2$$
$$3^2 + 7 = 4^2$$
$$\ldots$$
$$n^2 + (2n + 1) = (n + 1)^2$$
$$\ldots$$

they drew it as in the diagram opposite and as a simplex or stars in the sky.

This reproduces Euclid's definition without any significant differences: odd numbers make the set square around the inner square and reproduce with it, indefinitely, an outer square clearly resembling the first. With a diagram showing the right-angled

Each symbol placed in an angled arrangement includes the odd numbers which must be added successively to construct a new square. We find the bands of the set square on the numbers. Compare with p. 102.

bends, triangular, pentagonal and, in general, poly-gonal numbers can be produced. Theon of Smyrna calls them gnomonic numbers. Through these proce-dures we gain access to arrangements which are the forerunners of Pascal's triangle.

1
1 1
1 2 1
1 3 3 1
1 4 6 4 1
1 5 10 10 5 1
etc...

Pascal's triangle

The shaft of the sundial, the gnomon becomes a set square: an instrument or artefact in both cases. The former plots certain stations of the Sun on the sand while a ruler, *rectus* in Latin, right angle or straight line, like the set square, can describe them on the page. Geometry can be defined as the sci-ence which only allows itself the ruler and the com-pass. What should we think of the status, the place and the function of these artefacts in a perfectly pure knowledge? Second, set square or gnomon, lateral angled bands, two-sided complementary forms, enlarge or reduce and reproduce at will square or parallelogram, preserving their similitude intact. Thales' story can be turned round in both senses: the solar gnomon made him discover homothesis or through homothesis, gnomonic growth made him go from the post, a small-scale model, to the giant pyramid. Finally, the gnomon aligns sets of numbers. How can it be defined otherwise than a law of series? Add an odd number, add up the total of odd numbers, and you will obtain the successive squares. Or juxtapose the complementary band, and the similar parallelogram will appear. The gnomon is defined as a law of construction, like the rule of a series or its cause. An automatic rule, working by itself, recording the chain at will or each link without our intervention. This operation happens without an active or thinking subject, just as the shaft of the axle writes on the ground in our absence.

Everybody recognizes two sorts of artefact: those which do not depend on us and those which do. Only the first type functions relentlessly, or better, never ceases to be an artefact. Examples: the wall and the roof always protect us, even when we sleep, but when we lay down the spade and the pen, they sleep, useless and annihilated; intelligent only in our waking moments. In truth, real tools do not depend on us, the others rest too often to be genuinely entitled to that name. So to give the same name, expressing knowledge, to three automatisms, that of the post standing in the sun, that of the set square or the lateral band that is added or removed, and that of the operation whose repetition constructs series of numbers, brings us back to artificial intelligence. We see the avatars of it, the evolution in these three states: first thing, post or shaft, a speculative tool, then a rule capable of reproducing at will right angles, angles, ideal polygons, extracts or better, abstracts of this rule, lastly a formal operation on numbers, an automatic rule, an alogorithm.

Perpendicular and Automaton

When the ancients said 'according to the gnomon', they meant 'vertically'. We translate that as perpendicular to, for this word, in our languages and practises, refers to the plumb-line, that line which the Greeks called στάθμη (stathme). Here, the equipment of the mason is described with a word whose root implies stability, balance, like that of the word *episteme*, science itself. In this object, this artefact, are gathered, for an admirable coherence and support, the static origin of geometry which I found on reading Euclid's *Definitions* (cf. p. 114), in the *North-west Passage*, and its statuary foundation: epistemology and anthropology, linguistics and history. No longer just the earth and the sky, but knowledge and the thing. Shadows and light, the most ideal, abstract and formal laws and the most carnally human conspire wonderfully in this simple and easy plumb-line. Stable for mechanics, mass or heavy, dense stone, a straight statue pointing towards the low ground, a fine ruler which traces on the facing an almost perfect line as long as it is dyed with liquid colour (it then writes like the gnomon), this thing never makes mistakes and works automatically.

According to the plumb-line: perpendicularly. This adverb which we use without thinking should be re-thought or weighed up. What? The vertical gnomon means both intelligence and artefact? But so does the perpendicular. Certainly, it hangs like the mason's string and weighs like his weight, enjoys, of course, the maximum inclination, as much as the arm supporting the two dishes on a pair of scales, suspended like a pendulum. But it thinks. The verb 'think' in contemporary Romance languages – *penser, pensare, pensar* – has no other origin than the verb to weigh – *peser, pesare, pesar* – or to hang – *pendre, pendere, pender*. We may strive to make the connection between the hard literal and the soft figurative meaning, through evaluation or esteem, but in the decision on the scales concerning the gold content of a coin or of an ingot, even the anxiety bordering on fear or expectation, the reference remains the scales, the pendulum, still the plumb line or *stathme*: yes, the perpendicular thinks, or rather, the gnomon has the same connection or relation to knowledge, the same relation as that of the perpendicular to thought. Artificial intelligence is not new. From the beginnings of science there have been things or states of affairs that the history of our languages associates with mental activities, as if these artefacts, gnomon, plumb-line, ruler or compass, set-square, were considered as the subjects of thought.

That is not the same as restating the pragmatist theory of the origin of the pure sciences, according to which practice always precedes knowledge, things constructed by human hand holding or containing the secret of abstract speculations to come, as if the succession and the system of theorems unfolded, imitated, sublimated and rearranged a prior and obscure history of actions and gestures: deeds before the law. Ancestors, skilful but crude, did without knowing. We will never falsify or verify these judgements on the past, false and true at will like all laws of history, the misfortune being that education was founded on such arbitrariness. Nothing will ever either prove or invalidate pragmatism, the theory of teachers who believe that inventing consists of making an excellent copy of a text written by calloused hands or that discovery is reduced to interpretation. No, theory does not always come down to explanation of what is implied by the manual work. Sometimes yes, often no. A thousand manipulations do not lead to the rigour of the one who has already found it. But no matter. Profound linguists claim that the French word *baratin* (patter) comes from practice, or from the Greek verb corresponding to the French for 'to do', since the favourite discourse of intellectuals consists of extolling action, from which they refrain, to the detriment of abstraction, from which they never part. The height of *baratin* consists of speaking of doing while in fact merely holding forth, in short.

So, when it comes to knowledge, our languages bring us back to artefacts as simple and primitive as the plumb-line, and the gnomon tells us only that the human subject of thought is only recent. Artificial intelligence is more ancient than just intelligence, thought of as a faculty of the mind, itself being reduced, as the word expressly indicates, to a possibility of doing. *I think* is 300 years old, while the gnomon has been saying that it knows for more than 3000 years. And I find it harder to conceive of a virtual authority, internal to the individual, a transcendental condition of intellectual operations, than to see the line or the shaft of the sundial write automatically.

We use this last adverb carelessly. For us, an automatic action is accomplished without the participation of the will or human intention. Now the whole family from which this word comes is descended from an Indo-European root – *men* – where, on the contrary, mental activity is found: vehement, demented, commentary, mention, mendacity, memory, monument, monster, demonstration, the French word for watch: *montre*, money, come under the Latin subset from this root, while the words anamnesis, mania and automatic belong to a Greek relative. We use a word of understanding to describe something which we would like to be devoid of understanding. It is sufficient, within the family, to associate certain parents to obtain some lovely effects of meaning. For example, like a watch (*montre*), the automaton comments or demonstrates thanks to its memory and monstrously imitates mental activities. Here is a sentence which seems to meditate or decide on the seemingly bold questions which we ask ourselves concerning artificial intelligence, while it is reduced, in the eyes and to the ear of the craftsman of language, to the monotonous repetition of the same unit of meaning, to a sort of tautology or rather redundancy. The sundial probably owes its comparison to our watches to this. Our languages have known for a long time that automatons think, at least they said so before the Greeks, Arabs and modern and classical civilizations erected mobile statues for the ornamentation or the torment of their contemporaries.

In short, the automaton bears the same relation to mental activity as the gnomon has to knowledge, as the perpendicular or the pendulum to thought or the *stathme* or plumb-line to episteme, the stable statue to epistemology. Straight science, thought, knowledge, memory, mental acts, dementia or mania ... the philosophy we have learned prompts us to distribute them, like faculties, functioning well or badly, around a transcendental subject, compartment by compartment or like a crown, but the language which has recorded or spoken this philosophy for several thousand years recalls their origin, the shaft of the sundial, the set square, the string and the scales ... as if it were describing an objective intelligence. If there is a regulation, or several, for the governing of the mind, and if language still notes some redundancy between the orientation that this mind must follow and the thing that indicates it, since regulation and direction echo the Latin word *rectus* which means straight line, then the subject, in the third position, is only imitating an objective form. Does the mind, first, already reside in this objective form? And why fight against the refined pleasure of discovering the etymology, which is highly scientific, of *poêle* (French for frying-pan): a word which comes from the Latin *balnea pensilia*, hanging baths. What else is to be done in a frying-pan other than to say I think? (Descartes is said to have composed his *Discourse*, including the statement 'cogito ergo sum', by another kind of *poêle*, a stove.)

The philosophy taught today in the classroom, from which the lessons of things have disappeared, places the subject in language so that only those who hold forth acquire a noble status. They stop, timid, half-way towards this return to the objects of the world, since language re-sides in us, mouth, throat and movements of the body, and outside us, in libraries and signs, sound bands and radio receivers: internal and external, artificial and natural, and we are unable to make up our minds. The sub-ject there hesitates between a quasi-subject (from the collective culture to the individual unconscious) and a quasi-object (from books to codes). But what does such a phrase mean, when a word – subject – is elusive and cannot be pinned down between its literal meaning and its counter meaning? Constructed by us while we find ourselves constructed by it, collectively and in the course of a long history, used by us, individually and in groups, language, exercised in daily practice or in rare and stylized experience, teaches us immediately that it behaves like a thinking artefact. Its artisan is often led by it. In other words, language is part of artificial intelligence, like currency.

Gnomon also means set square and the perpendicular. A stele from the tomb of a Roman engineer showing the tools of his trade: the plumb line, compass, set square, level and ruler. (First century AD. Museo Capitolino, Rome. Photograph, G. Vasari.)

Matter and Form

The vertical gnomon, the angled set square, ruler, compass, perpendicular and pendulum assume a constant form: straight vertical, or horizontal in the case of the scales, normal or round, depending. Not only does form mean shape, face, edges, definition and determination in the literal sense, but also the object's principle of organization. The right angle describes both the appearance of the set square and its constructive framework, its construction. Thus form can be taken both as phenomenon and essence, appearance and reality. It is of no consequence whether stone, marble, iron or bronze are used as the raw materials for the shaft or the sundial, as long as it rises from the ground in the normal way. The information it shows or gives corresponds to its form and varies with it. According to its form, the information changes. Knowledge lies within the form. Language, again, assimilates form and information. The latter resides in the former.

The techniques of long ago informed matter: the potter modelled clay to extract the urn from the circle and from his tangential hands. Likewise, from a heap of stone the mason erected a house according to the architect's drawing, and the blacksmith twice did violence to the peaceful metal, with

95

fire and the hammer. Industry brought additional levels to the crafts, but in the same ways. We have changed all that. Our techniques today tend rather to explore or to recognize first of all the fine and complex forms scattered among the things of the world and to choose one of them or mix several when they correspond to our designs and to the constraints of the planned production: they even precede it sometimes. Certainly, we still make clocks out of metal, as in the past, but this crystal, that molecule, even that atom or isotope, now make better watches which are automatic and reliable, and this other crystal functions as a valve or semiconductor. The unformed forms lie in the things themselves where it is sufficient to retrieve them. Thus our works reverse ancient processes where information came only from our skilful hands or from expert understanding. Narcissistic idealism found in the world only its own image, which it stamped on with great effort. Science and technology reduce the real to their representations. Now the soft clay earth, the stone before the instrument, metal in its matrix, crystalline, in themselves and by themselves conceal hundreds of artefacts as in a cornucopia which ancient hands and desires were unaware of as they blocked it up. Our intelligence, our slightly foolish, violent, crude enterprise closed the doors of the treasury, while the world hides a thousand times more wonders than our decisions. The meaning, the direction, the work project are reversed. In this dawn of technology, we recognize first of all that the Universe has already created a great deal: here is the fount of information.

There is no matter in the Universe. Otherwise the physical sciences would have ended up encountering limits to their progress or their history, limits anticipated and set by materialist metaphysics. Now the latter vanishes as the former progress, ceaselessly finding forms without ever coming across a matter which they do not name, only recognizing the mass. Matter does not exist, only forms are found, like atoms, and right down to the tiniest particle, with or without mass, of countless forms, plus their mixing, chaotic or ordered, a system or clashing which shakes and agitates in its countless multiplicity as if in a basket. There is only information, the enormous store of which in the world, can probably be expressed by a very large number, mathematically finite but physically infinite, leaving science within an open history. Even weight codes a field of force, even any aggregate, colloid or organism supercodes a subset of coded forms. Only mixture and disorder, wrangle and chaos give the illusion of matter.

It follows that intelligence is immanent and, probably, coextensive with the Universe. The world provides a huge store of forms. Ours does not stand out from its black surroundings presumably to wait passively until we inform them. There is a vast objective intelligence of which the artificial and the subjective constitute small subsets. For us, knowing consists of putting ourselves into a form similar to what we know. The object we construct, we create in a fashion which is similar to certain things of the world, ultimately our guides. Intelligent, the gnomon intercepts the flow descending from the Sun and together, unaided, they draw on the ground, from which comes that erect statue, the objective and partial information of the shadow which speaks partly of the spherical form of the world.

Geometry was sleeping in the ground or dreaming in the radiance of the sun: the gnomon of the ancient Greeks or the Babylonians gradually wakened it through the singular forms common to shadow and light.

Meno

MENO: Yes, Socrates, but what do you mean by saying that we do not learn, and that what we call learning is recollection? Can you instruct me that this is so?

SOCRATES: I remarked just now, Meno, that you are a rogue: and so here you are asking if I can instruct you, when I say there is no teaching but only recollection: you hope that I may be caught contradicting myself forthwith.

MENO: I assure you, Socrates, that was not my intention: I only spoke from habit. But if you can somehow prove to me that it is as you say, pray do so.

SOCRATES: It is no easy matter, but still I am willing to try my best for your sake. Just call one of your own troop of attendants there, whichever one you please, that he may serve for my demonstration.

MENO: Certainly. You, I say, come here.

SOCRATES: He is a Greek, I suppose, and speaks Greek?

MENO: Oh yes, to be sure – born in the house.

SOCRATES: Now observe closely whether he strikes you as recollecting or as learning from me.

MENO: I will.

SOCRATES: Tell me, boy, do you know that a square figure is like this? [He draws in the sand.]

BOY: I do.

SOCRATES: Now, a square figure has these lines, four in number, all equal?

BOY: Certainly.

SOCRATES: And these, drawn through the middle [of each side of the square], are equal too, are they not?

BOY: Yes.

SOCRATES: And a figure of this sort may be larger or smaller?

BOY: To be sure.

SOCRATES: Now if this side were two feet and that also two, how many feet would the whole be? Or let me put it thus: if one way it were two feet, and only one foot the other, of course the space would be two feet taken once?

BOY: Yes.

SOCRATES: But as it is two feet also on that side, it must be twice two feet?

BOY: It is.

SOCRATES: Then the space is twice two feet?

BOY: Yes.

SOCRATES: Well, how many are twice two feet? Count and tell me.

BOY: Four, Socrates.

SOCRATES: And might there not be another figure twice the size of this, but of the same sort, with all its sides equal like this one?

BOY: Yes.

SOCRATES: Then how many feet will it be?

BOY: Eight.

SOCRATES: Come now, try and tell me how long will each side of that figure be. This one is two feet long: what will be the side of the other, which is double in size?

BOY: Clearly, Socrates, double.

SOCRATES: Do you observe, Meno, that I am not teaching the boy anything, but merely asking him each time? And now he supposes that he knows about the line required to make a figure of eight square feet; or do you not think he does?

MENO: I do.

SOCRATES: Well, does he know?

MENO: Certainly not.

SOCRATES: He just supposes it, from the double size required?

MENO: Yes.

SOCRATES: Now watch his progress in recollecting, by the proper use of memory. Tell me, boy, do you say we get the double space from the double line? The space I speak of is not long one way and short the other, but must be equal each way like this one, while being double its size – eight square feet. Now see if you still think we get this from a double length of line.

BOY: I do.

SOCRATES: Well, this line is doubled, if we add here another of the same length?

BOY: Certainly.

SOCRATES: And you say we shall get our eight-foot space from four lines of this length?

BOY: Yes.

SOCRATES: Then let us describe the square, drawing four equal lines of that length. This will be what you say is the eight-foot figure, will it not?

BOY: Certainly.

SOCRATES: And here, contained in it, have we not four squares, each of which is equal to this space of four feet?

BOY: Yes.

SOCRATES: Then how large is the whole? Four times that space, is it not?

BOY: It must be.

SOCRATES: And is four times equal to double?

BOY: No, to be sure.

SOCRATES: But how much is it?

BOY: Fourfold.

SOCRATES: Thus, from the double-sized line, boy, we get a space, not of double, but of fourfold size.

BOY: That is true.

SOCRATES: And if it is four times four, it is sixteen, is it not?

BOY: Yes.

SOCRATES: What line will give us a space of eight feet? This one gives us a fourfold space, does it not?

BOY: It does.

SOCRATES: And a space of four feet is made form this line of half the length?

BOY: Yes.

SOCRATES: Very well; and is not a space of eight feet double the size of this one, and half the size of this other?

BOY: Yes.

SOCRATES: Will it not be made from a line longer than the one of these, and shorter than the other?

BOY: I think so.

SOCRATES: Excellent: always answer just what you think. Now tell me, did we not draw this line two feet and that four?

BOY: Yes.

SOCRATES: Then the line on the side of the eight-foot figure should be more than this of two feet, and less than the other of four?

BOY: It should.

SOCRATES: Try and tell me how much you would think it is.

BOY: Three feet.

SOCRATES: Then if it is to be three feet, we shall add on a half to this one, and so make it three feet? For here we have two, and here one more, and so again on that side there are two, and another one; and that makes the figure of which you speak.

BOY: Yes.

SOCRATES: Now if it be three this way and three that way, the whole space will be thrice three feet, will it not?

BOY: So it seems.

SOCRATES: And thrice three feet are how many?

BOY: Nine.

SOCRATES: And how many feet was that double one to be?

BOY: Eight.

SOCRATES: So we fail to get our eight-foot figure from this three-foot line.

BOY: Yes indeed.

SOCRATES: But from what line shall we get it? Try and tell us exactly: and if you would rather not reckon it out, just show what line it is.

BOY: Well, on my word, Socrates, I for one do not know.

SOCRATES: There now, Meno, do you observe what progress he has already made in his recollection? At first he did not know what is the line that forms the figure of eight feet, and he does not know even now: but at any rate he thought he knew then, and confidently answered as though he knew, and was aware of no difficulty; whereas now he feels the difficulty he is in, and besides not knowing does not think he knows.

MENO: That is true.

SOCRATES: And is he not better off in respect of the matter which he did not know?

MENO: I think that too is so.

SOCRATES: Now by causing him to doubt and giving him the torpedo's shock, have we done him any harm?

MENO: I think not.

SOCRATES: And we have certainly given him some assistance, it would seem, towards finding out the truth of the matter: for now he will push on in the search gladly, as lacking knowledge; whereas then he would have been only too ready to suppose he was right in saying, before any number of people any number of times, that the double space must have a line of double the length for its side.

MENO: It seems so.

SOCRATES: Now do you imagine he would have attempted to inquire or learn what he thought he knew, when he did not know it, until he had been reduced to the perplexity of realising that he did not know, and had felt a craving to know?

MENO: I think not, Socrates.

SOCRATES: Then the torpedo's shock was of advantage to him?

MENO: I think so.

SOCRATES: Now you should note how, as a result of this perplexity, he will go on and discover something by joint inquiry with me, while I merely ask questions and do not teach him: and be on the watch to see if at any point you find me teaching him or expounding to him, instead of questioning him on his opinions. Tell me, boy: here we have a square of four feet, have we not? You understand?

BOY: Yes.

SOCRATES: And here we add another square, equal to it?

BOY: Yes.

SOCRATES: And here a third, equal to either of them?

BOY: Yes.

SOCRATES: Now shall we fill up this vacant space in the corner?

BOY: By all means.

SOCRATES: So here we must have four equal spaces?

BOY: Yes.

SOCRATES: Well now, how many times larger is this whole space than this other?

BOY: Four times.

SOCRATES: But it was to have been only twice, you remember?

BOY: To be sure.

SOCRATES: And does this line, drawn from corner to corner, cut in two each of these spaces?

BOY: Yes.

SOCRATES: And here have we four equal lines containing this space?

BOY: We have.

SOCRATES: Now consider how large this space is.

BOY: I do not understand.

SOCRATES: Has not each of the inside lines cut off half of each of these four spaces?

BOY: Yes.

SOCRATES: And how many spaces of that size are there in this part?

BOY: Four.

SOCRATES: And how many in this?

BOY: Two.

SOCRATES: And four is how many times two?

BOY: Twice.

SOCRATES: And how many feet in this space?

BOY: Eight feet.

SOCRATES: From what line do we get this figure?

BOY: From this.

SOCRATES: From the line drawn corner-wise across the four-foot figure?

BOY: Yes.

SOCRATES: The professors call it the diagonal: so if the diagonal is its name, then according to you, Meno's boy, the double space is the square of the diagonal.

BOY: Yes, it certainly is, Socrates.

The slave in Plato's *Meno* is evidence of a forgotten world which he re-calls before us, through an exercise in remembering.[8] Similarly, we must believe that Plato and Socrates consciously evoke the rhythms inspired by the poets who take them back to those lost times. But it is also necessary to describe precisely these worlds and those times which re-emerge during the demonstration.

When historians of science return to the problem of the duplication[9] of the square discussed here, they seek, in this section of *Meno* and in the figure, traces or evidence of fifth-century Greek geometry, forgotten today by every-body other than themselves, because only rare fragments have been pre-served, including this one. To reconstruct the diagram and demonstrate the relation between the side and the diagonal makes it possible to reconstruct lost knowledge and the past: a work of remembering. Now the history of science generally makes no reference to this theory which prompted Socrates to call upon an ignorant servant and raise for him this problem. Equally, the history of philosophy dealing with remembrance makes no reference to the duplication of the square itself. And if by chance the two memories were to make themselves known? Were Socrates and the slave engaged in the same task as ourselves, trying to recover forgotten knowledge? What relationship can science be said to have with memory?

Let there be a square whose surface we wish to double: how many feet will the side of the new square measure? Whatever the answer, we have to extend the two sides of the first square. We are back to the gnomon, the old form of the angled set square whose outline traces a space in the form of the initial square and whose wooden or iron extensions show the supplement added to it. Duplicating the given surface consists of building the set square: here is the problem of the gnomon. Can it be resolved thus? At least the young slave begins, with Socrates, to work it out like this: there is no doubt that he goes wrong because of such a drawing, since the real solution begins when he abandons it. His mistake is therefore a result of his asking this question of the gnomon. Hero of Alexandria says that everything which, added to a number or a figure, gives a whole similar to that to which it has been added, must be called thus. Duplication provides a crude individual example of such a similarity.

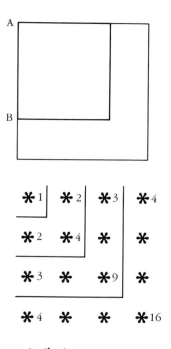

Now the mistake happens twice and twice through excess. Why? Starting from one side AB, which is two feet long, therefore with a surface of four feet, the slave extends the first side by doubling it, making it four feet long, and ends up with a surface of sixteen, when he was asked to make a surface of eight, the double of four. He then works backwards and chooses a side of three feet in length and obtains a square of nine square feet. This overshooting of the mark can again be explained by the problem of the gnomon. This word means the set square but also, let us repeat, the Pythagorean table which shows perfect squares, odd numbers and series of integers: the former on the diagonal, the latter along the sides. The odd numbers make the gnomon, on the remaining set square.

Now the ignorant young slave jumped from two to four and went down from four to three: so he followed the sides of the square in integers in the sense of the geometrical algebra of the ancient Pythagoreans. In other words, the gnomon still precedes him.

Reminiscences

He remembers. He recalls first of all attempts at definition in the dialogue: he must have heard it, hidden away in some corner. Let us remember that the similar exercise in the defininition of virtue consisted of that of the figure and that they agreed initially to reject the first two results: the figure is neither form nor colour; but they agreed to say that it is the limit where a solid ends. The line denotes the edge of the figure as the figure traces that of a body. So the slave is mistaken because he follows the edge, that of the square drawn by Socrates and that of the numeric table. But from the line to the surface area and from the surface area to the volume, that is from the limit to the variety which it surrounds or defines, the conclusion is not valid. The slave is mistaken because he remembers definition by the edge; short-term memory.

He remembers, secondly, the state of Greek geometry before the discovery of the diagonal, a forgotten world. He remembers geometric algebra, old Pythagoreans, the reign of whole numbers. The mathematical world of Plato, Theodorus, Theaetetus and Eudoxus is completely divorced from this one. In those days, people trusted the gnomon, whose job it was to know. The new school had lost that knowledge, which had become contemptible and only fit for slaves. And the young man knows it, says it and represents it. He knows the gnomon table well. Really? We witness, as we read and hear the dialogue, 2000 years later, that he knows his multiplication tables, since to the question: 'how many are twice two feet?' he replies without hesitation 'four'. And he confirms with ease that four times four make sixteen and that three times three make nine. But for Socrates and his school, this numeral and tabular knowledge amounts to ignorance. To know one's numbers is to know nothing. But we read that the slave recites his table. What is a table in fact, if not memory, the easiest to retrieve? The slave follows the table and the chart and the gnomon: he remembers. He remembers a knowledge which Platonism conceals and despises. In other words, behind geometry, precisely what determines a double square by the diagonal of the initial square, arithmetic and geometric algebra hide in oblivion, remembered by he who is despised. Suddenly, through his body, his language and above all his condition, he testifies to the rank to which the ancient science has fallen: into the order of ignorance and servitude, into the camp of the concrete in relation to the abstract. The philosopher reserves for himself the metalanguage in which this new relationship of the pure and the concrete is defined and so from now on can judge knowledge and its history as he pleases, by making them both begin with himself.

But Socrates also remembers when he says that he does not know. It remains true that he does not know. He doubts and seeks. And questions. And above all cuts up the great rhapsodical phrases and sections of encyclopaedia into elements and pieces. An infantryman, a pedestrian, he wants to walk step by step. First this, then that. Let us first make this unquestionable before moving on to that, which we will examine in the same way. Let us cut in two and proceed through dichotomies. Socrates only knows these procedures, which constitute a prudent, circumspect method or path. But let us take seriously again the divine theory he had just borrowed from Pindar, supposing he too remembered ancient knowledge Socrates remembers the step-by-step procedures of algorithmic[10] thought and he represents it through his character and his condition of a man who speaks but does not write. Since the dawn of time in the Fertile Crescent, division by two, which was favoured, makes it possible to work things out in one's head more easily. The little slave and Socrates walk together at the same pace towards the vanished world of which they are the personification. The elderly master rhetorician questions the ignoramus who can neither read nor write, according to the ancient and precise procedures which the latter is aware of, without ever taking his eyes off the previous link when he moves on to the next link and going backwards at once if he misses one – returning therefore to three after the abrupt leap from two to four.

The game no longer has two players but three: not Socrates, Meno and the slave, since the last two are interchangeable, but Plato, Socrates and the ignoramus. The Paedeia, education and history, goes through three stages: the philosopher-king, the foot-soldier and the manservant or worker in the fields, according to the ancient division. Plato thinks in the universe of geometry, pure space, rigorous metric theory, controlled irrationality: here is the diagonal, the alogos linked to the logos and mixed up with it, here is the Royal Weaver whose portrait ends the *Politicus*. As for the slave, he mentally counts the whole numbers in the traditional algorithm, the despised logistics of shopkeepers and farmers, while Socrates, still reasoning in the former state, without writing, discovers the new world of the square carrying the diagonal round its neck. He makes the link between the two kingdoms, like a messenger.

Plato haunts our thoughts, which we are unable to shake off, or rather we inhabit those which he conceived of while the little slave has not left the ancient Pythagoreans still concerned with Babylonian tables. Socrates knows nothing, like the child; nor does he write, like the slave; they both follow the ancient ways through which we and Plato remember, an ancient moment immersed in oral methods and step-by-step procedures, awestruck and hand in hand, but suddenly gaining access to a new, abstract world.

Algorithmic thought is engulfed by oblivion and no longer constitutes, through its counting rhymes, anything but the prehistory of science. The young slave remembers the gnomon and its tabular laws because it functions as a memory, like the multiplication tables. Algorithmic thought, which can be made artificial, was probably reduced to such memories. Let us not say: artificial intelligence, but rather: artificial memory. In the past, let us remember, knowledge was perhaps reduced to a memory. But the new geometry reveals its gaps: no number is to be found on the gnomon between 3 and 4 on the sides, nor between 4, 9 and 16 on the diagonal. Geometry compensates for its failures, invalidates a knowledge linked to memory. It invents another world teeming between the numbers, of which one soon loses count. A temporary end to the struggle opposes abstraction and the memory, both considered as economies of thought: here one has the upper hand while the other flees. Although it has lost the Greek battle, it carries on the fight among the Arabs in the Middle Ages, among the greatest classical mathematicians, such as Pascal and Leibniz, architects of algorithms rather than of geometries, and finally into the contemporary era. We have just learned to economize thought, thus to win, on the two tables: the one where the light of the Platonic sun still shines, pure mathematics, but also where memory has subjugated the very speed of that light. Objective slaves work at the heart of computers: the entire ancient dialogue follows procedures which can easily be written as software.

Measurement and Position

The discussion has suddenly branched off from arithmetic into geometry: hands up if you prefer not having to make calculations! Socrates is cheating, apparently. He asks the length of the side. The slave loyally replies four or three feet. He is required to give a measurement but he gives a quantity. But when the diagonal arrives as a side of the doubled square, it is only quality that is mentioned: on which line is the square with twice the surface built? On this one. Interrogative and demonstrative have now abandoned quantification to qualify what is being shown. Nobody asks the asker: how long? He questions the ignorant slave about a content about which nobody, however, bothers him. He found the side all right, but he did not measure it. Socrates is cheating: he knows that he will not find the exact length.

The slave erred twice on the side of excess because he measured the side of the square using whole numbers: the slave counted four and found sixteen, came down to three and found nine. First attempting an even number, then an odd, he twice overshot the mark. The number sought therefore will be neither this even nor this odd number.

Torpor and Narcosis

With an impasse, a problem, the dialogue ceases and Socrates, in the meantime, reminds Meno of his comparison to a torpedo. The metaphor expresses the contradiction and confusion in which the philosopher's inter-locutor finds himself. But we ourselves understand nothing before remembering the origin of the torpedo: this fish is so called because it plunges us not into stupor but torpor. Its touch causes us to lose consciousness, as if asleep. But again, we understand nothing if beyond its Latin origin we do not recall that in Greek, the torpedo is called νάρκη (*narke*), which associates it with narcosis and our narcotics. Here is a curious pharmacology. The shock received on contact with the creature seems electrochemical to us today. We elucidate this experience through various sciences, electrostatics, biochemistry, neurology, a whole range of sophisticated means at our disposal. Now our medicine chest of narcotics brings us back to the torpedo as if language, through its history, had followed the same path as science itself which, for at least two centuries, has carried out countless experiments around this extraordinary fish. It seems as if there were two parallel histories of science: one which tells of the manipulations of physiology and one which remembers the Latin torpedo and the Greek narcosis, narcotic sleep and the strange torpor into which the shock plunges us. We understand through our science something which is akin to electricity, about which Plato knew little, but Plato names a creature so that we might understand something akin to our chemistry, to our pharmacology, but also to his. The torpedo causes drowsiness like a narcotic. Narcissus was fascinated with himself to the point of falling asleep, totally wrapped up in himself, on seeing his image reflected in the smooth waters of a stream. Narcissus-narcosis bears the name of a fish, or bears that fish within him, and strikes himself down, totally alone like a *pharmakos*,[11] with neither a society nor an environment. Narcosis maintains with the lone individual the same relationship which the archaic victim called *pharmakos* by the Greeks had to the group. In the game of self-knowledge, are the Is going to kill the ego as the raging crowds put the *pharmakos* to death? Know yourself! Strike yourself down, narcissistic subject of thought! Philosophy of the subject, that suicidal drug ...

When we retrace its history, our knowledge developed in series – electricity, chemistry, pharmacology, neurology, psychopathology – closing the way a fan is folded, and our language alone, transmitted, links us to the past like a black line. Contemporary scientists appear proud, quite rightly, to have discovered a biochemical origin for the electric shock. Of course. But language already knew it, and had done for a long time. Sometimes the history of science requires only a certain memory, an artificial memory of the language.

Odd and Even: the apagonical demonstration, via the absurd

Let there be a square with a side of 1 and its diagonal b. From Pythagoras' theorem we know that $b^2 = 1^2 + 1^2 = 2$ hence $b = \sqrt{2}$. Since $1^2 = 1$ and 2^2

= 4, *b* has a value between 1 and 2. Let us express this value as m/n, supposing that this 'fraction' is reduced to its simplest expression. Therefore $\sqrt{2}$ = m/n from which we deduce: $m^2 = 2n^2$. Now m^2 is even: therefore so is m. First conclusion: *n is an odd number.*

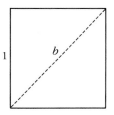

Now an even square is divisible by four, which is the case of m^2; therefore $2n^2$ is also divisible by four. So n^2 is even and: *n is even.*

Consequently, n is odd and even, which is impossible. The square root of 2 cannot be expressed as m/n. Since Antiquity, this first demonstration via the absurd has been known as apagogical.

It shows the failure of primitive Pythagorean arithmetic which only took into account whole numbers or, to be precise, rational numbers. Suddenly space shows lengths which calculation no longer includes. 'If you would rather not reckon it out, just show what line it is': this phrase of Socrates', more adroit and profound than meets the eye, clearly illustrates the bifurcation.

The apagogical demonstration shows that numbers make impossible that which space, obviously, makes possible.

Socrates' demonstration in *Meno* says that space makes possible that which numbers make impossible.

And they both go through even and odd numbers.

The dialogue remembers the apagogical demonstration and follows it, so to speak, in reverse. And the torpedo strikes through contradiction or absurdity. Apagogical means led away from the correct path, diverted or seduced. I certainly did say bifurcation. Seduced: fascinated by torpor.

Now the gnomon is formed by whole numbers, odds and evens: the young slave followed them. Show, now, do not count any more, show the diagonal! Here it is: it passes through 1, 4, 9, 16 . . . the numbers which will be called perfect squares from now on. Come, now show the diagonal side of a square with an area of eight feet! It is not there. Not demonstrable, indemonstrable.

The gnomon knows only perfect squares, the perfect science of the logos, unaware of the irrationals; the archaic and highly imperfect science of the perfect logos. Mathematics in its demonstrative authenticity is born, as a result, outside the logos, when it distances itself from the logos and can rigorously measure this distance. Science begins outside language. The gnomon therefore does not know at all. One can ask for or invent knowledge unknown to this memory which bears the name of that which knows. That is the torpedo's thunderbolt. The existence of knowledge outside the gnomon authorizes the search for what is not known – that which knowledge itself does not know.

The torpedoing of the gnomon, torpedoing of old practices, of their memory, of counting by space, of logos through the alogos, of the expressible through the inexpressible, of language through science, the torpedoing of artifice, of linguistic and artificial memory, of algorithmic thought.

Formerly judge, assayer, touchstone, the gnomon no longer decides or knows; ignorant as a boy slave, twice a fool. Deliverance! There is knowledge outside the memory!

There is no demonstration before the Greeks, before apagogical demonstration, before geometry, before the irrational. Certainly, there is only counting. If you would rather not reckon it out, just show what line it is! There's an original phrase. Show, demonstrate! To invent geometry and demonstration consists of filling the gaps of the gnomon, those of knowledge, of artificial intelligence, of algorithmic thought. The latter does not demonstrate. It only counted.

The Emergence of Ideal Figures

Faithful and refined as the reconstruction by algorithms of Greek mathematics in its infancy may be, the fact remains that a kind of infinitely withdrawn other world differentiates itself from this enterprise through the geometry of lines and solids, abstract space or ideal objects.

Algorithmic thought or practice takes into account the theory of numbers, of measurement, of variable and profound thoughts on rationals and irrational numbers resulting from the duplication of the square or that of the cube, but assumes, cube or square, sides without thickness and rigorous, transparent or perfect solids, which did not exist before the Greek dawn. We now have to understand the emergence of these idealities.

It can however go further than arithmetic, formally speaking, for its step-by-step procedures are a constant testimony to the intentional, controlled safety of its stages. It knows where it is going and does not go about it in a haphazard fashion. It is possible therefore to imagine a method, in the etymological sense of a designated path, which extends its process to more complex or more general rules which would make it possible to advance what had been planned in a given programme beforehand and only what one would find there to the exclusion of all else. The algorithmic procedure would then present a simple preliminary sample of what would subsequently become a demonstration in form. From the step-by-step process towards forbidding any unforeseen steps is only a short step. In other words, the theory and the practice of demonstration suppose an algorithm. In history, the one prepares the way for the other.

Zeno of Elea

Once again I imagine that the Elean school must have made a decisive contribution towards bridging the gap between the formula of rigour and the usual space of the ideal expanse where new objects manifested their appearance.

The paradoxes of Zeno[12] allow the *mise en scène* (setting) to be forgotten in favour of their *mise en forme* (formation). And supposing they led us infinitely from the one to the other? The arrow which flies from bow to target, or Achilles who strove to catch up with the tortoise, like the hare in Aesop's fable, both without hope of success, each taking a path; in other words, a method. Observe the precision with which all the elements of an algorithm fall into place: a path or method by which to reach a goal, the practical and simple aim of a device, the exact measurement of the segment covered, the breaking down of the process into elements, step-by-step procedure. There's no mistake about that, repetition echoed in the figure and the form, the scene and the number, the same gesture to be made after the same gesture, derived most probably from a fable. Observe again a certain imitation of the antipheresis, an alternative algorithmic subtraction, born of tradition, and which here takes away half of everything, then half of what is left, then again, half of what is left, and so on, as if Achilles or the arrow practised subtraction by moving. Observe finally, turning the clock forwards, how the infinitesimal algorithm still to be born, either in Abdera in a century or in the classical age of 2000 years ago, will bring little new compared to these procedures. The whole scenario then, starting with form, reveals algorithmic thinking.

Achilles runs or walks, the arrow is shot and flies, the whole formula fails. Neither the champion runner nor the tip of the arrow reach their goal. For the first time, a process spreads which is certain of its result and a good measuring formula, due precisely to their perfect workings and an illuminating and excellent example. Repetition only leads to repetition, marking time step by step with no possibility of stopping. The courageous hero will be mocked, a derisory image of the cowardly beast, speed no longer being of any use to them. In canonic fashion, Zeno assassinates traditional metrology: the age-old algorithm from the Fertile Crescent died in Elea.

The path of the arrow or of Achilles no longer goes towards the prescribed goal, but bifurcates, suddenly attracted to a very different target. Running, flying, the two vectors sink into the narrow rut of the segment, like an abyss linked to the tenacious algorithm, but together aiming for a single point at the limit of all the points actually travelled or possible, filtered through all the stations passed. That means that the places through which one passes or can pass are eliminated or subtracted, the places that have been reached or can be reached are disqualified, all the resting or occupied places are discredited, in favour of the only point towards which we travel without ever arriving. We can already hear Platonic rumblings. The process, which is in fact simple, that distinguishes this point from all the others, divides the segment through a single dichotomy; in other words, all points and one single point. On one hand, we can see, touch and tread on concrete ground, actually or virtually, remain there, gain access to it, pass through and leave the world or the path of these places remaining accessible to running or flight; on the other hand, a point emerges, intangible, impassable, inaccessible, which Achilles will never see, which the tip of the arrow will not pierce, and which will never be inhabited. It emerges from the vast sea of other points. The world which we can measure as much as we like, either approximately or even exactly, borders immediately on another world which is infinitely remote, without dimensions since metric theory is exhausted trying to reach it: an absent hole in the drawing. So draw the path of the hero or the flight of the vector in the sand, but you will not mark on their orbit the point towards which they are all hurtling: nobody can write or draw it. If you prick it on the page or in the sand, if Achilles or the arrow tip were here, it cannot be the point. You are holding the stylus, in other words, the spear itself, the flying line with which you write on the page, and it cannot inscribe the point after which it is always running. Science was born outside language; it is born outside writing. Here is the first intelligible, atopic place, at the end of this short path equal to the longest possible path. Geometric abstraction becomes the limit of the infinite sum of algorithmic subtractions.

Here is someone, of such and such an appearance or such and such an age, alive and well and distinctive, with a hundred characteristic marks; to think of him, says Plato, you have to imagine, in another world quite separate from this one, an idea of man or the ideal man. One has something of the nature of the other. How is it possible to imagine them both, the theoretical and the concrete together, says Aristotle, without forming an abstract idea of a third, of which they are a part? And again, how is it possible to imagine the three, the man of this world, the man of the first theory and the second man of the second theory, without a fourth who . . . and so on *ad infinitum.* This argument applied to the third man, far from criticizing or destroying the intelligible abstract place of ideas or forms, contributes to describing and founding it. Likewise Zeno's scenario leads

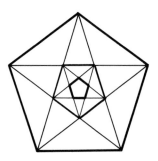

The line of the diagonals of the pentagon reproduce a pentagon whose diagonals create another pentagon and so on ad infinitum.

infinitely from the concrete representation or the metric formula to the surroundings, to the limit of non-representable ideality which cannot be drawn or written, subtracted to the point of exhaustion from all apprehension: visited or visitable points with an invisible, inaccessible goal. The abstract lies at the heart of this abyss, infinitely far, but infinitely near.

Here is the ignorant young slave, alive and well, who, under Socrates' torpedo, demonstrates the duplication of the square by building the irrational diagonal. Plato claims that he remembers a forgotten world since he knows without having learned. Without the intervention of Aristotle, we bring back into the picture the other slave in the other world in the process of calculating the area of another square, a scene which in turn brings to the abyss, in its echoed square, an infinite implication of diagonals, of sides or of aggravating Socrates. We have just remembered, by inventing it, the whole body of questions and problems, mathematical and philosophical, which enter the category of the third man. Interminably, a figure which constantly repeats or fits into itself, worlds within worlds disappearing upwards and oblivion rooted in missing memories, a young slave being reborn from his ignorance, calculates, counts, doubles a length then subtracts part, incapable of reaching unaided the diagonal which can and cannot be drawn, present there but irrational. The slave thinks algorithmically, the master does not forget geometry.

The definition of the geometric abstract, the model of the theoretical abstract required by Plato to think or exist or perceive, emerges from an infinite method or path down which Achilles and the arrow precede and guide us, leaving indefinitely behind them the algorithms which have become stuck.

Generalization

Zeno's reasoning is repeated: before arriving at the goal, it must pass through the middle of the segment, but before that it must pass through the quarter and again before that, the eighth, and so on, indefinitely, so that Achilles cannot start. The initial point therefore takes on the same status as the end point. The paradox, by the same step-by-step processes, is applicable to the middle point and then to any other point, so all segments become ideal.

This whole series of demonstrations has to be called a paradox because the elements which emerge from it are far removed from common belief. The term *Elements*, which translates into Latin and our modern languages the title used by Euclid and probably Hippocrates of Chios before him, originates from the letters L,M,N, in the same way as the alphabet spells the first Greek letters: alpha, beta, and the sol-fa sings the notes: sol, fa. The authentic title Στοιχεῖα *Stoicheia* does indeed mean letters, understood as elements of the syllable or of the word.

But as well as these elements of written language, it means those of the

Take a square with a side of 1 and its diagonal $\sqrt{2}$.

$$\sqrt{2} = 1 + \cfrac{1}{2 + \cfrac{1}{1 + \cfrac{1}{\sqrt{2}+1} + 1}} = 1 + \cfrac{1}{2 + \cfrac{1}{2 + \cfrac{1}{\sqrt{2}+1}}},$$

If we subtract them from each other we get $\sqrt{2} - 1$.
If we begin the subtraction again we get: $1 - (\sqrt{2} - 1)$.

Now the ratio of the diagonal to the side $\dfrac{\sqrt{2}}{1}$
is equal to the ratio between the two

remaining sides: $\dfrac{1 - \left(\sqrt{2} - 1\right)}{\sqrt{2} - 1}$ and so on:

$$\sqrt{2} = 1 + \cfrac{1}{2 + \cfrac{1}{2 + \cfrac{1}{2 + \cfrac{1}{2 \ldots}}}}$$

Which is verified by multiplying the means and the limits.
So everything starts all over again at infinity.
Let us find the value of $\sqrt{2}$ compared to unity.

$$\sqrt{2} = 1 + \left(\sqrt{2} - 1\right),$$

$$\sqrt{2} = 1 + \cfrac{1}{\sqrt{2} + 1}$$

It is enough to replace $\sqrt{2}$ by its proper value:

$$\sqrt{2} = 1 + \cfrac{1}{1 + \cfrac{1}{\sqrt{2}+1} + 1} = 1 + \cfrac{1}{2 + \cfrac{1}{\sqrt{2}+1}},$$

Theodurus? Thaetetus?

world, water, earth, in the manner of Empedocles who uses, from the term *rhizo*:[13] root, the radical origin of things; elements of the universe, stars, planets; of grammar, nouns, verbs; of logic, of rhetoric, of geometry . . . In this list or table no one discipline seems to have sought supremacy; neither language nor science prevails over things; the objects themselves do not precede signs. The night sky spreads out a series of points; the atoms, the punctual elements of things, are often presented as letters or numbers, unanalysable and to be combined. Moreover, both Proclus and Aristotle say of the elements of geometry that they constitute the subject-matter of geometry, as those who teach or learn it begin with these: foundations or rudiments, it depends. And so it does not appear that the Ancients sought or thought of elements absolutely first or last: there are elements everywhere, in local tables.

The corresponding verb, στείχω (*steicho*), designates the act of advancing in ranks, like an army in battle formation, so that the name corresponds to the line, the column, the row. Of this family, English has only retained certain technical words: distich, a unit of two verse lines or a couplet of a hexameter or pentameter; stichomythia, in Greek tragedy, dialogue in alternate lines, employed in sharp disputation and characterized by antithesis and rhetorical repetition. Prosody,[14] which uses these terms, counts the long and short syllables in dactyls, trochees and anapaests, with dots and lines as in Morse code. Atom, element: point, line. Again, what is an element? This mark, this line, the dash, the hyphen, in general the note, as these words were used by Liebniz. And in the plural, a series of these notes, a series generally grouped in a table or a chart of points and lines, in well-ordered lines and columns. As far as I know, the *Elements* of geometry also consisted of points and lines which we have to learn to draw.

Today, as in the past, everywhere we see similar tables: the letters of the different alphabets, numbers in all bases, axioms, simple bodies, the planets, markings in the sky, forces and corpuscles, the functions of truth, amino acids . . . Our memory preserves them so easily that they themselves constitute memories: objective, artificial, formal, in precisely the same sense as the ancient tablets of the law. What then does the term elements refer to? A table, open to every conceivable table; memory, in general: whatever a body of knowledge constantly refers to. Thus Euclid's *Elements* construct a system in the ordinary, logical sense, inferred and well-founded, but they also constitute a memory in the triple sense of history – hence the commentaries – of automation and of algorithms.

Now a unique and remarkable meaning emerges from this collection of meanings which are so coherent: στοιχεῖον (*stoicheion*) means the pointer that marks the shadow on the sundial; the gnomon, perhaps, but above all the local mark which records the hour. This hour, which popular etymology associates with the horizon, understood as a limit, the mark or line of the extreme boundary, finds itself here at the border between light and dark. Wondrous depth, time is defined as the common boundary between light and dark. Here is the ungula, the bronze or gold dash which preserves the memory of a fleeting moment. Here is the series of those stabilized marks, a range of elements along the substylar line, spelling in turn the longest day and the shortest day, the average night, the solstices and the equinoxes, the angle of the ecliptic, the axis of the world and the latitude of place . . . a range or table of elements for the map of the Universe; again an artificial memory sculpted on the sundial, elements of cosmology around the gnomon and marked by it. In semantic terms, an original trace can be discerned in the word element. Among the scattered or orderly lines on the table, we suddenly feel we can read who drew or traced them, as if, in this new sundial, a light, a shadow, a date were being revealed.

Statics

Has anything generated more literature, over the centuries, than Euclid's books, and especially his beginnings, *Definitions, Postulates* and *Axioms*, elements of *Elements*? Almost as much has been written on Euclid as on the Bible, and sometimes in similar terms, those of history and logic. And in the same way that a commentary on the Bible in turn becomes a Bible, so does a commentary on *Elements* of geometry become a system of geometry or of logic.

A description of the same order as the one before, analysing the meaning of the words used, led me in the past to say that beneath the statement of pure and formal idealities, recorded under the heading *Definitions*, lay a static base. The Greek terms which we translate as plane and trapeze mean first either a table or something at ground level. Similarly the verbs mean rest or equilibrium. Everything then happens as if the geometrically well-defined list were secretly building increasingly complex stabilities starting from the very simplest, the top spinning on one point, at the end of the list, from the lowest point, at the top. These things at rest become troubled or broken by successive angles to uncover, gradually, increasingly refined equilibria, as if the prolegomena of mechanics were concealed beneath the preliminaries of geometry. And this space of statics harks back to the Greek term ἐπιστήμη (*episteme*) which, meaning knowledge, still retains the trace of a stable invariancy.

I am convinced of this origin, particularly as the analyses of *Statues* have since relayed and reinforced this positive foundation with an anthropological foundation, expressing similar contents in a completely different language. The *episteme* will engender statics whereas it came from statues. Here I am weaving together the two preceding books and showing that the history of science follows on naturally from the anthropology of science. Such a transition was to be expected. The *Statues* emerge from the earth, slowly, as do *Definitions* and geometry. In short: at last we understand its name.

Who traces the elementary mark, who draws the line? The Sun on the earth, thing writing on thing, or is it the gnomon, standing like a statue made of earth, on the sundial's face, artefact drawing on artefact? The *Elements*, by their very title, seem to acknowledge an astronomical origin; the *Definitions* then, through their verbs and their nouns, enable us to surmise on a static root – or rather I would now say a statue's pedestal.

The Gnomon or Set Square breaks down into Ruler and Compass

Although the Ancients say nothing on the subject, the *Postulates* make it possible to trace the *Elements* of geometry, literally its lines, with the ruler and compass: the straight line, finite or infinite, the circle, the parallels, the right angle. In the case of the right angle, the set square disappears for it is enough to draw, using the two classic implements, a right-angled triangle and a semicircle. Formerly called the gnomon, the set square is thus broken down into two components, appropriate for drawing lines or elements: the ruler and the compass, which carry within them and preserve, invariably in a form in wood, bronze or marble, the possibility, the capacity of constructing or drawing dashes, lines, marks, points, short lines or curves, the real and intellectual elements of geometry.

The *Definitions* and *Postulates* construct the table or chart of elements or lines, in the formal, linguistic, pure or abstract sense of these terms, the sense in which they been understood ever since. But the compass and the ruler (or their sum, the gnomon) show the concrete table of these elements. For they make it possible to construct, draw, or lead them, they contain or imply in some way an infinity of straight lines, circles, points, right angles, parallels and possible figures: they truly constitute the memory in which they are enveloped and from which they can be extracted, abstracted, at will. Abstract: draw a line from the said table. The abstract line without any other dimension than its own is extracted from the wooden or marble ruler, is drawn from it in every sense. How is it possible to say otherwise that this element was included in it? Why does the theory of abstraction unfold its splendours in an imaginary space, separating the coarse senses from pure understanding? What place do feelings and faculties of the soul have here, when it is simply a matter of drawing lines with a ruler or a model canon, of a rigid form, when we can draw incessantly on this artificial memory as from a cornucopia which never runs dry? Yes, the verb abstract does have this truly elementary meaning.

We are still astonished at the interpretation of these things by souls and bodies. Who writes, in fact? The gnomon, upright like a statue; in other words, the element. What does it write? Lines, points and circles; in other words, elements. Where are these elements? In the ruler and the compass or in their derivative the set square, that is the gnomon; in other words, the element. The element writes elements, abstracts elements. Here is the beginning. The subjects, like the subject-matter of the discipline, haunt the intelligence or the artificial memory from which they are abstracted.

Present in the title like the pointer on the sundial, in *Definitions* like a statue in equilibrium, the artefact dogs us, haunts the *Postulates* and makes them possible. Euclid's geometry as a system or the development of a series of theorems from designated beginnings can thus be taken for an automatism. And that does not mean that its results remain finite.

Equality, Community

What should we think, now, to form a community? Equality. That nobody has the advantage over anybody else and that exchanges are reciprocated. Careful: 'you have failed to observe the great power of geometrical equality among both gods and men: you hold that self-advantage is what one ought to practise, because you neglect geometry,' thunders Socrates against Gorgias, an ambitious, well-born, young executive, newly graduated from the leading Schools, hungry for bloodshed and power, vain and competitive. Socrates shows him the surprising parallel between geometry and equality. No knowledge without the equals sign. No knowledge without this invariance. Now, once again, this notion and this operation also amount to order, justice, harmony, to the social bond. Equality is a precondition for community. Those who opt in favour of the invariance vote for social order.

So the term *Axioms* is the worst possible translation of Euclid's original title: Κοιναι ἔννοιαι (*Common Notions*), which deals with equality. One would have to be a blind believer in an individual subject of thought to imagine that we are talking about the inborn notions that each person carries or possesses from birth, innately, genetically, by right or by a miracle. One does not need much human or social experience to learn, on the contrary, that equality, in comparisons, roles or exchanges, is the least shared thing in the world – in the sense that it is the least common. If by chance you come across it, praise the heavens. People do not see themselves as one person just like any other, and do not know how to act as if they were. Perhaps we should only speak of people in general in this sense.

Having said that, equality is essential if one wants to found a community. It does not come from each individual, but from the project itself. From this point, common no longer means the usual or common denominator, but that which characterizes the public. The whole body of descriptions or implications of equality, its attributes, operations or properties, constitute indispensable notions for the setting up of the said community. Hence the title *Common Notions*. To understand this *koine*, we have to take leave of the individual subject of thought to think of a collective subject, which, in particular, constitutes and serves as a basis for a scientific community, which normal or elementary science develops by deducing and demonstrating from these beginnings, developing itself in the same way.

116

First Principles

Overall, Euclid's beginnings imply their own anthropology. The title itself recalls the gnomon, as well as the lines drawn by the Sun and the shaft on the primal earth evoked by geometry. From this earth rise the equilibria or repose of a fine statics described by the *Definitions*, through a series of successive inclinations, statues brought forth from the earth, upright like the shaft: the episteme begins. The *Postulates* describe the use of the gnomon, the set square which has given way to the ruler and compass, and how it works. Thus they indicate who draws the lines, or rather in which objects these lines are implicit and from where they are extracted or abstracted: artificial objects for the memory of the elements and their intelligence. Language itself leads us to describe as abstract the lines drawn or built from these artefacts as if they were abstracted from them. And lastly, *Common Notions* describes the conditions in the thought of establishing a community, that whole of which each individual is only a part. To sum up, there is the objective and the collective, in the absence of any subject in the modern sense.

There is something of the transcendental in Euclid's beginnings which echo the beginnings of geometry or which express and continue them; there is something of the conditional there, of the fundamental, of the elementary to be precise. But they do not lie in the subjective or in the *a priori*, nor in the formal or the pure, in the Cartesian or Kantian senses. They reside in the world, Sun, Earth, in the artificial, shaft, table, compass, ruler, statue, in short in the community, in the ill-named and therefore ill-conceived intersubjectivity starting with the individual subject. If the transcendental only adds an empty and sterile abstraction to the constructive idealities of geometry or subjective foundations to its formal bases, nothing distinguishes it from a fable, a fairy tale, a cosmetic ornament. If and when it exists, to know when the conditions it isolates, more than necessary become sufficient, it encounters anthropology. The origins of *Elements* can be located in the things of the world and in the collective culture.

The special – epistemological – conditions of science lie in the general – gnosiological[15] – conditions of knowledge which in turn lie in the hitherto obscure and unrecognized anthropological relations between the collective and the objects of the world, culture and nature. Does the group as a group itself become a thing? If yes, how?

Our philosophical tradition dictates that only the individual subject either perceives or thinks and constitutes the objective. As for the collective, it only constructs itself: the only object of our relations is our relations. The further we live from the world, the more we look after each other. This division which gives the loner the heroic role of an encounter with things in the silence of communication probably reflects the customary and tragic experience of the monumental human feats of history, but does not in any way reflect the real novelty of the exercise of science in relation to those acts. The control and the consensus of the community defined by that exercise constitute the subject-matter of science. It thinks collectively. The subject of this thinking only becomes individual in very rare moments of crisis: when the group under threat takes in someone who had been excluded while pretending to believe that they had been sent on ahead to reconnoitre, rather than having been excluded by the group.

Something which can be seen as a curiosity of history is the paradoxical fact that at the very time when science began to constitute itself as a group, if not as an actual profession, devoted to things themselves, to nature, to physics, a philosophy of the knowing individual subject emerged, as if this philosophy underlined the exception by patently ignoring what was becoming common law or the regulation of the community. And yet, only the tribunal of the scientific assembly, only the Church of experts keeping a check on each other, decides whether the Earth rotates, not the isolated hero. For if this subject were alone in thinking so, the Earth would not rotate and there would not be any science. Everything happens as if the Galileo affair had led the philosophers of knowledge to a misinterpretation, as if one of history's founding myths or the hagiography of science had caused them to forget that science thinks as a group, like a tribunal and a Church, and functions similarly. And so the history of science evolves, both in the detail and the collective laws, in a way that mirrors the history of religion. Religions advance through heretics, the sciences through inventors, fairly regularly expelled. There is nothing paradoxical in this comparison: religion gives us the first example of a collective subject thinking an object which transcends community relations.

In science, the egalitarian group of experts acknowledging each other does indeed constitute the subject of knowledge, as if one of the operating conditions of this knowledge is the reciprocal recognition of individuals thus made equal; science thinks in this way and offers, as well, guarantees that it thinks the objects of the world transcending its relations. Here is the exception, which surely does not concern the individual but the group, for the group in general behaves as if its relations were sufficient, as if there were no world. There is no object outside its own closed ambit. The corpus of its relations constitutes its definition and the redefinition of each relation constitutes the nourishment on which it feeds, its continuation and its replacement. The idealism which affirms that the world equals our representation is appropriate to certain serious mental illnesses and to all societies, without exception, whose relations are projected on to the environment. Sociologists are right to claim that groups recognize only their own laws: that is how hordes of animals and political animals behave, little puppets which are only able to move through the strings which connect them to each other.

The movement of one expresses or sums up the unrest of its social environment, from a particular angle which defines it. This kind of music box requires neither a spring nor a programme since each movement, dependent on the whole, immediately comes back into play as the cause of the next movement. There is nothing holding the strings and so the sociologist is always right to insist on the autonomy of his science because the whole is closed in on itself and is self-generating. In sum, this produces a few temporal fluctuations which are sometimes called history.

Suddenly, over two millennia, in this hazardous and monotonous time, a paradoxical school appears which devotes itself to thinking of some worldly object which exists independently of the networks, strands and knots which bind human beings to each other, as if it transcended them. In Lucretius' writings, when physics begins, the chaos, the world and atoms are accompanied by a transcendent[16] God who takes no notice of them. There is nothing paradoxical in this double affirmation: an absent God, indifferent to human relations, has the same status as the cloud of atoms, in that their actions, as individuals or as a group, remain eternally independent of whatever drives peoples. The natural object takes the place of God, can even coexist with him in the same place, the vital thing being to have a clear understanding of that place. Scientists believe in the existence of the outside world as the monk believes in God: neither can prove it, but they cannot exercise their faith or their science without this fundamental belief. In the Galileo affair, the stakes reside in that very place. A court only sits to dictate its law and only speaks performatively. Therefore for the court, this place does not exist: there are causes, not things, *causes* not *choses*. Let someone stand up in the middle and testify that the Earth rotates and there is still no science – after all it happens every day that someone in a group behaves abnormally. But a Church that has become united is ready for this. Only a religious tribunal could, on that occasion, waver. Condemn but make possible. Someone in its midst stands up and testifies that the Earth rotates and the jury reacts as if it is in the presence of a fanatic ranting on about his mystical intuition. Certainly, there is still no science, but a possibility opens up; there is a chance that, despite their claims, the members of the assembly might be converted to the astronomical revolution, accustomed as they are to debating real arguments, arguments of the Real Presence totally unconnected to their own relations. An ordinary tribunal, limited to causes, lacks such a place and cannot give in; the religious tribunal does not give in, but can give in, but will give in, open to this place. Suddenly, things exist, not just causes. Religion shuts its eyes to what links people but opens itself to the direct experience of God: in religion the learned clash with the mystics. The Galileo affair continues this canonic struggle. But it gives the idea of establishing a commission of experts responsible for things themselves, another tribunal beside the old tribunal: here is science, which does not speak performatively and in which the same debate is perpetuated. In short, there is one object or there are objects for us, for the collective, for this society whose brazen law of wages usually forces to behave as if they did not exist. Science forms a realistic, paradoxical group in the community which tends to be idealistic. Through this knowledge, we together have a relationship with a thing whose laws have no relation to our relations. No philosophy, to this day, as far as I am aware, enables us to think of such an event, since tradition dictates that there is no object of knowledge other than an individual subject and that the group cannot know objectively since its sole object is its relations. Certainly the difficult philosophy that would make it possible to think this would require imagining this transcendent place where God coexists with the objects of the world, mystic experience and experimentation.

It is no quirk of history, that tragic and fatal fact that, just when the death of God was announced, the objective world lowered its barriers, removed its obstacles, lightened cruel and ancient necessity, began to lose its battles against our agressive and triumphant technologies and retreated, humiliated, behind our representations, in short, entered its death throes. The bomb was the death knoll of the world, barely half a century after the death of God. The two transcendencies leave the same place at around the same time. Now we have to write a philosophy of the death of transcendental objectivity.

Foundations

Mathematics are the foundation of physics: a broad and vague statement, since they are equally the foundation of all the other sciences. In this crude sense, we mean that physics only becomes a science if it is expressed in mathematical language. So be it. But a foundation goes deeper.

To give an elemental reading of the dawn of mathematics, its beginnings in history and the preliminaries of the system, to read the elements of Euclid's *Elements* thus, we discover a forgotten world which has been lost from memory: a Sun and an Earth, shadow and light, the mark of time in space; heavy, ponderous objects, coming out of the earth slowly like statues brought back to life from the dead; artefacts, canons, rules or lines, objective memories implying the elements or lines we draw or abstract from them; the conditions for the setting up of a community, of a consensus. Agreement on the truth cannot be reached without equality, in short, a world and a group.

Here are the conditions or foundations of science: there is a transcendental[17] us whose object is a transcendental Earth. These are the foundations of scientific knowledge in general, be it abstract or concrete, concrete as the world and things. This geometry is the foundation of physics because it is a physics, because the world is its transcendental condition, as well as the object, as it is or manufactured. Similarly it is the foundation of technology

120

because it is a technology. But, on the other hand, it can be as abstract as one likes because it is the generator of abstraction. Pure geometry is born of the canon, of the ruler or of the compass, just as geometries, which are purer and even more abstract, will later come out of Euclid's geometry and its beginnings. Abstraction traces a continuous path which here quite simply resembles history. The first is drawn from artifice and the second from that which comes from it, and so on and so forth, in a fan which we gently open out.

Physics

Why, though, didn't the Greeks invent mathematical physics? The answer sometimes given is: because they had slaves. Those who do not use their arms have no need for tools and are content to contemplate. This is all very well. Do people believe that during the Renaissance, when physics did emerge, serfs had disappeared from Italy, Holland and France? Do they believe that in the nineteenth century the steam engine and thermodynamics appeared when the exploitation of men by those who do not believe they are their equals had ceased?

The Greeks would have hesitated before laws of physics because minor gods waged battles in space, each in charge of their own department. When a wood-nymph guards every tree and each fountain has a water-nymph to watch over the gushing waters, when the sea is swarming with sirens and the meadows with fauns, a thousand exceptions prevent the adoption of the general rule. It is not until the advent of the one God that the expanse suddenly becomes empty and no locality stands out against the homogeneous Universe. A Being beyond beings, here is a smooth universal which makes the natural and technical sciences possible. Openness and oneness abolish all peculiarity. The alliance, in other words, of a formula with phenomena open to experiment implies that the dogma of incarnation has been accepted. Conditions of a religious or metaphysical kind can appear more decisive than economic or social arguments.

But above all, the Greeks did not invent physics because of the human sciences, for human sciences preceded physical sciences. Anterior both in time and the conditions of the physical sciences, the human sciences prevented the latter from emerging. This conflict, outside the universities, haunts our early knowledge. We were concerned with our own relations well before we worried about the world. Humanity, a sociologist first, needed all its history before becoming a physicist. Conversely, history is this slow catching up of the world.

We have interpreted religions and mythologies in terms of the natural sciences for so long, a misinterpretation imposed by our modernity, that we still believe firmly that our ancestors were primarily afraid of lightning, meteors and the dark, of the sterility of fallow land. No, they feared others and the group, their enemies. All mythologies and religions are human sciences in an exquisite way, infinitely more precise, effective and sensible than what we call by that name today. To reach the world and then physics, we had to cross that particular barrier, erected by the groups themselves.

Numbers, first of all, regulate taxes, commerce and earnings: no known problem of measurement throughout the entire Fertile Crescent is about nature, as if bodies did not fall yet. All, on the other hand, quantified what depended on our relations. Even the assessment of cultivable fields whose boundaries had been washed away by the Nile floods sought to end disputes between neighbours through the power of the State and to re-establish the cadastral register in its integrity, which was the basis for taxation. This early geometry did not measure just any land, but rather weighed up credit and debit, and its constant errors of approximation were always in favour of the interests of the pharoah or of the stronger party.

A strange thing full of water, that was the inauguration of Thales. The Ionian physicists discovered objects – air, fire, earth – totally independently of our relations of will or power, things without human causes. There is a world beyond closed societies, where things are born, from fire, from water or from atoms, without rules or laws imposed by a king or a god. There is no god of weight. When the logos becomes a proportion it nullifies, through the effect of its relationship, rather as a fraction is reduced to its simplest expression, the mouths which say it and the orders which impose it, so that all that is preserved are the relations of the world to the world and those of the thing to itself. The new logos becomes the relationship between two former logoi or utterances. There are objects whose appearance and birth are independent of us and which develop by themselves in relation to other objects of the world. The rational logos which repeats twice, in Greek and in Latin, the proportion or the ratio, speaks without a human mouth like a law outside the law, from this transcendence. Among the infant physicists, what was taken for a voluntary affirmation of atheism and which remained one, for all that, indeed consisted of getting away from religions and mythology, but in that they express and sanction social relations. The world appears, is born, takes places, goes outside the city, without it: can the ancient city (*polis*) bear such 'apoliticism'? No. Another transcendence is required to shoulder it, that is, a religion which forces the move away from the sacred, from the crushing constraints of society.

The Pharoah Cheops, divine, all-powerful, representing the social body, had his pyramid built stone by stone by the people and Thales measured it without the proportion discovered taking into account in any way the king, his order, his tomb or that political relationship of the one to the many. The logos-proportion replaces the logos-discourse, there is a law or an order which does not recognize or is not recognized by order or social law, and the pharoah dies once again. What remains is the empty polyhedron, a transparent form.

Misfortune had it and still has it that this logos which kings, societies and language cannot bear, irresistibly returned to the mouths and the wills of power: a return to the almost inevitable archaism which the Greeks witnessed or suffered as we suffer it. The logos-proportion reappears in speech and social hermeticism: irrational or rational, it commands the weaving of the *Politicus*, educates the guardians of the *Republic* and Socrates crushes Callicles in *Gorgias*, through the all-powerful geometric equality among gods and men. It becomes mathematics for human sciences. Despite or because of the effort of *Timaeus*, the inaugural invention of an object-world independent from ourselves collapses again in the collective. Politics, human sciences and myths together and quite plainly prevented mathematical physics from developing.

> And wise men tell us, Callicles, that heaven and earth and gods and men are held together by communion and friendship, orderliness, temperance, and justice; and that is the reason, my friend, why they call the whole of this world by the name of order, not of disorder or dissoluteness. Now you, as it seems to me, do not give proper attention to this, for all your cleverness, but have failed to observe the great power of geometrical equality amongst both gods and men: you hold that self-advantage is what one ought to practise, because you neglect geometry. (Plato, *Gorgias*)[18]

4

Archimedes:
The Scientist's Canon

Michel Authier

In which we see that the history of science isn't new and that it has sometimes been the occasion of an edifying tale, and how Plutarch starts with a
famous mathematician, physicist and engineer and ends up with the cliché
of the ideal scientist.

Moscow 1980. After fighting for a reduction in the arms he had helped
to develop, Sakharov, the blue-eyed boy of Soviet physics and winner of the
Nobel peace prize, was stripped of office and honour and placed under
house arrest. Now he has been rehabilitated.

Washington 1954. After struggling hard to end research on the arms of
which from 1943 to 1945 he had been the chief architect, Oppenheimer,
'father' of the United States' atomic victory over Japan, was accused of being
in league with Communism. He was dismissed from his post as president of
the General Advisory Committee of the Atomic Energy Commission and sent
back to teaching. Now he has been rehabilitated.

Paris 1950. A winner of the Nobel prize for chemistry, responsible for the
first experiments in artificial radioactivity, which enabled the design and
manufacture of atomic weapons, Frédéric Joliot-Curie was sacked as High
Commissioner for Atomic Energy after opposing the military use of the atom.
Now he has been rehabilitated.

Should we conclude then that the relationship between scientists and

124

political power has always been difficult, irrespective of country or of political system? Is it really new, this logic which leads from scientific ambition, through success, guilt and partisan resistance, and on to political repression? We are going to try and discover the canonical narrative which established this relation between science and power.

One thousand three hundred years before the beginning of the Christian era, it is said that Daedalus, a wily Athenian and a contemporary of Aegeus, came and entered the service of Minos, king of Crete. By building a hollow wooden cow in which she concealed herself he enabled the queen Pasiphaë to conceive a monstrous creature by the sacred bull. This was the Minotaur, the symbol of Cretan power over the other cities of the region. It inspired such horror that the king insisted that Daedalus, the 'sorcerer's apprentice', should build a fortress to contain the abomination: this was the Labyrinth at Knossos. Horrified by the sacrifice of young Athenians to his creation, Daedalus enlisted the help of Ariadne to provide Theseus with the secret of the Labyrinth and the monster was killed. Minos then locked up the traitor in the trap he himself had devised . . . And the myth goes on to unfold the diverse figures of flight, pursuit and revenge between 'scientist' and sovereign.

Three centuries later, according to legend, Palamedes, the father of number, countered Ulysses' feigned madness with a clever trick, forcing him to go to war against Troy. He was later prosecuted for treason and stoned to death by those he'd helped, for having brought Ulysses into their camp.

Thanks to his genius, in 510 BC Pythagoras helped Crotona, his adoptive city, to crush its rival Sybaris . . . He was banished, a victim of his own success.

As we can see, there is no happy myth of the relationship between science and power. And yet, despite so many setbacks, the belief in an essential harmony remains, even though in every generation scientists who have served the State have been sacrificed for political ends. Rehabilitation generally comes later. Is this 'sacrifice' the apotheosis of a scientific career, making the scientist more than a scientific hero, a hero for all humanity? What is the model for such a Passion?

An Exemplary Scientist

In the *Calendar for the Advancement of the Human Race*, appended to his *Catechism of Positive Religion*, Auguste Comte gives to the fourth month, consecrated to the science of Antiquity, the name of Archimedes (287–212 BC; see p. 447). In the exact sciences, only he receives such an honour. Equal to Homer, Moses and Caesar, the great scientist of Syracuse stands higher in Comte's estimation than Newton, Kepler, Galileo and the rest.

In his famous meditation on the three orders, Pascal presents only two people: Archimedes and Jesus Christ. The first has his place in the order of intelligence in the same way as the second does in the order of charity. Few scientists have been treated with such veneration. And – a signal honour! – one of his discoveries has been made into a poem. In fact, the execrable poem which begins:

> Que j'aime à faire apprendre un nombre utile aux sages.
> Immortel Archimède, artiste, ingénieur . . .

gives the value of π in decimal notation, taking the number of letters in each word successively (so that que = 3, j' = 1, aime = 4, etc). With this mnemonic, the schoolteacher stamps upon the mind not only a mathematical constant, but also the name of the man who first discovered a way to measure the circle. In a short treatise, probably incomplete, entitled *The Meas-*

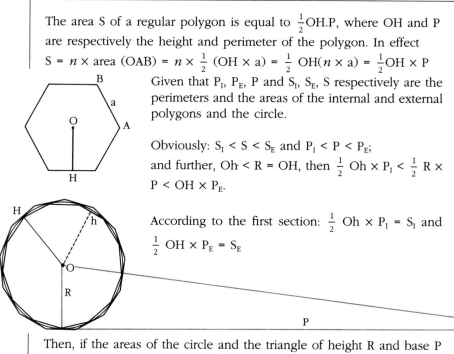

The area S of a regular polygon is equal to $\frac{1}{2}$OH.P, where OH and P are respectively the height and perimeter of the polygon. In effect

$$S = n \times \text{area (OAB)} = n \times \frac{1}{2}(OH \times a) = \frac{1}{2}OH(n \times a) = \frac{1}{2}OH \times P$$

Given that P_I, P_E, P and S_I, S_E, S respectively are the perimeters and the areas of the internal and external polygons and the circle.

Obviously: $S_I < S < S_E$ and $P_I < P < P_E$;
and further, Oh < R = OH, then $\frac{1}{2}$ Oh × P_I < $\frac{1}{2}$ R × P < OH × P_E.

According to the first section: $\frac{1}{2}$ Oh × P_I = S_I and $\frac{1}{2}$ OH × P_E = S_E

Then, if the areas of the circle and the triangle of height R and base P are different, their difference will always be between S_I and S_E, which is absurd because, in multiplying the number of sides of the polygons, the difference in their area may be made as small as one wishes (after Euclid, XII, 2).

urement of the Circle, Archimedes puts forward and proves three propositions, each important in its own way for the history of science.

First proposition: 'The area of any circle is equal to a right-angled triangle in which one of the sides about the right angle is equal to the radius and the other to the circumference of the circle.'

For us, brought up from childhood to use formulae relating to the circle, this proposition can seem tautological. Because its proof is also simple, many historians of science pass over it without noticing how innovative it is. The science of the time didn't wait for Archimedes before developing procedures which allowed the calculation of the area and circumference of the circle. Both the Egyptians and the Babylonians had already proposed effective algorithms for these calculations (see pp. 44–72).

What Archimedes is implicitly stating here is the relationship between the two problems of the squaring and the rectification of the circle (i.e. to discover a square having the same area as a given circle, or a line segment having the same length as the circumference). This means for us (we must guard against anachronism!) that the constants which enable the calculation of the area and the circumference of a circle are linked to the same number (which since the seventeenth century has been called π). He continues this work in the treatise *On Method*, where he writes: 'Judging from the fact that any circle is equal to a triangle with base equal to the circumference and height equal to the radius of the circle, I apprehended that in like manner, any sphere is equal to a cone with base equal to the surface of the sphere and height equal to the radius.' He thus proves, implicitly, that the measurement of the circumference governs that of the area of the circle and of the area and the volume of the sphere.[1]

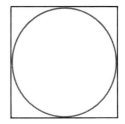

Second proposition: 'The area of a circle is to the square on its diameter as 11 to 14.' This result might seem banal, but it has to be understood in the context of its period, when the calculation for the area of the circle was related to that for the square. This was the practice of the Egyptians, the Babylonians and of the Greeks before Archimedes. To them he gives a response in the same manner, and he demonstrates its truth and greater precision. If we often underestimate this proposition, it's because it isn't addressed to us but to Archimedes' predecessors, whom he has now superseded.

If the square has a value of 14, then the circle has a value of very nearly 11; to establish this result, Archimedes uses his calculation for the circumference, which reinforces the idea of the reduction of squaring to rectification.

Third proposition: 'The ratio of the circumference of any circle to its diameter is less than $3\frac{1}{7}$ but greater than $3 \frac{10}{71}$.' Here Archimedes commands our admiration with a series of calculations which still astonishes those who know the Greek system of writing numerals. This result gave our schools the famous $\frac{22}{7}$ as an approximate value for π to within $\frac{1}{2500}$. The French fifteenth-century mathematician Nicholas Chuquet said: 'This is something that cannot be proved by demonstration' and this is the measure of Archimedes' genius. His technique for the approximation of the circumference consists of compressing the circle between two polygons, the number of whose sides can be increased as much as one wishes to obtain the required precision. This technique was still in use in the seventeenth century.

By reducing the cubing of the sphere, and the squaring and rectification of the circle to one single problem; by replacing the previous algorithms with his own calculations and by creating a method of successive approximation, Archimedes made his name as the present, past and future master of geometry's central symbol. This performance, itself enough to win great fame, is usually left unmentioned in the accounts of the historians who wrote about him: Polybius, Livy, Cicero, Silius Italicus, Plutarch and so many others.

These writers, then, whose works nourished the best minds of the sixteenth, seventeenth, eighteenth and even of the nineteenth centuries, and thus provided the cultural matrix for the growth of Western science, made no attempt to give a faithful portrait of Archimedes. As we follow Plutarch's text in particular, we will see how in these accounts of Greek and Roman history our authors do not give a faithful portrait of the man, but develop the archetypical image of the scientist.

A Famous Historian

In his *Life of Timoleon*, Plutarch writes: 'The history of great men is like a mirror in which I look so as to try to some extent to model my life on theirs and form it in the image of their virtues.' Born at Chaeronea around the year AD 50, he died there nearly eighty years later. The son of one of the town's foremost families, Plutarch settled there permanently after several journeys to Rome as a young man. As he said, he wanted the good opinion and high repute of his name to redound upon his town and upon his fellow citizens. A prolific author, he wrote many lives – called parallel, because the life of a Greek is always compared to that of a Roman – long treatises on morality and varied essays.

In what he does and what he writes, Plutarch is always concerned with the exemplary as the necessary foundation of human relations. His historical, philosophical and scientific learning are placed at the service of this morality. The pupil of a Platonist, he is convinced of the absolute reality of the world of ideas and its primacy over the world of things. So we can understand how Plutarch, in a world dominated by pragmatism and organized by legal authority, drew strength from his ideas, going to live far away from the powerful centre of the Pax Romana.

Though he wasn't a scientist himself, he certainly seems to have had a wide scientific culture, and not without originality. Evidence of this is his astonishing little work, *On the Face that Appears on the Surface of the Moon*, in which he describes certain cosmological theories and puts forward ideas about universal attraction and the theory of tides which follows from it.

His works were a source of themes for drama and reflection in Montaigne, Montesquieu, Rousseau and Shakespeare, as well as many others. As Jean Montucla, the historian of mathematics, puts it, they were 'in everybody's hands'. We can say then that there were very many scientists of the baroque and classical periods who would have come across the ten pages or so devoted to Archimedes while reading Plutarch's life of the general Marcellus. In 1654, Father Taquet observed of Archimedes that 'there are more who praise him than read him, and more who read him than understand him.' With this he suggests that Archimedes' fame extended beyond the narrow circle of the scientists. Following Plutarch's account step by step, we shall see how an exceptional mathematician could become the archetypical figure of the 'scientific genius'.

Syracuse and the Rest of the World

During the eighth century, the Dorians of Corinth came and settled on an island half a kilometre square, very close to the Ionian coast of Sicily. They called it Ortygia, the old name for Delos, so placing it under the protection of Apollo and Artemis. Later the island was joined to Sicily, whose fertile plains and thriving trade brought riches. From the fifth century BC, under the rule of the tyrants Hiero I and Gelon, the city expanded considerably and became one of the richest in Greece, with many hundreds of thousands of inhabitants.

Through three centuries, for the most part free and mistress of her own destiny, she took part in all the great Mediterranean conflicts. From near and far, Etruscans, Medes, Phoenicians, Carthaginians, Athenians, Spartans, Macedonians and Romans, all fought against Syracuse or rallied to her cause. Alliances and sieges followed one after another, monarchy and tyranny alternated with democracy. Despite everything, the city became richer and richer, endowing itself with temples and monuments crowded with works of art. Patrons attracted the greatest minds of Antiquity. Pythagoreans fleeing the south of Italy took refuge there; Plato came, and despaired of his dream of the Republic. Artists and scientists brought their talents, and contributed greatly to the expansion and prosperity of the city.

Thanks to its power, Syracuse controlled much of the Sicilian plain, whose fertility is legendary. Wheat grew there abundantly, and there was pasture for countless cattle. 'Measure for me, my friend, the number of Helios cattle which graze on the plains of Sicily, the Trinacrian isle.' So wrote Archimedes at the beginning of his curious *Problem of the Cattle.* Here he set out in the form of a poem a puzzle upon which Eratosthenes (librarian and geographer of Alexandria, who calculated the diameter of the Earth, and a correspond-ent of Archimedes) might test his mathematical skills. 'When you have solved it . . . you should know that you will be judged to have arrived at perfection in this science.' If we believe a study by a nineteenth-century mathematician, the solution in decimal notation would take 600 pages of figures. This is humorous testimony to the fertility of the Sicilian countryside; but it also underlines Archimedes' tremendous ability in calculation, unless, of course, unable wholly to solve his problem, he had turned, as he often did when faced with difficult problems, to one of the rare mathematicians capable of helping him. At that time they were few in number, and he had known almost all of them when he stayed in Alexandria in his youth. With its famous library, its museum and its celebrated scholars, it was *the* scientific centre, situated at the confluence of Greek, Egyptian, Babylonian and Jewish thought.

Reading the letters from Archimedes which preface many of his works, we discover the little world of mathematicians of the third century BC, who maintained a scientific exchange among themselves despite the thousands of miles which separated them. We should not forget what was necessary in the way of the stability and reliability of seaborne traffic for the safe con-veyance of these fragile scholarly scrolls. This is an indication of how much even in those days Archimedes' mathematical work depended on his coun-try's power. The hundreds of thousands of cattle inscribed in lists upon the papyrus crossed the seas in safety thanks to the thousands of artisans, war-riors and peasants who fed on the real cattle that grazed on the Sicilian plains. Plutarch shows us Archimedes' effort, with his airy intelligence, to do justice to the weighty masses of the world.

As the story begins, Syracuse is in crisis. King Hieron II had died the year before, after ruling for fifty-four years. His death followed that of Gelon, his son and right-hand man. After a disastrous alliance with Carthage at the beginning of the First Punic War, in 263 BC the Syracusans had entered an alliance with Rome, and there followed almost half a century of peace. Forty-seven years later, Hieronymus, Hieron's grandson and heir, made an alliance with Carthage, whose armies, under Hannibal, had just invaded the whole of Italy; this was the Second Punic War.

A few months later the assassination of the young king left the kingdom without a ruler, at war against its old ally and in alliance with its old enemy. Political instability was at its worst, and Plutarch's account of the siege of Syracuse opens in this politically leaderless city. For its defence it had the fortresses constructed on its perimeter during the previous decades and centuries; but it also had the astonishing engines of war which had been partly responsible for the city's wealth, being exported, from the fourth century on, to every corner of the Mediterranean, to Carthage, Rome, Athens and Rhodes. In control of these forces, there is one man: Archimedes. The scientist has been put on the spot, the crisis is at its very worst: what will he be able to do? This is the problem that Plutarch will examine.

Facing the Syracusans are the forces of Rome, enormous, well-armed and well-trained. They have just crushed the neighbouring cities, and at their head, as Virgil describes him, 'Marcellus advances, proud of his *spolia opima*.'[2] For Plutarch, despite the crowd of citizens, the hordes of soldiers, the terrifying machinery, the impregnable fortresses, the triremes and quinquiremes, the situation is archetypically simple. In the struggle between Roman power and the last great free city of Greece, the greatest of soldiers is face to face with the greatest of scientists. What will happen?

The Siege of Syracuse

The answer to this question belongs to the realm of ideas and not to the realm of military relationships between men. In this study we are working with Plutarch's text itself. This is what organizes the imaginative construction whose foundations we are investigating, and so we're not concerned here with the re-creation of some putative historical truth. What I am going to do is to try and show how the author has constructed his narrative; to show how he uses his philosophical and scientific knowledge to draw the picture he wants of the relations between pure science and material forces; to show how he emphasizes, de-emphasizes and distorts as he reviews the various episodes of the story and the different aspects of Archimedes' work, and develops what we may call the archetype of the scientist.

Plutarch's highly entertaining account alternates episodes of the greatest excitement with very intelligent digressions on the development of the sciences and on the work of Archimedes. The discussion which follows will divide up the text up according to its main episodes, and additional information will be provided when needed to bring out its assumptions and implications. Archimedes' principal works and discoveries will be presented as Plutarch refers to them.

131

In which we are to be convinced that the history of science is a necessary prelude to the history of battles

Plutarch's attitude is surprising. At sea, Marcellus is leading his sixty quinquiremes towards the little port near the ramparts of Acradina, while on land, the enormous Roman army approaches the fortifications of Epipolae. His only comment: 'But he had reckoned without Archimedes, and the Roman machines turned out to be insignificant not only in the philosopher's estimation, but also by comparison with those he had constructed himself.' Then he leaves the battle and begins a long digression.

Starting with the military context, Plutarch raises his account by stages to the level at which he situates his hero. He says first of all that, for Archimedes, military inventions in themselves have no importance. They are only re-creations of abstract ideas, making these perceptible. 'Archimedes did not regard his military inventions as an achievement of any importance, but merely as a by-product, which he occasionally pursued for his own amusement, of his serious work, namely the study of geometry. He had done this in the past because Hiero, the former ruler of Syracuse, had often pressed and finally persuaded him to divert his studies from the pursuit of abstract principles to the solution of practical problems, and to make his theories intelligible to the majority of mankind by applying them through the medium of the senses to the needs of everyday life.'

This separation, however, is not that simple. Even at the heart of this abstract world there existed an essential and unresolved mathematical problem which was already soluble by machine. But the use of machines was forbidden by the principle of the primacy of geometry put forward by Plato, doubtless because the material could not be a means of access to the transcendental laws of mathematics.

'Plato was indignant and attacked both men, Archytas and Eudoxus, who used machines for having corrupted and destroyed the ideal purity of geometry. He complained that they had caused her to forsake the realm of disembodied and abstract thought for that of material objects, and to employ instruments which required much base and manual labour. For this reason mechanics came to be separated from geometry, and as the subject was for a long time disregarded by philosophers, it took its place among the military arts.'

Here, probably for the first time, we see put forward in a historical text the too familiar separation between 'pure' science, abstract and intelligible, and 'applied' science, which is tangible, instrumental and coarse. This is the first outbreak of a conflict which will accompany the whole development of Western science. Thus condemned, mechanics fell into the field of military engineering. In emphasizing this problem, Plutarch passes in a few lines from the noise of battle to the music of the spheres. His plan is to persuade us that he who hears the second will rule the first. Ideas rule the world!

Let us follow his almost epistemological argument. First of all, technical inventions are no more than applications of geometry. Theory rules over practice, the abstract over the concrete and the artist over the engineer. If the scientist has to get involved with material objects, it is only out of duty to the monarch and from a desire to educate the masses. The scientist must respect his or her social context. Note in passing his exceptional status: it is the king who calls him to work, 'reducing' him to mechanical trivialities.

Nor was Archimedes the first to take on these technical problems. Archytas, who was once military commander of Tarentum during the fourth century BC, and his pupil Eudoxus, mathematician, astronomer and discoverer of methods later used by Archimedes, had already applied themselves to such matters. 'It was Eudoxus and Archytas who were the originators of the now celebrated and highly prized art of mechanics. They wanted to give geometry a certain grace, and to support by means of mechanical demonstrations easily grasped by the sense, propositions which are too intricate for proof by word or diagram.

'For example, to solve the problem of finding two mean proportional lines, which are necessary for the construction of many other geometrical figures, both mathematicians resorted to mechanical means, and adapted to their purposes certain instruments named mesolabes taken from conic sections'.

Here we catch a glimpse of Plutarch's mathematical knowledge. This is in fact one of the three great problems of Greek mathematics, the two others being the squaring of the circle (we have seen how Archimedes dealt with this) and the trisection of the angle (which he tried to solve in the eighth proposition of his book *The Lemmas*, if we accept this as authentic). The third great problem in which we are interested is a generalization of the problem of the doubling of the cube, a particular case of the problem of multiplying volumes.

Legend tells of the inhabitants of Delos, who had to ask the mathematicians for their opinion on these questions, because the oracle at the temple of Apollo (one of the tutelary deities of Syracuse) had enjoined them to double the volume of the altar. In his commentary on the works of Archimedes, Eutocius, a mathematician of the sixth century AD, cites a letter written by Eratosthenes to Ptolemy, the king of Egypt. Having recalled the legend, and having rapidly outlined various attempts at a solution, the geographer argues for the usefulness of his own discovery, a particular kind of mesolabe, stating: 'My invention is also useful to those who wish to increase the size of catapults and ballistae, for all needs to be increased in proportion if we wish to increase the throw in proportion, which is not possible without the discovery of means.'

The Mesolabe

Let us remember that a figure F will have double the surface of a figure *f* if each side A equals the diagonal of a square on the side *a*. There is no such simple process for volumes. Some Greeks, such as Eratosthenes, used a machine called a *mesolabe* for doubling volumes.

So to increase side *a* by a volume to obtain a similar side of double the volume.

Take three rectangular tablets of length 2*a* on which 3 diagonals have been drawn.

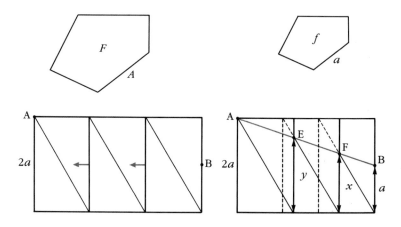

Slide the central tablet under the left-hand one and the right-hand tablet under the middle one so that points A, E, F and B are aligned. You then obtain the propositions:

$$\frac{2a}{y} = \frac{y}{x} = \frac{x}{a}, \text{ therefore } y^2 = 2ax \text{ and } x^2 = ay;$$

Therefore $x^4 = a^2y^2 = a^2(2ax)$; consequently $x^3 = 2a^3$.
The volume created by *x* proves double that created by *a*.

The references to the mesolabe, to curved lines and to sections show how well Plutarch was acquainted with the most subtle of mathematical solutions. But he doesn't point out the military relevance of this mathematical discovery, despite the fact that Syracuse too had its temple of Apollo, its catapults and ballistae, nor does he mention Archimedes' studies of the double mean, which must indeed have been known to him. In his treatise *On the Sphere and the Cylinder*, in fact, the mathematician supposes his problem solved, when for him it was a matter of constructing a cone or a cylinder one and a half times as large as a given cone or cylinder.

Did Archimedes need this knowledge for his war-like engines? Plutarch gives us no answer, because for him the scientist had to be on the right side of the divide established by Plato. In the meantime, Archimedes is white-washed. His science derives not from armaments, but from the will of the prince: it is the armament which is derived from science. The scientist him-self counts for nothing and, to show this, Plutarch abandons epistemology and gives us a little lecture on 'the sociology of science'.

Where we see how, as early as the third century BC, the scientist has to impress the king

'Then one day Einstein wrote to President Roosevelt.' Was this really an atypical episode? We should emphasize the honesty of Plutarch the 'sociolo-gist', who, having stated that the king came to look for Archimedes, men-tions just the same the latter's attempt to approach the king and his extraordinary boasting. 'Archimedes, in writing to Hiero, who was both a relative and a friend of his . . . declared that if he were given another world to stand on, he could move the earth.' First of all we should note the com-plete knowledge that Plutarch has of his hero (perhaps he had read the biography of Archimedes by a certain Heraclides, now lost, but still available in the fourth century). Behind the reference to his 'relative and friend' the king, and to the Earth he could move, there stands the figure of Phidias, Archimedes' father, who was, it is said, related to the royal family and at-tached to the palace as astronomer. And indeed, Archimedes makes refer-ence to his father's work in his astronomical treatise, *The Sand-Reckoner*.

In this strange little book, addressed to King Gelon, the son of King Hiero II, Archimedes attacks the idea of the infinite and proposes to evaluate the number of grains of sand (hence the name) required to fill a sphere the size of the Universe. To solve one problem, he has to deal with two. First, he has to give the dimensions of the Universe. Here he presents the astronomical knowledge of his time, and in particular the heliocentric system of Aristarchus,[3] and describes an experimental arrangement for the measurement of the diameter of the Sun. Second, he has to find a way of expressing the colossal number he is looking for. We should remember here that the Greek system of writing allowed only the expression of numbers less than the myriad of myriads (100,000,000). The system Archimedes proposes goes well beyond his own requirements.

Having once satisfied the young prince with his measurement of the universe in grains of sand, that is to say, measuring the greatest in terms of the smallest, Archimedes begins again with the king, suggesting that the heaviest may be moved by the lightest, that one man can move the Earth, that a small force can move a great mass, and that he himself is capable of moving a ship hauled up on the sand. My parallel isn't new, for at the end of the first century AD, it was written by Silius Italicus, the historian of the Second Punic War: 'He has counted the grains of sand on the earth, he who knew how to launch a galley with a woman's hand.'

Let us watch the performance: 'Archimedes chose for his demonstration a three-masted merchantman of the royal fleet, which had been hauled ashore with immense labour by a large gang of men, and he proceeded to have the ship loaded with her usual freight and embarked a large number of passengers. He then seated himself some distance away and without using any noticeable force, but merely exerting traction with his hand through a complex system of pulleys, he drew the vessel towards him with as smooth and even a motion as if she were sliding through the water.' According to other historians he 'simply' launched an enormous ship built for King Ptolemy in the yards of Syracuse. But whatever it was he actually did, this exploit was such that, according to Proclus, a fifth-century commentator on Euclid, 'from this moment on, Archimedes could be believed, whatever he proposed.'

In this story, the important thing seems to be the simplicity of the situation. The galley, an image of the world that Galileo would later use in his dialogues, is packed to the gunwales. The only people left on the beach are Archimedes, and doubtless the king looking on, unless, as Proclus relates, he himself is helping in the operation.

According to the historian of philosophy Emile Bréhier, 'Plutarch's Platonism is related to a very strong reaction in favour of Greek traditions, and also to a fierce critique of the grand post-Aristotelian dogmatisms.' As regards the situation we are concerned with here, we can follow Charles Mugler, a translator of Archimedes and a historian of Greek science, in seeing the experiment on the beach of Syracuse as intended to controvert in spectacular fashion one of the founding principles of Aristotelian mechanics.

So the incident is not just a bit of scientific politics, to get the king interested in Archimedes' activities. Looked at more closely, we can see how theoretical problems are inextricably entangled with the social problems that science is called upon to solve. Let me be more precise. In his physics, Aristotle affirms that the speed of an object is proportional to the relation between the force applied and the resistance to movement. According to the commonly accepted history, it is only with Galileo that this error is denounced (though what is at issue here is not whether medieval students of mechanics had got there before him). For Aristotle himself there was one exception to this law, which was dealt with in his second law of mechanics: 'If the force is weak and the resistance great, then the speed is zero' and no movement will occur. This should give us an idea of the king's astonishment.

This thesis of Aristotle's was universally accepted, all the more for its seeming to correspond to common sense; who in our own days would believe that a single child could move a stationary lorry? But this is what was

136

at stake. At a time when Aristotle's science was supported by 'common sense', Archimedes' attempt could only be considered an act of folly. This is why his success swept away all doubts, commanding recognition of the 'power of science' because, no indeed, he wasn't mad, but wise. According to Plutarch's text, this was the moment of Archimedes' triumph. In the same way as at Los Alamos twenty-two centuries later (see p. 636), 'The king . . . persuaded Archimedes to construct for him a number of engines designed for attack or defence, which could be employed in any kind of siege warfare. Hiero himself never had occasion to use these.' A fine argument for a policy of deterrence!

Perhaps it is time to life the veil a little on the foundations of Archimedes' powers. Obviously, the mechanics of levers he had theorized in his book *On the Equilibrium of Planes* confirms the possibility of moving enormous masses with the help of beams, fixed or mobile pulleys, or endless screws. But the Earth isn't really a simple geometrical figure. Of course the new system of numeration enabled the numbers in the *Sand-Reckoner* and the *Problem of the Cattle* to be written down, but what aspect of the real had to be hidden for all these theoretical possibilities to become credible? 'Give me a fixed point and I will lift up the earth.' Adam Ferguson, an eighteenth-century Scottish philosopher of the 'common sense' school, took Archimedes 'promise' literally. He calculated the time it would take a man pushing against the Earth at the end of a lever and moving at the speed of a cannonball, to move the Earth one inch. The result: 44,963,540,000,000 years! Everything now becomes clear, and the historian Montucla had also glimpsed it: to take advantage of the superpowers that Archimedes' theories seem to offer, you need time, plenty of time, time that neither Archimedes nor his hagiographer have talked about. In the spectacular effects produced by the suppression of the temporal dimension, there disappears that magnitude which is exchanged against those of these great masses and quantities. More generally, it is this conjuring away of the temporal which grounds the power of abstraction, and without it Aristotelian mechanics would be with us still.

The arrangements Plutarch tells us about make Archimedes an illusionist. This is the role the scientist is obliged to play to make an impression on the powerful. Here is an unexpected stage-setting for the play of relations between the temporal and the spiritual which Plato, a century earlier and in this same city, had attempted to master in his dealings with the tyrants Dionysius and Dyon. And with what lack of success! (Plato's *Letter VII*).

Archimedes, on the other hand, did succeed. For Plutarch, science was able to offer evidence beyond dispute. The defence of the city was entrusted to Archimedes. On this point though, he needs to make a slight correction to history. Herodotus could not have been unknown to Plutarch, and so he is being quite deliberate when he says nothing about the fortifications begun under Hiero I. He knew his Thucydides,[4] too, so why does he say nothing about the defences organized during the siege of Syracuse, in the course of the Peloponnesian Wars? So as to aggrandize his hero, he says nothing about previous defensive efforts, nor about the traditional production of armaments. Other historians, like Diodorus of Sicily, writing before Plutarch, had paid attention to the armament of those centuries. *Elepoles*, for example, used, as their name suggests, in the capture of towns, might be gigantic wheeled towers, with sides of 25 metres and a height of 50, and they might need 3000 men to move them. The same author tells us that at the siege of Rhodes, a century before Hiero II, Demetrius employed 30,000 engineers and workmen to build the necessary machines. Despite possible exaggeration, this indicates the considerable scale of the arms industry of the time, and when one knows the place of this industry in Syracuse, one can understand why Archimedes was so careful to win the confidence of the king.

Even though Plutarch says nothing, we can fix him better by noting what he doesn't say. Let's imagine for a moment that Archimedes was able to introduce very real innovations in military technology. For a time they would have guaranteed the safety of the city, but later, with exports, couldn't the machines the Romans brought forward be seen as the 'sisters' of the ones that Archimedes himself had invented? This could only increase his fame because, in this dramatic situation, Syracuse could at least boast of still having 'the inventor of the machines'.

The Disequilibrium of Terror

Having assessed the character of his protagonist, Plutarch returns to the battle. As a good moralist, he highlights the mental state of the troops being brought into play. No wounds, no deaths are described, and physical disasters are always merely understood. On the other hand, remarks about fear are frequent 'the Syracusans were struck dumb with terror . . . So frightened were the Romans, they ran away, shouting . . . Marcellus had the trumpets sound, which provoked panic and a desperate flight.' Behind the hurtling stones and the barrage of missiles, what was being exchanged was terror, fright, astonishment and horror . . . The important thing for Plutarch is what is going on in the minds of the combatants.

At the moment Syracuse's position, under attack from two sides at the same time, is hardly promising. What follows will show how the powers of Archimedes' machines will make fear change sides. Up till now, mind has mastered matter; now Plutarch will show us the specificity of battle, where matter in action in its most violent modes will rock the order of the mind. The state of war should be understood as symmetrical with what was described above. The distance between the world of things and the world of the spirit is as clear as ever; it is the relations of dominance which have been reversed.

The battle begins: here we are plunged into a world entirely mechanical. Archimedes, like a great clock-maker, sets his machines to work and, until the retreat of the Romans, there will be no other actor on the stage. The terrifying machinery is described as entirely autonomous, and follows the logic of its destructive function without any human intervention. In military science this is called the fatal logic of escalation. It can't be doubted that, taken as a whole, Archimedes' system is an infernal machine. Reading the descriptions of Livy and Polybius, one can more easily grasp Plutarch's originality. For him, the machines act by themselves, while for the two others it is men, and Archimedes and Marcellus in particular, who remain responsible agents.

This underlines how much our author's project was only incidentally historical. Described like this, the merely mechanical world throws the world of ideas into sharp relief. The variety of mechanisms allows us to glimpse the complexity of the technology of the period and Archimedes' skill. The battle begins: 'Archimedes brought his engines to bear and launched a tremendous barrage against the Roman army. This consisted of a variety of missiles, including a great volley of stones.' These are Appius' troops, attacking in the north, under the ramparts of Epipolae. A mass of stones and projectiles soon settles the situation. The description of the sea-battle is more detailed. The fleet has come in and attacked the city under the walls of the Acradina, a luxurious residential quarter, where, according to Polybius and Livy, Archimedes lived himself.

We must think of the boats right up against the walls, so as to make sense of the falling beams and in particular the operation of the iron grabs, which were very precisely described by Livy. 'The swing-beam projected over the wall and an iron grapnel was attached to it on a heavy chain; the grapnel was lowered onto a vessel's bows, and the beam was then swung up, the other arm being brought down to the ground by the shifting of a leaden weight; the result was to stand the ship, so to speak, on her tail, bows in air. Then the whole contraption was let go and the ship falling smash as it were from the wall into the water (to the great alarm of the crew) was more or less swamped.' Apart from its precision, this passage underlines the importance of men in the operation of arms, while there is nothing of this in Plutarch: 'At the same time beams were run out from the walls so as to project over the Roman ships . . . while others were seized at the bows by iron claws or by beaks like those of cranes, hauled into the air by means of counterweights until they stood upright on their sterns. Often there would be seen the terrifying spectacle of a ship being lifted clean out of the water into the air.'

Livy's account, which takes up that of Polybius and refines it, makes clear the technique of the lever just mentioned. By these means, not only was Archimedes able to move the ships, but if we are to believe Plutarch, he could make them fly. We should note here that the 'enormous beams' which were run out over the ships could well be those which served as levers to lift them up. In the sixteenth century Simon Stevin, a student of mechanics, would take up Archimedes' work on levers to produce a theory of the equilibrium of beams.

Mathematical Recreation

With the lever we come to the experimental apparatus which, in establishing the equivalence of the ratios of forces and of lengths, perfectly illustrates the central notion of proportion around which Greek mathematics was developed. It is therefore not surprising to see the greatest of the mathematicians of Antiquity attempting to master this fundamental concept in extreme contexts. The context of the infinitely great, in the world to be moved; the infinitely spectacular when there were monarchs to impress; the infinitely chaotic when there were battles to be won; and the infinitely small in the subtlest of mathematics. No, there can be no doubt that it is the same man who capsizes boats and measures the area of a section of a parabola.

Archimedes himself makes this connection between mechanics and mathematics in his treatise *On the Method relating to Mechanical Propositions*, where he evaluates the area of a section of a parabola. He writes to his friend Eratosthenes: 'I thought it fit to write out for you and explain in detail in the same book the peculiarity of a certain method, by which it will be possible for you to get a start to enable you to investigate some of the problems in mathematics by means of mechanics.'

The method in question is worth describing, and a quick plunge into Archimedes' mathematical universe will certainly help us to understand him better. We should recall that a parabola is a geometric curve obtained by the intersection of a right cone with a plane (see p. 184). It is called a conic section, like those others referred to above in Plutarch's text in connection with the problem of double proportional means. In our own day, the parabola is better known as the trajectory of an object which is launched from the ground and then falls back towards it.

What this 'exploration' shows us (see pp. 141–3), is the diversity of registers in use: the theory of conic sections, then in its infancy, the theory of ratios and proportions, the heart of Greek mathematics, theories of levers and of centres of equilibrium.

Yet Archimedes concludes: 'Now the fact here stated is not actually demonstrated by the argument used; but that argument has given a sort of indication that the conclusion is true.' This is why he qualifies his method as exploratory, and in *On the Quadrature of the Parabola* he will put forward two proofs which he thinks of as authentically mathematical.

To understand the dangers attending 'the method', which consists of identifying a surface with the sum of segments which compose it, we should note that two triangles of the same base are each composed of an infinity of segments parallel to the base, equal to those of the other, without their having the same area. Here we meet the paradox of infinite divisibility denounced by Zeno (see p. 109). Moreover, behind the vocabulary of weights, levers, equilibrium and centre of gravity there lie the experimental methods which, with the help of plates of material, allow a result to be obtained. The type of argument used in the *Method* thus demonstrates how much Archimedes' thought is organized by the model of the balance. In presenting 'a method by which it will be possible . . . to investigate some of the problems in mathematics by means of mechanics', he demonstrates a double break with mathematical orthodoxy, first in reintroducing a material dimension into the identities and then in playing with the taboo on infinite divisibility. At this level, the *Method* is a microcosm of Archimedes' relation to the world as it is expressed in his willingness to incarnate the mathematical essence in machinery or evaluate the finitude of the cosmos by means of its smallest element.

Archimedes' method applied to the area of a parabola

Archimedes idea, as a mechanical engineer, is to consider the equality between the two ratios $a/b = c/d$ as analogous to the equilibrium state of a lever bearing two weights such that

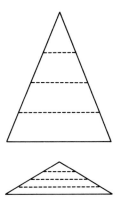

$$\frac{GB}{GA} = \frac{P_2}{P_1} \qquad \rightarrow$$

If he wants to evaluate a section of parabola AB he must find a figure where such a proportion exists.

This proportion he finds in the geometrical figure opposite, where AD is the tangent to the parabola AB, and BD is parallel to the axis. Making EFG parallel to BD, he can prove that

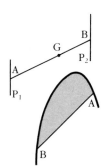

$$\leftarrow \qquad \frac{BE}{BA} = \frac{EF}{EG} \quad (1).$$

The principle of 'the method' is to consider (1) as an

equality between the ratio of length $\dfrac{BE}{BA}$ and the ratio of weight $\dfrac{EF}{EG}$. The section of the parabola and the triangle ABD are made up of all possible segments EF and EG respectively. Carrying through his analogy, Archimedes will construct an abstract lever which will satisfy the equality (1). This transforms a geometrical problem into a problem of mechanics.

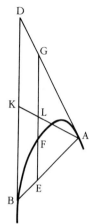

As the centres of the segments EG are all situated on the median AK, the 'weight' EG is therefore suspended at its centre of gravity L at the end of the lever-arm KL.
Since BD and EG are parallel, then

$$\frac{BE}{BA} = \frac{KL}{KA} \quad (2) \quad \text{and} \quad \frac{KL}{KA} = \frac{EF}{EG} \quad (1'). \qquad \rightarrow$$

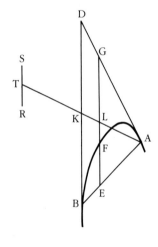

As KL is the lever-arm supporting EG, there has to be an arm of length KA to support EF. So Archimedes prolongs KL and constructs a segment KT = KA (3) at the end of which he places the weight RS = EF (4), so that T is the centre of RS, that is to say, its centre of gravity.

$\leftarrow$

Using the previous equalities, we can deduce

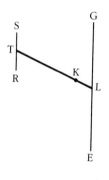

$$\frac{RS}{EG} = \frac{KL}{KT}, \qquad \rightarrow$$

that is to say the equilibrium about K of the weight and lever system.

Consequently, all the segments which constitute the triangle, in their own place, will balance those of the parabola transported to T. K is still the centre of gravity.

K is thus the centre of gravity of the system made up of the triangle and of the parabola (in the form of its constitutive segments transported to T).

In addition, X, such that KX = $\frac{1}{3}$ KA is the centre of gravity of the triangle (see the treatise *On the Equilibrium of Planes*) and so

$$\frac{KX}{KT} = \frac{1}{3}.$$

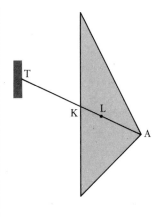

In the same treatise Archimedes shows that the ratio of weights is the inverse of that of the distances between the individual centres of gravity (X and T) and the common centre of gravity (K), so

$$\leftarrow \quad \frac{\text{parabola}}{\text{triangle}} = \frac{KX}{KT} = \frac{1}{3}.$$

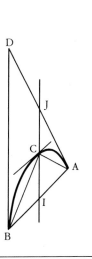

We have just seen how Archimedes finds that the section of a parabola is equal to a third of the triangle ABD. In fact, he expresses his result by saying that it is equal to four-thirds of the triangle ABC (the greatest triangle inscribed in the section). This is because of the properties of the parabola, from which it can be proved that I and C are the centres of AB and IJ, and thus that ABC is equal to $\frac{1}{4}$ of ABD, which completes the proof.

$\rightarrow$

This investigation and Archimedes' commentaries on it let us see his own position in the atomist-elementarist[5] conflict which pervaded Greek thought. His direct reference to Democritus, to whom, he says, a notable part of mathematics must be attributed, thanks to his illuminating intuitions, such as, for example, the manner of calculating the volume of a cone; the cosmology which underlies the *Sand-Reckoner*, which measures the world in terms of the smallest grain of matter, sand; his systematic study of the operation of the *stomachion*, a game in which a group of pieces having elementary geometrical forms are combined to produce a multitude of figures; all this confirms the interest he has in atomist theory. Michel Serres' demonstration that beneath the *De Natura Rerum* of the atomist Lucretius there lies the body of Archimedes' mathematics provides us with one more argument, *a posteriori*.

However, the most surprising thing is that in Archimedes' scientific practice, the two theories do not come into conflict. Atomism had a certain investigative quality which allowed one to 'approach certain propositions', unlike elementarism which, on the other hand, could serve as the philosophical foundation for the mathematical truths proved by Euclidean demonstration.

It is impossible to end this digression on the parabola without referring to the famous mirrors which Archimedes used to burn Marcellus' ships – especially as we know that the parabolic form is the one which best concentrates the rays of the Sun. The fact that Plutarch says nothing about them would be excuse enough to leave them out. Nor indeed do Livy or Polybius make any reference to them; the theme appears much later in Diocles and the Byzantine historians. Every reader of Descartes will remember that he thought them improbable, though it's true that he knew nothing of Archimedes' work in optics. Researchers have recently confirmed that the *Catoptrics*, a work hitherto attributed to Euclid, must in fact have been inspired by the thought of Archimedes.

After Descartes, the Jesuit Kircher and Buffon carried out experiments which made conceivable the technical possibility of setting fire to Marcellus' ships from a distance. To finish with fire, we may add that certain thinkers of the Renaissance even maintained the hypothesis that these fires, whose occurrence has never been confirmed, were produced by artillery pieces, which would make Archimedes the inventor of gunpowder and cannon. To him that hath shall be given!

The Defeat of Marcellus' Machine?

Facing all the machines in Syracusan hands, Plutarch mentions only one machine on the Roman side. He doesn't tell us where it came from, but he writes that it was erected by Marcellus 'on a huge platform supported by eight galleys lashed together'. What is the significance of this single *sambuca* in Plutarch's text? To understand it, we must go back to his sources. We have already seen what Diodorus of Sicily had said about these machines. As for Livy, he writes about the great number the Romans constructed 'with enormous effort', with many pairs of quinquiremes 'lashed together in pairs . . ., each pair was propelled, like a single vessel, by the outer banks of oars, and they were employed to carry towers, several storeys high, and other devices for breaching defences.'

The usually laconic Polybius has left a detailed description of the *sambuca*: 'As well as these vessels he had eight quinquiremes in pairs. Each pair had their oars removed, one on the larboard the other on the starboard side, and they had been lashed together on the sides thus left bare. On these double vessels, rowed by the outer oars of each pair, they brought up under the walls some engines called "sambucae", the construction of which was as follows:- a ladder was made . . . of a height to reach the top of the wall . . . each side of the ladder was protected by a railing and a covering or penthouse was added overhead. It was then placed so that its foot rested across the sides of the lashed-together vessels . . . with its other extremity protruding a considerable way beyond the prows. On the tops of the masts pulleys were fixed with ropes: and when the engines were about to be used, men standing on the stems of the vessels drew the ropes tied to the head of the ladder, while others standing on the prows assisted the raising of the machine and kept it steady with long poles. This construction has got the name of "sambuca" or harp for the natural reason that when it is raised the combination of the ship and ladder has very much the appearance of such an instrument.'

Reading this long extract, it is impossible to miss the fact that the techniques described derive from discoveries attributed to Archimedes. We shouldn't be surprised then at the ease with which Archimedes' own machines succeeded, according to Plutarch, in eliminating Marcellus' only *sambuca*. It is as if there were two different generations of armaments facing each other here. As for the accuracy of the shooting, almost miraculous according to Plutarch, Polybius, always trying to be exact, gives details that rouse a certain echo in us: 'But Archimedes had constructed catapults for every range; . . . he . . . wounded the enemy with stones and darts from tighter-wound and longer engines . . . and when far away, he used smaller engines graduated according to the range required from time to time.'

The proud power of the Romans had first of all terrified the Syracusans. Faced with Archimedes' science, the Roman fleet and army were forced to withdraw. Courageous, they returned during the night and tried to bring their attack as close as possible, to escape from the long-range shooting. 'However, Archimedes, it seems, had long ago foreseen such a possibility.' As clever close to as he had been at a distance, he had developed machines which shot at close range, the 'scorpions', otherwise called *manubalista*, which allowed shooting in rapid succession.

Once again, it should be pointed out that these little machines were, in some sense, scaled-down versions of larger models. For Plutarch the Platonist, the law of proportions allows no exception. He who knew its secrets could master the most varied of situations. Throughout the fighting, the machines crushed, raked and harried the Romans as they got closer and closer, and we have seen the proportions continuously being reduced. We won't be surprised then when, in the last stages of the struggle, Archimedes' ultimate form of defence is like 'an invisible hand which rained a thousand evils upon the Romans: it might have been a combat against the gods.'

The moralist is able to conclude: even in the most terrifying material chaos, in the greatest disorder of battle, he who holds the truth is as a god against his enemy. Having passed through all the stages of horror, at the denouement the situation is reversed and terror has changed sides. The Romans are dominated by fear, they are terrified by the least bit of rope, fighting is no longer possible. Marcellus must alter his strategy and find a new field of manoeuvre.

When he writes that, after the siege of Syracuse, Marcellus won great victories at Megara and near Aciles, Plutarch proves that the general's abilities are not an issue: it is Archimedes himself who presents a specific problem. But should we believe him, when we know that Syracuse had remained more or less impregnable through all the wars which came before, of which there were many?

Right through the story the machines have been reduced in size until they became almost invisible. This progressive reduction of the material prepares for a change of register heralded by the reference to combat against the gods. The landscape changes and at the Roman defeat we pass from a material world to a universe of logic. Marcellus, escaped from danger, is the first to strike the new tone: giving up the struggle, he puts down his arms, and starts to abuse his labourers and engineers. There's no intellectual content here, but just the same, we're well away from the fighting and entering, at its very lowest level, into the battle of ideas.

The Battle of Ideas

After such tribulations, we should hardly expect Marcellus to remain silent. At any rate, this is what Plutarch imagines, because no author before him concerned himself with the origin of the general's new strategy. So the defeated Marcellus speaks, 'making fun of his siege experts and engineers'. He said, 'We may as well give up fighting this geometrical Briareus (a mythical giant) who uses our ships like cups to ladle water out of the sea, who has whipped our sambuca and driven it off in disgrace.' In a note the translator Flacelière explains the allusion to 'cups' and 'after drinking'. After drinking well, the Greeks would sometimes play at cottabus. This was a game in which players would fling the contents of a goblet into a great basin full of water, in which there floated little earthenware cups representing ships, and the aim was to sink as many as possible.

The reasons why Plutarch ascribes such statements to the defeated general are clear. Having failed to master 'harsh reality', Marcellus attempts to dominate it by making an abstract image of it. The game is the classical image of transition from the concrete to the ideal, and it is through this aquatic image, probably very common at the time, that Plutarch imagines the general caught up in thought.

In this way Marcellus places himself in the same sphere as his enemy, whose role in Syracuse he finally understands. 'The whole population of Syracuse was the body, and Archimedes alone was the soul.' This is the classically Roman metaphor of the head and the members of the body. It is at the basis of many discourses on the legitimacy of power. Understanding the extreme dependence of the city on its 'saviour', a dependence heightened by the fact that non-Archimedean weaponry was not used – 'all other weapons were discarded' – the general has to find the weak point of his single adversary. I pointed out a while ago the blind spot which founds the overweening power of theory over the world. It is through this hidden dimension that Marcellus regains control of the situation: 'When Marcellus saw this, he abandoned all attempts to capture the city by assault, and settled down to reduce it by blockade.' Why should we be surprised at such a strategy? Isn't time precisely what is left unsaid in Archimedes? And do its 'necessities' not justify the supremacy of politics over the scientist and the philosopher?

Marcellus leaves the battle then to establish a long-term siege; on this point, all the historians agree. Taking advantage of the silence of arms, Plutarch draws a portrait of the victor.

The Portrait of the Scientist

Through the magic of this account, the man who so often saved Syracuse would become the archetypal figure of all mathematicians to come. Having described a battle of machines in which humans were practically absent, the author moves from one extreme to the other, rising now to the higher spheres of 'almost superhuman intellectual power' from which most people are excluded.

In giving a superhuman status to the world in which Archimedes moves, the philosopher historian underscores his Platonist convictions. He gives ideas an independence and an autonomy in relation to the world of things, while the all-conquering power of the scientist proves the dominance of the former over the latter.

In this way Plutarch, who has given up the honours of empire, uses his account of the siege of Syracuse to establish the terms of the struggle between Greece and Rome, between thought and power. In this perspective, we must see Syracuse as a metaphor for the spiritual which masters material contingency, untouched by the aggressions of the world and efficiently distributing, on the other hand, the materialized effects of its intelligible truths.

The thesis is simple: the forces of the spirit can materialize themselves and act upon the world. Confined to his small Greek city, where four centuries earlier Philip of Macedon had crushed continental Greece's last hopes of political autonomy, Plutarch never lays down his arms. Stylus in hand, he defends the superiority of Greek thought over the Roman imperium. It is from this struggle that the archetype of the Western scientist is born.

All the conventional elements are present in these few lines. First of all the disdain for 'all arts which serve the needs of life', underlined by the the attitude of Archimedes, who 'forgetting to drink, to eat, or care for his body', boasts of his indifference towards material circumstances. How can we not think of all the legends, well founded or ill, hawked about by the hagiographers. So-and-so worked for three days in a row without getting up from his chair, others forgot to dress, never wore socks, wrote out solutions which travelled far on the panels of a stage-coach, and abandoned the remains of uneaten meals among heaps of paper covered in formulae . . . A large book would not be enough to contain these anecdotes, which have in a certain way established the 'scientific character'. If we read, for example, a work by Bell entitled *The Great Mathematicians*, we see, spread through more than 500 pages, repeated throughout history, transposed into twenty countries and fixed on a hundred faces, the features of Archimedes as set down by Plutarch. It might seem that each scientist of the cohort brings his own particular trait to the portrait, this one bringing distraction, another stoical indifference, others powers of concentration, persuasiveness, shrewdness, and yet others joy in discovery, the reputation of genius, insolent ease, perseverance in difficulty, the mischievousness of riddles . . . All the way to the almost divine face of the aged Leonardo, there are no elements which are not present in the portrait painted by Plutarch. One might think that in analysing his hero, Plutarch had thrown all these elements across the centuries, into the faces of scientists yet to come. In this portrait the author's own intentions are clear. The spiritualization of the world must be made to appear necessary and irresistible, and the spirit must be installed at the apex of the hierarchy of values. And then by rejection of the needs of life and by praise of beauty, a field must be established where demonstrative science reigns above all. And on this pedestal there stands absolute excellence, Archimedes and his mathematical corpus.

'As for Archimedes, he was a man possessed of such exalted ideals, such profound spiritual vision, and such a wealth of scientific knowledge that, although his inventions had earned him a reputation for almost superhuman intellectual power, he would not deign to leave behind him any writings on his mechanical discoveries. He regarded the business of engineering and indeed of every art which ministers to the material needs of life, as an ignoble and sordid activity, and he concentrated his ambition exclusively upon those speculations whose beauty and subtlety are untainted by the claims of necessity. These studies, he believed, are incomparably superior to others, since here the grandeur and beauty of the subject-matter vie for our admiration with the cogency and precision of the methods of proof.'

Anxious to persuade his reader, Plutarch asks what the reasons were for this tremendous success: 'Some writers attribute this to his natural genius. Others maintain that a phenomenal industry lay behind the apparently effortless ease.' This is the alternative suggested; it is still with us, and Hugo's 'Five per cent genius and ninety-five per cent hard work' has not put an end to the debate. On the other hand, Plutarch makes no reference to the facilities available as a result of Archimedes' closeness to political power, the education he received from his astronomer father, the voyages to Egypt, the great scientist's conditions of life and work. Did he have a wife, children, friends, servants, slaves, assistants or premises? We will never know.

To reinforce his alternative between genius and hard work in favour of the first of the terms, Plutarch uses a procedure which never fails in works on the history of science: 'Look for the proof, you won't find it by yourself.' It works terribly well: if the reader doesn't want to look like an idiot, he has to award Archimedes that little bit extra we call genius. Here Plutarch differs from his master Plato, who in the *Meno* shows a slave rediscovering forgotten knowledge, with only Socrates' questions to show the way (see p. 97). The work of the genius is very different from Platonic anamnesis; in one case the rediscovery of a forgotten memory, in the other invention and creation.

The water flows from right to left; turning the screw by pushing against the vanes G, E and C beneath the surface of the water, the water enters through the mouth L. Trapped in the screw, it rises up, constantly attracted by the lowest point which changes with every turn, it leaves at K when this mouth points towards the right. (Engraving in J. Cardan, De Subtilitate, 1560. Bibliothèque Nationale, Paris.)

This procedure, which forcibly implicates the reader in the myth of genius, is at the origin of certain difficulties in the relationship between science and society. Individuals are thrown into a problem previously solved while all the circumstances of its solution are kept hidden from them. In the worst cases this has traumatized entire generations of individuals, forcing them to make an act of allegiance across the gulf of centuries, or has even led them to 'literary' disdain for the science from which they were expelled.

But the most violent of Plutarch's arguments is to come: 'The fact is that no amount of mental effort of his own would enable a man to hit upon the proof of one of Archimedes' theorems, and yet as soon as it is explained to him, he feels as if he might have discovered it himself, so smooth and rapid is the path by which he leads us to the required conclusion.' This is one of the disastrous commonplaces that mathematics has to drag around behind it. In all times, the very best mathematicians (Pascal, Leibniz, Chuquet) and the most competent specialists (Van Eecke, Itard, Mugler) who have studied Archimedes, have recognized the extreme difficulty of his works. It is, however, Plutarch's emollient version that wins out. Mathematics consequently appears as the archetype of limpid knowledge, a paradise of crystalline verities. Those who do not see behind their teachers' demonstrations the blinding clarity of the obvious exclude themselves from the kingdom, wander in the darkness of the forest, far from the light of intelligence. One can see that these few lines, read and read again over the centuries by teachers and their pupils, have forged an ideal which has channelled the greatest part of our intellectual energies towards science. What great scientist of past centuries has not seen in Archimedes the emblematic figure of excellence?

Continuing his work as propagandist, Plutarch shows that such glory should allow one to escape from lowly material worries, such as food and cleanliness, and reach a state of stoical indifference to material care. When we discover the scientist 'tracing geometrical figures in the ashes and drawing diagrams with his finger in the oil which had been rubbed over his skin', we grasp the root of the prejudice which insists that mathematics needs no funding to get done, that it needs, as the well-known expression has it, 'no more than paper and pencil'. So when, at the beginning of this study, Archimedes was writing to the king, it was to offer him material services. Now that it is a question of mathematics there is no question of petitioning, and the scientist's autonomy has become total.

How then should we not be surprised when in this hagiography of a scientist free of all constraint, the only external pressure to which he is subject is related to taking a bath? 'Carried by force, as he often was, to the bath,' our author writes. This is a serious point. Two centuries earlier, Vitruvius had written: 'One day when Archimedes was taking a bath . . .' What is it that Plutarch wants to be understood when he changes the version given by his predecessor, dispossessing Archimedes of his desire to wash and passing over in silence the episode of which Vitruvius speaks?

Let us remember that it was also 'unwillingly' that Archimedes was led to occupy himself with the material problems faced by the kingdom. If Plutarch has his scientist pushed into the bath, it's because he can't help it, for there is a story here that cannot be left untold. There has at least to be some reference to it, even if it does nothing for the stature of the hero. Let us return then to Vitruvius: 'One day when Archimedes was taking a bath, he saw, by chance, that the further he got into the bath of water, the more the water overflowed. This observation made him realize the solution he was looking for, and without waiting any longer, he was so carried away with joy that he got out of the bath and running naked through the house he shouted "Eureka! Eureka!" ' (meaning 'I have found it! I have found it!') Here we see the original joy, the beatific trance which comes over certain scientists after their discovery: those who have too much mind seem likely to lose it. But what, indeed, was he was doing in the bath?

The king had ordered a wreath and he suspected, without being able to prove it, that some of the gold had been replaced with silver. He called Archimedes in to solve the problem, which would be very simple if the weight and the volume of the wreath were known. We have mentioned often enough that Archimedes was the master of the beam-balance. However, the complicated shape of the wreath made it difficult to determine its volume. A method would have to be discovered: the bath provided the necessary apparatus.

The mass and capacity of a volume of water being equivalent, the bath becomes so to speak a balance for measuring volumes. In fact, the quantity of water displaced (what spills out of the bath) is equal to the volume of the object put into it. It is generally admitted that this measurement of the volume of a solid by means of the volume of a liquid is the basis of the treatise *On Floating Bodies*, in which Archimedes deals with the statics of fluids and the conditions of flotation of certain solids. Here he gains a universal reputation, proving that the surface of every liquid has the same curvature as that of the Earth and stating the famous principle which still bears his name.

Having mastered the numbers which regulate the Universe and the solids which dominate the real by their weightiness, the scientist now rules the liquids, whose laws he discovers in his bath. It is now easier to understand Archimedes' confidence in the ships which carried his scientific communications beyond the seas and across the centuries. The chief interest of the episode, however, probably doesn't lie here.

Nowhere in his text does Plutarch refer to finance, but when it is sacked the riches of the city will be plain to see. From the origins of merchant societies, the determination of the composition of an alloy has been essential to the trustworthiness of means of exchange. Syracuse had its own coinage; and even if Archimedes' method is unusable in practice, it remains the best theory for the testing of an alloy. The episode should be re-read in this perspective.

The relationship between solid and liquid is the primary source of the wealth of Syracuse. A city built on rock at the edge of the sea, sending its solid shipping out upon the waves, she melts her gold and silver to found her currency. The episode of the crown tells us this: the King's crown and Archimedes, symbols of power and of intelligence, are both plunged into the liquid. The first of the Greeks to theorize refraction finds himself above the overflowing water, sees a crown beneath it, compressed by the refraction of light (see pp. 315–43), and understands . . .

He understands that he alone controls the power structure, for he alone knows the falsity of the crown and thus the inauthenticity of kingly glory. Now all power is in his hands. The kingdom might fall, the Roman troops gather round the ramparts, the best of generals attempt the most subtle of strategies: 'Such was Archimedes' character, and in so far as it rested with him, he kept himself and his city unconquered.'

Archimedes' Power shown to be False

Impotent in the face of one who wields his intelligence to control space and matter, Marcellus relies on time, organizing a blockade and leaving for other conquests. 'Time passed,' writes Plutarch; more than two years in fact. We should halt a moment at this laconic ellipsis.

Yet how much went on! What extraordinary organization the Romans needed to blockade a city of more than 500,000 inhabitants for more than two years, surrounded, according to the Greek geographer Strabo, by a wall

180 stades in length – more than 30 kilometres. How many men had to be mobilized? What command structure had to be developed? Plutarch's silence is a measure of his intention to ignore what was profoundly original in the Roman republic, which was its discovery of arrangements enabling control to be exercised as long as was necessary over a space or a situation, arrangements which resisted the passing of time and the growth of disorder. Plutarch passes over the establishment and solidification of the siege of Syracuse, an episode characteristic of Roman genius, because nothing must detract from the pre-eminence of the Greek mind.

Mass, logistics, discipline: for Plutarch, these count for nothing in the fall of Syracuse. If mind governs matter, then if the Romans are to win the day, Plutarch must define other dimensions than those of things in which, thanks to science, Archimedes is omnipotent. All his finesse is in the way he brings these out these other dimensions, ignoring certain events where necessary.

The three causes of the defeat of the city are the three enemies of science throughout its development. These are:

- Time, which changes situations, disturbs stabilities, decomposes objects, multiplies singularities and transforms problems, preventing science from bringing to light its fixed principles. It is the essential enemy, so much so that science as the scientific community understands it will only be born with the invention of linear, immutable, indifferent time, emptied of all its turbulence, the undoubted heir of Rome.
- Lies and falsehood, treachery which destroys internal coherence and disorders perception of the world of things. This is the normative enemy, because it makes all statements equal and so forbids distinction between propositions, the authentification of practices and the autonomy of science.
- Superstitious religion, which plunges man into an irrational system, allowing truths not founded in the intelligence and challenging the power of science to master the world. This is the institutional enemy which proclaims a globalizing prophecy, competing with science's claim to be the very spirit of human development.

Plutarch will make these three forces (time, falsehood and religion) the allies of Marcellus, a member of the Roman college of augurs and the founder of a temple. Time having passed, the city was taken, delivered by treason, during a festival in honour of Artemis. The tutelary deity has betrayed the besieged, already in the grip of those two other enemies of the intelligence: drink and entertainment. Is it surprising that the huntress, the goddess of forests and the dark sister of Apollo, god of Light and Truth, should put herself at the service of Rome, this city born of the woods? The tragic end of Syracuse represents more than the fall of a kingdom. The Romans first captured the two neighbourhoods called Nea and Tyche: the Syracusans were losing both novelty and fortune.

With the fall of the city Plutarch presents an extraordinary defeat, the defeat of Greece by Rome, and more profoundly yet, of Greek intelligence by Roman power; at the highest level of abstraction this is the defeat of embodied intelligence by institutionalized power. Marcellus is moreover presented as the first Roman to understand the necessity of safeguarding the Greek inheritance. So we can understand it when the general, looking 'down from the heights upon the great and magnificent city below . . . wept as he thought of its impending fate . . . much against his will, he allowed his men to carry off property and slaves'. We have already seen the meaning of this 'much against his will': it is always the mark of the weight of material necessity as it confronts the intelligence.

For the purpose of the passage, Plutarch creates this image of Marcellus from scratch. In the comparison which follows the parallel lives of Pelopidas and Marcellus, the author admits that the Romans accused Marcellus of responsibility for the sack of Syracuse and the savagery with which it was carried out. This contradiction only a few pages later is one additional proof that the story of the siege of Syracuse is only incidentally a historical account.

In making the victor the most Greek of the Romans and describing the defeat as the doing of one traitor among half a million subjects, Plutarch imposes on the story the logic of fate which lies at the heart of Greek thought. Syracuse loses, despite everything, and in a ridiculous manner, because it was inevitable. Daughter of the gods of Delos, Syracuse has lost because the goddess of the night has betrayed her and the god of the Sun stood aside (and this perhaps explains the absence of the burning mirrors from Plutarch's myth).

The fall of 'the greatest city of Greece', for which Plato wrote his *Republic*, was determined from the moment of its foundation. This truth of Plutarch's is not at all historical. Its function is rather to resolve the fundamental problem of the worldly institutionalization of intellectual truths. If one considers the bounded universe of the Roman empire that Plutarch lived in, pessimism is not a surprise. 'The city of the mind' cannot last long in this world, it collapses of itself, betrayed by what it has established. This bitter regret will accompany the development of Western science, the reasonable daughter of a synthesis between Greek minds and Roman institutions. There is no happy and lasting institutionalization of science.

'The city was given over to plunder, except for the royal property, which was handed over to the Roman treasury.' In the midst of immense rejoicing, Marcellus showed his sympathy and compassion for the conquered: he wanted to see Archimedes. On this, his unforgettable day of glory, his wish would not be granted.

Multiple deaths are the privilege of the greatest of heroes. How many stories there are of the deaths of Ulysses, of Aeneas and of Romulus! Among disappearances, apotheoses and group murders, one doesn't know which death to choose. As for Archimedes, there is no end to the various versions. It would seem that the variety of stories makes the death go on for ever, giving the spirit time to set itelf free.

The Death

'As fate would have it, the philosopher was by himself, engrossed in working out some calculations by means of a diagram, and his eyes and his thoughts were so intent upon the problem that he was completely unaware that the Romans had broken through the defences, or that the city had been captured. Suddenly a soldier came upon him and ordered him to accompany him to Marcellus. Archimedes refused to move until he had worked out his problem . . . whereupon the soldier flew into a rage, drew his sword and killed him.'

We shouldn't be misled by the unlikelihood of the tale, with an enormous city captured during a spell of geometric thought. In a few words, this little scene emphasizes the gulf which lies between the subtle world of the mind and the vulgar world of arms. The mind is elsewhere: this is the essence of Plutarch's thesis. Our intention of marking in this text the characteristic features of the image of mathematics being developed should allow us to note that one of them is clearly absent here. It is commonly said that good mathematics is done by young geniuses, but this idea, probably the product of the Romantic nineteenth century, is absent from this text. Archimedes dies an old man, still at his work, and the letters which accompany his treatises afford further evidence of continuing scientific production. And finally, we should note in this death the scientist's extreme detachment. The death is mechanically described, an unfortunate but inevitable consequence of the situation.

The Execution

'According to another account, the Roman came up with drawn sword and threatened to cut his throat him there and then: when Archimedes saw him, he begged him to stay his hand for a moment so that he should not leave his theorem imperfect and without its demonstration, but the soldier paid no attention but despatched him at once.'

The pathos of this description is striking. There can be no illusions: here it is a question of the execution of the man who with his inventions had delayed the capture of the city and killed so many Romans. One can understand why the soldier is in such a hurry to kill him. What terrifying response might the scientist not be working on? And the humanity of the description is touching. Archimedes appears for the first time as a human being, passion-ately involved in science, of course, but really alive, as is emphasized by the final cutting of the throat. For one moment, we see the man behind the scientist. Then everything disappears.

The Murder

'There is yet a third story to the effect that Archimedes was on his way to Marcellus, bringing some of his instruments such as sundials and spheres and quadrants, with the help of which the dimensions of the Sun could be measured by the naked eye, when some soldiers met him and believing he was carrying gold in the box promptly killed him.'

The social realism of the scene is astonishing. This version is to be found in no other historian, while the two preceding ones are contained implicitly in Livy and Maximus Valerius. We should need several pages to bring out all its implications. The cosmographical instruments, for example, are not listed by chance, for each has a place in Archimedes' works.

There was perhaps among these spheres the one mentioned by Cicero in his *Tusculanes*: 'When Archimedes fastened on a globe the movements of the Moon, Sun and five wandering stars, he, just like Plato's God who built the world in the *Timaeus*, made one revolution of the sphere control several movements utterly unlike in slowness and speed. Now if in this world of ours phenomena cannot take place without an act of God, neither could Archimedes have reproduced the same movements upon a globe without divine genius.' Marcellus must have taken this sphere, the ancestor of plan-etariums and automata. In fact, Cicero tells us in his *Republic* that he saw it at the house of a friend of one of Marcellus' descendants: 'I had heard a great deal of this sphere . . . but did not admire the construction of it so much. But subsequently . . . the Sicilian appeared to me to possess more genius than human nature would seem to be capable of . . . the same setting of the sun was produced on the sphere as in the heavens, and the moon fell on the very point, where it met the shadow of the earth.' But we must resist the charm of Cicero's lengthy description and return to the text.

The Archimedes who dies in this third description is clearly close to the one we have had before us for the whole length of the story. With him there is completed the construction of the archetype of the scientist which Plutarch has being trying to develop throughout his account. Having understood that power and riches are no longer to be found in Syracuse, the scientist has packed his bags and is carrying away his secrets. In the same way as those scientists who, twenty-two centuries later would quit their ruined country to join the victors, Archimedes wants to join the camp of Marcellus. Plutarch is not indignant: he seems already to be telling us that science has no home-land other than that of the victors and of wealth. The scientist's death is a

155

typical accident, a result of the incomprehension of the ignorant and irresponsible, who know not what they do.

We shouldn't be misled by Plutarch's rhetoric: these three deaths only appear to be distinct. Taken together they present the whole drama, made visible in this triptych, where on each panel we may can see the abstracted intellectual faced by incomprehension, the victimized body whose throat has been cut by vengeance, and the turncoat scientist lost through the stupidity of his new allies. The juxtaposition of these three perspectives fuses the three figures into a single person; the trinity of mind, body and social individual is unified in the establishment of 'the archetype of the scientist', the hypostatization of the archetypal scientist. Once this is accomplished, the separate parts can disappear. The man, his intelligence and his work are combined in this new kind of man.

At the threshold of his house, as night is falling, an old man is scratching in the dust. He is thinking . . . Silent and absorbed, his head is full of the fury and the glory of his machines. In the lines, the circles and the curves with which he decorates the soil, he is looking for forgetfulness, crossing out the earth. The goddess has betrayed, the kingdom has fallen, the city has been destroyed and science has tasted the bitterness of power . . . But he must perpetuate his work, wipe out his dishonour and establish forever the 'practice' of his thought . . . Then, in the darkness, the executioner advances, the man stretches out his neck 'and the blood of Archimedes was confused with his scientific work'.

'Marcellus was deeply affected by his death . . . he abhorred the man who had killed him as if he had committed an act of sacrilege, and . . . he sought out Archimedes' relatives and treated them with honour.' It was probably he who commissioned the tomb. The scientist 'had asked his friends and relatives to place on his tomb, after his death, a cylinder enclosing a sphere, and as an inscription, the ratio of the solid containing to the solid contained'. Marcellus did what was necessary, thus inaugurating a fine series of mathematical epitaphs. Later, a seventeen-sided polygon would be carved on Gauss's tomb, and on Jacques Bernoulli's an Archimedes' spiral – he chose a logarithmic one. The German mathematician chose a memory of youth, the fruit of his arithmetical research, a result which inaugurated a profound relationship between geometry and algebra. The Swiss chose a symbol of rebirth despite transformation. What was the meaning of the symbol on Archimedes' tomb?

Let us listen to him one last time, in Book I of *On the Sphere and the Cylinder*. After thirty-four long and difficult propositions, a simple corollary: 'It is clear that every cylinder whose base is the greatest circle in a sphere and whose height is equal to the diameter of the sphere is $\frac{2}{3}$ of the sphere, and its surface together with its bases is $\frac{3}{2}$ of the surface of the sphere.' There follow a few lines of demonstration drawing conclusions from the preceding eighty pages. And so the result to which all this work is directed is described as a corollary. We shouldn't be taken in by this litotes: it is very much a mathematician's affectation to treat as a simple inference the end result of laborious researches.

Though it is difficult to reconstruct what Archimedes might have found fascinating in this proposition, we shall make the attempt. In the first place there is the very simplicity of the ratio, quite astonishing at first sight, especially when one thinks of the transcendental ratio between the sphere and its circumscribed cube. Then, it must surely be important that this result, already glimpsed thanks to the *Method* discussed above, allows the problem of the cubing of the sphere to be reduced to that of the squaring of its great circle, the base of the cylinder. And finally, perhaps the most astonishing thing is the utterly improbable observation that the ratios of the volumes and the ratios of the areas of the two solids are the same. Here there is an equilibrium so subtle that only mathematical reason can persuade us of it. These are some reasons for the power of this result.

As for the figure itself, it is a receptacle for all of Archimedes' work. Investigated in the treatise *On Method*, which makes abundant use of the concept of the lever, it was studied according to Euclidean orthodoxy in another book which establishes the relations between all possible figures (cones, truncated cones, spheres, cylinders, segments of the sphere and of the circle) and the circle. This figure is a memory in which is inscribed, implicitly, the whole series of formulae discovered by Archimedes, which deal with the calculation of volumes, areas and lengths, by means of a unit, incommensurable with the unit square, the area of a circle of radius 1. One may say of this unit circle that it is the Archimedean form of our number π.

There is no end to the number of references to Archimedes' works that can be pulled out from this hat: conoids, spheroids and paraboloids (whose volumes and centres of gravity are discovered in the books *On Method* and *On Conoids and Spheroids*) appear from their matrices: the cone, the sphere and the cylinder. And then with a turning and ascending movement we may obtain the famous screw, which, leaning with one end in the water, becomes, in Montucla's words 'a singular machine, in which the very tendency of weight to fall seems to be employed to make it rise'. It is a technical marvel whose path describes a helix in the cylinder.

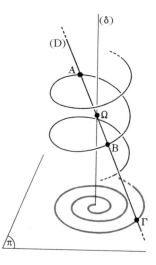

If we take a close look at this last object, we can see that all straight lines (D) which rest on points A and B and cutting the axis δ at a point

Ω traverse a plane perpendicular to the axis II at point Γ, inscribing in it a spiral of Archimedes. This relationship of construction between the screw, inherited from the Egyptians, and the spiral could make the latter the object of the theoretical study of the former. In work on the spiral Archimedes developed some of his most beautiful proofs: the first investigation of the tangent to a curve other than the circle, the area of each rotation once more being reducible to areas of the circle.

Let us here abandon the inexhaustible theory of Archimedes' works and allow the weeds and brambles to grow over the monument erected by the Roman general, who must have understood mathematics so badly and yet suffered its most mortal consequences.

Two centuries have passed. A biographer, an architect and a few historians remember the defender of the Sicilian metropolis. Like a wave weakening as it spreads, his fame travelled through the Mediterranean basin as it grew fainter at its point of origin. The man who fell in the dust as the Roman soldier struck was forgotten by his fellow citizens. 'I tracked down his grave . . . and found it enclosed all round and covered with brambles and thickets,' says Cicero in Book V of the *Tusculanes*, where he claims to have rediscovered the scientist's tomb. 'Accordingly, after taking a good look round (for there are a great quantity of graves at the Agrigentine gate), I noticed a small column rising a little above the bushes, on which there was the figure of a sphere and a cylinder . . . Slaves were sent with sickles who cleared the ground of obstacles . . . the epigram was traceable.'

So, despite the passage of time, the obliteration was not complete, and Archimedes' 'arms' allowed Cicero to rediscover the tomb. Thanks to this, he is able to conclude: 'One of the most famous cities of Greece, once a great school of learning as well, would have been ignorant of the monument of its most ingenious citizen, had not a man of Arpinum pointed it out.' Of course, we should understand this term 'monument' in its widest sense and discount Cicero's pretended modesty, seeing the 'monument' as a metaphor for Archimedes' monumental life's work. This hypothesis is confirmed by the passage in which these lines are found, where Cicero anticipates Plutarch in praising the happiness of thinkers above that of tyrants.

Beyond death, Cicero's account once more allows us to see the fundamental framework of Archimedes' thought, that of the effect of the minimum on the maximum: near the tomb we witness a socio-historical adaptation of an experiment of Archimedes': all by himself, a simple inhabitant of Arpinum is responsible for the rediscovery of the monument to the extraordinary genius. This is undoubtedly the true resurrection of the man who, on the evening of defeat, fell in the dust with a compass in his hand. Cicero has already warned us, when he compares the happiness of a tyrant like Dionysius to the imperishable happiness of a genius like Plato. At the beginning of the story of his discovery at Syracuse, he writes, 'Ex eadem urbe humilem homunculum a pulvere et radio excitabe, qui multis annis post fuit, Archimedem' which might be translated as: 'From this same city, from dust and compass I will conjure a humble mortal, Archimedes, who lived many years later (than Dionysius).' We have been warned: subjugated to the incantatory power of the Ciceronian text the reader is going to witness a veritable palingenesis, in which Archimedes will be resurrected from his circle of dust, brandishing the instrument symbolic of his science. Thus Plutarch must share with Cicero the credit for having brought to life from dust and forgetfulness the man who, thanks to their lustrous and legendary accounts, many years later became the very paragon of scientists.

Two thousand three hundred years have passed. Later, in 1988, a humble historian has come to breathe the light and sparkling air which witnessed so long ago the fall of the city and the death of the scientist. In front of him the sea fades into the sky, casting a blue light on the chalky hull of the peninsula of Ortygia. Around him, flies buzz along the contours of the rubbish tip that stands above the town, and here and there a stony plinth stands up against the rolling wave of rubbish. A few feet away the dreamer sees a regular shape, a block of white marble standing out among the filth. He approaches, bends down to touch it, feels his heart grow warm as he feels it smooth and cold beneath his hand.

The column is on its side on a clear patch of ground and, at the top, the symbol is still visible: the sphere and cylinder of stone. Taking hold of both of them he tries to stand the column up again. Worn by centuries of rain, decades of fumes and years of decay, the marble gives way at the bottom of the sphere. The cannon-ball remains in his hand ... Stunned by the accident, he puts it down without thinking, in the mouth of the stone cylinder – then he sees the light and everything becomes clear ...

Night is beginning to fall; the historian's thoughts become clearer, something in the fine image put forward by Plutarch is turned upside down. Is it so far between the scientist and power, between science and war? Didn't Archimedes himself want to be buried beneath this symbol of power? And then a general putting up a monument to pure mathematics, is it really credible? Lost in his thoughts, the man strokes the stone ... At the base of the column, the part which was earlier attached to the pedestal, his hand can feel the regular indentations which probably mean the presence of a carved inscription.

Impatient, careless and in a state of exaltation, he lies down in the filth. The palaeographer's eye has already detected the name of Marcellus as a signature, and above it a Latin sentence partly hidden under the remains of mortar ... Retracing in the depths of the night the stonecutter's lengthy labours, he uncovers the letters one by one. When all is done, his fingers will already have been able to read ... 'Qui tollis peccata scientiae' ('who bears away the sins of science').

5

Stories of the Circle

Catherine Goldstein

Certain mathematical problems topics are so old they seem natural. Here we will see that, like every other human creation, the idea of the circle is in fact highly dependent on its historical and cultural context.

In the beginning was the circle. Everywhere; hanging in the sky on nights of full moon; carrying chariots and, later, motor-cars; surrounding cities with stone; decorating vases, whose traces in the sand depict it still. Think of it, shut your eyes, open them, look for it: you will find it even in the O's on this page. And in your head no doubt, the everyday furniture of lessons in maths.

For four thousand years. Unchanging.

Twenty-three centuries ago, in his *Elements*, the mathematician Euclid put forward the following definition: 'A circle is a plane figure contained by one line . . . such that all the straight lines from one point among those lying within the figure are equal to one another. And the point is called the centre of the circle.' This definition corresponds almost word for word with the one found in modern dictionaries such as the *Shorter Oxford*: 'Circle . . . a plane figure contained by one line such that all the straight lines falling upon it from one point among those lying within the figure are equal to one another.'

Eternally identical to itself and present long before writing, the circle could therefore be one of those extra-historical objects favoured by mathematics, which provides us, as we have been taught, with eternal truths. Astronomers, philosophers, architects, mystics, surveyors, geographers and poets have studied, measured and described it, nourished their imaginations and polemics on it, drawn their inspiration from it or cradled their results within it without radically changing its image of serenity. A unique and recognizable figure rolling through time and space, 'the' circle makes its home in different disciplines and different cultures only so that they may illuminate its previously hidden properties or imagine new uses for it.

But what if it wasn't like that at all? If, unravelling this seamless tissue of world and time, its rounded curves were caught up in it and the circle lost its identity? If it sometimes even became difficult to recognize? If, in the end, it appeared as multiple and scattered, the descendant of a long line whose complex interrelationships required patient decipherment? If at each instant the human gaze did not so much uncover it as (re)create it? Attempting to capture what is common to everything we call a 'game', the philosopher Ludwig Wittgenstein compared the way different things are collected together under one name to spinning, in which the fibres are twisted together: 'and the strength of the thread does not come from one of the fibres running along its whole length but from many of them being combined together.'

To unwind all the fibres of this so solid thread we call the 'circle' would be, of course, impossible. It is not a matter of exhaustively examining all its different manifestations, but of understanding some of the stormy debates, disguised by the standard definition and seeing the reappearance of the objects it hides, attempting to explain, if possible, the overlapping of some of the threads, some of the metamorphoses. There will be some obvious gaps. Chinese scholars studied the circle for more than two thousand years, but they will receive no mention here; from Antiquity until our own day, the circle's property of enclosing the greatest area possible for a given perimeter has been debated, refined and generalized, yet I will not speak of it. Other omissions are less glaring perhaps, but they are innumerable and, of course, arbitrary; in a landscape so fertile and so civilized there are no inevitable oases, but journeys are chosen according to the wishes of the guide. Many civilizations beyond the Mediterranean – in Mesopotamia, in southern or eastern Asia, for example – saw the extensive development of scientific tools and problematics: I have chosen to begin our tour with an Indian wheel.

Taming some Circles: From India to Egypt, 2000–100 BC

'This is what must be known: he who has rivals should construct an altar fire in the form of a chariot wheel.'

Between the eighth and fourth centuries before the Christian era, the *Śulvasūtras,* whose name means 'treatise on cord (or rope)' were compiled in India. Within the framework of Vedic ritual they codified the rules for the construction and orientation of altars (*vedi*) and fireplaces (*agni*) intended for sacrifices. Just as in the saying of a *mantra,*[1] the effectiveness of the ritual depends on the exact observation of these rules. These treatises, then, develop and combine together techniques which we would call architectural, geometrical or calculatory, depending on the case. The basic figure for constructing these edifices, made of bricks of a fixed size, is a quadrilateral, in principle a square, on the basis of which all the precise forms and proportions required by the cult may be produced, in accordance with the objective to be attained:

> This is what must be known:
> He who desires the heavens must construct an *agni* in the form of a falcon. It has curved wings and a spread tail . . .
> He who has rivals must construct an *agni* in the form of a chariot wheel.

Then, at the cost of lengthy fractional calculations, there follow the details of construction, in terms of the basic brick. The essential instrument is the measure (the *sulva* or *rajju*), made of hemp or bamboo. Was it used to draw circles? And of what kind?

The answer is less easily obtainable than one might think. The texts have been passed down to us in many versions, (the main ones being those of Apastamba, Baudhayana, Katyayana and Manu), often accompanied by more recent commentary. They were composed in Sanskrit verses of sometimes elliptical form; in a context of oral transmission in which techniques of memorization were essential, they permit the coexistence of many synonyms, depending on the requirements of metre. Thus *vrtta, mandala* (which usually means a figure), *parimandala* (round figure) and *rathacakracit* (chariot wheel) all designate round things. In addition, circular lines are also drawn which have an auxiliary function in construction; they are not designated by any name, but indicated by a verb ('to trace in turning about'). These lines may be more or less complete, a peg and a cord (or a piece of bamboo) being used as a compass. Here, for example, is the construction of a square of 1 *purusa* (the basic measure), from Apastamba:

> One measures (the *agni*) by the measure of a man (he who performs the sacrifice). One measures with a stem of bamboo. This is what must be known. At a distance like that of the sacrificer with upraised arms, one makes two holes in a bamboo and a third in the middle. Having placed the bamboo along the *prsthya* (the east-west line of the altar) to

162

the west of the holes of the sacrificial posts, and having placed pegs in the holes:

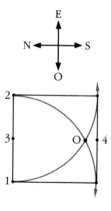

Having freed it at the west you must trace in turning about with the end (1) towards the south-east; freeing it at the east and fixing it at the west you must trace in a round with the end (2) towards the south-west;

Freeing the bamboo, fixing the terminal hole of the bamboo on the median peg (3), placing the bamboo to the south of the point of intersection of the lines (O), one fixes a peg in the hole at the end (4);

By fixing the median hole of the bamboo and placing the others on the ends of the lines, pegs are put in the holes.

This is the square of 1 *purusa*.

This is, as we can see, the detailed description of an operational procedure, of a manipulation, and the best way to absorb it and to understand the text is perhaps to provide oneself with pegs and with a stalk of bamboo (or with pins and a matchstick!) and to recreate the proposed construction *by hand*. This same approach is used for the construction of other edifices: the texts explain how to convert a square into other figures, and vice versa, all the while keeping the same area required by the rite. 'If you wish to eliminate your rivals you must construct a *rathacakracit*': the reason why and in particular the religious or concrete significance of the wheel is buried in the Vedic texts. The question of how, with which we are concerned here, receives several answers. Here is Apastamba's:

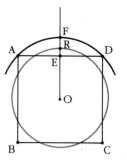

If one wishes to transform a square into a circle, one stretches (a cord) from the middle (of the square) to a corner; one turns around the side, one draws a circle with a third remaining. This gives the necessary circle, given that it is divided.

If one wishes to transform a circle into a square, one divides the diameter into fifteen parts, one takes away two, this gives the necessary square.

Thanks to later commentaries, which sometimes also contain justifications of the proposed constructions, one can elucidate the procedures suggested in the texts: the second consists of producing a square the length of whose side is thirteen-fifteenths of the diameter of the circle one starts with. The first is illustrated by the diagram on the previous page (OR is the radius of the circle sought and FR two-thirds of FE).

It is remarkable that these conversions are reciprocal neither in results nor in type of procedure: the circle obtained from the first construction by conversion of a square itself obtained from a circle by means of the second is significantly different from the original circle (that is to say more different than might be expected from the precision of other calculations in the text). Another version, Baudhayana's, proposes a more complicated conversion procedure which answers better to our desire for reciprocal conversion, but the mathematical, religious and technological origins of these preferences are not at all clear. To discuss them, however, is not a waste of time. Elsewhere in this book (see pp. 44–72), it has been shown how two civilizations contemporary with each other and much earlier than Vedic India, those of Mesopotamia and of Egypt, were interested in their time in our favoured figure. The circles drawn on the Rhind Papyrus and those engraved on Babylonian clay tablets do no more than illustrate the procedures related to them: and no more than in the India of the *Śulvasūtras* do they open the way to a theoretical exploration of the circle for its own sake. In all three cases, the interest is elsewhere. Already, however, differences have appeared. The fundamental topic dealt with by the Near-Eastern texts is the calculation of the area of the circle; for the Egyptians, directly from its diameter; for the Babylonians, through an intermediate calculation of its circumference – no explicit calculation of the circumference of the circle appears in the small number of Egyptian papyri we have. Apastamba's text, however, is concerned only with the reconstruction of a circle from a square while retaining the same area, that is to say, concretely, the same basic bricks. We can, of course, deduce from this a calculation for the area of a circle of given diameter, and Apastamba's two procedures give different values; but it is important, above all, to note that this question appears to be of no interest to the users of the *Śulvasūtras*. Thus, the treatment of the circle, discreet as it might be, provides information about the relative importance of matters connected with it, be they the exigent prescriptions of ritual architecture or the economic calculations relating to the construction of a granary. Given the architectural and religious evidence available to us, it is unlikely that texts like the *Śulvasūtras* would have been composed in Mesopotamia: one would, on the other hand, be very willing to believe that calculations of area were performed in Vedic India. There is nothing, however, to indicate any direct connection with those one might deduce from the *Śulvasūtras*. In 325 BC Alexander the Great tried to conquer India: this is the period when contacts, always very selective, with Greek science became common. But the Vedic influence had already diminished and with it the importance of ritual constructions; the Jain religion, born at the same time as Buddhism in the sixth century BC, developed a cosmography based on the circle. In it the

In Jain cosmography the madhyaloka, *the middle world, has the form of a disc on which is deployed a series of concentric rings representing alternately the oceans and the continents. This bas-relief from the temple of Ramakpur dates from the seventeenth century and shows the Nandisvaradvipa, the continent of rejoicing, with its fifty-two sanctuaries. (Photograph J.-L. Nou.)*

Sun and the Moon are supposed to revolve in concentric circles in planes parallel to the Earth, with Mount Meru at the centre. Between the second century BC and the second century AD, the mensuration of the circle, in its connection with the religio-astronomical sphere, became an essential part of Jain mathematical literature. Though there is no reason to suppose that the *Śulvasūtras* and their content were unknown in this period, it is elsewhere that one must look for explanations for the change in style, to the influence of late Babylonian astronomy, for example, which had been transmitted by Greek intermediaries and welcomed all the more enthusiastically for its concerns seeming very similar. Geometrical manipulations were replaced *en masse* by purely calculatory procedures; scrupulous precision was abandoned, the explicit requirements of the knowledge of celestial motions being – how paradoxical it seems to us! – less exigent in this respect than the necessities of ritual.

Formulae, stated in rhetorical form, give approximate relations between chord, arc, distance to the centre and diameter, as well as expressing the area of a circle as the product of its circumference and a quarter of its diameter; the circumference itself is calculated as the product of the diameter and $\sqrt{10}$. References are made to other traditions which used 3 instead of $\sqrt{10}$. Whether this value 3 derives from a transcription of Babylonian values into Indian figures or from a previous properly Indian metrological tradition, it is remarkable that no mention is made of the values more accurate than 3 which our own too well-informed eye can sometimes see in the *Śulvasūtras* (see 'The Measurement of the Circle', p. 167).

Prudence, then, is always necessary in advancing across this terrain, where our modern knowledge, even what seems to us to be the most natural, dangerously smooths out distinct steps, and makes differences invisible or reduces them to simple variants or even to blind spots. Three different civilizations (four, if one distinguishes between the Vedic and the Jain cultures) were constructed round objects which did not always bear a specific name, which do not appear in the same context, which are not treated in the same way and which do not produce identical calculations. From these texts there already emerge a host of themes, which are spun together in any quest for *the* circle: metrological questions connected with the calculation of its surface and circumference, on to which will soon be grafted the problem of squaring the circle; morphological questions, dealing with its more naïve aspects (in what verbal or graphic forms is the circle represented?) or its metamorphoses (to what other figures is the circle compared? Which of these are compared with the circle? What technical procedures do these conversions rely on, in what cultural context are they inscribed?). Even wider questions concern its habitat (in what sort of texts is it studied and why? In what kind of classification is it placed?). The section which follows will give other answers to some of these questions. New questions will also arise, which we, perhaps disorientated by the reading of more exotic texts, will address to Euclid's *Elements*, set up by our own education as the archetype of all mathematics.

The Circle is Defined: From Athens to Alexandria, Sixth Century BC to Second Century AD

Why Euclid again? To a great extent, because there is no choice. We have no earlier complete text, even if the philosophers of the fourth century BC, Plato and Aristotle among others, as well as later commentators such as Proclus or Simplicius, speak of mathematicians in Greece from the sixth century BC, and of a tradition of *Elements* presented in deductive and autonomous form from the fifth century BC. Of them there remain no more than fragments and references whose reliability is often difficult to assess. The other reason, perhaps connected to the first, is the fame of Euclid's text up to our own day and its historical impact, both as a formal model and as a reservoir of basic knowledge and problems. What then are we told about the circle by this (collective of) mathematician(s), who worked in Alexandria around 300 BC, at the beginning of the Hellenistic period?

The measurement of the circle

Here are some of the methods of calculating the area of a circle and their results from several civilizations. Also indicated (in italics) are results of a non-metrological nature, but relevant to the discussion.

Origin	Problem	Procedure and result (modernized)	Comments
Rhind Papyrus, Egypt, beginning of second millennium BC.	Calculation of area of round surfaces.	$(\text{diameter} - \frac{1}{9}\,\text{diameter})^2$	Exclusively numerical procedure; no specific name for the circle; no calculation for the circumference.
Palaeo-Babylonian tablet, Mesopotamia, beginning of second millennium BC.	Calculation of area of round surfaces.	$(\text{triple the diameter})^2 \times \frac{1}{12}$	Exclusively numerical procedure; via calculation of the circumference; the $\frac{1}{12}$ identified as a constant of the circle.
Śulvasūtra by Apastamba, India, first millennium BC.	*Conversion of circle into a square.* Conversion of square into circle.		*The area of the square obtained is about 3.004 times the square of the half-diameter of the circle; the area of the circle obtained is about 3.008 times the square of its half-diameter.*
Jyotiskarandaka, Jain text, India, second century BC–second century AD.	Formulae for the circle.	Circumference: $\sqrt{10} \times \text{diameter}$	Other formulae also existed.

Origin	Problem	Procedure and result (modernized)	Comments
Aryabhata, India, sixth century AD.	Circumference of the circle.	Approximate value of the circumference for a diameter of 20,000 is 62,832	So the ratio is 3.1416 (!); the area is calculated as in the Jain texts; method of approximation by polygons.
Nine Chapters on the Mathematical Art, China, first century AD (?)	Measurement of fields.	Area of the circle: $\frac{1}{2}$ circumference × $\frac{1}{2}$ diameter; or $\frac{3}{4}$ diameter^2 (*); or $\frac{11}{12}$ circumference^2.	Numerous later commentaries (among them that of Lui Hui in the third century AD) give justifications and more precise values, often with the help of polygons; in the fifth century AD, Zu Chongzhi gives a ratio of 3.141592614 for (*).
Euclid, Alexandria, 300 BC.	*The ratio of the area of a circle to the square of its diameter is constant.*		*No numerical values, of course.*
The Measurement of the Circle, Archimedes, Alexandria, third century BC.	Measurement of the circle.	The area is equal to that of a triangle, whose base is equal to the half-diameter and whose height is the circumference; the ratio of the circumference to the diameter is between $3\frac{1}{7}$ and $3\frac{10}{71}$.	The first known demonstrations on these points; method of approximation by polygons.
François Viète, France, sixteenth century.	Calculation of the ratio of the area of a circle to the square of its half-diameter.	Value of π correct to 9 places: 3.141592653.	Decimal numeration is available; polygon of 393,216 sides (the method in the text giving an infinite series is less accurate).
Ludolph van Ceulen, Germany, sixteenth century.	Calculation of the ratio of the area of a circle to the square of its half-diameter.	Value of π correct to 35 decimal places.	One of the most accurate determinations by the method of polygons!

Origin	Problem	Procedure and result (modernized)	Comments
From the eighteenth century on.	Area of the circle.	Area of the circle $A = \pi r^2$.	The value of π is determined analytically; today it is known to many thousands of places!

Book I begins with twenty-three definitions, and the fifteenth, the definition of the circle, comes after those of the point, the line, the straight line, the area and the angles; the following definitions relate to it:

15. A circle is a plane figure contained by one line called the circumference such that all the straight lines from one point among those lying within the figure are equal to one another;
16. And the point is called the centre of the circle.
17. A diameter of the circle is any straight line drawn through the centre and terminated in both directions by the circumference of the circle, and such a straight line also bisects the circle.
18. A semicircle is the figure contained by the diameter and the circumference cut off by it. And the centre of the semicircle is the same as that of the circle.

Then follow the definitions of rectilineal figures (enclosed by straight lines), of the different triangles and quadrilaterals, and finally that of parallel lines. Then, clearly distinguished from the definitions, there are five demands – postulates, as long as this word is not accorded an anachronistic technical significance and five common notions; we are concerned with only one of the postulates, which allows the author of the *Elements* to construct circles of given centre and radius: 'Let [it] be postulated . . . To describe a circle with any centre and distance.'

The circle appears in the propositions only in the third book. Euclid deals first of all with rectilineal figures and how they may be squared, that is to say the method for the geometrical construction of a square equal in area to a given figure. Circles are, however, referred to within certain demonstrations. As an example, here is the first proposition in Book I:

On a given finite straight line to construct an equilateral triangle.
Let AB be the given finite straight line.
Thus it is required to construct an equilateral triangle on the straight line AB.
With centre A and distance AB let the circle BCD be described; again, with centre B and distance BA let the circle ACE be described;

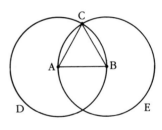

and from the point C, in which the circles
  cut one another, to the points A, B let the
  straight lines CA, CB be joined.
Now since the point A is the centre of the
  circle DCB,
AC is equal to AB.
Again, since the point B is the centre of the
  circle CDB,
BC is equal to BA.
But CA was also proved equal to AB;
therefore each of the straight lines CA, CB is equal to AB.
And things which are equal to the same thing are also equal to one
  another;
therefore CA is also equal to CB.
Therefore the three straight lines CA, AB, BC are equal to one another.
Therefore the triangle ABC is equilateral; and it has been constructed on
  the given finite straight line AB.
(Being) what it was required to do.

This construction is used later to draw a perpendicular to a straight line
at a point and thus to construct a square of a given side (propositions 11 and
46). Even if the equilateral triangle makes no explicit appearance in the
Śulvasūtras, the analogy between the constructions invites a comparison
between the two texts. They share the same type of question and the same
constraints, because the postulates authorize only the drawing of straight
lines and circles, and therefore the use of only 'ruler and compass', as we
would say, or bamboo cord and peg, as Apastamba would have it.

Where else than at the very beginning of the *Elements* can we see so clearly
the cultural gap which lies between Apastamba and Euclid? The Indian sac-
rificer, arms stretched out to give the measure of the altar he will construct,
points out to us to the disappearance in Euclid of the world beyond the text.
The Egyptian papyri and Babylonian clay tablets, in few words, none the
less often mention granaries, wells and cities and lay down for a designated
addressee a procedure that must be followed; however artificial these concrete
details may be, like the so-called practical problems of the old-fashioned
school-room, they involve the pupil and teacher in daily life. We must look
more closely at this absence in Euclid.

The first difference, and nothing could be more innocent, concerns the
word *circle*. It is used systematically: it is used in auxiliary constructions such
as the one described above, as well as where the Sanskrit would have been
satisfied with a verb. Now in Euclid a circle is a plane figure and not a simple
line; it is necessary to draw the complete figure before using it (a comparison of the drawings on p. 163 above and on p. 170 is, in this respect,
illuminating). What is more, bamboos, pegs and even compasses have all
disappeared together with the image they produced: the postulate which

states that one can and may draw a circle does not say with what. The verbal form ignores the possible mistakes that students might make, completely ignores the fact that there are students at all: who will draw the circle?

The constructions occupy an important place in the *Elements*. Like all the propositions they are accompanied, and here is another important difference from the *Śulvasūtras*, by demonstrations intended to support them *a posteriori*. Proved to whom? The ideal Greek mathematician is not as isolated nor as contemplative as one might have believed: he or she acts, constructs and justifies himself ... It isn't by a miracle, then, or by some internal requirement of the circle itself, that a so-called natural development should finally be disengaged from its material matrix, that the concrete universe of everyday life should disappear: the world is deliberately held at a distance, blocked within the definitions which describe it, fixed at precise points (the postulates) and within forms which contain its ambiguities and specify its practice. The Euclidian circle is born from this effort which condenses everything into a single figure, including the tool which has just drawn it. The relationship of this process to a conception of mathematics and of the educational system, to the presence of a professional environment, to a particular temperament or to philosophical issues is another topic whose connection with ours I only want to point out in passing (see pp. 344–71).

Later on in the *Elements* the circle appears as an object of study in itself. Book III studies its intrinsic properties (construction of the centre, comparison of angles intercepting the same arc), Book IV the inscription and circumscription of rectilineal figures, mainly the regular polygons (including the pentadecagon). Other propositions relating to the circle are scattered through the other books, for example, in those devoted to ratios or areas:

XII, 1: the areas of similar inscribed polygons are as the squares of the
         diameters of the circumscribed circles ...
XII, 2: the areas of the circles are as the squares of their diameters.

If the definition of the circle comes before that of the rectilineal figures, its study comes afterwards. Its relations with the straight line and itself are then explained (angles, intersections of straight lines and of circles), then its relations with polygonal figures classified by the increasing number of their sides. In these classifications we find traces of the double nature of its definition and status, the same word designating the auxiliary line associated with the straight line in all the constructions, and also the plane figure which comes at the end of all the others, whose properties are used to demonstrate its own by the method of exhaustion. The circle is approximated by means of inscribed and/or circumscribed polygons with an increasing number of sides, as in the demonstration of proposition XII, 2 (see also p. 126).

In this connection, though Euclid inscribes a square in a circle (IV, 6), circumscribes a square with a circle (IV, 7) inscribes a circle in a square (IV, 8) and circumscribes a circle with a square (IV, 9), with him there is never any question of squaring the circle, that is to say, constructing a square equal (in area) to a given circle, as he does for the rectilineal figures. For him this could only have been a matter of constructing it 'with ruler and compass', with straight lines and circles, and we know today that this is impossible. It is interesting to note that more or less identical rudimentary instruments, bamboo and ruler, peg and cord and compass, appear independently, as one may suppose in the absence of any evidence to the contrary, in such different civilizations, and that they should leave such durable traces. The reasons for which Euclid limited himself to constructions using only straight lines and circles are not clear: perhaps a respectful reference to their ancient and venerable use, in astronomy, for example; a desire to prove that it was possible to reduce all movement to a combination of two kinds, rectilinear and circular, at that time regarded as fundamental; or, indeed, simple distrust born of the theoretical and technological imprecision of other instruments. Many explanations have been put forward, but the present state of the evidence does not allow a final conclusion to be drawn. However, in its normative impact the Euclidian text would fix this obligation for centuries, or at least draw its successors' attention to its rigorous requirements. Here the *Elements* display their didactic side, explaining, reorganizing and even refining a mature and developed science, but without necessarily doing justice to other active research. The attempts at squaring the circle reported by Aristotle; the success of the mathematician Hippocrates of Chios in squaring the 'lune', that is to say the crescent-shaped figures produced by the intersection of two circles; the use of other instruments than the ruler and compass, and therefore of other curves than the straight line and the circle, such as the spiral: there is much evidence of the energy and imagination deployed in attempts to resolve the question. It was a question of interest not only to mathematicians.

In *The Birds*, performed in 414 BC, Aristophanes himself introduces our favourite subject:

> METON: . . . so all I have to do is to attach this flexible rod to the upper extremity, take the compasses, insert the point here, and – you see what I mean?
> PEISTHETAERUS: No.
> METON: Well, I now apply the straight rod – so – thus squaring the circle; and there you are.

Should we see here striking evidence of the ancient popularity of the squaring of the circle? It is more likely that the mockery is aimed at recent architectural projects: the round or rounded plan of ancient time as against the quadrilateral plan of the Hellene cities. And, with this concrete issue twisting a fibre of urban planning into our discussions, do they not become even more instructive for us?

Meton uses in passing a word we have almost forgotten: *measure*. Though in Greece it was not possible to square the circle geometrically, could one not at least measure its area? In his *Measurement of the Circle*, Archimedes proposes a solution the more interesting for its combining several threads essential to my discussion. Here is what was said by Eutocius of Ascalon, the author in the fifth century BC of a *Commentary on the Treatises of Archimedes*: 'Archimedes wishes to demonstrate what rectilineal area is equal to the circle, a problem for which the celebrated philosophers before Archimedes had for a long time sought the solution.' Archimedes shows in fact (see p. 126) that every circle is equal to a right-angled triangle in which one of the sides of the right angle is equal to the half-diameter of the circle and whose base is equal to the circumference of the circle: this result belongs to the purest Euclidian tradition, and indeed makes use of proposition XII, 2 on the area of the circle. This is in fact the first attested demonstration of a relationship between the half-diameter, the circumference and the area of a circle. This does not, however, resolve the question of squaring the circle because there is no construction of the base of the triangle (that is to say of a segment of straight line equal to the circumference of the circle) with circles and straight lines alone. The same Euclidian tendency perhaps led Archimedes to inscribe the circle within a rectilineal figure with 96 sides, as he did later. There is no explicit calculation in the *Elements*: but Archimedes shows that the ratio of the diameter of the circle to the perimeter of the 96-sided polygon is greater than the ratio of $4,673\frac{1}{2}$ to 14,688 (in modern figures): those who have seen how the Greeks calculated and wrote their numbers, will realize the remarkable labour involved here. It relies on fractional approximations of certain roots, for instance, on the approximate value of 265/153 for $\sqrt{3}$; one then proceeds by successive divisions by two. In this way Archimedes also deduced that the ratio of the area of a circle to the square of its diameter is approximately 11 to 14. There is evidence too of another logistical current of thought, more common perhaps, for after all one has to live, and therefore measure distances and build wells. Eutocius goes on: 'This book is necessary for the needs of life because it demonstrates that the circumference of the circle is equal to the triple of the diameter augmented by a segment taken between ten seventy-firsts and a seventh of the diameter.'

Between problems dear to philosophers and the needs of life, the line of demarcation is no more strict in Archimedes than it is in Aristophanes. Even when Plato wishes to draw the line, does this indicate an effective specialization or a model that ought to be followed? From before the Archimedean period we have no more than fragments, remodelled by the commentaries of the logistical tradition, and it is not easy to be sure to which of these currents (if not to both?) such people as Archytas and Thales belonged. It is only in the Hellenistic period that more precise traces appear in school textbooks or in practical texts. The most sophisticated rely upon or are accompanied by theoretical results of an Archimedean stamp: Hero of Alexandria (first century AD) composed, for example, in addition to his *Definitions* on the Euclidian model, treatises on *Pneumatics* and on mensuration, among them the *Metrics*, in which he combines, for the calculation of areas, geometrical demonstrations and numerical approximations: for the circle he uses Archimedes' results. The astronomers too, like Ptolemy, who in the second century AD calculated the chords of circular sections in his *Mathematical Syntax* (known to the medieval world as the *Almagest*), generally represented a mixed tradition in which the circle occupied an important place. In the complex theory of epicycles developed by the Alexandrians, from Apollonius to Ptolemy, the planets describe circular orbits about points which themselves move in circles about the Earth. From this example, or from that of the Jain cosmology mentioned earlier, one should not deduce too hastily a link between the circle and the heavens: Babylonian astronomy, which made very precise observations, did not organize them within a geometric schema and it had no use for the circle.

But the circle invaded the Greek landscape: for the philosophers it was a metaphor for eternity or for the world; in cosmological theory it was the form of the heavenly bodies. The circle was even taken as the basis for the lever and the balance in the third- or second-century *Mechanics* (for a long time wrongly attributed to Aristotle). The circle, word or image, the line or area combined together by the Euclidian purpose, the inevitable object of a whole culture brought together innumerable different paths: 'and the strength of the thread does not come from one of its fibres running throughout its length but from many of them being wound around each other.'

In which we follow more Modest Strands: Medieval Europe, Fifth–Fifteenth Centuries

It is 1484. A bachelor of medicine, now a 'teacher of algorithm' at Lyons, wrote 'a little treatise on the practice of geometry containing the manner of measuring all things'. A jump of a thousand years, a jump across the whole of the Mediterranean basin and beyond, from the delta of the Nile to the banks of the Rhône. In what we are going to discover we shall need to determine the contribution of independent local development, as in the cases of Egypt and Mesopotamia, which were already looked at briefly above, and the part played by transmission and even by patient recopying.

The Romans showed hardly any interest in the Greek mathematicians: traces of the Latin heritage must be sought above all in the treatises on field measurement. Compilations such as the *Ars gromatica* (from *groma*, surveying) brought together the most useful Greek results with numerical examples accompanying the rhetorical expression of the desired relationships: here, the area of the circle is equal to $\frac{11}{14}$ of the square of its diameter, as it was in the Heronian texts. Of the Euclidian thread, properly speaking, medieval Europe received first of all only fragments, perhaps fifth-century translations by Boethius of certain books of the *Elements*, a few allusions in texts by the Fathers of the Church and commentators on Aristotle. It was hardly ready to receive any more: while education, essentially classical, was spreading little by little, and arithmetic was taught seriously, geometry, on the other hand, remained ignored: the situation is well illustrated by an example from the eleventh century analysed by the historian Paul Tannery.

Francon, a schoolmaster at Liège, was aware of the problem of squaring the circle from a treatise on logic; for him the surveyor's formula mentioned above was an exact one: and it was easy to construct a *rectangle* whose area was $\frac{11}{14}$ of the square of the diameter of a circle. For him, therefore, the whole question was reduced to the transformation of this rectangle into a square and he struggled with it for a long time. As for the 'surveyor's formula' itself, he imagined, and he was far from being alone in this, that it derived from very carefully cutting up a piece of parchment! As highlighted by this example, the problems posed by the Greek mathematicians were not at all natural: they had no meaning except within the implicit culture which had given them birth. Medieval letters, which propose other expressions for the area of the circle, show too that the results given in technical texts were not known by all the learned or, at least, that the theoretical reasons for their validity were not so easily understood.

The situation changed when others were added to this Latin strand: the trail of the circle, which we left at Alexandria, crossed the Islamic world in the wake of Heron's works and Ptolemy's tables: the details of this journey will be found elsewhere (see pp. 191–216). For our own purposes we need only note that trigonometry was developed and refined at the confluence of Greek, Indian and Arabic texts.[2] Euclid's treatise played its role as a source of geometrical truths enriching astronomy, optics and metrology. There also appears to be more detailed evidence of a public which consumed this mathematical work: beside his *Commentaries on Euclid*, Abu'l-Wafā' (tenth century) wrote a treatise *On the Geometrical Constructions Necessary to Artisans*. Commercial and scientific contacts developed as Greek and Arab works were translated into Latin, and this work was assimilated little by little in the West. Of the travelling circles, one of the most representative is perhaps the astrolabe (see p. 204), both a measuring instrument and a system of stereographic projection used in the calculation of celestial

Is God a geometer? The supreme architect, Christ, here redraws with a compass the hitherto unformed sphere of the world. ('L'Architecte de l'Univers', a miniature from an Old Testament manuscript made in mid-thirteenth-century France. Österreichische Nationalbibliothek, Vienna.)

motions, constructed by Arab scientists on the basis of Ptolemaic descriptions. The use of this and similar instruments became widespread in Europe at the same time as the earlier astronomical works were assimilated: the operation of these instruments was taught in treatises which included the geometrical results necessary to understand it. Timidly and then more and more frequently, demonstrations reappeared: the road is marked by the *Practica Geometriae* by Leonardo Fibonacci around 1220, the *Treatise on the*

Planisphere by Jordanus de Nemore (in the thirteenth century) and Dominicus of Clavasio's *Practica Geometriae* in 1346, giving new life to the tradition. In 1484 a bachelor of medicine, now a teacher of algorithm at Lyons, also wrote a practical geometry: he was called Nicolas Chuquet. Though his work was not the most influential, it is none the less completely representative of its time.

> Here begins a little treatise on the practice of geometry containing the manner of measuring all things; of which some are measured by one dimension, some by two dimensions and the others by three . . .
>
> The straight line may be measured in two ways: the one naturally by fathoms or by feet, or by another measure, the one after the other; the other manner of measuring it is by the quadrant or by the back of the Astrolabe as will appear later when I deal with altimetry. The circular line is measured, as will soon be told in the mensuration of circular areas. The other curved lines can be reduced as much as is possible to the straight line or the circular line.

The tone is set from the very beginning: Chuquet writes not in Latin but in French, not for the clerks of the university but for the merchants whose mathematical training was based on calculation and the use of the abacus. Hence the title which precedes the first part of the work: *How the Science of Numbers can be Applied to Geometrical Measurement*; hence its style and its savoury combination of theoretical considerations and technical description.

In the section on the *Measurement of Areas*, the circle is presented before the rectilineal figures, an order uncommon in medieval practical geometries, which often retain an order according to the increasing number of sides, with the circle at the end. In Euclid the circle's place betrays its double function as linear tool and as a plane figure for study. Here the distinction is no longer valid, and Chuquet freely uses the straight line and even the compass; the order adopted seems to express respect for and knowledge of a tradition rather than an internal need of the text. Does the bachelor student peep out from behind the master of calculation? Just a tiny bit, perhaps.

> To measure (circular areas) one ought to know that in a circle there are three terms, that is to say, centre, circumference and diameter; the centre is the point equally distant from all parts of the circumference; the circumference is the circular line which encloses within itself the extremities of the circular figure; the diameter is a straight line passing through the centre, dividing the circle into two equal parts.

The whole is accompanied by an explanatory drawing; it is less a question of defining a circle than of recalling useful technical terms.

To measure and to know how to reduce all circular figures to a true square in accordance with what we have received from the ancients, there are many ways of which one is as follows: multiply the circumference by itself, and of the multiplication take $\frac{7}{88}$, and it will be done; example: let us say that the circumference is 22 feet; so multiply 22 by 22 giving the multiplication 484, which must again be multiplied by 7 and then divided by 88, and one will find 38 square feet and $\frac{1}{2}$, and this figure contains that much.

What is being put forward here, then, is the calculation of the area of a circle from its circumference; the 'other ways' mentioned are calculation from the diameter, from the circumference and the diameter, from half the circumference and half the diameter. Each time the presentation is identical, a general rule being accompanied by an example, traditionally 7 for the diameter and 22 for the circumference, which simplifies the numerical calculation. I have already traced some of the roads which brought into Europe, while sometimes erasing its theoretical subtleties, this body of knowledge 'received from the ancients' which features here only under its metrological aspect. And Chuquet adds: 'One should in any case understand that all these rules of the circular figure set down earlier are conjectural and very close to the truth: the ancients used them and the moderns use them still for want of better rules, though the squaring of the circle is an undiscovered art.'

The 'want of better rules'? Since the adoption of decimal numeration, Islamic science, and those who were aware of it like Fibonacci, had improved Archimedes' calculations by using his approach with polygons of an increasing number of sides; but communication was still uncertain and the metrological tradition had not, perhaps, assimilated these complicated calculations. Above all, it must be kept in mind that the intended public had a better knowledge of arithmetic than of demonstrations *more geometrico*, which would hardly have impressed or convinced the merchants of Lyons. On the other hand, there was immense progress in calculation, square and cube roots being manipulated without particular explanation and fractional calculations proliferating, as if to prove the mental agility of those who performed them. The calculation of a useful result for a mason or paver is also a playful exercise demanding speed and skill. The science of numbers does not appear in these disguises only, but helps too in the rereading of problems as well established as the inscription of rectilineal figures in a circle: construction by means of 'ruler and compass', the mark of Greek geometry, has disappeared. Chuquet provides himself *a priori* with a figure whose sides, for example, are of known length, and he tries to find numerically the diameter of the circle in which it may be inscribed. To do this he chooses a particular line as an unknown line to be determined, and transcribes all the geometrical relationships of the figure as a function of this line in algebraic form, more particularly in the algebraic form he developed himself in his *Triparty en la science des nombres* (1484).

Inscription of a regular pentagon in a circle by Nicolas Chuquet (1484)

We wish to inscribe in a circle a regular pentagon of side $ae = 4$: what should be the diameter of the circle?

Chuquet first of all calculates de from $de \times bc + db \times ce = dc \times be = de^2$;

If de is designated x (1^1 in Chuquet's notation), we have $4x + 16 = x^2$ ($4^1 \ \bar{p} \ 16 = 1^2$), and so $x = 2 + \sqrt{20}$.

We then calculate ag, for $ag^2 = ac^2 - gc^2$, whence $ag = \sqrt{20 + \sqrt{320}}$;

then gh from $gh \times ag = bg \times gc$, and hc from $hc^2 = gh^2 + gc^2$.

Finally, the diameter ah is given by $ah^2 = ac^2 + ch^2$ and thus

$$ah = \sqrt{32 + \sqrt{204 + \tfrac{4}{5}}} \ .$$

One should note the mixture of algebraic calculation and geometrical results, Pythagoras' theorem and Ptolemy's ratio between the sides and diameters of a quadrilateral inscribed in a circle.

At the end of the fifteenth century then, the circle seems to have stabilized: metrological, architectural and astronomical practice had assimilated the Euclidean demonstrations as operational properties. But another algebraic approach, born in the Arab world and passed on to the West, where the ground was far more favourable to it than to the Euclidean tradition, began a renewal of old themes. It was still only a matter of looking at the same objects with a new eye. Chuquet's circles, however, whether they schematize masonry work or serve as a pretext for a game of calculation, would no longer order space . . .

From Trivial Pursuit to Useful Art: Sixteenth–Seventeenth Centuries

Translations of Greek and Arab works meanwhile continued to pour into the West. Accompanied by commentaries, completed, even hypothetically reconstructed from scraps of information left by the editors of ancient compilations, they put the heritage of Antiquity at the disposal of a more and more educated, better and better organized public, with its network of correspondence, its exchanges of books and its discussion meetings (see p. 347). The trail of the circle now has to be followed through this new environment. To establish the relationship between the area of a circle and the square of its diameter, or to calculate it, the Alexandrians approximated the circle by means of polygons of a sufficient number of sides for the area remaining to be smaller than any area given in advance: they then deduced the properties of the circle from the properties of rectilinear areas enclosed by polygons. I have already mentioned that, thanks to the progress of decimal numeration, it had been possible to improve the calculations during the medieval period. In the century that followed, the search for metrological precision seems to have given way, little by little, to sheer prowess, in what some have called 'the race for decimal places': 15 places for Adrien Romain (1561–1615), 35 for Ludolph van Ceulen in 1609. Does this have anything to do with 'the needs of life'?

This enthusiasm for calculation accompanies and illuminates the progressive abandonment of Euclidian precautions: the demonstration of proposition XII, 2 in the *Elements* uses a clumsy argument to reduce the study of the circle to that of the polygonal figures. In 1615 Johannes Kepler simply identified the circle as 'a polygonal figure of an infinite number of sides' and its surface as an infinity of triangles. In his *Dialogue on Two New Sciences* Galileo re-examined the classical questions on the simultaneous revolution of two concentric circles, studying first of all the case of concentric hexagons and then the polygons, because, he says, 'circles (are) polygons with an infinity of sides.' Of course, Euclid, who dealt with the figure of the circle after all the others, testifies to a similar view: but his explicit procedure on the other hand was entirely different and attempted to avoid any hasty extrapolation. Here at the end of the Renaissance, supported by the ease of algebraic calculation, the passage from the finite to the infinite was hardly worrying. A long list of works expresses the area of the circle by means of formulae with an infinite number of terms.

Many similar developments were obtained during the following decades by the use of similar methods, whose origin was less and less geometrical. They did not all supply, far from it, more decimal places more rapidly than the Archimedean approach. Another desire than that for sheer numerical adventure was now at work: it may perhaps have relied on the same sources, worked on the same problems, but it did so for another purpose. François Viète, for example, claimed to have rediscovered the true method of discovery, the analysis of problems which had been obscured by the synthetic Euclidian style of exposition. Confronting a geometry still surrounded by its classical prestige, a new approach to mathematics, with its own favoured

The expression for the area of a circle of unit radius by François Viète (1593)

The area of the triangle OAB is the product of the lengths OH × HA, therefore the area of a polygon of n sides is $n \sin \alpha \cos \alpha$, and in the same way the area of the polygon of $2n$ sides is $2n \sin \alpha/2 \cos \alpha/2$. Using the formulae for the duplication of circular lines

$$\sin \alpha = 2 \sin \alpha/2 \cos \alpha/2 \text{ and } \cos \alpha = 2 \cos^2 \alpha/2 - 1,$$

the ratio of the areas of polygons of n and $2n$ sides is therefore $\cos \alpha$. Repeating the procedure, one finds that the ratio of the area of a polygon of n sides to the area of the circle (a polygon of infinite number of sides) is $\cos \alpha \cos \alpha/2 \cos \alpha/4 \cos \alpha/8 \ldots$ Viète chooses a square (so $n = 4$ and $\alpha = \pi/8$) of area 2 to begin his calculation and therefore states that the ratio of 2 to the area of a circle of radius 1 is:

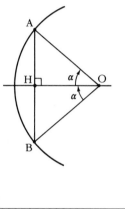

$$\sqrt{\frac{1}{2} + \frac{1}{2} \sqrt{\frac{1}{2} + \frac{1}{2} \sqrt{\frac{1}{2} + \frac{1}{2} \sqrt{\ldots\ldots}}}} \text{ , since } \frac{\pi}{4} = \sqrt{\frac{1}{2}}.$$

tool, algebra, developed in the wake of this work. The Greek inheritance adapted to the conditions of the period was revised and subordinated to arithmetic.

As Descartes puts it:

All the problems of geometry may easily be reduced to such terms, so that afterwards, it is only necessary to know the length of certain straight lines in order to construct them.

And as all of Arithmetic is composed of only four or five operations, which are Addition, Subtraction, Multiplication, Division and the Extraction of roots, which may be taken as a kind of division; thus one has nothing more to do in Geometry, with regard to the lines one is looking for, to prepare them to be known, than to add to them others, or take them away; or indeed, having one I shall call unity to relate it all the better to the numbers, and which can ordinarily be taken as one wills, and then in having two others again, to find a fourth, which is to one of these two as the other is to unity, which is the same as Multiplication. And I will not fear to introduce these terms of Arithmetic into Geometry, so as to make myself more intelligible . . .

But often one has no need to thus trace the lines on paper, and it is enough to designate them by letters, each by a single one. So that to add the line BD to GH, I call one a and the other b and write $a + b$ . . .

The geometry which accompanies Descartes' *Discourse on Method*, then, is not so innovative in its methods as in its systematization . . .

> Such are the plane problems.
> And if (a question) can be solved by ordinary Geometry, that is, by the use of straight lines and circles traced on a plane surface, when the last equation shall have been entirely solved there will remain at most only the square of an unknown quantity, increased or diminished by some other quantity also known . . . For example, *zz = az + bb.*

A first step: every construction with ruler and compass (but Descartes, who has read his classics, says 'with straight and circular lines') provides at the most an equation of the second degree; a point can be constructed through ordinary geometry by a series of equations of this type, and it is now a question of studying the equations themselves.

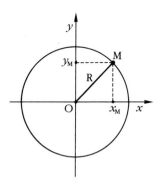

> The manner of distinguishing curved lines of certain genera . . .
> To understand together all (the curved lines) which are in nature, and distinguish them by order into certain genera, I do not know anything better than to say that all the points of those one can call Geometric, that is to say which fall under some precise and exact measure, necessarily have

How to find the equation of a circle. Because of the connection with Pythagoras, $x^2_m + y^2_m = R^2$ where M crosses the circle with centre O and radius R.

some relationship to all the points of a straight line, which is perhaps expressed by some equation, in all by the same. And that, when this quantity amounts to no more than the rectangle of two indeterminate quantities, or to the square of the same, the curved line is of the first and simplest genus, in which only the circle, the parabola, the hyperbola and the ellipse are included.

A second step: the curve itself may be entirely described by a second-degree equation for the circle. With the disappearance of the geometrical support, it was on the basis of this equation that the properties of the curve would be definitively described and calculated. Here the circle is no more than the circular line. I have stressed the effort made in the *Elements* to mention the circle only as a plane figure; the tools of algebra and those who used them gave priority to what they could deal with and thus to the line which could be defined by an equation. In addition, this form itself intersects with other categories than does the geometric approach. In his classification Proclus distinguished between plane lines (straight lines and the circumference of circles) and solid lines (the conics, obtained by cutting a cone with a plane). The algebraic classification, for its part, identifies the circle with the conics, whose equations are of the same genus.

The circle as plane figure is a polygon with an infinite number of sides, the circle as line is an equation, and the tension which unified the Euclidian definition no longer exists, even if habit retained the use of the same word. Other loosenings correspond to the unravelling of these fibres and to their being retwisted elsewhere: metrology and calculation had already been split apart. After very many efforts to make theoretical models coincide with Tycho Brahè's very precise astronomical observations, Kepler stated that the trajectory of a planet followed an ellipse. The circle no longer ruled the sky. Galileo, studying falling bodies, described them with the help of parabolas: the circular line was no longer the only alternative to the straight line, necessary in the final analysis for an account of non-rectilinear movement. But above all the circle, a geometrical drawing, the traces of which may still be deciphered everywhere by the educated eye, was, in a parallel process, expelled from view.

In which our Sight deceives us: Sixteenth–Seventeenth Century

During the Renaissance many works were devoted to perspective. It was seen at first as the exchange of one form into another whose rules had to be established, essential as they were for both painting and cartography, whose mastery was made even more necessary by voyages and expeditions. The treatises on commercial arithmetic have their counterparts in the works of Filippo Brunelleschi, Piero della Francesca, Leon Battista Alberti, Albrecht Dürer and Leonardo da Vinci. In Euclid's *Treatise on Optics*, a circle seen from the side 'has unequal diameters', and there was no question of identifying it with an ellipse. As painters have taught us, the knowledge which we have very often reconstructs what we see, and our intuitions the connections we make. The necessity of representing three dimensions in two was frequently mentioned at the beginning of the treatises, and little by little in the seventeenth century, the eye, the point of projection, was integrated into the plane of the drawing.

In 1639 Girard Desargues, an architect familiar with the Greek treatises (in particular, Apollonius' *Conic Sections*) and with scientific groups, had fifty copies printed of a book which was part of his plan to provide a universal method to unify graphic techniques: this was the *Brouillon Project* ('Rough Draft on Conics'). The necessary technical terms are expressed by floral metaphors, perhaps in a deliberate attempt to purify mathematical language

of its imprecisions, but this disconcerting style and
the book's restricted distribution did not help its
reception among mathematical practitioners.
Desargues's work interested only mathematicians
like Pascal and Philippe de La Hire and the en-
graver Abraham Bosse, who took on the defence
of the master's ideas. The most fundamental is
that the conic sections (called 'sections of a roll'),
'formed from the ways in which one cuts a circular cone, must partake in the properties of the
circle'. It is no longer a matter of studying the way
in which things are changed by projection, but,
on the contrary, of discovering the properties
which remain invariant. As all the conic sections
can be reduced to the circle under a suitable projection, the separate examination of each type
may be replaced by a general theory which will
deduce from those of the circle alone, often
more easily established, the analogous properties
for each conic section. Did this restore to the cir-

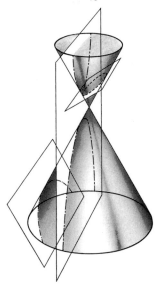

———— circle
-------- ellipse
·—·—· parabola
—·—··· section of a hyperbola

cle a power of unification it had lost? Not really. Now the accent has been
shifted away from characterizing a figure by its particularities, towards distinguishing among its features those which are shared with a greater class of
objects, which it will thus be possible to grasp and to comprehend. The
main thread is no longer the circle, but what it is that makes the circle a
generator of all the conic sections.

It is in its transformations, its shadows, that the circle drawn here is particularly interesting. What is most significant is the change that has been
produced even in the mental representations that organize the way we see
things. A circle viewed from the side is now an ellipse and not a deformed
circle. Was it the geographer, the painter, the astronomer or the mathematician who prepared the way? It isn't easy to say, for the movement gained
in strength from every success it encountered: the circle is a conic section
and it is very hard not to believe that it had not always been so. Looking
at the lost and rediscovered threads we are investigating may help to convince us that there is nothing inevitable or natural in these identifications;
but it is surely even more striking to recall the resistance of the very people
whose contributions to this change in point of view have already been mentioned. Kepler wandered a long time in a search for ovals, and not ellipses,
to replace circles in accounting for the movement of heavenly bodies. Galileo, for his part, continued to use circular trajectories in his cosmological
works.

In which we lose a few Threads: Eighteenth Century

In mathematics, under the leadership of the Cartesians, the success of the analytico-algebraic point of view led to the progressive elimination of drawings, since now it was a matter of basing effectiveness on the accuracy and mechanical certainty of a calculus. Once again, the questions relating to the squaring of the circle changed in appearance: as soon as curves were classed according to the type of their equation, the restrictive use of the ruler and compass for converting the circle to a square hardly made sense any longer. The new techniques of differential and integral calculus allowed quadrature,[3] which now meant to calculate the areas enclosed by different curves on the basis of their equations. In his *Histoire des recherches sur la quadrature du cercle* of 1754, the historian Jean Étienne Montucla summed up contemporary opinion: 'on what grounds, indeed, do we look on the circle as more simple than the other figures? (It is) only a particular kind of ellipse ... The equality of the diameters has no influence at all on the relationship of its ordinates to the abscissae, nor on that of the inscribed or circumscribed polygons which limit it. The curves in which these relationships are simpler, like the parabola, are squarable, though less regular to our own eyes.' What was considered interesting, then, was the determination of the relationship between the area of a circle and the square of its radius, which is also that of the circumference to the diameter. It was in this period, finally, that it received a name, $\pi = 3.14159$. Compared to the millennial adventures of the circle, the most famous number in the history of mathematics is then a mere child. The manner of calculating it will leave this account to follow other independent paths, its determination being more related to the infinite series of the sine, cosine, and tangent functions and, more recently, to the development of computers. Metrology will retain the formula $A = \pi r^2$ and more decimals for π than it would ever be able to use. But its properties keep it for one moment on our own road: in 1766, the mathematician Jean Henri Lambert showed that π was not a rational number. Numerous false solutions were still being proposed for the squaring of the circle, but by 1775 the French Académie des Sciences decided not to inspect them any longer. A profession had been created which standardized scientific activities, identifying the problems recognized as its own and the valid methods by which they ought to be be approached. What mathematicians now attempted to prove was the impossibility of squaring the circle 'with ruler and compass', or rather the impossibility of obtaining π as the solution to a series of at most second-degree equations. The specialists interested themselves particularly in the solution of these equations and in the properties of the solutions. It was only in 1882 that Ferdinand von Lindemann, carrying on the work of Charles Hermite, proved that π was not the solution of any algebraic equation: the quest for the impossible square came to an end almost in the absence of the figure of the circle.

Almost, because the techniques which were being developed, differential or infinitesimal calculus, had difficulty in dealing with the problems abundantly generated by their initial triumphs and almost anarchic proliferation. Were the solutions to the equations really numbers? Did these infinite series have a meaning? Throughout the eighteenth century, traditional geometry offered a disturbing model of certainty. In 1798 a book was published by Lorenzo Mascheroni, *The Geometry of the Compass*, 'which determines the points by means of the compass alone, and without the help of the rule'. The author mentions that he wants to return to the sources of geometry to see if it is not possible to simplify them further, and that he has been encouraged by contemporary work in astronomy, for which the more precise compass was an important aid. Until the end of the nineteenth century and beyond, new results on the circle continued to accumulate in hundreds of articles. New branches of mathematics and fields of study in formation, different schools, points of view and distinct objectives appropriated extracts from them according to their needs or remade the circle in their own image. An ordinary mathematical topic, taught in the schools, it remained a useful example since it furnished easily available models or counter-examples. It remained present in mathematics through the survival of problems connected with it which would receive a solution only later when they were submerged in other currents, combined with other threads, as had been the case with the squaring of the circle. Two examples among many will explain what I mean.

In which we scatter a few Circles: Nineteenth Century

The first example is taken from the *Disquisitiones Arithmeticae* published by Carl Friedrich Gauss in 1801, dealing with integers and equations with integer coefficients; included in it was a chapter on 'the equations which determine the divisions of the circle'. Gauss begins by underlining the importance of the circular functions in all the branches of mathematics and then he looks at the old problem of the inscription of regular polygons in the circle 'with ruler and compass'. Euclid had already dealt with the case or the pentadecagon, but the inscription of a polygon of 17 sides resisted the attention of researchers until the nineteenth century. Gauss sets up the problem in these terms: 'we designate by P the circumference of the circle or four right angles . . . we will limit our investigations to the case in which one has to divide the circle into an odd prime number of sides.' To inscribe a regular polygon of p sides is to distribute p points regularly along the circumference, and therefore to divide P into p parts; it is also to divide the angle at the centre (of 360 degrees or *4 right angles*) into p parts: the vertices of the polygon correspond to the angles P/p, $2P/p$, ... $(p-1)$ P/p whose circular functions Gauss then states. He shows that they satisfied equations of degree $p-1$ which he studies in detail. If $p-1$ is divisible only by 2 and its powers, the solutions can be obtained from a sequence of equations of second degree and can therefore be constructed geometrically 'with ruler and compass': this is the case with the polygon of 17 sides, as $17-1=16=2\times2\times2\times2$, but not that of the 19-sided polygon since $19-1=18=2\times3\times3$.

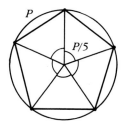

The procedure is exemplary: a classical problem is expressed and solved in algebraic terms, and the definition of the circle (*4 right angles*) retains only that part which is relevant to the calculation. The circle only appears effectively in the form of circular functions. What is more, in the very investigation of the equations, the accent is put upon the relationships between the different solutions, relationships which are geometrically self-evident (the vertices of a regular polygon being interchangeable by rotation), but here transcribed in algebraic form. The issue at stake is also exemplary: it was by solving this problem that Gauss became famous. It is connected to the circle only by a thin thread; the power of the method derives from a general theory born elsewhere.

The second example I have chosen is that of Gaspard Monge. Made responsible in the Year III of the Revolution for the introduction of descriptive geometry into the programme of the École Normale, he had to establish the theoretical basis for graphical procedures useful in stone-cutting, in perspective and the surveying of fortifications: in brief, to realize Desargues's old dream. But Monge's public was not made up of reluctant practitioners or supportive friends: there were now students to be trained and thus the possibility of impressing upon them as natural, in its turn, another way of approaching the subject. In 1822 the geometer Jean-Victor Poncelet published a *Treatise on the Projective Properties of Figures* which generalized the work of Desargues and drew on the lectures given by Monge and Lazare Carnot. He highlighted the properties which 'remain constant under the effect of projection'. Metric properties connected to distance and to angles that are not conserved in projection were carefully distinguished from those which were. In the transformation under projection of a circle into a conic, the equality of the radii, a metric property, is not conserved. But Jean Poncelet also wanted to free pure geometry from the restrictions imposed on it by the particularities of a given figure. The issue is the possibility of giving geometry the unifying power possessed by analysis. From the point of view of their equations, circle and ellipse may be assimilated; but if in general ellipses could intersect at four points, two circles could have only two points of intersection at most. Desargues had already ran into analogous problems: to enable a uniform treatment of the different types of projection (central and parallel) he had introduced the 'point at infinity' where parallel lines met – a projection parallel to a given direction could thus be considered as a central projection whose vertex was at infinity. In the same way, Poncelet introduced the 'cyclic points', fictive points where all circles of the plane were supposed to meet; two circles intersected at these two points as well as at their ordinary points of intersection. Poncelet's circles, subject to other constraints, included other points than the Euclidian circle did.

To complete the work, it now had to be stripped of all reference to metric notions: what should be done with the circle? The fact of the question being asked shows that it was not yet conceivable to do geometry without its familiar objects, though they might be deformed or modified by comparison with the Euclidian corpus: but it is the field which gives the object its acceptable form. It was necessary to show the efficacy of an approach and to radicalize it, and it was therefore necessary to give the circle a definition without distance or measure of length. In this way the relationship which connects it projectively with the conic sections could easily be expressed.

Tossed about in the mathematical currents of the nineteenth century, was the circle still everywhere? It was rather nowhere: it served those who wanted it, as they wanted it: the threads which formed it loosened, twisted together with other threads and no longer fitted together so closely. We have measured the fragility of its position in the last place in which it remained as a crucial if not central object, in pure geometry. In 1872, when he joined the University of Erlangen, the mathematician Felix Klein laid down a new programme for geometry. The examples mentioned above underline the growing role of transformations in mathematics, from those which exchanged the roots of an equation to those which retained metric properties or did not. For Klein it was these transformations which defined geometry: 'Let us abstract from the material figure which, from the mathematical point of view, is not essential,' he said, 'geometrical properties are characterized by their invariance relative to transformations.'[4] If one allows more transformations, only a part of the properties will be conserved: if one allows projections, metric properties will have to be abandoned. And conversely, if one imposes on transformations the fixing of the cyclic points described above, they will automatically conserve the metric properties and will distinguish circles from other conics. Figures will now be classified according to the transformations which change or conserve them. One of the most spectacular instances of this point of view is the interchange of straight lines and circles by inversion: better yet, in the nineteenth century, articulated systems were constructed which converted the trace of a circle into the trace of a straight line: with the impossibility of squaring the circle with ruler and compass and the transformation of circular movements into rectilinear, the nineteenth century extracted from their circles very different responses from those of the Alexandrians.

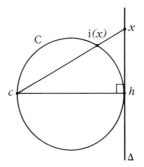

The transformation x → i(x) converts the straight line Δ into the circle C. The point h is its own transform.

Magnifying glass or astrolabe, ellipse or straight line, the circle does not have the inert self-evidence of a natural object. It was from priests, mathematicians and painters that the regard and the word which created it had to be borrowed. It might seem convenient to combine together in a univocal entity the partly overlapping heterogenous threads, some of whose unravellings and some of whose intertwinings we have just examined, but there is sometimes nothing in common between the realities uncovered by these multiple and even contradictory definitions, developed as they are from distinct fibres spun by architects, poets or geometers. The circle now shelters much more than the precise trace of an archaic mirage.

Some modern images of the circle

The circle as equation is the ensemble of solutions to $x^2 + y^2 = r^2$ with an appropriate choice of coordinates. If $r = 0$ one obtains no more than a point; if one allows only integer values of x and y, one obtains several different cases depending on the values of r. And if one extends the possible numbers to the complex numbers $a + b\sqrt{-1}$ (see p. 362) there is no difference between $x^2 + y^2 = r^2$ and $x^2 - y^2 = x^2 + (\sqrt{-1}y)^2 = r^2$: among the circles now also appear the hyperbolas.

In topology one is interested only in the properties of objects under continuous deformation: a circle as surface, a disc, is identical from this point of view with any flat surface *without holes*. The circle as line is identical with a closed line of any form.

One may note that the circle has a constant radius of curvature: it shares this property with the straight line and the helix.

One of the models of non-Euclidian geometry is the sphere: in this model, straight lines, in the sense of geodesics (the shortest path from one point to another), are circles passing through the two poles.

When David Hilbert published his *Foundations of Geometry* around 1900, he declared that he would not define a straight line or point; these could be what one liked on condition that they satisfied rules governing their mutual relationships, which he stated carefully, as 'every *straight line* contains at least two *points.*' Here a circle is still the set of points such that the segments drawn from them to a fixed extremity are equal (Hilbert says 'congruent'): but 2000 years apart, it is in the gaps and absences of the Euclidian definition that Hilbert's must be deciphered: *equal* and *intersect,* which Euclid did not define, now receive their definitions. To an insistence on the description of things, point, line and area, there corresponds the insistence, the heir of the 'Erlangen Programme', on the relations between them, *to be situated, to be equal, to enclose.* The notion of distance in particular must be specified: the circle, once the *ensemble of points at an equal distance from a centre,* will now only be what the metric permits it to be, quite round or square if it likes, or a tree or worse.[5]

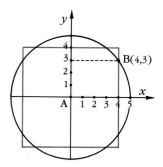

How a circle can be a square: the distance from A to B can be expressed as $\sqrt{(x^2 + y^2)}$, the circle of centre A and radius AB will then be a proper 'round' of radius 5. But it can also be expressed as the greater of the pair ($\sqrt{x^2}$, $\sqrt{y^2}$). For this other distance, just as lawful mathematically, the 'circle' of centre A and radius AB will then be a square.

6

The Arab Intermediary

Paul Benoît and Françoise Micheau

In which it will be seen that the ten centuries of the Arab and Latin Middle Ages play an essential role in the emergence of European science, although this has been denied for a long time. There follows a description of Greek science's journey to the West, with Arab science acting as an intermediary, together with a discussion of the specific contribution of Arab science.

In a lecture on 'Islam and Science', given at the Sorbonne on 29 March 1883, Ernest Renan said: 'Such is the great body of philosophical thought which is customarily called Arab, because it is written in Arabic, but which is in reality Graeco-Sassanid. It would be more precise to say Greek; for the fruitful element in all this came from Greece ... Greece was the sole source of science and rigorous thought.'

A century ago Renan was at the height of his fame and authority, and this was how he resolved the problem of origins, or more precisely, the problem of the influences and filiations, which allowed the emergence of the European science from which there developed the world science of the present day. Everything came from Greece; the Arabs had no more than an intermediary role. We might think that the brutality of Renan's statements might indicate some anti-Semitism, if it were not the case that the same themes appear, though less forcefully, in the discourse of more recent historians of science, some of whom are loud in proclaiming their anti-racism. Such a position is related not only to the cultural Eurocentrism of Western historians, but also to the belief, very strongly held by most of them, that there exists only one science, classical science, true and positive. Although it has now been challenged, particularly by non-European historians of science, many people still share this position. Can it survive a scrutiny of the historical facts?

A refusal to follow Renan uncritically leads to a whole series of questions about the origins of European science. In its development, its filiations and ruptures, we have a relatively coherent view of the history of science from Galileo or Copernicus to our own day. We are now able to trace the links between classical science and the science of the late Middle Ages, even if in this area a great deal of work remains to be done. What still needs to be determined are the factors which allowed Christian Europe to develop its own science between the thirteenth and the fifteenth centuries. Among these factors, there is the assimilation of the heritage of Antiquity. But by what means did the Greek inheritance reach the West? At what period? In what form? Was it not modified as it was transferred? At its birth, was the Greek legacy all that European science had?

To try and answer these questions is on one hand to provide the data to deal with a historical topic located in time and space – the emergence of European science; but it is also to take on the more general problem of the transfer of scientific knowledge.

All these questions will receive here only a limited and inadequate response, which is bound to leave the reader unsatisfied. Uncertainties will be more numerous here than elsewhere in the book, for the history of science is first and foremost history and it is founded on scholarship, on the establishment of texts and facts, and on a chronology. In our own field these foundations are lacking. This may perhaps be due to a lack of documentation, but there is also the lack of researchers, historians and linguists. The documents lie in libraries and archives in the Middle East and in India, in Western Europe and the United States, but the means of access are lacking. Language, training, political constraints and the absence or inadequacy of funding are inhibiting factors which are added to the lack of interest in the history of science displayed by too many specialists in the ancient East. Take Syriac, for example: the theological and hagiographical literature and the chronicles have been translated and studied, while scientific texts are ignored. All these factors come together to prevent the proper development of a history of scientific cultural transfer in the Middle Ages. The history we present is only provisional.

What was the State of Fifth-century Greek Science?

What was the state of Greek science at the collapse of the Western Roman Empire in the fifth century AD? To speak of a unitary Greek science in this period is risky. Between Thales and the Milesians of the seventh and sixth centuries BC to Diophantus of Alexandria, who probably lived in the fourth century AD, there is not much less than a thousand years. Greek science, many-sided, developed in very differing contexts. Born in cities jealous of their independence, it developed in empires, a great part of whose population wasn't Greek at all. Having emerged under an undemanding polytheism, in its latter centuries it experienced the sway of monotheistic religions of totalitarian inclination, and in 529 the Emperor Justinian closed the schools of Athens on account of their paganism. Greek science none the less presents a certain unity, a unity of language above all, for Greek had spread beyond Hellas and Asia Minor, and was the bearer of a tradition whose works were highly respected: Aristotle and Euclid, Ptolemy and Hippocrates.

In the West, just before the fall of the Empire, scientific knowledge was in retreat. With the weakening of public authority, the decline of the town and the return to the countryside, the schools closed, classical culture diminished and the use of Greek was lost. Though Christianity did not systematically reject Greek science, it limited its application. St Augustine, whose thought influenced the Middle Ages before the thirteenth century more than anyone else's, insisted that to accede to Christian science, that is to say, to the understanding of the divine word, it was necessary to have a basic education, which meant the kind received by the sons of the aristocracy of the Late Roman Empire, at the hands of the teacher of rhetoric. This is described in the treatise by Martianus Capella entitled *The Wedding of Mercury and the Sun*, which was current throughout the Middle Ages. It is concerned with the liberal arts, which are divided into the *trivium* and the *quadrivium*. The *trivium*, an introduction to the art of reading and of correctly interpreting texts, is made up of grammar, rhetoric and logic, while the *quadrivium*, which has a more scientific air, is made up of arithmetic, geometry, music and astronomy. One shouldn't be deceived by the terminology. Arithmetic wasn't practical calculation, as understood in the schools of our own day, but a speculative arithmetic, dealing with numbers and their nature, starting off with odd and even. There was nothing instrumental either in music, the science of harmonies. Boethius gives us a glimpse of what science remained in Rome at the beginning of the sixth century. Born into the Roman upper aristocracy, he occupied posts of the first importance at the court of Theodoric I, the Ostrogoth King of Italy. Boethius still knew Greek and translated Aristotle's *Categories* and *On Interpretation*. He also left an arithmetic and a geometry, which show that Euclid and Ptolemy were not completely forgotten, but taken as a whole they remain at a very elementary level. Limited as it is, Boethius' work was essential, for until the twelfth century it constituted the principal means of access to the science of the Greeks. The vestiges preserved are miserable, compared to the scale of the edifice, but for the men of the Middle Ages they were evidence for the existence of a true science the Ancients had once possessed. Cassiodorus, a disciple of Boethius, took up the work of his predecessor and sanctified the division of the liberal arts in his *Institutiones*, written for the monks of the monastery he founded. Under cover of philology, the *Etymologies* of Isidore, Bishop of Seville (560–636), bring together knowledge of every kind, scientific in particular. To our eyes a mediocre work, this compilation met with great success throughout the first centuries of the Middle Ages.

The scientific life of the West at the time of the barbarian invasions had therefore been reduced to very little. The essence of the antique inheritance was sunk in forgetfulness, fallen from the memory of populations incapable of comprehending it. Boethius' translations had no successors, but they were copied and recopied throughout the Middle Ages. The decline of science accompanied the decline of the city. The West was going through a period of profound change and there emerged a rural world, where culture was based in the monasteries. This was a religious and literary culture, not at all scientific. The Early Middle Ages had no need of scientists.

In the East there survived a Roman Empire which had died in the West. It was a world still urban, with a continuing State, a world where Greek, now the official tongue, was spoken by the urban elites, a world where there were still schools: different in every way from the West. But as well as Greek, other languages, and Syriac in particular, a Semitic language, were becoming languages of culture. Greek was, however, still known by all the scientists. Despite the lacunae in the sources and the inadequacies of research, one can still distinguish a certain scientific activity. In the fourth century, the most probable date, Pappus of Alexandria wrote a *Mathematical Collection*, a vast compilation on which the Middle Ages could draw for one part of the Greek inheritance. In the following century Alexandria saw the arrival of students from all over the Near East, who came for teaching in law, medicine, mathematics and philosophy. A few particular scientists stand out: the female mathematician Hypatia, killed in a riot in 415, John Philoponus, a convert to Christianity, the author of a treatise against the pagans, of commentaries on Aristotle and works on optics and mathematics, or again, Paul of Aegina, who edited a collection of Galen's writing under the title *The Seven Books of Medicine*. Still in the fifth century, Proclus made a commentary on Euclid at Athens. Mediocre no doubt, and certainly badly known, a scientific life based on commentary and compilation survived in the great metropolitan cities of the Eastern Empire: Athens, the capital Constantinople, Alexandria in Africa, Antioch in Asia.

At the eastern margins of the empire, at Edessa, at Harran, at Ras el-Ain, and on the routes of communication between Asia Minor and the Persian Gulf, an intellectual life was developing in these cities of Semitic culture. Translators were active; Sergius, a physician and a Syrian Monophysite priest, who was educated in Alexandria, worked at Ras el-Ain, where he translated a great many works from Greek into Syriac, before dying in Constantinople in 536. At Edessa in the fifth century there developed a school, called 'the Persian School'; there Aristotle and his commentators were translated into Syriac by Proba, Cumas and the bishop Ibas. But the Emperor of Constantinople, angered by the school's Nestorian orientation, ordered its destruction. Activity at Harran, badly known and difficult to establish, seems essentially to have been turned towards Platonist philosophy. But this centre of Hellenism must have maintained its vitality; Thabit ibn Qurra, the great Arab mathematician and astronomer of the ninth century, was born in Harran.

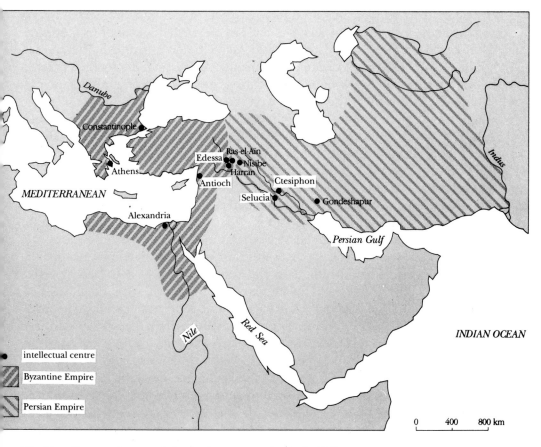

The Near East on the eve of the Arab conquest (towards 630).

Confronting the Graeco-Roman world, the Sassanid Empire included lands of ancient culture, the plateaux of Iran and Mesopotamia in particular. For a time the conquests of Alexander had made them Greek, or Hellenistic, as the historians say. Here there are enormous gaps in our knowledge; Persian science and its debts to the various traditions remain very badly understood.

195

But there did exist a scientific milieu. The Nestorians exiled from Edessa after the closure of its school in 489 and the philosophers expelled from Athens in 529 found refuge in the Sassanid Empire. During the reign of the emperor Khosroes Anushirwan (531–578), the capital Ctesiphon, on the banks of the Tigris, enjoyed a brilliant intellectual life: men of letters, scientists and astrologers thronged the court. Nisibe, in Upper Mesopotamia, benefited from the transfer of students and teachers from Edessa; at the beginning of the sixth century, on the eve of the Arab conquest, this centre had more than 300 students. Gondeshapur, a city in south-east Iran, founded by Shapur I, from whom it took its name, became a great scientific centre thanks to the Nestorian scientists. They brought with them translations in Syriac, among them those of Sergius of Ras el-Ain. They organized a school of medicine, modelled on those of Alexandria and Antioch, and a hospital. The city became a very important centre and remained active until the tenth century, a place where the Greek heritage, Indian scientific tradition and Semitic culture all met together. The Greeks were not the only visitors to Persia. At the time of Khosroes it was equally influenced by the civilization of India, though literary contacts are better known to us than the transfer of medical and astronomical traditions. However, Khosroes himself sent his physician Burzoe to India to copy manuscripts, and he ordered the revision of the great astronomical tables in accordance with data given in the Sanskrit treatises.

Alexander's conquests, which extended as far as the Indus, had developed the connections between India and the Graeco-Persian world. In the second century AD, on the border between India and Persia, there developed in Bactria a civilization still strongly affected by Greek influence. Under the Gupta Empire, from the third to the sixth centuries AD, India enjoyed a particularly brilliant period. It is impossible here to deal with Indian science, but we can indicate certain features. Greek influences appeared in medicine in the art of diagnosis, and in astronomy, where the Ptolemaic system (see p. 285) gradually replaced the traditional model of the Universe. On the other hand, the Indians excelled at observational astronomy. For their measurements they used the sine of the angle, and the earliest known table of sines is Indian, dating from the fifth century, while the Greeks employed the chord of the double angle (see p. 83). Such astronomy favoured the development of trigonometry. The Indians were calculators; under the term *ganita*, which means calculation, they included what we would call arithmetic and algebra. At the beginning of the sixth century Aryabhata used decimal places to represent numbers – his system later developed into the Arabic numerals we use today. He devoted a treatise to elementary operations: he started with addition, subtraction, multiplication and division, and continued with squaring and the extraction of the root, to finish with the cube and its own root. But he could also solve equations of the second degree, and he wrote on indeterminate equations. In the following century Brahmagupta used letters in equations to designate unknowns and calculated with negative numbers.

In the seventh century Greek scientific thought, widely diffused even if it no longer produced original work, had already been confronted for a long time with other traditions, other cultures and other sciences.

The Arab Conquest and the Foundations of Scientific Development

This was the world into which the Arabs erupted. The Arab conquest, a brutal event, is still surprising; in a few decades, a people until then unknown destroyed the great empires which dominated the Middle East and spread a new religion as far as India. An explanation of the spread of Islam belongs to another history, but it is necessary here to grasp its main stages and to see what influence it had on scientific life.

In 632, at the death of the prophet Muhammad, Islam had united the western part of the Arabian peninsula. The years that followed would see astonishing victories: in 635 Damascus fell into the hands of the Muslims, then Jerusalem in 637; they were then the masters of all of Syria and Palestine. At the same time they pushed towards the north-east, taking Ctesiphon, capital of the Persian Empire, in 637; all of Mesopotamia was conquered. The Arab expansion extended as far as Armenia and reached the plateaux of Iraq and Iran. On a third front, Egypt was subjugated between 639 and 646. In less than fifteen years the Sassanid Empire had disappeared, while the Byzantine Empire survived, but considerably diminished. From this time on, all the great intellectual centres of the Middle East, the bearers of a great scientific tradition, passed into the hands of a people ignorant of science.

After a period of stasis the expansion continued, towards the end of the seventh and the beginning of the eighth centuries, under the leadership of the Umayyid caliphs. A drive towards the west brought the Arabs into North Africa, where in 670 they founded Kairouan, which was to become one of the great intellectual centres of Islam. Once their domination of the Maghreb was secure, they crossed the Straits of Gibraltar in 711, conquered Spain, and made many excursions into Gaul. A far-reaching raid was halted by the Frankish cavalry at Poitiers in 732. There was a drive to the east as well: Bukhara was conquered in 709, Samarkand in 712. Muslim domination extended as far as Sogdiana and Transoxania, on the edges of central Asia. Islam reached India, but halted at its gates. Direct contact with Indian science then became possible.

As in many other fields, the conquerors respected the existing situation. The burning of the library of Alexandria by the Muslims is a myth that needs to be exposed as such. During its first period, the Arab conquest hardly changed the conditions of scientific life at all, but it gave the Arabs access to Greek texts, because the principal libraries of the Graeco-Roman world were now in their hands. The work of translation continued and expanded. At the end of the eighth century Theophilus of Edessa was still translating Aristotle's works from Greek into Syriac, like his contemporary Yahya ibn al-Batriq. At the beginning of the next century Job of Edessa, a physician attached to the court of the Caliph al-Ma'mun, owed his fame to his translations of Galen into Syriac and to his great encyclopaedia of the natural sciences. All these scientists were Christians. As the decades passed, and especially from the eighth century on, there were important changes. Arabic developed into a language of culture and administration. The reign of Abd el-Malik (685–705) was decisive; by his order Arabic replaced Greek, Pahlavi and Syriac in all official documents. Thus Arabic, later to be called Classical, spread; in all the countries dominated by Islam the scientists, whether Muslim, Jewish or Christian, Arab or Persian, all spoke, thought and wrote in Arabic. Two of the greatest names in the science of the Islamic countries, al-Biruni and Ibn Sina, the Avicenna of the West, were Persians, and their scientific work was in Arabic. A scientific and intellectual community was born, encouraged by a society feverish for knowledge and supported by wealthy rulers.

The succession of the Prophet proved complicated, conflicts broke out and the capital passed from Medina to Damascus. The coming to power of the Abbasids in 750, which corresponds to the desire to create a Muslim rather than an Arab empire, which would offer a place to believers of any national origin, found expression in the establishment of a new capital, Baghdad, by the Caliph al-Mansur in 762. He attracted scientists to this city, wishing to secure their services and to establish the prestige of his new regime. Affected by indigestion, he sent for Jurjis, a famous physician of Gondeshapur, and once cured, he tried to keep him at his court. And so men of science settled in Baghdad, bringing with them books, knowledge and experience. The great Abbasid caliphs who succeeded al-Mansur continued this policy; it was a descendant of Jurjis who founded in Baghdad at the beginning of the eighth century the first hospital worthy of the name, modelled on that of Gondeshapur.

Al-Ma'mun, who reigned from 813 to 833, is the model of these enlightened princes, friends of the arts and sciences. If he did not found it, he certainly gave a very strong impulse to the activity of the House of Wisdom. In this vast library, astronomers, mathematicians, thinkers, scholars and translators worked and met together. Their work was encouraged and financed by the caliph. The book-stock was enriched with works of Greek philosophy and science, obtained from the Byzantine Empire, but mainly collected in the libraries of the Near East; copies and translations put them at the disposal of a cultivated elite. Paper-making, a process of Chinese origin, was introduced to Baghdad at the end of the eighth century; paper was more solid than papyrus and less costly than parchment, and it allowed the growth of a real market for books. The Caliph al-Ma'mun was also a patron of the first great projects of astronomical observation; a team of scientists put together at his expense was commissioned to verify the data in Ptolemy's *Almagest*, which led to the establishment of new *Tables*. Exemplary, but not at all isolated, al-Ma'mun's policy was pursued by numerous caliphs, viziers, emirs and sultans.

From the eighth century the first signs of the fragmentation of the empire established by the Umayyids and the Abbasids began to make themselves felt. The regional chiefs, emirs who had become masters of political power, the caliphs of Cairo and of Cordoba, began to assert themselves against the Abbasids. Their capitals attempted to rival Baghdad, which once had been able to claim the monopoly of intellectual and cultural life. If you mark the location of scientists on the map, then you discover the great phases of Islamic history: at the high point of the Abbasid caliphate (eighth to tenth centuries), Baghdad was the great pole of attraction towards which most scientists made their way. Then other cities took over: in the west, Cairo and Cordoba, capitals of caliphs who were the rivals of the Abbasids; in the east there were the great cities of Persia where, in the tenth and eleventh centuries, Turkish and Persian principalities were founded. Conquered by the Seljuk Turks, who occupied Baghdad in 1058, Iraq became less and less important. In the thirteenth and fourteenth centuries, the great local dynasty of the Ayyubids, founded by Saladin, and that of the Mamelukes ensured the strength of Syria and of Egypt, faced first by the Crusades and then by the Mongols. All through the Middle Ages Arabia was absent from the historical scene, while Andalusia, a distant province, enjoyed an original and lively economic and cultural life, providing a bridgehead to the Christian world. Scientific activity continued throughout the period and continued long after the ascendancy of Baghdad. The break-up of empire and the growth of regional metropolises favoured the geographical diffusion of culture.

Astronomers, physicians, mathematicians and philosophers often lived at the court of a prince, dependent on his good will and generosity. Avicenna's whole career was played out among the Persian courts of the tenth and eleventh centuries: Bukhara, Isfahan, Samarkand, Hamadhan and Rayy. He was much appreciated for his advice in medicine and in politics. Now coveted and sought after, now pursued by his enemies, always entangled in the political intrigue of the day, Avicenna suffered tribulations of every kind, but also experienced long periods of security at the courts of his patron princes, at Rayy and at Hamadhan, where he died in 1037.

For intellectual life depended, directly or otherwise, on the prince's favour. Science never comes free. No library could be founded, no hospital built, no translation undertaken, no programme of observation completed, without funding from sovereign, vizier, influential courtier, emir or local notability. In one sense, the chief scientific institution of the Arab world was patronage. This attraction that science had for princes was not the result of a passing infatuation, its motivations more or less obscure. It was part of a concrete political plan: to increase their power and prestige by giving Islam the means to elevate itself to the level of the civilizations already conquered, and even to overtake them, allowing the development of a scientific culture which might reach the heights attained by Greek, Syriac and Indian science.

And so there developed a science, usually called Arab science. The expression can be considered inaccurate: a great many of its practitioners were not Arabs. But to speak of Islamic science is heavy with ambiguity, giving scientific activity a religious connotation it didn't have, for the scientists were Muslims, but also Jews and Christians. The common feature providing these scientific activities with an external unity is simple. The language, the chief factor making for unity, may legitimately serve to characterize the science of the Islamic countries of the Middle Ages, for it is the vehicle of the ideas, the concepts and the knowledge. And indeed, if one wished to argue against the expression 'Arab science', it would then become necessary, for the same reasons of ethnicity and religion, to argue against 'Greek science' as well.

The earliest scientific texts in the Arabic language are translations, from Greek, Syriac, Sanskrit and Pahlavi, which continue and extend the labours of the Hellenistic and Persian centres of learning. These works fostered the development of a scientific vocabulary, which did not exist in the language of the Bedouins of pre-Islamic Arabia, which the Koran had made the language of the Revelation. To give one example: the Greek word ἐπιληφία, which was the term for epilepsy, was first of all just transcribed in Syriac, and then from Syriac into Arabic: *ibilimsiya* (the b replacing the p which is absent in Semitic languages). An Arabic treatise of the ninth century gives as the title of one of its chapters: '*Fi ibilimsiya*, which is to say, *al-sar*', this last word being a derivative of an Arabic verb meaning 'to throw someone to the ground'. Some decades later, the word transliterated from the Greek had fallen into disuse; the properly Arab word was enough and it was then the only one employed; from now on it belonged to a unified terminology, well known to all.

In the chain which leads from Greek to Arabic, Syriac is an essential link. In fact most of the translators were Christians, fluent users of Greek and Syriac, the learned languages of the period, who looked down a little on Arabic as lacking an adequate vocabulary. They worked in Arabic only when commissioned to do so, by Muslim scholars or notabilities, and later on, when Arabic was imposed on everyone. Hunayn ibn Ishaq was one of these translators: his activity is a good example of the conditions of the transfer of antique science to the countries of Islam. Born in 808 to a family of Nestorian Christians at Hira, on the lower Euphrates, he was bilingual: Arabic was the language spoken in his native city, while Syriac was his mother tongue. Attracted like so many others by the prestige of the Abbasid capital, he came there to study medicine. He then learnt Greek, perhaps at Alexandria. Returning to Baghdad, he worked as a physician and as a translator. We owe him several treatises on medicine and an impressive number of translations of works by Plato, Aristotle, Hippocrates, Ptolemy, Porphyry, Rufus of Ephesus, Paul of Aegina and Galen. During a spell in prison in 856, he composed a short work in which he tells how he translated 129 treatises of Galen's. He knew but criticized the previous translations, for he re-did a great number of them. He had a preference for Syriac, hardly ever translating directly into Arabic, and often gave one of his students the task of translating the Syriac text into Arabic. The activity of Hunayn and his fellows gave the Arabs access to the knowledge of Antiquity: the great works of Greek philosophy, logic, medicine, astronomy, mathematics, botany and mechanics were now accessible.

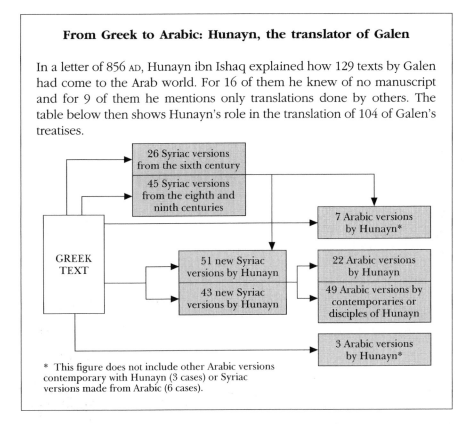

From Greek to Arabic: Hunayn, the translator of Galen

In a letter of 856 AD, Hunayn ibn Ishaq explained how 129 texts by Galen had come to the Arab world. For 16 of them he knew of no manuscript and for 9 of them he mentions only translations done by others. The table below then shows Hunayn's role in the translation of 104 of Galen's treatises.

GREEK TEXT

26 Syriac versions from the sixth century

45 Syriac versions from the eighth and ninth centuries

7 Arabic versions by Hunayn*

51 new Syriac versions by Hunayn

43 new Syriac versions by Hunayn

22 Arabic versions by Hunayn

49 Arabic versions by contemporaries or disciples of Hunayn

3 Arabic versions by Hunayn*

* This figure does not include other Arabic versions contemporary with Hunayn (3 cases) or Syriac versions made from Arabic (6 cases).

There was translation from Greek into Arabic, but also into Arabic from Sanskrit and from Pahlavi. This is less well-known and surely not on the same enormous scale, but it can no longer be left out of account. One of the most important astronomical treatises of the Indian tradition, the *Mahasiddhanta*, was translated from the Sanskrit in the decade following 770, at the instigation of the Caliph al-Mansur. Together with other Indo-Persian works, it was the basis of a school of Arab astronomy which survived for a long time in Andalusia.

Originality and Inheritance

The Arabs were conscious of the originality of their science and the inheritance from which it had benefited. In eleventh-century Spain, Sa'id al-Andalusi singles out in his book *The Categories of Nations* the peoples who had cultivated the sciences. He recognized eight: the Indians, the Persians, the Chaldeans, the Greeks, the Byzantines, the Egyptians, the Arabs and the Jews. In the second half of the ninth century al-Razi, the Rhazes of Latin authors, a very well-known physician, a physicist and alchemist, and the author of around fifty works, declared that scientific progress was only possible by following in the steps of the Ancients, because 'the most recent benefits from the gains made by his predecessors and adds to them by his own studies.' He puts science back into history. For al-Razi this consciousness of inheritance did not in the least negate the claims of contemporary science.

What are the main features of this science? What does it owe to the civilizations that came before it? What is original in it? What is its role in the development and transfer of scientific knowledge? The degree of originality, the scope of innovation and creation among the scientists of the Islamic countries varied according to the period. The early centuries, the time of translations, were those of a science gathered in conquered lands or in neighbouring countries. We have to wait until the ninth century, with al-Khwarizmi the mathematician and astronomer and al-Kyndi the philosopher and geometer, for the development of a properly Arab science. All branches of science, physics, mathematics, astronomy and medicine presented common characteristics, explained by their political and economic context, the surrounding culture and the scientific institutions; but they drew on varied sources and they developed according to their own methods. A study of Arab science has to give some account of the great fields of scientific study.

Physics

The basis of the work of the pre-Socratics, and essential to Aristotle, physics played a predominant role in Greek science until Archimedes and beyond. The Arabs called it 'the science of natural things' and took it up on their own account. Commentators on Aristotle were interested in the physics, and scientists in their turn believed that the world was composed of four elements, fire, air, water and earth; but this physics derived from the Greeks did not lead to any specific developments as regards the structure of the universe.

On the other hand, certain disciplines which we include in physics, optics (see p. 323), statics or the science of balances, had an important place in the Islamic countries. More practical and also connected to the realities of the economic and social context, they opened up new paths without challenging the Aristotelian edifice.

Astronomy

There exists an edition of the Latin translation of the Arabic translation of Ptolemy, annotated in the hand of Copernicus, a fine example which demonstrates the place of the Arabs as intermediaries between Greek science and the science of the West. The very name of *Almagest* which Westerners applied to Ptolemy's *Great Composition* itself came from the Arabic. However, the Islamic world knew of an astronomy other than the one which came from the Greeks. The earliest Arab astronomy developed first of all on the basis of Indian and Persian sources, and beyond these, Mesopotamian. The *Zij al-Sindhind* (the 'Table of Sindhind'), composed around 830 by al-Khwarizmi, is designed according to an Indian model, which is particularly clear when he explains his calculations for the positions of the planets. The *Almagest* was translated into Arabic at the beginning of the ninth century; from then on Ptolemy's work dominated the history of astronomy. Nearly all the models and interpretations developed by the Arabs start off from Ptolemy, whose model of the Universe dominated the Arab world as it had done the Hellenistic. By means of a complex combination of circular motions, it gave a much better account of the movement of the stars than had previous models. This recourse to perfect uniform circular movement, the only one conceivable in the superlunary world, allowed astronomy to be given its place in the currently dominant physics and cosmology, those of Aristotle. None the less, this need to use circular movements to render non-circular orbits led Ptolemy to a system whose compatibility with physics was questionable, and which had indeed been questioned during the Hellenistic period. From the second half of the ninth century al-Battani made a commentary on the *Almagest*, but he also corrected and completed it, relying on new observations. He was the first of the great Arabs to continue Ptolemy's work. The work of the Alexandrian astronomer, a descriptive mathematical model, was also criticized, particularly in terms of a return to Aristotle. The end result of this discussion, the work of the School of Maragha (Iran) in the thirteenth century, and particularly that of Nasir al-Tusi, was at the origin of Copernicus' conception of the planets.

On the other hand, Arab astronomers were able to use more powerful means of observation and of calculation than the Greeks had at their disposal. Drawing upon the Indian tradition, they used decimal numeration and they developed trigonometry; they built large observatories; they produced astronomical tables – more than a hundred have been catalogued – for

Made in Iraq in the ninth century by Ahmad ibn Khalaf, this astrolabe once belonged to the son of the Caliph al-Muktafi (902–908). On the lower disc one can see the lines showing the projection of the Earth on to the plane of use, surmounted by the rete which gives, at the end of each pointer, the position of a star. (Bibliothèque Nationale, Paris. Photograph, Giraudon.)

superior to those of their predecessors. Widely distributed in the West, they were surpassed in accuracy only by those of Tycho Brahe in the second half of the sixteenth century. None the less the Ptolemaic influence, even through the controversies it gave rise to, remained considerable. From the Greeks, Arab astronomy had inherited cosmological conceptions and a mathematical model which effectively accounted for the appearances. It took up many questions which had arisen in the Hellenistic period and discovered original and innovative responses, relying on new methods of calculation and observation.

Why was there this growth in observational astronomy? Scientific curiosity no doubt, but this was linked to social and cultural facts. Astronomical measurements are indispensable to the practice of Islam. Worship requires knowledge of the hours of sunrise and sunset, and the lunar Islamic calendar requires the calculation of the beginnings and ends of months, of Ramadan in particular. The appearance of the crescent moon marking the beginning of the month presents a practical problem to which Thabit ibn Qurra devoted a scientific treatise dealing with questions of astronomy, physics and optics. Mosques need to be oriented towards Mecca. Everything leads to the multiplication of observations and to the search for greater and greater precision. To a great extent, Islam developed in countries which had practised worship of the stars, as witnessed by the ziggurats of Mesopotamia and the Temple of Artemis at Ephesus. Astrology, part of the heritage of the Hellenistic world as well as that of the East, was very popular. Here too, one needed to predict the movements of the planets and their positions relative to each other and to the Sun. Islam appeared as a world of observation, from the simple quadrant to the princely observatory. This is attested by the number of treatises on the astrolabe and the number of instruments surviving into our own day. The astrolabe wasn't used to observe the heavens but to interpret them, and its manufacture required a thorough knowledge of the stars and of their apparent movement. It was made up of a circular plate against which there turned another disc, mostly cut away, resembling a spider. The plate bore a projection of the earth from a particular place, while the spider was a map of the sky with the principal fixed stars, among them the Sun. It allowed the determination of the azimuth and the hours of sunrise and sunset, the position of the stars on the horizon and much other information indispensable to astrology.

Alchemy

Astronomy and astrology, alchemy and chemistry, distinctions which in the twentieth century draw the line between science and charlatanism, were not distinctions at all for East or West in the Middle Ages. In all the classifications, chemistry or alchemy – for there was only the one Arab word – was one of the sciences of nature, just like medicine. Just as Ptolemy provided

The observatories of medieval Islam

The works of the Arab astronomers depend as much on observation as on theoretical models and the art of calculation. The production of copper instruments of great precision and the construction of observatories was therefore indispensable to them. Close study allows a division into two periods.

Until the eleventh century, the astronomical work supported by the rulers of Baghdad, Shiraz, Isfahan and Cairo was conducted with simple observation posts set up for programmes of limited duration. So the greatest scientists of the time were brought together by the Caliph al-Ma'mun to verify the data of Ptolemy's *Almagest.* They took up positions with their instruments in the north of Baghdad and on the hill which rises above Damascus. The death of the caliph in 833 prevented them from making all the observations planned.

Only the last centuries of the medieval period saw the construction of real observatories which functioned continuously through the years. The magnificence of the buildings and the reputations of the scientists who worked there made these establishments famous, though it is true they were few in number. At Maragha, a city of Azerbaijan, the grandson of Genghis Khan built an observatory provided with a rich library and with instruments of great technical perfection: the armillary sphere, quadrants of different kinds, parallax rules, the sextant, the celestial globe, the astrolabe and the sundial. This great scientific centre of the late thirteenth century attracted both scientists and students, and one can speak of a 'School of Maragha' whose most illustrious representative is Nasir al-Tusi. Today only the foundations of the complex survive. The observatory of Samarkand, on the other hand, founded in the mid-fifteenth century by the grandson of Timur – whom we know as Tamburlaine – has been the object of several archaeological investigations. Some impressive remains have been restored: part of a gnomon of 40 m radius, used to determine the height of the Sun by means of the length of the shadow, and a vast cylindrical building more than 30 m in height, sumptuously decorated with glazed tiles, which had doubtless borne a dome pierced at its highest point to allow the Sun's rays to pass through.

But the dearth of written records, the difficulty of interpreting the few archaeological remains, and the absence of monographs leave the organization and operation of these establishments very much in the dark.

True scientific institutions, they did however play a decisive role in the development of the exact sciences in Islam, with the drawing up of *Tables* constantly revised, the training of students, the concentration of scientists and the financing of series of observations. They had a certain influence on similar constructions in Istanbul and in Mogul India, and perhaps, though more questionably, on Tycho Brahe's installations.

The elements in Aristotle and in Arab Science

The elementary qualities are four, and any four terms can be combined in six couples. Contraries, however, refuse to be coupled: for it is impossible for the same thing to be hot and cold, or moist and dry. Hence it is evident that the 'couplings' of the elementary qualities will be four: hot with dry and moist with hot and again cold with dry and cold with moist. And these four couples have attached themselves to the *apparently* 'simple' bodies (Fire, Air, Water and Earth) in a manner consonant with theory. For Fire is hot and dry, whereas Air is hot and moist (Air being a sort of aqueous vapour); and Water is cold and moist, while Earth is cold and dry. (Aristotle, *On Generation and Corruption*, fourth century BC)

   To the partisans of the two doctrines, I would reply that Aristotle had already put them forward in his treatise entitled the *Logic*, which is one of the most marvellous of works; he divided it into four parts called: the *Categories, On Interpretation*, the *Analytics* and the *Topics*. He preceded them with an introduction, and so he gave us the first treatise on proofs, and in this he was the first among philosophers. And so philosophers prostrate themselves before Aristotle who created this science . . . As for the two luminaries, the Sun and the Moon, having created all things from the four elements, fire, water, air and earth, God made emerge from the ancient worlds the four qualities: heat, cold, wetness and dryness. The combination of these elements produced fire, which contains heat and dryness; the earth which has cold and dryness; the air, which had heat and wetness, and the earth which has cold and wetness. It was with the help of these elements that God created the upper world and the lower world. When there is equilibrium between their natures, things continue to exist, despite the passage of time, without being consumed by the two luminaries, nor rusted by the water of the pools; such is gold which nature has cooked and purified in all its parts, without having drugs, or analysis or refining. (Jabir ibn Hayyan, *The Book of Balances*, tenth century)

the principal model on the basis of which Arab astronomy developed its own interpretations, it was the Greeks' conception of the world and of matter, and Aristotle's in particular, which provided the theoretical framework for Arab chemistry and medicine.

   The sublunary world, the world of generation and corruption, as opposed to the immutable world of the fixed stars, was composed of fire, air, water and earth, from whence came the 'qualities', Aristotle's 'elements', which are heat, cold, the wet and the dry. The combination of these elements gives rise to all substances which exist in nature. Natural substances could be modified. There existed the theoretical bases for a chemistry. In the Greek world, the Aristotelian theory had already come up against another tradition, which we now call the alchemical. Its origins have been sought in Egypt or Meso-potamia, in China too. It is likely that all these hypotheses have an element of truth: men have always dreamed of powers over matter by means which have extended from magic to experimentation. The Arab tradition of alchemy was able to draw on many sources, but essentially, it would seem, it was influenced by the Alexandrian alchemists who had already absorbed some of Aristotle's ideas.

The early tenth-century compilation of texts attributed to Jabir ibn Hayyan, very well known in the West under the name of Geber, demonstrates a new conception of alchemy. From the Greeks it took a physics of elements and believed in the transmutation of metals, but it had its own classification of the minerals, which it divided into three great categories. These were the spirits, 'volatile substances', metals, 'fusible substances which may be hammered' and bodies, 'substances fusible or not, which may not be ham-mered and which turn to powder'. There were five spirits, sulphur, arsenic, mercury, ammoniac and camphor. The metals were made of sulphur and mercury. In one of the major collections of the alchemical corpus, *The Book of Balances*, Jabir defined the principles of the art: the reduction of all the phenomena of nature to the laws of quantity and of measure. His alchemy is not cut off from the world, and like a modern chemist he looks for applications in metallurgy, in pharmacy, in dye manufacture. Organized by a conceptual framework entirely foreign to us, the chemistry of the Arab Middle Ages is little known and difficult to understand.

Medicine

The duality of theory and practice is found in Arab medicine as well. Inher-ited from Aristotle, but also and especially from Hippocrates and Galen, its theoretical foundations owed much to Greece. From Hippocrates Arab medicine took its theory of humours. Fire, the product of heat and dryness, produced bile, which was to be found in the gall-bladder. Blood came from air, the combination of heat and wetness, and resided in the liver. Phlegm came from water which was cold and wet, and was situated in the lungs. Black bile came from earth, and so from cold and dryness, and occupied the spleen. Health depended on an equilibrium of the humours, an equilibrium which varied with the individual. Cure therefore depended on an exact diagnosis pinpointing the disequilibrium and on a treatment which took individuals into account, together with their temperaments, their dietary habits and their previous illnesses. Besides this omnipresent theory, the Arabs drew on other sources. The physicians of Islam augmented their pharmacopoeia by contact with India. In his medical encyclopaedia, *The Book which Con-tains All*, al-Razi states for each disease the opinions of the Greeks, the Syrians, the Indians, the Pesians and the Arabs, which he rounds off and critiques with his own remarks and observations. It isn't any longer just a question of theory. Al-Razi was also the director of the hospital at Baghdad. On the basis of a complex inheritance, the Arabs developed a medicine of practice and observation. The precise descriptions of smallpox and of measles by al-Razi, the discovery of the lymphatic system by the Egyptian Ibn al-Nafis at the end of the thirteenth century, the constant progress in the anatomy and physiology of vision are the product of a thought which had its bases first of all in practice and observation. In the tenth century Ibn Sina, the Avicenna of the West, a physician, but also a great commentator on Aristotle, composed a major work entitled the *Canon of Medicine*,[1] which remained one of the bases of medical culture both in the Islamic countries and in the Christian West.

Canon of medicine: *an encyclopaedic compilation offering a broad, complete and methodical synthesis of Graeco-Arab medical knowledge. It owes its fame to its rigorous construction, to the scale its theoretical contribution and to its demand for rationality.*

The theory of humours in the Arab world

The first humour is the bile. It is derived from fire, which is the product of heat and of dryness. The bile resides in man's body near the liver, in the gall bladder. The second humour is the blood. It is derived from air, which comes from the combination of heat and wetness. Its seat in man is the liver. The third humour, the *pituita* (or phlegm or lymph) derives from water, which was created from the combination of coldness and wetness. It resides in the lungs. The fourth humour, the atrabile (or black bile), derives from earth, which is composed of cold and dryness. It occupies the spleen. These four humours make up the materials of the body, they determine its state of well-being or ill-health. (Al-Suyuti, *The Book of Mercy in the Art of Curing Illness*)

Geometry

Geometry is considered to be the quintessential science of Classical Greece. 'Let no one enter here who is not a geometer' says the tradition, speaking of Plato's Academy. It witnessed considerable development during the Hellenistic period, the time of Euclid and of Apollonius of Perga. Many Arab mathematicians taught the books of their predecessors and made commentaries upon them and in certain areas they were not lacking in originality. The application of calculation to geometry developed within the context of highly innovative work in algebra. It allowed the calculation of π correct to the sixteenth decimal place. From the ninth century Arab geometry was marked by the importance of research on parallel lines, which had already received attention from Euclid's Hellenistic successors. Al-Khayyam devoted part of his work to these questions and influenced Nasir al-Tusi's *Exposition of Euclid*. Neither of these claimed to challenge Euclid's postulates, but their work opened up the way in this direction. Even when Greek geometry was challenged on certain points, the argument was carried forward in the framework set down by the Ancients.

The Sciences of Calculation

The sciences of calculation, which is what we may call algebra, arithmetic and trigonometry, owed much less to Greece and much more to the East. Sciences of the practical and of the concrete, which were needed to calculate the area of a field, to share out an inheritance or to forecast a profit, they developed very early, outside the Hellenic tradition. In the ninth century AD, al-Khwarizmi wrote an *Arithmetic*, as his work on calculation is usually called, based on the use of decimal numbers on the Indian model. With this work, and probably other writings and other contacts, the Indian system of number spread through the Arab world and thence to the West and then to the whole world. Al-Khwarizmi's treatise gives a lot of room to the practical. Besides this arithmetic he left many other works, among them the treatise *Al-Jabr wa l-muqabala*, which achieved great fame through the centuries. A new discipline, dealing with calculation by means of unknowns, appeared for the first time under its own name: the term 'algebra' comes from the very title of this treatise. It is true that algebraic technique remained elementary; reliance on solutions using geometrical figures made it impossible to conceive of negative roots, and above all, there existed no algebraic symbolism, everything being done by means of figure or rhetorical expression. It is true none the less that the systematization of argument, and the grouping of different equations into a certain number of canonical equations, whose solutions were found later, opened the way to an algebra that owed nothing to the Greeks. Al-Khwarizmi's work belongs to a tradition of algorithmic calculation which may go back as far as the Mesopotamian civilizations. The type of calculation used to solve equations of the second degree also leads to comparison with India.

Al-Khwarizmi's *Kitab al-Jabr wa l-muqabala*

The 'Little book of *al-jabr* and *al-muqabala*' was written in the first half of the ninth century by Mohammed ibn Musa al-Khwarizmi, born at Khiva, in the Khwarizm, in Central Asia. He worked and wrote in Baghdad. His life remains little known. He composed astronomical tables. His arithmetic, translated into Latin in the twelfth century by Robert of Chester and by Adelard of Bath, made a great contribution to making 'Arabic' numbers known in the West, together with the calculations that they

---

made possible. He was the first to write a treatise having 'algebra' in its title, the *Kitab al-jabr wa l-muqabala*.

In his introduction al-Khwarizmi gives the reasons which impelled him to write the book: 'The imam and emir of the believers, al-Ma'mun, encouraged me to write a concise work on the calculations *al-jabr* and *al-muqabala*, confined to a pleasant and interesting art of calculation, which people constantly have need of for their inheritances, their wills, their judgements and their transactions, and in all the things they have to do together, notably, the measurement of land, the digging of canals, geometry and other things of that kind.'

The terms *al-jabr* and *al-muqabala* signify reduction and comparison respectively; they define the method used by the author. The first thing to be done is to relate an equation to one known and soluble. Al-Khwarizmi gives six standard or canonical equations, which, translated into modern terms, are the following:

$$ax^2 = bx,$$
$$ax^2 = c,$$
$$bx = c,$$
$$ax^2 + bx = c,$$
$$ax^2 + c = bx,$$
$$bx + c = ax.$$

To arrive at these kinds of equations, which all contain only positive numbers, it is first of all necessary to proceed by means of *al-jabr*, which means to balance the terms of the equation. And so an equation such as $2x^2 + 100 - 20x = 58$ will give by *al-jabr* $2x^2 + 100 = 58 + 20x$, which is transformed by *al-muqabala*, that is, by simplification, into $x^2 + 21 = 10x$.

Solutions to equations are furnished by algorithms based on geometrical proofs. The example used to find the value of the algorithm which permits the solution of an equation of the type $x^2 + bx + c$ is the following:

'Let the square and ten roots be equal to 39', which is to say in rhetorical language, that the square of the unknown and ten times the unknown equal 39, or in our contemporary mathematical language: $x^2 + 10x = 39$, an equation of the fourth canonical type.

'The rule is that you divide the roots into two halves, here we get 5, which you multiply by itself, we have 25, to which you add 39, and we get 64. You take the root which is 8, and you reduce it by half the number of roots, which is 5, giving 3, which is the root of the square for which you are looking, the square is 9.'

This algorithm we can express in the form:

$$x = \sqrt{\left(\frac{b}{2}\right)^2 + c} - \frac{b}{2}.$$

211

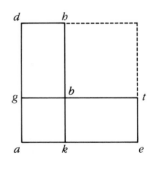

Al-Khwarizmi gives the proof by positing a square *ab*, which represents the square of the unknown, whose side is therefore the unknown. Two perpendicular sides of the square are prolonged by a length which is the half of 10, which is 5. One can then construct on the sides of this square two rectangles, rectangles *g*, *b*, *t*, *k*, one of whose sides is equal to the unknown and the other to 5. The total of the two rectangles is therefore equal to ten times the unknown, and that of the figure formed by the square posited at the beginning and the two rectangles together is 39, because it corresponds to the value of the unknown and ten times the unknown. If one then completes this figure to make a square with a square of side 5, the area of this latter square will be 52, which is 25, which is added to 39 to give $\sqrt{64}$. The side of the big square *de* is therefore equal to 64, which is 8. To discover the unknown, one then has to subtract 5 from 8, which gives 3. The algorithm which allows the answer to be found is therefore expressible, in modern terms, in the form:

$$x = \sqrt{\left(\frac{b}{2}\right)^2 + c} - \frac{b}{2}.$$

a solution which implies positive roots only.

The algorithms giving solutions to the other canonical equations are proved by methods of the same kind.

The search for al-Khwarizmi's sources is, however, far from being over. A striking character in the history of calculation, al-Khwarizmi inaugurated a current of mathematical research which has continued uninterrupted since his day. After him treatises on calculation multiplied, always in the same tradition.

Was a geometry based on the Greeks confronted by a novel science of calculation which took advantage of Eastern learning? This is too simple. In his *Recherches sur l'histoire des mathématiques arabes*, Roshdi Rashed says that, after al-Khwarizmi, algebra developed in two directions: geometry and arithmetic. Al-Khayyam based his researches on cubic equations on a geometrical conception, but in his algebra figures play only an auxiliary role, and he considers the solutions to equations as the intersections of curves. According to Roshdi Rashed, the 'arithmetization of algebra' marks a decisive step forward. 'By arithmetization,' he says, 'one means the transposition and extension of the elementary arithmetical operations, algorithms like Euclidean division, or the extraction of the root, to algebraic expressions, and to polynomials in particular.' Even if this idea is fiercely rejected by some specialists, it is none the less the case that at the end of the eleventh century al-Karaji's book presented a theory of algebraic calculation characterized by the extension of arithmetical operations to the polynomials. It relies on al-Khwarizmi's algebra as it was developed by his successors and on the translation of Diophantus' *Arithmetic*, from which he drew a good many of his problems. His successor, al-Samawal, stated unhesitatingly that it was necessary 'to operate on the unknowns with all the arithmetical instruments, even as the arithmetician operates on his knowns'.

The outline presented here tends to obscure the considerable lacunae which remain and which make it difficult to grasp the specific character of Arab science, what it derived from antecedent cultures and what it developed from its own resources. It would seem, however, that the Greek world, Ancient Greece and the Hellenistic world, provided the Arabs with a decisive part of the foundations of their scientific life. In the first place, there was a model of the Universe. In the works of Aristotle and Ptolemy Islam's scientists found the idea of a Universe divided into two parts, the sublunary world of generation and corruption and the immutable world of the fixed stars. From Aristotle they also took a logic and, in particular, a way of thinking: abstraction came from Greece.

It is much more difficult to distinguish and classify the influences of India and Iran, the stratified traditions of the Middle East, with their origins in the Mesopotamian world. They are there none the less and indisputable, particularly where the Arabs made an enormous contribution, in the field of calculation. Where did the types of algorithm used by Al-Khwarizmi come from? Was it from Mesopotamia or India, or might they derive from a broader tradition of calculation? Even if Greek science, and Hellenistic science particularly, was often more concerned with practice than is usually said, Arab science drew on traditions clearly more pragmatic, more utilitarian and more interested in calculation. It was probably on these different foundations that the Arabs constructed their sciences of observation: astronomy, medicine, natural science and chemistry. These were the multiple, complex and still somewhat indeterminate influences which gave Arab science some of its specific character.

The Way to the Latin West

While Arab science, the science of the Islamic countries, had reached its apogee, what was happening to the science of Christian Europe? It was vegetating and, according to Guy Beaujouan, it didn't really take off until the last quarter of the thirteenth century. But a desire for knowledge found expression long before this. Before it gave rise to its own scientific developments, the West came to rely heavily on knowledge from the East.

The first contacts between Arab science and the Western world were marginal, because they occurred at the very edges of Islam and of Christendom, because they concerned only very few people. Much remains unknown and, despite any efforts that might be made, it will remain so for lack of adequate sources. The tenth century, when the main structures of the future Europe were being established, is the least known century of our history, in which writings are the most scarce. There are none the less certain events and certain men to whom attention can be paid, in order that we may retrace this history in the absence of any more extensive documentation. The works of Gerbert d'Aurillac show that, by the second half of the tenth century, Arab science had already penetrated to the West. Gerbert, born around 940–950, was a monk and then director of the episcopal school at Reims; he became Archbishop of the same city, and in 999 became Pope under the name of Sylvester II. This remarkable character travelled to Spain, or at least to Catalonia, where he met certain people, with whom he later maintained contact by letter. If the mathematical work attributed to him may be disputed, his letters show that he asked his correspondents for books, among them one with the evocative title *Of Multiplication and Division*. At the same time, 'Arabic' numerals were appearing in Latin manuscripts. The oldest known and dated was copied in 976 at a monastery in northern Spain. From the beginning of the eleventh century, they become more and more

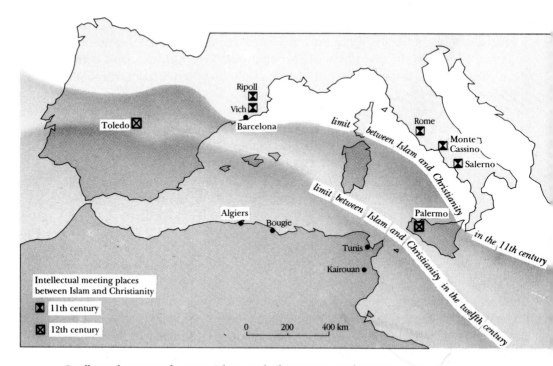

Intellectual contacts between Islam and Christianity in the West

214

common, notably in copies of Boethius' *Geometry*. Could these be the first signs of the influence of Arab science? They probably are, as is suggested by other evidence. Astrolabes, unknown in the West until then, are mentioned at the end of the tenth century and at the beginning of the eleventh; Gerbert brought one back from Spain.

Spain was doubtless the main conduit of Arab influence, but not the only one. At Salerno, near Naples, tradition had it that the School of Medicine was founded by four professors, one Greek, one Latin, one Jewish and one Arab, each one teaching in his own language. In fact this isn't true, but the legend is significant. Southern Italy was at the crossroads of the Mediterranean, marked by many different influences. In the first medical works from the School of Salerno the Arab influence is imperceptible, but this soon changed with Constantine Africanus. Born at Tunis in North Africa, he underwent many trials – as a traveller in the East, an envied scientist, a merchant on the Mediterranean – before reaching Italy, bringing with him many manuscripts. He was converted to Christianity and became a monk at the abbey of Monte Cassino. Though Constantine's biography isn't entirely clear, tradition has it that he was a most productive translator. He left a voluminous corpus, produced between 1065 and 1085, essentially composed of translations of Greek and Arab texts taught at Kairouan. Disseminated through the School of Salerno, these texts formed the basis of European medical education for centuries.

In the twelfth century these first contacts were succeeded by the period of the great translations. The sites remained the same: the Latin East, the outcome of the First Crusade, has left practically no evidence of scientific contact between Crusaders and Muslims. Everything happened on Islam's western borders. Sicily, in the hands of the Muslims since the ninth century, had previously experienced centuries of Byzantine domination. Conquered by the Normans between 1061 and 1091, in 1194 it came into the possession of the Emperor Frederick II. Situated between the Latin, Greek and Arab worlds, the island became a meeting-point for products, men and ideas. As well as the majority of translations done from Arabic into Latin, other texts were translated directly from Greek. In Spain, Catalonia remained a focus for exchange, but after its reconquest other centres developed. At the instigation of the bishop Raimond, from the moment of its capture by the Christians in 1085, Toledo became a great centre of translation. There were Mozarabians, Spaniards of Arab language and culture, Jews, some converted to Christianity, others not, and foreigners who came from all over Europe to study. The most prolific was the Italian, Gerard of Cremona, who translated at Toledo around eighty works in most of the scientific disciplines. The role of the Jews was fundamental. Few Christians knew Arabic and they therefore had recourse to an intermediary, often a Jew. The translation was then done in two moves and was the work of two men. The first would translate from

Arabic into a European language and the second would then translate this into Latin. Wanting to enrich their own communities, the Jews translated into Hebrew and wrote in it; some of their works were retranslated into Latin. Translation developed rapidly in the course of the twelfth century. Certain texts were translated in successive versions: Euclid's *Elements* were done by Adelard of Bath, Hermann of Carinthia and Gerard of Cremona. Ptolemy's *Almagest* by Gerard of Cremona and Eugene of Palermo. There were many adaptations of al-Khwarizmi's *Arithmetic*, translated by Adelard of Bath: John of Seville's *Alghoarismi de arismetrice*, Sacrobosco's *Algorismus* and finally, the *Carmen de algorismo* by Alexandre de Villedieu, composed in verse so as to be easier to remember. The very origins of the translators mentioned above go to show that if the chief sites of the meeting between Islamic science and the West were on the borders of Christendom, the effort of translation and adaptation had mobilized intellectuals from all over Western Europe in a large-scale and sometimes systematic endeavour contrasting with the limited relationships that obtained during the preceding centuries.

The *Elements* from West to East, or Euclid in parallel

Euclid's *Elements* is at the beginning of the history of mathematics in Europe. How did it arrive? The history of this transfer goes back very far. There survives no manuscript of the *Elements* dating from the period of its composition. There is no agreement about who Euclid was. For some the author is an individual, for others the *Elements* are the product of a school. The first text must have been composed at the end of the fourth or the beginning of the third century BC. The first known commentator is Hero of Alexandria, two centuries later, but it seems certain that Apollonius of Perga knew Euclid. The founding texts of the Euclidean tradition are the product of late Latinity. The commentaries by Pappus (fl. AD 300–350) and by Proclus (410–485) are made on the edition by Theon of Alexandria, who wrote in the second half of the fourth century. Theon's text, the Theonian edition, provided the basis for most editions up to the nineteenth century.

The Arabs very soon obtained the *Elements*. They had made it an affair of state, the Caliph al-Mansur obtaining a Greek version from the Byzantine Emperor. Some decades later, in the reign of Harun al-Rashid (786–809), Hajjaj made the first translation, on the basis of a Syriac text, and in response to a request from the Caliph al-Ma'mun (813–833) he produced an abridged version for teaching purposes. In the first third of the ninth century the Arab world had a version of the *Elements*, thanks chiefly to princely intervention. This text can't have seemed good enough, for Ishaq ibn Hunayn, the grandson of the great translator Hunayn ibn Ishaq, got together with a mathematician, Thabit ibn Qurra, to produce a new version closer to the Greek text. The Arab world produced a

---

216

considerable number of works relating to Euclid's *Elements*. There were translations, résumés, like the one Avicenna incorporated in his *Kitab al-Shifa*, and commentaries. The best known, by Nasir al-Tusi and Muhyi l-Din al-Maghribi, date from the thirteenth century.

At the same time Euclid had not totally disappeared in the West. He is cited by authors of late Antiquity, such as Martianus Capella and Cassiodorus, as well as in a compendium for the use of surveyors, and especially by Boethius. The tradition was not lost, and a translation from the Arabic by Adelard of Bath shows signs of Boethius' Euclid. An Italian manuscript of the tenth century gives a fragment of an unknown translation of Euclid. Its mediocrity and the incompetence of its author may explain why it remained unknown; on the other hand, another most precise, word-for-word translation of a Theonian manuscript, done at Salerno around 1160, seems to have been hardly any better known.

The translations which established the *Elements* in the West came from the Arabs. The two traditions meet again, Gerard of Cremona situating himself in the line of Ishaq and Thabit, the line of the Arab text closest to the original Greek. Though less faithful, the three translations and adaptations by Adelard of Bath, based on the version by Hajjaj, had a far greater influence. The second version, an abridged commentary, met with great success. Adelard's work was the basis for that of Campanus of Novara, who made his own version of the *Elements* in the 1250s.

Adelard and Campanus were the basis for all later work. In 1482 the *Elements* were printed at Venice; it was the Campanus translation. It was only in 1505 that there was printed a version newly translated from the Greek by Bartolomeo Zamberti, and it was in 1533 that the Greek text itself was printed, at Basel. However, the translation by Campanus remained important until 1572, when Federico Commandino published a Latin translation from the Greek which was regarded as authoritative until the nineteenth century.

The story of the *Elements* allows us to provide our history with detail and nuance. The influence of sovereigns was decisive. The various translations produced at Baghdad only gave rise to original work after the passage of many centuries. The most important were the work of a Persian, Nasir al-Tusi, and of a North African, Muhyi al-Maghribi. Although the *Elements* had never been forgotten, and despite the translations from the Greek made in southern Italy, Arab science played an essential role in the transmission of Euclid to the West. It was only in the second half of the sixteenth century that a new translation from the Greek allowed the Arabo-Latin tradition to be forgotten. When Gerard of Cremona and Adelard of Bath were translating the work of the Arab translators, Euclid's work was a living tradition for the mathematicians of the Islamic countries, and the great commentators al-Tusi and al-Maghribi had not yet been born. From the two models, one living, the other dead, the West chose life.

At the end of the twelfth century, the West found itself the inheritor of a very important part of Graeco-Arab science and philosophy: most of Aristotle, with his legions of commentators, Euclid's *Elements*, Apollonius' *Conics*, the *Almagest*, translated from the Arabic but also from the Greek, Hippocrates and Galen, as well as al-Khwarizmi, the Alhazen of the Latins, Jabir ibn Hayyan, called Geber, Avicenna and Averroes. This considerable mass should not be allowed to hide the gaps; certain works remained, sometimes for centuries, unknown to the West. Diophantus, known to the Arabs, is cited for the first time by Bombelli in his *Algebra* of 1572, and was translated four years later. Wallis discovered Nasir al-Tusi only in the seventeenth century, and it was only the nineteenth century which saw al-Khayyam's *Algebra* become available in Europe.

The Latin work of an Italian trained in Arab mathematics remained for the most part unknown. This man, Leonardo Fibonacci, called Leonard of Pisa, is considered to be the greatest mathematician of medieval Europe. The son of an officer of the Republic of Pisa, made responsible around 1192 for the colony of Bougie, he received instruction from 'an admirable master', to use his own description. A merchant, he travelled on business to Egypt, Syria, Greece, Sicily and Provence, bringing back many manuscripts. Returning to Pisa in 1202, he began to write. His *Liber abbaci* was widely disseminated. A treatise on arithmetic in the Arab manner, taking account of the whole of al-Khwarizmi's contribution, the book is divided into fifteen chapters, extending from the Arabic system of numbers to problems of algebra and geometry, though the greater part is devoted to operational arithmetic and its commercial applications. The *Liber abbaci* would be one of the essential sources for the European commercial arithmetic of the late Middle Ages (see p. 275). His geometry, the *Practica geometriae*, was also extensively reproduced in the Latin countries, but Fibonacci's other works remained largely unknown in the West until Prince Baldessare Boncompagni exhumed them in the 1850s. Living at Pisa, Leonardo maintained very close relations with the court of Frederick II in Sicily. Two of the problems in the *Flos Leonardi* are solutions offered in response to a challenge issued by John of Palermo in the presence of the Emperor himself, just as the *Book of Squares* responds to another challenge: to find a square number, which, augmented or diminished by 5 would give a perfect square. The connections with Arab mathematics, through his education in the first place, through the Sicilian milieu he frequented and then in the problems he dealt with, are clear. But probably because it was too learned, Fibonacci's work remained for the most part unknown to his contemporaries.

The Arab line of transmission was not the only one to give access to Greek thought. The Greek language had never been completely lost in southern Italy, and certain works of Antiquity reached the West directly and without an Arab intermediary. Part of Aristotle's writings, transmitted through Boethius, had always been known in the Latin countries, and many of Euclid's minor works came through adaptations of the Greek. But the tradition deriving from these, measured by the numbers of manuscripts and of commentaries it inspired, could never have had the importance of the Graeco-Arab line.

From the thirteenth century direct translations become more frequent, doubtless from a concern for accuracy. William of Moerbecke (c. 1215– c. 1286) made a Latin version of the essentials of Aristotle's work, and also translated Archimedes, Hero of Alexandria and Galen. Symptomatic of a new rigour, William of Moerbecke's work is situated in the period when Western science is beginning to take off, having assimilated a decisive part of the Graeco-Arab contribution, the translations, commentaries and original works by the scientists of the Islamic world.

The West wasn't able to absorb all the science the Islamic world was able to offer it. Why should this be? At the time when translations were being produced in great numbers, Europe was going through an exceptional expansion after centuries of stagnation. The first sign of this was an increase in population. With demographic growth there came an extension of territory. The clearing of forests and the draining of marshes increased the area of cultivable land.

Spatial expansion took advantage of technological changes, which, if they were not directly related to scientific development, no less completely modified the means of production. The swing-plough was replaced by a plough with an asymmetric share. Air and water now circulated much more freely in the ploughed earth, heavy soils became far easier to work and yields increased. There were innovations too in the field of energy with the appearance of the rigid collar, which allowed a more rational exploitation of the horse's strength, and the construction of water-mills. Hydraulic power was used to mill grain and for many other purposes, particularly for the hammering of iron blooms produced in reduction furnaces. An essential innovation, still too little understood, the application of water power to the production of iron totally changed the conditions of production of the metal, which was used to make ploughshares and iron parts for mills as well as arms.

The territorial growth of the West also put it into contact with the Islamic countries. The spectacular failure of the distant Crusades was of little moment compared to the expansion of western Christianity at its frontiers. In the East the pagan Slavs gave way before the Germanic thrust, opening up new lands for colonization. In the south the Normans, younger sons of noble families settled in Normandy, arrived in southern Italy at the beginning of the eleventh century.

They expelled the Byzantines from Bari, their last stronghold, which fell in 1071. Moving on to Sicily, then an Islamic country, the Normans completed its conquest in thirty years. Thus was born the Norman kingdom of Sicily inherited by Frederick II.

The eleventh century also saw the beginnings of the Reconquista in Spain. This suffered various setbacks and took centuries to complete. In its history the eleventh and twelfth centuries have a special place. In 1085 Toledo, the old capital of the Visigothic kingdom, fell into the hands of Alfonso VI, King of Castile, and in 1212 an army made up of knights from all the kingdoms of which Spain was then composed, but also of Crusaders from all over western Christendom, inflicted a crushing defeat on the Muslims at Las Navas de Tolosa. This was a decisive victory, which hemmed in Iberian Islam within the confines of the tiny kingdom of Granada. In Spain as in Sicily, the Christians were confronted with civilizations of a much higher intellectual and scientific level than anything they can have known. At these cultural crossroads, the West found the scientific resources that it lacked.

Rural growth, increase in population and an opening towards other worlds: everything favoured urban development in the West. Towns were encircled by new walls, bigger than the old, cathedrals were built, new parishes were formed. These were all signs of growth. The people of the cities had new needs. At Reims, at Chartres and at Laon episcopal schools were set up. In Paris the canons of Saint-Victor and the monks of Sainte-Geneviève provided education, in addition to that offered by the canons of the cathedral. The restored papacy favoured the education of priests. In 1079 Gregory VII instructed all the bishops to ensure the study of the 'literary arts'. A century later the Third Lateran Council of 1179 required that each cathedral should possess its own school, supervised by a priest specifically responsible for it, the schoolmaster.

But what should this learning be built on? The memory of antique science remained; it was there in the schools of the twelfth century, but whole tracts remained unknown. The physics at Chartres was based on Plato's *Timaeus*, and logic at Paris was based on a limited part of Aristotle's *Organon*[2] and its commentators. Compared to the monumentally compendious works of Aristotle or of the great Arab thinkers, those of the masters of the twelfth century are insubstantial. The explanation of the world is to be found elsewhere, tempting but at the same time threatening, for it is being expressed by infidels and by pagans.

The need for knowledge is present in the victorious West of the twelfth century as it was in the victorious Islam of the seventh and eighth centuries. The West needed not a religion, for Christianity dominated its intellectual life, nor a tradition of jurisprudence, which it found in Roman law; what it needed were philosophical and scientific foundations: an explanation of the physical universe, physics, astronomy and mathematics, the riches possessed by the Arabs.

From the tenth to the thirteenth centuries, the Arabs acted as intermediaries between Greek science and the West. Through them came the first stirrings in the tenth and eleventh centuries, through them too the great mass of texts which in the twelfth century provided the foundation for the intellectual renewal of the West. This transfer affected all the disciplines: mathematics and physics, astronomy and medicine, chemistry and optics. The role of direct transmission from Greek to Latin was minor, even if later the Latins found it convenient to turn to the original texts.

But through translation and through direct encounters the West also came into contact with sciences which did not come from the Greek world. Just as there were translations of the Arab versions of Greek texts, there were also translations of the works of Arab scientists. They commented on or developed the works of Antiquity, but they also presented novel researches, quite different from the science of Greece. This was particularly the case in the field of calculation, with decimal numeration, trigonometry and algebraic procedures. Should we therefore assume the existence of a Mediterranean science? One must go further. Through their writings the Arabs transmitted knowledge which came from the East, and from India in particular. The Arab intermediary appears then as an agent who puts several clients in touch with each other, more than two in any case. But this role is not enough. The Arabs modified, improved and transformed; they created, on the basis of learning acquired elsewhere, the resources of their own civilization. The science passed on by the Arabs to the West has its own identity, profoundly original in relation to its Greek and Indian contributions, just as the science of the medieval West would itself be different.

If too much is missing to allow a study of the passage of Eastern science towards Islam, the two transfers, from Greece to the Arab world and from the Arab world to the West, deserve comparison. They do not at all resemble the contemporary transfers which have imposed European science across the world. Empires in crisis furnished their knowledge to conquering and victorious civilizations. The Arabs had just taken over a large part of the known world when they came to master the knowledge of Antiquity. In the twelfth century Europe was enlarging its own space and extending itself to the detriment of its neighbours. At the most we can say that an antique science which repeated itself more than it renewed itself was confronted by an Arab science which was still creative in the period when it nourished the West. There is no question here of a knowledge imposed but rather of a science captured, as if the dynamics of civilizations lacking the foundations of a scientific culture led them to search elsewhere to make up their deficiencies. None the less, the science transferred has to be understood in order to be adopted, and it took many centuries for Islam and for Christian Europe to construct their own sciences on the bases of these external contributions.

There was some transfer, of course, but in a manner more complex than has been believed. The Arabs were much more than simple intermediaries. Their science should not be considered as a simple relay-station, but as a period in the history of Euro-Asiatic science.

7

Theology in the Thirteenth Century: A Science Unlike the Others

Paul Benoît

In which it will be seen that, thanks to Aristotle, theology acquired the status of a science and then lost it at the moment when a new science was being born in medieval Europe; in which it will also be seen that the definition of science can change.

Utrum sacra doctrina sit scientia? Is theology a science? For most of us in the twentieth century the question sounds strange, but this is a question that Thomas Aquinas asks in the first few pages of his greatest work, the *Summa Theologica*. He answers it in the affirmative, putting theology at the summit of the hierarchy of knowledge, and his whole approach is based on this position. How was it that in the thirteenth century theology was thought of as a science, and the dominant science at that? And after it had been accorded this status, how did it lose it, despite the fact that we generally think of the development of science as the extension of its field with the emergence of new disciplines? To put it briefly, can we determine what entitles a discipline to be called a science at any particular time? The concept has varied in different periods, the term has covered very different realities and its meaning has sometimes been more or less precise. It has always been difficult to define. In a text dealing with thirteenth-century theology, it can't be a question of thinking about science in terms of our contemporary norms, but rather of

Theology: the word and the thing

The term appears in Plato's *Republic* (II, 379a), with the general meaning of stories about the gods, which might be false and immoral, such as those of mythology, or reasonable and edifying, such as those that Plato wished to impose on the priests. In the *Meteors* (II, 1) Aristotle uses it when speaking of mythology.

In the Church Fathers of late Antiquity the term 'theology' finds only a limited use. In his *De Doctrina Christiana*, Augustine presents his conception of what we would call 'theology' but never uses the word himself in this context. In the *City of God*, however, he uses 'theology' in speaking of the conceptions of the Platonists or even of pagan mythology. It is only with Abelard that there emerges a sense rather closer to the current meaning of the word.

During the twelfth and thirteenth centuries, when it was accepted into the classification of the disciplines, theology split into two parts. Natural theology, more or less merged with metaphysics, dealt with what might be known of God by the exercise of reason alone, while a revealed theology expounded the articles of faith. To designate this second theology thirteenth-century authors, including Aquinas himself, preferred to use the expression *sacra doctrina*, reserving the term 'theology' for the theology of the philosophers. Only at the end of this century and in the course of the next did the word 'theology' take on its current meaning.

determining the different conceptions that were held by the people of the Middle Ages.

Sociology, psychology, linguistics and literary criticism have each in turn claimed and been granted their status as sciences. The idea of the 'human sciences', an institutional distinction within faculties of arts, expresses this extension of the meaning of the term. Is this an irreversible fact, or can we imagine that a discipline might lose its status as a science? By virtue of what epistemological, but also political, social and institutional criteria, is a science recognized as such? The history of medieval theology offers the student an exemplary case of a discipline that won its claim to scientificity and then practically abandoned it. By seeing how Thomas Aquinas (1224–74) and his contemporaries dealt with the question, 'Is theology a science?' we can try and grasp what the people of the Middle Ages understood by 'science' at a time when Western Christendom had come into possession of a good part of the Greek and Arab inheritance, and was also in the process of developing new social structures. The thirteenth century, the century of Saint Louis and of the French hegemony, also witnessed the triumph of papacy over empire, the development of the papal monarchy and the creation of the mendicant orders. A time of urban growth, it was the century of the cathedrals, of Dante and of the *Roman de la Rose*.

During the dimly apprehended tenth and eleventh centuries, when Europe's economic and demographic growth was just beginning, culture was a matter for churchmen, clerks in holy orders; it had taken refuge in monasteries and in cathedral cloisters. For a religion of the book, it was necessary to train a clergy that wasn't illiterate. To a great extent, the West of the tenth and eleventh centuries was ignorant of writing and, at the beginning of the twelfth century, in certain areas it was almost enough to know how to read and write to be considered a scholar. This explains the importance attached to correct language and to grammar, the first branch of learning. Still in 1231, the Pope commanded that Priscian, the Latin grammarian *par excellence*, be taught at the University of Paris. This also explains the veneration for the culture of Antiquity, for a golden age of knowledge now beyond recovery. Manuscripts were conserved and copied with great respect.

Schools in the Twelfth Century

Circumstances, however, were changing rapidly. In the course of the twelfth century there was a proliferation of discourse on language and on its exactitude and significance. The eleventh and twelfth centuries are the centuries of dialectic, of the logic of language. Great changes happened not in the monasteries but in the towns. The twelfth century saw an unprecedented phase of urban growth, which came to an end in the thirteenth. The influence of the episcopal schools of Laon, Rheims and Chartres, which flourished in the first half of the twelfth century, was followed by the supremacy of Paris. Under Philippe Auguste, Paris, the city of the Capetians, became the capital, seat of the king and of his services. The expansion of the kingdom and of its monarchy gave the city a dynamism which attracted teachers and students. Studying or teaching in Paris were Englishmen like John of Salisbury and Italians like Peter the Lombard, but there were also Germans and Scandinavians. The monastery schools and those of the regular chapters of Sainte-Geneviève and Saint-Victor, which had shone in the first half of the century, now gave way to another kind of education. The cathedral school, the 'Cloister of Notre-Dame', meaning the canons' residences around the cathedral, became the chief centre for studies, which were mainly theological. The bishop and chapter controlled the teaching by means which, in the twelfth century, are largely unknown to us. The growth of the schools led teachers to install themselves on the bridges joining the Île de la Cité to the Left Bank, and then on the Left Bank itself among the vineyards. In this way they escaped the overcrowding of the Île de la Cité and also the constraints imposed by the hierarchy.

In order to express his opinions free of the jurisdiction of the bishop of Paris, Pierre Abelard (1079–1142) taught for a while on land belonging to the abbey of Sainte-Geneviève. He was a son of the Breton petty nobility, as touchy as he was impetuous, and tradition remembers him especially for his unhappiness in love. He was the herald of a new age. A paid intellectual, he lived from his teaching, receiving fees from his pupils. A dialectician, his work seems to have played a decisive role in the reception of the new Aristotle. He subjected the texts he studied to the tests of formal logic and he applied the same methods to Scripture. No other late thirteenth-century teacher was as famous as Abelard, but schools were developing. They were schools which first taught the liberal arts, grouped together, as they were in Antiquity, in the *trivium* and *quadrivium* (cf. p. 192). Indeed, the scientific disciplines, in the current sense of the word, were ignored in favour of the study of language, the study of grammar, rhetoric and dialectic sanctioned by the then available texts of Aristotle's *Logic*. This was an initial training which opened the way to higher studies, including theology. Rich and complex, twelfth-century theology sought above all to determine the facts of Revelation and to illuminate them by means of the work of the Church Fathers. It was more a matter of ordering and organizing the facts of Scripture than of doctrinal synthesis. Commentary, discourse on the sacred text, led to the exposition of the questions which arose, to their presentation and to an attempt to order them. It was the time of the collections of maxims and of the first summae. Their appearance was the result of the intellectual activity which burgeoned in Paris during the last years of the twelfth century.

The University

In the midst of all this activity there emerged the university. *Universitas* – an assembly, union or community: the term was commonly used in the Middle Ages to designate different kinds of associations. At Paris the *Universitas magistrorum et scolarium*, together with Bologna the earliest of the universities, brought together teachers and students, those who had knowledge and those who sought it. It was a professional association, a guild, like those which were beginning to unite the craftsmen of the medieval town, those who were, in Rutebeuf's words, 'workers with their hands'. It was a new urban association, finding its identity in opposition, as the free towns often did themselves.

We have no exact date for the formation of the University of Paris, but the latest work puts it between 1170 and 1190. At this time it wasn't an organized institution but an unrecognized association, opposed by the public authorities and unintegrated into the life of the city. In 1200, as the result of a brawl, five students were killed by the sergeants, police officers of the provost of

Paris, the representative of royal power in the City. Philippe Auguste very quickly disowned his provost, punishing him and granting the *scolares*[1] a privilege which removed members of the University from the provost's jurisdiction. They were now subject only to the ecclesiastical authority of the bishop. By doing this, the king gave the scholarly world independence from his own power and from the city, with which relations were often strained.

The guild became organized. The University's main adversaries were the bishop of Paris and the chancellor of the chapter of Notre-Dame. In accordance with the decisions of the Third Lateran Council (1179), they had the right to grant the *licencia docendi*, the licence to teach, which allowed them control of what was studied. Faced with this degree-awarding monopoly, the teachers wished to impose their own authority. The course of the conflict is still not very well known, but it ended with victory for the teachers. In 1212–13 the chancellor was forced to award the licence to candidates judged worthy by the teachers, and in 1215 the University received its statutes from Robert de Courçon, the papal legate. The last great crisis affecting the freedom of the University erupted in 1229. Once again it began with a brawl, with students killed in a confrontation with the king's officers. The University replied to this violence by going on strike and dispersing itself. Teachers and students left Paris for Orléans, Oxford, Toulouse and the cities of northern France. This reaction can't simply be ascribed to a desire to make amends for the deaths of the unfortunate students. The conflict was ended by the intervention of Pope Gregory VI, proof of the affair's importance. The papal bull, *Parens scientiarum*, confirmed the victory of the University. There followed a return to Paris. The contents of the bull show the issues very clearly. The privileges granted in 1212–13 and the years that followed were confirmed. The University became autonomous, awarding degrees and becoming responsible for its own recruitment. It could make its own regulations and elect its own representatives. These privileges, to which others must be added, both jurisdictional and fiscal, were not at all out of the ordinary. The craft guilds also had the right to award the degree of master, to elect their own tribunals and, though their statutes were granted to them by the public authorities, in reality they were often themselves their authors. But there was a great difference between the guilds and the University. The latter was composed entirely of churchmen, clerks in holy orders. Even if they were for the most part only in minor orders, all were subject, at least for the duration of their studies, to the authority of the Church. This clerical status gave the University great independence in relation to the civil power and provided it with guarantees in the face of an urban population with which it was imperfectly integrated. The struggle against bishop and chancellor did not estrange the University from the Church but simply brought it under the authority of the Pope, a distant and well-intentioned power.

Three centuries of new universities

At the end of the twelfth and the beginning of the thirteenth centuries, teachers and students were slowly organizing themselves in the different regions of western Europe. It is impossible to give a precise date for the formation of the first universities, for official documents do no more than confirm an already existing reality. In the twelfth century the law students of Bologna, many of whom would already have been practising as lawyers, formed an association. They gave what would be their University its characteristic features: training in Roman and canon law, authority over its students. At Paris, Oxford and Valencia, under the pressure of teachers and students, university structures were established at the beginning of the thirteenth century.

These spontaneously created institutions, often born from struggle with local powers, were succeeded by establishments set up by the authorities. In 1224 the University of Naples was created by Frederick II, who wished to provide his States with civil servants. When the University of Toulouse was set up by the Papacy in 1229, it was intended to provide the manpower and the arguments to win back to Catholicism the areas taken over by Catharism. From the second half of the fourteenth century on, the number of universities increased greatly. They were necessary to the prestige of a ruler, providing him with a literate staff, capable of operating the machinery of the State in the process of formation. At the end of three centuries, around 1500, from Uppsala to Valencia and from Cracow to Coimbra, Europe had more than fifty universities. Paris's domination over theology and even the arts and Bologna's over law had disappeared. Christendom's centres of learning were succeeded by national universities.

Part of the movement towards guild organization, the rapid establishment of the University presents the historian with many questions. How could a group of intellectuals, without statutes, with no organized power and no economic strength, have succeeded in bringing to an end the pretensions of the bishop of Paris and the cathedral chapter? How was it able to assert its independence in the face of expanding royal power in a city which was becoming the capital of the kingdom? There are many factors which go to explain this state of affairs. First of all, the ever-increasing number of teachers and students, a sign of Paris's reputation and a manifestation of the need for learning in a France in full expansion. More directly, the University received the support of the king of France: by training intellectuals, it provided a staff for the State then being born. There is a direct connection between the development of learning and the development of administration, even if to the eyes of an inhabitant of the twentieth century it may seem very faint.

The mendicant orders and the university

At the beginning of the thirteenth century new religious orders appeared in the West. Known as the mendicant orders, they were the Franciscans and the Dominicans. Although born in very different circumstances, they answered a need for a renewal of the Church, faced with a changing world which could not be satisfied by traditional monasticism. The mendicants were to live in the poverty of the apostles, possessing nothing, either personally or collectively, which distinguishes them from the monastic orders. These monks were subject to a Rule and therefore regular, but they none the less lived in touch with the world and with secular life, for which they were to provide a good example and which they were to convert. Preachers and missionaries, they needed a good intellectual training.

The Dominicans established themselves at Paris in 1217 and the Franciscans in 1219. Very soon the friars began to follow the courses of study provided by the secular teachers. They obtained university degrees: in 1229, the Dominican Roland of Cremona became regent. Secular teachers joined the orders, Jean de Saint-Gilles joining the Dominicans in 1230 and Alexander of Hales going to the Franciscans in 1231. Around 1240, out of twelve chairs of theology, the Dominicans had two, the Franciscans certainly one and very likely two. The success of the mendicant orders within the University, their being answerable to the Pope alone, the manner of nomination of teachers by the order, all these led to a confrontation between the secular clergy on one hand and the Franciscans and the Dominicans on the other. The quarrel came to a head in the 1250s. The most famous of the regular teachers got involved in the conflict, Aquinas for the Dominicans and Bonaventure for the Franciscans. Finally the Pope decided in favour of the mendicants.

The mendicant orders had a decisive role in the intellectual history of Europe during the thirteenth century. They dominate the history of theology and philosophy at Paris. The Dominicans represented the moderate Aristotelian current, with Albertus Magnus and Thomas Aquinas, while the Franciscans with Bonaventure continued the Augustinian tradition. At Oxford the Franciscans, particularly Robert Grosseteste and Roger Bacon, also representatives of Augustinian Platonism, were the initiators of a scientific current more mathematically inclined and strongly orientated towards optics.

The decisive interventions of Popes Innocent III and Gregory IX were those of popes who had developed the papal monarchy, who had given the Church a more efficient administration of finance and justice extending throughout western Christendom.

This period when the universities were being established sees a move from crusading to teaching, with a Church which preaches and instructs, its new identity reflected by the *Beau Dieu*, the statue of the teaching Christ at the west portal of Amiens cathedral. One needed to know how to convert the infidel, to triumph over the heretic who reasoned, to bring the people of the cities back to the strait and narrow. So, while the Papacy developed a monarchical centralism, the Church tried to develop the Christian faith and also control all the new forces, the dynamism of a western Europe in expansion. Like the creation of the Franciscans and the Dominicans, the mendicant orders, the Pope's support for the emerging universities is a part of this project, and the scientific policy of the Papacy arises within this context. This last expression might seem anachronistic but, in his bull *Parens scientiarum*, Gregory IX affirms the scientific nature of sacred learning: 'Paris, mother of the sciences . . . Here the iron is taken from the ground, while terrestrial weakness is strengthened by moral force, and with it is made the armour of faith for Christ's soldiers, the sword of the spirit and the other arms, powerful in the face of the powers of brass' (H. Denifle and E. Chatelain, *Chartularium Universitatis Parisiensis*, 1889).

For the Pope the science of divine things is the science *par excellence* and not at all gratuitous. It must exalt the greatness of God and prepare for the struggle against the forces of evil. The political will of papal power made theology a science. This was not a sufficient but a necessary condition for its being accepted as such. It is in the university, a product of papal power, that this status may be granted to it. The ideas that society has of science depend mostly on institutional recognition.

In the thirteenth century the University of Paris and the other universities often shared a similar form of organization by faculty. A faculty of arts, providing a kind of preparatory study, opened the way to three higher faculties: medicine, law and theology. At Paris the faculty of law taught only canon law, the law of the Church, civil law being taught by the University of Orléans. The faculty of theology, on the other hand, exercised its influence over the whole of western Christendom, for if papal power had its seat in Rome, it was in Paris, 'the most noble city of learning', that doctrine was defined. All the great theologians of the century were at Paris. It was 'the city of philosophers' in the words of Albertus Magnus, or indeed, according to a Dominican of the late thirteenth century, 'the new Athens'. The faculty of arts also enjoyed an exceptional reputation.

The students began their studies very young, sometimes between the ages of twelve and fourteen, attending the Faculty of Arts for six to eight years and obtaining degrees. Once they had received the baccalaureate (Bachelor of Arts degree), they could obtain the licence, the *licencia docendi*, which authorized them to teach. Then they might become masters, a degree attainable only at the age of twenty. Theological study lasted much longer: it needed fifteen years to obtain a doctorate, which meant one was thirty-five at the youngest, if all went well. Theology was the concern of professionals, charged with the establishment and transmission of knowledge. According to Aristotle. an essential feature of a science is the possibility of its being taught, and in this theology acquired one of the characteristics of being a science.

The career of Brother Thomas, the Angelic Doctor

Thomas Aquinas was born at the end of the year 1224 or at the beginning of 1225 into a family of the feudal nobility, in the fortress of Roccasecca, near Naples. He was the youngest son of Landolph of Aquino, who in 1230 offered him as an oblate to the great neighbouring abbey of Monte Cassino, the cradle of the Benedictine order. Political conflict led him to leave the abbey in 1239, theoretically just for a while, but in practice for nearly all his life. He was then sent to Naples, where he began his university studies, and spent five years there. In 1244, despite the violent opposition of his family, which did not hesitate to have him locked up, he entered the Order of Preachers, the Dominican order. In 1245 he arrived in Paris, at the priory of Saint-Jacques, and pursued his studies in theology, under the direction of Albertus Magnus in particular. He followed his teacher when in 1248 he was sent to Cologne to head the *Studium generale* established by the order. Thomas stayed until 1252, when he returned to Paris to teach and to prepare for his master's degree in theology. He made commentaries on the Bible and on the *Sententiae* of Peter of Lombardy and received his degree in 1259, when the Pope called him to the Curia as a teacher and he stayed there until 1268. This period in Italy was particularly fruitful: he began the composition of the *Summa Theologica*, wrote commentaries on several works of Aristotle's and some of his commentaries on the Bible.

He was already well known in the world of the schools when in 1268 he returned to Paris, where he returned to his chair in theology. He had been called on by his order at a time when the University was going through a serious crisis, which arose from contrasting interpretations of Aristotle and from a fundamental challenge to his ideas. From 1269 to 1272 he was very active, fighting both the radical Aristotelians and the conservatives who were hostile to Aristotle altogether. He finished the *Summa Theologica*, published various commentaries on Aristotle and on the Bible, and various other works, among them the *De unitate intellectus contra averroistas parisienses*. In 1272 he once again left for Italy, to take charge of theological studies at Naples. He died on his way to the Council of Lyons on 7 March 1274. His ideas, partly condemned in 1277, rapidly returned to importance. He was canonized in 1323.

The Scholastics

Theology acceded to the ranks of the sciences by virtue of its language and its methods too. University teachers, at least in their professional life, read, wrote and thought in Latin, in a world where vernacular tongues were now respectable. French was the language of daily life, and poets, *romanciers* and moralists, even historians, expressed themselves in French. The renaissance of the theatre found expression in the common tongue. The century of Aquinas was also that of Rutebeuf, of Joinville and of the *Roman de la Rose*. Latin became the language of the schools, a scholastic Latin which had to be useful and technical, suitable for scientific discourse and therefore, above all, precise. It was essential to determine the proper meaning of words. Without care for literary style, identical phrases served to introduce different parts of a text. *Sed contra*, for example, usually served to announce the third part of a question opened by *utrum*. It was a scholarly language, the work of the teachers of arts as much as the theologians, more suitable for demonstration than for emotional effect. In the thirteenth century, monotonous scholastic Latin seemed an efficient tool. It was far from being a dead language fixated on an antique model. It was the language of translators who lacked the equivalents for terms in Greek or Arabic, the language of authors who moved in a universe very different from that of Cicero, and it had to develop its own vocabulary. The instrument of a learning based on the analysis of texts, scholastic Latin was the scientific language of the thirteenth century.

In a profoundly Christian world, where the source of all religious truth was a book, the Bible, in a world where science was the inheritor of a prestigious and venerated past, one can understand that the first steps of scientific learning were founded on the study of the *auctoritates*, a term difficult to translate, meaning both authors and authorities. The *auctoritas* is a work which serves as a reference, from which one quotes, upon which one writes commentaries. For commentary was multiplying: commentaries on the Bible and on Aristotle, on the works of the Fathers of the Church and of Ptolemy, or on those of Peter of Lombardy, teacher at Paris in the twelfth century and author of a *Book of Maxims* before becoming bishop of Paris, or of John of Halifax, without doubt the most famous of mathematicians and astronomers in thirteenth-century Europe. The Middle Ages created its own authorities. The new works, the treatises and summas are themselves full of references which serve as a basis for argument.

For the Scholastics, the study of texts was the foundation of knowledge and of science. The same methods were applied to Scripture as to profane works. Thomas Aquinas produced commentaries on Aristotle's physical works, the *Physics*, the *Treatise on the Sky* and the *Treatise on Generation and Corruption*, as well as commentaries on the Bible.

In *Towards Understanding Saint Thomas*, Marie-Dominique Chenu notes that in the first twelve questions of the *Summa Theologica*, Thomas uses 160 citations: fifty-five from Aristotle, forty-four from Augustine, twenty-five from Dionysius, twenty-three from the Latin Fathers, four from the Greek Fathers and nine from profane authors.

This kind of intellectual approach must be understood in the context of a civilization without its own scientific culture, the meeting-point of two traditions founded on the book. The supreme authority was still the Bible, the Word of God which one had to be able to understand and to explain; Saint Augustine himself had recognized the need to have recourse to Classical culture in order to have access to Scripture. The scriptural heritage of the Church Fathers met with another just as literary.

Scholastics

The method of teaching and of exposition used in the schools, scholastics began with the *lectio*, the reading, a term which then had a specific technical meaning. The teacher, who might be a simple Bachelor of Arts, read the text, but also commented on it.

In this reading three levels of interpretation were distinguished: according to the *littera*, the letter, a simple explanation of the words and phrases; according to the *sensus*, the sense, in which one looked for the meaning of the text, which might even mean translating it into a language those being taught might more easily understand; and finally the discovery of the deeper meaning and the acquisition of a true understanding in the *sententia*. The first indispensable stage of scholastic learning, the reading gave rise to the *quaestio*, the question. It was born from difficulties encountered in the text and then became the framework, the method for dealing with a topic under discussion, not because there were real doubts, but because the procedure allowed better development of an argument. Beyond this, the question was transformed into a *disputatio*, a rarer exercise, during which the teacher, always a doctor, would launch the debate with an exposition. Then the propositions would be discussed by the other teachers, masters and bachelors in turn, followed by the students. The bachelor-student of the teacher who had made the exposition would reply to the questions, while the teacher kept silent. In a second session, the teacher would 'determine' the issue himself; he took up the arguments and sorted them out, drawing his own conclusions.

Aristotle

In the twelfth century the Christian West had been hit by the weight of Arab and Antique science, a mass of learning which in great part consisted of commentary. The period which saw the establishment of the universities saw 'the growth of Aristotelianism'. At the beginning of the thirteenth century Europe possessed most of Aristotle's writings. First of all the logical works, the works of the *Organon*, the *Analytics*, the *Topics* and the *Refutation of the Sophists* being added to the *logica vetus*. Then the *Libri naturales*: the *Physics*, the *Treatise on Generation and Corruption*, the *Treatise on the Sky*, the *Treatise on Meteors* and the *Parva naturalia*. Large parts of the *Metaphysics* and of the *Nichomachaean Ethics* had also been translated. The work of translation continued throughout the century. Aristotle did not arrive alone: he was accompanied by Greek, Jewish and Arab commentators. There were translations of Porphyry's *Isagoges*, and of the works of al-Farabi, Avicenna and Averroes.

Among the Greek and Arab works which invaded the West, that of Aristotle had a special place. It furnished a coherent system of explanation of the world, based on a scientific method founded on the *Organon*.[2] There was an enormous gap between the works known to twelfth-century Europe and the monumental work of the Greek philosopher. For the thought of the West in full effervescence, the discovery of Aristotle was a revelation, a scientific but also a philosophical revelation which brought new tensions, posing in an entirely novel manner the question of the relationship between faith and reason. Aristotle's physical works, the *Libri naturales*, in principle posited an eternal uncreated world radically different from that of the Christians.

Aristotle's four causes

Aristotle's work constitutes a real philosophical encyclopaedia which deals with all fields of knowledge: morals, politics, poetics, logic, natural history, physics and metaphysics. One of the dominant themes is the theology of the four causes. In the creation of an object, there are involved

- the material cause: the material;
- the efficient cause: the craftsman who acts on the material;
- the formal cause: the form given to an object;
- the final cause: the use for which the object is intended.

These four causes are found in all natural processes: substances, made up of matter and form, are subject to all kinds of changes. The heavenly bodies, for their part, do not undergo change, and are animated by a regular and circular movement. Beyond, at the edge of the world, the first cause, God, the prime mover, remains motionless and eternal.

The excited fascination with Aristotle experienced by young Masters of Arts who often opposed the theologians led to conflicts within the University.

233

Not much is known about the reception of Aristotle at Paris in the first decades of the thirteenth century. The Masters of Arts left far fewer writings than the theologians; younger, occupying a transitional status, their courses, it seems, were rarely followed by publication. It is through the statements and accounts of those hostile to it that we are best able to sense the progress of Aristotelianism within the University of Paris. In 1210 a provincial council held at Paris forbade the 'reading of the natural books of Aristotle or their commentaries'. There are almost identical terms in the Statutes granted by Robert de Courçon (1215). 'Reading' in its technical sense meant teaching. Only the teaching of the texts was forbidden, not the use made of them by University teachers to support their own writings. These steps seem to have been taken in response to the danger to the faith represented by the physical and metaphysical works of Aristotle and his Arab commentators. These measures were taken to defend a certain tradition in the study of the sacred texts but could not match the threat represented by the arrival of this new science and this new philosophy. In Aristotle's eternal and uncreated world God is the prime mover, the ultimate cause of all movement and of all change, with nothing to indicate that he has an interest in Man, nothing to suggest a relation of fatherhood. As far as we know, these multiple inter-dictions had their effects. Until the 1240s Parisian arts students only heard lectures on the *Organon* and the first three books of the *Nichomachaean Ethics.* Paris is a particular case where it took longer than elsewhere to overcome the intransigence of the authorities, perhaps because of the intensity of intellectual activity which developed there, which was surely the result of the weight of the faculty of theology, the guarantor of sacred doctrine. At Toulouse, where the Pope had created a university just after the Albigensian Crusade, 'to purge the country of heretical depravity', a little work praising the institution declared: 'The works of natural philosophy forbidden at Paris are read here, and those who wish to study the secrets of nature may hear the lecture.' There is some discussion as to whether this is a propaganda circular or a piece of school-work, but it says loud and clear that the works forbidden at Paris were studied elsewhere. The novelty is less striking when viewed from Oxford where, with the agreement of the hierarchy, the *Libri naturales* had been taught since the beginning of the century.

The earliest evidence for the teaching of the new Aristotle at the faculty of Arts in Paris comes from a young teacher trained at Oxford, Roger Bacon, who obtained the degree of Master of Arts at Paris around the year 1240. He taught there until about 1245, and commented on various works of Aristotle's until then forbidden, among them the *Physics* and the *Treatise on Generation and Corruption.* Roger Bacon's time in Paris overlaps partly with that of Albertus Magnus, the future Universal Doctor, who came to get his degrees in theology. Partly qualified from 1240 to 1242, he became one of the Dominican Master regents from 1242 to 1248. His great commentaries on Aristotle would come later, but he was already making much use of the *Libri naturales* in his theological works. Albert is not an isolated case among the theologians. This turnabout can be explained: the hostility of the theologians diminished across the decades and rejection of a work to which indeed everyone had access was followed by the need to come to terms with it and perhaps the simple attraction of an exceptional body of thought.

Theology as a Science

In the middle of the thirteenth century, when Aristotle's work was beginning to be accepted almost in its entirety, theology dominated the structures of the University of Paris, standing at the summit of scientific education. It had its place too in the 'classifications of science', a literary form which flourished in the twelfth and thirteenth centuries. It seemed necessary to introduce some order into knowledge. The Christians of the West had examples of these 'classifications' at their disposal in the translations of Aristotle or al-Farabi. There also survived a tradition deriving from late Antiquity, for Boethius (*c.*480–524) had already proposed a classification of the sciences which served as a model for a long period. He already placed theology at the summit of the hierarchy of knowledge. At Paris, as early as the first half of the twelfth century, Hugues de Saint-Victor gave a classification in his *Didascalicon.* For him, 'philosophy is divided into theory, practice, mechanics and logics, and these four branches comprehend all scientific knowledge.' And crowning theoretical knowledge was theology. Around 1150 Dominicus Gundissalinus, archdeacon of Segovia, wrote a *Treatise on the Division of Philosophy* which gives a classification of the sciences showing the influences both of Greek thought and of Arabic examples. In it theology has its place at the summit but is clearly distinguished from human knowledge. The last of the medieval treatises devoted to this topic is that of Robert Kilwardby, *On the Origin of the Sciences.* This English Dominican, Master of Arts at Paris around 1250, was writing to educate the students of his order; like his predecessors, he put theology at the head of the hierarchy.

In all these classifications, neither the concept of science nor the concept of theology is clear. The scientific thought that Hugues de Saint-Victor wished to grasp in its entirety included the construction of arms and the arts of the theatre, and Kilwardby included cooking and building. More than a classification of the sciences, what these works offered was more of a prospectus of human knowledge. Theology was without difficulty ranged among the sciences, but with greater rigour came a more precise concept of science, and the place of theology was then open to challenge.

Thomas Aquinas

Theology possessed certain scientific features, but was it really a science? In 1245, when Brother Thomas Aquinas arrived at the priory of Saint-Jacques at Paris, aged about twenty or so, this question was the order of the day. Apart from the status of theology itself, it posed the problem of the definition of science, which until then doesn't seem to have been dealt with in the Christian West. Alexander of Hales, a secular teacher who had become a Franciscan, had made his point of view quite clear: theology wasn't a science.

> It should be noticed that there is a science of the cause and a science of that which is caused. The science of the cause has no end but itself, while the science of that which is caused does not exist of itself, seeing that caused things relate to the cause of causes and are dependent on it. From this it follows that theology, which is the science of God, and has for its object the cause of causes, is only directed upon itself; and that consequently the name science is properly applied only to the science of caused things, while that of wisdom should be reserved to the science of the cause of causes. This is why Aristotle himself affirms that the first philosophy, which is directed only upon itself and treats of the cause of causes, should be called wisdom. For the same reason the doctrine of theology, which transcends all sciences, certainly merits the same title. (Alexander of Hales, *Summa Theologica*, the Theology, v. 1266)

Alexander the Franciscan had used the authority of Aristotle to deny theology the status of a science and was thus led to suggest limits to the concept of science, distinguished from that of wisdom. Basing himself on the same authority, Thomas proposes the opposite, arguing that theology is indeed a science.

He develops his argument on three different occasions, in his first work, the commentary on the *Sententiae* of Peter of Lombardy, the basis of his teaching as a Bachelor between 1254 and 1256, and then in his commentary on Boethius' *De trinitate*, also dating from 1256. He takes up the argument for the third time in the first articles of the first question of the *Summa Theologica*. Although the argument is elaborated at greater length in the commentary on *De trinitate*, we will look at the text of the *Summa*, the first part of which was composed between 1266 and 1268, because it shows the fruits of later reflection, and the more influential phase of his thought.

Already for his predecessors, the science of divine things was divided into two parts. First of all there was the Greek inheritance, now called 'natural theology', a part of philosophy which reasoned from tangible effects to divine causes. Aristotle's metaphysics is one example. Then there was sacred doctrine, as it was called at the time, a new discipline which could draw on the facts of natural theology, as is shown by the *Summa Theologica* itself, but whose essence was to argue from articles of faith taken as axioms. This division had two epistemological consequences. To justify the existence of two sciences dealing with the same subject, in this case God, Thomas Aquinas has to argue that the sciences are not distinguished so much by the diversity of their objects but rather by the difference of their principles. The cosmographer, *astrologus* in Thomas's text, and the physicist, the *naturalis*, both demonstrate the spherical shape of the Earth, the first by mathematics and the second by arguments taken from the nature of things.

Once this point has been accepted, it still remains to be shown that theology – sacred doctrine – is a science. Thomas Aquinas then develops a thesis to explain the paradoxical situation of a scientific discipline whose principles are given by revelation. He takes it for granted that a theology based on Revelation defines, reasons and proves on the basis of articles of faith taken as principles; theology cannot possibly ground these itself, otherwise it would be no different from natural philosophy and Revelation would have been unnecessary. These principles are not self-evident or all the world would accept them. The solution must therefore be to

---

Utrum sacra doctrina sit scientia? Is theology a science?

Thomas Aquinas presents a certain number of arguments for and against, and then replies:

> I answer that sacred doctrine is a science. We must bear in mind that there are two kinds of sciences. There are some which proceed from a principle known by the natural light of the intellect, such as arithmetic, geometry and the like. There are some which proceed from principles known by the light of a higher science. Thus the science of perspective proceeds from principles established by geometry, and music from principles established by arithmetic. And in this way sacred doctrine is a science, because it proceeds from principles established by the light of a higher science, namely the science of God and the blessed. Hence just as the musician accepts on authority the principles taught him by the mathematician, so sacred science believes the principles revealed to it by God. (St Thomas Aquinas, *Summa Theologica*, First Part, Q1, Art. 2)

---

consider articles of faith as being founded by a superior science, the science of the blessed. The beatific vision, an intellectualized version of Paradise, is not only the ultimate goal of theological speculation, but is also its epistemological foundation. Aquinas explains his position by means of a significant analogy: the relationship between revealed theology and the science of the blessed is the same as that between music and arithmetic, for music does not demonstrate its basic principles, which are provided by mathematics.

Sacred doctrine therefore does not have to argue for its principles; they are a matter of faith. On the contrary, it argues so as to demonstrate other truths that follow from them: 'I answer that, as other sciences do not argue in proof of their principles, but argue from their principles to demonstrate other truths in these sciences, so this doctrine does not argue in proof of its principles, which are the articles of faith, but from them it goes on to prove something else, as the Apostle from the resurrection of Christ argues in proof of the general resurrection' (St Thomas Aquinas, *Summa Theologica*, First Part, Q1, Art. 8).

This same solution, called subalternation,[3] integrated theology into the body of the Aristotelian sciences while still maintaining its revealed character. On the basis of a few lines of Aristotle, Aquinas was led to develop a whole theory which defined the status of the sciences through this concept of subalternation. It was a brilliant but fragile solution, which could be challenged by any change in the intellectual landscape, as much in the field of theology and of epistemology, and therefore of philosophy, as in the field of the sciences.

The Thomist theologians of the first generation, the pupils of the Angelic Doctor,[4] followed his teaching faithfully, except on one point. For them theology was not a science in the exact sense of the term. How can one explain this reversal, this abandonment of the master's thought on a point which he had established with so much care? No research has been done that would allow one to give an altogether satisfactory answer. Too many points remain badly understood, too many texts unpublished, too many sources unexplored. One can only put forward the most likely factors.

First of all there were the uncertainties in Aquinas' thought. Theology's status depended in particular on the fact that it drew its principles from a superior form of knowledge, the science of the blessed, which is to say Revelation. But if theology did take its principles from the science of the blessed, a point accepted by all, then these principles were not a consequence of Revelation, they were part of it. There was therefore no subalternation in the exact sense. This weakness is important, and explains why Aquinas' thought wasn't followed, though it doesn't tell us when his point of view was actually abandoned.

The Condemnation of 1277

In addition to the internal weakness of Aquinas' thought, there were external factors. In a decree of March 1277 Étienne Tempier, the Bishop of Paris, condemned nineteen theological and philosophical propositions described as 'execrable errors which certain students of the faculty are not afraid to treat and discuss in the schools'. The holders of the doctrines condemned had a week to recognize and abandon their errors, on pain of excommunication. Composed in haste, the document mixes together all sorts of propositions, set out in no particular order, but in its confusion it expresses a violent hostility to the Aristotelianism which had become widespread since the middle of the century. The period of condemnation of the *Libri naturales* had been followed by the triumph of a Christianized Aristotelianism. Using Aquinas' work, some were attempting to integrate a biblical vision of man and salvation with an Aristotelian conception of the material world. In the 1260s a more radical Aristotelianism developed within the Faculty of Arts, whose most remarkable representative is Siger de Brabant. There is still discussion about how much these heterodox Aristotelians were influenced by the commentaries of Ibn Roch, known to them as Averroes. Starting from a very precise reading of Aristotle, in the technical sense of the term, certain members of the Faculty of Arts believed they saw contradictions between philosophical truth and revealed truth. Even though for Siger the limitations of the human mind prevented it from reaching the truth, and despite the fact that Revelation still always won out when revealed, truth was faced by the truth of reason, the dangers to the faith quickly became clear. The subjects most discussed were the eternity of the world and especially the individuality of the human soul. Aquinas vigorously attacked the positions of the heterodox in his treatises on the *Unity of the Intellect* and the *Eternity of the World*. The threat roused the anti-Aristotelian party, and Étienne Tempier's decree was aimed at all Aristotelians, the moderates as well as the radicals. It attacked certain positions which had been held by Aquinas, who had died three years to the day before the publication of the decree. None of the articles condemned dealt with the question of whether theology was a science, but the condemnation of Aristotle certainly damaged the credibility of the system Aquinas had constructed. To defend it, reliance could be placed only on what was solid. Above all, the episcopal condemnation marked a rupture: institutionally, theology was no longer at the side of the other sciences of the Aristotelian corpus.

This event was part of an older conflict and followed others, but it was remarkable for its violence and the memory it left behind. The physicist and historian of science, the Catholic Pierre Duhem, considered the condemnation to be the birth certificate of modern science, the act which definitively broke the links between science and theology and freed them one from the

239

other. The Church could be thought of as the initiator of a movement which started within medieval thought and would end with classical and contemporary science. Duhem's argument is attractive, but for it really to be convincing, it would have to be shown that Western science was born on the morrow of the condemnation of 1277.

Pierre Duhem and the birth of modern science

Pierre Duhem (1861–1916), a French physicist and chemist, gained recognition in his own lifetime with work in the fields of thermodynamics, chemistry, physics and hydrodynamics. He taught at the universities of Lille, Rennes and Bordeaux. The intransigence of his positions and his conflicts with Marcellin Berthelot kept him far from the capital. In our own day Duhem is better known for his historical and philosophical work than for his work in physics and chemistry. His most important book, in ten large volumes, is *Le Système du monde: histoire des doctrines cosmologiques de Plato à Copernic*, which is still a standard work, even though marked by the author's ideological positions. Clerical, nationalist and conservative, Duhem opposed the dominant scientism of his day with a form of Christian positivism. For him, scientific theories could not claim an absolute truth, but should limit their ambition to the provision of a rational picture which could explain the phenomena. By limiting science's realist claims and forbidding it to restrict God's creative power, Christian metaphysics could then be seen as permitting the growth of real science. The interventions of the ecclesiastical authorities, usually presented as examples of obscurantism, now appear, in Duhem's account, as a rather paradoxical anticipation of positivism.

This is how he writes of the condemnation of 1277:

> If we had to assign a date to the birth of modern science, we should certainly choose the year 1277. Understood as a condemnation of Greek necessitarianism, the condemnation led many theologians to affirm as possible by virtue of the omnipotence of the Christian God, philosophical and scientific positions hitherto held to be impossible by virtue of the essence of things. By allowing new mental experiments, the theological conception of an infinitely powerful God freed the mind from the finished framework in which Greek thought had imprisoned the Universe. (P. Duhem, *Études sur Léonard de Vinci*, 1906–9)

Though Duhem's views have been challenged and are no longer accepted as they stand, he vigorously demonstrated the importance of medieval science and posed the question of the origins of modern science in a novel manner.

The 'Other' Science

During the third quarter of the thirteenth century, when Thomism was at its apogee and events were preparing for Étienne Tempier's decree, Western Europe developed a science. Having plagiarized and more or less assimilated the Graeco-Arab models, original scientific thinking sprang up in different places and under very different conditions. From 1254 to 1270 Albertus Magnus undertook the immense work which brought him the title of 'Universal Doctor'. It was based on a commentary on Aristotle, whose knowledge he wished to put at the disposal of Christians. But rather than a literal commentary, such as Thomas Aquinas produced, he paraphrased Aristotle's text and introduced the remarks of other commentators as well as the results of his own observations. His own original scientific work was as a naturalist. A great observer, he wrote a *Treatise on Plants*, based on an apocryphal treatise attributed to Aristotle, in which he attempted to classify plants and to understand their physiology, while also giving practical advice on vinification and the storage of manure. His *Treatise on Animals* begins with a commentary on Aristotle, but he develops much more individual views. He performed dissections on the eye of the mole and on the scorpion, whose nervous system he studied, and compared the eggs of birds and fishes.

Albert was one of the great theologians of the thirteenth century. He taught at Paris, but, unlike the situation in theology, Paris was not the centre

---

Roger Bacon (*c*.1216 – *c*.1292)

Roger Bacon began his studies at Oxford, where he obtained his MA degree, then went to teach at Paris. He returned to Oxford around 1247 and joined the Franciscans in 1257. A disciple of Robert Grosseteste, he left a body of scientific work essentially devoted to optics, but he is known above all for his statements in favour of science. He gave experience a predominant place, and said that man could make machines that would propel themselves on land, on the sea and in the air. He thought that all human knowledge should be placed at the service of the Church to conquer the infidel and the Antichrist, whose coming he thought was near. In fact Bacon's thought was less innovative than some of his formulations might lead one to think. His idea of experience is not that of present-day science. According to him the machines he mentioned, apart from the flying machine, had already existed and surely in his own time existed in other places which he didn't know. Extremely open to innovation, he remained profoundly anchored in his own time. He was not the visionary genius that some have seen in him.

---

of the world in the case of science. Refusing Aristotle, the Franciscans of Oxford had affirmed their confidence in experiment. The most famous of all was Roger Bacon, who announced that 'reasoning proves nothing, everything depends on experience.'

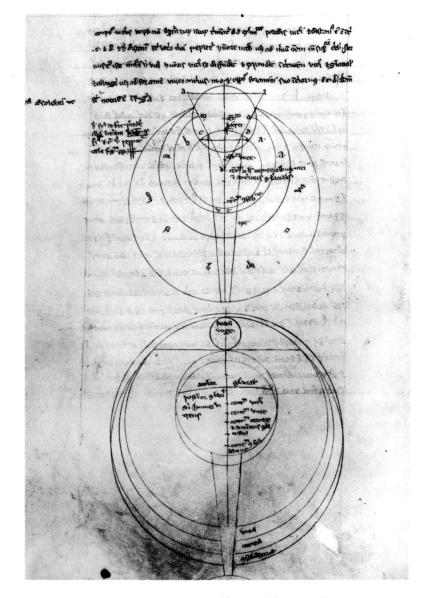

Roger Bacon's optics. With the help of geometrical diagrams Bacon reconstructs the passage of light through the eye, allowing him to show the curvature of the refractive layers, that is to say the parts where the light undergoes the phenomena of refraction (cf. pp. 315–43). (Drawings in R. Bacon, Opus Majus, thirteenth century. British Library, London.)

His work proves to be less original than his legend leads one to think. He had a profound admiration for a certain Pierre de Maricourt, whom he considered to be one of the greatest mathematicians of the time. Apart from this praise by Bacon, we know of Pierre de Maricourt's life only what he put into the two known texts by him, the *Epistola de Magnete* and a treatise on the astrolabe, which was published only in part. Pierre de Maricourt wrote his letter in 1269, beneath the walls of Lucera, then besieged by Charles of Anjou, the brother of St Louis. Many think he was a military engineer in the service of the prince, one of the *ingeniatores* that all great armies then had. It is a plausible hypothesis, for no trace of him is to be found in the world of the universities. The *De Magnete* is a small work, whose printed version comes to no more than fifteen pages, but its importance has nothing to do with its length. Its first original feature is the absence of all reference to *auctoritates*. The explanation of this is that Pierre de Maricourt's account relies essentially on observation and on experience. Nor is this experience

---

Pierre de Maricourt's *De Magnete* (1269)

The northern part of the stone attracts the southern and vice versa, as has been stated, in which attraction the stone of the stronger virtue is the agent and that of the weaker the patient. I think the cause of this phenomenon may be explained in this way; for the active agent strives not only to join the patient to itself but to unite with it, so that out of the agent and the patient there may be made one. And this you can find out in the case of this marvellous stone in this manner. Take one stone which you may call by AD, in which A is the north and D the south point. Divide it in two parts so that two stones are made from it. After this place the stone which contains A on water so that it may float; you will see that A turns towards the north as before. For breaking does not take away the properties of the parts of the stone if it is homogeneous. Hence the part of this stone which is at the point of fracture which is B, must be the south. Let, then, this stone regarding which we have been speaking be represented by AB; as to the other stone which contains D, if it is placed on water you will see that D is south as at first, because it turns towards the south, if placed on water. But the other part near the fracture, which may be designated by C, will be the northern; this stone will therefore be CD; let the first stone AB be the agent, CD the patient, and thus you see that the two parts of the two stones which, before the separation, were continuous in one stone, after the separation were found to be, one the northern and one the southern part. But if the same parts are again brought together, one will attract the other until they are joined at the point BC, where the break took place. (In H. D. Harradon, 1943)

---

in the Aristotelian sense of the term, meaning the confirmation of a fact by its reproduction in Nature, but true scientific experiment, produced and repeated at will. The act of breaking a magnet to examine the position of the poles in the two fragments is an experimental operation, a new break-through which appeared outside the University.

There were profound changes too in the field of optics. The Silesian, Vitello (c.1220–75), who studied at Paris and at Padua, seems to have worked mostly in Italy and probably in Bohemia. Well read in the works of the Ancient and Arab worlds, Vitello made experiments. He built his own para-bolic mirrors and, on the basis of work by Ptolemy and al-Hazen, he de-veloped a method to measure angles of refraction (see p. 328) in air, water and glass while varying the angle of incidence of the rays. He tried to give this a mathematical expression.

To the same period must be assigned the *Liber Jordani de ratione ponderis*, which developed work done by Jordanus de Nemore during the first half of the century. He was particularly interested in bent levers and in the inclined plane. Medieval statics was born, and these works of the mid-thirteenth century influenced sixteenth-century students of mechanics, Simon Stevin (1548–1620) in particular.

Another novelty was the *Theoria planetarum* of Campanus of Novara, a character of whom little is known, except that he wrote his work in the third quarter of the thirteenth century, probably at the Papal Court. As distinct from earlier treatises 'On the Sphere', which often presented highly simpli-fied versions, his work was the first to present in Latin a highly detailed exposition of Ptolemy's system, and on certain points, such as its way of calculating the size and distances of the planets, it shows improvements on the *Almagest*. Campanus of Novara was the first to offer a model of an 'equator', a complicated apparatus intended to reproduce the movement of the planets and of the stars on the basis of the Ptolemaic theory.

For astrologers it was essential to know the positions of the heavenly bodies in the sky, and the equator was one solution, the other being given by astronomical tables. The progress of observational astronomy had allowed new ones to be drawn up.

These examples express the emergence of a new science. It referred often to the Ancients and some texts, such as those of Albertus Magnus, are still stuffed with references to Antiquity. Others would ignore them, like Pierre de Maricourt, but with observation and experiment and with their recourse to mathematics, they went beyond what had come from the Greeks and Arabs. Science was no longer satisfied with commentary or with a résumé of the learning inherited from the past. Of course the language remained Latin, of course reverence for Aristotle survived, but the ambience changed and rea-soning on the *auctores* was succeeded by reasoning on the data acquired by the scientist.

Thomas Aquinas' attempt to make theology a science ended in failure. While his theological and philosophical work was generally rehabilitated after the events of 1277, for he was canonized in 1323, his theory of the scientific nature of theology was not retained. A separate discipline, theology, which finds its basic givens in faith, cannot be a part of a science which is grounded in observation and in reason. Alexander of Hales enjoyed a posthumous triumph. The epistemological foundations of Thomas's position were too weak in themselves to withstand the attack launched by his adversaries on his whole system of thought. In addition, his idea of science rested on a very literary and scholastic conception of knowledge, which was challenged by the emergence of the new science.

Rather than looking for the reasons for this failure, which seem obvious, it is doubtless more important to ask why an intelligence as remarkable as that of Thomas Aquinas should have done everything possible to make theology a science. He never explained this. It would seem to correspond to the need to oppose to the organized and scientific knowledge of Aristotle and his commentators, then considered as a model, another knowledge just as organized, just as scientific but more true, which depended on the divine Word. It was possible just so long as science was based on discourse, defining itself on the scholastic model, the one familiar to Thomas Aquinas. To be a scientist was to reason on the Bible as on Aristotle, and in both cases to employ the same scientific language.

Aquinas' intention matched that of the religious authorities, to make theology a science, something that was taught, something that could be proved. To make theology a science was to guarantee effective means for conversion. The position of the Church, and of the Pope in particular, but of many other churchmen as well, depended on an ambiguity of vocabulary and an absence of definition in the concept of science, as is shown by the different 'classifications of the sciences'.

To constitute theology as a science, Thomas Aquinas had himself been obliged to give the concept of science a far stricter definition, defining it in terms of its object, its methods and its relations with other sciences. At the same time he gave a special status to theology, so as to distinguish it from the other disciplines and from philosophy. He helped to define science as it was being born in the West.

But this definition ran up against two obstacles. The first was a too Aristotelian and therefore dangerous definition of science (which explains the attitude of the authorities), with scientific truth threatening to collide with revealed truth and making a separation necessary. The second was the appearance alongside the science defined by Aquinas of the premises of another, founded on observation and experiment, which brought out the epistemological weaknesses of the Thomist argument.[5]

245

8

Algebra, Commerce and Calculation

Paul Benoît

In the late Middle Ages mathematicians in Florence and elsewhere had to teach merchants' children how to calculate. We shall see how they did their algebra and what followed from it.

Greek science was geometry: its physics reasoned and deduced but hardly ever calculated. In our own day calculation is the essential foundation not only of the sciences but also of technology and economic activity. Although this is a commonplace in the age of computers, the urge to put the whole world into equations goes much further back in history. First of all, though, you need to know how to solve an equation, to formulate it indeed. One can have long discussions on the value of Greek methods of calculation, on the possible antecedents of Diophantus' work; one can show how Archimedes and the Alexandrian mechanists were able to calculate; but it is none the less certain that algebraic calculation only developed in Christian Europe at the end of the Middle Ages and at the beginning of the Modern period. According to the commonly accepted view, algebra was born in Western Europe with Viète, the first it is said to use letters to stand for the unknown. Viète was able to make his own advances thanks to the rediscovery in the sixteenth century of the work of Diophantus of Alexandria. Can it have been as simple as all that? Should we call someone the father of algebra just because it was his system of symbolism that served as a model? Before

François Viète, others had already sought to express algebraic realities by means of symbols, among them Jérôme Cardan, Raffaele Bombelli and Nicolas Chuquet. With Nicolas Chuquet, a French mathematician of the second half of the fifteenth century, we enter a world very different from Viète's. Chuquet wrote not in Latin, the scholarly language of the Middle Ages and of the Classical period, but in French, and he was one of those mathematicians whose job it was, at the end of the Middle Ages, to try and drive some maths into the heads of merchants' sons. Chuquet was not an isolated case. There were mathematicians like him living and working in Italy, in Venice and Florence in particular, and they produced works which will be of undoubted significance for the history of arithmetical and algebraic calculation. 'Will be' because this history is still at an early stage. Work is being done by Italian, German, American, English and French researchers. It is an ungrateful task: the texts are numerous, and often very like each other, mostly in manuscript and sometimes badly written. Certain authors' thinking is sometimes confused and mistakes are frequent. For one Chuquet or Benedetto of Florence, there are many anonymous unoriginal authors – whose works may none the less contain clues to evolutionary development. This is a history which doesn't show us the archetypical great scientists; it is rather the history of a scientific milieu. We are not looking for the origins of algebra in Western Europe, but discovering the conditions in which a science of calculation was able to develop in a particular period – the last centuries of the Middle Ages – and in a particular milieu, that of the merchants involved in big business. Afterwards these conclusions will have to be integrated into a wider framework, the history of calculation in the Mediterranean world. Even if the conclusions are only partial, they may allow the problem of the relationship between scientific and socio-economic development to be put in clearer terms.

Mediterranean Trade

The growth of medieval Europe, with the expansion of agricultural production, the development of the towns and the spread of money, was accompanied by commercial expansion at every level. The increase in the numbers of local markets went hand in hand with the establishment of an international trading network. Western Europe traded in particular with the Byzantine and Muslim East, exporting both metals and money, receiving in return luxuries such as silk and spices from the Far East and certain items essential for its own textile industry, such as alum,[1] a mordant indispensable for the production of cloth and other dyed products. Western Europe also traded with the Muslim West: there was wheat from North Africa or Sicily, wool, leather and coral from the Maghreb, cloth from Italy and Catalonia, gold and spices from Africa, silver from Europe. Thanks to its traditions and its geographical situation, Italy had a central position in this international trade, which made the fortunes of cities like Genoa, Venice, Pisa and Florence.

In the twelfth century Venetians and Genoese came together and organized themselves so as to be able better to undertake their overseas operations. A capitalist would provide funds to an itinerant merchant, who would contribute his work and sometimes put in some capital of his own. Already you would have to know how to calculate, so as to share the profits and losses as agreed in the contract. During the thirteenth century, in the inland cities, great companies were set up on a permanent basis. The members of a family group and their associates would provide the *corpo*, the capital. They shared profits in proportion to their investments, and they would be responsible for any losses in the same way. The company would also accept investments from individuals, who received a fixed rate of return. Associations of this kind might grow to a considerable size, like those of the Bonsignori of Siena in the thirteenth century, and those of the Bardi and Peruzzi of Florence in the fourteenth. The great Florentine concerns had branch offices throughout the Mediterranean and the countries of Western Europe. For these, trade was accompanied by very substantial banking activity which brought them into contact with the greatest people of the age, with the Pope and other rulers who called upon their services.

A bill of exchange

'In the name of God, 18th December 1399, you will pay upon this letter at usance, to Brunacio di Guido and Co., CCCCLXXII pounds and X shillings of Barcelona, which 472 pounds and 10 shillings being worth 900 écus at 10 shillings and 6 pence the écu have been delivered to me here by Ricardo degl'Alberti and Co. Pay them in due form and put them to my account. May God keep you. Ghuighlielmo Barbieri. Greeting from Bruges' (in J. le Goff, 1986).

Usance was the usual period allowed for the payment of a bill of exchange between one place and another. Between Bruges and Barcelona in the fifteenth century, the usance was thirty days. Such a bill of exchange could cover money-changing, transfer and credit operations. It became one of the essential instruments of Italian commerce in the late Middle Ages.

In the middle of the fourteenth century Europe was struck by a very wide-ranging crisis. There was a demographic crisis: after centuries of growth population stagnated or began to decline as a result of the Black Death (1348). According to the best estimates, Europe lost around half its population in the course of a century. This took place, however, in a context of economic

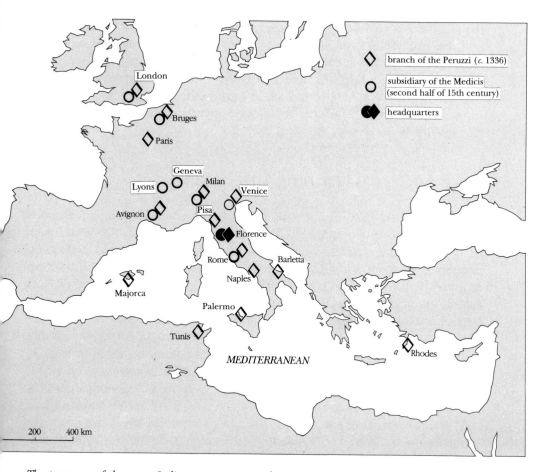

branch of the Peruzzi (*c.* 1336)

subsidiary of the Medicis
(second half of 15th century)

headquarters

The presence of the great Italian companies in fourteenth- and fifteenth-century Europe

depression and political conflict. Prices and production collapsed while the continent was ravaged by wars, most notably the Hundred Years War. The phenomenon was complex. Even before the Black Death had reached Florence, the Bardi and Peruzzi had gone bankrupt, victims of loans made to princes unable to repay them, among them the King of England. During the second half of the fifteenth century, business was reorganized along new lines. Companies with branches were replaced by companies with subsidiaries, that is to say one single capitalist group, a family group as before, controlling other companies, which were legally distinct: the family group was a holding company, as we should say now. The system was more flexible and local problems would not necessarily bring about the collapse of the whole enterprise. The commercial and industrial empire of Cosimo de' Medici (1389–1464) was built according to this model.

This scale of business demanded letter-writing and book-keeping, while the amount of capital involved demanded insurance. The last centuries of the Middle Ages witnessed the emergence of 'double-entry book-keeping'[2] and the proliferation of insurance contracts. Means of payment were also needed and the big Italian firms were also bankers. They developed the bill of exchange, a brief document in which one person promised to repay the sum of money advanced to him, at a later date, in some other place and in another currency. This was therefore a means of credit, of transfer and of exchange.

Learning the Business

'You mustn't be lazy about writing.' This statement by an anonymous fourteenth-century Florentine merchant is very understandable. With letters or figures, the merchant's work was done in writing. This was a business you needed to learn. It is possible to trace the basic outlines of the education received by the future Italian merchants of Pisa, Venice, Genoa and of Florence in particular. About the age of seven, the children went to an elementary school for two or three years; there they learnt reading, writing, and the rudiments of grammar. A fourteenth-century Florentine, Donato Velluti, wrote of his son: 'Having learnt to read in very little time, he became a good grammarian . . . then he went on to the abacus.' 'Abacus' was a word which had first meant the table on which counters were moved about to perform mathematical operations, but now it had taken on a wider meaning and was used in the sense of 'calculation'. Besides arithmetic, the masters taught 'things useful for business'. Pupils might be taught by a tutor or by a schoolmaster, who would have only a few pupils. So in the tax records of Lyons, Nicolas Chuquet, before appearing as an '*algoriste*' (an arithmetician), was described as a an '*escrivain*' (a writer or scrivener), the name given in Lyons to those who taught the children of patricians and great merchants. Luca Pacioli (*c.* 1445–*c.* 1517), the author of a celebrated *Summa arithmetica* printed at Venice in 1494, began his career as a tutor employed by Antonio Rompiani, a rich Venetian merchant. In the Italian cities, however, nearly all the little merchants-to-be were sent to school. In Florence in 1338, according to the chronicler Giovani Vilani, 'the children learning abacus and algorism in six schools numbered between a thousand and a thousand two hundred.' These are impressive figures for a city of 100,000 inhabitants, which was perhaps unusual on account of its importance as a trading city and centre of intellectual activity. In 1345, however, there were public schools for the teaching of the abacus at Lucca. In Milan in 1452, thirty-seven businessmen sent a petition to the Duke, asking him to finance the teaching of book-keeping to their children. In Genoa in 1486 it was the Wool Guild, the guild of woollen cloth manufacturers and traders, which opened a school.

The Florentine schools are the best-known, no doubt because of the importance of the city, but also because there the teaching of mathematics occupied a special place. Even the Venetians, who were Florence's competitors and often its enemies, recognized the Tuscan city's superiority in this field. It seems that the Florentine schools, the *botteghe dell'abbaco*, literally the calculation shops, were all private. In the middle of the fourteenth century Master Paolo dell'Abbaco owned his own school, and he left it to a colleague and friend. The bequest included the premises and all the equipment needed for teaching. More than any other source, his will throws light on the life of the Florentine mathematician in the fourteenth century. Drawn up in 1367, probably very shortly before its author's death, it shows us a well-off man, the owner of two houses in town and another in the country, who had a capital of 1000 florins at a time when a servant earned 10 florins a year, a master mason 40 and a notary around 300. This was a considerable fortune. His executors included a teacher of abacus but also a rich silk-merchant. What can be gleaned from the documents about his colleagues' situation show that Paolo was not an exceptional case. Less well-off than the great merchants with whom they came into contact, well-known teachers of abacus earned more than craftsmen, and this put them among the richer members of the middle class.

Others, on the other hand, had a lower standard of living, appearing in building accounts as measuring the work done and calculating the volumes of material used. Their pay, rather low, made up what they earned from teaching. A contract of 1517 gives the conditions of employment for a young teacher taken on by Francesco Galigai, a more famous schoolmaster who needed an assistant. This beginner was very badly off, and the minimal salary he received was comparable to a workman's pay. In Florence there was a group of teachers, professionals who lived from their mathematics and, more particularly, from calculation. They had a recognized and valued place in the life of the city. Towards the end of the fifteenth century, Luca Landucci, a Florentine, giving an account of 'the most noble and most valorous' men of his city, listed with them, and with Cosimo de' Medici, seven artists and two teachers of mathematics.

The Treatises

The textbooks give us some idea of what was taught in the schools. At Florence in the 1430s Paolo dell'Abbaco wrote an arithmetic for the use of merchants. Works of this kind proliferated in Italy, in Florence and Venice in particular. They were taken up by the printers; the first book of commercial arithmetic was printed at Treviso in 1478, and another three years later in Florence. Some of them were very successful: the *Nobel opera de arithmetica* by Piero Borghi, a Venetian, went through sixteen editions between 1484 and 1577. Luca Pacioli's *Summa de arithmetica, geometria, proportioni et proportionalita*, an enormous work which included commercial arithmetic, was printed at Venice in 1494. The first German commercial arithmetic was printed at Bamberg in 1482, a few years before the work of the most famous German teacher of calculation, Johannes Widman. At Nice in 1497, Francès Pellos published the *Compendion de l'Abaco* in the Nissard language. Francesch Sanct Climent had his practical arithmetic printed in Catalan in 1482. There is also a Provençal manuscript from the mid-fifteenth century. If Italy produced the most numerous and best-known works, the trend extended far beyond Italy and even beyond Mediterranean Europe.

In the present state of research, we know five French manuscripts belonging to this tradition, dating from the second half of the fifteenth century. The *Kadran aux marchans* was written by Jehan Certain in 1485, a year after Nicolas Chuquet finished his *Triparty en la science des nombres*, whose last part is entitled: *How the science of numbers may serve in doing business.*

All these works were written in the vernacular tongue and not in Latin, the language of most earlier and contemporary scientific literature. This clearly shows that these treatises were aimed at a different public, intended neither for the University nor for nascent humanism, but for a public for whom knowledge was not to be confused with the culture inherited from Antiquity. Their language, derived from both popular and scholarly languages, was not yet fixed: terms varied from one textbook to another, testimony to the youth of a discipline which had not yet defined its vocabulary.

They were written for a practical purpose. Chuquet wanted to apply the science of numbers to business, and Jehan Certain wanted his book to be 'guide, instruction and declaration to all merchants as to how count properly.' As for Borghi, he wrote his book for 'young people who will go into business'. Most of these treatises make explicit mention of their practical educational purpose.

The usefulness of mathematics for merchants, according to Jehan Certain's *Kadran aux marchans* (1485)

I would compare my treatise to a sundial, yet I would call it the merchants' sundial, because just as the sundial is the leader, guide and way of all manner of people to know the measure of time and of the day. So will this little treatise be guide, instruction and declaration to all merchants as to how to count properly, so as to take and to give in buying and selling to each according to his lawful right . . . this second part will speak of weights, measures, companies, exchanges and other contracts and is therefore necessary to anyone who would do business.

A New Art of Calculation

Another of their common features was another innovation: all made use of calculations on paper. The method was just becoming common in certain circles, among the astronomers as well as the merchants. The quasi-impossibility of calculating in Roman figures had for centuries made it necessary to use the abacus and to calculate with counters. These were still, at the Renaissance, the techniques used for the public accounts. The arrival of Arabic numerals (see p. 210) had already made a noticeable difference. The calculations were put down, but written in wax, in sand or dust, with the intermediate results being rubbed out as one went along, and the workings were lost. The spread of paper in the West, a much cheaper medium for writing than parchment, not only completely altered the material conditions of calculation but also changed the way it was done. The figures were written down and intermediate results were retained: new ways of laying out mathematical operations became possible, and with them, new methods.

A Common Fund of Knowledge

The treatises all begin with arithmetic and, with the exception of Nicolas Chuquet's book, go straight into commercial problems. Except for a few details, they are all planned in the same way. We can take the example of the *Kadran aux marchans*. It is divided into four parts. The first deals with Arabic numbers, continues with addition and subtraction and their proofs, followed by multiplication and subtraction and the proofs by 7 and by 9. He then goes on to fractions, which he calls 'broken numbers: reduction, addition, subtraction, multiplication and division, and he deals with simplification at the end of the chapter. Then there begins the second part, devoted, in the words of the author 'to weights, measures, companies and exchanges' (evidence of his practical aims), which deals with the rule of three and its applications. This part is divided into chapters following practical rather than mathematical criteria. Here the author deals with problems related to the compound rule of three (see p. 259). Then he writes, but much more briefly, about problems of simple and double false positions[3] and what he calls 'the rule of opposition and remotion'[4] before finishing, even more rapidly, with arithmetical and geometric progression. His economic concerns find expression in the title of the third part, 'Money and copper coin, gold and silver'. In it we find the methods of calculation a merchant must know so as to be able to deal with all the problems raised by the handling of money and precious metals. The fourth part, specific to this book, entitled 'Alloys and assaying', has a clearly technical character, giving methods for the refining of precious metals, and therefore falls outside the scope of our account.

The plan of Borghi's book is in many ways similar to that of Jehan Certain's: the numbers, the four elementary arithmetical operations, fractions,

The Abacus

The Middle Ages had several types of abacus, which were tables or tablets used for calculation, like the counting frame still used today in the Soviet Union and the Far East. The most common was the linear abacus, which took the form of a board on which there were lines which represented the units and powers of 10. An intermediary counter might be put between the lines, worth 5 between units and tens, 50 between tens and hundreds, etc. Counters placed on a line were worth one of what that line represented: three counters on the units line were worth 3, five on the thousands line 5000, etc. The cross indicates the thousands line, so 5807 looks like this:

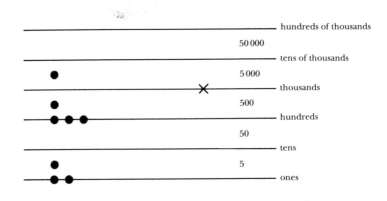

Addition and subtraction were easy, but the other operations were much more difficult. The addition of 17,617 and 4861, whose result is 22,478 looks like this:

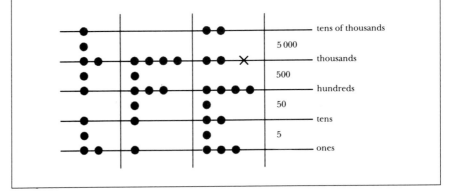

the rule of three and then metal alloys, *De ligar metalli.* The book ends with a series of problems, some relating to business, but others belonging to a different tradition. One such is the problem of the snake, which climbs up a tower during the day, getting a certain distance, but then falls back again at night, but not as far as it has climbed: the question is to know how many days it will need to reach the top. There are a few rare differences in presentation: multiplication and division come before addition and subtraction, and the paragraphs on false positions and 'opposition and remotion' are absent from Borghi, as is the final technological chapter.

The similarities, however, are greater, and they are to be found in all the other treatises. While influenced by differences in education, environment and interests, the authors constructed their books with little modification on the basis of a body of mathematical knowledge relating to similar commercial operations, which constituted a common fund of knowledge considered to be indispensable for the training of the future merchant.

The Operations

Positional notation,[5] a topic sometimes accompanied by a brief discussion of the different ways of writing numbers, does not receive a lengthy exposition, being treated rather as something already understood. The same goes for addition and subtraction. The problems are posed and solved in the same way as today, with only very minor differences. The examples given usually relate to money, which was then a little more difficult. Decimals were unknown and the currencies in circulation were split up in different ways. In the dominant system, one pound was worth twenty shillings and each shilling was worth twelve pence. The addition of two sums therefore might require the use of division and subtraction as well. Addition was proved by subtraction and vice versa.

Authors sometimes had difficulty in defining multiplication, but they performed it with ease. Many ways of multiplying existed; their names differ from manuscript to manuscript. One could multiply sequentially, a method which required many partial results to be carried over and which therefore risked more mistakes, or by *gelosia* or *carrat*, putting the intermediate results in a grille in such a way that there was no need to carry over, or finally one could use a technique very similar to the one in use today. Other methods existed, for the Italian abacists were men of fruitful imaginations.

Sequential multiplication seems to have been an outmoded survival from the days when calculation was done in sand with the intermediate results crossed out. The trace of the past is even more visible in division, the most difficult of all the operations. Until around 1460 in Italy or 1485 in France, this was done by successive divisions, striking out what had been divided and keeping the remainders. This is what Jehan Certain, in his *Kadran aux marchans*, called '*partir par gallée*' ('galley division'). The intermediate results were crossed out, rather than being wiped away as with sand.

The newer form of division, which like our own conserves the results of successive subtractions, completely transformed the way the operation was carried out. It was clearer and therefore safer, reducing error considerably. Multiplication and division were verified by the rules of 7 and 9.

Multiplication

The multiplication of 578 by 76,589 would have been done with the help of a matrix (on the left), and worked out as shown on the right. All one has to do is to multiply each figure of the multiplicand by each figure of the multiplier and place the result in the box located at the intersection of the columns at the head of which the figures are found. As there is no carrying over, the operation can be done in any order. The result is obtained by adding the figures along the diagonals. The result, read from top to bottom and left to right, is 44,268,442.

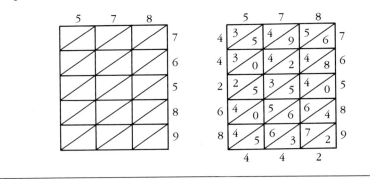

Old style division

Division done in the ancient manner, striking out the intermediate results (Bibliothèque Nationale, Manuscrit français 2050, v. 1460).

Remainder 17

Number to be divided ....
Number arising from the division ... 133
The product ....

leaves $34\frac{1}{2}$

The operation breaks down like this:

11
4̶5̶39
1
3̶4̶,

so 45 − 34 = 11.

Then divide 113 by 34.

1
7̶
1̶1̶1
4̶5̶39
13
3̶4̶4
3̶,

thus 3 × 3 = 9, taken from 11, leaves 2 giving 23; then 3 × 4 = 12, taken from 23, leaves 11.

The operation ends in the same way:

1̶1
2̶2̶
1̶1̶1̶7
4̶5̶3̶9̶
133
3̶4̶4̶4
3̶3

The calculation wasn't difficult, but the way it was laid out could lead to many mistakes. If the dividend and the divisor were large numbers, it took up a lot of space. It was often called a 'galley' division, because its shape resembled that of a ship.

Division in the modern style

Division in the modern style from the *Kadran aux marchans* (1485): 'One wants to divide 6753 pounds between 12 persons . . .

6753   67
0562   60
  12   75
            72
            33
            24
             9.'

Which gives a result of $562\frac{9}{12}$ or $562\frac{3}{4}$.

Only the layout and the absence of decimals distinguishes this from the method used in French- and English-speaking schools.

Fractions

After operations on integers, the treatises move on to fractions, *rotti* in the Italian texts, *nombres routz* or *nombres rompus* (broken numbers) in the French ones. The authors understood fractions even if they found them a little difficult to define. Reduction to a common denominator was commonplace, as was simplification. The addition, subtraction and multiplication of

---

The division of fractions

Jehan Certain in the *Kadran aux marchans* (1485): 'To divide a broken number (fraction), it is necessary first of all to reduce and to know the value of what you want to divide and what you want to divide it by and then divide as if they were whole.' The example given is that of the division of $\frac{1}{2}$ by $\frac{1}{3}$. He reduces them to the same denominator and arrives at the result 'this gives $\frac{1}{3}$, with a half left over', which is wrong.

Another manuscript (Bibliothèque Nationale, Manuscrit français 2050), from the south of France, around 1460:

$$
\begin{array}{r|r}
8 & 9 \\
\text{Divided} \dotfill \quad 2 & 3 \\
\text{by} \dotfill \quad 3 & 4 \\
\end{array}
$$

which gives $\dotfill \frac{8}{9}$

$$12$$

Here $\frac{2}{3}$ is to be divided by $\frac{3}{4}$. The author reduces the two fractions to the same denominator, giving $\frac{8}{12}$ and $\frac{9}{12}$. He shows the value of the denominator under the vertical line, and the numerators from the reduction are placed above the original numerators. One then divides one numerator by the other, which is the same thing as multiplying the fraction to be divided by the inverse of the fraction which is the divisor.
From another (Manuscrit français 1339) of around 1460:

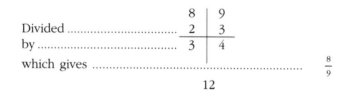

$$
\begin{array}{ccccccc}
 & 3 & & & & & \\
28 & & 8 & & & & \\
\frac{7}{8} & & \text{by} & & \frac{1}{4} & \text{(which gives)} & 3\frac{4}{8} \\
\end{array}
$$

---

fractions hardly seem to have presented any difficulties, but division remained a source of error for a long time. At the close of the fifteenth century, a good calculator like Jehan Certain still didn't know that he could divide by multiplying by the inverse fraction, although several of his contemporaries were indeed using this method.

The Rule of Three

With the exception of division, and particularly the division of fractions, the authors had mastered the elementary operations. With much experimentation and hesitation, they were creating something new. In very simple mathematical language, they produced the first formalization of the basic arithmetical operations as they have been done until our own day. But the greater part of these treatises is devoted to the rule of three, which Chuquet calls 'lady and mistress . . . of the proportions of numbers'. Most of the problems make use of it, because of its many applications in commerce: it helps calculate prices, value currencies, share the profits and losses of a company and many other things useful for merchants. Though they might all apply it to the same problems, not all of them state it, and therefore they do not apply it in the same way.

'The rule of three is called the rule of three because there are always three numbers, that is two similars and a contrary. And if there are more, they must be reduced to these three, and one must multiply the thing one wants to know by its contrary and then divide by its similar.' This is how the rule appears in the *Kadran aux marchans*. Jehan Certain continues with an example: 'How to say if 3 florins are worth 2 royal francs, how much 20 florins of Avignon are worth.' According to the rule, the 20 florins, 'the thing one wants to know' should be multiplied by 2 francs, 'its contrary' and divided by 3, 'its similar'. He writes the operation, and its solution:

If 3 ff 2 fr 20 ff
are worth 13 fr 6 s 8 d

This expresses an attempt at presentation and clarification making an abridged representation of the discursive statement, but the rule of three isn't stated as a ratio, which would have helped to simplify it.

More developed forms also correspond to these techniques, and Francès Pellos' *Compendion de l'Abaco* uses terms which would be familiar to a school-child of the twentieth century: 'If 4 are worth 9, how many will 5 be worth . . . multiply 5 by 9, which gives 45, which you divide by 4 to get 11 and a quarter.' Having presented the principle and practice of the rule of

259

three in discursive form, he demonstrates a method 'by which you may rapidly find what you are looking for', and the method here is different: 'If three and a half are worth 6, how many are four worth? Lay out your operation in the following way:

$$
\begin{array}{c}
\quad 12 \qquad 48 \\
7 \diagdown \quad 6 \text{——— } 4 \text{ the number to be divided is } \quad 48 \\
\diagup\times \\
2 \diagup \quad 1 \text{——— } 1 \\
\quad 7 \qquad 7 \text{ the divisor is} \qquad\qquad 7.'
\end{array}
$$

While today we would write $\dfrac{6 \times 4}{3.5}$ = Pellos writes $\dfrac{6}{3.5}$ as $\dfrac{12}{7}$; then he does the multiplication $\dfrac{12 \times 4}{7}$.

By reducing the fraction to an integer, he avoids calculations which might give rise to errors.

From the simple rule of three one moves on to the compound rule of three with the usual problems of exchanging the most diverse currencies. 'If 100 Modena pounds are worth 150 of Venice, and 180 of Venice 150 of Corfu, and 250 of Corfu 360 of Negrepont, how many Modena pounds are worth 850 of Negrepont?' This example from Borghi uses cities where Venetian trade was established. Chuquet has a similar example, but he chooses Paris, Lyons and Geneva.

The rule of three was, however, unable to solve all the problems faced by mathematicians. In most cases the treatises devote a chapter or two to the rule of single and double false position, and there might be another on the extraction of square and cube roots; but though these topics are evidence of the mathematical knowledge of the teacher who wrote the textbook, they were not directly relevant to commercial practice. They are given very little space.

It is this mathematical knowledge, rather scanty to our eyes, which constitutes the fund of knowledge common to all the treatises: elementary operations on whole numbers and fractions and the rule of three. Teaching this, however, poses certain problems, some of them difficult to appreciate, related to the novelty of what was being said. Ignorance of the signs, +, − and = doesn't seem to have been a major problem for teaching. On the other hand, the absence of decimals necessitated the constant use of fractions which were very difficult to handle. The pedagogical effort put into the treatises is plain. The chapters begin with the exposition of a 'rule', that is to say they provide a method to solve a certain kind of problem. The rule is followed by numerical examples, starting with simple ones and going on to the complex. The idea of a proof was entirely unknown to men who were above all concerned to provide an effective algorithm.

Despite our authors' efforts, not all merchants seem to have reached the level of mathematical skill required to apply their textbook lessons to business. Nicolas Chuquet often insists on the difficulties of calculation, and puts forward 'short and simple' rules, methods which might replace division by a series of mediations[6] or find simple ratios between units of money and of measure. Francès Pellos recommends the use of simplifications which bring about 'less fatigue'.

Arithmetic and Commercial Practice

The practical side comes out even more clearly in the exercises and problems. Examples are given in money of account – pounds, shillings and pence – or in real money, such as florins, ducats or écus. Many of the problems concern prices: how to find a total price when one knows the price of a unit, or the other way round, how to calculate a cost price or a profit. Other exercises deal with the value of a product as a function of its dimensions, such as material at so many pennies a yard; concrete problems which aren't just school-room exercises.

Barter

Certain problems relating to typical commercial operations of the late Middle Ages appear in all the books, barter first of all: *Troques et changement de marchandises* in Chuquet, *baratti* in Borghi and again *De barati et usso in fra mercanti de baratare* in Gori in 1571. This presents a simple problem: how to exchange products against each other using money as an accounting unit to quantify the prices. An example in the *Compendion de l'Abaco* involves exchanging cloth worth 3 florins the cane against wool at 16 florins the quintal. But the question is complicated by the fact that the merchant's quoted price is for cash, and he will want a higher price if he is going to barter.[7]

In a period when coin was rare, and replacements such as bills of exchange were used only in large-scale international trade, you had to pay to use cash. Nicolas Chuquet, a true mathematician, none the less reviews every case, hoping that 'no one will be deceived.' He sometimes puts so much stress on practical concerns that he advises against the operation in certain cases.

Associations

Other concrete problems examined by all the authors have to do with associations of merchants. They look at several types of company, all with simple structures: associations of merchants with one bringing the capital, the other the work, or merchants putting in different amounts of capital, or the withdrawal of one partner during the course of a commercial enterprise. All these provide plenty of examples to be solved with the rule of three. The practical side is always in view. Chuquet, finding several ways of sharing out the profits of a business, provides several precise mathematical solutions, and then declares 'each should take the one which seems to him the most just.'

A problem of association, from Nicolas Chuquet (1484)

There is a merchant who agreed to give his factor 500 pounds to manage and direct an enterprise, the factor to take $\frac{2}{5}$ of the profit. It happens that above and beyond this agreement and the consent of his master, he has put 100 pounds in company with his master. What has to be known is what part of the profit the factor should have by the first and uncorrupted agreement. Answer: for the first the factor by virtue of the agreement made for his services should take $\frac{2}{5}$ of the profit on the 500 pounds put in by his master. Now the 500 pounds are $\frac{5}{6}$ of the whole capital of the company so the factor by virtue of his service should take of the profit the $\frac{2}{5}$ of $\frac{5}{6}$ which is $\frac{1}{3}$ of the whole profit. And after for the 100 pounds he has put in, which are $\frac{1}{6}$ of the company, he should take a sixth part and thus the factor should take $\frac{1}{3}$ and $\frac{1}{6}$ of the profit which is $\frac{1}{2}$. And one should understand that in this manner (the merchant) neither gains nor loses from the 100 pounds which his factor put into the company. Thus, seeing that the factor was responsible for all and that the merchant was not at all involved, neither should he have a share at all.

Currency

The Middle Ages used metallic coin for currency, alloys of gold and silver, defined by their weight, composition and type. From the thirteenth century the gold coins of Florence were florins, those of Venice ducats, and the écu was a type often struck by the kings of France. These coins, like the far more common silver coin, were called real money. They bore no indication of value, which was given in pounds, shillings and pence, a system of money of account inherited from Charlemagne. A pound was worth twenty shillings and a shilling worth twelve pence. The relationship between real money and money of account was fixed by the authorities.

Currency questions came into nearly all problems. The period had very specific difficulties in currency exchange. Currency was metallic, made up of gold and silver coin; accounts were reckoned in money of account, pounds, shillings and pence. Each coin had its own value or values, because its commercial rate might not coincide with the official rate fixed by the authorities. Many different coins were in circulation, and foreign coins, though in theory demonetized, were always acceptable in practice. Currency exchange was part of the daily work of any merchant operating on a certain scale. But business also needed other kinds of skill. Under different headings, 'ligar', for example, meaning 'to alloy' gold and silver, or *fair le sou de fin*, or some other term, the textbooks of calculation described highly practical activities relating to precious metals. The minting of new coins required that precious metal be taken to the mint to be melted down. The public authorities, in France the king and in Italy the city, offered to buy metal at a certain price. It was very important for a merchant to know whether it would pay to sell his gold and silver, which coins he ought to send to be melted down and which coins he should keep. He had to know how to calculate the composition of a coin, how to determine the quantity of precious metal in a coin when the composition was known, and how to make an alloy of a given composition. Some even went further, for the state of technique and gradual wear meant that not all coins weighed the same. Chuquet shows the way to calculate the weight of coins which ought to be sent to the mint while keeping the lighter ones for one's own use: a totally illegal operation, but undoubtedly very common.

The Lacunae

The treatises taught simple mathematics corresponding to very basic commercial operations. By the fourteenth century, double-entry book-keeping (see p. 249) was known to the Italians. It might have been of interest to the teachers of calculation, but as far as we know at the moment, it isn't until Pacioli that we find a mathematician adding a treatise on accounting to his arithmetic. Pacioli's is a special case, in any event, for the *Summa* contains much more information than other contemporary mathematical treatises. Nor was compound interest studied as such in the arithmetics, although lending at interest was common.

From Commerce to Algebra

Despite their highly practical character, these textbooks did not contain all the knowledge needed by the merchant-to-be. He completed his education by practical experience gained through an apprenticeship in a shop or a company's warehouse. There he would be initiated in particular into the subtleties of double-entry book-keeping. If the teaching had to provide an education both technical and general, it is no surprise that alongside the dominant practical arithmetic, certain authors should have added chapters to develop the powers of reasoning. Many textbooks contained problems belonging to a very ancient tradition, such as the one about the snake which climbs a tower during the day and falls back a lesser distance at night: how long does it take to reach the top? One sometimes even finds questions which can be answered without recourse to calculation, like the one about the man who has to cross the river in a boat with a cabbage, a goat and a wolf, who can take only two out of three with him at any one time, and who knows that he cannot leave the wolf alone with the goat, nor the goat alone with the cabbage.

A Computational Geometry

But alongside these games there appeared chapters of innovative mathematics, geometries first of all, even though they might be limited to a few pages. They share many characteristics with the arithmetics. They were not concerned with demonstration, but rather with furnishing mathematical solutions to very concrete problems: to find the area of a field or a piece of fabric, the volume of a well or barrel. Possible business applications were very clear. They used well-established pedagogical methods, simple examples being succeeded by more complicated ones, such as flooring a building or digging a square well in the middle of a courtyard. Calculation was dominant and always expressed in numerical examples.

Lists of Equations

In the Italian texts there is sometimes an algebra, included under various names: *algibra* or *argibra*. As a rule, the Italian algebras of the late Middle Ages are only chapters added to treatises on abacus, arithmetic or algorism, depending on the author's terminology, treatises on commercial arithmetic, in any case. So in the *Trattato di praticha d'arismeticha* by Master Benedetto of Florence, a large volume of 506 folios written in 1463, algebra has three books out of sixteen. It was often no more than a few pages. From the fourteenth century onward, however, there appeared in Tuscany works which were concerned only with algebra, such as the *Aliabra argibra* attributed to Dardi of Pisa, a manuscript of 112 folios composed at the very end of the fourteenth century.

These algebras were primarily lists of equations and of the algorithms which provided the solution to each. There was no general rule for the solution of equations, but a series of cases, to which, as in al-Khwarizmi, a problem had to be reduced. The mathematicians of the late Middle Ages were for a long time ignorant of the possibility of reducing the number of equations.

In these lists there are often equations of the following type:

$$ax^3 + bx^2 = cx \text{ and } ax^2 + bx = c$$

each one with its algorithmic solution, seemingly without their identity having been recognized.

Al-Khwarizmi had given six typical linear and quadratic equations (see p. 210). In the fourteenth century the search for solutions to problems of higher degree led to a considerable lengthening of the lists. In 1328, in the first known Italian algebra, Paolo Gherardi gave a list of fifteen equations, six quadratic and nine cubic. Dardi of Pisa reached 198, though Piero della Francesca around 1480 had only 61, one of them in the sixth degree. According to the algorithms inherited from al-Khwarizmi, the solutions to equations of the second degree were positive and could not be zero. Beyond this, the Italian algebrists had pursued their researches into the problems of the higher degrees. The *Aliabra argibra* ends its list of 198 equations with a case of the type $ax^4 + bx^2 = \sqrt{c}$, whose proposed solution is:

$$x = \sqrt{\sqrt{\left(\frac{b}{2a}\right)^2 + \frac{\sqrt{c}}{a}} - \frac{b}{2a}}.$$

The authors never prove the validity of the algorithms they propose. Even if they were known and included by certain authors, like Master Benedetto in the 1460s, geometrical proofs for equations in the tradition of al-Khwarizmi and of Leonardo of Pisa were more and more being forgotten. They were only applicable to the second degree. Algebraic calculation on the other hand was undergoing development. Few texts give it much room, but there is a fourteenth-century manuscript which explains the multiplication of monomials. The chapter begins with a statement of the rule of signs: 'First I say that plus times plus makes plus and minus times minus makes plus and minus times plus makes minus and plus times minus makes minus' (Riccardiana of Florence).

The Development of Algebraic Calculation

In the fourteenth century, two authors gave a systematic exposition of calculations involving monomials and polynomials. Multiplication was done according to the rules used in arithmetic. Despite the use of abbreviations, the absence of a real algebraic symbolism made the operations difficult, and the division of polynomials was very briefly treated.

The algebra of the fourteenth and most of the fifteenth century was written rhetorically, most often in Tuscan. The unknown was called *cosa*, the 'thing', its square *zenso* and its cube *qubo*. The higher powers were expressed by a combination of these basic terms, which wasn't without its own problems. In one author the phrase 'square of the cube' meant x^5 and in another x^6, in the first case x^{2+3} and in the other $x^3 \times x^3$.

During the fifteenth century, Italian algebra developed within the same framework. In his *Summa* Master Benedetto of Florence took entire chapters from his fourteenth-century predecessors, and his language had very little relation to our own. Just as in the arithmetics his examples were always numerical, the problems posed and the solutions stated rhetorically. He doesn't write $x^2 + c = bx$ nor even $x^2 + 21 = 10x$, but 'the square (*zenso*) plus 21 units (*dramme*) equals 10 roots' or '10 things'. Despite this handicap, which makes the development of calculation difficult, Master Benedetto of Florence managed to deal with many problems relating to the second and higher degrees.

The thirty-six typical equations of Master Benedetto of Florence (1463)

(The equations have been converted into the algebraic language of our own day)

1. $x^2 = bx$ $x = b$

2. $x^2 = c$ $x = \sqrt{c}$

3. $x = c$ $x = c$

4. $x^2 + bx = c$ $x = \sqrt{\left(\dfrac{b}{2}\right)^2 + c} - \dfrac{b}{2}$

5. $x^2 = bx + c$ $x = \sqrt{\left(\dfrac{b}{2}\right)^2 + c} + \dfrac{b}{2}$

6. $x^2 + c = bx$ $x = \dfrac{b}{2} \pm \sqrt{\left(\dfrac{b}{2}\right)^2 - c}$

7. $x^3 = c$ $x = \sqrt[3]{b}$

8. $x^3 = bx^2$ $x = b$

9. $x^3 = bx$ $x = \sqrt{b}$

10. $x^3 + bx^2 = cx$ $x = \sqrt{\left(\dfrac{b}{x}\right)^2 + c} - \dfrac{b}{2}$

11. $x^3 + cx = bx^2$ $x = \dfrac{b}{2} \pm \sqrt{\left(\dfrac{b}{2}\right)^2 - c}$

266

12. $x^3 = bx^2 + cx$ $\quad x = \sqrt{\left(\dfrac{b}{2}\right)^2 + c} + \dfrac{b}{2}$

13. $x^4 = bx^3$ $\quad x = b$

14. $x^4 = cx^2$ $\quad x = \sqrt{c}$

15. $x^4 = dx$ $\quad x = \sqrt[3]{d}$

16. $x^4 = e$ $\quad x = \sqrt[4]{e}$

17. $x^4 + bx^3 = cx^2$ $\quad x = \sqrt{\left(\dfrac{b}{2}\right)^2 + c} - \dfrac{b}{2}$

18. $x^4 + cx^2 = bx^3$ $\quad x = \dfrac{b}{2} \pm \sqrt{\left(\dfrac{b}{2}\right)^2 - c}$

19. $x^4 = bx^3 + cx^2$ $\quad x = \sqrt{\left(\dfrac{b}{2}\right)^2 + c} + \dfrac{b}{2}$

20. $x^5 = bx^4$ $\quad x = b$

21. $x^5 = cx^3$ $\quad x = \sqrt{c}$

22. $x^5 = dx^2$ $\quad x = \sqrt[3]{d}$

23. $x^5 = ex$ $\quad x = \sqrt[4]{e}$

24. $x^5 = f$ $\quad x = \sqrt[5]{f}$

25. $x^5 + bx^4 = cx^3$ $\quad x = \sqrt{\left(\dfrac{b}{2}\right)^2 + c} - \dfrac{b}{2}$

26. $x^5 + cx^3 = bx^4$ $\quad x = \dfrac{b}{2} \pm \sqrt{\left(\dfrac{b}{2}\right)^2 - c}$

27. $x^5 = bx^4 + cx^3$ $\quad x = \sqrt{\left(\dfrac{b}{2}\right)^2 + c} + \dfrac{b}{2}$

28. $x^6 = bx^5$ $\quad x = b$

29. $x^6 = bx^4$ $\quad x = \sqrt{b}$

30. $x^6 = bx^3$ $\quad x = \sqrt[3]{b}$

31. $x^6 = bx^2$ $\quad x = \sqrt[4]{b}$

32. $x^6 = bx$ $\quad x = \sqrt[5]{b}$

33. $x^6 = b$ $\quad x = \sqrt[6]{b}$

34. $x^6 + bx^5 = cx^4$ $\quad x = \sqrt{\left(\dfrac{b}{2}\right)^2 + c} - \dfrac{b}{2}$

267

35. $x^6 + cx^4 = bx^5$ $\quad x = \dfrac{b}{2} \pm \sqrt{\left(\dfrac{b}{2}\right)^2 - c}$

36. $x^6 = bx^5 + cx^4$ $\quad x = \sqrt{\left(\dfrac{b}{2}\right)^2 + c} + \dfrac{b}{2}$

(from L. Salomone, 1982)

Geometrical proof of an equation after Master Benedetto of Florence (c. 1460)

In his *Trattato di praticha arismetricha*, Master Benedetto of Florence devotes a section to '*la reghola de Algebra Almuchabale*', borrowing the Arab words directly. 'One square (*zenso*) plus 21 units (*dramme*) equal 10 of its roots, that is to say, 10 things.' We would put this as $x^2 +$

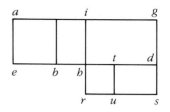

$21 = 10x$. On the basis of this numerical example, Master Benedetto looks for the algorithm that will allow him to solve equations of the type $x^2 + c = bx$.

The author draws a square, a geometric figure, whose side is equal to the thing, the unknown, which he calls *ab*, after the names of the two opposing angles. He adds a rectangle *bg* which has one side in common with the square and whose area is 21. The area of the rectangle obtained by adding *bg* to *ab* is equal to the square and 21 units or, according to the terms of the equation, 10 things, that is to say 10 unknowns. One of the sides of the big rectangle is therefore equal to 10 and the other to the unknown. The equation has been put in a geometric form.

One then puts a point *i* at the middle of the line *ag*, and drops a perpendicular *ih*. The line *hd* is equal to the line *he*, and the line *ih* is equal to the line *ae*, the side of the square and therefore equal to *be*.

One then augments the line *ih* by a length equal to *bh*, which will be *rh*. The square *rg* (also named from the two opposing angles) has an area of 25, because its side is equal to half of *ag*, which is 10.

The area of the square *rg* is therefore equal to 25 and that of the rectangle *bg* is 21. The difference between the two areas is therefore 4. One puts on the line *hd* a point *t*, such that the length *td* is equal to *ih*, and from *t* one drops a perpendicular to *rs* at *u*. The rectangle *du* is therefore equal to the rectangle *bi*. The difference of 4 between *rg* and *bg* therefore corresponds to the area of *rt*, which is a square, because *ri* is equal to *hd* and one has taken away *be* which is equal to *hi*; *ht* remains, which is equal to *hr*.

The area of this square is equal to 4, and its side (its root) is equal to 2. Now the length of *ri* is 5, half of 10, and that of *hi* therefore 5 less 2,

268

which is 3, that is to say the root of the original square, which is what was to be found.

'By this rule, take the half of the roots and multiply it by itself (which for us would be $(\frac{b}{2})^2$), and take away the number (for us c); take this root from half of the roots, and this gives the thing (the unknown)' (L. Salomone, 1982).

Or as we would put it today:

$$x = \frac{b}{2} - \sqrt{\left(\frac{b}{2}\right)^2 - c}$$

No doubt exceptional, but none the less reflecting the efforts of the Italian algebrists, the *De radice de' numeri e metodo di trovarla* was the work of an unknown author, probably a Tuscan writing in the second half of the fifteenth century. He developed a system of writing, which, through abbreviation, almost became set of symbols:

cossa, the unknown:	C,
zenso, its square:	Z,
qubo, its cube	Q.

To express higher powers he multiplied his symbols, so x^8 was written ZZZ, $2 \times 2 \times 2$, and ZQ was 2×3. For the first odd powers he combined addition and multiplication: x^5 was written CZZ, $1 + (2 \times 2)$ and x^7 was CZQ, $1 + (2 \times 3)$.

But the *Radice de' numeri*'s most novel contribution is, however, something else. Instead of giving an endless list of possible equations, the author divides them into eighteen basic types and provides a solution for each of them. He groups together equations such as these (in our language):

$$ax^2 + bx = c$$
$$ax^3 + bx^2 = cx$$
$$ax^4 + bx^3 = cx^2$$
$$ax^5 + bx^4 = cx^3,$$
all equations of the type
$$ax^{n+2} + bx^{n+1} = cx^n.$$

This is a considerable advance, not only in practice and in convenience of expression, but also as a new conception of the equation. The abandonment of the geometrical tradition, the development of algebraic calculation and of a kind of symbolism, the effort, doubtless isolated, to define and to treat equations more simply, the search for solutions to equations of higher degree than the second, are the chief characteristics of the Italian algebra in the vernacular which was developed, particularly in Tuscany, during the fourteenth and fifteenth centuries.

The Case of Chuquet

French algebra of the late Middle Ages was much less prolific and Nicolas Chuquet was altogether more isolated. He worked in areas comparable to the Italians', but he does show a certain originality. He created a symbolic algebra which was able to express the unknown, the powers and the roots; like many of his contemporaries he used scribes' abbreviations ($\bar{p}$ and $\bar{m}$) less

---

The rule of firsts, or the beauties of algebra, from Nicolas Chuquet (1484)

As Boethius says in his first book, where the first chapter the science of numbers is very great, among the sciences of the quadrivium is one about which every man ought to be diligent in inquiring into it. And elsewhere he says that the science of numbers ought to be preferred as an acquisition before all others, by reason of its necessity and all the great secrets and other mysteries which are the properties of numbers. All sciences have some part of it, and none are without it. Yet it is a science of great utility and also of great necessity in so much as it is convenient and useful to clerks and to laymen. Many wise men have studied to attain the great marvels and subtleties of it, many rules have been made, of which one is the rule of three which is lady and mistress of the proportions of numbers, and so highly to be recommended that it is called by some philosophers the golden rule. Similarly the rule of position, by which are done so many beautiful and delightful calculations that they cannot be counted. Also the rule of two positions which serves to enquire into things so deep and of such great subtlety that none of the above-mentioned rules can attain to them. And similarly there is the rule of apposition and remotion. There is also the rule of mean numbers, of which I was the discoverer, by which I made any calculation which by two positions I was not able to do. Mention of all these rules is made in the first part of this book. But above all these above-mentioned rules marvellous beyond measure is this rule of firsts which does what the others do and so does above and beyond this innumerable calculations of unimaginable profundity. This rule is the key, and entrance and the gateway to the depths which are in the science of numbers. (From A. Marre, 1880).

The algebraic language of Nicolas Chuquet and his successors

The use of letters to designate mathematical magnitudes was not unknown in the Middle Ages. Around 1225, Jordanus de Nemore wrote; 'one gives a number a which is divided into b, c, d, ...' but there was no algebraic symbolism properly speaking, for these values were not used in calculation.

Luca Pacioli (1494), like most of the algebrists of his day, could write: '1 number added to its square equals 12', which would be translated into our current algebraic language as: $x + x^2 = 12$.

Nicolas Chuquet, on the other hand, already had a symbolism which allowed him partly to escape from rhetorical language. To designate the unknown, which he called a first, Chuquet used the notation 1^1, which we would call x, and to designate the square of the unknown x^2, he would write 1^2, etc. In the same way, he wrote 34 where we would write $3x^4$. He didn't have the signs +, − and =, but he replaced the first two with the abbreviations $\bar{p}$ and $\bar{m}$, which were already becoming symbols. He was therefore able to write: '12 $\bar{p}$ 3^1 equal to 4^2', meaning $12 + 3x = 4x^2$.

To designate the root he used a letter with an exponent, writing R^2 and R^3 where we would write $\sqrt{}$ or $\sqrt[3]{}$. His algebraic symbolism allowed him to use negative exponents: 12^{2m} for $12x^{-2}$.

He would write, for example: 'If you divide 84^2 by 7^{3m} the number divided by the number gives 12. Then afterwards you must subtract $3m$ from 2 leaving five for the denomination thus giving 12^5', which is, in our own language $\dfrac{84x^2}{7x^{-3}}$ which gives $12x^5$.

Jérôme Cardan, on the other hand, in 1545, put his equations in a much more literary manner: 'A cube p: 6 things equal 20' or, as we should say, $x^3 + 6x = 20$.

With Raffaele Bombelli in 1572, a notation very similar to Chuquet's: '1^6 $\bar{p}$ 8^3 equal 20', which is $x^6 + 8x^3 = 20$.

And around 1590, François Viète wrote:
'1 QC − 15 QQ + 85 C + 225 Q + 274 N equals 120', which if C means cube, and so x^3, Q square, and so x^2, and N number, x, we get the following equation: $x^6 − 15x^4 + 85x^3 + 225x^2 + 274x = 120$.

and less to express addition and subtraction, he juxtaposed terms to be multiplied to show the product, he expressed division by placing the divisor underneath the bar of a fraction. He created his own symbols to designate unknowns, powers and roots, and he didn't hesitate to use negative exponents. With this symbolism, for the creation of which he didn't himself claim responsibility, Chuquet was able to develop a calculus of monomials and polynomials and solve a number of equations. It was what he called the rule of firsts.

Like the author of the *De radice de' numeri*, he groups equations under a certain number of major types. He never uses geometry as a proof of the solutions he puts forward. Without sufficiently accurate information, it is difficult to tell what Chuquet owed to his predecessors and contemporaries. It seems that, in addition to the education he received in Parisian university circles, he knew the tradition of the Midi, and also that he also got to know the work of the Italian algebrists at Lyons or even perhaps in Italy itself, during the course of a journey which he very probably undertook. Though an isolated figure in France, a mathematician original in his vocabulary as he was in his methods and the solutions he proposed, he represents a significant example of the nascent algebra of the West.

Algebra in Europe

The men who developed this early algebra were those who taught calculation to merchants' children. If many mathematicians showed themselves to be to some extent algebrists, all the algebrists mentioned wrote commercial arithmetics, both the Florentine masters and Nicolas Chuquet. Even the *De radice de' numeri*, a treatise on algebra and geometry and not a commercial arithmetic, contains a chapter on loans at interest. Apart from this evidence, necessary but not sufficient, the interconnections are difficult to establish.

Commercial Practice

One might wonder whether commercial practice itself did not require recourse to algebraic operations. The merchants' need to be able to fix prices, to share out profit and loss, to deal in currency and precious metals was satisfied by the rule of three. However, there were certain complex problems which were difficult to solve without algebra. In his application of the science of numbers to commerce, Chuquet uses it only once, to calculate compound interest; he only gives the result, saying that he found it by the application of the 'rule of firsts'.

By the fourteenth century the Italian authors offer many more examples: 'A man lends 100 pounds to another and after three years he receives 150 pounds for the capital and for the compound interest. I ask you what was the monthly rate interest per pound.' This problem, as it appears in the *Aliabra argibra*, can arise in practice. It can be expressed in terms of an equation of the third degree.

As well as problems of interest, there are questions related to other commercial operations. When in the fifteenth century Master Benedetto put forward 26 problems in commercial algebra, 8 concerned currency exchange,

8 were about journeys, 6 were price calculations or quantities of goods, while alloys, wage calculations and currency exchange each had one problem devoted to them. Commercial practice was always present to some degree. In one manuscript 20 out of 39 problems are commercial, while in another the figure is 20 out of 44.

Besides this direct relationship, new business techniques provided the merchant and the mathematician with new circumstances which had to be dealt with. The development of book-keeping, of double-entry book-keeping in particular, which requires that every credit be balanced by a debit and vice versa, broadly favoured the development of the concept of negative numbers. As a general rule, commercial arithmetics were ignorant of negative numbers, as there was no need of them in explaining the mathematical foundations of simple commercial operations. But the mathematician was confronted by their existence by the development of algebraic calculation, and in particular of equations of the second degree. Leonardo of Pisa refused to accept negative roots, but they appear in an algorithm in Provençal composed at Pamiers around 1430. Pellos uses them only once in the *Compendion de l'Abaco*, but they are used systematically by Nicolas Chuquet. He suggests that the negative root of an equation should be considered as a debt. A pedagogic trick, no doubt, a sense of the concrete and of education's very close connection to practice, but the author of the *Triparty* has a conception of negative numbers which goes much further than this. He manipulates them like positive numbers, multiplies and divides them, uses them as exponents. Commercial practice and the development of mathematical thought were connected, many examples confirming what seems to be obvious, but the relationships remain complex, and in many cases still very difficult to determine.

Traditions and Exchanges

The models used by the writers of arithmetics and algebras provide additional information on the connections which might exist between commerce and mathematics. In their writings many teachers of calculation recognized a debt to their predecessors or to their great forerunners. Aristotle is sometimes cited in the introductions, evidence of the importance of the myth of the universal scholar at the end of the Middle Ages, but Boethius and al-Khwarizmi are also mentioned. It was Boethius who had initiated into arithmetic generations of pupils educated in the liberal arts. As part of the *quadrivium* alongside geometry, astronomy and music, it wasn't a calculatory science. A speculative arithmetic, as it was called at the time, it reasoned on odd and even numbers, on triangular numbers and square numbers, on perfect numbers. The tradition went far back into Antiquity, one of its best-known exponents being Nicomachus of Gerasa (end of the first or beginning

273

of the second century AD). Borghi demonstrates his knowledge by stating that 'there are several kinds of numbers, as Boethius declares in his arithmetic' but he intends to treat only those which are useful to merchants.

During the thirteenth century a new arithmetic spread through the West; this had come from the Arab world. Its great exponent was Sacrobosco, the author of an *Algorism* which was very successful throughout the Middle Ages. He used Arabic numbers and positional notation, and in his calculations he obliterated intermediate results (see p. 000). Extensively influenced by Arab models, Sacrobosco's arithmetical work seems poorer than that of al-Khwarizmi: nothing on fractions, the rule of three, or false position, nor anything on how calculation might be used by merchants. Less orientated towards practice, Sacrobosco's *Algorism* is the work of an academic, written for academics. Details of construction and a particular vocabulary indicate that this model influenced the fifteenth-century commercial arithmetics of northern France, evidence of the influence of the scholastic model in areas close to Paris. But if the tradition of Sacrobosco made the West familiar with new techniques of calculation, it contained nothing which could lead directly to algebra.

Another source of innovation in mathematics was astronomy. Here again, the French manuscripts give evidence of the development of fractional calculation among the astronomers, and Paolo dell'Abbaco is cited by his contemporaries as a famous astronomer. Skill in handling fractions would have helped develop calculatory practice, but this didn't require recourse to algebra.

In the thirteenth century the influence of Arab mathematicians followed a more direct route. Some years before Sacrobosco, Leonardo of Pisa, in his Latin *Liber abbaci*, had gone through the bases of calculation, but had developed his work in a very different manner. The practical part was well developed and the book ought properly to be regarded as the principal inspiration of the European treatises on commercial arithmetic of the fourteenth and fifteenth centuries. Alongside the commercial questions, the *Liber abbaci* devotes its fifteenth book to geometry and algebra. In the direct tradition of al-Khwarizmi, it presents different types of equations and their geometrical solutions. Among Leonard of Pisa's algebraic work, only what was contained in the *Liber abbaci* was widely diffused in the West. His more scholarly work remained unknown for a long time.

As well as this definite Italian route, there are other ways this knowledge could have reached Western Europe, particularly through Spain. The arithmetics of southern France, whether in French or in Occitan, have characteristics absent from the Italian texts, such as calculation by apposition and remotion, an arithmetical method for the solution of indeterminate problems of two equations with three unknowns. Chuquet himself had drawn on this

---

Table of Contents from Leonard of Pisa's *Liber abbaci*

Leonardo Fibonacci, or Leonard of Pisa, wrote his *Liber abbaci* in 1202. The work, whose contents are very similar to those of the fourteenth- and fifteenth-century treatises, was written in Latin.

1. The nine figures of the Indies, numbers and numeration.
2. Multiplication of whole numbers.
3. Addition.
4. Subtraction.
5. Division.
6. Multiplication of whole numbers with fractions, multiplication of fractions.
7. Addition, subtraction and division of whole numbers and fractions, and reduction to a common denominator.
8. Buying and selling.
9. Barter.
10. Companies.
11. Changing money.
12. Solutions of multiple problems.
13. The rule of *chatayn* which allows the solution of many problems.
14. The extraction of square and cube roots and operations on roots.
15. Geometry and questions of algebra.

(From G. Libri, 1838–41)

---

source before coming into contact with the Italian sources. This southern French tradition is the subject of current research. A well-grounded hypothesis relates it to the development of Jewish mathematics in Spain and in the South of France. Its history has distant origins, for the first work about algebra written in the West dates from the first half of the twelfth century, composed in Hebrew by Abraham bar Hiyya Ha-Nasi, and very soon translated into Latin. Still poorly known, this tradition influenced Spanish mathematicians as well as the French. It is in the Provençal treatise from Pamiers (*c.* 1430) that for the first time a negative solution to an equation was accepted in the West. This text also contains a new approach to irrational numbers, and this treatise too is above all a commercial arithmetic.

The search for origins shows that the routes followed by the algebra of al-Khwarizmi and his successors as it was diffused were often the same as those followed by the commercial arithmetics. It is more difficult to establish the influences which may have been at work at the end of the Middle Ages. Arab science, even if not as exciting as it had been in the preceding centuries, continued to exist and to produce. The effort made by the great Eastern algebrists of the eleventh and twelfth centuries, to free algebra from the grip of geometry, was continued in the Muslim West, in the Maghreb and in Spain, during the thirteenth and fourteenth centuries. Alongside the lists of algorithms which circulated as they did in the Christian world, mathematicians like Ibn Badr in Andalusia or Ibn al-Banna at Bougie had developed a symbolic script, had rethought the classification of equations and had proposed new solutions. It is still impossible to establish the precise relationship, but in general algebra progressed in comparable directions on both sides of the western Mediterranean: first of all in the Islamic countries during the thirteenth and fourteenth centuries, and then in the Christian countries during the fourteenth and fifteenth centuries. The shores of the Mediterranean were the site of continuous commercial exchange.

The Weight of Environment

The commercial world was certainly an intermediary in the transmission of the mathematics of the Islamic world, and of algebra as the Arabs understood it, to Western Christendom. It was also a site of mathematical innovation. This fact, now well established, leads to several questions, questions of chronology first of all. Why was it only in the fourteenth century that the work of al-Khwarizmi became widely diffused? Why did algebraic innovation, which we have begun to see was connected particularly to the development of calculation and the abandonment of argument based on geometry, only begin to develop in Italy in the fourteenth century, and even later in other countries? One has to come back to the initial facts: it was the same men in the same social situation who produced the commercial arithmetics and the first treatises on algebraic calculation in the vernacular. The same people, that is to say, a group of professional mathematicians, the first in Western Europe, for whom mathematics was neither geometry nor reflection on the nature of number, but calculation and algorithms. It was a necessity of their profession: they had to teach calculation to men whose first concern was for commercial effectiveness, men for whom it was a daily need to be able to calculate quickly and correctly. This helps us understand the place of calculation in a city like Florence, which in the fourteenth century was one of the most important centres of international trade, and the home city of companies whose activities extended throughout the Mediterranean basin and Western Europe.

Present in daily life and present in art, financed by the very people whose wealth was founded on trade, and therefore on calculation, mathematics was everywhere. For the art historian Michael Baxandall, the education received by the Florentines is a major element necessary to an understanding of the art of the Tuscan Quattrocento. For him, measuring barrels on the basis of calculation, the daily use of the rule of three and thus the determination of proportions, created 'a very particular intellectual world'. A symbol of this world, Piero della Francesca, one of the creators of the new pictorial space, was also the author of an arithmetical treatise. The place of the mathematician and calculator was recognized in the city, and economic and social conditions favoured the development of a new science on the basis of external contributions now assimilated. While it had been able to take on whole chunks of Graeco-Arab culture, in the fields of astronomy and physics in particular, the University had been able to assimilate only a small part of the mathematics developed in the Islamic countries, which had not been very useful for the creation of a model of the Universe. The merchants on the other hand helped bring about the emergence of a science of calculation by founding an institution, the vocational school, which provided mathematicians with the means of life, and by creating an innovatory milieu, in the area of business technique as well as in art, where mathematics occupied an essential place. A scientific milieu had been born.[8]

The role of the mathematician was recognized in the city, which provided the social foundation of his activity. The connections between the history of art, the history of science and the city can be seen in the chronology. In the sixteenth century, when Florence ceased to be a great centre of artistic activity, when Rome and the princely courts attracted the artists of greatest renown, this is when the great Italian mathematicians who continued the development of algebra ceased to be Florentines or belong to the world of the merchants. Another dynamic had been created. It wasn't that mathematics abandoned calculation, as is proved by Bombelli, Fontana, Tartaglia, Cardan and Viète, but there was a change in the milieu. Business gave way to humanism and the merchant gave way to the prince. In Paris in 1495, Pedro Sanchez Ceruello published the *Arithmetica speculativa de Bradwardine*, including a practical arithmetic very close to the treatises on arithmetic written for the merchants, but the text was in Latin. The humanities assimilated the contribution of commercial mathematics, but cut it off from its roots. In the sixteenth century the great names of algebraic history belonged to a milieu very different from that frequented by Paolo dell'Abbaco and Chuquet. Tartaglia (1499–1557) began his career as a 'teacher of abacus' and wrote in Italian. He is different from his predecessors in the breadth of his interests, which included mathematics, but also geography, astronomy, architecture and optics. Cardan (1501–76) also taught, but he began in a humanist school where the pupils learnt Greek, dialectic and astronomy alongside mathematics, and later he became professor of medicine at Padua. His essential works, the *Practica arithmetice* and the *Ars magna*, were written in Latin. Bombelli (1526–72) wrote in Italian, but he was an engineer and an architect, in the service of an important Roman nobleman. Outside Italy, Viète (1540–1603) was a councillor of the French Parliament, who undertook many missions for Henri IV; he too wrote in Latin.

The teaching of calculation to young merchants none the less continued, and for centuries, but it became repetitive and no longer creative. Models existed which proved to be adequate.

There should be no illusions about the conclusions which can be drawn from this chapter: they are no more than provisional. The uncertainties are great and the lacunae considerable. Too many texts remain unknown or little studied. Relationships between the mercantile world and the development of mathematics are only seen really clearly in Florence, a very special milieu, and the French examples seem to impose a greater need for nuances. The research must be extended to Germany, where trade flourished in the late Middle Ages, and where mathematicians and teachers of calculation also wrote commercial arithmetics. Greater knowledge is also needed of the links established between the mathematics of the Islamic countries, chiefly of the Maghreb, and Christian Europe within a commercial space in which trading was constant. The fate of this mathematics needs to be traced through the sixteenth century, to know to what extent it contributed to the formation of the algebrists of the Renaissance, who received the essential core of Diophantus' work and paved the way for classical algebra. Finally, it will be necessary to look at the world in which the merchants lived. From the fourteenth century the Parisian physicians, in their search for a first mathematization of the world, were led, like Nicole d'Oresme, to develop a new mathematical language. None the less, the great adventure of European algebra in the last centuries of the Middle Ages remains essentially mercantile and Mediterranean.

At the end of the Middle Ages the sciences of calculation took a decisive step in Christian Europe. The form of what later came to be called elementary mathematics was fixed for centuries. The chapters to be found in books intended for the training of young merchants of the fifteenth century are still there in the books of primary-school pupils of the Third Republic in France: numbers, addition, subtraction, multiplication, division with their proofs, fractions, the rule of three, simple and double false position. Methods of calculation hardly changed either, the diffusion of paper having made it possible to write down the intermediate results and conserve them. The present forms were being established, very quickly for addition and subtraction, a little less so for multiplication and especially for division. Even the division of fractions began to be dealt with by multiplication by the inverse fraction. Only the rule of three, essential as it was for the solution of problems, was not yet stated in the same way as today.

The counterpart to this well-established calculatory arithmetic was an algebraic calculus in full mutation. The differences from the algebra taught in the secondary schools are considerable. The language is often still rhetorical, and when a symbol does exist, it varies with every mathematician. Demonstration was unknown, and the algebrist developed algorithms whose validity was measured by their efficacity. Important innovations none the less occurred. Algebra abandoned the geometric tradition going back to al-Khwarizmi and transmitted, in particular, by Leonardo of Pisa. Using different methods, it developed a calculus of powers and of roots, of monomials and polynomials, and was therefore able to tackle solutions to equations of greater degree than the second, by far. The centre of development changed for centuries, and the sciences of calculation now developed in Western Europe, and in Italy in particular. In the sixteenth century the Italians played a determining role; at the end of the century Viète, more than any other, fixed the rules of classical algebra. The impact of the translations of Diophantus on the construction of classical algebra must be reconsidered from this perspective.

In the fourteenth and fifteenth centuries the essential innovations appeared outside the University, in which mathematics had a very limited role, being orientated towards kinematics, the study of movement, rather than towards algebra. There mathematics was more the means to establish the laws of physics than a science in its own right. Nor did algebra accord with the model still present in the minds of the academics: the model of Greek science symbolized above all by Aristotle. Diophantus of Alexandria, the only great algebrist to have written in Greek, remained unknown in medieval Europe. The chief architects of the expansion of algebra were, however, teachers, professional mathematicians who gave access to knowledge rather than to culture. Intended for the training of merchants' children, their teaching had an indisputably practical character. As the merchant had above all to know how to count, the teachers had to be calculators. Rather than being the consequence of a specific demand on the part of the merchants, innovation arose from new conditions. Algebra developed first in Florence, in a world where calculation was ever-present, in daily life as in art, where calculation was recognized as indispensable to the life of the city; it was also a milieu in which mathematicians were numerous and were able to live from their science. The critical mass was reached. A milieu developed which was conducive to research, a scientific milieu within which mathematical problems were posed and went far beyond the practical requirements of business.

The Galileo Affair

Isabelle Stengers

In which we find several different affairs involving different actors, each a response to the question 'What was it that Galileo did?' It is suggested that this multiplicity tells us something about both history and historians.

In his book *Les Atomes*, Jean Perrin points out a difference between the maps one can make of the coast of Brittany and the suds that one makes with soap in water. The first presents a problem of representation: at a given scale, a map traces the outline of the coast as series of regular curves, at each point of which a tangent can be constructed. Drawn at a larger scale, the continuous segments will be replaced by a more complicated contour, though this itself will also be made up of regular curves. Uncertainty about the choice of tangents is therefore connected to the choice of scale, the choice of the map to be used.

> This is because the map is a conventional drawing, in which, by virtue of its own construction, every line does have a tangent. It is an essential feature of our soap-suds, however, that whatever the degree of magnification, one *suspects*, without being able to see them altogether clearly, the presence of details which absolutely forbid any attempt to construct a tangent (just like the coastline itself, in fact, if instead of looking at it on the map one looked at the thing itself from whatever distance).

Can Perrin's comparison also be applied to the relationship between concrete historical situations and the descriptions given of them? One thing has to be noted first of all: the differences between historical accounts can't be understood in terms as simple as differences of scale. We are going to discover several 'Galileo affairs' and this multiplicity expresses not so much a difference in precision or in degree of resolution, as a difference in points of view between the historians and philosophers who give their accounts.

Should we therefore conclude that these accounts are worse than Perrin's maps, that the real is inaccessible here not only on account of the infinity which blurs the least of its details, but also because of the bias of those who describe it? I want to show that the interest each brings to the Galileo affair shouldn't be thought of as a screen which hides the truth from us and leaves us with nothing but subjective projections. Of course the different versions stand in a polemical relationship to each other, and those who put them forward may believe their truth will beat all others. However, none of them can be considered as the final, neutral, disinterested version of the 'affair'. This is why I shall proceed by the exploration of successive bifurcations. At each bifurcation a new landscape comes into view, not more or less detailed, as in Perrin's maps, but different, organized by the questions and the reasons which brought each historian or philosopher to the 'Galileo affair'.

In this sense, each of the Galileo affairs examined here can be compared to a chemical reaction. Looked at like this, the historian's or philosopher's interest in what is seen as the outcome of the affair acts as a reagent. The interest which relates past to present is not an obstacle, but an operator which actively produces new perspectives and interesting problems, constructing narratives whose very divergence gives the past its depth, just as a chemical substance gets its identity from its multiple possibilities of reaction.

Historians, however, were not the first to get themselves involved. Galileo was recognized as the greatest scientist of his age; his books were written not in Latin, as was the custom, but in Italian, the language of the ordinary people.[1] They were clearly addressed not only to his colleagues but to all reasonable men, to inform them of a radical transformation of their model of the Universe, the ending of the tradition which rested on the authority of Aristotle, and the practice of a science which respected the 'facts' and not the texts. Everyone knew the historical significance of Galileo's condemnation, which would echo all over Europe. In fact, what is significant about the Galileo affair is that those involved knew that they were part of an 'affair', that they were making history. As a result, the interests of present-day narrators are precisely comparable to those of the historical actors. Unlike the rocks and grains of sand, unlike the streaky suds which are indifferent to being neglected or otherwise by the cartographer, these people addressed themselves to the future in the same way that historians address themselves to the past. They did not live a history whose narrative would be invented by a

281

later historian. They themselves tried to determine the perspective, arguing for a particular cartography, constructing the regular curves, the narrative which governed their own position, in the name of which their adversaries could be condemned.

This chapter will be no exception to the rule. As we shall see, I also have my version of the Galileo affair, which doesn't cancel out the others but provides a different perspective. And this is no accident. One doesn't come across Galileo by accident, as just one episode among others. One turns towards Galileo, as he turned to his own public, in order to establish a thesis on the new science, to discuss its novelty and the 'rationality' to which it corresponds.

The Earliest Affairs: 'And yet it moves!'

'*Eppur, si muove!*' Galileo is supposed to have murmured, at the moment of his condemnation. The power of the Church cannot fight against the power of facts. This is certainly the message of Galileo's works before the condemnation: two powers confront each other, the power of a tradition supported by literal interpretation of Scripture, and the power of the facts, of which Galileo is merely the humble representative. The Church had to recognize and exercise its only legitimate role as guardian of the faith and abandon the rest to human reason. So, in his *Letter to Castelli* (1613), Galileo writes:

> I would willingly believe that the authority of the holy Scriptures is in-tended only to teach men those articles and propositions, which, neces-sary to their salvation and lying beyond human reason, can be taught and made believable only by the very voice of the Holy Spirit. But that God, who endowed us with sense, reason and intellect, intended that we should neglect to use them, that he intended to give us another means of knowing what we might grasp through these, this I do not think it necessary to believe.

And Galileo summons his adversaries onto the terrain of the facts:

> But if they truly believe that they have the true meaning of a particular passage of Scripture, and are therefore assured of having to hand the absolute truth on the point under discussion, let them just simply tell me, whether they believe that in a dispute on a question of natural science, he who happens to be upholding the truth has a great advantage over the other who has to defend the false. I know that to such a question they will reply yes; he who upholds the truth, they will say, will have on his side a thousand experiences, a thousand necessary demonstrations, while the other will have no more than sophisms, paralogisms and

fallacies. But if, according to them, within the limits of natural reason and without the use of other arms than those of philosophy, they are conscious of being so far superior to their adversary, why then, when the moment comes to face up to him, do they not immediately take in hand that invincible and redoubtable weapon the very sight of which would terrify the cleverest and best trained of combatants.

Galileo was just pretending to be writing to Father Castelli, but he knew the letter would be passed on to his protector, Duke Cosimo II de' Medici, and to his mother, Catherine de' Medici. He knew it would be circulated and that it represented a real challenge: those who opposed him should gather together the facts which proved that he was wrong. On his side Galileo had thousands of experiments, while his adversaries had only the displaced authority of the Scriptures.

In 1616, after a denunciation of the *Letter to Castelli*, Cardinal Bellarmine informed Galileo of the prohibition on his publicly asserting the truth of the heliocentric doctrine. He submitted to this prohibition, but in a way which ridiculed those who had imposed it on him, even as he expressed his obedience. In fact, the *Dialogue on the Two Chief World Systems* (1632) presents three speakers: Salviati, representing Galileo, Sagredo, a cultivated and enlightened man, capable not only of being completely convinced by Salviati's arguments but also of carrying them forward or corroborating them on the basis of his own 'common-sense' reflections, and finally Simplicio, a narrow-minded Aristotelian, endlessly convicted of error and blind submission to tradition. It is, however, neither Salviati nor Sagredo who concludes the dialogue, but Simplicio, who invokes the authority of an 'eminent and learned' person, faced with which all must fall silent. According to this eminent person's teaching, everything hitherto presented as certain demonstration is neither true nor necessary, because God has the power to produce these apparently conclusive 'facts' by means which are beyond our imagination. It made no difference that Simplicio's interlocutors, Salviati and Sagredo, actually accepted this 'angelic and admirable' doctrine: it's said that Pope Urban VIII, the author of the argument, still felt himself to have been held up to ridicule. The way was now open for Galileo's enemies, and it led to the condemnation, one year after the appearance of the *Dialogue*.

Here then is a classic version of the affair, pitting Galileo against the Church, a version which corresponds to Galileo's own ideas, making him out as a hero of free thought, the target of obscurantist persecution. Other historians, however, have listened to other actors. So Arthur Koestler makes Cardinal Bellarmine a model of tolerance. What he forbade Galileo to do was to claim absolute truth for the heliocentric doctrine, while he recognized his right to use it as a scientific hypothesis. Shouldn't it have been enough for a mathematician to show that appearances were better saved by the

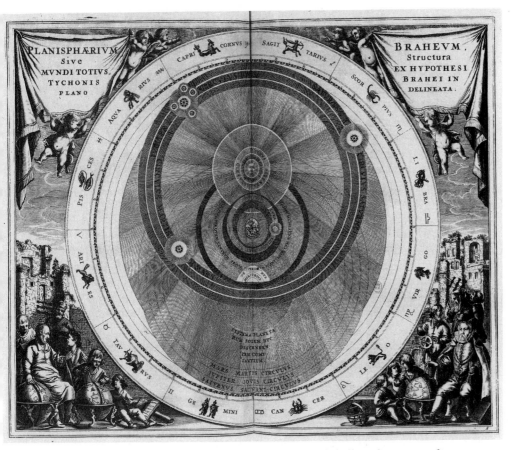

*The three rival models of the Universe: the systems of Ptolemy (c. 90–c. 168
AD), Copernicus (1473–1543) and Tycho Brahe (1546–1601).*

*In Ptolemy's geocentric system the Earth is immobile at the centre of the
Universe; the first seven spheres carry the 'planets' (including the Sun and
the Moon) while the eighth carries the stars.*

*For Copernicus and his heliocentric system the Sun is immobile at the centre
of the Universe; the Earth and its satellite, the Moon, as well as the other
planets, revolve around it. The movement of the stars is only apparent, the
result of the rotation of the Earth.*

*Lastly, Tycho Brahe's system is intermediary between the two others: the Earth
is again immobile at the centre of the Universe and the Sun revolves around
it. The difference is that the planets revolve around the Sun, accompanying it
as it travels around the Earth.*

*These engravings of 1661 are a good illustration of the conflict for
precedence and magnificence between the Earth and the Sun. Rather than
orbits, the planets have zones of influence. Jupiter's satellites, discovered by
Galileo, are also shown. (Engravings in A. Cellarius,* Harmonia
macrocosmica, seu Atlas universalis et novus, *1661. Bibliothèque Nationale,
Paris.)*

The heliocentric theory

The first 'heliocentrist', the first who dared think that the Earth might be moving around the Sun, was Aristarchus of Samos, said to have been born in 310 BC. Of his work, there survives only a treatise on the size and distance of the Sun and the Moon. The first edition of Copernicus' *Six Books on the Revolutions of the Celestial Orbs* was published in 1543, the year of his death. While it is commonly accepted that the glaring discrepancies between astronomical observations and predictions based on the system of the Alexandrian astronomer Ptolemy played a role in the 'Copernican revolution', it is less commonly realized that Copernicus' system, which also utilized circular orbits, was just as complicated and imperfect. The system which saved the phenomena in the most satisfying way, until Kepler broke the circle and dared to make the planetary orbits elliptical (1605), was that of his teacher, Tycho Brahe. The Earth was at the centre, but the (other) planets revolved around the Sun, which itself revolved around the Earth. However, the Copernican revolution gave rise to passions which went beyond the technical issues. In this sense, it was much more of a cultural than a scientific revolution. Galileo, who remained ignorant of Kepler's ellipses, made himself the spokesman for this other revolution.

---

hypothesis of the Earth's movement, rather than needing to assert that the Sun was truly at the centre of the Universe?

As Koestler reads it, the *Letter to Castelli* changes its meaning. Galileo was in fact attempting, entirely for his own benefit, to make a unilateral declaration of the rules which should govern all who would contribute to the long history of discussions on heliocentrism. Not only did he assert the absolute truth of his thesis, but he then sent the ball back to the others; it was for the others to try and disprove his truth. Koestler points out that if Galileo precipitated the final outcome of the conflict by refusing the agreement offered him by Bellarmine, it was surely because he had no way of showing that he saved the appearances better than the old astronomy of Ptolemy. The Copernican system, which like its predecessor postulated the circularity of celestial movements, was, as Koestler points out, more complicated again than the Ptolemaic.

Koestler thus questions the role of Galileo the astronomer: his discovery of the satellites of Jupiter and of sunspots made him a propagandist for the Copernican hypothesis, but more than twenty years after Kepler's *Astronomia Nova* (1609), the *Dialogue* maintained the postulate of circular orbits, and kept quiet about the fact that these didn't allow one to save all the facts. Koestler concludes:

> For almost fifty years of his life, he had held his tongue about Copernicus, not out of fear to be burnt at the stake, but to avoid academic unpopularity. When, carried away by sudden fame, he had at last committed himself, it became at once a matter of prestige to him. He had said that Copernicus was right, and whosoever said otherwise was belittling his authority as the foremost scholar of his time ... This was the central motivation of Galileo's fight ... It does not exonerate his opponents, but it is relevant to the problem whether the conflict was historically inevitable or not.

Koestler thinks that it wasn't, and that without Galileo's pride, the conflict organized by Galileo himself, between an obscurantist Church and a science henceforth identified with freedom of thought, might have been avoided. Perhaps another history might have been possible, which did not fix science in a heroic role to which it was unfitted.

For Koestler, this is what is at stake in the Galileo affair: the dramatic split between 'rationality' and 'value', which today allows men to wield the atomic bomb without having learnt the implications of its overwhelming power. Science has made Galileo the symbol of its liberty, but this symbol primarily expresses a vehemently asserted incompatibility between its own conditions of development and the values of society now identified as obstacles, as resistance and obscurantism.

Should Galilean science be identified with liberty of thought, with the light that struggles against the dark? Other performers enter on stage. Up to now, our historians have accepted the identifications of Galileo and Bellarmine themselves. But isn't the truth something more secret than this? The historian Pietro Redondi suggests that the hidden origin of Galileo's condemnation was a serious accusation of an entirely different kind, the clues to which he believes himself to have uncovered. It was an accusation relating to the faith, and not to a 'question of natural science'. If the Galileo affair was indeed set in motion by the denunciation of the atomist doctrine presented in *Il Saggiatore* (1623), if Galileo was accused of contradicting the dogma of the Eucharist, the issue isn't any longer what was presented by Galileo as the distinction between questions of faith and questions of natural science. On the other hand, if we follow Lerner and Gosselin, his contemporaries did not accept the image he put forward of himself, and saw him as another Giordano Bruno. In this version, Galileo's condemnation has to be understood in the complex context of international politics and of the Vatican's oscillation between alliances with France or Spain. In 1632 a gesture towards Spain was necessary, in this case the condemnation of a supporter of an alliance with relatively tolerant France, which, after Bruno, was promoted by the Hermetics. Was Galileo condemned instead of Campanella, who was himself too dangerous, who knew too much about the Hermetic sympathies of the Pope?

The atomist doctrine and the Eucharist

The dogma of transubstantiation in the Eucharist was proclaimed in the sixteenth century at the Council of Trent (that is to say in the context of the Counter-Reformation, for the Protestants did not accept that in the hands of the priest the bread and the wine really became the body of Christ). This dogma included a reference to the scholastic notion of 'substance', which didn't allow it to be explained, but at least allowed it to be said how, when by a miracle the 'substance' was transformed, the 'accidents' of the bread and wine (texture, taste and colour) remained. In *Il Saggiatore* (1623), Galileo deprived bodies of their substantial reality, making a distinction between what had to be attributed to bodies – a geometric shape, a position and a speed – and the tastes, smells and colours which he attributed to the subject or to the animal whose senses were excited by tiny particles emitted by the body. And if we perceive the taste of bread, it is therefore because, despite transubstantiation, the host is still giving off the same tiny particles as the bread.

Hermeticism

Hermeticism was a mystical philosophy peculiar to the Renaissance. For its adepts, the author of the doctrine was Hermes Trismegistos, a semi-divine figure from Ancient Egypt. In degraded form, they claimed to discover in the Judaeo-Christian tradition the 'secret knowledge' of Egypt, passed on by the 'initiated': Zoroaster, Orpheus, Plato and Pythagoras. Hermeticism also had a political dimension: it was necessary not only to liberate the divine essence found in every person, to rediscover the transparent analogy between microcosm and macrocosm, but also, by returning to the 'true philosophy', to reunite the Churches torn apart by the Wars of Religion. The Hermeticists wanted to convert an 'enlightened monarch', who might guide the world towards a golden age. Giordano Bruno (1548–1600), who described an infinite universe containing an infinite number of inhabited worlds, and who died in the flames of the Inquisition, adhered to this philosophy, as did Tommaso Campanella (1568–1639), who defended Galileo (1616), and attracted the favour of Urban VIII by his reputation as a magician and astrologer, arguing with him for a 'natural' and 'reformed' Catholicism and for an alliance with the France of Louis XIII and the Edict of Nantes.

We have a choice between a denunciation kept secret because its revelation would have led Galileo to the stake, or machinations at a high level of international politics. Here the historians are claiming or are trying to claim independence from the narratives of the actors themselves, and making the

Galileo affair the result of a combination of circumstances of which the victim himself was unaware. The stage is populated with new relationships and new problems, with characters pulling strings and turning the official narratives into so many booby-traps. And the boobies, of course, are primarily those today who see the Galileo affair as an affair of our own modern culture, and who project onto it the concerns by which they themselves are preoccupied. The affair is, and can only be, the concern of professional historians. What is at stake in the Galileo affair is the autonomy of the historian, which cannot allow itself to be caught up in the biased narratives of those who believe themselves to be actors, which searches for clues to truths carefully dissimulated by other more hidden actors whose roles will prove to be decisive. Everyone must be lying, there is a conspiracy of silence, and the historians are in their element, until the discovery of a new document makes the whole construction crumble. So, after a searching analysis of the bias underlying Redondi's 'police investigation', two other Italian Galileo specialists, V. Ferrone and M. Firpo, produce a letter of Galileo's *from after his condemnation.* Galileo is replying to a correspondent who has innocently suggested to him that it would be interesting to study the relations between the atomist doctrine and the doctrine . . . of the Eucharist. Rather than recoiling in horror, Galileo feels that the idea is very interesting . . .

Whether or not one accepts these theories offered as examples, they do throw light on the meaning of the first bifurcation, a first choice among the interests which set the past in relation to the present: either we stick to a professional, disinterested history, considered as the only one which purifies the Galileo affair of all parasitic issues, or, as we shall do, we can follow those who are interested above all in the question, posed at the time of the Galileo affair, of this scientific truth for whose sake Galileo believed, at the very least, he entered into conflict with the Church. A first bifurcation and a change of scene.

The Question of Modern Science

Here is a second Galileo affair, then, which isn't some left-over from the first, revealed by a more detailed approach, but the result of different questions which weren't asked by Galileo's contemporaries. The affair is no longer situated in the seventeenth century but in the twentieth, and the actors opposing each other are the historians of philosophy who recognise in Galileo's scientific texts the first expression of the question in which they are interested themselves, the question of the specificity of modern science.

But isn't this Galileo affair simply the affair of philosophers, using Galileo to discuss modern science in general? Perhaps; but such an affair is not foreign to the history of science, in the sense that the way in which historians read Galileo's work depends on what these historians feel entitled to think of the modern science whose founder Galileo is thought to be. So Pierre Duhem, who attempted to limit the implications of the event which actually constituted 'Galileo's foundation of modern science', was refused a chair in the history of science at the Collège de France and passed over in favour of more respectful scientists.

Can one describe Galileo's position and those of his scientific adversaries in a symmetrical fashion? Is it possible to read Galileo without thinking the facts proved him right? We're speaking here not of astronomy but of the theory of falling bodies which won him the splendid title of the 'founder of modern science'. This is what was attempted, in their different ways, by the philosopher Alfred North Whitehead sixty years ago, and by the contemporary epistemologist Paul Feyerabend, among others.

In his *Science and the Modern World*, Whitehead accepts Galileo's claims at face value: the 'historic revolt' which is the foundation of modern science is based on the facts.

> Galileo keeps harping on how things happen, whereas his adversaries had a complete theory as to why things happen. Unfortunately the two theories did not bring out the same results. Galileo insists upon 'irreducible and stubborn facts', and Simplicius, his opponent, brings forward reasons, completely satisfactory, at least to himself. It is a great mistake to conceive this historic revolt as an appeal to reason. On the contrary, it was through and through an anti-intellectualist movement. It was the return to the contemplation of brute fact; and it was based on a recoil from the inflexible rationality of mediaeval thought.

Poor Italian theologians, Whitehead concludes, poor belated medieval scholars, attacked by the Protestants, held up to ridicule by Galileo, held in contempt by the bishops themselves, who had recommended at the Council of Trent that they should avoid useless and superfluous discussion! Poor theologians who defended the lost cause of overweening rationalism in the middle of a world at war! Like Koestler, who agrees with him on this point, Whitehead is arguing for a reconciliation between science and reason, for a science which gets away from naïve faith in the facts, from an intolerant application of abstraction, a science which doesn't deny that to which it is unable to assign a meaning, as a result of the very constraints of its own method. Galileo is the founder of a method blind to its own limits, and his unfortunate opponents may expect greater justice from the future that Whitehead is working for.

For his part, in his book *Against Method*, Feyerabend challenges the idea that Galileo respected the facts. He wants to show that modern science cannot legitimately claim a truth essentially different from that of other forms of thought, whether mythic or religious. Not only does Galileo interpret the facts, but, and this is the vice that Feyerabend wishes to purge from modern science, he uses psychological tricks and propagandistic manoeuvres to disguise the fact that he is interpreting, to make people believe that the facts unequivocally prove him right. How could the 'facts' prove that the Earth is in motion, when intuitive observation shows it to be immobile? For the facts to speak in favour of Galileo, he has first of all to teach his readers and interlocutors to redefine them, to describe them in a new language. Feyerabend discusses Galileo's use of the famous example of the ship. Here he needs to show that a sailor on board a ship, if he has no point of reference in the sea or on land, has no way of telling whether or not he is in movement; and further, he has to convince Sagredo, who in the dialogues stands for the man of good sense, that everyone had always known this but had forgotten it, and that he, Galileo, was doing no more than pointing out the obvious. But what he is trying to argue is not in the least self-evident, but implies a profound transformation of the idea of movement. Galileo's theory implies that a stone falling from the top of a ship's mast will fall at the foot of the mast, whether the ship is moving or not. The stone, separated from the movement of the boat, must therefore maintain this movement at the same time as it falls towards the deck. Movement maintained by itself, this is what the hypothesis of a moving Earth forced one to believe, against all the empirical evidence (all the movements we observe having a tendency to come to an end) and against all tradition. This is what Galileo presents as common sense, and succeeds in having it accepted by Sagredo.

Feyerabend does not criticize Galileo for having introduced a new observational language, impregnated with theory. What he wants to do is point out that none of these observational languages, whether scientific, theological, mythical, etc., can rely on an unequivocal relationship to the facts which would allow it to pronounce judgement on the others. He says:

> A science that insists on possessing the only correct method and the only acceptable results is ideology and must be separated from the state, and especially from the process of education. One may teach it, but only to those who have decided to make this particular superstition their own . . . Of course, every business has the right to demand that its practitioners be prepared in a special way, and it may even demand acceptance of a certain ideology . . . That is true of physics, just as it is true of religion, or of prostitution. But such special ideologies, such special skills have no room in the process of *general education* that prepares a citizen for his role in society. A mature citizen is not a man who has been *instructed* in a special ideology . . . a mature citizen is a person who has learned how to make up his mind and who has then *decided* in favour of what he thinks suits him best.

The science with which Whitehead and Feyerabend are concerned receives its overall definition in terms of its (difficult) relations with philosophy, mythic thought, etc. Other paths could be followed which would highlight other relationships, for example, those between scientists, State policy and 'intellectuals' (see Brecht's *Life of Galileo*). But another possible landscape also opens up, and this is what we shall go on to explore. Here the question is no longer, 'What is modern science in general?' but 'What is this particular discipline we call mathematical physics?'

The Question of Mathematical Physics

A new affair, and a new way of approaching the texts. Here we no longer have a confrontation between Galileo and the Church, nor again between science and other forms of thought. The adversaries depart from the field, or only appear there when necessary to demonstrate the singularity of what Galileo put forward and the incomprehension it must have met. So Alexandre Koyré calls upon Descartes and Mersenne, who represent 'modern thought', unlike Galileo's theological opponents, but still did not accept Galilean physics.

The field is now therefore purified of any context. Only the texts count, but the texts in their turn are at the centre of an affair concerned with the question of theoretical physics, of the new relationships Galileo is supposed to have established between mathematical and empirical descriptions. Even before Feyerabend, Koyré had shown in his *Études galiléennes* that Galileo did not 'respect' the facts. But for him this isn't a question of propaganda, justifying a political critique of scientific pretensions. Galileo's physics wasn't 'impregnated with interpretation', just like every other description: what it did was invent a new, experimental relation to the facts. It was because Galileo was convinced that he had to go beyond the phenomena in order to grasp their essence, and because he believed that only mathematics could express this essence, that he was able to formulate the laws of falling bodies. And it is this formulation which distinguishes Galileo from Giordano Bruno, who had also said that because the Earth moved, stones would fall to the foot of the mast of a ship in motion, and distinguishes him too from Descartes, who formulated the principle of inertia – which Galileo never did – but rejected the idea that a falling body might be following a law.

> It is very easy to understand Descartes, who 'denied' all of Galileo's experiments! How right he was! because all of Galileo's experiments, at least all of the real ones, those which resulted in measurements, in precise values, were falsified by his contemporaries. In spite of this it is Galileo who is in the right. For as we have seen above, he was not at all looking to found his theory on facts gained in the realm of experience: he knew perfectly well that this is impossible. And he was also aware that concrete

observation – even experiment – performed, for example in the air and not in a vacuum, or on a smooth plane and not on a geometrical plane, and so on *cannot* produce results as predicted by the analysis of the abstract case. Therefore he does not expect this of them. The abstract case is an assumption. Experiment can confirm that it is a good assumption. It can do this within its limited means, or rather within the limits of our means.

Galilean physics deals with abstract examples: the concept of an absolutely smooth plane, of an absolutely spherical sphere, of a perfect vacuum doesn't come from the experiment, but allows this experiment to be judged in the name of a mathematical ideal. 'And so one shouldn't be surprised to see that "experience" cannot be in entire agreement with the deduction. But it is the latter which is right. It is this, and its "fictive" concepts which allows us to understand and explain nature, to ask questions of it and to understand its replies.'

Koyré feels that these arguments are enough to triumph over two kinds of opponents: the 'Marxist' historians who see in Galilean physics an expression of the new technical and economic imperatives of nascent capitalism, and the empiricists who explain it by the subordination of judgement to observation. There remains a third opponent, the more redoubtable, as we shall see, as Koyré himself relied on his analyses: this is Pierre Duhem. Koyré's *Études galiléennes* regularly refer to the third part of Pierre Duhem's *Études sur Léonard de Vinci*, which bears the subtitle '*The Parisian Precursors of Galileo*'. But these references are not enough to suggest how carefully Koyré has to go forward, how close his argument about the fictive, abstract character of Galilean concepts comes to Pierre Duhem and his theory of precursors.

The interest which directs Duhem's work is clear and explicit: Galilean science does not mark a rupture with medieval Christian science. The effort which enabled modern physics to replace the physics of Aristotle

> was based in the most ancient and the most illustrious of the medieval universities, the University of Paris. How could a Parisian not be proud of it? In the fourteenth century, its most eminent promoters were Jean Buridan from Picardy and Nicolas Oresme from Normandy. How could a French person not feel a legitimate pride? It was the result of the stubborn struggle by the University of Pairs, in those days the true guardian of Catholic orthodoxy, against neo-Platonic and Peripatetic paganism. How could a Christian not render thanks to God?

The landscape opens up suddenly. Brought alive again by Duhem's interest, these medieval theological controversies threaten to engulf us. Here we must resist and limit ourselves to the point in dispute between Duhem and

Koyré: at the end of the Middle Ages, was the 'fruit', as Duhem says, so ripe that the lightest touch – Galileo's – was sufficient to detach it? Or was it rather the case, as Koyré argues, that Galileo had no proper precursors and was indeed the inventor of mathematical physics?

Here at last we have to deal with the content of Galilean physics, and more precisely with the laws of naturally accelerated motion which Duhem discovers in the work of the fourteenth-century thinkers. This is the most dangerous point for every history of science, when the scene, empty now of characters, is occupied by a problem similar to those dealt with by beginners in physics, threatening to have not a historical but a pedagogical character. In fact, however, the concept of a 'uniformly deformed quality' invented by medieval scientists does not send us back to our school benches but returns us to that speculative moment when heat and cold, dry and wet, ceased to stand in the relation of mutual opposition which governed their use in the Aristotelian system.

What is a quality, for Oresme? Charity may be a quality, as may speed or heat. A quality may be characterized by its degree, by its intensity. A Christian isn't charitable, but more charitable or less. A body is not hot or cold, it is more cold or less cold, and in the same way, more or less hot. But a quality also has extension, it qualifies a space and a time. Charity qualifies the life of the Christian. A certain degree of heat qualifies a region of the body, or indeed such a region of the body for such a period of time. How can one represent the 'becoming' of a quality, the way in which it increases or decreases from moment to moment and from point to point? It is this development that Oresme represents by a graph in two dimensions. The horizontal line, or longitude, represents an extension of the quality, in space or time. At each point of the horizontal a vertical, the latitude, may be constructed: its length represents the intensity of the quality at this moment or at this point. The succession of intensities is therefore expressed by a plane figure. A triangular figure (or a trapezoidal, if the initial intensity is not of zero value), represents a 'uniformly deformed' quality such as heat, which diminishes in a linear manner through time. A rectangle represents a uniform quality.

According to Oresme, this graphic representation allows a more rapid and perfect grasp of the properties of the quality. In particular, it allows him to give a geometrical demonstration of a rule put forward by teachers at Merton College, Oxford (notably Thomas Bradwardine, Richard Swineshead and William Heytesbury, who were philosophers, theologians and mathematicians at the same time) during the first half of the fourteenth century.

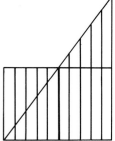

'Every quality uniformly deformed has the same total quantity as if it affected the subject uniformly according to the degree of its middle point': the quantity of the quality is measured by the area formed by the latitudes which succeed each other in an interval of time or in a given space. It is clear that the area of the triangle representing the quantity of the uniformly deformed quality will be equal to that of a rectangle representing the quantity of a uniform quality which has the same extension and an intensity equal to that of the uniformly deformed quality at its middle point.

Oresme explicitly deals with the case where the quality is in fact speed, increasing or decreasing in a uniform manner in the course of time. And in this case the Merton rule leads to the establishment of a relationship of equivalence between the quantity of an accelerated movement and that of a uniform movement. It corresponds in this case to one of the rules about uniformly accelerated movement taught today at school: the average speed of such a movement is the mean of its initial and final speeds. Galileo for his part expressed it as follows: 'The time it takes for a certain space to be traversed by a moving object beginning at rest with a uniformly accelerated motion, is equal to the time it takes for the same space to be traversed by the same moving object with a uniform motion whose degree of speed is half the last and greatest degree of speed attained in the course of the accelerated motion.'

Equivalence of the medieval and modern formulations presupposes that when the quality is speed, its quantity, the area produced by the longitude and latitude, is nothing other than the distance covered during the time measured along the longitude. Duhem is forced to recognize that Oresme nowhere explicitly identifies quantity with distance. But, he supposes, this is because for Oresme this was self-evident. And in any case, he points out, Oresme's disciples understood without any problem that the 'common quantity' shared by uniformly accelerated movement and uniform movement of mean speed during the same interval of time was indeed nothing other than distance.

When we look at Oresme's diagram, we moderns can't help but read off the Galilean laws of accelerated motion, the descrip-
tion of a speed varying in a linear manner through time. And the question immediately presents itself: why didn't Oresme go further? Why didn't he understand that he had the means to establish measurable physical relations between the distance travelled by such a movement (the area of the triangle) and time? Isn't it obvious that the area increases as the square of time? And doesn't the application of the Merton rule to different sections of the figure allow one to put forward an equivalent relationship between

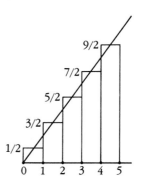

distance travelled and the square of the time taken to travel through it; in uniformly accelerated motion, the distances travelled during successive equal intervals of time are like the series of uneven numbers. In fact, Clagett points out that in his *Questions on Euclid's Elements*, when examining the general case of a uniformly deformed quality, Oresme put forward a rule according to which if one divides the longitude into equal parts, the successive quantities produced by this division will be like the series of uneven numbers. Why was it necessary to wait for Galileo to learn that the distance travelled by bodies in free fall in fact increases with the square of the time of falling?

In putting these questions, we presuppose precisely what Koyré is contesting when he denies that the *physical* theory of the uniformly accelerated motion was born in the Middle Ages. Certainly, he admits, Duhem was able to locate a few cases where a medieval scholar illustrated the abstract notion of a uniformly deformed quality by the example of a falling body. But nobody thought that the purely mathematical analysis of a concept could take the place of physical explanation in providing an account of falling bodies. Galileo had no precursor, Koyré explains, and the Middle Ages were not pregnant with the theory of falling bodies, because none of the mediaeval theorists had taken the decisive step which Galileo himself would accomplish in 1604: to abandon any physical explanation of the movement, no longer searching for the reason why the rising body slows down and the falling body accelerates. No one identified the explanation with the mathematical essence. Galileo had no precursor because he was the first to conceive the search for laws in Platonic terms: mathematical law rather than physical explanation. According to Koyré, it is a philosophical influence, Plato's influence, which explains the particularity of Galilean physics, and through this the invention of the mathematical physics which explains the observable empirical with the unobservable mathematical.

However, in its essentials Koyré's response concedes the thesis he is trying to fight. If the birth of mathematical physics derives from a philosophical decision, then there's no need to ask how Galileo invented the abstract mathematical essence of accelerated motion from which he deduced the phenomenal laws which describe falling bodies (the two rules quoted above). It would seem then that Koyré is implicitly admitting that Galileo got them from the medieval tradition. Duhem and Koyré discuss the same story, one stressing the continuity of the mathematical analysis and the other the radical innovation constituted by the decision to identify the understanding of actual physical motion with an abstract mathematical essence. The theory of rupture needs and presupposes the theory of continuity all the more, because Koyré, as we have seen, also denies any role for the experiments Galileo carried out on the subject of falling bodies.[2]

But did Galileo inherit the concept of accelerated motion from medieval scholars? Here we reach the last bifurcation in this account. Again the land-scape changes, as well as the texts. We are no longer listening to the Galileo of the *Dialogue* or of the *Discourses and Mathematical Demonstrations*, Galileo demonstrating to the public the discoveries he had made decades before. Now we must look at the Galileo who took the decision, if decision there was. We have to go back to the fateful date of 1604, when, in a letter to Paolo Sarpi, Galileo explained the principle to which the motion of falling bodies corresponds. According to Koyré, this is the date at which Galileo, then forty years of age, becomes our Galileo, the one who abandons all physical hypotheses in order to state the mathematical essence of acceler-ated motion.

But in 1604 Galileo made a mistake. Put in modern terms, he related acceleration not to time but to distance. The question is whether, as Koyré claims, this error confirms the theory of 'philosophical influence', Galileo having allowed himself to be carried away by his geometrical passion – but apart from this detail being the same Galileo when mistaken in 1604 as he would be in 1609, when historians believe he wrote the note showing the correct version, which would then be taken up practically word for word in the *Discourse* published in 1638.

The Affair of 1604

The history of science doesn't follow the logic of geographical representa-tions discussed by Jean Perrin. Here we are dealing with five years: the time-scale is reduced, while the few texts which are to be investigated, the scribbled notes in which Galileo attempted to refine his ideas and their implications, become more detailed. The problem itself, however, doesn't become more detailed, but is again transformed, as it has been each time we have raised the question, 'What were the consequences of Galileo?'

We have gone from freedom of thought to modern science, and then on to mathematical physics. As we shall see, what is at issue, for me, in the contrast between the Galileo of 1604 and the Galileo of the *Dialogue* and the *Discourses* is the question of the specificity of rational mechanics in relation to mathematical physics. This is my Galileo affair, but to approach it, we need to know first of all how to read the Galileo of 1604.

> I postulate (and perhaps I will be able to prove this) that a heavy body falling naturally moves with a speed which continually increases as the distance from the point at which it started grows larger. For example, if the body starts from point A and falls through the line AB then I postulate that the degree of speed at the point D will be as much greater than the degree of speed at the point C as the distance DA is greater than CA; thus

the degree of speed at C will be to the degree of speed at D as CA is to DA, and similarly at each point on the line AB the body will have a degree of speed which is proportional to the distance of that point from the starting point A. This principle seems to me to be very natural, and it is in agreement with experiments on machines and instruments which work by impact, in which the effect of the impact is greater to the degree that the fall is from a greater height. If this principle is assumed then everything else can be demonstrated.

Before we go any further, let's look at this way of approaching the question. Galileo is certainly 'mistaken', because he describes a speed which increases with distance and not with time. But it is easy to understand Koyré, who thinks that 'everything' is there already: the formulation of an abstract mathematical principle from which will follow the observable, pheno-menological properties of the motion of falling bodies. Should we take seriously Galileo's reference to a purely empirical experiment with machines which act by shock, or does this not rather represent a simple search for plausibility clearly foreign to the line of reasoning? We should note too, as another problem, that it is only when Galileo speaks of 'degrees of veloc-ity' that he states not only that they increase with distance but also that this increase is a *linear* function of distance.

Make a line AK at any angle with AF, and draw parallels CG, DH, EI, FK through the points C, D, E, F. Since the lines FK, EI, DH, CG are to each other as the lines FA, EA, DA, CA, it follows that the speeds at the points F, E, D, C are as the lines FK, EI, DH, CG. Thus the degrees of speed increase at each point on the line AF in the same proportions as the increase in the parallels drawn from these same points.[3]

Apparently Galileo has just taught us to draw a diagram very similar to Oresme's: the degrees of velocity, the 'intensities' successively achieved by the accelerated motion are the par-allels drawn at each point of the 'longitude'. What then is represented by the area of the triangle, the 'quantity' of the medieval motion? This cannot be distance, as it is in Oresme's diagram, because distance is already represented by the vertical line. And this is where things begin to go wrong.

Moreover, since the speed with which the moving body goes from A to D is composed of all the degrees of speed which it has acquired at each of the points on the line AD, and the speed with which it has covered the line AC is composed of all the degrees of speed acquired at each of the points of the line AC, it follows that the speed with which it has

covered the line AD is to the speed with which it has covered the line
AC in the same proportion as all the parallels drawn from all the points
of the line AD up to AH are to all the parallels drawn from the line AC
up to the line AG: and this proportion is that of the triangle ADH to the
triangle ACG, i.e., that of the square of AD to the square of AC. Thus the
speed with which the line AD has been covered is to the speed with
which the line AC has been covered as the double proportion of that of
DA to CA.

Here, suddenly, the double resemblance between the Galileo of 1604 and
our Galileo on one hand and Oresme on the other disappears. The quantity
of the accelerated motion, the area of the triangle, is nothing other than the
velocity itself! What then are these degrees of velocity whose sum makes up
the velocity? If we insist that Galileo thought of a uniform increase in velocity as a function of distance travelled, how can the degrees of velocity which
increase as a function of distance be summed to give a velocity? What is
summed to give a velocity, for us and for our (future) Galileo, are the
accelerations. But then the 1604 degrees of velocity cannot be our acceleration,
because the degrees increase with distance, whereas for us the acceleration
is uniform. Koyré's discussion here is so indulgent as to be almost blind, and
shows how important it is for him that all the oddities of Galileo's texts
should follow from the 'initial mistake'. He refers to Duhem's analysis, implying that he agrees with him in understanding the triangle and Galileo's
mistake on the basis of the medieval diagrams. Strangely, he accepts that the
total velocity of the moving body is the sum of the instantaneous velocities
(of the medieval intensities?) which it acquires at each point, and limits
himself to the remark that the sums of velocities which increase as a linear
function of distance cannot be represented by triangles.

We should note, that as regards velocities proportional to distances travelled,
Galileo doesn't speak of velocity at a point but of the velocity with which
the body has reached this, and then let us go on to what follows, which
is, if it is possible, even worse.

And since the ratio between the speeds is the inverse of the ratio of the
times (for to increase the speed is the same as to decrease the time) it
follows that the time of the motion through AD is to the time of the
motion through AC as the sub-double proportion of that of the distance
AD to the distance AC. The distances from the point of departure are
thus as the squares of the times, and consequently the spaces covered in
equal times are to each other as the odd numbers *ab unitate*: this is in
agreement with what I have always said and with observed experiments.
Thus all the truths are in agreement.

Here again, Koyré shows himself to be extraordinarily indulgent. He reproaches Galileo for not having seen that the ratio of the velocities is the

inverse of that of the times only in the case where the distances travelled are identical, but he doesn't point out that this relation is limited to uniform motion. Moreover, he doesn't notice that *even if* this relationship is accepted, *it would not allow* one to pass on to the relationships which follow, between distances and times, which correspond, according to Galileo, to observed experiment. Here it isn't possible to speak of a 'conceptual' error. Elementary mathematics are sufficient to conclude that the truths are not in agreement.

We are faced with a choice: either that, error or not, Galileo didn't know at the time how to reason, or indeed how to pass from one mathematical relation to another, or perhaps the reading is wrong, and the Galileo of 1604 was not the Galileo of the *Dialogue* and of the *Discourses*.

Here the professional historian again appears upon the scene, not to challenge the stories of the actors themselves, but because in 1604 Galileo said nothing about himself, or very little. Study of the texts published by Galileo doesn't allow one to say who Galileo was in 1604.

The historian Stillman Drake is well-known for having contradicted Koyré on one essential point. He has shown that with the resources he described, Galileo would have been able to make far more precise experiments than Koyré thought, experiments whose results, though they might not be exact, would certainly have been significant. In particular, the relationship according to which the distances travelled in equal times increase like the series of odd numbers could well have an empirical origin, as Galileo claimed.

Drake's demonstration on the 1604 problem is based on a reconstruction of experimental arrangements described by Galileo, but also on Galileo's notes. These 200 sheets show Galileo passing from one problem to another, scribbling strings of figures, diagrams, outline arguments. Galileo writes, 'I say that . . .' but he is not addressing the public. Perplexed, he is thinking for himself. And these sheets are not dated, which is why we need a professional historian capable of the real detective work needed to put them into chronological order.

As a result of this detective work, Drake offers us a very different picture of Galileo in 1604. At this time he wouldn't have tried to deduce the observable consequences of an *a priori* mathematical definition at all. This would imply a homogeneous field of meaning, in which the articulation of the different terms is fixed. It was this articulation, however, that Galileo was trying to construct. Galileo's 'demonstration' is nothing other than a search for 'agreement' between the different relationships he had grasped concerning motion. In other words, Koyré and Duhem were deceived by the similarity between medieval and modern definitions, and didn't understand how far the Galileo of 1604 was from this line of continuity, the extent to which the 'degrees of velocity' indicated by the parallel lines of his triangle were different from medieval intensities.

Let us return to the medieval definition of uniformly deformed motion. It implies that, like any other quality, velocity must be defined along two dimensions, longitude and latitude, extension and intensity. Velocity is therefore defined as a magnitude relative to space or time (its extensions), and it is measurable in terms of the distance travelled and the time taken to travel it. Space and time, for medieval scientists as for ourselves, were continuous magnitudes: one can say that such a body is at such a point at such an instant. But this did not mean that velocity is a continuous magnitude, that a body can be said at any particular instant to have such an (instantaneous) velocity. The medieval degree of velocity expressed a logical consequence of the conceptual definition of accelerated, 'deformed' motion, so it could be posited *a priori*. It was not, however, a measurable physical magnitude: it had no extension and therefore no spatio-temporal measure. Oresme and his successors, when they describe how the velocity of a movement increases, always decompose it into a succession of uniform motions, the intensity of the velocity varying in a discontinuous manner after a determinate interval of time: during each interval, the velocity can be defined by the relationship between the distance travelled and the time taken to do so. For medieval scholars, the measurement of a velocity, even at the level of conceptual abstraction, involved relating the movement characterized by this velocity to a uniform movement of the same extension. The velocity of a uniformly deformed movement could only be quantified by using a uniform movement as an intermediary, by the equivalence between the quantity of the uniformly accelerated movement and the corresponding quantity of a uniform movement.

According to Drake, Galileo cannot be considered as the faithful heir of the medieval tradition (which was, he points out, in any case held up to ridicule in the Italy of Galileo's time). The medieval definition produces uniformly deformed motion as a concept, posited *a priori*. Galileo, on the other hand, tried to characterize the accelerated motions of falling bodies as they happen. He tried to give a measurable physical meaning to the idea that such movements gain in velocity. He attempted to describe these movements not as decomposable *a posteriori* in terms of a succession of motions of increasing velocity, but as the accumulation of what the body gains at each point in its fall. This question implies that, unlike the medieval scientists who defined the quantity of velocity of an accelerated motion by its equivalence to that of a uniform motion, Galileo conceived of it as the product of an accumulation of quantities which should themselves be measurable.

The 1604 text then becomes clear, at least in its first part. Galileo is proposing a local measure, by means of shock, of the degree of velocity at a point. Koyré didn't bother to discuss this empirical argument, although it can't be intended to convince any public at all. He did not take seriously that for himself and for himself alone, Galileo notes that, in the case of machines which work by percussion, the effect depends on the distance travelled by the falling weight. Here at least is a measure which does not refer to a motion produced across a distance, during an interval of time; the shock is produced at a point, just as a degree of velocity is reached at a point. There follows this 'principle': the shock is the measure of the degree of velocity. The degrees of velocity therefore increase proportionately to distance. The idea that the sum of degrees of velocity can be identified with spatio-temporal velocity, the velocity of a body travelling a certain distance, then loses the glaring absurdity it would have if Galileo had known that his degrees of velocity were no other than the medieval intensities and his future instant-aneous velocities. Contrary to what Koyré thinks, Galileo's degrees of ve-locity in 1604 are not therefore instantaneous velocities. Velocity relates to a distance travelled, the degree of velocity to a body at a given point. Galileo is proposing a definition concerning two ideas whose articulation is not fixed *a priori*: the velocity with which a body travels a given distance results from the accumulation of degrees of velocity which the body acquired at the different points of its fall.

Will the velocity with which a body falls through a given distance therefore increase with the square of this distance? For us, such a relationship has to be verified (or rather refuted) by observation. But how, without the idea of instantaneous velocity, can one establish the way in which the velocity of a non-uniform motion varies with varying distances and times? The velocities vary with the distances, perhaps, but the only magnitude capable of direct measurement is the time taken to travel the distances. Galileo measured the distances travelled successively during equal intervals of time. He deduced from these observations that the ratio of the distances to each other was the ratio of the squares of the times taken to travel the distance. How could this relationship between distance and time be articulated with the relationship between distance and velocity that had to be established? It is here that Galileo bluffs his way through, seemingly positing a relationship between velocity and time, while, in fact, it is the agreement that he wishes to pro-duce between the two relationships that he already has – the relationships between the velocities are as the square of the distances travelled, the re-lationships between the times are as the square roots of the distances travelled – which will give a content to the 'inverse proportion' between velocity and time. As Drake reminds us, for Galileo the notion of 'inverse proportion' has no technical meaning determined *a priori*. It allows one to give a common-sense idea – for the same distance travelled, to increase the velocity means to diminish the time – the appearance of a premise which allows Galileo to say what he needs: that in relation to distance, velocity and time play as it were contrary roles, the square and the square root.

In 1604, says Drake then, Galileo is not primarily a mathematician but a perplexed physicist. He is trying to articulate among each other the different measurements which may be made of an accelerated motion, and particu-larly the measurement by shock and the rule of the odd numbers *ab unitate* which establishes the relationship between distances travelled and the time taken. He is trying to get these measurements to accord with each other, that is to say, to understand in each case what is being measured. The deduction will only follow later, in 1607 – according to Drake – when Galileo knows what he is measuring. And this is not as a result of a philosophical decision, but because he takes the step which is still so difficult for schoolchildren of our own day, dissociating velocity from its spatio-temporal measurement. As Salviati says on the third day of the *Dialogue*, in accelerated motion the moving body does not remain for one instant at any degree of velocity. And yet, at each moment and each point of its fall, this body has an instantaneous velocity. The degree of velocity is now the instantaneous velocity, a velocity which does not characterize any actual motion, a velocity at which the body travels no distance in no time. The velocity is no longer the attribute of a movement but of a body at a particular instant (or a particular point).

In 1608 Galileo made experiments in which the (instantaneous) final velocity of a body was converted into uniform motion. A grooved inclined plane was connected to a grooved horizontal surface, and the uniform velocity with which the moving body rolled along the horizontal surface was the velocity it had at the point where the two grooves joined together. Measurement of the variation in velocity of a fall no longer presented a problem. It was enough to vary the starting-point of the moving body on the inclined plane to obtain the relationship between distance travelled on this plane and the velocity resulting from the descent through this distance (the ratio of the velocities to each other is that of the square roots of the distances travelled). Galileo even went on to reconvert the uniform motion (in an experiment whose possibility has been verified by Drake and his colleagues): the horizontal groove led to the edge of the table, the ball fell and a numerical relationship was established between the distance from the edge of the table to the point of impact and the starting height in the inclined plane. Galileo the experimentalist had been born. Now he knew what he was measuring.

Drake concludes, with Koyré in mind: 'It is a mistake to think that he assumed from the outset that mathematics governed nature and physics must conform to it; rather mathematics gradually forced his hand in this thorny question of literally continuous change. Galileo became our Galileo not as the result of a philosophical decision, not because he decided to think about abstract bodies, spheres perfectly spherical and hard, perfectly smooth planes. Nor was it enough simply to abandon a physical explanation of falling bodies in favour of a definition on the basis of mathematical law. This law still had to have a physical meaning. One still had to know how measurable and observable velocity could be considered under the purely mathematical concept of the 'degree of velocity'. Koyré and Duhem therefore underestimated the problem that Galileo had to deal with: to conceive of a velocity independently of the movement which allowed it to be measured, which also meant to invent a physical meaning, a way of characterizing and of measuring a velocity to which no movement corresponded.

The Measurement of Accelerated Motion

What have we learnt from this new Galileo affair, his 'mistake' of 1604? How does it affect, as others do, the way we read what follows from Galileo? At first glance one can take Stillman Drake to be one of those empirical historians criticized by Koyré. Doesn't he revive the idea of a Galileo measuring and establishing empirical relations? But perhaps the other thing we might learn from his analysis is the common unspoken presupposition of the opposed philosophical categories of empiricism and rationalism: they underestimate the problem of naming the facts, of knowing what a relationship characterizes, whether empirical or theoretical. For this is precisely what Galileo didn't know in 1604. In other words, before the philosophers – including Galileo himself addressing the public and discussing what he had done – could argue among themselves, there had to be a solution to an apparently far more humble problem. How could one measure accelerated motion, without subordinating it to the categories of uniform motion, the distance travelled and the time needed to do it?

Galileo, who had submitted observation and physical reasoning to the ideal conditions of mathematics, was acclaimed by Koyré as the inventor of mathematical and experimental physics. Koyré couldn't see the problem which led him to this invention. How could one give physical sense to the logico-mathematical concept of accelerated motion? It wasn't enough to imagine a physical instantaneous velocity: it also had to be linked to observable spatio-temporal magnitudes. It was therefore necessary to redefine space and time in such a way as to make sense of something other than the velocity of an actual movement, so that they allowed motion to be measured in terms other than those suited to uniform motion, the distance travelled and the time taken to do so.

In 1604 Galileo brought new kind of measurement into play, the measurement of motion accelerated by distance, independently of time. He referred to the effect of his percussive machines, which depended only on the height from which the weight fell. I now have to show that this type of measurement – which I shall call causal, because it characterizes what a body has gained by the effect this body is capable of producing as a result, because it transforms what requires quantitative evaluation into the cause of an effect which allows it to be evaluated – is at work after 1604 in the texts which won Galileo recognition as the 'founder' of physics in the modern sense of the word.

Let us move on now thirty years after the episode we have just examined, to the time when Galileo, condemned by the Vatican, is writing his last great work, a true scientific testament, the *Dialogues Concerning Two New Sciences*. During the course of the third day of the imagined discussion, Galileo deals with the question of 'naturally accelerated motion'. And it really is dealt with: Salviati, Sagredo and Simplicio are there, just as in the *Dialogue*, but the text is no longer structured by their discussions, which serve only as a commentary on points which Galileo particularly wants to draw out. When everything has become clear, the theorems, propositions, corollaries and scholia are all linked together in a necessary and impersonal argument, as in a textbook. So Galileo uses Sagredo to mark the distance between what he is putting forward and a 'pure abstraction' of the medieval type. Salviati has just come out with an abstract definition:

> I say that a motion is equally or uniformly accelerated when, starting from rest, it receives in equal times equal moments (*momenta*) of velocity [and Sagredo remarks] Although I have nothing against this definition, rationally speaking, nor against any other, whoever might produce it, because they are all arbitrary, I am still doubtful, if I may say so without offence, whether such a definition, elaborated and accepted in the abstract, is fitting or suitable for the type of accelerated motion obeyed by naturally falling bodies.

There follows an exposition of the difficulty which in any case the definition gives rise to: if time is infinitely divisible, the degrees of velocity corresponding to the 'instants closest to the starting-point' would, if the body then began a uniform motion, correspond to a velocity infinitely small. And Salviati responds by passing from measurement by uniform motion to measurement by the effect of shock: he considers the effect of a hammer dropped on a pile from less and less height. 'Finally, if one drops it from the height of a finger, what more will it do than if it had just been placed there without percussion? Very little, you may be sure, and the effect would be imperceptible if it were lifted only the thickness of a leaf. But because the effect of the percussion depends on the velocity of the moving object, who can doubt that the motion would be very slow and the velocity absolutely minimal when the effect of the shock is imperceptible.' It is this relation of cause and effect which gives plausibility to the instantaneous velocities as quantities increasing in a strictly continuous manner, in the same way as time or height.

It isn't yet a question of a quantitative measurement. In Galileo's abstract definition, it is the time of fall which serves to measure the velocity attained, and the other possible articulation between the velocity and the distance of the fall is only used to calm down the disconcerted Sagredo. However, after other exchanges during which Salviati recounts the hesitations and even the errors of 'the Author', when the definition has been properly explained and considered as 'settled', it is time to demonstrate the 'rigorous agreement' between the properties which may be demonstrated on the basis of the definition and the results of experiment, an agreement which 'makes the difference' with arbitrary abstract definitions. Now Salviati asks that one single principle be accepted as true: 'The degrees of velocity acquired by the same body on different inclined planes are equal, provided that the heights of these planes are equal.'

Salviati will try to demonstrate the probability of this principle by argument from analogy. A few pages later, when he wants to use it, he offers a demonstration; this is a version in dialogue form produced by his pupil Viviani for the text of 1656, from a text dictated by Galileo in 1639. This addition confirms the essential importance of the principle. In fact, the definition first put forward by Galileo – the body receiving equal degrees of velocity in equal times – has no operational application except in the comparison of motions characterized by the same acceleration, while the inclined plane, the experimental apparatus *par excellence*, allows the degree of inclination, and therefore the acceleration to be varied. How could a velocity be measured in such a way that this measurement included as a variable the acceleration which the inclined plane allowed one to manipulate? How could one compare the accelerated motion of two bodies rolling along different inclined planes?

It is here that the measurement of velocity attained by the height of the descent during which it was acquired will play an essential role. With inclined planes, the time of descent means nothing by itself; it doesn't allow any relationship to be established because it varies with the angle of inclination. The height of the descent, on the other hand, if it allowed the determination of the velocity reached at the end of it, would allow the deduction of the time taken by the descent if the length of the plane is known. The measurement of acquired velocity given by the height of the descent alone will effectively give Galileo the means, in theorems and propositions and scholia, to relate the times of fall for inclined planes of varying length, varying height, and finally variations in both length and height.

How does Galileo justify the unequivocal relationship between the degrees of velocity acquired during the course of a descent and the simple height of the descent? In two different ways, as I have said, but in the two cases the justification makes it clear that motion is no longer characterized in terms of space and time. Instantaneous velocity is characterized in terms of what it makes the body capable of at the given moment, which is to say through a causal measurement, which assumes an equivalence of the cause to be measured *with* the effect which makes it possible to measure it.

Galileo's first justification appeals to the movement of the pendulum. Galileo

describes an experiment showing that the height
h reached by a pendulum does not depend on
the course of its swing but only on the height
from which it started. A nail placed in the path
of the wire changes the curve of ascent but leaves
the height unchanged. The experiment proves that
the momentum acquired during the descent, which
is 'obviously' the same as that which makes the
pendulum capable of its matching re-ascent, makes

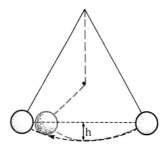

it capable of any ascent of the same height. And so neither does the moment
acquired during the descent depend on the curve but only on the height.

The 1639 demonstration brings in statics, the science of the equilibrium
of bodies. It identifies equilibrium as a measure of
the relative propensity to move (or of the *impeto*,
or of the energy, or of the momentum of descent)
of two bodies in equilibrium with each other. Their
rest signifies that the two bodies are halted, one
by the other, i.e. that the propensity to move of
each makes it capable of resisting the propensity
to move of the other. Now Galileo, who studied

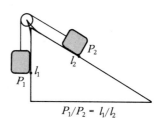

$$P_1/P_2 = l_1/l_2$$

mechanics at Padua (though one doesn't know whether he knew the more
or less contemporary work of the Dutch mathematician and engineer, Simon
Stevin), could say that these propensities to move were proportional to the
degree of inclination of the planes on which the bodies to be put into
equilibrium were resting. We will not consider the details of the demonstration. What is important here is that equilibrium, the balancing of one body
by another, was conceived by Galileo as a mutual cancellation of movement,
and therefore became the instrument of a causal measurement. Each body,
so far as it is characterized by a propensity to move, can be described as the
cause of an effect which is precisely the cancellation of the other body's
propensity to move.

What is measured by the ascent of the pendulum to the same height as
it had at the beginning? What is measured by the beam-balance? In neither
of these cases did Galileo explain what was equal. In the first case it was
Christiaan Huyghens who would specify the precise relationship: it is the
ratio of the squares of the velocities which is equal to the ratio of the heights
of the fall (we would say, $mv^2/2 = mgh$, where v is the velocity, h is the height
and g is gravitational acceleration). In the second case it was Gottfried Wilhelm
Leibniz who would systematically and explicitly identify the 'propensity' or
'momentum of descent' to the 'first movement', to the velocity taken by a
body at the first instant of its fall, beginning again and cancelled at each
instant by the equilibrium, that is to say, what the physicists would then
begin to call acceleration.

Galileo, for his part, had no need to explain what was equal in the measurement of cause by effect. It is the '=' sign that counts, the situation in which a cause can be measured by its effect in such a way that one can be sure that the measure is exhaustive, that is lets nothing get away. The identify of cause and effect, that is to say, of what is conserved, need not precede the measurement but can follow it, as with Huyghens and Leibniz. What is important to Galileo is that, in the situations presented, the cause is completely exhausted in the effect which it produces; the pendulum returning to its starting height has exhausted the velocity which it acquired in the course of its descent, the propensities to move are cancelled in equilibrium. And as a result of this, each situation allows the measurement, which is to say the putting into relationship, of magnitudes which cannot be characterized by actual movement happening in space and time. The first allows the measurement of instantaneous velocity, the second, the 'propensity to move', the first physical definition of what we call acceleration.

The Question of Rational Mechanics

After Koyré many people tried to place Galileo in relation to both Aristotle and Newton. He is thought to mark the destruction of the Aristotelian cosmos and of the questions authorized by this cosmic order. For Aristotelian reasons Galileo substitutes the mathematical question of knowing how a body falls. But Galileo is thought to have halted on the threshold of the Promised Land. He did not conceive of homogeneous and isotropic space in which an isolated body might continue its uniform rectilinear motion to infinity. Galilean physics is a science of bodies with weight in a space under the effect of gravity. With the notion of 'force' Newton, on the other hand, would give a generalized mathematical explanation of what Galileo had merely described in its effects, that is to say, gravity. Galileo then might have purified physics of its antique causality, but it was Newton who invented the new type of causality which made it possible to progress from (kinematic) mathematical description to mathematical explanation.

Why do bodies accelerate? In fact, Galileo never asked this question. Nor did he postulate the homogeneous and isotropic space corresponding to rectilinear and uniform inertial motion. But should we therefore follow Koyré when he describes Galilean space, waiting for Newton, as still bound up with a physics of bodies with weight? Of course, one isn't going to deny the importance of the Newtonian idea of force or of the idea of reciprocal interaction which it introduces into physics. That the Sun does not attract the Earth without the Earth exerting an equal force of attraction on the Sun, this is a radically new idea. None the less, it is not enough to say that Galilean space is 'not' isotropic and homogenous: it must be stressed that it too is radically new. It is a space knitted together by the sign '=', by the equality of cause and effect which allows the characterization of velocity. Tell me where you are coming from and whatever the course you followed, and I will tell you what velocity you have attained, and therefore also, with this velocity, where you can go. The '=' sign between cause and effect, which allows the definition of instantaneous velocity, articulates a determinate past with the ensemble of futures which this past makes possible for the body; it balances its past and its possible futures. Galileo might have destroyed the Aristotelian cosmos, but he replaced it with another, just as rational, subordinated to what Leibniz, a reader of Galileo, would call the 'principle of sufficient reason'.

Leibniz recognized the key role of the equality of cause and effect, and the name he gave his principle expresses very well the rational character that can be claimed by causal measurement. Who but a fool could deny that the effect is equal to the cause? But he also grasped its operational implications: the application of the principle of sufficient reason makes the moving body a measurable object, and determines the way in which it may be interrogated – whether the experiment be actual or conceptual. In the statement 'The cause is equal to the effect' it is the '=' sign, as we have seen, which precedes the definition of the cause and the effect themselves. The measurement of cause by effect therefore guarantees that causes and effects are defined completely, and independently of the particular choices or point of view of the one who describes them. Their 'objective' definition lets nothing get away. The principle of sufficient reason defines motion as subordinated to its own reasons, as self-determining.

The subordination of physical magnitudes to a '=' posited *a priori* defines not mathematical physics, but a particular tradition in physics called 'rational mechanics'. With this, I arrive at the end of my own Galileo affair. For me Galileo is not the one who prepares for Newton, Moses leading thought to the edge of the Promised Land, but the inventor of rational mechanics, of space redefined through the equality of cause and effect to which in the eighteenth century mechanical physicists such as Euler, d'Alembert and Lagrange would subject Newtonian forces.

If I am not to start on a Newton affair here, I will have to restrict myself to mere assertion. It ought to be remembered first of all that it was only in 1737 that Euler wrote what we call Newton's second law, $f = ma$ (where f is the force acting on a body of mass m, producing an acceleration a). Euler's definition of force subordinates this, whatever might be its phenomenological value (for example $f = mm'/r^2$, where f is the force of attraction between two bodies of mass m and m', the distance between them being r), to an *a priori* identity. As a cause, a force is equal to its effect, the acceleration. The '=' in $f = ma$ is the rational equality which one can posit *a priori* before identifying the cause and the effect it leads to define, the '=' invented by

Galileo. Maupertuis, who was a contemporary of Euler's, was not mistaken; he traced Euler's principle back to Galileo and not to Newton.

What can we say then about force in Newton's sense? Newton, as we know, was treated as a hero of positivist thought because he had written: 'I do not make hypotheses. I keep to the phenomena.' Many studies have since revealed a speculative Newton. But, very strangely, these studies allow us to understand why Newton had to keep to the phenomena, why the world, as he conceived of it, could not be made intelligible on the basis of a rational principle posited *a priori*. In other words, Newton did not belong to the field of rational mechanics whose discoverer was Galileo. For him, the forces express and give effect to the present activity of God in the Universe, and they can, phenomenologically, be recognized by observation, in so far as they determine an acceleration which permits the mathematical reproduction of observable motions. But their identities and their reasons are not related to their effects, but to what is beyond the reach of the physicist, God. The '=' which figures in the $f = mm'/r^2$ is not the '=' in $f = ma$. The second is posited *a priori*, making the physicist a judge giving his questions an *a priori* rational object, while the first is a phenomenological definition, which makes the physicist the reader of a world whose author is a God free of all rational constraint. Of course, rational mechanics is also the child of Newton, in the sense that the 'causal' space which now governs its calculations is not given once and for all, like the space of Galilean heavy bodies. The causality defining Galilean space was a function of uniform gravity; in rational mechanics it has to be redefined at every instant because the relative distances between objects and therefore the forces of interaction between these bodies are changing at every moment. None the less this space of variable links, described after Lagrange by a 'potential' function,[4] is not Newton's homogenous and isotropic space. Homogeneous and isotropic space goes with uniform motion: the space of rational mechanics is fitted to the idea of an instantaneous velocity freed from the categories of uniform motion and determined by what it makes the body potentially capable of doing.

Why am I so interested in the distinction between the general category of mathematical physics and the particular tradition of rational mechanics that I want to provoke a new Galileo affair on account of it? I could argue, prudently, that it seems to me to give a better account of the history of mechanics in the eighteenth century, a better account of the problem which had to be solved by Euler, Lagrange, d'Alembert and the others: the need to articulate together two types of causality, one discovered by Galileo and the other introduced by Newton's forces. But my interest isn't solely historical, no more than were the interests of Koestler, Whitehead, Feyerabend, Koyré and Duhem. What Koyré was trying to understand was the origin of a typically post-Einsteinian physics, a physics which had abandoned its image of science proceeding by generalization on the basis of the facts in order to

310

announce itself as a conceptual science, inventing the meaning of the observed facts on the basis of a theoretical hypothesis. My own problem is that within mathematical physics itself, not all the laws are equal, not all of them have the same status. Fourier's Law for the diffusion of heat is a mathematical law,[5] but it is, for most physicists, merely phenomenological; it bears no relationship to the causal measurement invented by Galileo.

Here then are new actors who appear on the scene of my story. My own Galileo affair seems to be a matter of what one calls 'internal history': no more Jesuits, no more Wars of Religion, no more diplomats, no more Pope, no more medieval thinkers, even no more Platonic tradition. Here is a man faced with a moving body, which he learns to define objectively, and how to articulate coherently its measurable variables. Yet the internal–external distinction doesn't work. Rational mechanics might well come from an idea, the operational application of the equality of causes and effects, but it is inscribed not in the heaven of ideas but in a concrete historical field which it electrifies, giving rise to new interests, new actors, new relationships between these actors, who all, engineers, physicists and philosophers – including me – are or were interested in the relevance of the equality of cause and effect which gives its identity to rational mechanics.

The Relevance of Rational Mechanics

How can one measure movement? By the quantity of movement, mv, as Descartes suggested, or by its 'living force', mv^2, as was suggested by Leibniz, following Galileo and Huygens? How can one justify this velocity to the power of two, foreign to the clear ideas of geometry? This confronts us with the dispute over *vis viva*, which lasted for decades. We can't go into the history of this dispute, which, until Kant and Lagrange, mixed together the histories of physics and of philosophy. Let us simply point out how it illuminates the particularity of rational mechanics. Descartes certainly admits that the effect is equal to the cause. But he wants to give the cause a rational definition, a clear and distinct meaning. The formula 'mv^2' has no rational meaning; the Cartesians that Leibniz had to deal with felt that the effect should be identified on the basis of a rational cause, and it could not therefore be that designated as a cause by mv^2. 'Mr Leibniz is mistaken,' said the Cartesian Abbé Catelan in 1686: he measures the 'force' of a moving body only by the distance this force enables the body to travel and neglects the time it takes to do it. And Leibniz replies that time has nothing to do with it, that it was like saying a man was richer because he took more time to gain his money.

How can one measure an effect? If we double the velocity with which a body is thrown upwards, is the effect that it goes four times as high, as Leibniz thought, or does it go twice as far, as was argued by Samuel Clarke, a disciple of Newton's who took the side of the Cartesians on this occasion, because it takes twice the time?

How can one measure an effect? This question wasn't asked only by philosophers but also by engineers. A typical question: at what velocity, that is to say, in how much time, should one raise a body so as to minimize the cost of the operation? This question mixes up what rational mechanics, since Galileo, had separated. Certainly, from the Galilean point of view, the time of fall or of lifting up 'counts', but not in evaluating the 'force' of a moving body from the point of view of its potential effect, that is to say of what its velocity makes it capable of at any instant. From this point of view, the change in height is enough: the course followed by the moving body in going through this change in height, the time which it takes are completely irrelevant. Time counts when it is a matter of characterizing the particular path along which the effect the body is made capable of by its velocity becomes actual. Depending on whether the body rises vertically, or along a plane of such an inclination, or follows such and such a curve, it will re-ascend in such and such a time. Here we rediscover the constitution of Galilean space: one must choose, either to describe an accelerated motion in space and time, i.e. in terms of its successive accelerations, or to describe it from the point of view of height alone, i.e. through the equality of the linkages which allows causes and effects to act as measures for each other. Space-time or height – one has to choose and the two can't be mixed together.

When one leaves the ideal world of rational mechanics for the world of the engineers, where bodies are subject to friction, time obviously begins to count in any case. Galileo's approach therefore makes no sense of the question the engineers were interested in. The reason defined by causal measurement doesn't only link up space, it also imposes an *a priori* grid on the field defined by the problem shared by engineers and specialists in rational mechanics: what can a movement do? It brings with it a judgement which situates the preoccupations of the engineers and defines them as merely relative to the gap which separates their world from the rational ideal of the mathematical specialists in mechanics. 'To the extent that bodies are subject to friction, the mechanical effect is always less than its cause. How then can the loss be minimized?' This is a question for engineers only.

With regard to the philosophers, the principle of sufficient reason meant an intrusion of history, so far as it challenged the geometrical self-evidence of the quantity of movement. As for the engineers, it raises the historical issue of how these situated themselves in relation to this 'rationalization'. As a result their problems were relegated, if not to the irrational, then at least to the status of side-issues in relation to the rational ideal that characterized the world within which such problems arose. Rational mechanics therefore gave

312

rise to a professional and political problem, that of the possible subordination of the engineers to a judgement which situated their practice. One can do no more than indicate a few historical signposts: in 1775 the Academy of Sciences in Paris brought upon itself the hatred of 'inventors', because it decided from then on to refuse, *a priori* and without examination, any design for a perpetual motion machine, whose aim was necessarily in contradiction to the conservation of the cause in its effect. At the beginning of the nineteenth century the professional education of French engineers was based on the principles of rational mechanics; the effective functioning of machines was judged in terms of the gap between the real and the ideal. The inventors had more or less disappeared, and the art of the engineer was a matter of applied mechanics.

But the story wasn't over. Fourier's law, stated in 1822, describes the way in which a difference of temperature levels itself out in the course of time. In the nineteenth century the diffusion of heat would become the typical example of an irreversible process, in the sense of the Second Principle of Thermodynamics, a process which annihilates its own cause with no possibility of return: the difference in temperature disappears without producing an effect that might recreate it. The Second Principle defines processes which are irrational in terms of the principle of sufficient reason, in terms of the equality of cause and effect.

The Second Law of Thermodynamics

The Second Law has been formulated in very many different ways, which reflect the complexity of the history from which it arose (the reinterpretation in 1850 by Rudolph Clausius and William Thomson (the future Lord Kelvin) of the optimal output of the transformation of heat into mechanical movement stated by Sadi Carnot in 1824). Its most universal expression we owe to Clausius (1865): 'The entropy of the Universe increases towards a maximum.' The Second Law of Thermodynamics defines the class of processes called 'irreversible' by the increase in a function, the entropy. No natural process can be characterized by a spontaneous decrease in entropy, which means that if, starting from a given state, a physico-chemical system has undergone a change which increases entropy, no natural process can restore it to its former state.

This is why philosophers and physicists have since been discussing the status of the Second Law within physics: is it a product of our approximate descriptions of a world 'objectively' governed by the principle of sufficient reason, and therefore 'merely phenomenological', or does it put the principle of sufficient reason in question? The Cartesian defenders of the quantity of movement, like the early engineers before the École Polytechnique, belong to the past. But it was these contemporary discussions, which are at the same time scientific, speculative and political (for to speak of status is to speak of hierarchy, of domination and subordination), about the difference between fundamental and phenomenological physical laws, which led me, after so many others, to the Galileo affair.

10

Refraction and Cartesian 'Forgetfulness'

Michel Authier

We travel the continents and the centuries, in order to tell the story of the emergence of a law of physics, uncovering, layer by layer, everything that lies behind it, though without ignoring the novelty it represents.

'It wasn't yesterday that the air came to lie over the surface of the globe. It is a law of nature, which it is logical to think of as having existed from the beginning of the world to our own day. So it would be reasonable to think that there has never been a time without refraction.' This is Kepler, in his *Paralipomena to Vitellion*, written at the dawn of the seventeenth century, and this is the way he presents the universality of refraction, the phenomenon responsible for so many natural wonders. Poets have been enchanted by 'rosy-fingered dawn', by crimsoned dusk, by haloes, glories, by multiple suns and rainbows. Astronomers have long noticed heavenly bodies becoming visible before the time they should appear above the horizon and abnormal angular distances between stars. They have seen coloured moons. Nomads and sailors have always been familiar with the mirage: oases, palm-trees and cities float above the dunes; and when the weather is very hot, coasts, lighthouses and boats become visible from very great distances.

Responsible for significant discrepancies in astronomical calculations (because observations of the relative position of heavenly bodies are affected by the angle of incidence of their light upon the atmosphere), refraction was from the earliest days a matter of concern to scientists. How often did Archimedes, himself the son of an astronomer and a brilliant observer, stand on the coast of Alexandria or of Syracuse and see the light bending at the edge of the sky, see it refracted in the sea? It is to him, probably, that we owe the first experimental description of the phenomenon: 'If you put an object in the bottom of a vessel, and if you move this vessel away until you can no longer see the object, you will see it reappear at this same distance as soon as you fill the vessel with water.' Thanks to the extreme simplicity of the arrangement, we can see clearly what the experiment was intended to do. Substituting water for the atmosphere, the vessel for the sky, and the object for the Sun, we have a small-scale model of the astronomical phenomenon of refraction. Later, the development of geometrical models and their numerization became sufficiently sophisticated for scientists to talk about refraction without making reference to any phenomenon in particular. In the following pages we will try and reconstruct certain threads in this single cord which joins together the Homeric dawn and the Sine Law.

It is quite common in the history of science for a problem to appear a long time before the emergence of what one customarily calls its solution (which corresponds often enough to the dissolution of its emotional connotations). This gap may of course be due to the intrinsic difficulties, though this isn't saying very much, because the difficulty of a problem is itself most commonly measured by the time it takes to solve it. In other cases, the problem does not remain in continuous existence, but disappears and then reappears a long time later, only to disappear again, without any self-evident continuity and even without any obvious logic. What is interesting about refraction, on the other hand, is the enduring character of the problems it gave rise to. From the Greeks to the scientists of the Enlightenment, from the Persian Gulf to medieval England, in all these very different societies we will discover people who were concerned to understand it.

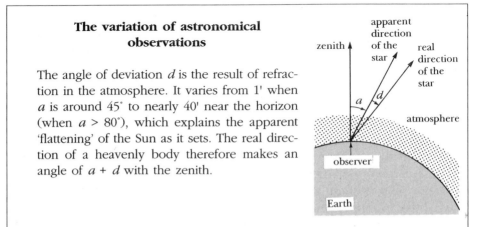

The variation of astronomical observations

The angle of deviation d is the result of refraction in the atmosphere. It varies from 1' when a is around 45° to nearly 40' near the horizon (when $a > 80°$), which explains the apparent 'flattening' of the Sun as it sets. The real direction of a heavenly body therefore makes an angle of $a + d$ with the zenith.

Before becoming a scientific object, light was studied by philosophers and artists. Despite their effectiveness, the electromagnetic or quantum solutions offered by modern and contemporary physics cannot dispel the original questions. Indeed, though we may not challenge the existence of the radiation (*lumen*), we can still wonder whether without *psyche*, light (*lux*), identified as the necessary and sufficient condition for the visibility of things, would still exist.

We shouldn't forget Hesiod's *Theogony*, where the couples Erebos/Nux and Aether/Hemera join together to give the absolute principles of darkness and light. Nor should we forget Dante, who wrote in his *Convivio*: 'It is the habit of philosophers to call "clarity" light as it is in its originating principle; to call it "ray" in so far as it traverses space; to call it "splendour" so far as it is reflected in another place which it illuminates.' And finally we shouldn't forget Kepler, who wrote that 'The ray of light is not at all the light itself which emanates.' Yet it seems difficult to escape the slow displacement of *lux* by *lumen*. The physical description of phenomena affects our perception of them and sensations are transformed by the knowledge of theory. It has become difficult to understand the birth of *psyche* and of *lux*, the essential conditions of a humanity capable of understanding and a world conceived as visible.

Light, the precondition of sight, is in Greek thought the most important medium of the relation between mind and the world. In the *Timaeus* Plato wrote: 'It follows from my argument that sight is for us the cause of very great benefit, for of the present theories we have about the Universe, none could have been held if we had seen neither the heavenly bodies, nor the Sun, nor the sky.' Later he goes on to say: 'From this we have drawn a kind of philosophy which is the greatest good that has come or will ever come to the race of mortals from the kindness of the gods.' We see here how thought, the daughter of light, now tries to understand what has engendered it.

One of the benefits of the history of science may be that it makes us understand how the obviousness of the 'natural' is in fact dictated to us by our scientific and cultural environment. If today it seems to us certain that light travels towards our eyes, four centuries ago this was far from being the case. Wishing to exclude humanity from the explanation of nature, the Atomists Leucippus, Democritus, Epicurus and Lucretius (see p. 143) were the only thinkers of Antiquity to believe that it was objects that made their own presence manifest. According to them, they did this by sending *eidola*, a certain kind of shadow, very thin peelings off the surface, simulacra which travelled through space in an extremely short time (the word really stresses simultaneity) to enter the eyes and imprint themselves on the retina with their forms and colours.

At the time, and indeed until the recent past, this theory was systematically

317

ridiculed. However, Book IV of Lucretius' *De natura rerum*, where the theory receives a lengthy exposition, repays an attentive reading: 'So, I re-peat, you are obliged to recognize these emanations of simulacra which strike the eyes and produce in us the sensation of sight . . . It is true that all bodies continually emit emanations of every kind, which go out in every direction, without ever stopping or coming to an end . . . The surface of all bodies being provided with a multitude of imperceptible corpuscles which can detach themselves without losing their order and original form and spring forward the more quickly the fewer obstacles they have to overcome.' We can see that this theory resolves the problem of the preservation of form in vision. In addition, it provides an explanation of the act of seeing which maintains a total independence of subject and object. This was far from the case in other Greek theories.

For the historian of science, Plato is the filter through which percolates the little we know of those who went before him. We shall not deal here with the problems that this raises. The reader should, however, keep this simple fact in mind: there exist practically no primary sources for Greek science, and everything we know derives from secondary documentation. Whatever their virtues, Plato's texts are of this kind. In them we discover that in the sixth century BC the Pythagoreans postulated a *quid* which left the eye to travel towards the object and to touch it. Sight, like touch, was an active sense, unlike hearing and smell, and later Aristotle would justify this by remarking on the convex form of the eye, which contrasts with the concave form of the nostrils and the ears. With various modifications, this was the idea that remained dominant for nearly two thousand years.

Empedocles, for his part, anxious to adapt his theory of similars to the problem of vision, developed a system of double emanations, originating in both organ and object, and meeting in the air between. 'When then there is the light of day all around the visual ray, this pours out, like to like, and combines with it. A single substance, adapted to our own, constitutes itself along the line issuing from our eyes, in such a direction that the fire which springs from the interior comes to strike against that which comes from the external objects . . . It transmits the movements through the whole body as far as the soul, and procures for it the sensation thanks to which we say we see' (Plato, *Timaeus.*)

But Plato did not restrict himself to these wonderfully poetical expositions of others' thoughts. In Book VI of the *Republic* we can find a primitive optics in the myth of the cave. Here light is separated from sight, and the Sun becomes its universal source. On the philosophical level, this allows Plato to define the relationships between knowledge, the real and the good. It also leads to a conception of light independent of the Sun and of sight. In this separation there is the beginning of a renewal of the physics of light, and this exercised a considerable influence on Western thought.

We know that, when he approached the Medusa, Perseus provided himself with a mirror, which avoided his being struck directly by her fiery glance. Thanks to this trick, he escaped the malefic fluid and was able to kill the hideous creature. This symbolism of the look is preserved in a good many poetical and popular expressions and it also allows the establishment of a simple relationship between visual and solar rays. It was Archytas of Tarentum (430–348 BC) who gave this a systematic expression: a fire left the eye in a straight line and went to touch the objects looked upon. The fire, the straight line and the eye-object direction of travel were characteristic of theories to come, apart from the theories of the Atomists, as we have seen, and also, as we shall shortly discover, Aristotle's theories.

It is very difficult to give a brief account of Aristotle's theory as it appears, for example, in the second and third chapters of his little treatise on sensation and on the sensitive. Here he systematically attacks all previous positions. For him nothing emanates from the eye or from the object, for in one case, any emission would make night vision possible, and in the second case, a needle would be visible in a haystack. For Aristotle, there is only an alteration of the intervening medium, which conveys a pressure onto the eye and which disappears with darkness. 'As has been explained in this work [*On the Soul*, II, 6–11], light is the colour of the diaphanous medium by accident . . . But what we call the diaphanous medium does not properly belong to the air or to water or to other substances so named, but it is a certain nature, a certain common force, which does not exist separately, but which is in the substance . . . In the diaphanous medium, the nature of light is therefore indeterminate.' At the beginning of this short treatise, Aristotle stresses the difficulty of reconciling the five senses with four elements. Is the diaphanous medium not perhaps another element in gestation? Until the beginning of the twentieth century, physicists of light would always be tempted to associate with light a substrate unlike any other.

In conclusion, we can say that the Greek philosophers had no stable or unanimously shared ideas about the nature of light. The primacy accorded to vision above all other senses (to the extent that for Aristotle, imagination takes its name, *phantasia*, from that of light, *phaos*) ensures that the problems presented by vision are among the central concerns of the great Greek 'scientists'.

The First Physics of Light

To see in the Lucretian *eidola* or in the Aristotelian diaphanous and its alteration the distant ancestors of photons, of ether or of waves, is not so much a misinterpretation as a nonsense, for this would be to impose a physics of matter on a theory of sensation. The first steps towards such a physics would be taken only in the Hellenistic period.

Although an Aristotelian, when he came to develop his first geometrical optics, Euclid based his work on Archytas' model. He did this by removing any reference to the theory of elements, and thus any reference to the order of sensation. There remained only the straight line and the direction of propagation. Following the example of his *Elements of Mathematics*, his treatise is developed on the basis of a series of postulates:

I.   Suppose that the straight lines which emanate from the eye are propagated divergently from the magnitudes;

II.   That the figure comprised by the visual rays is a cone having its point at the eye and its base at the limits of the magnitudes that are being looked at;

III. That the magnitudes upon which the visual rays fall are seen; while those upon which they do not fall, are not;

IV.   That the magnitudes seen under greater (smaller or equal) angles appear greater (smaller or equal);

V.   and VI. That the magnitudes seen under higher angles (or lower, or more to the right, or more to the left) will appear higher (or lower, or more to the right, or more to the left);

VII. And finally, that magnitudes seen under more numerous angles appear more distinctly.

This list clearly shows how close the physicist of light was to the mathematician, so much so that the concept of the straight line and that of the ray of light should be thought of as twins. The relationship between the angle of the cone and the size of the objects also leads one to suspect the existence of a close relationship between the then emergent optics and the theory of geometrical proportions which already existed in astronomy. It is worth remarking that the cones leaving the eye are of great geometrical simplicity compared to bundles of rays springing out from every point of the objects. We should also note Euclid's neglect of the fact that we have two eyes.

This optics, no more and no less than a system of perspective, was obviously not born *ex nihilo* in the city of Alexandria, where there was a concentration of men and materials indispensable to the 'scientific policy' of Alexander the Great, who had himself been Aristotle's pupil. The art of perspective, indeed, had existed for some time already. In his *Architecture*, Vitruvius recalls that since the time of Aeschylus it had been possible to create an illusion of reality in theatrical scenery, by the use of vanishing lines, and anyone who has seen a Greek temple knows how Greek architects enlarged the ends of the pediments so as to create an impression of equilibrium in their constructions.

The results brought together by Euclid appear to constitute a minimal basis for optical study. Light loses all substance, the rays are governed by an elementary geometry, vision is a matter of an eye at a single point and the world is reduced to a representation which may be observed from the left or the right, from above or from below. Paradoxically, this text, which is no more than a synthesis obtained by the impoverishment of other sciences, draws its richness from this poverty, establishing, for optical phenomena, a disembodied space and a set of abstract rules. Here we probably have the first elements of mathematical physics. Optics would never really lose this position as the most abstract of the physical sciences, even though it was closely followed by the mechanics it carried along in its wake.

The first treatise we know of in catoptrics (the science of mirrors) was attributed for a long time to Euclid, but is probably a recension of a lost work by Archimedes, who also made a systematic study of refraction. We should remember that this great mathematician was also an astronomer (see pp. 124–59) and that refraction is of considerable importance in astronomical observation. Unfortunately, no work of his relevant to our topic has survived, except for the experimental arrangement of vessel and object mentioned above (see p. 316), which appears in the pseudo-Euclidean catoptrics. According to Apuleius, other works contained an explanation of the rainbow and of refraction. Although we are unable to check the accuracy of this information, we should at least be struck by this juxtaposition of the two phenomena.

In the centuries which followed, the mechanical physicist and mathematician, Hero of Alexandria, and Claudius Ptolemy, the great astronomer, drew up tables of measurements of refraction. They have come down to us through the Arabs, who made their own improvements to them. In the fourth century AD, when Hellenistic science had already been in decline for a long time, Damianus was the last of the Greeks we know to have added to the edifice of Alexandrian optics. According to the historian Vasco Ronchi, it was Damianus who compared the rays of the Sun with the visual rays and demonstrated the identity of their properties. He also postulated that sight should attain the object to be seen as quickly as possible and, relying on this 'unitary principle', he demonstrated the law of reflection.

For eight or nine centuries, then, light was a central preoccupation in the mythological, philosophical and scientific thought of Greece, or perhaps, more properly, of the Mediterranean. The origin of the world's genesis in Babylonian, Egyptian, Biblical and Hesiodic accounts, light gradually lost substance and unity. Through the centuries this process of separation split optics into different parts: psychology, physiology and physics. In identifying it with its straight-line model, mechanical physicists and astronomers, usually Alexandrian, freed thought about light from the considerable problems posed by its nature. Above all, they wanted to develop a geometry of the movement of the visual ray to which light had now been reduced.

As a result, there arose the habit of dividing this physical optics into three parts: perspective, catoptrics (the study of reflection) and dioptrics (the study of refraction) (see p. 151). In this way a certain stability was found, and just as in the case of astronomy, there was no more to do than perfect the tables of measurements. Broken up like this, geometrized and emptied of substance, the Alexandrian theory of light was able, after a fashion, to solve the problems put to it by astronomy.

Primary and secondary rainbows

It was very soon noticed that primary and secondary rainbows always appeared at the same angle (around 42° and around 50° respectively). This explains why:

– the Sun is always at the observer's back;
– the rainbows are visible (depending on the latitude) only in the morning or in the late afternoon; as soon as the Sun rises above 42°, the rays of the primary rainbow pass above the surface of the Earth; similarly above 50° for the secondary bow;
– contrary to its appearance (visible to the observer only), the rainbow is not an arc of a circle in a plane, but results from all the raindrops passing through the space encompassed by two cones (shown here by lines) originating at the observer's eye and having their axes parallel to the direction of the Sun's rays, and with an angular radius of approximately 42° for the primary and 50° for the secondary bow.

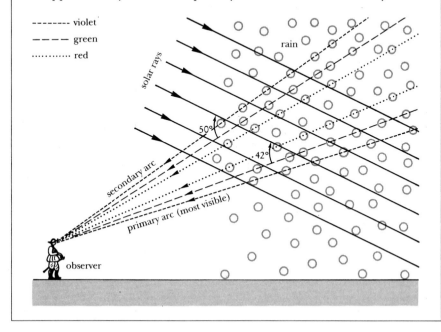

Arab Optics

Far from being mere intermediaries (see pp. 222–45), the Arabs entirely recast the science of optics. It is difficult to explain the occurrence of such an important change in a society which had renounced representational images. There is a hypothesis, difficult to verify, that it was this renunciation which allowed the abandonment of the perspective cone in favour of a ray of light which started from the object and entered into the eye. On the basis of this radically new idea, Arab physicists not only improved the quality of optical tables, but, most important, they restored optics to its previous unity. The most important figure is al-Haytham, known to the West as Alhazen, who was born at Basra in 965 and died at Cairo in 1039.

By reversing the direction of propagation of the ray, Arab optics was able to develop new answers, but more than this, it was able to abolish old problems and create new ones. In fact, for an eye which was now a receptor, the problem of emission no longer existed, nor the one presented by the ability to see objects at very different distances at the same time. On the other hand the problem of perception returned, though now greatly complicated by a punctiform decomposition of the object into a multitude of rays which must be reconstituted by the eye. This difficulty placed the eye at the centre of the problems treated and for more than six centuries raised it to the first rank among optical mechanisms.

Was this conception of a light which radiated from every point visible rapidly accepted by Arab scientists? It is difficult to reply to such a question, given the scarcity of early documentation. A prosperous city from the seventh to the ninth centuries, when al-Haytham was born Basra had been in decline since the weakening of Abbassid power. When he came to Cairo, which, with its mosque university, was enjoying a full-scale renaissance, he must have got to know the work of the Alexandrian School, the ancestral model of the great Arab scientific centres. Alexandrian geometrical optics, dealing above all with measurement and emptied of any concern with substance, would not have opposed the new conception. On the other hand, the highly argumentative and sometimes polemical tone of al-Haytham's work suggests that his ideas were not the dominant ones of the time.

In making the eye an optical apparatus, al-Haytham forcefully stresses the idea, already present in Aristotle that 'in vision, refraction is all.' Though he doesn't consider the inversion of the retinal image, nor the focusing effect of the crystalline lens, it is in his work that the general structure of the mechanism of vision appears for the first time as it is still taught in our own day.

From now on, refraction would be the key problem not only of physiological but also of geometrical optics. In this field al-Haytham's conception had very creative consequences. As the eye was unaware of any particular effort at the moment of emission, the Greeks had found it difficult to conceive of the visual ray according to a mechanical model, despite the efforts of Hero and of Damianus. The reversal performed by Arab science, however, allowed the thorough exploration of the mechanical metaphor, now justified by the fatigue felt by the eye at an excess of light.

In al-Haytham's work each ray is now considered as an 'arrow-ball' travelling at very high speed, and this model explains the linearity of propagation and the equality of the angles of reflection. More astonishing yet, when he considers refraction between two different media, such as water and air, the movement of the ray is broken down into two components, one parallel and the other perpendicular to the plane of separation of the media.

In his *Discourse on Light*, the Arab scientist explains the bending of the light-ray in refraction as follows: 'Lights which propagate themselves in transparent bodies propagate themselves with a very rapid motion, imperceptible on account of its rapidity. However, their motion in thin substances, that is to say those that are diaphanous, is more rapid than their motion in thick substances . . . In fact, when light travels through it, every diaphanous substance opposes to it a slight resistance which depends on its structure.' This is the argument in terms of speed which would preoccupy Western science for decades. With al-Haytham, optics freed itself from a static geometry and became the spearhead of the mechanics for whose inauguration it was itself chiefly responsible.

Another original point in al-Haytham is the way in which he describes phenomena by rules which are not, as they were with the Alexandrians, essentially numerical. As we can see from what follows, they describe the relations between the different angles in refraction:

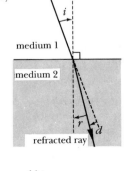

- when the angle of incidence i increases, the angles of refraction r and deviation d increase (if $i_1 < i_2$, then $d_1 < d_2$ and $r_1 < r_2$);
- but this variation is greater for angles of incidence i than it is for the angles of deviation d ($d_2 - d_1 < i_2 - i_1$);
- the ratio of the angle of deviation to the angle of incidence d/i increases as the angle of incidence i increases ($d_2/i_2 > d_1/i_1$);
- for a refraction from a less dense (medium 1) to a denser medium (medium 2), $d < 0.5$ i; conversely, for a refraction from a denser to a less dense medium, $d < 0.5$ $(i + d)$;
- a less dense medium deflects the light further away from the normal n;
- a more dense medium deflects the light nearer to the normal n.

As one can see, this is a law of refraction; reference to the 'natural' phenomenon is completely suppressed, and there remains no more than a series
of principles or rules. Although it would only be printed in the West in 1572,
the work of this mathematician and physicist exercised a considerable influence on those few individuals who contributed to the advance of optics
in the Middle Ages. Until the fourteenth century, in any case, Arab optics
remained extremely lively, while other sciences entered into decline much
earlier.

Divine Optics

While Antiquity and Islam saw optics flourish in the great centres of intellectual and political activity, the thirteenth century begins a period when the
study of light developed within Europe, but at some distance from its dominant centre. At Paris, the intellectual and temporal capital of the Western
world, faithful to the antique conception of vision and absorbed in the study
of the *auctores*, and in particular Aristotle, the logician and metaphysician,
the Scholastics of the Sorbonne neglected the study of optics (see pp. 222–
45). The scientists of the Oxford school, on the other hand, and more particularly its founder, Robert Grosseteste, Bishop of Lincoln (1168–1253),
put optics at the very centre of their search for truth. Taking up the Augustinian conception of light as analogous to divine grace, their reading of
Aristotle differed from that of the Parisian Schoolmen.

By making a distinction with in science between knowledge of facts and
knowledge of causes, Grosseteste points towards three fundamental aspects
of scientific research: the inductive, the experimental and the mathematical.
He advocates the validation of hypotheses and consequents by experiment,
and to justify his methods he puts forward a principle of the economy of
nature: 'Every operation of nature is accomplished in the most definite, the
most quick and the most perfect manner possible.' Applied to light, this
gives: 'Nature acts according to the shortest possible path.'

In these circumstances, the study of light was central to the conception of
the physical world, but in addition, because he establishes a close parallel
with his metaphysics, the key to which is the emanation of beings from
unity, Grosseteste makes light the central problem of all knowledge: 'All is
one, issued from the perfection of a single light, and things are multiple only
thanks to the multiplication of light itself.' This is important because, as
optics is inseparable from geometry, philosophy is impossible without
mathematics, 'for all the causes of natural effects ought to be expressed by
means of lines, of angles and of figures, for otherwise it would be impossible
to know the reason of these effects.' Within the framework of this optics,
conceived as the first of the sciences, in which light is the 'elementary form'

and 'first principle of the movement of efficient causality', the Oxford scientists were particularly interested in two phenomena: the rainbow and spherical lenses. This choice of objects, of course, wasn't accidental. The first, a felicitous natural wonder and symbol of the covenant with God, had its model in the second, the sophisticated products of a glass technology stimulated by the art of the stained-glass window. Through these two phenomena refraction, the key to both, became associated with the problem of colour.

'The function of optics is to discover what the rainbow is, because in doing this, it shows the reason of it, so far as there is added to the description of the rainbow the manner in which this kind of concentration can be produced in the light which goes out from a luminous heavenly body and through a cloud to a determinate place, and which then, by particular refractions and reflections is directed from this place to the eye.' So wrote Thierry de Freiberg in his *De iride*, where for the first time the rainbow is explained (though without the mathematical law of refraction).

In the thirteenth century, Roger Bacon, and later Thierry de Freiberg came to think that rainbows were produced by splitting and reflection within raindrops – glass balls filled with water allowed them to make accurate measurements. Within each ball (an extremely magnified raindrop) one could see how the rays of the edge colours of the primary and secondary rainbows were formed. The rays of the primary bow undergo one reflection and those of the secondary two, which explains the inversion of the colours. The coloration is due to the refraction undergone by the rays of white light entering into each drop. The size of the drops is important: when they are less than a tenth of a millimetre the coloration isn't visible and a white rainbow may be seen.

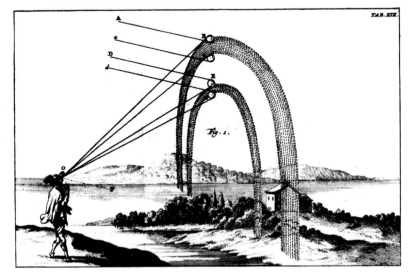

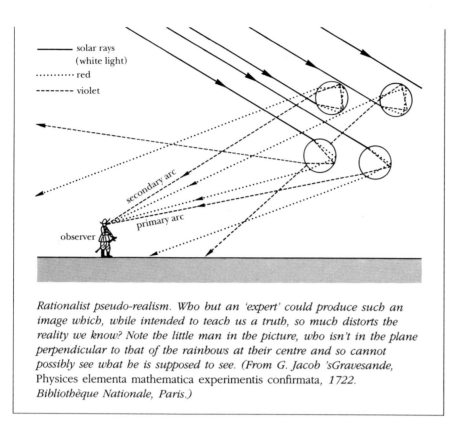

Rationalist pseudo-realism. Who but an 'expert' could produce such an image which, while intended to teach us a truth, so much distorts the reality we know? Note the little man in the picture, who isn't in the plane perpendicular to that of the rainbows at their centre and so cannot possibly see what he is supposed to see. (From G. Jacob 'sGravesande, Physices elementa mathematica experimentis confirmata, *1722. Bibliothèque Nationale, Paris.)*

A person of the twentieth century cannot fail to be surprised that so much importance should be attached to a phenomenon, however majestic it might be, which we regard as a simple optical illusion. We should, however, remember the fundamental place of theology in the social relationships of the thirteenth century, and remind ourselves once again that the rainbow, the symbol of God's covenant with man, had been the subject of disagreements among the theologians.

We know that Aristotle thought of the rainbow as the result of the reflection of light in a cloud. Grosseteste made it depend on refraction, whose principles had then to be investigated. This is why in his own *De iride* Grosseteste formulates a law of refraction whereby the refracted ray r follows the line which bisects the angle formed by the normal n and the incident ray i.

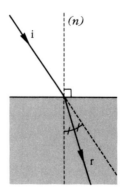

Here we can see the cavalier manner in which experiment was treated by these early practitioners. Several centuries before, al-Haytham had said that the angles made by r and i with n were constant. By fixing this relationship as $\frac{1}{2}$, Grosseteste demonstrates incontestably that the imperatives of experimental rationality were overridden by divine geometry's concern for harmony.

Roger Bacon, the 'Admirable Doctor' (1219–92) (see pp. 241 and 242) continued his master's work. He judged the speed of light to be very high, comparing it to the sound of the cannon, which is heard long after the smoke is seen, and he refined the explanation of the rainbow. Using the work that the Arabs al-Haytham and al-Kyndi had done on lenses, he gave a geometrical description of the rainbow's position in the sky, understanding it to be composed of a multitude of droplets.

It was in this same period that the Polish philosopher and physicist Vitello (c. 1230–75), also known as Vitellion, a Dominican born in Silesia, made a double impact on the science of optics. In establishing, thanks to careful experimental work, precise tables of refraction between different media for the different colours, he provided a precious tool for the construction of optical instruments. The care with which he did his experiments led him to suppose, as al-Haytham had done, that the deviation of a ray of light was greater when the medium was more dense. Despite this, his work was little known; it is true that at the time the diffusion of scientific work was uncommon and very much subject to chance. It was only in 1572 that it was published, together with the work of al-Haytham. With Kepler, we shall see the considerable effect of this double publication.

Another Dominican, Thierry de Freiberg, deserves to be considered the father of the theory of the rainbow. It isn't important that he didn't formulate the correct law of refraction (which came only three centuries later). In detecting the whole sequence of optical effects (overall composition by a multitude of droplets, rectilinear course of the Sun's rays, reflections, refractions, angles of appearance . . .) which produce it, he can be said to have explained it.

He also added an explanation of the secondary bow which one sometimes sees above the first. Following on from the work of Bacon, it was by lifting flasks of water above his head that he was able to observe the variation in colour. With this apparatus he was able to fix precisely the angles at which the colours appeared.

Al-Shirazi and al-Farizi produced a similar explanation at around the same time. This is evidence of the fertility of al-Haytham's ideas and of the fact that Middle Eastern scientific production hadn't stopped even after parts of the area had been seized by the West. At the end of the fourteenth century, Arab optics was far from being moribund: al-Farizi continued his illustrious predecessor's work on the camera obscura, later to be taken up by Leonardo da Vinci, and he put forward an explanation for refraction, suggesting that the speed of light was proportional to the density of the medium.

Because they had made of sight the sense which gave first access to the truth of the world, the Aristotelians at the Sorbonne distrusted everything they thought of as mere appearance. This led them to an attitude which might to us seem paradoxical. Although they were fascinated by optical illusions, already analysed by the Arabs, medieval scientists, with the exception of those we have mentioned, for three centuries refused to consider ocular lenses as an object of study. 'The things which they cause to be seen are not objects of the creation,' they said. A residue of this resistance was maintained until the beginning of the seventeenth century, when Galileo tried to have the telescope, or rather the reality that it made visible, accepted.

Probably discovered by glass-workers, converging lenses which improved long sight were used from the beginning of the thirteenth century. While they would have offered a good model for the study of the crystalline lens of the eye, it was only in the sixteenth century that serious research was undertaken. It was a strange situation, in which men wearing glasses wrote page after page in thick volumes on vision, without seeing that the key to the solution was at the end of their noses. One has to believe that for them the problem was not posed in the same terms as it is for us. We shouldn't forget that for the most part Arab optics remained unknown, and that the Greek idea of the *quid* which left the eye avoided any questions about the mediating function of the ocular lens.

But this should not be enough to convince us that this lack of thought about spectacles sprang simply from a particular theory of the propagation of the visual ray. Dante, in fact, who abandons most of Archytas' theory in favour of Aristotle's, little regarded at the time, declares in the *Convivio*: 'Visible things come into the interior of the eye not really but intentionally.' After much reflection on long sight and the means to remedy it, he says that the object must be made more distant from the eye 'so that the image may enter more light and more subtle'. Yet he makes no reference to corrective lenses.

So with the exception of a few people in contact with Arab optics (in a way not very well understood), medieval optics maintained until the fifteenth century a Scholastic conception of the world in which light, the common medium of the celestial and sublunary world and the higher form of all communication, could admit of no interference and no intermediary. This was a matter of a direct relationship with God and, from this point of view, sight was the highest of the senses.

Market Optics

Leaving the churches and the monasteries for the merchants, bankers and *condottieri*, painting was the main agent in the transformation of optics, and perhaps even more important in changing its relationship to society. By popularizing the problems of perspective and its distorted forms. Anamorphoses in particular, painting took optics away from the philosophers and theologians and made it an object of profane consideration.

Around 1500 Leonardo repeated al-Haytham's experiments on the camera obscura, although we do not know whether he knew of his predecessor. On the basis of this work, he identified its operation with that of the eye and noted the inversion of the image on the retina. An inspired forerunner in many fields, at the beginning of the sixteenth century the work of this great painter and engineer marks a renewal of optical study which began in Italy and spread through the whole of Europe. Confined until now to the *studii* of a few churchmen, in the second half of the sixteenth century it was the subject of enormous popularization.

In 1558 the *Magia Naturalis* of Giovanni Battista Della Porta (1534–1615) had its first printing. It was a great success! Endlessly enlarged, the book went through very many editions: twenty-three in Latin, and then more than twenty in the vernacular languages (Italian, French Spanish, Dutch, Arabic). As one might suspect from its title, the *Magia Naturalis* is a work in which the spectacular often takes precedence over strictly scientific discourse. This excessive popularization was perhaps necessary, for the book had a decisive effect. What is significant is that this exhibition of the wonderful and the 'magical' authorized the inclusion of numerous discussions on lenses. This changed the scientific attitude towards these objects and they rapidly became the central problem of optics. And it was through the study of every kind of glass that the study of refraction was renewed.

In 1593 Della Porta published his second great work on optics, the *De refractione*, in which he brought together all the work done on various phenomena: the formation of the retinal image, concave, convex and spherical lenses, together with the rainbow. But his explanations often mark a step backwards in relation to those of the Oxford school or of the Germans of the thirteenth century. This underlines, if it were needed, the extremely confidential character of their works. In Della Porta the concept of the light ray is very unclear, and that of the punctiform composition of the image, as explained by al-Haytham, has not been completely accepted, despite the publication of the latter's work, together with that of Vitello, more than twenty years before in Basel.

In general, this half-century saw the optical landscape considerably transformed. Problems, solutions, 'marvels' and apparatus left their confinement to become public. In 1590 the first divergent lens was made in Italy. And in 1571 the founder of modern optics, Johannes Kepler, was born in a little village in Württemberg.

Kepler and Baroque Optics

In 1604 the great German astronomer dedicated to Rudolph II, Archduke of Austria, King of Bohemia and of Hungary, the work he had just finished, in the composition of which he had been engaged for several years. This was the *Paralipomena to Vitellion*. In it he dealt with the optical part of astronomy, where there were two problems: refraction by the atmosphere of the light from heavenly bodies, and the diminution of the diameter of the Moon during eclipses of the Sun. In these concerns we can see a revival of those of the Alexandrians.

An attentive reader of al-Haytham and of Vitello, and a collaborator of the Dane Tycho Brahe (1546–1601), Kepler had understood the considerable influence of optical phenomena on astronomical observation. We should note that it was the Tycho Brahe's observations, remarkable for their precision, which allowed Kepler to establish his astronomical laws on absolutely sure foundations. One cannot fail to be struck by this recurrent connection between optics and astronomy. Refraction, for which Brahe had just drawn up new tables, was once again the key problem for observation.

In his big book, modestly entitled 'a supplement to Vitello', Kepler put forward his overall conception of optical science, through a long sequence of definitions, propositions, demonstrations, digressions and descriptions of experimental arrangements. Occasionally, he gives an account of the ancient theories, in order to criticize them, recalls the theories of Alhazen and Vitello, returns to these authors' refractive tables and to those of Brahe. Begun in 1600 as a study of the camera obscura as a simulation of the difficult problem of the eclipse, the *Paralipomena* overflowed into all the domains of physical and mathematical optics, not to mention metaphysics.

Making the sphere the image of the divine trinity, which then becomes the archetype of light as it diffuses from the centre towards the surface, by means of its rays which propagate themselves instantaneously towards infinity, Kepler demonstrates his conception of the nature of light. He adds some observations on the relationship between light and heat, and makes the Sun, the 'body in which resides the faculty of communicating itself to everything', the centre of the Universe.

In his *Astronomia Nova*, published in 1609, in which he develops a theory of the mutual attraction of bodies, he says that the Sun 'emits into the fulness of the Universe an immaterial species of its substance, analogous to the immaterial species of light'. Many historians of science have been astonished that, provided with this comparison, and having recognized in the propagation of light the inverse square law of distances, Kepler did not go on to propose an analogous law of attraction. At the risk of seeming sententious, it should be pointed out that *a posteriori* evidence is the historian's most deceptive guide. If the road followed by Newton today seems 'natural', trodden as it has been by thousands of footsteps, in Kepler's time it did not exist. When Auguste Comte wrote of Kepler that 'metaphysical considerations considerably delayed his progress,' one is right to ask what a delay could possibly be when one doesn't know where one is going.

In any case, Kepler himself considered this analogy which he finally rejected. For him the photometric law could not be applied to the motive effect of the Sun on the other planets. He believed that this *vis motrix*, applied tangentially, can only diminish linearly (that is to say in proportion to $1/r$, where r is the distance between the Sun and the planet in question), while light shining on a surface diminishes geometrically (that is to say in proportion to $1/r^2$). This was just 'obvious' to the astronomer at the dawn of the seventeenth century. Strangely, the profound connections between astronomy and optics, which assert themselves when experiment and observation are in question, seem to dissolve as soon as the problems concern the nature of light and of attraction!

The discoveries presented in the *Paralipomena* are legion. About heat, we learn that light heats bodies more or less depending on whether they are black or white, and that this heat is not material. Binocular vision is comprehensively treated in all its aspects as a topic of geometrical optics. Thanks to the measuring triangle, the position of images in the mirror receives its definitive explanation. Refraction alone, or almost so, resists this flood-tide of explanation in which all the old questions receive an answer. Exhibiting a rare intellectual generosity, Kepler is profuse in his explanations, declaring in his dedication to Rudolph II: 'As it was necessary to give a complete explanation of the way the vision works in refraction, in the images of objects seen and in the colours, one should not be astonished if I have made digressions . . . on the subject of conic sections . . . on optical wonders . . . on the nature of light and the colours as well as on other subjects. Even if these questions have nothing to do with astronomy, they deserve to be studied in their own right.'

The book is a huge cauldron in which all traditions are mixed together: 'I have accomplished this vast and stringent labour by going deep to the heart of questions neglected for centuries, any one of which would have been worthy of a book to itself.' Then he adds, still addressing the king: each of these works would have brought 'as much patronage to other people as to myself'. For he is very much aware of the size of the task he has taken on, and comparing himself to the king in his war against the Turks, he calls upon his generosity: 'I shall therefore no longer need be anxious that indigence, the most pernicious enemy of science, should oblige me, constrained by hunger, to abandon this task of mine, this fortress entrusted to my honour; nor need I doubt at all that Your Majesty will provide me when they are needed with the help and supplies whose dispatch would allow me to withstand the siege.'

This whole long strategic metaphor helps us understand the change in the status of the scientist since the time of the thirteenth-century Franciscans and Dominicans. To flatter the prince in this century of patronage, the scientist has to support himself with 'this one thought, worthy of a German, that it would be fair to die for so great a Prince'. Later, in his *Astronomia Nova*, he again begs 'His Majesty to consider that money is the sinews of war, and to graciously command his Treasurer to allow to his general the sums necessary to raise new troops'.

Bolstered by his success in explaining reflection, and led to it by an etymological analysis of the Latin and Greek terms in the closely related vocabularies of reflection and refraction, he attempts to identify the latter with reflection in mirrors with very special surfaces, which, however, he does not succeed in defining precisely. Then, on the basis of the degree of refraction, he makes a systematic investigation of the relationship between one medium and another. Finally, resuming the work of Alhazen, he concludes that the two angles are directly proportional, an almost exact approximation in the case of small angles and therefore appropriate to astronomy.

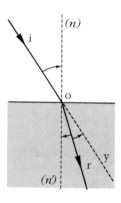

But this did not explain the other phenomena, and refraction still resisted this impressive effort with which, 'sweating and out of breath, he followed in the tracks of the Creator.' Despite the vast size of the work, his review of optical phenomena wasn't exhaustive, and in all this, Kepler didn't say one work about convergent lenses to compare them with the crystalline lens of the eye.

In this same period, Dutch lens-grinders started to manufacture astronomical telescopes after the Italian model of 1590. Five years later Galileo made of this instrument, a simple curiosity of 'natural magic', the final lever to bring down the edifice of Aristotelian science. There is no need to stress further the considerable influence of the *Starry Messenger*: in publishing the discoveries made with the help of the telescope, Galileo uncovered in this book an unknown heaven in which new stars appeared.

We know that, despite his successive alliances with the merchants, bourgeois, nobility and certain dignitaries of the Church, Galileo had above all to fight against the Aristotelianism present in the Church and all-powerful within the University. There followed the summons of 1616 and the condemnation of 1633, but that is a

Kepler did not arrive at the law of refraction because instead of taking the angle n' o r and comparing it to i o n, he looked at r o y, which is the angle of deviation the astronomers were interested in. Astronomy, which had led Kepler into the investigation, now obscured the way when he was close to his goal.

different affair (see pp. 280–314). In the history of refraction itself Galileo's contribution is rather minor; none the less the use of the telescope, which allowed the discovery of Jupiter's satellites, had an unexpected effect on theories of light. In fact, it is thanks to the satellites discovered by Galileo that Olaius Romer was able in 1675 to give the first measurement of the speed of light, given as 308,000 km/s, thus ending a 2000-year-old dispute on the instantaneity of propagation.

Despite certain reservations, Kepler rapidly understood the importance of Galileo's telescope. In 1611 he wrote his *Dioptrics*, a treatise on lenses, in which he explained for the first time the principle of the telescope. This is a simple and lucid work which benefited from the difficult gestation of the *Paralipomena* and in which Kepler took full advantage of the geometrical optics and principles of reflection and refraction developed seven years earlier. While the law of refraction had still not been established, the working of all the optical apparatus is perfectly described.

Here one might well be tempted to wonder, a little impertinently, what the use of a law of refraction is, when all the mysteries it was supposed to have illuminated a few decades later in Descartes' writings had already been solved. One might equally extend this impertinence to the 'revolution' which is supposedly represented by the collapse of the ancient optical conceptions. Was it the result of Galileo urging opinion-formers to have a look through his telescope? Was it the victory of Kepler's 'baroque genius', which was able to confound together all traditions and come up with a new paradigm? Or is the victory Della Porta's, the successful popularizer, who allowed the development of new ideas? Or perhaps there never was a rupture, either ideological, epistemological or sociological.

In his *Histoire de la lumière* Ronchi writes that, in 1523, a little less than a century before the appearance of Kepler's *Dioptrics*, in a monastery lost in the Apulian mountains, Francesco Maurolico, a priest from Messina, finished a little optical work which would only be published in 1611. It is strange to find, eighty years earlier, an account of many of the discoveries made by Kepler, as well as a fairly closely related conception of light and a similar manner of argument.

It isn't a question of Kepler's honesty: he frequently mentions his predecessors, even attributing to them merits they didn't always have. Maurolico's work wasn't published, and according to Ronchi, who devotes several pages to it, only a few Church officials knew about the book. Given the lack of any direct contact, there only remains the hypothesis of a common filiation.

We may recall that that, in the middle of the fourteenth century, Middle Eastern optics, which shared in the fortunes of the great urban centres of the time, was still very active. There is one fact connecting the southern Italian priest to the German astronomer: at the end of the fifteenth century Maurolico's father fled Constantinople as it was invaded by the Turks, and a century later the whole of eastern Europe, where Kepler lived, was struggling against these same Turks. Can one imagine that this flourishing Islamic optics, sharing in the varying fortunes of the great cities of the East, arrived in the West through the marches of Turkey? We would then have to recognize, behind the argument for rupture, the effect of our ignorance. Under this risky hypothesis, the fall of Constantinople would take on a new meaning, and this would not be the least of historical ironies.

The Cartesian 'Inversion'

You need to be someone living in twentieth-century France not to know that in about 1620 the Dutchman Willebrord Snell, called Snellius (1591–1626), put the crowning touch to the Keplerian system by stating the law of refraction, which today bears his name in all other countries of the world. According to the latest news, the war still goes on. In the article 'Descartes' Law' in the *Grand Larousse encyclopédique*, Snell's name is absent, as is the name of René Descartes in the article on 'Snell's Law' in the *Encyclopedia Britannica*.

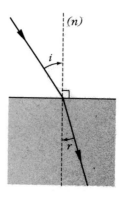

So the first part of the history of refraction comes to an end at the beginning of the seventeenth century. As we have seen, for almost all the problems of elementary and instrumental optics for which refraction provides the basic explanation solutions had already been found. And strangely, it was at the moment that the tool of refraction was no longer useful that there appeared the formulation of the law of refraction that would henceforth remain unchanged: the ratio of the sines sin r/sin i is constant for all angles i.

As the consequences of this law were all known before its formulation, thanks to the availability of very precise tables of refraction, the question that remained was the reason why it should be so. In the first third of the century there came about a long-term reversal: it was no longer a question of knowing what refraction produced, but of knowing what produced refraction. As a result, the problem of the nature of light returned to the forefront. In this change of perspective, Snell's Law, which had been the culmination of the history of optics, became in Descartes' account the cornerstone of an optics stripped of all ideas of genesis and put into direct contact with nature by the agency of reason alone.

If barbarism may be characterized by the systematic refusal of history, we should not be afraid to say that the appearance, 350 years ago, of the *Discourse on Method, followed by Three Essays: on Dioptrics, on Meteors and on Geometry*, is the act of barbarism (saying so being no judgement on its necessity) which founds scientific rationalism. All his predecessors' work was wiped out, its usefulness denied. In the sixth part of the *Discourse*, for example, one can read: 'I do not boast of having been the first discoverer of any of them [scientific discoveries], but rather that I have never accepted them because they were held by others, nor because they weren't, but simply because reason has persuaded me of their truth.' Glory be to pure reason, Descartes only has to reason a little, and 2000 years of work and scientific discovery are obliterated in order to be rewritten in accordance with the dictates of the new order.

> In the Eighth Discourse, *On Meteors*, the second essay in the *Discourse on Method*, Descartes attributes to his method 'knowledge that was never possessed by the authors of the writings we have'. We have already pointed out this attitude of Descartes, who discounts the discoveries of all those who came before him. The illustration of his rainbow also tells us a great deal about his way of thinking. The scene here is completely unrealistic.
>
> On one hand the two rainbows, primary and secondary, seen by the same observer could not possibly be produced by the same raindrop. And on the other, and more important, it is quite impossible to see from one side the rainbow seen by another person. It seems to me that this picture's objectifying tendency tells us a great deal about the Cartesian method, which invites us to take the place of the observing subject so that we may be persuaded by reason of the rightness of his opinions. 'As for the opinions which are mine, I need make no apology for them as being new, for if their reasons are well considered, I am sure that they will be found so simple and so much in accord with common sense that they will seem less extraordinary.'

And a few lines later, he writes: 'If I write in French, which is the language of my country, rather than in Latin, which is the language of my teachers, it is because I hope that those who use nothing but their natural reason may be better judges of my opinions than those who believe only in their ancient books.' Let there be no mistake, Descartes is anything but ignorant, and he knows Kepler, Vitello and al-Haytham perfectly, as well as De Dominicis' theories of the rainbow as they are given by Thierry de Freiberg. What's more, at that time text-books on atmospheric phenomena such as Froidmont's were studied in all schools ... And yet these scientists, and many others, have no mention in the *Discourse* and the *Essays*, though their discoveries are to be found everywhere within them. Descartes' strategy is clear: he will not recognize any inheritance, and the only ones to be mentioned by name are 'a certain Jacques Métius, who never did study', and Father Maurolico, who is mentioned only to be criticized.

It is therefore impossible to think of the *Discourse* as anything but a manifesto whose slogan might be: 'Forget the past!' (The reader will also find this strategy in other great 'founding geniuses', Lavoisier, for example (see pp. 455–82.) What I have tried to show, however, is that in so far as it was the work of people, done in particular contexts, on the basis of a scientific heritage more or less well known, there was no revolution in the history of optics.

What appears with Descartes is not a scientific revolution but rather a revolution in the manner of presenting results. All the discoveries are there, but the actors have disappeared. In this new account history is dissolved and nature becomes the only referent. Experience is governed by reason, scholarship is governed by common sense. The language of the middle classes takes over from that of the Church and of the University, universalizing the right to scientific judgement. That nature can be represented according to the rules of reason is a mystery, just like the history of those who made possible the work of Reason. They are completely hidden by the new philosophy and lie buried in the foundations of the new science.

In one single movement, epistemological foundations, social alliances and the meaning of interpretation have all been turned upside-down. Nature, common sense and intuition will specify the nature of light, and from this will follow a law of refraction which will govern, by deduction, the interpretation of all optical phenomena. In this same period the Aristotelian University brought about its own dissolution, taking part in fierce witch-hunts – never had so many been burnt at the stake – young scientists received their training at the hands of the Jesuits, and one of the first networks of scientific communication in Europe was developing around Father Marin Mersenne (1588–1648).

The transformation of scientific practice was considerable. This was of course partly the result of the establishment of new institutions, which brought money, security and information, but the Cartesian discourse also played its part. In leaving out of consideration the genesis of scientific results Descartes invented a new kind of scholar, who could be ignorant of the history of science, who practised it like a logical game, whose pieces were produced by nature and organized by reason alone. This new man, who was often more ignorant and arrogant than scholarly, was so much the slave of his practice that he bore its name: he was the modern scientist.

At the end of the *Discourse* we find the 'Ten Commandments' of this new practice:

> The whole should be read with patience and attention, and I believe the reader will be satisfied: for it seems to me that the reasons follow on from one another in such a way that as the last are demonstrated by the first, which are their causes, these first are also demonstrated by the last, which are their effects. Nor should one imagine that I am here committing

the fallacy the logicians call a circle; for as experience renders the greater
part of these effects most certain, the causes from which I deduce them
do not so much serve to prove them as to explain them; on the contrary,
the causes are proved by their effects.

Here we have the ancestor of the sequence – hypotheses/logical principles/
conclusions/experimental control/validation of hypotheses – to which today
it is claimed that all scientific knowledge may be reduced. Without any
reference at all to those who preceded him on the road to this method,
Descartes produces *ex nihilo* the magic circle of science from which history
and the world are permanently banished.

At the risk of being boring, it must be repeated that Descartes' great genius
consists in his total lack of scruple with regard to earlier theorists. He steals,
splices, glues together, twists and turns his pieces, fragments and ideas, to
make up his own coat, cuts off what gets in the way, changes the meanings
of words, stretching them or cutting them down as suits him best, dissolving
300 years of work in three lines of text, and then goes on for twenty pages
about some trifle. His *Dioptrics* is a good example. What has been most
criticized in this essay, both by Cartesians and by historians of science, is the
first discourse on the nature of light. This is also the only one of Descartes'
contributions that has some claim to originality. Three conceptions are
successively presented. Light is a stick, a highly rarefied fluid, and a little
spinning ball, all at the same time. These conceptions, which it would be
inappropriate to think of as contradictory, because a concern for overall
coherence does not appear to be among Descartes' major preoccupations,
serve in turn to provide solutions for different problems.

Sensation? Nothing simpler: 'It must sometimes have happened to you
when walking at night without a torch, that when the going was difficult,
you had to use a stick to guide you on your way.' A marvellous explanation,
in which the blackest night illuminates the problem of light! The Pythagorean *quid* isn't very far away. But the air between the eye and the object is
very far from having the consistency of wood.

What substance then is the medium? 'Look at a vat at the time of the grape
harvest, full to the brim with half-crushed grapes . . . Consider that there is
no vacuum in nature . . . the pores must be filled with some highly rarefied
and highly fluid matter . . . compare with the wine in this vat . . . you must
judge that the rays are nothing but the lines along which this action happens.' This is Aristotle's theory of the diaphanous medium, even to the very
metaphor, the wine which clouds the spirits. Descartes says nothing of this,
and passes on to another level, for it must not be forgotten: the fate of a ray
is always to hit upon an obstacle, or as Descartes suggests, to curve under
the influence of a gravitational field.

This then is the question why 'when they meet bodies, they are liable to be deflected by them, or absorbed, in the same way as the motion of a musket-ball.' Here we have al-Haytham's idea. With this, light becomes a little ball and obeys the laws of motion which were an object of study for the whole of the seventeenth century. Strangely influenced by the models of the stick and the spirits of wine, the ball does not actually move, but simply has an inclination to do so. It is an astonishing paradox to see the laws of motion called upon to explain the effect of an immobile light.

Should we be surprised? Descartes, who pretended to owe nothing to the past, has just looted three solutions from the scientific heritage. He insisted on founding everything on reason, and then put forward an idea of light as solid like a stick, liquid like wine and discontinuous like mobile immobile balls. But finally, it doesn't matter at all that these hypotheses on the nature of light are mutually contradictory, for here, at least, Descartes has been a pathfinder, the most modern physics having taught us not to be astonished at the supposed contradictions of appearances. What is important is to have enough models so as to be able to deduce from some of them the effects that experiment can validate. If, for example, it is a question of explaining the colours, the light that was a very small ball can become 'for the purpose in hand' a package of little balls rolling around each other.

If the law of refraction has to be demonstrated, the mechanical model will be used. The demonstration has to be followed in detail, to persuade us of the omnipotence and the 'ahistoricity' of reason. In refraction light, which doesn't move, is like a ball projected towards water at very high speed by a tennis player. A ball 'whose course is changed neither by weight nor

---

Diagrammatic representation of Descartes' demonstration

This circle, which has its place in each step of the explanation, here reduced to its bare bones, allows one to argue by an equalization of the distance travelled in each medium, so that speed (or the tendency to move) becomes the essential variable in the phenomenon.

A ball leaves A in the air at a speed which is increased by a third when it enters the water. The horizontal component remains unchanged. As the distances AB and BI are the same (that is why we have the circle), the times of travel along AB and BI are in inverse proportion to the speeds, so GI will be one third smaller than AH, and the ratio CB/BE will be constant, wherever the point A may be. (If the radius of the circle is 1, CB and BE are the sines of the angles i and r.)

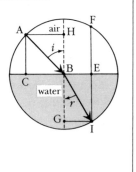

---

lightness, neither by its size nor by its shape, nor by any other external cause.'

Starting with this postulate, Descartes explains refraction by supposing that light's tendency to move is less impeded in water than in air. 'Which you will no longer find strange, if you remember the nature that I have attributed to light . . . a movement or action occurring in a highly rarefied substance . . . and if you consider that like a ball it loses less of its movement to a soft substance (air) than to a hard substance (water).' This is why Descartes' successors would say that the speed of light is greater in water than it is in air, a statement that is meaningless, according to the Cartesian orthodoxy, in which light has only an inclination to move.

Given Descartes' influence on the development of science, knowing that in the end the law of refraction is probably the only scientific result that he could lay claim to, it needs to be considered whether his contribution was really original. Is this really the demonstration of a law or is it, as Huygens believed, the product of a manipulation intended to disguise the theft of Snell's results? From this point of view, we need to have a look at the very heart of his 'reasoning':

> Finally, so far as light in this follows the same laws as the movement of the ball, it must be said that when the rays pass obliquely from one transparent substance to another, which receives them more or less easily than the first, they are there deflected in such a way that they find themselves always less inclined to the surface of these substances on the side on which there is the one that receives it the more easily, than on the side where the other one is. One must only be careful that this inclination is measured by the quantity of the straight lines like CB or AH, and EB or IG . . . For the ratio or proportion which exists between the lines AH and IG or others like them, remains the same in all refractions caused by the same substances.

Everyone may judge for themselves whether this may be called a demonstration established 'by reason alone', or whether a quick glance at Snell's rough draft in the course of a journey to Holland did not fix the correct ratio in Descartes' mind as an end to be attained by any means possible: by sophistry, metaphor or contradiction. The whole business disguises the tautology: 'The law of refraction is demonstrated by the fact that the ratio of AH to IG remains the same in all refractions' (which is nothing but a restatement of the law itself).

Can we laugh carelessly today at this idea of light, a mixture of stick, of new wine and of little balls? If one can be very critical of Descartes' physical arguments, should one not be more prudent about his metaphysics? In the three comparisons we have seen three totally contradictory natures, but have we looked properly? Might it not be possible to keep all three, but think of them as corresponding to different degrees of magnification? And finally, if we look at them in this way, don't we have here a history of light?

If we read Descartes' *Dioptrics* as a metaphysics, we can attempt some tentative identifications. The stick first of all in its rectilinearity may be considered as governing the whole of geometrical optics before the seventeenth century. Then the rarefied fluid corresponds to the ideas of the ether and of waves which would govern the explanation of optical phenomena in the two centuries which followed, while the spinning corpuscles foreshadow the photons which appeared at the beginning of the twentieth century, and which, associated with frequency, spin and probability, now underlie the explanation of all phenomena connected with light.

If, however, Descartes' basic suppositions are of a metaphysical order, their value depends on finding their true meaning in his work taken as a whole. A contemporary philosopher has attempted to solve this problem. We know that Descartes distinguished two operations in thought: intuition and deduction. Noting that *intueor* in Latin means to observe, to consider, to look at attentively, Michel Serres reminds us that in the ninth rule of his *Rules for the Direction of the Mind* (1628), Descartes symbolizes intuition, 'which is born of reason alone' by a blind person's stick. What's more, intuition is not only the basis that makes the work of deduction possible, but it is also the result of it. For Descartes, in fact, the frequent exercise of reason in a deductive chain which is travelled rapidly enough and often enough, 'exercising the mind as one exercises ones sight', converts the movement of reason along the chain into immediate comprehension: intuition. It is this which allows Michel Serres to conclude that in certain conditions the stick which is intuition is a chain of deduction in which the links no longer exist as independent units.

This ninth rule, which was written ten years before the *Discourse* which united deduction and intuition, allows one to understand how the three natures of light, certainly contradictory in a physical sense, form within Cartesian thought a coherent whole. Perhaps it is thanks to the work of this thought that few today consider it scandalous that light can be thought of under an apparently contradictory double aspect, as waves and as particles.

Of course, Descartes' 'discoveries' did not bring to an end the questions raised by refraction. In the same way as it can be said that the formula which links the volume of a solid to its mass tells us nothing about space or about matter, Snell's Law tells us nothing about the nature of light or about the different structures of the transparent media traversed by its rays. None the less, its existence did make a lot of difference. Physicists, taking over from craftsmen, were able to go further with the work that Kepler, Galileo and others had done on telescopes and microscopes. Then, with a precision entirely due to theory, and with far greater efficiency, they were able to

341

design optical instruments which would increase the visible world, from the infinitely large to the infinitely small.

Mastered in this way, not in its nature but in its measurement, refraction, which for centuries had been an obstacle to astronomy, became the phenomenon that governed the principles of construction of apparatus which allowed more and more precise observation. In the matter of efficiency, it isn't accidental that the *Essay on Dioptrics* from the *Discourse on Method* should end with the a study of 'Figures which transparent bodies must have to deflect rays by refraction in all the ways useful to sight' (Discourse 8), 'The Description of Lenses', (Discourse 9) and 'The Manner of Cutting Glasses' (Discourse 10).

Inaugurating an expanding dialectic between comprehension and phenomena, the refinement of physical laws and the improvement of instruments, the law of refraction, with lenses, microscopes and telescopes would bring into sight new phenomena like the coloration of thin films, Grimaldi's fringes, Newton's rings, birefringent crystals . . . and new concepts like diffraction, double refraction, interference and polarization.

None the less, quite apart from the technical successes due to its metrological validity, the law tells us nothing about the 'reasons' of the phenomenon. Whatever the reproaches one might make to Descartes, it has to be said that in looking for a 'reasonable' explanation of refraction, he wanted to understand the 'why'. That he failed is probably not so important, because in his attempt to answer the question, he defined the field of thought in which his successors would have to operate, whether they were enemies or supporters. For one has to recognize that Descartes' explanations didn't even satisfy those who, like Maupertuis, based their work on his hypotheses.

Some twenty-five years after the appearance of the *Discourse* there broke out an important polemic between the Cartesians (who were anxious to defend one of their master's finest productions) and Pierre de Fermat, one of the great mathematicians of the time, co-inventor with Pascal of the differential calculus, of the calculus of probability, and famous above all for number theory (see p. 346). Using the principle that light takes the shortest possible time to travel from one point to another, and postulating that the speed of light is greater in less dense substances, the mathematician proved the correctness of the law of refraction. He did this at the end of a difficult geometrical proof whose principles were all present in Euclid but whose general direction foreshadows the calculus of maxima and minima.

Noting that the 'master's' conclusions had been obtained by means of such contradictory hypotheses, the Cartesians 'went into raptures that the same truth should appear at the end of two roads entirely opposed to each other', Fermat wrote in a letter which was published after his death. But that is another story, and its subject would be not exactitude, but relevance. We should then see that depending on whether light is governed by its supposed nature (wave or corpuscle), or by a quasi-divine principle of minimization (of time, resistance or of action), scientists like Huygens, Newton and Grimaldi on the one hand and Fermat, Leibniz or Maupertuis on the other, all found the explanation of the phenomenon in efficient or final causes. These disputes would last a century before the triumph of the wave model put forward by Huygens . . . One has to wait for the twentieth century and quantum physics to see the dissolution of the dilemma between efficiency and finality.

The wave model proposed by Huygens

The plane wave front AD meets the contact surface between the media 1 and 2. By the time the wavelets D_1 arrive at D_2 the wavelets A_1 have been propagated along a ray R; the front of the plane wave front A_2D_2 is tangent to the wavelets, so A_1A_2 is perpendicular to A_2D_2; one sees that

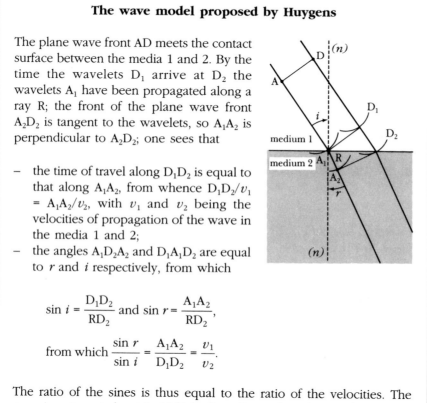

- the time of travel along D_1D_2 is equal to that along A_1A_2, from whence D_1D_2/v_1 = A_1A_2/v_2, with v_1 and v_2 being the velocities of propagation of the wave in the media 1 and 2;
- the angles $A_1D_2A_2$ and $D_1A_1D_2$ are equal to r and i respectively, from which

$$\sin i = \frac{D_1D_2}{RD_2} \text{ and } \sin r = \frac{A_1A_2}{RD_2},$$

from which $\dfrac{\sin r}{\sin i} = \dfrac{A_1A_2}{D_1D_2} = \dfrac{v_1}{v_2}.$

The ratio of the sines is thus equal to the ratio of the velocities. The experiment carried out in 1849 by Jean Foucault and Hippolyte Fizeau showed that the speed of light is proportional to the refractive index of the medium, as suggested by Huygens, finally closing a problem inaugurated almost a century before by the great Arab scientist al-Haytham.

This wave explanation also explains other phenomena such as the refraction of waves in the neighbourhood of river banks, and that of sound-waves when there is a variation in the temperature between adjacent strata of air.

11

Working with Numbers in the Seventeenth and Nineteenth Centuries

Catherine Goldstein

Certain fields of mathematics, number theory in particular, seem never to have been anything more than intellectual games. Here we will discover, in the internal changes in the discipline and in its economic and social integration, what it was that both underlay and demonstrated the transformation of amateurs into professionals.

In the first half of the seventeenth century a *conseiller* at the *parlement* of Toulouse, a lover of Greek and Latin poetry, noted in the margin of a Latin translation of a Greek treatise on mathematics that it was not possible 'for a cube to be written as a sum of two cubes or a fourth power to be written as a sum of two fourth powers or, in general, for any number which is a power greater than the second to be written as a sum of two like powers' and that he had discovered 'a truly marvellous demonstration of this proposition which this margin is too narrow to contain'. In 1850 a professor at the University of Breslau published in volume XL of the *Journal für Reine und Angewandte Mathematik* a general demonstration that the 'equation $x^n + y^n = z^n$ has no whole number solution' for exponents n subject to certain technical hypotheses; for this work the author was nominated for a Gold Medal from the French Academy of Sciences.

Amateur or professional, playing or working: it isn't always easy to label precisely those who practise science, with the exception of the mythical but attractive case of the isolated, unknown and uneducated individual who makes the discovery of the century quite outside official circles. Depending on the period and the state of their discipline, very different kinds of people have been described as amateurs; occasional readers of journals of scientific popularization or collections of puzzles, the enlightened and or enthusiastic consumers of the science produced by others; Saturday-night astronomers and Sunday-afternoon botanists exploring the heavens and the undergrowth in search of original observations or new species to record, the collectors of data for more official investigations; full-time researchers whose personal resources permit them to pursue unpaid the same daily work, and to be judged according to the same criteria, as scientists paid by the universities. They are equally interested and trained and have access to the same publications. The relationship between these 'amateurs' and the neighbouring 'professionals' varies from the pure and simple absorption of one category by the other through relations of hierarchically organized mutual assistance to complete mutual indifference.

The double example with which this chapter opened illustrates a particular type of interaction (others are also discussed in this book: see pp. 18, 124, 401 and 455), in which the questions raised by amateurs provide the raw material for professional work some centuries later. But how should these two stages be characterized and distinguished? How do they succeed each other or overlap? The choice of mathematics and more particularly number theory as a field of study is in this respect more interesting for being an extreme case: among the sciences mathematics, and the theory of numbers in particular, retains, rightly or wrongly, a reputation for splendid and unchanging isolation which discourages over-hasty explanation of its professional development. What exactly is the use of a proof that a cube is not the sum of two cubes?

It is, *a priori*, of such little use that it is tempting to forget it, to turn to a history of people, of institutions and of money. The marginal note and the theorem of pure mathematics are then considered relevant only to another kind of history, which deals only with them, and in which the place where they appear and the people who produce them are no longer relevant at all: the margin would be famous only for the dreams it gave rise to ('if only it had been bigger and the proof had been written there!') and the journal no more than an occasion for an amusing anecdote about its founder.

But these two possible histories, each of which serves the other as its background, its reservoir of picturesque detail, its sunny side of the street, are in fact only one: connections exist and they can be discovered between a discipline, the social forms which mould it, the problems it generates, the sources it draws on, the means of expression, communication and protection it adopts.

345

Beyond all idiosyncrasy, *Homo ludens* and *Homo faber* must therefore be grasped just as firmly in time as in their mathematics, in order that they may be distinguished one from the other.

Homo ludens

What can we say then about this manuscript note scribbled in the margin? Rather than the necessarily incomplete answers that it offers, what concerns us are the questions it gives rise to: who wrote or spoke of numbers in the seventeenth century? What were the sources used? In what form and in what context did one express oneself? And finally, how and to what extent does mathematics as a game reflect or produce these particular answers?

Men and Mathematics

Pierre de Fermat, a *conseiller* at the *parlement* of Toulouse, annotated a copy of Diophantus' *Arithmetic*, translated with an abundant commentary by Claude Gaspard Bachet de Méziriac. Fermat is known in our day as a mathematician, the inventor, just as much as Descartes, of analytic geometry; but he was also a member of the Commission de l'Édit at Castres and of the Conseil Général of Beaumont. The nobleman Bachet de Méziriac, grandson of one of Henri II's *conseillers*, owed to his talents as a mythologist the place he was offered in the Académie Française, shortly after its creation by Richelieu in 1635. In the sixteenth and seventeenth centuries it is among diplomats like Kenelm Digby, *conseillers* at the *parlements* like Pierre de Carcavy or François Viète, or at the Cour des Monnaies, like Bernard Frenicle de Bessy, among engineers in the service of kings or princes like Raffaele Bombelli or Philippe de Girard, tutors and secretaries to noblemen, like Jean Beaugrand or Simon Stevin, or even soldiers like René Descartes or monks like Antoine de Lalouvère, Marin de Mersenne or Jacques de Billy that one has to look for the authors or recorders of investigations of number. The irregularity of their mathematical researches itself offers occasional evidence of the imperatives of their social life and of political events.

Nor in the seventeenth century did the forms of mathematical activity make up a unity: it is indeed its multiplicity, even its contradictions that best express its specificity. We have just encountered one of its most important manifestations, the translation of the books of Antiquity (cf. pp. 191–221). Continuing work initiated in the Mediterranean world, scholars deciphered and translated into Latin Euclid, Apollonius, Archimedes and Diophantus, the main Greek mathematicians, sometimes reconstructing their works from fragments and from the muddled evidence of later compilations. It was a question, of course, of re-appropriating the heritage of Antiquity, assimilating and understanding it, but also, more and more, of criticizing it and rewriting it in accordance with the standards of the day. This marvellous source of problems was by no means the only one: engineers, artillerymen, navigators and specialists in fortification used the results of mathematics and sometimes developed them. Another current was provided by the 'Cosist' tradition (from Italian *cosa*, a thing, the unknown), known today as algebra, inherited from the Arabs and well established first of all in Italy and Germany. The Cosists, who were consulted by merchants, prioritized effective methods, presenting them in the form of concrete problems considered as examples. An entirely different passion for numbers found expression in esoteric compilations, whose hidden numerical properties were thought to reflect the magic secrets of the world . . . As well as learned commentaries and textbooks of every kind there were collections of puzzles, full of numerical problems dressed up in different disguises, depending on the fashion and the public they were aimed at. In the preface to his *Recreation mathématique, composée de plusieurs problèmes plaisants et facetieux, en faict d'Arithmeticque, Geometrie, Mechanicque, Opticque et autres parties de ces belles sciences* ('Mathematical Recreations, composed of several pleasant and amusing problems, in Arithmetic, Geometry, Mechanics, Optics and other parts of these fine sciences'), the Jesuit Lerrechon states that 'the nobility does not study mathematics to swell its purse or for any gain it might look for, but to satisfy the mind, to honestly occupy the time and to have the means to entertain company with beneficial yet recreational discourse.' Molière's *Femmes savantes* were not far away.

The pleasures of skilful and rapid calculation, the subtle decipherment of ancient manuscripts, amusement, and the attraction of hermetic mysteries: the mixture would hardly surprise us if it wasn't often to be found in the same libraries, even inside the same head. John Dee, deeply versed in alchemy and numerology, also gave advice on navigation and voyages of exploration. Bachet de Méziriac did not content himself with a scholarly translation of Diophantus: his *Problèmes plaisans et delectables qui se font par les nombres* ('Pleasant and Delightful Problems with Numbers'), a mine of parlour games and numerical puzzles published in 1612, was republished as late as 1959!

Coming from such different social and intellectual backgrounds, impelled by such different concerns, were these men able to understand each other? Did they even try?

Networks and Margins

There are already two signs of mathematical exchanges, at first sight contradictory. 'The margin is too narrow to hold [the proof],' wrote Fermat. Rather than look for the buried proof, historians wanted to know how and why it

Some of the essential names in the story

Diophantus (possibly third century AD), Alexandria

Ramus, Pierre La Ramée, called (1515–72), Paris

Bombelli, Raffaele (1522–72), Italy

Dee, John (1527–1607), England

Holzmann, Wilhelm, called Xylander (1532–76), Heidelberg

Viète, François (1540–1603), Bordeaux

Bachet de Méziriac, Claude-Gaspard (1581–1638), Lyons

Mersenne, Marin de (1588–1648), Paris

Descartes, René (1596–1650), Paris, Holland

Leibniz, Gottfried Wilhelm (1646–1716), Hannover

Bernoulli family, Basle

Goldbach, Christian (1690–1762), St. Petersburg

Euler, Leonhard (1707–1783), Berlin, St. Petersburg

Lagrange, Louis de (1736–1813), Turin, Berlin, Paris

Legendre, Adrien-Marie (1752–1833), Paris

Gauss, Karl Friedrich (1779–1855), Göttingen

Humboldt, Alexander von (1769–1859), Prussia

Carcavy, Pierre de (1600–84), Toulouse, Paris

Fermat, Pierre de (1601–65), Toulouse

Roberval, Gilles Personne de (1602–75), Paris

Billy, Jacques de (1602–79), Champagne

Frenicle de Bessy, Bernard (1605–1675), Paris

Wallis, John (1616–1703), Oxford

Pascal, Blaise (1623–62), Paris

Huygens, Christiaan (1629–95), Paris, Holland

Crelle, August Leopold (1780–1855), Prussia

Jacobi, Carl (1804–1851), Berlin, Königsberg

Dirichlet, Gustav (1805–1859), Berlin, Göttingen

Kummer, Ernst Eduard (1810–1893), Breslau, Berlin

Borchardt, Carl Wilhelm (1817–1880), Berlin

Eisenstein, Ferdinand Gotthold Max (1823–1852), Heidelberg

Kronecker, Leopold (1823–1891), Berlin

had been lost. What absolute solitude or even solipsism allowed a mathematician to refrain from writing down anywhere such a 'marvellous proof'? The second trail, on the contrary, is the evidence of that copy of Diophantus, published in 1621, in the hands of Fermat. At least writing is handed on. How?

Publication in the seventeenth century was neither easy nor very widespread. Often it was vital to supervise the printing or to arrange for a friend to oversee the process, paid for by the author, in order to produce an acceptable result. The works of Viète were not published until long after his death, and the marginal notes of Fermat have come down to us thanks to his son Samuel, who republished the *Arithmetic* in 1670, supplemented by the famous annotations. Given these difficulties, in general only whole treatises were published or sometimes a commentary on a translation. The precarious conditions surrounding the acquisition and transmission of information are easily to be seen in letters, like this one from Fermat to Mersenne: 'I should be grateful if you could arrange for me to receive any new treatises or books on mathematics which have appeared in the last five or six years . . . And I must tell you, however, that I have completely reconstructed the treatise of Apollonius, *De locis planis*. Six years ago I gave Monsieur Prades, whom perhaps you know, the only copy that I had written out.'

Here we can also see the importance of personal relationships, either face to face or in letters. The absence of a unified tradition of mathematics and, in particular, of research on numbers, already show that education, reflection, research were essentially private activities. If the Jesuit schools, among them La Flèche, where Descartes was educated, considered mathematics important, French universities remained on the fringes of new trends for a long time. It was then through personal relationships that problems were studied, and especially in new books. It was probably through Viète's students, for example, that Fermat was made aware of the work of this man, and it was through his former colleague in the *parlement* of Toulouse, Carcavi, that he was able to begin an acquaintance with Mersenne.

Exchanging letters on all sorts of philosophical and scientific subjects with correspondents scattered across the whole of Europe and even as far as Turkey, Mersenne maintained an invaluable network of knowledge. He also brought together, in the Minim monastery in the Place des Vosges an astonishing company, among them Gilles Personne de Roberval, Pascal, Hobbes, Descartes and Gassendi. Such academies flourished all over Europe at this time: inspired by their Italian equivalents, set up by private initiative during the sixteenth century, in the wake of the Platonic revival, and directed against the Aristotle of Church and University, little by little these cultivated societies began to specialize, some in poetry, others in art and still others in the exploration of natural phenomena. Depending on the personality of the host, meetings might be accompanied by dinners or concerts, by dissections or astronomical observations. In France, the most famous and most productive in the exchanges they made possible were those organized around Nicolas Fabri de Peiresc, the brothers Dupuy, Mersenne, and later Pierre Rémond de Montmor.

These vast networks of correspondence and the nodes of transmission represented by meetings of the academies wove the fabric of mathematical communication across the country and far beyond: letters were carefully copied and sent on elsewhere, sometimes with great effort. Given a letter for Fermat by Frenicle, Digby added this note to the consignment: 'I have had it copied by my secretary, for you would not have been able to read it; [Frenicle] ordinarily writes on scraps of paper and so quickly that he's the only one able to read the writing.' On occasion, correspondents would publish rewritten versions of letters received, of which they are sometimes today the only evidence. In this way the *Inventum Novum* was composed, the 'New Discoveries in the Science of Analysis Collected by Father Jacques de Billy, Priest of the Society of Jesus, from the Various Letters Sent to Him at Different Times by Monsieur Pierre de Fermat, Counsellor at the *Parlement* of Toulouse', a work with a discursive title which contained some of Fermat's work on problems from Diophantus.

It is therefore tempting to speak of a community in formation; but if this was indeed the dream of Mersenne and others, the reality was less idyllic, especially as regards questions of number. In 1640 Roberval wrote to Fermat that, in order to discover the 'great mysteries' of number 'there would have to be several together, without jealousy and of one accord, whose genius is naturally inclined to this speculation, which is very difficult to find.' Beneath the protestations of friendship and mutual admiration, the normal mode of communication was in fact the challenge; evidence of a half-serious, half-playful combat in which one demanded proof of the others' skills, the letters often pose problems while very carefully hiding any hints at a solution. Let the cleverest solve the problems presented. One of the incidents most characteristic of this way of thinking is that of the 'Challenges to Mathematicians' issued by Fermat in 1657 (in the hope perhaps of finding an able correspondent), and addressed in particular to Frenicle and the Englishmen, John Wallis and William Brouncker. Having set out the problems, Fermat concludes: 'I await the solution of these questions: if it is furnished neither by England nor by Belgian or Celtic Gaul, it will be provided by the country of Narbonne,' meaning by Fermat himself. In other words, it wasn't usually a question of shared problems on which one consulted other specialists, but rather of sophisticated riddles, to which the questioner already knew the answer. What is more, even if, in the currents of the seventeenth century, the movements and techniques which would soon characterize the scientific community were beginning to coagulate, questions regarding number were hardly well-regarded. To Digby, who had passed Fermat's challenge on to Wallis (and although he would eventually meet it!), the latter wrote: '[Fermat] seems to have a singular fondness for [questions of number]; but I admit that for my part at least they do not have so powerful an attraction that I am led to devote much time or effort to them, and I do not think them sufficiently important for me to turn to these speculations on number, neglecting the other investigations in Geometry which please me more.'

Why were numerical problems seen like this?

Of Numbers and their Power(s)

Fermat's note is a commentary on a problem in Diophantus on the decomposition of a square number into two squares, as in 25 = 16 + 9. This is a typical question from the *Arithmetic*, in which the problems generally involve finding (whole or mixed) numbers fulfilling certain conditions or relations, such as: finding two numbers whose sum and whose product are given; finding three numbers such that the product of any two of them added to a given number is a square, etc. For particular numerical values an explicit solution is provided, sometimes accompanied by a procedure for obtaining it.

But other kinds of questions about numbers were debated in the seventeenth century: the readers of Books VII, VIII and IX of Euclid's *Elements* had at their disposal a definition of the odd and even numbers, a study of divisibility, including prime numbers, which have no factors other than themselves and 1, such as 5 or 7, and perfect numbers, which are the sum of their factors, such as 6 = 1 + 2 + 3. Each of these two mathematical traditions brought with it its stock of favoured questions, which they sometimes had in common under different disguises. At the beginning of their correspondence Fermat and Frenicle discovered a shared passion for magic squares, tabular arrangements of numbers the sum of whose rows and columns has in each case the same value. These magic squares were also used as talismans.

Today, many of these questions seem to belong to very different levels of difficulty, and not the least surprise for us is to find them side by side. Any classification evidently has to be based on the formulation offered or the tools used to solve the problems. The problem of decomposition into a sum of squares can be put into an algebraic perspective; its generalization will then be concerned, as in Fermat's commentary, with higher powers, cubes and quartics (fourth powers) etc. But it can also be expressed in an entirely different form: according to the theorem of Pythagoras, the relationship $a^2 + b^2 = c^2$ defines a right-angled triangle with sides a, b and c (c being the hypotenuse); one would then be interested in triangles satisfying particular properties, such as that the area added to the height is a square and the perimeter a cube. In the absence of a shared training in the new algebraic methods, this old type of presentation was prevalent. Frenicle, who

The magic square. All the rows, columns and diagonals add up to 34. 'I hardly know anything more beautiful in mathematics,' Fermat wrote to Mersenne about them in 1640. (Engraving, a detail from Melancholia, A. Dürer, 1514. Bibliothèque Nationale, Paris.)

351

had written a *Treatise on Right-angled Triangles*, and who played a not inconsiderable role as an interlocutor for Fermat, knew very little, even nothing at all, of algebraic techniques; they were therefore avoided in their correspondence. This discordance of formation was even more marked in

---

Aliquot parts

In the seventeenth century the 'aliquot parts' were the factors of a whole number. A number was said to be 'perfect' if it was the sum of its aliquot parts (for example $6 = 1 + 2 + 3$), 'sub-multiple' if the sum divides it (in general in a given ratio like 2, 3 or 5). Two numbers were said to be 'friendly' if each was the sum of the aliquot parts of the other (for example 220 and 284, because $220 = 1 + 2 + 4 + 71 + 142$ and $284 = 1 + 2 + 4 + 5 + 10 + 11 + 20 + 22 + 22 + 40 + 44 + 50 + 55 + 110$).

In Euclid's *Elements* there is already a proof that any number of the form $2^n(2^{n+1} - 1)$ is perfect, provided that the term $2^{n+1} - 1$ is a prime number, for example, $6 = 2 \times 3 = 2 \times (4 - 1)$. In the seventeenth century and even before, it was known that the numbers $2^{n+1}(18 \times 2^{2n} - 1)$ and $2^{n+1}(3 \times 2^{2n} - 1)(6 \times 2^{2n} - 1)$ were friendly provided that all the terms in brackets were prime numbers: for $n = 3$, for example, one finds the pair $18,416 = 2^4(18 \times 2^6 - 1)$ and $17,296 = 2^4 \times 23 \times 47$.

If $n = p^r$, p being a prime number, the factors of n are $1, p, p^2 \ldots p^{r-1}$, and thus the sum of the aliquot parts of n is $1 + p + \ldots + p^{r-1}$; since $(p - 1)(1 + p + \ldots + p^{r-1}) = p^{r-1}$, one can see that the study of aliquot parts leads to that of the factors of numbers of the form $d - 1$. Fermat stated his 'fundamental proposition of aliquot parts': if q is a prime and a any number, then q is necessarily a factor of one of the numbers $a - 1$, $a^2 - 1$, $a^3 - 1$ etc. for which r is a factor of $q - 1$; q is also then a factor of $a^R - 1$ for every multiple R of the smallest r possible. Thus, for $a = 2$ and $q = 7$: 7 is a factor of $2^3 - 1$, $r = 3$ is indeed a factor of $q - 1 = 7 - 1 = 6$; what is more, $q = 7$ is a factor of all $2^R - 1$, R being a multiple of $r = 3$, so that $2^6 - 1 = 63$, $2^9 - 1 = 511 \ldots$ A factor q of $2^{37} - 1 = 137,438,953,471$ must be such that 37 is a factor of $q - 1$: the first candidate is therefore 149 (which doesn't work), the second 223 (which does!): this proposition thus allows the saving of much time and calculation in testing the possible factors of number $a^r - 1$, and therefore in searching for perfect and friendly numbers.

Fermat was also interested in the factors of $a^r + 1$, conjecturing on several occasions that $2^{2^r} + 1$ was always prime like 3, 5, 17 and 257. Unfortunately for him, $2^{2^5} + 1$ is divisible by 641; this can be demonstrated by a simple variant of the method described above, but it seems that Fermat had made a simple error of calculation . . . This error was only discovered by Euler, a century later, a fact which throws a great deal of light on the state of the discipline.

---

the case of another of Fermat's interlocutors, Father Jacques de Billy. His text, written following Fermat's letters and mentioned earlier, is liberally sprinkled with errors; his mastery of the techniques under consideration is doubtful, and Billy often seems to be heaping up anodyne variations of the same procedure without explaining it in principle.

In accordance with the traditions from which the problems were derived, the solutions are nearly always put forward in purely numerical form: demonstrations still belong to geometry. In the case of numbers, they generally proceed by induction, that is to say, by studying and calculating example after example; they are often accompanied by a more general statement, but without any further justification: the proof of the author's competence lies in his ability to find all the numerical answers he might be asked for, and more particularly, in finding very large ones. Concerning the number of possible ways of making up a magic square of a given size, Fermat wrote to Mersenne: 'To show you the extent of the knowledge I have of it, the square of 8 which is 64 can be arranged in as many different ways as there are units in this number: 1,004,144,995,344, which will doubtless astonish you, because Bachet and the others that I have seen give only one.' And in the same way, to convince a sceptical correspondent that he possessed a general result which would help to generate sub-multiple numbers, he wrote: 'A long time ago, I discovered and sent out the propositions of the two numbers 17,296 and 18,416, and to do that I had to develop a general solution.'

What is specific to these numerical questions, then, is that unlike geometrical problems, it still seems enough to demonstrate a particular solution, or at best a recipe, in order to convince. As Descartes says, echoing Fermat: 'I have only to add the proof of it' (he is dealing with a numerical question) 'for I save time and in the matter of problems, it is enough to give the method; then those who have proposed it may examine whether it be well resolved or not.' 'I save time' reminds us of the margin too narrow to hold the proof. Seventeenth-century exchanges are traversed by the theme of lack of time, space and availability, and we have already mentioned the material conditions which lay behind it; it justifies a more sloppy, less precise and detailed presentation than the composition of a work in principle requires. Above all it justifies being satisfied with examples, because these are sufficient to solve the particular problem, given the terms in which it is posed: 'Find numbers such that . . .' Competition between correspondents and the challenges issued reinforce this tendency, which itself was perfectly adapted to them; for the potential adversary one chose the most extravagant numerical data to make the task more difficult.[1] On the other hand, this side of things repelled those who were apparently interested in more general studies. 'Not that the questions connected with Arithmetic are more difficult than those of Geometry,' wrote Descartes, 'but that they may sometimes be better found by a painstaking man who will doggedly study the series of numbers than by the skills of the greatest intelligence there might be.' A vicious circle then, in which the type of question and the answer expected put off exactly those who might have been in a position to change their nature.

Let us look again at our favourite note: 'it is impossible for a cube to be written as a sum of two other cubes or a fourth power to be written as a sum of two fourth powers or, more generally, for any number which is a power greater than the second to be written as a sum of two like powers.' Because it is presented in a negative form, it isn't possible to provide a numerical response to the problem raised.

Nowhere else in Fermat is there an example of a general case, and it is therefore reasonable to see in this another example of the hasty extrapolations made possible by the absence of full workings-out and rigorous discussion. But the example of the cubes and the biquadratics is repeated on several occasions, and we know at least the principle of a possible demonstration; Fermat calls it 'infinite descent'. Assuming that there is a solution to the problem, another solution may be deduced by various algebraic manipulations, which will be necessarily smaller: yet there cannot exist an infinitely decreasing series of integers. The initial hypothesis is therefore false, which is what one wanted to prove. If the principle can be stated fairly easily, its application isn't always obvious: and this is precisely what Fermat refused to give any further details of: 'I will not add the reason why I infer [that there would be a smaller solution] because the account would be too long, and it is the whole secret of my method. I will be happy for the Pascals and Robervals and so many other scholars to be searching for it following my clue.'

Even when its form seems the least appropriate, we find the question subjected to the ordinary conditions of seventeenth-century arithmetical practice. Moreover, the reaction of Fermat's correspondents was hardly

---

Infinite descent

Two properties of integers are necessary: first of all the fact that any integer may be uniquely decomposed into the product of prime numbers themselves undecomposable, for example $28 = 2 \times 2 \times 7$. A fundamental consequence of this, which we will use, is that if two numbers have no common factors, and if their product is a square, each of them must itself be a square. The other fact is that there cannot exist a strictly decreasing series of integers. We will now show, by 'infinite descent', that there are no squares which are the sum of two biquadratics, that is to say, no integers x, y and z such that $z^2 = x^4 + y^4$. This will show *a fortiori* that a biquadratic is not the sum of two biquadratics, which is a case of Fermat's 'theorem'.

We will need this characterization of right-angled triangles, which was known before the seventeenth century: if x, y, and z are the integer sides of a right-angled triangle, so that $x^2 + y^2 = z^2$, then they may be written as follows:

$$(1) \quad x = (a^2 - b^2)d, \; y = 2abd, \; z = (a^2 + b^2)d;$$

a, b, and d being integers and d being their highest common factor. If x is even and y not, then their positions are exchanged (they cannot both be odd or the sum of their squares would not be a square). Let us give an example: 12, 9, and 15 are the sides of a right-angled triangle, because $12^2 + 9^2 = 15^2 = 225$ and $x = 9 = (4 - 1) \times 3$, $y = 12 = 2 \times 4 \times 1 \times 3$, $z = 15 = (4 + 1) \times 3$, in the form (1), with $a = 2$, $b = 1$ and $d = 3$.

Let us suppose now that there exist integers x, y and z such that $z^2 = x^4 + y^4$. Rather than divide everything by their common factors, we will suppose, in order to simplify, that they do not have any. Now $(x^2)^2 + (y^2)^2 = z^2$, and therefore, according to characterization (1) they can be written:

$$x^2 = a^2 - b^2, \; y^2 = 2ab, \; z = a^2 + b^2 \; (d = 1).$$

If x, y and z have no common factors then neither do a and b: and as y is even then y^2 is divisible by 4 and either a or b must be even, let us say b. Following the first principle mentioned above, since $a \times (2b)$ is a square, a and $2b$ having no common factor, it must be the case that a and $2b$ are squares, therefore:

$$a = A^2, \; 2b = (2B)^2 \; (2b \text{ being even is the square of an even number}).$$

In particular, $x^2 = A^4 - (2B^2)^2$, that is to say: $x^2 + (2B^2)^2 = A^4 = (A^2)^2$.
Still in accordance with the characterization of right-angled triangles, there exist numbers a' and b' such that:

$$(1') \quad x = a'^2 - b'^2, \; 2B^2 = 2a'b', \; A^2 = a'^2 + b'^2$$

Once again, a' and b' have no common factors and their product B^2 is a square, and they are therefore themselves squares, so that $a' = A'^2$ and $b' = B'^2$, whence $A^2 = A'^4 + B'^4$.

In other words, starting with a solution x, y and z satisfying $z^2 = x^4 + y^4$, we have constructed another, A', B' and A. Since $z = A^4 + b^2 > A$, this solution is strictly smaller than the previous one. One could of course repeat the procedure and thus construct a strictly decreasing series of numbers z, A, etc.: We said at the beginning that this was impossible, and thus there cannot exist solutions x, y, and z with which to begin the process.

We can see here how the use of specific properties of integers coordinates the algebraic manipulations of the demonstration.

355

encouraging: working by tables of examples, they were disorientated by 'negative' propositions. They were hardly interested, and sometimes even overtly complained about them.

A bubbling mixture, perhaps, but one without harmony, a discordance of visions and approaches highlighted by the lacunae in the demonstrations and the material and intellectual chanciness of communication: how did this come to be transformed two centuries later?

Homo faber

Men and the Theory of Numbers in Everyday Life

Our point of departure this time is an article by a famous academic honoured by the French Academy of Science; if the fugitive traces of the seventeenth-century amateurs of number had to be sought in their letters and in the margins of their books, by the first half of the nineteenth century these matters were plainly published in the *Journal für Reine und Angewandte Mathematik*: more than 150 notes, articles, problems and reports appear under the heading 'Number theory' in the cumulative table of contents for the first fifty numbers.

We have left behind a '*conseiller* in the *parlement* of Toulouse', the noblemen and the diplomats, and here the authors are lined up monotonously and repetitively through the different issues: *Privatdozent*[2] at the University of Berlin, professor at the University of Breslau, professor of mathematics at Halle, Braunschweig, Brandenburg, Paris, Oxford, professor at the Karlsruhe Polytechnic Institute, etc. The authors are recruited from a closed milieu: the few who do not hold a university post, either as professor or as *Privatdozent*, are usually at least entitled doctor, which indicates a university training, or are teaching in secondary schools; in accordance with the regulations of the time, they would also have attended institutions of higher education.

From this point of view, Ernst Eduard Kummer, the author of the article mentioned at the beginning of this chapter, had an exemplary career. Having gone to university to study theology, he eventually turned to mathematics which, he wrote to his mother, 'would provide him with a living'. He taught first of all at the Gymnasium at Liegnitz, while writing mathematical articles. Made a member of the Academy of Science at Berlin in 1839, he was appointed professor at Breslau in 1842, and then at Berlin. He worked on number theory during the whole of this period, obtaining the results already mentioned on the problem posed by Fermat. A corresponding member of several European academies of science, he also became Rector of his University, giving lectures and running a research seminar (the first such seminar in mathematics to be organized at Berlin). He was the chief examiner on the jury for the theses of thirty-nine students, of whom seventeen would obtain university posts, among them Immanuel Lazarus Fuchs, Leopold Kronecker and Paul Bachman, who themselves also published in the *Journal für Reine und Angewandte Mathematik*. Filiations of researchers and teachers were being established, occupying the pages of the journal just as they occupied the posts available in the universities.

All this no doubt demanded energy, work and patience; in volume XL of the *Journal* alone there are three articles by Kummer, taking up more than forty large pages. These qualities were indeed clearly valued. In 1846 Kummer wrote to Kronecker: 'As you know, our mathematical literature is made up of treatises, large and small. This is why I would give you this friendly teacher's advice: from the very beginning, pursue your mathematical studies in such a way as to produce treatises, that is to say, you must work on certain subjects until they are sufficiently polished, in such a way that even if they offer, from certain points of view, material to be further developed, they may constitute in themselves a finished whole.' The mathematician Richard Dedekind, for his part, talked of his own *aurea mediocritas*, whose strength, he added, had no other origin than a stubborn perseverance.

Here teaching and research are closely connected, the first serving to provide new recruits for the latter; tasks as diverse as attending an administrative meeting, reading an article, looking for a new result, giving a course, all contributed to the maintenance or improvement of the organism which gave rise to them. The universities were at the heart of mathematical research, the creation of the University of Berlin in 1810 witnessing the new importance attached to it in Prussia, together with other reforms of the educational system. Gustave Dirichlet was a member of the last generation to travel to Paris because it was impossible to learn up-to-date advanced mathematics in Germany. After him (and in part, thanks to him!), students could find, if not everywhere, then at least in Berlin and Göttingen, the basic courses, the explanations and the models which would introduce them to the most recent methods. Visible evidence of this formative activity, the texts of the lectures, together with the short notes published in specialist journals, make up the other side of the mathematical literature. Dedekind and Dirichlet's *Vorlesungen über die Zahlentheorie* appeared in 1863, and well before that entire courses on advanced number theory were offered at the University of Berlin. This daily gathering together of mathematical forces in the same places did, of course, unify the field of problems dealt with. People were drawing on the same sources: Euler, Lagrange and the mathematicians of the French School (then considered the best in the world), and especially Gauss, who in 1801 had published his *Disquisitiones Arithmeticae*, the Bible of nineteenth-century number theory. Gauss's name appears everywhere in articles on the subject, which particularized or generalized his work, re-using his notations and his point of view. The unifying function of this educational process was also a normative function; a common culture was being born, with its rules and its language. We now know the men who made up this community, but how did they interact among themselves?

From Crelle's Journal *to the Academy of Sciences*

The first important journal devoted entirely to mathematics was published at Montpellier at the very beginning of the nineteenth century by the mathematician Joseph Gergonne: this was the *Annales de Mathématiques pures et appliquées*, which disappeared fairly quickly, but none the less formed an explicit point of reference for later mathematical journals, Joseph de Liouville's in France and Crelle's in Germany, whose titles were exact translations of each other: in English, *Journal of Pure and Applied Mathematics*.

August Leopold Crelle, who trained as an engineer, was employed in the Prussian administration. More or less self-taught in mathematics, he took his doctorate at Heidelberg, and from 1828 he worked at the Ministry of Education as a specialist in mathematics and mathematical education. In the preface to the first number of his journal, which appeared in December 1825, Crelle explained that the Germans loved mathematics and that it was necessary to meet the demands of an interested public by making accessible all recent work in the field 'independently of fashion, authority or school'; there would be, if necessary, translations into German, and commentaries on other articles. Correcting the presentation of certain texts, at least, recruiting both mathematicians for the universities and collaborators for his journal, introducing side by side articles by different authors on the same subject, Crelle offered a place where controversies between mathematicians about methods or priorities might be both neutralized and made public, and contributed greatly to the constitution of a better regulated and more unified field of study.

Of course, the exchange of letters between specialists or close colleagues continued, and we have seen an example between Kummer and Kronecker. But the journal, however limited its distribution, relied on the existence of anonymous readers who were none the less capable of understanding the articles published. Peremptory claims that 'It is well known that . . .' and notations without definitions were evidence of the implicit existence of a site, the university perhaps, where the necessary knowledge had been supplied. All science, indeed, makes what it says opaque for those who do not hold the key. Fermat used contractions and allusions in his correspondence, but he had personal knowledge of his correspondents and of the limits of their knowledge. In the nineteenth century, a community existed, at least a potential one, which could be distinguished by the fact that its members could read and/or publish in this journal. Like any other, this node of information marks the limits of a world governed by a conception of mathematics and the means of speaking or even living it. The changes in scientific organization can be measured both by the growing percentage of academics and the percentage of articles published in specialist journals. The different means of recruitment, the places available at Berlin or with Crelle, weave together the shared space of mathematical activity. The publication of Kummer's article was therefore both a banal event and a significant one in the author's life. And academic reward? This has a long and instructive history, an account of which we will borrow from the historian Harold Edwards.

Everything began far from Germany, at a meeting of the Academy of Sciences in Paris: Gabriel Lamé, a mathematician, announced that he had a general proof of Fermat's proposition. Let us recall that this was a matter of proving that it was impossible for integers x, y and z to satisfy a relationship of the type $x^n + y^n = z^n$ unless $n = 2$ or one of the integers is zero. The method of infinite descent allows the cases $n = 3$ and $n = 4$ to be settled; complicated and/or partial proofs had been found for other powers, but a general proof seemed far away. In his presentation Lamé said that he had used ideas from Lagrange and from Gauss, and he recognized the essential contribution made by a conversation with Liouville. We should note in passing that these references to colleagues, far from diminishing the merit of he who made them, now helped to ensure a favourable reception. Unfortunately for Lamé, Liouville himself declined the favour, passing on the responsibility for his modest contribution to other illustrious predecessors: he was justifiably anxious about some of the hasty generalizations that Lamé had made. The latter, as we shall see in a moment, had extended to other kinds of numbers the usual properties of integers, divisibility, decomposition into prime factors, etc. Gauss had certainly followed a similar procedure in a particular case, but had carefully justified these properties at each stage. With his lack of caution, had Lamé got himself onto dangerous ground?

Another person spoke up at the same decidedly eventful session of the Academy: this was Augustin-Louis Cauchy. He recalled that he had already spoken on the topic several months earlier; he had not had the time fully to develop his ideas, but it wouldn't be much longer . . . And indeed, on the French side, the following months were full of feverish activity: who would complete the proof?

It was then that Liouville received a letter from Kummer in Germany. The latter, having heard of the situation, probably from Dirichlet, answered in the negative the thorny question raised by Lamé's work: no, the properties of integers could not so easily be extended. In the same letter, however, Kummer announced two articles, one to appear soon, in which this problem would be resolved by the introduction of a new sort of number, which Kummer called an 'ideal number', and the other (which would appear in 1850) on the application of this new theory to the original question raised by Fermat.

The story might well finish here; all the ingredients are there to make it an eternal western, the victory of the good, from every point of view (Kummer), over the beast (Lamé) and the crook (Cauchy). And the whole thing happened under the eye of the wise old man (Liouville). All the ingredients are there to illustrate the operation of a professional milieu: problems drawn from the same sources, considered to be essential at the same time; local public sessions to announce, confirm, exchange and regulate these researches; contacts with other schools and other countries to stimulate and verify national activity; journals, of course, places of compromise and peace, very well typified by the letter written by Liouville to accompany the appearance of an article by Kummer in his journal. Cauchy, even if he continued to protest the value of his own work on the question, is a very good example of the ideal mathematician: 'If Mr Kummer has advanced the question some steps further, if indeed he had removed all the obstacles, I would be the first to applaud the success of his efforts, for what we ought to desire above all, is that the work of all friends of science should come together to make known and promote the truth.'

The story, which did not finish there, is even more exemplary.

In 1850 the Academy of Sciences at Paris decided to offer a prize for a complete proof of Fermat's 'theorem'. Kummer, in fact, had only succeeded in establishing one for integers n satisfying certain technical hypotheses (in particular for n up to a hundred, with the exception of 37, 59, 67 and 74). Seven years later, no satisfactory solution having been offered to the Academy, it was suggested that the prize be given to Kummer. The committee set up to look into the question included Cauchy and Lamé. The first expressed reservations about certain details of Kummer's proof; after a certain amount of wavering and an exchange of letters orchestrated once more by Dirichlet, the situation was saved by the appearance of a new article by Kummer in 1857, once again in the *Journal für Reine und Angewandte Mathematik*. This dealt with all the lacunae of the first article, but by following a very different path. This is a characteristic example of the sorts of pressures which lead to publication, of the procedures of control set up when an important result is at stake. This is what happened to those who were connected with numbers. Did the field itself offer evidence of what had happened?

The Law of Numbers

We have seen that the permanent re-reading of the past provided arithmetic with its daily bread. At the same time this re-reading was buried under new notations which expressed the new approach. In the seventeenth century, one looked for the prime factors of $a^r - 1$: for such a prime number q, the remainder of the division of a^r by q is therefore 1. The stress on the remainder from division is expressed by the notation $\equiv$, which we owe to Gauss: one says that two integers a and b are congruent modulo an integer q ($a \equiv b$ mod q) if a and b give the same remainder when divided by q. The 'fundamental proposition of aliquot parts' (see p. 352), that is, that a prime number q is a factor of one of the numbers $a^r - 1$ for a suitable r (a factor of $q - 1$) is then written: $a^r \equiv 1$ mod q for an r such that $q \equiv 1$ mod r. The remainders (and thus the congruences) can be added and multiplied together, thus facilitating the search for the smallest possible r: $25 \equiv 4$ mod 7, therefore $25^5 \equiv 4^5$ mod 7, and there is no need for the calculation of 25^5. There were also analogous results for modulus q not necessarily prime. Many numerical tables on these themes for different values of q also appeared in Crelle's *Journal*, several of which were produced by the founder himself: these numerical tables served as a reservoir of examples from which to draw in the search for properties to demonstrate. If the discourse is perhaps not so immediately decipherable, the inductive activity, the calculation of particular cases, a characteristic facet of the study of numbers in the seventeenth century, thus found a properly integrated place in the development of the common project. It was now a preliminary activity and a public service, no longer an individual game providing insoluble riddles for adversaries.

The decomposition of numbers into sums of squares (or, as was still said, the search for hypotenuses of right-angled triangles) became quite simply the study of the form $x^2 + y^2$: and the classification of forms of this type is one of the crucial problems raised by Euler, Lagrange and later Gauss. Algebra therefore became a language in which questions could be expressed, arithmetical ones. Dirichlet, for example, used infinite series and the analytical methods developed in the eighteenth century to prove that for two integers a and n without common factors, there exists an infinity of primes p such that $p \equiv a$ mod n. These infinite series in their turn became the subject of fundamental research in the field. But this growing complexity was managed through the unification of the goals aimed for: if languages, notations and techniques became more difficult to use, they organized with more certainty a corpus of privileged themes in the service of a discipline, whose mission was the study of the properties of integers. It is in this light that we need to re-read our favourite problem.

The new tools we shall need are the complex numbers. Introduced several centuries earlier into the study of algebraic equations, they served first of all as aids to calculation. This is how they were used by Lagrange and Euler, for example, in the study of expressions such as $a^2 + b^2$ or $a^3 + b^3$. Just as $a^2 - b^2 = (a - b) \times (a + b)$, one can write:

$$a^2 + b^2 = (a - b \sqrt{-1}) \times (a + b \sqrt{-1}),$$

$\sqrt{-1}$ being the complex (or 'imaginary' number) whose square is -1. Gauss went further by showing that complex numbers of the form $a + b \sqrt{-1}$, a and b being ordinary integers, behave very like integers: they can be multiplied, added and even uniquely decomposed into 'prime' complex numbers not themselves decomposable. These new integers (called Gaussian integers) could be useful intermediaries in the study of true integers: if p is the sum of two integers, it can be written:

$$p = a^2 + b^2 = (a + b \sqrt{-1}) \times (a - b \sqrt{-1}),$$

that is to say, it decomposes into products of Gaussian integers: the study of its decomposition into such products therefore replaced decomposition into the sum of squares. A prime number, the sum of two squares, is no longer prime as a Gaussian integer: for example $5 = (2 + \sqrt{-1}) \times (2 - \sqrt{-1})$ is decomposable, while 3 and 7 remain 'prime' as Gaussian integers. As one has already seen in other cases, these complex 'integers' can become a source of interest in themselves. Do those derived from other decompositions ($a + 5b^2$ for example) still have the usual properties of integers? What is it that characterizes an integer, essentially? These questions, and the search for the general laws of decomposition of these exotic numbers were indeed all the rage in nineteenth-century Germany. How does this relate to 'our theorem'?

What Lamé proposed was to decompose $x^n + y^n = z^n$ with the help of complex numbers in the form:

$$x^n + y^n = (x + \zeta_1 y) (x + \zeta_2 y) \ldots (x + \zeta_n y) = z^n,$$

where the ζ_i are the n complex numbers whose nth power is 1. Lamé's argument consisted of granting to the numbers $(x + \zeta_i y)$ the same properties of factorization as the usual integers: in particular, accepting that the product above could only be an nth power if each term was – this type of argument was always used in a classical 'infinite descent'. Unhappily for Lamé, this property, essentially true for natural and Gaussian integers, is not true for the general case: this is the specific problem which Kummer attacked and resolved in the articles mentioned above.

Ideal numbers

How could the creation of ideal numbers allow one to retain a fundamental property of integers, the unique factorization into prime factors? To try and understand this better, let us imagine for an instant that we are considering only numbers of the form $4k + 1$. The numbers can be decomposed into prime factors of the same kind, but not perhaps in a unique manner: for example $441 = 21 \times 21 = 9 \times 49$, and 21, 9, and 49 are undecomposable into numbers of the form $4k + 1$. If one wants to maintain the uniqueness of the decomposition, one has to 'invent' numbers of the form $4k + 3$: one then has $9 = 3 \times 3$, $21 = 3 \times 7$, $49 = 7^2$ and 441 decomposes uniquely into prime numbers $441 = 3^2 \times 7^2$. It was analogous phenomena which necessitated the introduction of Kummer's 'ideal numbers'.

Apart from placing a number of new tools and techniques at the service of theory, this example underlines how previously isolated questions found their place in the unified panorama the discipline now offered.

It was in this framework, broader than Lamé's, that Kummer worked: the article on Fermat's proposition made up only eight of the forty pages he published in volume XL; the partial solution provided was the result of a lucky detour from a more ambitious programme. This was an exemplary application of the advice he offered Kronecker; while aiming at a global question, day-to-day work ought to explore details and investigate possible applications. To the *Allgemeiner Beweis* (general proof) of the beginning of the title of the article in question there corresponds, as an echo, the technical restrictions at the end which specify its validity, here the powers effectively dealt with by the proof.

There had thus been a considerable change, both in the vision of the discipline and in the means of practising it. The field, whose prestige depended of course on that of the professionals who worked within it, did not content itself with using developments in other disciplines, but coordinated them with its own, or even stimulated them itself. The connections between the integration of functions such as $1/\sqrt{(1 - x^4)}$ and solutions to $y^2 = 1 - x^4$ opened new perspectives for research, and in his work on analytic functions, the German mathematician Karl Theodor Wilhelm Weierstrass was able to draw on the research on factorization whose importance we stressed earlier. As Gauss put it, number theory had become 'the queen of mathematics'.

At the same time, the queen defined her court: frontiers were fixed, which specified and made more rigorous the conditions of access to it. We have already seen that education alone allowed comprehension of the texts: Fermat would perhaps have had some difficulty in recognizing his problems but the nineteenth-century number theorists (without always having read him directly) claimed him as their distant ancestor. How and why was his inheritance, in particular, passed on? How and why did it become so fruitful? It is to his sources that we must now turn to understand this.

Putting Things into Perspective

Fermat, as we have said, must have reflected on the work and algebraic approach of François Viète. His *Challenge to Mathematicians* rings out like a declaration of faith, and it is worth a closer look:

> There is hardly anyone who proposes purely arithmetical questions, there is hardly anyone who knows how to resolve [problems about numbers]. Is this because arithmetic has until now been treated through the medium of geometry rather than for itself? This is the tendency which appears in works both ancient and modern, and in Diophantus himself. For though he distanced himself from geometry a little more than the others by restricting his analysis to the consideration of rational numbers only, he did not completely escape it, as is superabundantly proved by Viète's *Zetetics*,[3] in which Diophantus' method is extended to continuous quantities and then to geometry.
>
> Arithmetic, however, has a field proper to itself, the theory of whole numbers; this theory was only slightly dealt with by Euclid and has not been sufficiently cultivated by his successors ... the mathematicians, therefore, must develop or renew it.

Here then is a subject and a unity. Here there is a line of thought pretty firmly upheld in other places in his correspondence: with methods extensible to continuous ('geometrical') quantities the particularities of whole numbers disappear. So a more appropriate method must be developed. This is exactly the role of the famous 'infinite descent' we have mentioned (see p. 354): it shows how the specificity of the integers can be introduced into an argument based on algebraic formalism. We have already shown how this type of procedure, having become common, then led to a more thorough interrogation of the very idea of an integer and its characteristics. We can understand at the same time how a selection operating according to the professional criteria of the nineteenth century would have viewed Fermat.

Looking at other characteristics too, Fermat seems like the number theorists of the nineteenth century, whose behaviour we have already examined: is not his restoration of Greek treatises the typical product of a patient devotion to the advancement of science? Are not his attempts, towards the end of his life, to persuade Pascal or Carcavi to help him publish, or even to write and complete his works on number theory, an expression, though late, of a desire for diffusion among the mathematical community? In the *Challenge to Mathematicians* he gives precious information on his work in progress: 'It is known that Archimedes did not disdain to work on propositions of Conon's which were true but not proven, and that he was able to provide them with demonstrations of great subtlety. Why should I not hope for like succour from you eminent correspondents, why should I, a French Conon, not find an English Archimedes.' And he goes on to explain his conjectures on the prime nature of 2 to the power of 2 to the power of $n + 1$.

What this teaches us is that the conditions of validation of such an activity do not belong to the isolated individual. The mathematical and arithmetical fabric is too frayed: what is true, concretely, whatever Fermat's declarations of intent and whatever the multiple readings one can make of them, is that Fermat will not find among his contemporaries an English Archimedes to complete his conjectures or to refute them, nor even a Cauchy to express doubts about their validity. No one will help him compose nor even publish his work on numbers (Carcavi's attempts ran into the inertia of Huygens, and after Fermat's death, Carcavi himself seemed little inclined to offer Fermat's son copies of the precious letters he had received from his father). His enthusiastic supporters in this field, Father de Billy at their head, seem to have been more impressed by the number of solutions he had obtained rather than the manner in which he had done so. Above all, and much more seriously, his numerical investigations were met by a general disdain on the part of the most talented mathematicians of the time, those who were beginning to form what would become a scientific community. We have already mentioned Wallis's reaction and Descartes's. Frans van Schooten knocked another nail into the coffin, making fun of the tone in which Frenicle announced his own solution to Fermat's questions: 'And here Paris gives a solution to the problems which neither your English nor the Belgians have been able to find; Celtic Gaul is proud to take the palm from the Narbonnaise, etc. As if it were an affair of State to know these numbers, and that everyone should attach so much importance to this solution, that he could find no better way to pass his time.'

At the end of the seventeenth century, as one can detect in reading the correspondence, a milieu was being formed, the academies took on a capital A and received grants of royal funds. The Academy of Sciences was founded in France in 1666 and among its members were Carcavi, Frenicle and Roberval. In England the Royal Society published its *Philosophical Transactions* from 1665, from which it 'might appear that numerous hands and minds are in many places industriously employed' for 'the general benefit of humanity'. The public good was on the agenda, the progress of man would come through the progress of science put at the service of the State: the accent was on utility and on social convenience, sometimes seen in terms of the new classes and the priorities of the Industrial Revolution. If then a mathematical community was constituted, it did not include arithmetic among its essential themes. Highly symbolic in this regard, the first volume of the *Transactions* contained an obituary of Fermat, which mentions his principal works, his position as *conseiller*, but practically nothing of his work on numbers. How then, two centuries later, could they offer a prize for a proof of his 'theorem'? Where then are Fermat's inheritors?

A first answer is provided by the professionals themselves: it was Euler, then Lagrange, Adrien-Marie Legendre and Gauss whom they studied and thought about. Euler, indeed, provides an example of the ideal transitional case: he received his mathematical education from the Bernoullis, and he then obtained and filled with distinction several posts with the Academies of St Petersburg and Berlin; he published innumerable articles in their journals, for example, in the *Acta eruditorum* created by Leibniz on the model of the *Transactions*. But his contribution to number theory takes up only four volumes out of the seventy or so of his complete works. Frederick II certainly didn't pay him to apply most of his energy to it; his activities rather included advice to military engineers, work on ballistics and artillery; his sight was seriously damaged by the work he did on maps. It was through his correspondence (one of those nexus of scientific communication of which we have found traces in other periods) with Christian Goldbach that he discovered Fermat's conjecture on the powers of 2 and picked up the gauntlet. The greatest part of his arithmetical work reconstitutes or complements Fermat's affirmations, and he went so far as to search (in vain) for the remains of letters or other clues that Fermat might have left concerning his proofs. The flame, revitalized in exchanges with Lagrange, would never again be extinguished.

But these famous names allow us only to note the changes in method. If Goldbach insistently mentions Fermat's hypotheses in a letter to Euler, it is because throughout the eighteenth century there continued to be people who loved numbers, who transmitted the inheritance and prevented it from falling into complete oblivion. The arrival *en masse* of the professionals definitely put an end to the Billys and the Frenicles themselves, on two counts: a historical exclusion erased them from the filiations established for the determination of precursors, in this case Fermat (for other reflections on the question of precursors see pp. 73, 455 and 506); a second, sociological exclusion, some aspects of which have been dealt with above, pushed them outside the milieu in the process of being constituted. Elsewhere we find the true, undisciplined amateurs, in the journals of recreational mathematics, in letters to the academies, received with a little sigh of amusement or of boredom when they announce the squaring of the circle or a general proof of Fermat's theorem by elementary methods. Their origins, complex motivations and multiple activities offer few points of reference for characterization; their contours dissolve, their histories become more and more individual, the freezing up of a milieu once more having exiled them to the margins of our discussion. It is on the centre then that we shall fix our attention. Whatever one says, a university isn't made up only of professors: there are also students, secretaries, administrative and maintenance staff, money for courses, salaries, post and the building up of libraries. How, and to what extent was all this finally mobilized, deciphered and understood in terms of a property of numbers? In other words, to take up Schooten's expression, how did it become an 'affair of State'?

At the end of the eighteenth century, number theory was still no more than a flower-filled country lane, disdainfully ignored by the great mathematical roads. Jean-Étienne Montucla, the first historian of mathematics, was still able to write: 'Geometry is still the general and only key to mathematics.' A woman, Sophie Germain, prevented by her sex from following a course of higher education, was still able successfully to solve certain cases of Fermat's problem by elementary methods and to maintain a real exchange of ideas with Gauss. Gauss himself, after the publication of his *Disquisitiones*, whose importance was mentioned earlier, sought a post less dependent on the generosity of a patron; he needed, he said, to devote himself to more useful work, that is, astronomy. Whatever the always keen interest Gauss had in numbers, whatever the sophistication and difficulty of the techniques employed, one is still very far from Kummer, who began his researches in this field as soon as he was appointed to a university post. It took one generation, it seems, for the point of view to change; and in the meantime the world too had changed, in the storm of revolution.

Throughout the eighteenth century, France occupied a remarkable political and intellectual position in Europe (see pp. 422–54). It was not Euler but Maupertuis, a Frenchman, who was President of Frederick II's Academy. It was the ideas of the French Enlightenment that provided the ideological points of reference. It was the French one read, admired and detested. German defeats in the face of the Napoleonic armies saw the culmination of this state of mind. Gauss, still thought of, with some reason, as ensconced in his ivory tower, was horrified, and expressed himself forcefully on the situation. The generals too responded, with what was at the same time an explanation, an excuse and a cure. It was through the scientific training of its soldiers that France had won (see p. 428). It was thanks to such French institutions as the Polytechnique and the Écoles Centrales, established during the Revolution, that Prussia had lost. What strikes us now is that this discourse was widespread; the Humboldt brothers – Alexander, especially, who knew the French mathematicians and who favoured the development of the discipline in Germany – and Crelle made their own contributions.

Crelle was sent on a mission to France. His report, highly positive on the whole, expressed a most curious reservation, especially coming from a specialist in technical questions. It seemed to him that there was in France too great an insistence on immediate, concrete and practical problems, to the detriment of a real education suitable for the development of the whole person.

To understand this criticism better, it must be remembered that priorities in nineteenth-century France and Prussia were very different. Compared to France, and to the countries of Western Europe generally, Prussia suffered from considerable technological backwardness, but when the decline in French influence came, no 'industrial class' was yet ready to take over the lead. What is more, and here was another difference from France, intellectual activities had for a long time been regarded with disfavour. The newly emerging classes had to be provided with the means of their education and a feeling of integration into the society that was being constructed. The envious bitterness towards France, exacerbated by military defeat, encouraged refusal of a rationalism imposed according to the Enlightenment model, and it was from the 'neo-humanist' philosophy that nationalist sentiment took its themes and modes of expression. Postulating the fundamental unity of the essence, thought and development of the individual and the social good, in the nineteenth century this new way of thinking prevailed in many parts of Prussian life.

The same letter in which Kummer encourages his pupil to work carefully and go forward with small steps, continues with these significant words: 'What I say here is generally valid, it is applicable to everything that has to do with development, yes, just as much to the history of the world as to the life of States and of individuals.'

Kummer, the pastor's son who had become Rector of the University of Berlin, is a typical example of social ascent in nineteenth-century Germany, as indeed Fermat, the son of a well-off merchant who became a member of the *noblesse de robe*, was in seventeenth-century France. In this respect it is instructive to compare the composition of French mathematical society with what we know of the German mathematicians: the academic sons of pastors find their match in France in engineers and artillery officers.

It is easier to understand now how in 1810, at the newly created University of Berlin, it was the departments of philology which provided the scientific models thought of as universally valid, and also the highest salaries. Indeed Gauss in his youth had wanted to devote himself to this discipline, and this was the subject studied by many of our number theorists when they first entered university. Kummer several times asserted that mathematics and philosophy were only two forms of the same activity. But the same interpretation is to be found elsewhere, without regard for social origins or political differences: Carl Jacobi, the first professor of Jewish origin to be appointed in Germany, and as much of a political radical as Kummer was a conservative, wrote to the French mathematician Legendre: 'It is true that M. Fourier held the opinion that the principal goal of mathematics was public usefulness and the explanation of natural phenomena; but a philosopher such as he should have known that the only goal of science is the honour of the human mind, and in this respect, a question about numbers is worth just as much as a question about the Universe.'

To the economic requirements of the moment, to the necessity for professional education, for apprenticeship, and therefore the development of the universities, were added the constraints imposed by these groups themselves, who established themselves there, and the desire for disciplinary purity. Lecturers at the university or at the military academy, the theoreticians were able to devote themselves to what research they wished, since in any case a necessary harmony guaranteed its social relevance. The mandarin theme of academic freedom runs through official discourse like a leitmotiv: nothing must constrain the free working of the mind, seen less as a rational progress than as a constitutive element of the good order of the world. In universities in which what we today call the human sciences or even literary disciplines were pre-eminent, number theory and so-called 'pure' mathematics generally had a privileged position *vis-à-vis* other disciplines orientated towards more immediate practical applications. They were almost natural allies defending their progress and their professional status with the same arguments. The change in methods too made this adoption easier, since the most sophisticated techniques were now employed and could therefore be learnt there.

This way of thinking penetrates to the heart of Kummer's article on ideal numbers. His story has allowed us to understand better the behaviour of a professional in the field of number theory, and it is only right that it should also allow us to distinguish the particular conditions in which this discipline became a profession. Having explained why he had been obliged to introduce these ideal numbers, Kummer compares them to the radicals in chemistry, a rising science soon to become all-powerful in the second half of the nineteenth century. These radicals testify, says Kummer, to the presence of a physical phenomenon, without themselves being isolable. Then he adds:

> These analogies should not be thought of as amusing coincidences; on the contrary, we find in them a firm foundation for believing that chemistry, just like the theory of numbers we have been considering here, has, even if they are in two different spheres of Being, the same basic concept as their principle, that is to say that of composition . . . The chemistry of natural materials, and the chemistry of complex numbers treated here should be seen as realizations of the concept of composition and of the conceptual spheres that depend upon it: the one as physical [realization],

369

connected to the conditions of external existence, and in consequence more rich, the other as mathematical [realization], perfectly pure in its own necessity, but therefore poorer than the first.

Here then, we see at work, at the very heart of number theory, the principles which led to its institutionalization; we have already briefly noted the socio-economic conditions in which it took place. It is then easy to understand why it was in Germany in the nineteenth century, and not in France, that arithmetic really became a professional discipline.

To decompose a power into the sum of two similar powers ... To demonstrate that this problem has no solutions if the powers concerned satisfy certain conditions ... The apparent filiation has only made clearer the differences in the status of the two statements. A fragmentary science and the still lively attraction of amusing riddles offered up to the intelligence of the amateurs of numbers became progressively bound up with a technical language inaccessible to non-specialists, losing the hint of adventure in the academic dust, and intellectual play came to be underwritten by academic distinctions. They became serious, and gained the social recognition of professional mathematics.

The identification of the characteristic features of the amateur and the professional has enabled us to show how these distinctions only take on meaning to the extent that a recognizable social entity establishes the definitions, the relations and the rules. We have also seen that in order to develop as a profession a discipline has to mobilize substantial resources, which cannot be expressed in terms of strict utility. Or rather, the concrete fact of the preferential insertion of pure mathematics in Germany has led us to search for more unexpected forms of public utility. The professionalization of disciplines without immediate applications did not take place under the wing of those disciplines which did have them, even if, of course, the marginal interest of the professionals already in place contributed to the development, when the time came, of a discipline hitherto disdained. Neither is it particularly a question of a risk taken by a society sufficiently developed, a risk that a field might in the long term produce useful results; one would still have to say for whom and for what. Nor is it the inevitable luxury which accompanies 'progress' and which encourages with the same enthusiasm any disinterested activity in art, music or number theory. The movements of disciplinary structuring, of public recognition of activities are less often fortunate chances than the results of tensions, sometimes contradictory, sometimes complementary, which need to find resolution. The researcher, in front of the blank page, can and should forget that other worlds exist; but the necessity of taking one's seat at the *parlement* of Toulouse always threatens to dislodge one.

We are also the heirs of the nineteenth century and the discourse of their number theorists, deciphered here, still runs through our heads: pure mathematics versus applied, public utility, which must be re-read all the time to see whom it contains; the honour of the mind, yes, going as far as reconstructed versions of their own history. It is from them that we still borrow some of our ways of looking at mathematics and their strange relationship to the world. This has changed, however, just as the subtle alliances which cunningly bring together number theory, the public economy and social advancement have done. We have several times discarded explanations by chance or simple inertia. From the tarnished finery of the nineteenth century, what threads still constrain the development of modern number theory?

12

Ambiguous Affinity: The Newtonian Dream of Chemistry in the Eighteenth Century

Isabelle Stengers

In which we see how a concept can become outmoded, despite having organized the language, operations and reasoning of a science for a century; and how chemistry itself became 'modern', in two stages.

First, a problem of method: ought we to speak of the history of science or of the history of the sciences? Both these alternatives can be used in equally simplistic ways. 'History of science' can be taken as implying that there is properly only one single scientific project, and that this one project covers the whole series of different fields through a kind of division of labour. Behind the specificity of any one science, there is 'science' itself, the common identity that lies behind diversity. But 'history of the sciences' too can be just as loaded with *a priori* judgements; in this case each science is taken as developing its own individual project, which has to be identified in its purity, beyond the shared appearances and in particular the attempts at unification now thought of as artificial or ideological. The historian traces the dissolution of presuppositions, the appearance of a clear awareness of the specificity of the scientific object and of the conceptual and technical instruments corresponding to it.

The case of chemistry is very interesting from this point of view, because it allows us to displace the question and transform its meaning. For the choice between *science* or *sciences* is not primarily a methodological question for historians, but a question asked by chemists themselves throughout the history of their discipline. When did chemistry become entitled to the name of science? How did it 'discover' the nature of its relationship with physics? How did it establish its autonomy from craft and later, in the nineteenth century, industrial production? And finally, was chemistry a specific science, or should it be conceived as a specialized branch of 'science' in general? These seem to be questions for the historian, but they are in fact questions asked by some of the actors within the history the historian is trying to understand. Like the historians, these actors referred to the history of chemistry as they worked out their answers, arguing for a particular future and a particular identity for their science.

If then the historian's 'methodological' problem has to be reformulated, I would suggest that it is important to avoid becoming one actor among the others. That is to say, we must understand these questions – all of which imply value judgements – not as questions we have to answer ourselves, but as elements in the complex situation which has to be understood.

Is chemistry a specific science, and if so, what is the nature of its specificity? Let us hear one of the actors, Fontenelle, writing in 1699:

> By visible operations, Chemistry resolves substances into certain gross and palpable principles, salts, sulphurs, etc., but Physics, by subtle speculation, acts on principles as Chemistry acts on substances; it resolves these themselves into other principles yet more simple, into tiny moving bodies shaped in an infinity of ways ... The spirit of Chemistry is more shrouded and indistinct, it resembles the compounds, in which the principles are more confounded one with another; the spirit of Physics is more simple and more free, it goes back to first beginnings while the other does not go all the way.

Tiny many-shaped bodies: here is an allusion to Cartesian chemistry, but if we leave aside such detail, which helps date the text, and if, for example, we substitute 'chemical elements' for 'gross and palpable principles' and 'quantum wave functions' for 'tiny bodies', many contemporary physicists and even some chemists would agree with the description. They would say that there is no difference in theory between physics and chemistry. It is physics which allows us to understand the Periodic Table, which is both the outcome and foundation of the practice of chemistry. (See Mendeleyev, pp. 556–82). The real difference, which might justify a distinction no longer valid in theory, is summed up in this well-known saying; 'Physics you understand, chemistry you learn.' The 'spirit of the physicist' is compelled to elucidate the relationships between 'first principles' and their consequences, while chemists are interested in the compound, in the complicated cases in which these principles are in fact 'confounded', and where arguments are only a combination of theoretical intelligibility and approximation based on experience. Chemistry's specificity therefore derives from the fact that the operations it performs and the questions it puts are determined not by theory but by utility. Chemistry is a hybrid produced by the compromise between practical concern and physical intelligibility.

This characterization might be criticized by suggesting that the compromises chemists are reduced to are less the expression of a compromise with utilitarian interests than the consequence of the limitations of these famous 'first (quantum) principles' themselves, and there is no reason why this argument should not be pursued. But this can only be done at the price of becoming an actor oneself, and using, like the other actors, all the available theoretical, historical, epistemological and socio-political arguments in order to bring about a desired result: here, a transformation in the image of chemistry. Let us therefore resist the temptation and let us remain historians, though we should recognize that history as told by historians is itself part of the history of the different sciences. This is so both in the questions historians will ask, the choice necessarily made, always being reconsidered by the profession, of what is considered as 'needing explanation' against the background of what is seemingly natural, and in the use of the products of historical narrative in creating the images scientists have of themselves or of the controversies which divide them. I will therefore be considering a problem which is properly historical in the sense that all present-day scientists regard it as settled, but which allows a 're-historicization' of the contemporary idea of chemistry by showing that this is not a final conclusion, but rather the current solution to a problem which has dogged chemistry ever since scientificity became an issue. This problem is raised by the history of the concept of affinity.

Outmoded Affinity?

Today, when chemists use the notion of affinity, they do so in the context of 'chemical thermodynamics', established in the second half of the nineteenth century with the work of physical chemists such as Cato Guldberg, Peter Waage, August Horstmann, Henry le Chatelier, Jacobus van't Hoff, Josiah Willard Gibbs and Pierre Duhem. This designation refers to a formal extension of thermodynamics, a physical science, to the description of chemical reactions. In other words, like the interpretation of the Periodic Table from the first decade of the twentieth century (see p. 578), the history of the notion of affinity from 1860 to our own day has been a question of physics.

Here, however, we aren't concerned with an interpretation. As we shall see, for eighteenth-century chemists the concept of affinity combined the problem of the 'chemical bond' with the problem of the reaction in which bonds are formed and destroyed. Thermodynamic affinity, on the other hand, refers only to the direction of the chemical reaction. The reason for a reaction occurring is not a question of thermodynamics but of quantum analysis. Thermodynamic affinity treats the chemical reaction as a function of (thermodynamic) general conditions, while it is the task of quantum mechanics to understand it as an event, as the creation and destruction of bonds between atoms.

This separation of the reaction and its conditions is enough to show that the affinity of the present-day chemists is not that of their eighteenth-century counterparts. Far from being a 'first approximation' to modern affinity, eighteenth-century affinity died without offspring and, when they needed the term some fifty years later, nineteenth-century chemists borrowed it from a past that was over and done with.

Eighteenth-century affinity is then a classic example of what a philosophical historian like Gaston Bachelard would call a 'lapsed' concept, meaning a purely parasitic concept expressing an epistemological obstacle, which can be eliminated without loss from the true history of science, as it demonstrates the dynamics of scientific reason. In his *Matérialisme rationnel*, we can see how Bachelard judges Berthollet, one of our major protagonists: 'How could an experienced experimentalist, a great chemist like Berthollet, be satisfied with such a view as the following: "The powers which produce chemical phenomena are all derived from the mutual attraction of molecules, which has been called affinity, to distinguish it from astronomical attraction. It is probable that both are the same thing".'

Bachelard 'judges' Berthollet in the name of what became clear in the twentieth century: that the chemical bond has nothing to do with the gravitational attraction between masses. He redoubles his condemnation with a reference to the historical context: to 'demonstrate the vanity of such views, which merge together astronomy and chemistry with a stroke of the pen', Bachelard gives another example of 'gratuitous synthesis', the *rapprochement* between human relationships and 'chemical affinity or relationship' attempted at that time by the French writer Louis de Bonald. 'The *rapprochement* of two such widely different topics deprives them both of any cultural value.' De Bonald is but one example, and Bachelard might have mentioned others such as Schelling, Hegel and Nietzsche, who all thought the problem of the relationship between affinity and gravitational attraction worthy of discussion. Above all, he could have poured scorn on Goethe's *Elective Affinities*, a book to which we shall return, which puts 'chemical attraction' at the heart of a story where the storms of passion are compared to the lawful calm of marriage. In any case, affinity twice is doubly condemned, both by the progress of science and by the fact that its meaning did not remain narrowly circumscribed within a scientific discipline; proof that it was the vector of other interests than those of science.

But Bachelard's verdict poses a problem, which is the one we shall be dealing with here. Of course, at the end of the eighteenth and the beginning of the nineteenth century, the notion of affinity wasn't 'pure'. What in fact it represented within scientific culture was the problematic relationship between chemical linkage and the Newtonian force of interaction; this is also the problem of the specificity of chemistry's object, and it was therefore of interest to all those who were concerned with the scope and significance of the 'first principles' of physics. This being so, the 'irredeemably lapsed' character of this notion of affinity, despite its being used again in thermodynamics fifty years later, is part of the history of the question of the relationship between physics and chemistry. Whence the problem: what was it that made the issue represented by affinity outmoded, and made it outmoded in such a way that we can observe today this strange resonance between Fontenelle's description of the relationship and the dominant opinion of contemporary physicists, which is that chemistry is a branch of physics, characterized by the gross or approximate character of its intellectual and practical operations? 'Physics you understand, chemistry you learn.'

Newtonian Chemistry in the Land of the Cartesians

What was the origin of the idea of affinity? It is usually traced back to the *Table of the Different Relationships Observed between Different Substances*, published by Geoffroy in 1718. But relationship rather than affinity? The fact that one term was be replaced very quickly by the other is less important, as we shall see, than the absence of a third term, attraction.

At that time 'affinity' was part of the traditional vocabulary of chemistry as of alchemy. As Louis Guyton de Morveau wrote in the article on 'Affinity' written for the 1776 *Supplément* to the *Encyclopédie*: 'This term for a long time had no more than a vague and indeterminate sense, indicating a kind of sympathy, a veritably occult property, by virtue of which different substances more or less easily united with one another.' There is no need here to mention the older chemical or alchemical meanings of the word; these had been forgotten, the tradition they belonged to held in scorn by everyone who in the eighteenth century borrowed the term from what they thought of as the prehistory of their science. As far as they were concerned, the most important property of the word was its neutrality, its belonging to the tradition of chemistry itself; it did not prejudge the relationship between this science and Newtonian physics.

For what was indeed at issue was the importation of Newtonian chemistry into France: an importation problematic, at the very least, because intellectual France was officially Cartesian and had therefore rejected Newton's force as one of the occult powers – action at a distance – of which physical bodies had been divested by Cartesian physics. Geoffroy did not claim to be a Newtonian, but he had visited London in 1698 and had been made a member of the Royal Society. He later became the Academy of Science's official correspondent with the Royal Society, and between 1706 and 1707 he presented Newton's *Optics* to the Academy in a series of ten lectures. What is it he presents in the *Table of the Different Relationships Observed between Different Substances*? They are what are called 'displacement reactions', in which one substance substitutes for another in its relationship with a third. This 'third substance' is at the head of each column, followed by all the substances liable to combine with it, in the order determined by their mutual displacements; any substance displaces all those which come after it and is displaced by all those which come before it. In the *Optics* we can read: 'A solution of iron in aqua fortis dissolves the cadmium which is put into it, and abandons the iron; a solution of copper dissolves iron and abandons the copper; a solution of silver dissolves copper and abandons the silver; if one pours a solution of quicksilver in aqua fortis onto iron, copper, tin or lead, this metal is dissolved and the quicksilver is precipitated.'

If we look at the third column in Geoffroy's table, the one which deals with nitric acid (or aqua fortis), we see that Newton's observations are there in their essentials (iron displaces copper which displaces lead which displaces mercury which displaces silver), but not the conclusion Newton goes on to draw: 'Do these experiments not show that the acid particles of the aqua fortis are more strongly attracted by cadmium than by iron, more strongly by iron than by copper, by copper more than by silver; that they show a stronger attraction to iron, copper, tin and lead than to quicksilver?' Unlike the 'relationship' prudently put forward by Geoffroy, Newtonian attraction is an explanation: it explains both the chemical bond and the reaction in which bonds are altered.

Few were deceived by Geoffroy's agnostic presentation. In his *Éloge* of Geoffroy (1731), Fontenelle, a Cartesian, noted how these affinities 'distressed some, who feared that these were attractions disguised, all the more dangerous for clever people having already given them a seductive form'.

Here we will take the opportunity for a brief excursus. According to our current understanding of Newton's life, the association and dissociation of chemical substances was one of his main areas of research, and he spent more time in his laboratory, it has been said, than at his calculations. In this interpretation, the idea of a force acting between bodies was not the pure product of a hypothetico-deductive procedure intended to explain the movement of the planets. It was rather in order to mathematize the attraction between substances that Newton first turned his attention to astronomy, as a special case and, he hoped, a simpler one, of the general problem, and the anti-mechanical idea of attraction came to him from chemistry. But the mathematization of the heavens produced a great surprise: a single universal force was enough to account for all the movements. The attractions were not specific to each planet. One universal force, proportional only to the masses of the bodies which it joined and inversely proportional to the sphere of the distance that lay between them, was enough to account for the movements of the planets.

This interpretation of Newton's itinerary makes it easier to understand the abandonment of Kepler's problem presupposed by the idea of force. The classical Keplerian problem, which is also that of Huygens or Leibniz, implied that each of the planets turned all by itself around the Sun. It was then a question of explaining each orbit, and it is in this sense, first of all, that we must read Newton's idea that the Sun 'attracts' each planet. But for Newton, this is no more than an approximate description, for the forces are mutual: the planets attract each other and attract the Sun which attracts them. While the Keplerian problem was that of the movement of the different planets around the Sun, Newton poses the problem of the 'society' of heavenly bodies, the question of the system. Now if this idea of a system of masses mutually attracted to each other constitutes a radical innovation in astronomy, it is less surprising coming from a chemist, who knew that in a chemical reaction all the substances present play a role in relation to each other. So for Newton, the solvent, which enables the reaction between two substances, is an intermediary, a 'middle nature': the 'unsociable' particles are 'made sociable by the mediation of a third'.

If however, the 'chemical' hypothesis contributed to Newton's innovation in celestial mechanics, the latter also brought about a profound transformation in the traditional conception of the chemical substance. For the idea that chemical reactions could be understood on the basis of a 'Newtonian force' implied that the substances themselves were inert and without specific properties. In 1758 the natural philosopher Roger Boscovitch even proposed the reduction of inert mass to a point without extension. All chemical properties were relational. None was attributable to a substance in itself, all arose from relationships.

The idea that chemical properties are 'relational' runs counter to the whole of the chemical tradition since Aristotle, against what one might call 'the chemistry of substances'. There the chemical substance was the subject, whose reaction expressed a quality. The Newtonian chemical substance was no longer a substance except as a linguistic approximation. The only true subject was the ensemble of bodies present and in reciprocal interaction.

When a new word is created, its future often bears little relation to the intentions of those who created it. The word 'affinity', which at the beginning was officially only agnostic, was in fact adopted by all the protagonists, both by the Newtonians and by those who held that affinity had nothing to do with Newtonian attraction. None the less, if the term itself cannot serve as a distinguishing mark, its role in chemists' practice, on the other hand, expresses the profound ambiguity of the concept. We will see that it is as a supposedly 'purely empirical' and neutral notion organizing the practice of the 'table-makers' that affinity expresses its Newtonian side.

A Research Programme for Chemistry?

To all appearances, the tables of chemical affinity was indeed expressing a problematic proper to chemistry, the science of combination in both senses of the word: combination as reaction allows the comparison of combination as bond, a comparison of the strength of the bonds a substance can enter into with a series of others. Chemists should therefore have been satisfied with the tables, finding in them a tabulation of known reactions and a principle of organization for those yet to be discovered. The idea of affinity, however, also faced eighteenth-century chemists with the problem of its cause, and in this way it allowed the Newtonians to consider the progress of empirical chemistry from the point of view of its possible foundation as an explicitly Newtonian science. When the problem of the relationship between affinity and the Newtonian force of attraction was solved, the ensemble of chemical knowledge assembled according to the principle of the tables would be interpreted and understood.

The notion of affinity was therefore adopted by both Newtonians and anti-Newtonians, but only the Newtonian chemist would identify the very progress of chemistry with the construction of tables. Only he could make his goal the experimental investigation of every possible relationship between substances. For him, nothing at all could be said of substances taken independently of one another, no more than one could describe the behaviour of the Earth without reference to the Sun and the planets. The chemists who followed Georg Ernst Stahl, the author of the doctrine which rivalled Newtonian chemistry in the eighteenth century, believed on the other hand that chemical substances had an intrinsic power, and that the chemical reaction merely revealed this power, which was a property of the substance itself. There was therefore no need to investigate every possible relationship, because it was enough to look at those which demonstrated the power of a given substance in the most characteristic fashion. For those who made the elements responsible for chemical action, the tables were only a practical tool. It was therefore as an instrument of systematic empirical research that affinity expressed a Newtonian allegiance.

'Physics you understand, chemistry you learn': here then is one meaning of the distinction. The discovery of the universality of force, and of the uniformity of matter which it implied, brought Newton closer to mechanism, at least in means of expression; the different heavenly bodies could be considered entirely independently of their nature, in terms of the 'quantity of matter' measured by their mass. But Newtonian chemistry, when deprived of any possibility of referring to substances by their 'nature', had to learn. If they couldn't be deduced mathematically, in the same way as the movement of the Earth and the planets could be deduced from what one knew of the force of attraction, the possibilities of reaction between substances had to be subjected to systematic investigation.

For chemistry to escape empiricism, for it to become a deductive science, understood rather than learnt, the affinities would have to be interpreted, of course, and their specificity, so different from universal attraction, had to be explained. The Newtonian problem of affinity was this: what was the relationship between the force of attraction, which depended only on mass and distance, and affinity, which seemingly depended on the chemical specificity of the substances present?

There were two 'Newtonian' responses to this problem, one from Boscovitch in England and the other from Buffon in France.

In his *Theory of Natural Philosophy*, published in 1758, Boscovitch understands the diversity of affinities as a complication of the universal force. The formula $1/r^2$ is valid only for large distances,[1] while chemical phenomena occur at very short distances, where the force might be attractive or repulsive, depending on the distance. Here Boscovitch was taking up a position of Newton's: what we call a chemical substance is made up of a complicated cluster of particles, which may be decomposed and recomposed in reactions. The forces of attraction characterizing two substances are therefore specific because they are determined by the construction of the cluster, because they are the resultant of the forces exerted by each point of it. Part of a 'cluster-particle' can therefore exert an attraction on another particle, while another of its parts repels it. Within this framework, it is even possible to explain the role of substances which encourage a chemical reaction. With a certain particle, for example, an intermediary might cancel the repulsion which prevents this particle from attracting a third.

In a hypothesis published in 1765 (in volume XIII of his *Natural History*), Buffon, whom Voltaire called the 'leader of the Newtonian party in France', adopted the other logically possible solution. $1/r^2$ was indeed the only formula for Newtonian force, but chemical substances were much closer to each other than were the planets, and the approximation utilized in Newtonian astronomy, which reduced all masses to points, was no longer valid at this scale of magnitude. The specificity and diversity of chemical activity was therefore to be explained by the diversity in the shapes of the particles of chemical substances.

Both solutions were equally good from the logical point of view: one connected the diversity of affinities to a variation in interactive force dependent on the distance between the particles, the other related it to a straightforward force of interaction acting between bodies with different shapes. However, the two authors drew very different conclusions from their hypotheses. Boscovitch came to the conclusion that the theory of chemical operations would never allow the prediction of associations; the determination of the effects produced by different chemical constructions was far beyond the powers of the human mind. Buffon, on the other hand, declared that their 'grandchildren' would be able to calculate chemical reactions just like Newton calculated the course of the planets. Affinities could be deduced from the shapes of the constitutive particles and one would be able to predict the possibilities of reaction. Two logically equivalent propositions were therefore interpreted in opposite ways. While England saw one as a speculation with no direct relevance to the progress of chemistry, the French saw in the other the road chemistry would have to follow in order to become a real science.

Why was the 'Newtonian dream' of a quantified chemistry considered to be a mirage in England, but thought of in France as a programme for development? The historian Arnold Thackray has shown that in England the Industrial Revolution had produced a new kind of chemist, who wasn't worried by mathematical concerns or interested in speculative programmes. The new interest in chemistry also saw the emergence of teacher popularizers more concerned to spread the practical benefits of the science than to deal with its conceptual problems. According to Thackray, Dalton was one of this new class of chemist, and this explains the naïve character (from the Newtonian point of view) of his conception of chemical atoms. In England then, chemistry's specificity, autonomy and interest were based on its practical utility, and not, as in France, on the validity of its claims to the status of science. This reading is related to the more general theses of the historian Joseph Ben-David, according to which eighteenth-century England measured the value of science by its contribution to social, economic and technological development. As in Francis Bacon's model, this was a 'useful' science, a science whose prestige came from its service to society rather than to truth or pure knowledge's progress.

The model of Francis Bacon

Francis Bacon (1561–1626), was Chancellor to James I of England, and the author of several influential books (the best known being the *Novum Organum*, 1620) which put forward a new, anti-Aristotelian conception of knowledge and its role in human history. He has been valued (and particularly by his eighteenth-century readers) for his hostility to any kind of system, for his interest in efficient causes (as opposed to final and formal causes), the minute observation of individual phenomena and prudent generalization therefrom (Bacon has been called 'the father of the inductive method') and finally, for his view of the practical character of knowledge, which allows nature to be placed at the service of humanity.

In this sense, the fate of the notion of affinity reveals the difference in what was called 'science' in England and France (and those other countries where monarchs had created academies on the model of the French Academy of Science). In both cases, the 'scientists' certainly carried out a great deal of practical activity, but the continental academies made a distinction between these activities and the pursuit of a 'rational science', a science whose value lay in the promotion of reason as such. Members of the Royal Society, on the other hand, were satisfied with a science whose value was related to its usefulness, its role in the development of technology. The question whether chemistry could become a 'rational' science which escaped empiricism, which was understood rather than only being learnt, was in France an absolutely central issue, from the point of view of the status of the chemists and the prestige of their science; in England it was regarded as a speculation of little interest.

The first 'table of affinities' allowed Geoffroy to import Newtonian chemistry secretly into France. But the growing success of such 'tables of affinity' on the Continent can be taken as symptomatic of the specific form taken by Newtonianism there, a veritable pursuit of the 'Newtonian dream', quite different to what was happening in officially Newtonian England. Though there were only two new tables published between 1718 and 1750, there were three during the 1750s, four in the 1760s and five in the 1770s. The proliferation of tables from 1750 on may be linked to the publication of Pierre Joseph Macquer's *Éléments de Chimie* (1755), which contained the first systematic exposition of the theory of affinity. Macquer stressed the empirical character of the tables of affinity, a theory-free starting-point for interpretation. He alluded guardedly to the Newtonian theory, but also had recourse to Stahlian types of interpretation, referring to the notion of an element as the bearer of intrinsic qualitative properties. In 1766, doubtless because Macquer had adopted Buffon's position, his very influential *Dictionnaire de Chymie* suggested that the Newtonian theory of affinity was probably true, and closed with an appeal to those who had enough chemistry and mathematics to help resolve the problem, the key to the most hidden phenomena of chemistry.

However, the high point of research into affinity, as well as the most marked example of its Newtonian character, was the work of the Swedish chemist Torbern Bergman on what he unhesitatingly chose to call 'elective attraction'. The tables published by Bergman from 1775 to 1783 brought together the results of thousands of chemical reactions in twin tables (of reactions in solution, on one hand, and 'dry' reactions, 'brought on by heat', on the other) of 49 columns (showing 27 acids, 8 bases and 14 metals and other substances).

As the historian Maurice Daumas said, Bergman 'took very great pains with his task, a conscientious craftsman who nibbled away at the labour which lay before him. He seems to have intended to perform all the re-actions imaginable, so as to compare their results and thus classify sub-stances in relation to each other . . . He seems, however, not to have been very satisfied with his results: he thought another thirty thousand exact experiments would be necessary to bring his table to a certain degree of perfection.'

To Bergman's labour there corresponds the work of his contemporary, Guyton de Morveau. The first was developing 'learnt' chemistry to its fur-thest limits, while the other took the first steps towards 'understanding', particularly with his attempt to give a quantitative measurement of the force of affinity. Guyton de Morveau measured, for example, the forces necessary to separate different sheets of metal from the mercury on which they floated, and he discovered, much to his satisfaction, that these forces followed the same order as the chemical affinities.

The chemistry of the late eighteenth century therefore didn't bring to-gether chemical affinity and attraction 'by the stroke of a pen', as Bachelard had it. This relationship had already provoked long-term research, and it constituted a real programme of development. Chemistry was defined less as a territory (as it was defined, as we shall see, by the chemist Gabriel-François Venel) than as a perspective which required an enormous collective labour: to learn, so as eventually to understand: to accumulate 'data', which were indeed empirical, but which would, like astronomical observations before Newton, constitute the terrain which theory would then organize in deduc-tive form. This was, however, a very distant hope, and at the end of the century, the Newtonian model was not the only route to recognition as a science. It is well known that Lavoisier tried to find another foundation for chemistry. In his *Traité élémentaire de chimie* he wrote that he would leave the problem of affinity to his colleague Guyton de Morveau; he believed, however, that the science of affinity was to chemistry what 'transcendental' geometry was to elementary geometry. But it is clear enough that for Lavoisier chemistry didn't have to be able follow the model of geometry or of celestial mechanics in order to become a rational science.

However, the problem is far from being reduced to a choice between Guyton de Morveau's 'Newtonian' strategy and Lavoisier's, which was in-spired by the work of the French philosopher Étienne de Condillac. Apart from the ambiguity of the notion of affinity denounced by Buffon, when he wrote that many chemists used tables of affinity without understanding them, that is to say without understanding that affinity was only the effect of universal attraction, there is the question of whether the chemistry of affinity really was destined to be superseded by a rational science deducible from simple principles. Was empiricism, as both Guyton and Lavoisier would have agreed, the expression of a science still awaiting its formal basis?

Affinities and Circumstances

We can now turn to Goethe's *Elective Affinities* (1809). First the Captain describing chemical activity:

> You must see with your own eyes these apparently lifeless but actually very dynamic elements and observe with interest how they attract, seize, destroy, devour and absorb each other and then emerge out of that violent combination in renewed and unexpected form. Only then will you agree that they might be immortal or even capable of feeling and reasoning, because we feel our own senses to be almost insufficient for observing them properly, and our reason too limited to understand their nature.

How then should chemistry be understood as a science? Let us listen to what Mittler, the mediator, has to say about the subject. Mittler is active, he hurries about, intervening where people don't get on, and he helps the protagonists come together: Charlotte, her husband Edward, young Odile and the Captain. He refuses, however, to predict the outcomes of the meetings he brings about: each of them is an adventure. As soon as it is a question of contact between substances, one has to learn from what actually happens, one has to give up deduction and prediction. Charlotte seems to agree, when she says to Edward, at the moment they invite the Captain to share their life: 'That strange fellow, Mittler, may be right after all. All such undertakings are a gamble. No one can foresee how they will turn out. New combinations can have fortunate or unfortunate results; and we cannot even claim that the outcome is due to our own merit or our own guilt.' In another situation, however, Charlotte forgets the lesson: she and Edward believe they can conclude from a previous meeting, in different circumstances, that Edward can feel no attraction towards Odile. The adventure will prove them wrong, without its being the fault of anyone at all.

Was Goethe's a 'Newtonian' chemistry, then? Is the impossibility of knowing related to the relational character of affinity? What seems clear is that the realm of chemical activity is opposed to the realm of law, in which effects are regular and predictable. When, rather pompously, the Captain describes to her the chemist's art of separation and recombination, the way in which diluted sulphuric acid takes up the lime which is contained in a combined state in a chalky soil, while the second element of the combination, a weaker acid, is freed, Charlotte replies: 'But in this case I should never think of a choice but of a compelling force – and not even that. After all it may be merely a matter of opportunity. Opportunity makes connections as it makes thieves. As to your chemical substances, the choice seems to be exclusively in the hands of the chemist who brings the elements together. But once united and they *are* together, God have mercy on them!' Charlotte has understood that chemistry is not a science of laws but an art of circumstances. It remains for her to learn that the union between two creatures, sanctioned by the laws of marriage and the agreement of minds, of interest and of reason, can be swept away against all reason by the accidental intervention of a third. Charlotte, 'so wonderfully united with Edward', will feel herself ousted by the complicity which develops between Edward and the Captain, like the weak acid in the parable. And Odile, who is to come to console Charlotte for the distance which has grown between her and Edward, as a fourth substance unites itself with the third which has been left behind, will have an entirely different effect. No more than chemical combination and separation are human passions amenable to rational prediction.

Goethe's elective affinity is not so much the expression of a science waiting for its reason as it is a sign that another kind of science ought to be recognized. It will have to be accepted that knowledge through learning, on the basis of real experience, is not inferior to deductive knowledge based on laws, and corresponds to an inevitable necessity imposed by chemical activity, and by the passions and affinities of the subject-matter.

Should chemistry become a law-abiding science, following the model of celestial mechanics, or should it, 'the manipulative art of chemists', maintain the privileged connection with craft activity which had characterized it? Could 'rational progress' be the work of mathematicians contemptuous of the 'workers', as was the case in mechanics, or did it belong with a new kind of collaboration between 'systematizing intelligence' and 'craft knowledge'? These questions are implied in Goethe, but they had made an explicit appearance in Diderot's writings and in the *Encyclopédie*.

Learning Chemistry

Diderot's *Encyclopédie* or *Dictionnaire raisonné des science, des arts et des métiers* (1751–72) does not speak with a single authorial voice, and in it there are many articles devoted to chemistry or in which it is involved. In the article on 'Attraction', d'Alembert declares that Newtonian attraction should be able to give an account of chemical activity. But the article on 'Chemistry', on the other hand, expresses at every level the very different role which Diderot saw this science as playing. He had given this article to the chemist Gabriel-François Venel, an ex-pupil, as he was, of Hilaire Rouelle, who had introduced Stahl's chemistry (see pp. 462–3) into France. It is in this article that for the first time we see the identity of chemistry treated as a problem involving science, philosophy and politics all together.

The article on 'Chemistry' appeared in the third volume of the *Encyclopédie* in 1753, and in the same year Diderot published a very similar analysis in his

A chemistry laboratory, illustrating the article 'Chemistry' by Gabriel-François Venel in the Encyclopédie *of Diderot. The equipment is that of the former tradition of alchemy. There is no balance nor yet an enclosure to isolate gases. Most chemists during the eighteenth century worked at home, in a room of their houses. (From Diderot's* Encyclopédie, *facsimile of the first edition, 1751–80. Bibliothèque Nationale, Paris.)*

Pensées sur l'interprétation de la nature. He argued for an open scientific practice, in which those who 'think' would finally deign to associate themselves with those who 'act', in which 'those with lots of ideas and little apparatus' would learn to collaborate with those who have 'lots of apparatus and few ideas'. What he is denouncing is less the speculative hypotheses of the mathematicians than the contempt of those who think for those who learn on the basis of experience. In the same way he wrote in his *Principes philosophiques sur la matière et le mouvement* (1770):

> What does it matter to me what goes on inside your head! What does it matter to me whether you regard matter as homogeneous or heterogeneous! What does it matter to me that abstracting from its qualities, and considering it only in its existence, you see it as being at rest! What does it matter to me that you therefore look for the cause which moves it! You can do geometry or metaphysics as much as you like; but I am a physicist and a chemist, who takes bodies as they are in nature and not as they are in my head, and I see them existing, various, having properties and activities, and acting in the Universe as they do in the laboratory . . . If matter is to be moved, they say, there must be an action, a force: yes, either external to the molecule, or inward, essential, and internal to the

386

molecule, constituting its nature as an igneous, aqueous, nitrous, alkaline or sulphurous molecule. The force which acts on the molecule exhausts itself; the internal force is never exhausted. It is immutable and eternal.

While Goethe's Charlotte, who seems to defend a relational conception of affinity, none the less describes bodies internally ready for activity, Diderot, who adopts a thoroughly Stahlian approach to the question – with inherent and essential rather than relational properties – still uses the term 'force'. In the second half of the eighteenth century, relationships and principles were less opposed to each other than were the theses on the nature of chemistry as a science. This is the level at which Venel first of all discusses the specificity of chemistry and the difficulties it faces.

Like Diderot, Venel understands these difficulties politically, in the sense that the title of 'science' is first of all a title recognized or not by society. Certainly he looks forward to the 'revolution which will grant Chemistry the rank which it deserves, which will at least put it at the side of mathematical physics', but for him this revolution would not be the birth, at last, of scientific chemistry. The clever, enthusiastic and bold chemist who could bring about this revolution, will be the one who 'finding himself in a favourable position and taking advantage of fortunate circumstances will be able to attract the attention of scientists, first of all with noisy ostentation, a decisive and affirmative tone, and then with reasons, if his first weapons have broken through the prejudice'. Chemistry is suffering from the contempt of the 'scientists', and the 'new Paracelsus' must first of all be a propaganda agent.

Such a revolution would, however, transform the practice of chemistry not at all. The most learned, the most enlightened chemist will always have that 'genius' which among workers is called only 'common sense'. For only common sense will enable one to deal with the obstacles which halt the inexperienced chemist: the variety of chemical techniques and products, singularities, apparent oddities, isolated phenomena difficult to reproduce, etc. The power of deductive theory will never be able to take away from chemistry what it is that makes it a matter of long, arduous, laborious and patient practice. Abstract *a priori* truths will never reduce chemistry, an art of circumstances, to a science armed with laws which could dispense with 'the faculty of judging by feeling', the 'look' which understands the clues and is able to grasp the circumstances.

Chemistry demands an apprenticeship which is an apprenticeship of the senses, of the body (one must have the thermometer at the tips of one's fingers and the clock in one's head, says Venel) and of the mind. In this sense it is a passion. For Venel, this is what explains the fact that chemistry has been called a mad passion, that the chemist has been called eccentric or abnormal, devoured by a desire which loses him his fortune, his time and his life. (This is the story that would be told by Balzac in *La Recherche de l'absolu.*) But 'these difficulties and inconveniences should mean that the scientists who have the courage to brave them are regarded as citizens deserving of all our gratitude.'

Chemistry, the Science of the Heterogeneous

The specificity of chemistry as a passion, in its difference from the calm of deduction, expresses the fact that chemists are concerned with the intrinsic properties of substances, while the physicists stop at the superficial, at 'what can be seen by the donkeys and the oxen'. Venel does not attack the notion of affinity as such any more than Diderot, but he is concerned to demonstrate the difference in nature between the aggregate which is the object of physics, whether Newtonian or Cartesian, and the compounded unity which has to be thought in chemistry. The parts of the aggregate have with each other only relations of 'vicinity', and all change is a question of a change in spatial disposition, becoming nearer or further apart, without the parts undergoing internal change. The properties which may be understood through such changes are therefore 'exterior' or 'physical'. They can vary without any transformation of the corpuscles which make up the aggregate, and are therefore no help in answering the question, asked, for example, in investigating alchemical fraud, 'what is it that makes gold gold?' On the other hand, internal qualities 'properly specify the substance, constitute it as a particular substance, are what makes water, gold and nitre, etc. be water gold and nitre, etc.' These properties are intrinsic to the corpuscles, integral parts of the aggregate.

The proper object of chemistry is therefore the actions of corpuscles on each other. These actions 'depend on the interior qualities of corpuscles, among which homogeneity and heterogeneity are worth considering first, being essential conditions: for aggregation only takes place between homogeneous substances . . . heterogeneity of principles, on the contrary, is essential to compound (*mixtive*) union.' Unlike the aggregative union studied by the physicist, compound union, the knot which unites the principles of corpuscles, in fact produces the homogeneous, endowed with intrinsic properties, from the heterogeneous. 'Masses adhere to each other by reason of their vicinity, their size and their shape; corpuscles do not recognize this law: their unions are made by reason of their relationships or affinity; and in the same way, masses are not subject to the law of affinity . . . and a new homogeneous substance is never the result of the union of one mass with a mass of different nature.'

Venel thus defines affinity, and the compound union produced as a result of it, by means of a contrast with physics. He doesn't know what the 'knot' is in compound union, nor, in particular, does he claim that it is explained by Stahl's 'principles', but he knows that the union produced by affinity cannot be reduced to aggregation. And he refuses to believe that the absence of an intuitive explanation for this union, the absence of a mechanical agent which would be its cause, is a defect, for, as he says, those who use this ignorance to accuse the chemists of obscurity are simply reassuring themselves in the face of unintelligibility. The chemists, for their part, are brave enough to prefer obscurity to error. In this they are following the example of Newton and of all those before him who knew 'that nature brings about most of its effects by unknown means; that we cannot number its resources; and that the really ridiculous thing would be to try and limit it, by reducing it to a certain number of principles of action and means of operation: it is enough for them to have noticed a certain number of related effects of the same order to constitute a cause. Do the chemists do anything else?' Those who want to 'rationalize' chemistry are in fact demonstrating a most irrational horror of the unintelligible, a horror not shared by Newton, who had dared to introduce a force which acted at a distance as a 'cause' for the behaviour of the planets and of heavy bodies.

The 'Chemistry' article in the *Encyclopédie* is therefore a counter-attack against all those who see submission to the principles of physics as the only possible scientific destiny for chemistry. This counter-attack is remarkable for its being developed on several distinct levels articulated among themselves: there is the presentation of the social and practical specificity of the chemists' science, of their necessary, passionate apprenticeship, but also a demonstration of the specificity of chemical phenomena; and finally the announcement of a counter-theory of knowledge, in which the significance of the opposition between the rational intelligibility of mechanics and the obscurity of chemistry is reversed.

The distinction which Venel stresses, between what he calls 'aggregative' and compound ('*mixtive*') union was in fact generally accepted by all the eighteenth-century, chemists including those who, like Bergman and Guyton de Morveau, hoped to be able to explain affinity with the same kind of force as explained the aggregate. No one challenged the idea that a chemical reaction always had a clearly determined direction, which was determined by the stronger affinity, although certain physical conditions, such as the state of aggregation, might prevent it while certain others, such as heat, disaggregation, or putting into solution might remove the obstacle. This was the common ground of the very procedures employed in the trade chemistry of the period. Venel points out the double language referring to shared procedures.

> The worker says: overly concentrated nitric acid will not attack silver, but diluted with a certain quantity of water and excited by a certain degree of heat, it will dissolve it. Science says: the aggregative union of the

concentrated acid is greater than its relationship (its affinity) with silver, and the water added to the menstruum [the acid] loosens this aggregation, which the heat loosens even more, etc. The worker never generalizes; but here science would say more generally: in every act of dissolution, the tendency toward compound union overcomes aggregative union.

As we shall see, this common property, the double language bringing together the chemistry of 'scientists' and of 'workers' would find itself challenged at the beginning of the nineteenth century. The chemical knowledge of the eighteenth century, based as much on craft procedures as properly scientific, gave a meaning, as we have seen, to a qualitative distinction between compound and aggregative union, the latter having if necessary to be 'overcome' so that the former could take place. This distinction would be challenged at the very beginning of the next century in the name of the Newtonian theory of affinity. From then on, the harmony which had existed in France between Newtonian affinity and the science of the chemists would come to an end, and with it the programme of double complementary progress: the expansion of tables of affinity which organized empirical knowledge, together with the quantification of affinities, which might allow the deduction of the knowledge represented in the tables.

Elective or Functional Affinity?

A relational concept of affinity entailed the exhaustive study of all possible reactions, not only the 'interesting' reactions, the reactions which had interested the craft practitioners. It was however the 'interesting' reactions which had given rise to the qualitative distinction between (chemical) compound union and (physical) aggregative union. In fact, the favoured reactions corresponded to a purpose, which was to obtain a product in its most homogeneous and purest form.

It was the choice of such procedures which underlay one distinctive aspect of the idea of compound union: if a substance had a stronger affinity towards another than the one this other had towards a third with which it was combined, the first completely displaced the third. If one abstracted from the obstacles which might be presented by aggregative union, the chemical reaction was meant to be complete, the choice that a substance made for a particular substance rather than another following the qualitative law of 'all or nothing'. Bergman called this elective attraction, explaining the qualitative difference between compound and aggregative union accepted at that time.

Bergman's tables, which brought together a great number of reactions foreign to the traditional corpus of chemistry, brought more and more 'anomalies'. In numerous cases, he had to explain that physical factors interfered with 'truly chemical' affinity, and prevented the reaction from taking place completely, or even made it happen in the 'wrong' direction. He often had to note that he had to use many times the quantity which ought to have been necessary to bring about a complete reaction.

Contrary to what was supposed by Bachelard, who was apparently convinced, as we have seen, that any 'fine experimentalist' would have seen that affinity and attraction had nothing to do with each other, the increase in experimental knowledge did not lead to the abandonment of the identification of affinity and attraction, but at least first of all, to its reinforcement. Until then, critics of Newtonian chemistry like Venel had been alone in their stress on the fact that attraction could not explain the formation of a third homogeneous substance from two heterogeneous substances, nor account for the qualitative difference between aggregative and compound unions. As we shall see this incapacity became a virtue when Newtonian chemistry used it to deny the elective nature of affinity and the qualitative difference 'election' was intended to explain. The experimental questioning of the idea of the complete reaction, corresponding to a law of all or nothing, would confirm the implications, hitherto neglected by its defenders, of the Newtonian interpretation of affinity.

Here we come across, at the same time, the work of the chemist Louis Berthollet and the French Revolution, the latter being important here not for its ideas or for the death of Lavoisier (see pp. 422–54 and p. 478), but for the new problems presented to chemistry, and for the institutional transformations which it brought about with the creation of professorial chairs consecrated to the systematic teaching of chemistry.

During the Revolution, Berthollet devoted himself to a problem entirely foreign to the tradition of craft chemistry: he worked on rationalizing the production of the saltpetre necessary for the manufacture of gunpowder. Instead of gathering saltpetre where it lay, as had been done before, it was now a question of producing it in a controlled industrial environment. Berthollet realized that when nitrous rocks were washed, the more saltpetre there was already dissolved in the water, the less efficient was the washing; it was better to wash in several steps, using fresh water each time. At the same time, each wash dissolved a smaller quantity of saltpetre.

In a course which he gave at the École Normale in year III of the Revolutionary Calendar, Berthollet concluded that a substance's tendency to combine with another decreases in proportion to the degree of combination which has already occurred. This meant that affinity, rather than being a characteristic of one substance in relation to another, was a function of the physico-chemical state of the environment and, in particular, of the concentration of reactants present.

In 1800 Berthollet restated the same idea in more forceful terms. In the meantime he had taken part in Napoleon's expedition to Egypt, and the story has it that his beliefs about affinity derive from something he had seen on his travels: a 'sodium lake'. In contact with the calcium carbonate of the lake bottom the salt contained in the water produced sodium carbonate which was deposited on the shores of the lake. In the laboratory the reaction didn't happen in this direction. Berthollet explained this difference by two factors: the quantity of salt and of calcium carbonate, and the fact that the two reaction products are continuously eliminated from the reaction medium, calcium chloride being drained through the ground and the sodium carbonate being precipitated on the bank.[2] As soon as he got back to Paris Berthollet undertook a systematic reversal of the chemistry's categories of experimental judgement. The conditions of reaction (temperature, reactant concentrations) had been thought of as a source of interference which explained anomalies; now it was the complete reaction in which one substance was entirely displaced by another that would be anomalous, needing to be explained by particular factors (such as the elimination of products through volatilization or precipitation).

In this way Berthollet highlighted the particularity of the traditional procedures in craft chemistry, the fact that they had been chosen for the completeness of the reactions which they brought about. He showed that the irreversible order of displacements in Bergman's tables was nothing other than the order of solubility of the products. 'Complete reactions' were to be explained by the volatility or by the low solubility of one or other of the products, which therefore escaped from the reaction medium.

When he returned to France, Berthollet was well placed to ensure that his ideas had maximum impact. He was, together with Laplace, one of the founder members of the Société d'Arcueil, a private association which brought together the leading scientists of the age; it published a journal under the title *Mémoires de Physique et de Chimie de la Société d'Arcueil*. He was also a Senator in the Napoleonic legislature. Laplace and Berthollet exemplify the privileged position enjoyed by 'great scientists' under the Empire: they were involved in politics, they were responsible for the creation of new educational institutions, and they obtained the financial rewards which allowed them, in particular, to set up private laboratories at Arcueil. Newtonian physics provided the official doctrine of the Société d'Arcueil, and Berthollet, who was certainly an experienced experimentalist, demonstrated in his *Statique chimique* (1803) that the generally incomplete character of chemical reactions and their intrinsic dependence on 'circumstances' were the normal consequences of chemical affinity considered as a Newtonian force of attraction.

Venel had been right. Newtonian attraction did not allow a distinction between 'physical' and 'chemical' forces, nor between 'aggregative' and 'compound' union. Berthollet understood that it was necessary to choose between the chemical tradition and the hitherto unexamined implications of a Newtonian conception of affinity. For him, affinity was one factor among others, so that one could, by manipulating the other factors, produce a reaction in one direction or in the other. More precisely put, Berthollet sought to abolish any distinction between the 'natural' direction of a reaction and what interfered with this direction. For Berthollet, chemical reactions no longer had a 'natural direction'. The chemist no longer had to 'manipulate circumstances' in order to allow chemical affinities to be expressed. The chemist manipulated a function: the direction of a chemical reaction is a function of a purely chemical affinity, it depends on concentrations, but also on the temperature, on the cohesion of the reactants, etc.

In this function, 'chemical' affinity resulting from attractive forces no longer plays a privileged role to be opposed to the effect of 'circumstances', it is itself part of the ensemble of circumstances which the chemist must take into account. In the *Statique chimique* Berthollet writes:

> The chemical activity of a substance does not depend solely on quantity and on the affinity proper to the parts of which it is composed; it also depends on the state of these parts, resulting either from their current combination, which causes them to lose their affinity to a greater or lesser degree, or by their dilatation or condensation, which alter their distance from each other: these are conditions which modify the properties of the elementary parts of a substance and form what I call its constitution: to come to a correct analysis of chemical activity, one has not only to know all these conditions, but also all the circumstances to which they are in some way related.

But Berthollet goes further, confirming Venel's analysis: the force of attraction can only explain the mixture and its proportions, not the chemical combination which produces new homogeneous substances from heterogeneous substances. Berthollet therefore denies the specificity of compound union. The product of a chemical reaction is nothing but a state of equilibrium in which the different kinds of compounds coexist all mixed together. Nor do these compounds themselves have a well-defined identity. A substance does not completely expel from the neighbourhood of another the third substance which attracts this other less than it does itself; it simply exists in the neighbourhood of the other in greater quantity than its weaker rival. Every substance is therefore a mixture, and as a general rule, a chemical reaction cannot result in a pure product. And worse, the products of a reaction are not characterized by clearly defined proportions of the substances from which they were formed: the composition of a substance depends on its history, on the different factors characterizing the reactive milieu in which it was formed.

The Controversy

When Bachelard expressed astonishment that Berthollet, an experienced experimentalist, could identify affinity and attraction, he was adopting the commonly accepted point of view of the controversy provoked by Berthollet's theories. It is usually said that the chemist Joseph Louis Proust proved that Berthollet was wrong, with an experiment which showed that chemical substances do exhibit definite proportions. The experiment is treated as a triumph of experimental chemistry, the victory of 'fact' over eighteenth-century speculation. This view is doubly false.

First of all, the idea of definite proportions was well rooted in tradition. In his first article in 1799, before the beginning of the controversy, Proust drew general conclusions from a few experiments: he wanted to demonstrate that composition was independent of the origin of the product. He therefore worked on mercury oxide produced in the laboratory, and also on a mineral form from the mines of Peru. Just when Berthollet discovered, during his voyage, proof of the exceptional nature of what traditional chemistry considered as normal, Proust was sending his substances on voyages to prove the correctness of the traditional ideas. In the 1799 article, Proust also quotes Stahl, who spoke of the *pondus naturae*: definite proportions were signs of the invisible hand of nature, which worked in the same way, whether in the laboratory or in the far-flung reaches of the globe, fixing the invariable proportions of a substance's constituents. The chemist had no more power over the *pondus naturae* than he had over the law of election which governed all combinations. We tend to think of the law laid down by Proust as new because we think it true, but this law has its roots in the chemistry of the eighteenth century, and in the years 1792–1802 it had already been used by the German chemist Benjamin Richter in an attempt to produce numerical values for affinity.

What was novel in the controversy is above all the fact that finally Newtonian relational affinity and the Stahlian view of affinity as what revealed chemical principles had found their locus of disagreement, the point where, finally, a choice had to be made between them.

Between 1799 and 1807 experiments and counter-experiments followed one after another, but without either side recognizing that the other had 'established a fact' which entitled them to claim victory. Proust never ceased to claim that the facts proved him right: it was his only weapon against the authority of Newtonian theory and against Berthollet's prestige. Looking back, those who 'know' that he was right are impressed by his 'modern' tone; scientists, he never tired of repeating, cannot construct theories which go against the facts. But the facts themselves, as they could be discovered by the methods of the time, remained ambiguous. The chemistry of Proust and Berthollet was still the chemistry of the eighteenth century and not the analytical chemistry which was yet to be born. Ten years later, where Proust had used hundreds of kilos of oxide to perform an analysis, only a few grams would be needed. In retrospect, we know that experimental precision was in this case essential: most of the metallic compounds analysed by Proust and Berthollet were, as we know today, mixtures of several types of oxide, with the results that each of the protagonists could interpret the 'facts' to his own satisfaction, accusing the other of fudging and of introducing *ad hoc* hypotheses.

The controversy ended in 1807 without having reached a definite conclusion, each one considering that he had beaten the other. In 1832 Louis Jacques Thénard and Joseph Louis Gay-Lussac put forward an interpretation of Berthollet's chemical equilibrium which respected the principle of definite proportions, separating what the chemistry of the eighteenth century, Berthollet included, had confused in the problem of the 'cause' of reactions. Gay-Lussac and Thénard distinguished the question of the bond, that is to say the fixed proportions of pure substances making up a compound product, from the question of the reaction. Every reaction produces products exhibiting definite proportions, but reaction is generally incomplete and results in a mixture of reactants and reaction products, a mixture whose composition is, as Berthollet had shown, a function of the conditions of reaction.

This is the 'rational' conclusion of the controversy, and if we follow it we may say that the eighteenth-century notion of affinity proved to be a fruitful concept, which led to an unsuspected distinction, foreign to Newtonian physics, between the 'bond' which the chemist cannot control, and the process reaction in which these bonds are transformed, which he can manipulate through the conditions of reaction.

Nineteenth-century chemistry was to confirm definitely the specificity of the chemical bond. Those like the Englishmen, Humphry Davy and Michael Faraday, or the Swede Jöns Jacob Berzelius who would have liked to substitute a hypothetical 'electrical' bond for the Newtonian force of attraction were defeated by the development of the new organic chemistry. At the end of the 1830s Jean-Baptiste Dumas showed that chloracetic acid was formed from acetic acid by the substitution of chlorine, electro-negative, for hydrogen, supposedly positive. The theory of substitution, born of the need to understand the jungle of products introduced by organic chemistry, was the end of the last non-specific interpretation of chemical processes (despite Berzelius, who continued to insist that it was valid only for organic chemistry).

But what is surely even more interesting than the rational conclusion of the controversy is the fact that, at the time, very few people were interested in it; notably neither Gay-Lussac nor Thénard paid it much attention. This indifference would be maintained until the second half of the century, until a chemistry which now endeavoured to perform synthesis on an industrial scale came up against the problem of yield. Now chemists would again become interested in the possibility of adjusting concentrations and physical conditions of reaction so as to modify the proportions of reaction products. This is when the term 'affinity' reappeared as the name of the (thermodynamic) function on which the direction of the reaction depends, defining as a function of concentration, pressure, temperature, etc. what the old notion of qualitative affinity designated as the 'choice' that one substance made for another rather than for a third.

The controversy between Berthollet and Proust poses the question less of knowing who was right than of understanding why it ended amid general indifference. Berthollet lost not so much to Proust as to a change in the interests of chemists.

The Outdated Eighteenth Century

In their defence of chemistry Diderot and Venel brought together arguments on several levels: analysis of what was known of chemical combination, analysis of the relationship between theory and practice, issues in the philosophy of scientific knowledge. Their strategy might be considered as scientifically illegitimate: the value and significance of an idea like 'affinity' as promising a 'rationalization' of chemistry ought to have been decided entirely on the basis of considerations internal to the science. But it is in taking all these levels together that we can understand the fate of affinity, its becoming outmoded. Affinity didn't become a lapsed concept because it turned out to be absurd; in other possible histories it might have survived by transforming itself in such a way that the identification of affinity and attraction would seem as historically respectable as Dalton's atomic hypothesis, for example. It became a lapsed concept because chemistry changed on all levels at once, because the techniques of chemistry, the questions asked by chemists after the 1810s, and the chemists themselves all changed.

French historians of chemistry have seen in the development of analytic chemistry the normal consequences of the work done by Lavoisier. Others have preferred to stress the importance of the double discovery, by Dalton and by Gay-Lussac, of the 'laws' obeyed by all chemical combinations. After the confirmation of Dalton's law, the balance, the instrument to which Lavoisier had subjected chemical practice, became an effective instrument for the investigation of chemical combinations. It not only allowed a rigorous experimental control, ensuring that nothing had entered or left the site of the reaction – that is to say, making a clean sweep of the old chemistry.

It also enabled combinations to be described in terms of the weight of reactants combined, and the description of reactants themselves in terms of the ratios of the weights in which they entered into the various possible combinations.

Simple proportions

Around 1803, probably, John Dalton showed that chemical combinations were formed in accordance with definite ratios of the weight of reactants. In cases where different combinations may be produced from the same reactants, the proportions of one reactant to another in each combination bear a simple relation to each other such as 1:2:3 . . .

In 1805 Louis Joseph Gay-Lussac and Alexander von Humboldt showed that, when measured at the same temperature and the same pressure, two volumes of hydrogen and one volume of oxygen were needed to produce one volume of water. In 1809 Gay-Lussac generalized the observations; there is always a simple ratio between the volumes of gas entering into a chemical combination, as there is between the sum of these volumes and the volume of the gas resulting from their combination.

Dalton published his results in 1808, in the *System of Chemical Philosophy*, in which he presented his 'law of weights' as a proof of the atomist hypothesis. Dalton's atoms, which were intended as 'Newtonian', had nothing in common with the complex constructions Newton thought constituted reactive substances (and which had allowed him to understand the creation of gold from other metals by the composition and recomposition of the cluster). Dalton didn't use Gay-Lussac's law; Dalton's and Gay-Lussac's laws were unified by Amadeo Avogadro, whose hypothesis was not widely accepted until the International Chemical Congress at Karlsruhe in 1860.

The idea of indefinite proportions implied by Berthollet's notion of affinity therefore found itself contradicted by 'laws' which, for analytic chemistry, were not simple experimental laws among others, but laws which entail new experimental arrangements and the interpretation of these new results. If one followed Berthollet, the very conditions of analytic practice would have crumbled away; one could never have been sure, when analysing a product by making it react with another, that the experiment would be reproducible. No standard product could be reliable; each one, having been formed in different conditions, might have had a different composition.

Definite proportions, now discovered to be in simple ratios, took on a significance quite foreign to the Proust and eighteenth-century chemistry: not rules relating to the nature of the chemical reaction, but principles for analysing the constituent parts of a substance. Analytical chemistry wasn't a science of reactions but the analysis of combinations, of products; the reaction itself was of only instrumental interest, an analytical tool which gave access to the chemical composition of a product. It was a 'modern' chemistry, but in one sense it marked a return to the situation that prevailed before the development of tables of affinity, before what I have called the 'development programme of Newtonian chemistry'. A limited number of 'good' reactions were enough, using standardized and powerful reactants to separate compound substances. It was only during the second half of the nineteenth century that chemists returned to Berthollet's concern with 'incomplete' reactions. Indeed, as the chemist Wilhelm Ostwald remarked, synthetic chemistry, which lay behind this new interest, necessitated the use of the whole range of possible reactions. Newtonian physics, however, had nothing to do with the return of the incomplete reaction, which was now studied as a particular case in thermodynamics, the general physics of physico-chemical transformation.

This brief look at the transformation of chemistry at the beginning of the nineteenth century will have been enough to show that the relationships between a science and what one might describe as 'industrial development' need to be analysed with care. Analytical chemistry was, of course, of interest to industry, and the relationship is reciprocal. However, when he took the consequences of the Newtonian interpretation of affinity seriously, Berthollet was posing a problem crucial to the 'rationalized' production of saltpetre, and which was to become essential to the synthetic chemistry of the second half of the nineteenth century: what were the reaction conditions which would ensure the best yield? The analytical chemistry of the first part of the century, on the other hand, rediscovered the 'privileged reactions' characteristic of the craft tradition whose particularity had been brought out by Berthollet's work. We can see that 'industrial development', once having opened up Bethollet's perspective, was able to go in several directions: it could concentrate on 'good methods', which allowed the extraction of the desired product from the mixture resulting from the reaction, or, as in Ostwald's time, it could revive the problem of yield and the means of influencing the composition of the mixture.

These also, however, are *a posteriori* generalizations. Other factors and other circumstances have to be integrated into the account. One can't deduce the transformation of chemistry from the social and industrial context, nor from an *a priori* distinction between the 'speculative chemistry' of the eighteenth century and the 'positive chemistry' of the nineteenth, but one can follow it, a bit like Goethe's chemist, who can only learn from what happens when reactants are actually brought together. Then many different 'circumstances' become significant. Could analytical chemistry have played the role that it did, if it had not been for the opening, at the same time, of a new field of exploration, the whole 'new continent' of reactions and compounds represented by organic chemistry? But in the end, wasn't the real break that made eighteenth-century chemistry an outmoded science, the change in the practical identity of the chemist and of chemistry?

Venel, defending the chemist's' 'glance', the ability to 'read the clues' which made for an able chemist, had written that a chemist with a thermometer would be as ridiculous as a doctor with a thermometer. The generation of Liebig and Gay-Lussac would see a radical change in chemical practice – with the standardization of instruments and products allowing the reproduction of experiments and thus the establishment of experimental protocols – which in a few years would consign Venel's 'passionate' chemist to prehistory. Chemistry became more and more a matter of precise measurement, and the experienced chemist, educated by years of unremitting labour, was replaced by chemists trained in four years at Liebig's laboratory at Giessen, or in laboratories modelled on it, chemists who essentially had learnt to follow the protocols and to use more and more sophisticated instruments.

Nineteenth-century chemistry wasn't a science of experience but of experimentation. It was 'learnt' still, but it wasn't learnt in Bergman's way, a learning that waited for the deduction that would make it unnecessary, nor in Venel's, the passionate learning of body and mind. It was learnt in the way invented by Liebig, a systematic training in the use of instruments and of experimental protocols, the efficient education of chemists who shared the same 'facts', the same approach, the same methods and the same lectures. Chemists came from all over the world to 'learn' in Germany. Liebig's way was to become a model: chemistry was the first science to coordinate the production of research and the production of researchers, to train its students in a way which 'imitated' research, and it was the first effectively international science. Venel and Diderot had argued for an open science which respected the complications of the phenomena, which would prefer obscurity, or 'conjecture', as Diderot put it, to a false and reductive intelligibility, and bring together practitioners and theorists in a new way. Chemistry was no longer the science with the double language of scientists and workers. 'Chemical truth' was defined within the closed vessel of the academic and later the industrial laboratories. The chemists who worked there learnt neither the lapsed history of their science nor the practices of the craft chemistry of their time. They no longer learnt to play with many specific circumstances, for they had the means to overcome them, and as Marcellin Berthellot said, to 'create' new objects, related to the new instruments and protocols they were developing.

The triumphant chemistry of the nineteenth century gloried in its identity as an active science, which no longer submitted to a manifold and circumstantial nature, but mastered its workings. It also prided itself on being autonomous and disinterested, in short, academic. Liebig was the first to argue against the 'Baconian' conception still prevalent in England: chemistry shouldn't be at the service of industry, but should pursue its own questions,

399

and it would be the consequences of this 'pure' science which would in turn drive forward industrial development. Nineteenth-century chemistry therefore won its title of 'science' while following a course which was neither Venel's or Buffon's. It became the model of positive science, articulating pure and applied science, the supreme expression of Man's rational and creative mastery over Nature. Chemistry was neither deduction nor passion. It was action, practical and therefore rational, impassioned in its production of new products which transformed society and the life of the individual.

We started off with the question of the strange resonance between Fontenelle's scorn for chemistry and the attitude of contemporary physicists. Affinity didn't allow us to solve the problem, but only to clarify it. In the course of the nineteenth century, chemistry not only won the status of an autonomous science, but also became a science at the cutting edge, a queen of sciences and a model for positive science, exemplifying an effective conception and practice of science as experimental and pragmatic. It is for the history of the twentieth century to understand how this successful strategy eventually turned against itself, and how in the eyes of the public, and also in those of certain scientists, chemistry became 'merely practical' and thus beholden to social and economic interests.

13

From Linnaeus to Darwin: Naturalists and Travellers

Jean-Marc Drouin

In which we see how the naturalist travellers, who crossed the world to catalogue the variety of living species, brought back from their travels the materials for a geography of plants and animals and thus the foundations of a theory of evolution.

The coincidence of dates has often been noticed: there is exactly a century between the establishment of Linnaean nomenclature in the tenth edition of the *Systema Naturae* in 1758 and Darwin's first public presentation of his theory of evolution at a meeting of the Linnean Society of London. To some authors a century seems little enough, given all that separates these two highlights in the history of biology. For Linnaeus did not only put forward a code for the designation of plant and animal species and a system in which they might be classified; he was also inclined to make species the fundamental and invariant givens of creation. In reconstructing 'the origin of species' Darwin, however, put their fixity into question, so that it is tempting to see the passage from one to the other as a radical mutation.

One talks of the Darwinian revolution: this avoids losing sight of the novelty amid a welter of precursors, and at the same time suggests an analogy with the revolution associated with the names of Galileo and Copernicus, which had come about in cosmology 200 years before. Should pre-Darwinian natural history really be considered, though, as a sort of prehistory suddenly brought to an end by the emergence of the theory of evolution?

For quite a long time now, historians have stressed the role of the early nineteenth-century naturalists. The controversy between Lamarck and Cuvier, in particular, has generated an abundant literature. Certain authors have depicted Lamarck, the philosopher-naturalist, as 'a French precursor to Darwin', subjected to the sarcastic attacks of the highly conservative Baron Cuvier. Other authors, however, have insisted on the modernity of Cuvier, 'the founder of palaeontology', and recalled how, thanks to his knowledge of the inter-relationship of organs, he was able to reconstruct an entire skeleton on the basis of a few bones. Apart from their confrontation on the question of the transformation or fixity of species, Lamarck's most durable legacy to his successors was perhaps a new classification of the invertebrates, while Cuvier's essential contribution was surely his work on the comparative anatomy of vertebrates.

It is accepted, then, that if Darwin wrote the history of living beings, others had already begun to decipher the archives. However important they were, though, studies of fossils and comparative anatomy were not the only disciplines exploited by the theory of evolution. One cannot fail to be struck, on reading *The Origin of Species*, by the frequency of arguments taken from biogeography. The role played by this discipline, which studies the distribution of plants and animals, and isolation, barriers and migration, has in recent decades been noted and analysed by many historians of science. If one turns to Darwin's biography, one sees how important for his development as a naturalist were the years spent voyaging on the *Beagle* (1831–6). Significantly, Alfred Russel Wallace, who arrived independently at very similar conclusions about the role of natural selection, had spent many years exploring Amazonia and the Malay Archipelago. Beginning in the seventeenth century, the adventure of the traveller-naturalists reached its apogee at the end of the eighteenth century and the beginning of the nineteenth. Logically and chronologically, is it not one of the threads which connect Linnaeus' establishment of the classification of species to its transformation into genealogy by Darwin?

Travels and Travellers

Not all travellers were naturalists and not all naturalists were travellers, even taking these words in their widest sense; in every period there have been travellers indifferent to flora and fauna, and naturalists of the study or garden who travelled only in their thoughts. To make a career in the natural sciences, it was sometimes even preferable not to travel too far from the capital! Having said this, there are many travellers known for their contribution to

→ voyage of the *Beagle*

Invited by Captain Fitzroy to accompany him on his voyage round the world on board the Beagle, *the young Charles Darwin left England in December 1831 and did not return until October 1836.*

natural history. There can be no question of dealing with all of them here, but mention may be made of some of their travels. One finds collective enterprises sponsored by governments, as well as individual adventures, which indeed might well attach themselves to the former.

One of the first, and perhaps the archetype of all the others, were the travels of Joseph Pitton de Tournefort (1656–1708). Journeying for two years through Anatolia and the Greek islands with his two companions, this French botanist reveals himself in his letters to be a writer of great humour. These letters, which make up a real diary when taken together, were republished in 1982 in an abridged paperback edition, under the title *Voyage d'un botaniste*. They describe landscape and vegetation, but also the inhabitants and the political and religious situations of the countries traversed. This journey wasn't just three men's adventure, but also an enterprise financed and protected by royal power, as is clearly shown by the note from Pontchartrain, the Contrôleur-général des Finances, to the Abbé Bignon, Secretary of the Académie des Sciences, and dated 16 January 1700.

In this letter one discovers, condensed, and seen from an administrative point of view, all the political factors behind the naturalists' voyage. An itinerary is laid down, in this case a tour of the Mediterranean basin (though it would in fact be only partially covered). The objectives of the mission are clear: to discover the natural resources available within the Ottoman Empire. A team is put together: it was reduced to only three, but we should note that one of them had to be a draughtsman, who played an essential role on every scientific expedition before the invention of photography. In this case it was the painter Claude Aubriet; the third member of the team was a German botanist, André de Gundelsheimer. Funds are provided, up to a certain limit, and on condition that they should be accounted for; in the meantime, an advance payment will be made. Finally, assurances are given to Tournefort: his career will not suffer, quite the opposite, as a result of this long absence. And natural history is only one aspect of the voyage: ancient ruins, ways of life, and political and religious organization will be of as much interest to our travellers as rocks and vegetation.

Tournefort's mission

A note from Monsieur Phélypeaux, comte de Pontchartrain, to the Abbé Bignon, Secretary to the Academy of Science, dated 16 January, 1700, cited by Stéphane Yérasimos in his introduction to *Voyage d'un botaniste*: 'I have informed the King, Sir, of the proposal that has been made to send M. de Tournefort, a botanist of the Academy of Science, to Greece, Constantinople, Arabia, Egypt and the Barbary Coast, there to investigate the plants and the metals and minerals, to learn about the illnesses of these countries and the remedies which are used there, and about every-thing regarding medicine and natural history; His Majesty has strongly approved this design, desires that it be carried out, and has no doubt that it will be of great usefulness in the improvement of medicine and the advancement of the sciences; His Majesty thus commands me to write to you to tell him to prepare to leave as soon as possible, with a capable man whom the Academy will choose to work with him, and a draughts-man: His Majesty will pay him on his return all the expenses he will have incurred, in accordance with the accounts he will provide, on condition that he will incur these expenses with great economy; meanwhile I will send him today an order for 3000 pounds, which will be paid before his departure; I do not think it necessary to tell you that his pensions from the Academy will be continued and paid regularly during his absence, and even being away, he will have an even greater right to claim those increases and favours which His Majesty may accord to Academicians; he must come here so that I may present him to the King; I will also have sent to him all the passports and letters of recommendation he may need, so that he may make the voyage with all the safety and comfort we here may be able to afford him.

Thirty years later, and leaving from Sweden, a country which was at that time much less rich than France, Linnaeus' journey in Lapland was a more modest affair than Tournefort's. It was a summer tour which one of his recent biographers, Wilfrid Blunt, felt able to compare to 'the expeditions undertaken today by enterprising students in order to escape the boredom of the long vacation'. This summer tour, for which he had obtained a modest grant from the Royal Society of Sciences, would however play a not negligible role in the Swedish botanist's career, by allowing him to make himself better known. His journal, *A Tour in Lapland*, enables us to follow him step by step, and to appreciate the number, variety and precision of his observations.

The second half of the eighteenth century saw large-scale scientific expeditions. The one sent by Catherine the Great to explore Siberia, under the direction of the German zoologist Peter Simon Pallas, is still famous for having discovered the remains of mammoths preserved in the ice. Historians' attention has been particularly drawn by the rivalry between the French and the English. The great circumnavigations by Bougainville, Cook and La Pérouse all resulted in the discovery of new species of animals and plants, drawings of which, together with descriptions and dried, stuffed or living specimens, arrived in London and Paris at the same time as the two powers increased their knowledge of the sea routes which were also the paths to commercial and military hegemony.

The closest connection between political demonstration and scientific expedition was surely reached with the French expedition to Egypt. The army was accompanied by a whole council of scientists in an adventure which ended rather badly, but which none the less profoundly marked the France of the time. The naturalist's task was sometimes facilitated by these links between geographical discovery and imperialist acquisition. Even an isolated botanist with few financial resources – like some of Linnaeus' disciples, or the Frenchman Michel Adanson – could take advantage of merchant ships for transport and of trading posts for their accommodation. However, rivalry between European countries, and sometimes the distrust or hostility of the natives, increased the difficulties of voyages already full of danger.

From this point of view, the large expeditions were not always the safest. In 1788 La Pérouse's expedition disappeared in the Pacific. In 1791 France sent another expedition to search for him, under the command of Antoine d'Entrecasteaux, in which the botanist La Billardière took part. The expedition ended at Java in political disagreement, without finding any trace of La Pérouse, but not without its crop of new species. Ten years later, the expedition to the Southern Seas under the command of Commandant Nicolas Baudin, which brought together many young scientists, was marked by a succession of illnesses, withdrawals and deaths, as well as by bitter conflict between scientists and naval men.

The main relevant dates and expeditions by great travellers of the eighteenth and nineteenth centuries:

1700–2:   journey to the Levant (Greece and Turkey) by Joseph Pitton de Tournefort.

1732:   Linnaeus' travels in Lapland.

1735–70:   Joseph de Jussieu's time in South America; leaving with the expedition under Charles Marie de La Condamine, he stayed there for thirty-five years.

1749–54:   Michel Adanson lives in Senegal as an employee of the Compagnie des Indes.

1763–75:   the voyages of Captain James Cook, which allowed Joseph Banks, and then the Germans Johann and Georg Forster, to study the southern flora and fauna.

1767–71:   Louis Antoine de Bougainville's voyage around the world, Philibert Commerson being the expedition botanist.

1768–74:   expedition to Siberia, led by the German zoologist Peter Simon Pallas.

1785–9:   expedition under the command of Jean-François de La Pérouse, ending in the disappearance of the two ships, the *Boussole* and the *Astrolabe*.

1791–4:   expedition under Antoine d'Entrecasteaux, sent to find La Pérouse's ships.

1799–1804: voyage to Latin America by Alexander von Humboldt and Aimé Bonpland.

1800–4:   Commandant Nicolas Baudin's expedition to the Southern Seas.

1831–6:   Charles Darwin's voyage on the *Beagle*.

1832:   death of Victor Jacquemont in Bombay.

1848–52:   voyage to Amazonia by the Englishmen Alfred Russel Wallace and H. W. Bates.

At the same time as these great collective enterprises were taking place, a number of naturalists set off more or less alone on profitable travels. Mention might be made of the three best-known: Alexander von Humboldt, the German geographer and physicist, and Aimé Bonpland, the French botanist, who left for South America in 1799 and returned in 1804 after a particularly fruitful voyage; in the early years of the century, Jean-Jacques Audubon, the American painter and ornithologist of French extraction, who travelled in the United States; and in 1832 the Frenchman Victor Jacquemont spent four years studying the flora of India, before dying in Bombay at the age of thirty-one.

If a tendency is observed towards better-organized expeditions with more and more resources at their disposal, it is still true that until the middle of the nineteenth century at least, a great number of the traveller-naturalists were isolated individuals, among whom were a good number of missionaries and religious.

Beyond the often tragic, sometimes idyllic and always highly coloured image of these epic journeys, it is important to come to some assessment of the work that was actually done. First of all, to stick very closely to the adventures themselves, there are the accounts of the travels, which may be considered an important contribution to eighteenth- and nineteenth-century European culture. Not all the travellers write as well as Tournefort, in whose hand the least anecdote takes on the allure of a tale by Voltaire; but all these accounts, with their descriptions of distant peoples and landscapes, profoundly affected the picture of the world one gets in the literature of the age of the *philosophes* and then in that of the Romantic era. From 1748 onward the Abbé Prévost composed his *Histoire générale des voyages* ('A General History of Voyages'), collecting and summarizing the accounts of many

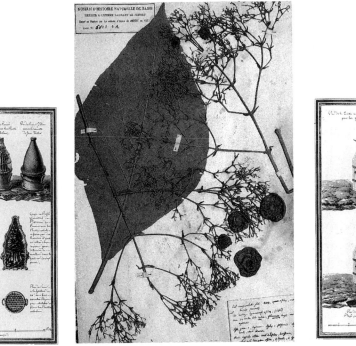
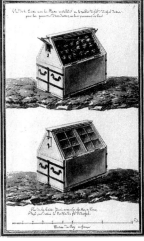

*This anonymous watercolour shows baskets for bringing back samples of living plants during the voyage of La Pérouse (1785–8). Another watercolour shows a case for living samples from the same voyage. (Archives Nationales, Paris.) A sheet (centre) from the herbarium of Philibert Commerson, the naturalist on Bougainville's voyage (1766–9). The plant is teak (*Tectona grandis*) from Mauritius, then called the Île de France. (Laboratoire de Phanérogamie, Muséum National d'Histoire Naturelle, Paris.)*

travellers. If he is today better known as the author of *Manon Lescaut*, his history, as demonstrated by Numa Broc, was none the less the source of an abundant literature whose influence is still to be found in the following century. At another level, Bougainville's voyage, and his stopover in Tahiti, the new Cythera, gave Diderot the idea and the occasion for writing the *Supplément*, in which ethnographic description is combined with amusing philosophical fiction.

The traveller-naturalists, like other 'educated' travellers, not only supplied literature with narrative elements, but also encouraged the success of philosophical ideas on the diversity and relativity of ways of thinking, and contributed to debates on the hypothetical state of nature, with arguments both favourable and unfavourable to 'the myth of the noble savage'. In a more direct way, through their descriptions of tropical vegetation, they enriched the European imagination with a theme whose success continues undiminished.

When they came back from their travels, however, the naturalists not only brought back their stories; they were also loaded with hitherto unknown species; hides of exotic animals, and especially the sheets of herbaria (collections of dried plants), together with seeds and sometimes plants in pots. One knows that a number of plants cultivated for food or ornament today were introduced to Europe from Asia or America. In three centuries, the scale of the living world, the idea one has of its diversity, has changed in order of magnitude. In the vegetable kingdom alone, the number of species known in the sixteenth century and described by the botanists was no more than a few thousand. At the end of the seventeenth century Tournefort was able to describe more than 10,000 plants. In 1833, in his inaugural botany lecture at Montpellier, Alire Raffenau-Delile, who had taken part in the expedition to Egypt, spoke 'of the discoveries of indefatigable observers, of travellers to new lands explored every day', and he added 'We owe to their zeal our knowledge of more than 50,000 species of plants, in the current state of science.' Today, it would probably be necessary to multiply this number by five at least. Let us note, for comparison, that the animal kingdom has considerably more than a million known species, most of them insects.

Naming and Classifying

Such figures didn't come from nowhere. They presuppose specimens collected, prepared, drawn and described, gathered together in museums, gardens, herbaria and natural-history laboratories, where everyone could come and see them, observe them and compare them . . . Voyages and collections

---

The Equipment

René Lesson (1794–1849), the French naturalist and Navy pharmacist who had taken part in the voyage of the *Coquille* (1822–5) under the leadership of Louis-Isidore Duperrey, was the author of the article on 'Taxidermy' in the *Dictionnaire des sciences naturelles* (Levrault, 1828). He begins with this definition: 'Taxidermy is the art of preparing and conserving for collections the objects of natural history', and he ends with a list of 'Objects necessary for the conservation of natural historical collections during voyages of discovery':

Before embarking on a campaign of discovery, whose presumed length will be three years, one will provide oneself with all the objects indispensable for ensuring the success of the enterprise:
Colourless spirits of wine, three hundred litres . . .
Jars of strong white glass, three hundred . . .
[The jars and the alcohol allowed the transport of small-sized animals.]
Mastic . . . twenty-five kilograms . . .
Corrosive sublimate, contained in a glass flask with a ground-glass stopper and always stowed in a medicine chest, five hundred grams.
['Corrosive sublimate', like the 'arsenical soap', was used to treat skins to avoid putrefaction.]
The other indispensable objects are:
1. Sheet lead the thickness of thin card, to make labels, three hundred square feet;
2. A punch, the size of a *sou*, with a series of stamps bearing little numbers. The numbers thus stamped into the lead will identify each jar, and each number will be repeated on a list on which are written the notes relating to object enclosed.
3. Three hunting guns, with their powder flasks . . .
4. Two slightly flattened tin boxes, for hunting and for botany;
5. Arsenical soap, twenty-five kilograms, in a little barrel;
6. Twelve boxes lined with cork and fitting each inside the other, for insects;
7. Fifteen reams of paper for plants, and fifty kilograms of waste paper for wrapping minerals.

---

thus appear as the two poles of natural history. But nothing could happen between these two poles if an effort was not made to name and classify all the specimens brought in. Between the adventure of the voyages and the poetry of the gardens, nomenclature and classification were not an obstruction or diversion, but an exchange connecting one to the other, determining the production of knowledge about living things.

Some people grasped this at the time, Jean-Jacques Rousseau in particular, whom we know devoted his leisure time to botany, and who in 1774 started to write a *Dictionnaire des termes d'usage en botanique* ('A Dictionary of Terms Used in Botany').

> I ask every intelligent reader how it is possible to pursue the study of plants, but reject the study of nomenclature. It is as if one wished to become learned in a language without wanting to learn the words ... It is a question of knowing whether three hundred years of study and observation should be lost to botany, if three hundred volumes of figures and descriptions should be thrown to the fire, if the knowledge gained by all the scientists who have devoted their money, their lives and their attention to immense, costly, difficult and dangerous voyages, are useless to their successors, and if each starting off again from zero can by himself attain to the same knowledge that a long series of investigations and studies have distributed through the mass of the human race ... To accept the study of botany but to reject that of nomenclature is therefore to fall into the most absurd contradiction.

In other words, botany, and this goes for zoology as well, although Rousseau doesn't talk about it, can only become a cumulative science if all of those concerned with it, travellers, amateurs, gardeners and collectors, adopt a common nomenclature. Well, says Rousseau, this already exists, proposed by Linnaeus; it has met with resistance deriving from 'national jealousies', but it will eventually win through everywhere, 'even in Paris, where M. de Jussieu has just adopted it in the Royal Garden, thus preferring public utility to the glory of a new production.'

Linnaeus' contribution to nomenclature was in two parts. First of all, he continued the work of his predecessors, of Tournefort in particular; he established rules of determination for genera and then for species, and he applied them first of all in drawing up the catalogue of the garden of George Clifford, a rich Anglo-Dutch amateur, and then in drawing up an inventory of all the animal and plant species he could discover. To each species he gave a generic name which it shared with related species, and a phrase which 'specified' it, allowing it to be distinguished from the others. Up till then, as Rousseau says, 'he had determined the greatest number of known plants, but he had not named them; for defining a thing is not to name it.' The second stage, the creation of names which were labels and not phrases, was done almost surreptitiously, and above all for pedagogical reasons. Gradually, Linnaeus came to divorce the name properly speaking from the description, still called the diagnosis.[1] It was in order to facilitate memorization and identification in the field that he gradually introduced the binomial names, still in use today, which attributed to each species a generic name and a specific adjective or noun. So the common oak was called *Quercus robur*, the holm oak *Quercus ilex*, the cork oak *Quercus suber*, etc. Linnaeus started to use this binomial nomenclature for certain species in 1745, generalizing it to the whole of the plant kingdom in his *Species plantarum* in 1753, and to the whole of the animal kingdom in the tenth edition of his *Systema naturae* in 1758.

Of course, not everything was solved as if by magic, and problems of synonymy continue to occur. On one hand, the principle which assigns to each species the name given it by the first naturalist to describe and name it in print according to Linnaean nomenclature can necessitate the investigation of priority, which can be a source of difficulty. On the other hand, certain Linnaean genera have been broken down into several, which automatically alters the name. Having said this, the conflicts and errors which may still exist are as nothing compared to the confusion which would reign without the nomenclature. The risk that the same plant or animal species will be 'discovered' several times by different travellers and listed in different museums under different names has not disappeared, but it has been reduced to such proportions that one can see Linnaean nomenclature as one of the decisive advances in the history of the natural sciences.

In botanical gardens, in herbaria, in natural-history laboratories, as in the books, it is now theoretically possible to decide whether the plant or animal before one belongs to a species already known. But it is not enough to label specimens, for they have to be arranged, and for this they have to be classified. The determination of the genus constitutes the beginning of classification, because several species may belong to the same genus. And so the donkey and the horse belong to the genus *Equus*, the cowslip and the primrose to the genus *Primula*. But this beginning is not enough; the very smallest collection presupposes a more comprehensive classification. Should animals be grouped together according to the element in which they live – on the ground, in the air or in the water – at the risk of putting the bats with the birds and the whales with the fishes, while both bear their young and suckle them in the same way as do mice and elephants? Should one divide plants into grasses, shrubs, bushes and trees, or group them together by habitat, or should we perhaps apply criteria based on their structure? This question occupied naturalists for a long time; it became extremely important in the eighteenth century, precisely because it affected the integration into gardens and collections of the plants brought back by travellers, and thus in its turn affected the production of catalogues, guides and floras which allowed each traveller to take advantage of the work of his predecessors.

In this matter too, Linnaeus undertook the work of legislation. Following the example of his predecessors, he put the animals in six great classes: Mammals, Birds, Amphibians, Fishes, Insects and Worms. These classes were themselves divided into orders. In the tenth edition of the *Systema naturae*,

The misfortunes of a naturalist

Commerson was a man of unflagging activity and of the profoundest knowledge. If he had himself published his collected notes, he would have been considered a naturalist of the first rank. Unfortunately, he died before being able to finish preparing his work for publication; and those to whom his manuscripts and his herbarium were confided neglected them in a culpable manner ... His herbarium fell first of all into the hands of his heirs; and then it arrived at the Jardin des Plantes where it still remains. Many new plants may perhaps still be found there, although it has in recent years been investigated by many clever botanists, such as Jussieu and Lamarck. The fishes Commerson collected stayed in their cases until nearly twenty years ago, when M. Duméril discovered them in an attic at Buffon's house. The manuscripts were handed to Lacepède, who drew on them extensively for his *Histoire des Poissons*, in which he did not actually include them, but combined them with his own work ... The descriptions are done in the Linnaean style, with the greatest detail and the greatest precision ... They are accompanied by drawings, some by Commerson and some by Sonnerat, and others by artists who went with Bougainville. All these drawings, also given to Lacepède, were also used in his *Histoire des Poissons*, for which they were engraved ... In addition, as Commerson had not settled on his names, it happened that a single creature might be multiplied by three, with one relying on the figure, the second on the characteristic phrase written on the figure and the third on the description. Lacepède, writing in the country, where he had been exiled by the Terror, and not having the original papers, but only notes, was not able to make the comparisons needed to be able to avoid these errors. Travellers who die, who have not sent back in good order the material they have collected, and whose works have been deposited for later use in public institutions, are exposed to the unhappy fate suffered by Commerson. (From Georges Cuvier and Magdeleine de Saint-Agy, 'Voyages scientifiques', *Histoire des sciences naturelles*, 1841–5.)

cited by William Stearn, the class of Amphibians includes three orders: the 'Reptiles', in which one finds our present-day Batrachia (such as the frogs and toads) and reptiles, with the exception of the snakes, which have an order of their own, while the third order, the 'Swimming Amphibia', corresponds more or less to our cartilaginous fishes. Many of these groups have been contested and reorganized, particularly to take account of comparative anatomy.

Buffon, for his part, planned to do without any classification in his *Histoire naturelle*: 'Is it not better to follow the horse which is soliped with the dog which is fissiped, and which in fact does usually follow it, than with a zebra which is little known to us, and which has perhaps no other relation to the horse than being soliped?'

In fact, there is nothing naïve about the anthropocentrism of this declaration: it expresses the importance Buffon accorded to geographical factors and the effect of climate in particular. In any case, the author of the *Histoire naturelle* can only afford the luxury of doing without a classification by restricting himself to groups such as the mammals and the birds, in which the species, limited in number, all have common names. Behind the dispute over methods there is an implicit hierarchy of disciplines and objects.

In the time of Linnaeus, however, disagreements about the classification of animals remained limited. The classification of plants, on the other hand, was the subject of a major controversy recounted by all historians of biology.

Concerned above all with logical rigour, and struck by the importance of recently discovered plant sexuality, Linnaeus put forward his 'sexual system': he divided the flowering plants into twenty-three classes, depending on the number of stamens, the male organs, and then redivided these into orders in accordance with the type of pistil, the female organ. If one takes the autumn crocus, for example, one sees six stamens surrounding three little columns, the styles, which stand upon the ovaries and support the stigmata where the pollen is deposited. Looking then at the crocus, an apparently very similar flower, one sees no more than one style in the middle of three stamens. In the Linnaean system the autumn crocus falls into the order of the *Hexandria Trigynia*, six husbands for three wives, and the crocus into the *Triandria Monogynia*, three husbands for one wife. At that time, this vaguely erotic ethnographical metaphor was not to everybody's taste; some were offended, while others made fun of it. But this was not the real problem, which lay in the too arbitrary character of the divisions. As noted by Antoine Laurent de Jussieu in 1773, in an article in the *Comptes rendus* of the Académie des Sciences, it needs only one aborted or supernumerary stamen to throw into confusion the 'followers of the sexual system'. What was more, the sexual system required the rejection of long-established groups, substituting others sometimes less well-founded.

Other botanists, such as Adanson, the Jussieus, and then Candolle, put up against the 'sexual system' and the others which had preceded it a 'method', more empirical perhaps, which consisted of grouping genera exhibiting affinities with each other into natural families: the Umbelliferae, the Compositae, the Rosaceae. These families in turn were arranged into classes. There was no longer a single criterion, such as the number of sexual organs in Linnaeus' system, but a combination of characteristics from several different parts of the plant: the number of cotyledons[2] (one or two), the mode of insertion of the stamens, the number of petals, etc. Linnaeus himself seems to have been aware of the advantages of a less artificial classification, and he made various sketches for a division into natural families.

413

The weakness of the Linnaean system, as of all the other systems, was no doubt the attempt to satisfy two incompatible requirements. On one hand, they had to make it possible to identify any particular species by means of a finite number of simple operations: multiple-choice questions and the counting of features. On the other hand, they were put forward as a means of grouping living beings together in accordance with their affinities. The first function presupposes characteristics easy to recognize and easy to combine; the second, characters which are of determinate significance for the structure of the organism; and these are not necessarily the same.

This is why Lamarck's publication of his *Flore française* in 1778 marked a decisive stage. The 'Preliminary Discourse' precisely distinguishes the two 'objects' of a classification, and then puts the question: 'Can one fulfil both objects at the same time? That is to say, is it possible that the means which ought to allow us to discover the names the botanists have given to the plants we wish to know about could offer us at the same time the degree of the particular relationships by which plants are connected to each other?'

In answering no to this question, Lamarck leaves himself entirely free to put forward a key which makes its artificiality plain. With this a plant can be identified by means of a series of questions with two alternative answers, which have no other justification than their convenience. This example would soon be followed by others. Freed from the need to identify, classification no longer had to do anything but group species together in the most 'natural' way possible, and systems like Linnaeus' were abandoned, while the natural method found expression in botanical gardens.

Distribution and Genealogy

Even more than the herbarium, the botanical garden is a place of deceptive simplicity, a place of discreet charm and quietly murmuring life. A label next to each plant indicates its scientific name, assigning to it the task of representing the whole species which bears this name. Around it, in the same bed, crowd plants which would usually live in completely different places, even in different continents. Most of these plants would never have come together in the same place if classification had not assigned them to the same family. Their meeting, however, leaves one question hanging in the air: where can these plants be found in nature? The classification is drawn up ignoring this kind of question, because it is concerned with morphological criteria; but this is one of the things that really strikes the travelling naturalist: one doesn't find any old species anywhere. Behind this question, and this almost self-evident observation, there lie in fact two distinct problems: on one hand, in what environment does this plant live (in water, in the desert, on high mountains, etc.), and on the other, in what regions of the world is it present and in which is it absent?

The key to the flora

To explain how to discover the name of a plant, in the 'Preliminary Discourse' to his *Flore française* (1788) Lamarck begins by imagining that there exist only eleven species of plants – which appear in the text under the Latin names given to them by Linnaeus – and which we call hawkweed, dog fennel, male fern, chickweed, meadow sage, field mushroom, the pear tree, a moss (*Bryum murale*), scarlet pimpernel, the yellow cep and the milk thistle. Then one takes a specimen of one of these plants, supposedly unknown, chickweed for example, and one answers a series of questions:

- a flower whose stamens and pistils may easily be distinguished;
- a flower whose stamens and pistils are nonexistent or cannot be distinguished.

The first answer must be chosen, which leads on to the following question:

- numerous florets joined together in a common calyx;
- independent flowers not joined together in a common calyx.

The second answer must be chosen, which leads to the following question:

- Monopetal corolla;
- Polypetal corolla.

One replies 'polypetal corolla' which leads to the last question:

- ten stamens or less
- eleven stamens or more?

The first answer is the correct one and gives us the name of the species: chickweed, which Linnaeus called *Alsina media* and which twentieth-century botanists call *Stellaria media*. In the *Flora* itself, which runs to hundreds of pages, this analysis is carried out through consulting a series of tables, at the end of which, one is assured, if one has not made a mistake, of finding any species which is described and named by the author. For the reader who finds this a rather long way of going about things, Lamarck evokes 'the nature of geometric progression. In effect, if you divide the figure 4096 continuously by 2, after the eleventh division you get to one.' In other words, it takes no more than about ten questions to cover thousands of species.

The practice of acclimatization, whose purpose is essentially utilitarian, relies entirely on this distinction, and it forms the equivalent of a whole series of experiments which bring out its nature. When one imports an exotic plant from a distant land, there are three possible outcomes: it may be unable to survive in the normal conditions of its new environment; it may become a cultivated plant; and finally, it may be naturalized and become part of the local flora. The first case is the simplest: a plant brought back from Amazonia, for example, cannot be grown in France except inside a greenhouse in which the heat and the humidity it needs are provided. The role of such important climatic factors is easily seen. The second case, of plants imported and then cultivated – fundamental to the material history of our societies – has, in addition, the advantage of demonstrating the effect of fairly subtle physical factors. In their work, the grower and gardener adapt the soil and the microclimate to the cultivated plant, and fight against more vigorous species which might compete against it. The third case, complete naturalization, has great theoretical interest. To take an example given by Linnaeus himself in 1744: *Erigeron canadensis* (horse-weed) transported from North America to France in the middle of the seventeenth century, and introduced into a few botanical gardens, had become a century later one of the most common wild plants of its adopted country. One could also mention the American cacti which were introduced into the Mediterranean flora, and all the European plants introduced to the United States. By their success in their new countries, these plants negate any exclusive explanation by the influence of environment: how can their initial absence from the local flora be explained when their naturalization clearly indicates that the environment was indeed appropriate to them? If the culture of exotic species reveals, in the difficulties it meets, the importance of physical determinants of the distribution of species, the successful naturalization of species introduced from elsewhere marks the limits of this kind of determinism and calls for another explanation.

Parallel to these deliberate or accidental plant migrations, there developed a new science, botanical geography, which deals precisely with the question: how are plant species distributed on the surface of the globe? Augustin Pyramus de Candolle, the Swiss descriptive and classificatory botanist, was interested in agronomy, and was one of those who contributed to the foundation of the new discipline. In the article on 'Botanical Geography' in the *Dictionnaire des sciences naturelles*, published in 1820, he lays out the problem. He deals first of all with factors which affect the distribution of the different species of plants, and then of the 'stations', that is to say the environments in which they are found, and then the 'habitats', by which he means the regions in which they naturally grow.

On many points this text resembles Alexander von Humboldt's *Essai sur la géographie des plantes*, published in 1807, in which he shows the influence of temperature on vegetation, using the observations he had made in the

Andes with Aimé Bonpland, in the course of his travels in Latin America. Candolle is not properly speaking a traveller-naturalist, even if he had journeyed a great deal in France and in Switzerland, but his article owes a great deal to materials brought back from long-distance voyages. This is particularly clear in the third part, which deals with the 'habitats', and of which he says at the very beginning that it touches on facts which escape 'all existing theories', because they touch 'on the very origin of organized beings, that is to say, on the most obscure topic of natural philosophy'.

Having demonstrated the influence of temperature, Candolle writes:

> Up till now I have sought to prove that the habitats considered as a whole appear to be determined by temperature. Of course, there must be combined with it those considerations deduced from their stations; for it is clear that the more sandy a country is, the more one will find there the plants of sandy soils, etc. But, even when one has granted to these causes as great an affect as may be attributed to them, does this enable one to give a complete account of the best-known facts? I think not, and this requires a new discussion.

He notes 'the small number of phanerogamous species common to the different continents'.[3] Thus one-eightieth of the plant species observed in New Holland – meaning Australia – are shared with Europe. He then studies the means of transport of seeds and then turns to the flora of islands. He proposes this formula: 'The plants on islands are part of the vegetation of the neighbouring continent more or less in inverse proportion to the distance between them.' He also talks of human action, which modifies insular floras, and he utters this warning, in the form of a research programme: 'Let us hurry, then, while there is still time, to make up the floras of distant countries; let us recommend to travellers especially those islands little frequented by Europeans; it is in studying these that there will be found the solutions to a host of questions of plant geography.'

These remarks give meaning to the concept of the botanical region, which the author defines in these terms: 'From all these facts taken together, one can deduce that there exist *botanical regions*; I designate by this name those areas, which if, one excepts species introduced from outside, offer a certain number of plants particular to them, which one might call truly *aboriginal*.'

Candolle points out the fact that many genera include one North American species and a European or Asiatic species, after which he enters into what for us is the substance of the matter, and which for him is the difficult point: 'The whole theory of botanical geography rests on the idea we have of the origin of organized beings and of the permanence of species.' He gives his own position: 'The whole article one has just read has been composed in accordance with the idea that the species of organized being are permanent, and that every individual comes from another individual like it.'

417

We find this creationist declaration somewhat astonishing in retrospect, because we know that the geography of plants and animals provided the theory of evolution with plenty of arguments. Explanation by religious prejudice, the panacea of the history of science, will not work here: nothing in this text or in Candolle's *Mémoires et souvenirs*, published by his son in 1862, suggests that he would have defended creationism in order to safeguard the literal meaning of Genesis. Candolle, a liberal Geneva Protestant, often seems closer to free thought than to religious fundamentalism. In reality, to understand his opposition, one has to look at what he opposed. Although he doesn't mention it explicitly, one can see that here he is getting at a conception which associates the transformation of species with spontaneous generation, that is to say a conception which grants to the environment total power to produce living beings and to mould them. To this mechanistic vision of evolution, attributed more or less justly to Lamarck, the particularity of the botanical regions becomes inexplicable. Some of these regions, indeed, have analogous climates, so how can one explain that they did not produce the same species? 'The partisans of spontaneous generation seem to be . . . completely unable to explain the common and incontestable fact that a great number of clearly determined species can be found only in one region, and are not to be found wild in countries in which all circumstances are favourable to them, and where they live well when once they have been introduced.'

Comparison with the chapters on 'Geographical Distribution' in *The Origin of Species* is illuminating. Darwin too notes the inadequacy of explanations based exclusively on physical factors – soil and climate – and offers as proof 'that the division of the Earth into the Old and New Worlds is one of the most fundamental divisions in geographical distribution', while there was 'no climate or condition in the Old World which does not have its equivalent in the New'.

The resemblances were no less astonishing than the differences. The species of equatorial South America have more affinity with those of temperate South America than with those of Africa. In fact, what one learns from the American and the Australian examples is 'the affinity which exists between the productions of the same continent'.

Until then, one could say that Darwin was doing no more than extend the problematic of phytogeography to the whole realm of living things, while refining its analyses in order better to grasp the role of topography: barriers, passages, islands and archipelagos. He compares the land which separates two marine faunas to the sea which separates two terrestrial faunas. It is here perhaps that Darwin most reveals himself as the naturalist-traveller he was first of all. The diary of his voyage on the *Beagle* is full of biogeographical annotations; one knows in particular of his interest in the different species of 'Galapagos finches', whose distribution he studied across the archipelago.

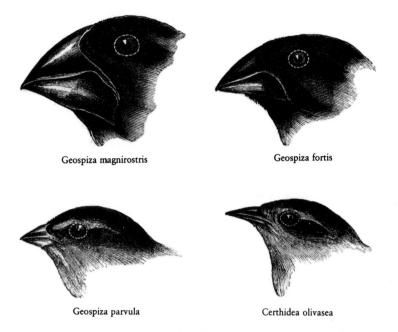

Geospiza magnirostris

Geospiza fortis

Geospiza parvula

Certhidea olivasea

Some species of Galapagos finches. (From C. Darwin, A Naturalist's Voyage round the World, *1875. Bibliothèque Nationale, Paris.)*

If afterwards he never left the English countryside, he did experiments on the conditions in which species might travel, examining the effects of seawater on seeds, and listing those which might be carried on bird's feet. But Darwin's real innovation occurs when he begins to interpret the distribution of flora and fauna: 'These facts denote the existence of some profound and intimate organic connection in time and space, across the same areas of sea and land, independently of physical conditions. A naturalist would have to be most indifferent not to be tempted to investigate this connection.' For him, 'this connection is quite simply heredity.' In other words, the affinity between the species of a same biogeographical region is to be explained by a common origin, and the differences arise above all from natural selection, which in different milieux favoured different variations. Island flora and fauna, related and at the same time different to those of the mainland, are a good illustration of this process. However, taxonomic and geographic proximity did not always go together; but it remained the case that 'the different species of the same genus, even if inhabiting points of the globe very distant from each other, must have the same origin.' It was therefore necessary to look for common ancestors and to rediscover their migrations. The geographer became the historian of the living world. In the experience of Darwin the traveller, biogeography meets palaeontology, as we are reminded by the first lines of introduction: 'When on board H.M.S. *Beagle*, as a naturalist, I was much struck with certain facts in the distribution of the

419

A means of dispersal

To understand the current distribution of flora and fauna, Darwin had to study the 'means of dispersal' of animal and plant species, and to do this, he had to carry out several experiments simulating the conditions of accidental transport. For example, he observed the effects of seawater on seeds. Of 87 seeds immersed for 28 days in seawater, 64 were still able to germinate. He also dried plants and saw how long they would float: 19 out of 94 (not all belonged to the same species as in the last experiment) floated for more than 28 days after drying. Combining the two results, he was able to suppose that '14 per 100 of the plants of a given country could be carried for twenty-eight days by marine currents without losing the capacity to germinate.' Taking account of the average speed of such currents, these seeds could be transported more than a thousand kilometres to a favourable shore. (C. Darwin, *The Origin of Species*, 1876)

---

inhabitants of South America, and in the geological relations of the present to the past inhabitants of that continent. These facts seemed to me to throw some light on the origin of the species.'

The connection between geography and classification, which until then had been a matter of statistics – the number of species or families present in a continent – was now assured by the 'theory of descent with modifications', or, as we call it today, the theory of evolution. At the very moment when biogeography was conceived as the end-point of a history, classification was defined as a genealogy. In the chapter on 'Mutual Affinities of Organic Beings', which follows the two chapters on geographical distribution, after a discussion of the different principles on which the different systems of classification were based, Darwin wrote:

> All the foregoing rules and aids and difficulties in classification are explained, if I do not greatly deceive myself, on the view that the natural system is founded on descent with modification; that the characters which naturalists consider as showing true affinity between any two or more species, are those which have been inherited from a common parent, and, in so far, all true classification is genealogical; that community of descent is the hidden bond which naturalists have been unconsciously seeking, and not some unknown plan of creation, or the enunciation of general propositions, and the mere putting together and separating objects more or less alike.

In presenting 'common descent' as the principle hidden behind systems of classification, and which his predecessors were unconsciously seeking, Darwin defines the way in which he intends to situate himself in relation to the natural-historical tradition.

In fact, like many voyages, the itinerary which leads from Linnaeus to Darwin ends where it began. At the beginning, a reform of nomenclature and classification was necessary in order to name and to classify the mass of specimens brought back from their voyages by naturalists, so that natural history could be a description of nature rather than a strange inventory continually being begun again. As soon as each voyage could contribute to the constitution of a cumulative science of living things, there appeared a new programme of research: how were plant and animal species distributed across the surface of the globe? At the very moment when Europeans were trying hard to modify this distribution by their efforts at acclimatization, they were attempting to explain it. The voyages and migrations that man imposed on other species established the limits of geographical determinism. The affinities between the natural productions of each region, just like the irregularities of geographical distribution, had to be referred to a process of evolution which combined chance and constraint. The geography of living things had to be seen as the current end-point of a history of living things. The classificatory table was no more than a transverse section across this tree at a given moment. The road from Linnaeus to Darwin travels through maps and collections, and it is to a great extent the road which was trodden by the traveller-naturalists.

14

Paris 1800

Michel Serres

The history of France meets the history of science, and they prove to have much in common: with the French Revolution, the scientists come to power.

The origin of mathematics, of Western mathematics at least, may be located on the map at a remarkable feature, that is to say along an axis, or perhaps a crevasse, which passes between Greece and Turkey and extends from Constantinople to Mecca. Ancient Egypt, Babylonia, Persia, and Jewish Palestine all lie alongside this line. These were the lands of the biblical prophets, of Christianity and of Islam, the lands of the great monotheistic religions as well as the home of Greek philosophy and of Greek and Arab science. They also witnessed the birth of writing, money, and the manufacture of iron and bronze. Could there be specific locations in space, dense as seeds, where discovery is intensified? How can the phenomenon be explained? Could it be the threefold contact between Africa, Europe and Asia at a meeting-point unique in the world? Or was it the fierce encounter between Semitic and Indo-European culture, leading to collision and to miscegenation?

Features in Space and Time

In the same way, might there exist singular moments in time just as remarkable and as powerfully productive? In the fifth century BC it took scarcely two generations to furnish Athens with a profusion of beautiful art and great literature: the very potters were visited by genius. In the same way, Paris of the Classical period or Western Europe of the late nineteenth century needed only twenty-five to forty years to give their all in the domains of artistic and intellectual creativity. These are like sudden volcanic eruptions which transform the landscape all around. How should they be explained, when we know that so many powerful and wealthy groups reached the very pinnacles of social life, but remained sterile and culturally impoverished in the midst of their economic wealth and military power? In such times great works of tragedy are created, which are absent in other places and other times. One notices too the great comic writers, witnesses to the health of their contemporaries, the like of which are not to be found in the periods before or after.

If it is social and not necessarily individual, does genius sometimes flash out meteor-like in particular places and at particular times?

Paris

In the period which falls on either side of 1800, from the end of the *ancien régime* to the restoration of the monarchy (1789–1814), there occurred an extraordinary succession of political, social, intellectual, scientific, religious and anthropological events. They were events of such violence and of such great consequence that, when they were over, there sprang up whole philosophies devoted to understanding them. The owl of Minerva flies only at dusk, said Hegel; his phenomenology is perhaps only an account of what happened then. In these years, indeed, Paris did in reality what Hegel and Comte would only speak about, as if there remained to the generations to come only the tasks of understanding and telling, the regretful jealousy of interpretation. But what did these philosopher-historians talk about? They were concerned with the totality of knowledge as it developed within the totality of history. What was it then that went on in Paris around 1800? Precisely that, for here a local and temporal singularity consciously bore universality within it. It was a volcanic eruption, an earthquake, a total transformation of the world.

Now we have forgotten the Paris of 1800, which can only be seen in Turgot's plan: one of the most astonishing accumulations of architectural beauty in history. The nineteenth century saw such frenetic demolition and construction that the Paris of today, which we call historic, is in fact a very

423

new city, and newer, for example, than New York, according to an accurate recent calculation of the average age of its buildings and monuments. On both sides of the river, but particularly on the Left Bank, there were churches and their squares, houses, esplanades and gardens. As elegant and as luminous as the Sainte-Chapelle, the Hôtel de Sens or the Place des Vosges, they clustered together or lay along the streets, dozens together, until they were overtaken by the destruction wrought by the Consulate and the two Empires. To get a clear impression of this loss, we should remember that when Haussmann constructed the Boulevard Saint-Germain, he demolished more than forty churches built on the same scale and in the same style as the one which remains on the Île de la Cité. We have no more memory of this city of beauty than we do of the sciences which were practised there. Was it vandalism or adaptation? Urban dynamism and the power of renewal? Every kind of thesis has been put forward, as usual. But the place was transformed like nowhere else.

Portraits

In a quarter of a century, France went through practically every possible form of constitution: from kingdom to republic to empire, moving from different varieties of chaos to every kind of tyranny, as if Paris was living through a universal history of institutions. Was this a thoroughgoing upheaval? Or was there an essential stability? As usual, every kind of thesis has been put forward.

In 1787, and in the middle of these transformations, Joseph-Louis Lagrange, who was born at Turin in 1736, succeeded Leonhard Euler at Frederick II's Academy in Berlin. He had already founded the Academy at Turin. He then accepted an invitation to Paris from Louis XVI and settled there. This superbly French Italo-German European was given lodgings in the Louvre, an apartment very near to the studio of Hubert Robert, the painter who would soon find himself carted to the scaffold, where he miraculously escaped death.[1] While his neighbour painted, Lagrange spent fifteen hours a day on mechanics and calculus. He would have seen the hats and cloaks passing in the courtyard below, nearly every season changing shape and colour with the fashion. Sometimes too there would be heads carried rather higher than was usual. Later on the artist would draw ruins in the palaces, while the mathematicians were working out the theory of invariants by variation. Fêted by the King, by the Terror and by the Emperor, decorated and ennobled, Lagrange died two years before Napoleon's Hundred Days. He thought science; he spoke many languages; he frequented every kind of ruler.

A marquis and an extremist, Marie Jean Nicolas Caritat de Condorcet frequented firstly those great by birth, but in the salons and the academies, being talented himself, he got to know the great thinkers and great oppositionists, more powerful even than power itself: Turgot, Voltaire and d'Alembert. He was elected to the Legislative Assembly and to the National Convention, where he moved among the great men of politics. He lived surrounded by every kind of greatness, but spoke with fervour about equality. A student of real analysis and of mechanics, an astronomer, the author of writings on the inclination of the ecliptic (see p. 82) and the three-body problem (see p. 435), he was also a statistician before statistics were invented, extending the theory of probability for applications in what we would call the social sciences. Condorcet covered all, or nearly all the rigorous mathematics of his time. As Secretary of the Academy before it was dissolved, he was at the head of science. When his close friend Turgot became Prime Minister, he became head of the administration. During the Revolution, he wrote up the Parliamentary debates for several influential newspapers, one of the leaders of the media of the time. He is a forerunner of modern political power, which is exercised through language and discourse. No one may resist the truths of scientific knowledge, nor the ever-present seduction of circulating information, nor may one oppose the administrators who organize the social order, without penalty of error, of silence, of illegality. Condorcet spoke fervently of liberty.

A warrant was issued for his arrest, and he was sentenced to the scaffold. He went into hiding in an obscure street, where he wrote the *Esquisse d'un tableau historique des progrès de l'esprit humain* (translated as *Sketch of a Historical Picture of the Progress of the Human Mind*), in which science and rational language are shown as directing and embracing the whole of history. Mind, or universal reason, which is embodied in science, in Condorcet's science, finally comes to power everywhere and for all time, dominating every culture. It speaks fervently of fraternity.

Under threat of death Condorcet fled for two days, wandering the roads south of Paris, near Bourg-l'Egalité, formerly Bourg-la-Reine, sleeping in the open, using a borrowed name, dirty and badly shaven, eating in public houses. There he encountered at last the people and the poverty he had described so often. Forty-eight hours of direct experience made up for a life of speeches; it killed him (1794).

The exact sciences

Mathematics

ARBOGAST, Louis (1759–1803).
Symbolic calculation.
ARGAND, Jean-Robert (1768–1822).
Imaginary numbers.
CARNOT, Lazare (1723–1853).

Infinitesimal calculus.
CAUCHY, Augustin-Louis (1789–1857). Real analysis.
CONDORCET, Marie-Jean Nicolas Caritat de (1743–94). Algebra.
FOURIER, Joseph (1768–1830).
Partial differential equations.

FRANÇAIS, Jacques-Frédéric (1775–1833). Imaginary numbers.

GERGONNE, Joseph-Diaz (1771–1859). Duality.

LACROIX, Sylvestre-François (1765–1843). Analytical geometry.

LAGRANGE, Joseph-Louis de (1736–1813). Real analysis, mechanics.

LAPLACE, Pierre-Simon de (1749–1827). Mathematics, physics, astronomy.

LEGENDRE, Adrien-Marie (1752–1833). Algebra.

MONGE, Gaspard (1746–1818). Descriptive geometry.

POINSOT, Louis (1777–1859). Statics.

POISSON, Siméon-Denis (1781–1840). Probability.

PONCELET, Jean-Victor (1788–1867). Projective geometry.

Astronomy

BAILLY, Jean Sylvain (1736–93).

BORDA, Jean Charles (1733–99).

DELAMBRE, Jean-Baptiste Joseph (1749–1822).

LALANDE, Joseph Jérôme François de (1732–1807).

MECHAIN, Pierre (1744–1804).

MESSIER, Charles (1730–1817).

Physics and chemistry

ARAGO, Pierre-François (1786–1853). Electricity.

BERTHOLLET, Claude-Louis (1748–1822). Chemistry.

CARNOT, Nicolas Léonard Sadi (1796–1832). Thermodynamics.

COMTE, Auguste (1798–1857). Mechanics, astronomy.

COULOMB, Charles (1736–1806). Electricity, magnetism.

DULONG, Pierre Louis (1785–1838). Acoustics.

FOURCROY, Antoine François de (1755–1809).. Mineral chemistry.

FOURIER, Joseph (1768–1830). Theory of heat.

GAY-LUSSAC, Louis-Joseph (1778–1850). Physics.

GERMAIN, Sophie (1776–1831). Acoustics, mathematics.

HASSENFRATZ, Jean-Henri (1755–1827). Chemistry.

HAÜY, René-Just (1743–1822). Mineralogy.

LAVOISIER, Antoine-Laurent de (1743–94). Chemistry.

MALUS, Etienne-Louis (1775–1812). Optics.

NIEPCE, Nicéphore (1766–1833). Photography.

PARMENTIER, Antoine Augustin (1737–1813). Agronomy, pharmacy.

PRONY, Gaspard Marie Riche de (1755–1839). Engineering.

PROUST, Louis (1754–1826). Chemistry.

ROME DE L'ISLE, Jean-Baptiste (1736–1790). Mineralogy, crystallography.

SAVART, Félix (1751–1857). Acoustics.

THÉNARD, Louis Jacques (1777–1857). Chemistry.

Biology and medicine

BICHAT, Marie-François-Xavier (1771–1802).. Histology.

BLAINVILLE, Henri Decrotay de (1777–1850). Natural history.

BRAVAIS, Louis (d. 1842). Botany.

BRONGNIART, Alexandre (1770–1847). Mineralogy.

BROUSSAIS, François Joseph Victor (1772–1838). Medicine.

CABANIS, Pierre Georges (1757–1808). Medicine.

CANDOLLE, Augustin Pyrame de (1778–1841). Botany.

CORVISART, Jean Nicolas (1755–1821). Medicine.

CUVIER, Frédéric (1773–1838). Geology.

CUVIER, Georges (1769–1832). Anatomy and palaeontology.

DAUBENTON, Louis (1716–1800). Natural history.

DUPUYTREN, Guillaume (1777–1835). Medicine.

DUTROCHET, Henri (1776–1835). Medicine, osmosis.

ESQUIROL, Jean (1772–1840). Neurology.

GALL, Franz Joseph (1758–1828). Anatomy.

ITARD, Jean Gaspard (1774–1838). Endocrinology.

JUSSIEU, Antoine Laurent de (1748–1836). Botany.

LACEPÈDE, Etienne (1756–1825). Natural history.

LAËNNEC, René Théophile Hyacinthe (1781–1826). Stethoscope.

LAMARCK, Jean-Baptiste de Monet de (1744–1829). Biology.

LATREILLE, Pierre-André (1762–1833). Entomology.

PINEL, Philippe (1745–1826). Psychiatry.

SAUSSURE, Théodore de (1767–1845). Chemistry of vegetation.

Lazare Carnot studied real analysis and was Lagrange's rival, the author of *Reflections on the Metaphysics of Infinitesimal Calculus*, in which he attempted to master differentials or so called vanishing quantities. He combined success in mathematics and in physics. Elected as a Deputy to the Convention, he not only escaped the fate of Condorcet, his colleague in mathematics and philosophy, but he became a ruler. He joined the Committee of Public Safety, where he embodied the spirit of war, creating fourteen armies for the republic, working out all the campaign plans, organizing victory and preparing the destiny of Napoleon. His was a thoroughgoing triumph in analysis and in reason, pure and applied, in politics and in strategy, crowned at the end by ideological martyrdom when he was exiled at the Restoration. Few towns in the country, even small ones, are without their rue Carnot; few names have gained greater glory. Carnot was science, Carnot was power, Carnot was victory and memorial plaques on the walls. Nothing or very little survives of his science, and we know what is left on the battlefields at the end of the day. A grandson of his would find fame as President of the Republic, only to be assassinated. He himself lived to the age of seventy, from 1753 to 1823.

His eldest son Nicolas Léonard Sadi was a physicist, who died mad at the asylum of Charenton, then under the direction of Esquirol, screaming in anguish, restrained in a strait-jacket. He had hardly reached the age of thirty-six. We know nothing of him, of his youth or of his genius. From the sad shipwreck of his life there survives only a manuscript on the power of heat-engines in which he discovers and founds the science of thermodynamics, describing the heat cycle by stating the Second Law (cf. p. 313). A new time really was being born, amid insults and poverty, in the straw, dirt and solitude, in the extraordinary suffering of the abandoned genius.

The political revolution passed, with its two great tides of speech and death, the one instead of and by reason of the other, with their ebb-tides of Empire, of *journées populaires* and of Restoration, reversals marked by hatreds and by sublime and irreversible ideals: this was war. The Industrial Revolution came about elsewhere: coal, capital, the recapitulation of money and of mines, steam engines and fierce exploitation of the poor: this too was war. The authentic scientific revolution took place in silence and in isolation, far from power, far from glory, far from fortune, in the cell of an asylum, in the midst of unforgiving unhappiness: was it far from war? The father was triumphant, chopping off heads; he was sterile and organized armies; he made history, as we say. The son, wretched, organized the future.

Science and Power

Positive science took power. Lazare Carnot and Condorcet were mathematicians, Bailly, the first mayor of Paris before he lost his head, an astronomer; Lagrange and Laplace were students of mechanics, Fourier and Arago physicists, Fourcroy and Berthollet, chemists. There were a few doctors, Cabanis among them, and it was a geometer who rose to the heights of military and political power. His name was Bonaparte, and he had solved what was called Napoleon's problem: to divide a circle into four equal parts by means of compasses alone, according to the elegant method of the Italian, Lorenzo Mascheroni. The Emperor not only embodied the first of the doomed struggles between Southern Europe and the triumphant North, but also the second of the victorious struggles of the sciences against the humiliated humanities. Chateaubriand fled abroad, Beaumarchais was imprisoned, Chamfort committed suicide, Chénier was beheaded, Madame de Staël in exile. Positive science took power and humanities lost it.

War within a war, or revolution within the Revolution, this battle of the faculties was the continuation of a conflict which had begun with the Enlightenment and which continues in our own day. Science put its hands on reason and became its exclusive possessor; outside science there was only the irrational. All other contents of knowledge and culture, even those which had paved the way for the emergence of rationalism, such as metaphysics and theology, and even those we now group together under the title of the human sciences, dedicated to myths and shadows, saw themselves expelled from the field of reason. The so-called Romantic movement confirmed and strengthened this unfair division by taking seriously what we now tend to think of as only *Sturm und Drang*. Since then we have lived with the self-evident fact that science and the rational form a single united domain, though in fact it is the first which has claimed the second for its own. As a result of the transfer of power, this triumph of advertising, already well prepared by the Enlightenment, reached its apogee in Paris around 1800. Society gave itself to reason, which handed itself over to science, which then expelled the cultural. The universal triumphed over the particular.

The Universal

Its title coming from a Greek word invented by Rabelais, or from a classical concept long considered by Leibniz, its aim adopted by every philosopher worthy of the name, from Aristotle to Auguste Comte, its inspiration drawn from the work of the Briton, Ephraim Chambers,[2] the *Encyclopédie* or *Dictionnaire raisonnée* was written over twenty years, from 1751, by d'Alembert and Diderot, with the help of collaborators both famous and obscure, among them Voltaire, Montesquieu, Rousseau and Galiani. It sums up the Age of Enlightenment: the whole of knowledge was recorded in it, in alphabetical order.

Their title deriving from the name of a mythical Attic hero, whose gardens served as a meeting-place for Plato and his disciples, brought into existence in nearly all the great capitals of Europe under the patronage of the kings of the Classical Age, the academies met according to a fixed calendar, assembling in a central location not books, but men, specialists in their respective disciplines. They dominated the Age of Enlightenment. The totality of knowledge could be brought together around a table or in a room, ordered in accordance with yet another convention.

During the Revolution, this totality constituted the equivalent of a Council of Ministers.

Science without Frontiers

Today we would happily think of Lagrange or Beccaria as Italians, Gauss German, Linnaeus Swedish, Benjamin Franklin American, d'Alembert French, Abel Norwegian, and Euler, the Bernoullis and Saussure as Swiss. The seventeenth century, however, knew neither the categories nor the adjectives. Scientific Europe existed, and Lagrange lived in the Louvre, just as Voltaire lived in Prussia and Diderot in Russia. In the period around this date, 1800, chosen as a round figure, Paris wasn't a centre like London, or later the United States, places where world decisions were taken: it was more like a clover-leaf junction in a space without frontiers. It has been said that the centre changes its location through time; but this is to believe in a homogeneous and isotropic space and time, which goes against all we know of history and its varying circumstances. The centre may change its nature and, at certain exceptional moments, it may no longer coincide with the site of power. What must be determined is the place where cosmopolitans could meet, and the language in which they spoke to each other. In this period Paris lost power and gained universality. In the same way, in the past, Athens had attained the second, though never having achieved the first.

What is the deeper meaning of the enormous battle against fanaticism and religion fought by the men of the Enlightenment during the century which had just come to an end? Certainly it was a struggle against power. But it also rejected particular cultures in favour of the emergence of a rational universal, a shared new language. At the end of the eighteenth century, Europe was not so much speaking French as this new language. What today, and precisely since the beginning of the nineteenth century, we call cultural identity, which is expressed in regional languages, founded on religions, recognized in anthropology and often embodied in what we call the humanities, seemed at that time to be a major obstacle to universal rationality, which therefore had to sweep away these particularities. This is what was really at stake when science dealt with universal history, as, for example, in Condorcet's *Sketch for a Historical Picture of the Progress of the Human Mind.* Even today, there are difficult issues which remain unresolved.

To try to impose a centre, then, is to place at the summit of an intolerable hierarchy a particular disguised as the universal. In those days, national meant against the centre, anti-royal. What individual man, what culture could call itself sovereign? None. So there was no umbilical point in Paris, nor in Ferney, Turin, Potsdam, Berlin or Saint Petersburg. Paris in 1800 was located not in France, but in the universal, if not in Europe. Whether it succeeded in the enterprise I cannot say, but it's clear that it tried. In any case, this is what everyone believed. Reason was embodied in an eccentric space. The proof: Paris committed suicide as a centre. It gave up its claim to power for the sake of universality, it gave up power for knowledge.

Reason in Power

Eudoxus and Aristotle in Ancient Greece, Sacrobosco and Aquinas in the Middle Ages, Descartes and Galileo in the Classical Age and Newton and Leibniz at its end stand out as exceptional individuals, but during the course of two millennia, never yet had science been tempted, nor had it the means to put itself at the centre of philosophy nor at the centre of the State, still less at the centre of society or of history, as it was lived or as it was recounted. It was a servant and it remained peripheral. Paris in 1800 wished to be at the centre of no space and of no empire, because it had granted or left this place to science taken as a whole. Now that the universality of reason was understood, now that scientific society was properly organized, the totality of knowledge as such, only recently defined in the *Encyclopédie* and by the academies, accumulations without order, set out to conquer and from now on attempted to occupy every post, all the room and the very centre of space. This is how it was understood in Europe. So *Paris 1800* is less the title of a chapter in the history of France than the marker of a time and place decisive for the history of the Western sciences and humanities. What is at issue is nothing less than the annexation of the whole of life by science, in the shape of the class of scientists. In the seventeenth century the master and possessor of nature, science now wanted to make itself master and owner of men. This plan, already visible then, met with a defeat which can only be considered temporary. Control, direction, even the drive were given, so that the movement begun then would never cease to approach its goal, a victory very near today. Certainly, fragments of individualized disciplines made their appearance there at this time, but what really emerged is what we might call the totality or order, or the collectivity, or even better, the sociology of science.

Before this date, science as a whole offered informal collections rather than systems: the conventional disorder of alphabetical order or of the meeting around the table. In order to take power, everything had to be classified. The *Encyclopédie* made a circle within which everything might be grasped and enclosed, so that its centre should be the one where all was decided. Will science take over political power and will reason take over history? Hegel *happened* here before it was written, and it was written here even before Hegel by Condorcet's dying hand.

Never had so many scientists come close to the centre of political power: the phenomenon has to be seen as a whole and not in its particular details. Jean Sylvain Bailly (1736–93) was an astronomer and historian of science: he became, as we know, the first President of the Constituent Assembly and also the first mayor of Paris. In this very book one can read of the intellectual and political career of Lavoisier (see pp. 455–82), a farmer-general (a privatized tax-collector) under the *ancien régime*, who ended up beheaded. Condorcet was at the Legislative Assembly, and then the Convention, as was Lacepède, who was advised by his friend Cabanis. Lazare Carnot presided over the Committee for Public Safety. Laplace was a Senator, Monge Minister of Marine, while Fourier was a prefect and Arago and Chaptal were ministers. Such examples are plentiful. When Bonaparte left geometry the solution to a problem, in this he matched Louis XIV, who left literature a translation of Julius Caesar. Nothing has been left out: from mathematics to economics, by way of physics, chemistry, natural history and medicine, all the sciences suddenly entered the political arena, not individually but as a whole.

It brought with it its own habits and its disputes. The jealousy of colleagues played its part in Lavoisier's death. Marat's behaviour is in part explained by his resentment at the condemnation in 1780 of his *Découvertes*, in a report by the Academy of Science, signed, among others, by Condorcet. Many accounts were settled by proscriptions and at the scaffold. But the cowardly prudence of these speculative minds allowed them to come out from the revolutionary torment rather better, taken as a whole, than those in many other professions, especially those who were concerned in public affairs. At the end, this caste was not only powerful but also safe and in good health. In this we may see a measure of its true power and its solidarity and concerted action despite all.

One can imagine that one or another, Condorcet, Bailly or Lavoisier, made his way compelled by ambition, interest or ideology. Of course. But the movement is so great that it can only be understood in its totality. All of a sudden, science made up a whole which advanced all alone towards political office. Scientists thought, lived and acted in a collective subject to its own laws. No one will deny that this process of consolidation developed slowly through two centuries, speeding up as it did so, but it was the French Revolution which provided the opportunity for its crystallization. Like the totality of knowledge, science tends to become a total social fact.

Recapitulation

Now one could look differently at what came before this battle and the temporary victories. Already well developed, the main disciplines, real analysis, mechanics, astronomy, physics, chemistry and history, went through a process of recapitulation, more or less all at the same moment. This was the age of the great systematic regional treatises signed with the names of those who just before were running after titles and office. The *Encyclopédie*, hitherto diffuse, became concentrated. One might say that a review was being carried out. These were local circles within a global circle. It can be understood in many ways. An inventory is, first of all, the precondition and necessary predecessor of invention and discovery. This is true, but recapitulation also builds up a capital. Or again, concentration can define and strengthen the centre. And finally, a review or recall are for memory and for knowledge what they are for soldiers under arms. Science becomes conscious within itself of its internal and external strengths; by concentrating, it could be said to be preparing. It did not come to power by chance.

Paris didn't aspire to be a centre, which it wanted to disperse everywhere through an *Encyclopédie* whose immense circumference would contain everything. The totality of knowledge, now mobile, went on its travels. Bonaparte put it in a boat and took it to Egypt: the academies would meet on the banks of the Nile, where the *Encyclopédie* looked for its ancestors. Once again, what would later only occur in the text or in writing was actually was done in reality: science as a whole went in search of its own history.

Another universal: the whole world, the Universe, the globe, both site and object of knowledge. The great journeys of exploration begun in the fourteenth century were concluded by Bougainville, Cook and d'Entrecasteaux.

The sailors who went about the seas sometimes knew nothing of the Revolution. The geographical discovery of new lands came to an end; these roads had all been travelled. These voyages had circled every part of the globe; they were begun again in the name of the *Encyclopédie*. The new explorers intended not so much to conquer as to know: to observe the stars at the Cape, to unfold the map of heaven, to triangulate an arc of meridian. Science inspected and prospected an Earth that was speculative and experimental, physical and astronomical, while waiting for ethnology to make men themselves its objects. The globe changed, now not so much scenery or something to be appropriated as a circle of objectified circles, concrete support for the *Encyclopédie*. From there, animals and plants were sent to zoological and botanical gardens and museums, where they were recapitulated: a central collection or inventory which could take place anywhere.

To the age of the inventory or of the great treatises there corresponds the one in which the museum was invented, to hold not only flora and fauna but also human productions. It is also the age of great cultural thefts from weak nations by the strong. Recapitulation required the theft of beautiful living monuments. From the Louvre, founded in 1791, they removed genuine producers like Fragonard and Robert to make room for the offices of curators such as Vivant Denon. The whole of the fine arts really needed management. When this same Robert drew the Louvre in ruins, was he mourning the expulsion of the artists for the benefit of the teachers of history and of the administrators? Beauty becomes no more than a heap of stone when one talks in terms of dates, references and files. This vainglorious science also killed culture stone dead.

The Great Treatises

The recognized successor of Euler and d'Alembert, Joseph-Louis de Lagrange (1736–1813) was the author of two treatises held up as models. His *Analytical Mechanics*, published in 1788, deduced from one single principle, the principle of virtual velocity, all the sciences of rest and of motion, the statics and dynamics of solids, liquids and gases. The author boasted of the complete absence of figures in his book, of never once having to have recourse to intuition. The geometer Sylvestre-François Lacroix invented, if not the thing itself, then the name at least of analytic geometry. In 1797 Lagrange wrote his *Theory of Analytic Functions*, in which he tried to introduce some order into differential and integral calculus, unfortunately around the idea of the derivative differential coefficient. The fortune of the adjective 'analytic' dates from then: to have at one's disposal a clear and perfectly mastered language which enables one to elucidate without ambiguity the questions with which one is dealing. The analytic ideal was born more or less at the same time as positivism, two very modern schools of thought 200 years old.

Each of these two great treatises twice constructs a local encyclopaedia, first of all by summing up the field with which Lagrange is dealing, but also by summarizing or assuming the whole of its history. Before constructing his building, the architect unrolls all the times which came before him, and refers, in mechanics for example, to the works of Archimedes and of his Classical predecessors such as Galileo, Stevin and Pascal. The totalization of history goes along with the totalization of knowledge. This is a specific characteristic of the period, just like the expedition of the academies to the land of the Nile. Hegel and Auguste Comte only had to copy this idea of double integration to be found in the great scientific treatises which preceded them. In science, there is only science and history. This is also a characteristic of that other universal called the university. The scientist who takes power will become the professor: a civil servant of history or of science, to the exclusion of all else.

The Three Revolutions

In 1800 Paris had hardly woken up from the myth of the Revolution before it immersed itself in the Napoleonic legend. Had France lived through a real political revolution or, quite the contrary, had it developed a concentration of power even greater than that of the royal state? This is a question still being discussed. In the same period, England was the theatre of the Industrial Revolution, after which nothing would be the same again. Can we in the same way distinguish a scientific revolution? And in what sense?

Before it was used for politics, science or industry, the word revolution concerned the heavens. In 1543 Copernicus published his *De revolutionibus caelestium libri VI*, in which he described the orbits of the planets about the Sun. When one revolution is accomplished, the heavenly bodies return to the same point, and it is impossible to tell the difference between the state of affairs at the beginning and at the end of the cycle. With this same word we say, without intending it, that France remained unchanged after the stirring and spectacular changes of the Consulate and of the Terror. Does history hold something of the reversible within the irreversibility of time?

In 1796 Pierre-Simon de Laplace published his *System of the World*, and then *A Treatise of Celestial Mechanics* between 1798 and 1825: another two great works. Let us repeat after him: the world constitutes a system, for three reasons. The first, mathematical, one might say even Euclidean, is that all figures and all real or apparent observable movements can be deduced without exception from the law of central forces, called Newton's Law. The world is a system by virtue of oneness, deduction and coherence: it follows from a principle. Laplace takes pains to demonstrate the validity of this principle in detail for local regions which seem to escape the rule, like the satellites of Jupiter or the rings of Saturn. The apparent exceptions are reduced to small-scale models of the principle itself.

The law of universal attraction now reigned without challenge. The second reason concerns determinism. Calculus deploys law in a system of differential equations (and we can notice in passing this new occurrence of the word 'system', and doubtless in the same sense) which combine together constants and variables. Before Laplace, d'Arcy had expressed doubts of the possibility of integrating these equations as soon as the planetary system had to take account of three bodies or more. A god, who quickly became famous, was brought in to define Laplace's determinism: supposing that at a given moment he knew all the parameters, then with the help of the equations he would be able to deduce all past and future positions. The world is a system first of all by mathematical deduction, and secondly because it can be known integrally.

But the third reason concerns revolution: it is enough to look at the Greek word 'planet' to recognize that there are aberrations to be seen in the heavens. Few things happen there in a regular manner; some heavenly bodies wander, others nutate, the Moon oscillates in libration; anomalies appear everywhere. How can these be deduced from a single law? By bringing them back to equilibrium. The nutations can be considered as vibrations, the wandering about as an oscillation, whose period can be calculated; some unevennesses take a year to be reabsorbed or resolved, others perhaps ten years or a century. Variations were even discovered with a period of almost a thousand years, but when everything was taken into account, stability returned. The term 'system', signifying a harmonious and concurrent ensemble of spinning-tops or wheels in individual or collective equilibrium because they spin about themselves or as part of a larger whole, was therefore entirely suited to a world invariant in its variations. Can one distinguish a difference between the state of affairs at the end of a cycle and that at the beginning? Not one. This state of affairs occurs for the nth time, has already occurred and will re-occur in exactly the same way. In this way a reversible time flows and returns, in which the past absorbs a future which one can only predict by memory. The memory of this cyclical machine is engraved in differential equations. It isn't yet a case of the eternal return, but simply an example of revolution in its first sense. The stable world turns and returns to itself, indefinitely.

Heavenly bodies, however, do not all have the same appearance. Take the Earth for example: a solid skin, mantled by the sea, wrapped round with an airy scarf. An absolutely certain mechanics held good for the first of these materials; at this period, the tides could be explained by harmonic equations with partial derivatives; but of gases, nothing was known. Fire, however, brought the three states together under its law, liquefying solids and evaporating liquids. In the preface to his *Analytical Theory of Heat* (another great

435

treatise), Fourier (1768–1830) writes that nothing escapes from heat because every substance contains it, receives it and diffuses it, and that it is therefore as universal as gravitation. He was right: before Laplace, the whole of science was presided over by Newton, because of the importance of force; after Fourier, it was dominated by clearly thermal phenomena, as was civilization. There was a shift from the watch to the boiler, if we take these two as cultural models. This is revolution in the second sense of the word: a break, a new state of affairs, not at all a moment in a repetitive cycle.

But the first sense always threatens to win out over the second, and in the following way. In note VII to *The System of the World*, Laplace ends his cosmology[3] with a cosmogonic[4] hypothesis. First, he objects to his previous reductions to equilibrium by raising certain points, some of a general and some of a residual order, where no compensating symmetry can be found. The stars and satellites move or spin from west to east, without any compensating translation or rotation from east to west. Although it is a weak effect, the eccentricity of orbits does exist, the central forces do not act at the centre and these elliptical orbits occur in planes all at slightly different inclinations. These are uncompensated discrepancies.

To understand them, Laplace changes time and leaves the closed stability of the system. Yes, the world vibrates and returns, but this oscillation is the result of a history. Let us note once again, in passing, the obligatory association of science and history. We had forgotten that the Sun lies at the focus, a word which signifies centre, but also flame. Here is fire again. In the beginning, everything was burning. In the centre there is also the origin. Everything begins with a hot and volatile nebula, turning in a spiral, as if Descartes came before Newton, as if turbulence comes before attraction. This magma cools with the passage of a new time, irreversible, along whose path nothing may return. Everything becomes more and more cold, nothing can reheat without an external input. This slow cooling produces the planets, detached from the mass, their conditions changed. Reversible time organises the symmetries to be seen in the present state of affairs, while the discrepancies can be explained by directional time. This is how these cosmological exceptions imply a cosmogony.

This is how the discrepancies, the slight asymmetries in the global symmetries of space lead to a new, irreversible, directional time. One might say a time exposed, in its etymological sense . . . Is this turbulence of the forces and of fire, this directionality, a structure common to time and to space? One might say that all things which become pendant rest initially and fundamentally on something inclined, as Lucretius said. Left and right warp space and begin time.

In one of the papers which conclude his *Elements of Statics*, Louis Poinsot (1777–1859) finishes Laplace's demonstration of the stability of the world, referring the whole of the solar system to a fixed plane, which one might almost call eternal, at the centre of which there acts the general couple which produces all the movements of the system, the wheel which moves all others. But what is a couple: two equal forces applied in opposite directions to the two ends of a rigid segment. Here again, in the final analysis or the beginning of a series, we again have an operator both symmetrical and asymmetrical, that is to say, directional. When in the *Metaphysics* Aristotle speaks of an unmoved mover, is this oxymoron[5] too intended to refer to directionality?

The new time passes without returning upon itself: this is the revolution which does not return to its starting-point. One has to keep both meanings in view: reversible and irreversible. The circle of the law of forces rolls along the straight line of the law of heat. The circle of cosmology rolls along the curve of cosmogony. This means that the stable world none the less continues its history. It is eternal but becoming. Will this tension be re-absorbed, will one side win out over the other?

For the Laplacean changed conditions inaugurate or imply a time, itself inclined, which may find its re-equilibration. The oceans rub against the skin of the Earth, the space traversed by the heavenly bodies is scattered with tiny particulate obstacles, and all of this exercises a braking effect. Little by little the movements slow down, bodies fall back towards the still flaming centre and the melting-pot begins anew. In its end state, the new nebula is exactly like the original, and we are back with the eternal return rediscovered by Auguste Comte in extrapolating Laplace, after Kant and long before Nietzsche. This is too easy to do in cosmology, where one only had to think of the cycles of the eclipses or of comets or just the Moon, but extraordinary in cosmogony because it takes reversibility from irreversible fire.

Should the directionality of the origin be seen as a kind of fossil? Chateaubriand wanted the world to be born old: already in ruins on the first morning. Is the starting-point the completion of a circle?

System and Evolution

Two scientific revolutions governed Paris around 1800: taking 'revolution' in in its second and its first sense, as a line or a circle. In mathematics, mechanics, astronomy, physics and chemistry, Lagrange, Laplace, Fourier, Lavoisier and Lamarck devoted themselves to recapitulation, this universal reassessment which we have already seen in politics. Each great scientist creates for his field a great system, universal in its way. We could call it x-ology: cosmology, thermology . . .

Each great treatise constructing the system begins, however, with a great preface which tells of everything that has gone before. Science has a history, like the world; it is a system in a state of becoming. Lagrange takes up the work of Archimedes, Stevin, Galileo and Pascal and the statics–dynamics opposition, whose synthesis he produces in his own work. Hegel will be able to translate, or rather, re-copy. Laplace perfected and Fourier transcended Newton's universality. The circle of the system rolls on the straight line of history. The encyclopaedic circle of circles rolls along the road of time, combining the reversible with the irreversible. We can look in a new way at Napoleon embarking the whole of knowledge onto a boat, in the shape of the academies; the scientific system was returning to its genesis, towards its Egyptian origin. Suddenly, history begins to make the same gestures as the sciences themselves, as if it was acquiring the same universality.

This said, it is the sciences themselves, not their history, their preface, which are on the irreversible road of time, along which there is no going back. Let us call them x-ogonies, like cosmogony. These days, geology and biology are wrong: they should clearly be geogony and biogony. How and where are they going, the Earth and its living creatures? Where do they come from, and where are they going?

Lamarck

One example from among thousands. Vertebrates have two eyes: the mole has tiny ones, hardly visible. The aspalax, a Persian mole, has none, no more than *Proteus*, a little aquatic reptile, which has taken refuge in deep waters, underneath the earth. Vertebrates have teeth, but not the whale with its flexible baleen, nor the anteater with its sticky tongue or the birds with their horny beaks. There are exceptions to certain laws. Others have none: all vertebrates have ears. Sound passes obstacles which bring light to a halt.

Lamarck said that life progressed according to an overall plan: irreversibly, time composes, complicates and perfects. Here and there external causes affect the execution of this plan, without actually destroying it, whence the gaps in the series, the blind moles, the hens without teeth. These causes arise from circumstances: climate, milieu, soil and weather, in other words the concrete, which resists, like a tangled chaos, the irresistible advance of the single plan, or as one might almost say, of the spirit or vital force. So it is necessary to rethink change as variegated and disrupted; Lamarck sees this as a second law which cuts across the overall plan, the first law. Living beings are on the move, even plants: they adapt. Needs change with the environment, new habits appear which, after a very long time, alter the physical organs. Here Lamarck is putting forward what is called the transformationist hypothesis, a favourite with the peasantry, while the Darwinian hypothesis tends to attract the scientists. You therefore find hidden teeth in the foetus of the whale and grooves which might have held them in the bird's beak. Under the skin of the aspalax and of *Proteus*, the vestiges of eyes are to be found: all marks of the straight line in dizzy Nature, traces of the universal in the singular. The term biogony is more applicable to such an evolution, as it is to Darwin's, than is the word biology, which remains the classical term despite the misconstruction.

The Irreversible

The revolution which returns upon itself makes for systems while the other sometimes dissolves into continuity: both senses were present in Paris around 1800. Let's move on to the second.

The mathematics laid down by Lagrange, Laplace, Lacroix, Monge and Comte now seem to us to be rather concrete and applied. Geometry describes forms and figures, calculus brings measurement to better approximations – this is theory for the ex-pupils of the Polytechnique. The French School had a horror of abstraction. No logic, despite the heritage of Condillac: it would come later, and what's more, only with Boole. Very little arithmetic, no theory of number, in which the German School of the nineteenth century would renew the work of Euler. No pure algebra, except the theory of equations. Our mathematics would be founded a little later by Abel and Galois, who practically forgot all their predecessors. The whole of mathematics was traversed by a deep gulf which separated the French style, highly descriptive and practical, from the theories and formalisms which were to make the mathematical future. The great treatises suddenly seem like tombstones.

It was the same in astronomy. Once the vision of the world offered by Laplace and Poinsot had been perfected, it remained only to step out into the Universe, beyond the solar system. Balzac's friend Savary showed that certain stars are binary. An extraordinary fault line separates Borda, Lalande and Cassini, astronomers of the Age of Enlightenment, from Messier, whose catalogue lists the blurred patches scattered randomly across the sky like exceptions, but whose organization would become the main preoccupation of modern astronomy: the galaxies. Our astrophysics is as detached from the world as astronomy in Paris in 1800 was cut off from the Universe. Here again, there is a clean break: the men of the Revolution were conservative scientists.

Before this period, physics was no more than geometry and mechanics: it described and explained figures, movements and forces. The optics which triumphed in the Classical period, until Maupertuis, is an excellent example. Electricity hardly went any further than magnetism, which with Charles Coulomb, was precisely reduced to Newton's Law. The latter triumphed on the macro-level with Laplace, and in close interactions in the attraction of magnetic masses. It was always mechanics, even in acoustics. What we call physics today didn't begin in the late Renaissance with Galileo and Bacon, but with Fourier's study of the phenomena of heat in both theory and experiment. Little by little it left geometry and dynamics behind and began to express its own originality. When Sadi Carnot founded thermodynamics, an even more revolutionary science in the second sense, this was as profound a break from Fourier as was Fourier's own in relation to the physicists who had preceded him. It wasn't any longer a question of dealing with the propagation of heat, but of treating heat itself as a motive power. Here is the intersection between scientific and industrial revolutions. Modern and contemporary physics were born here, together with our hot civilization. Irreversible in its own time, through its theory heat introduces a universal fault-divide in the time of history, in the history of technology and in the history of science. Bernard Belidor, a typical engineer before 1800, dealt exclusively with hydraulics and with architecture: no engineer, after Carnot, would be interested in fluids except in fire-powered machines.

The life sciences fit so well into this double model of x-ology and x-ogony that one might think it had been taken from them. It has been said that the nineteenth century thought above all about history and about time. If one thinks of the nomenclatures, taxonomies and systematizations of the Enlightenment, one will get some idea of what I have called recapitulation: in the age in which museums were founded, and Vivant Denon returned from Egypt, the Royal Zoo and the Museum of Natural History accumulated the living creatures of the earth, the flora and fauna. This is what history now described and it attempted to explain the advance, the progress, the transformation of living beings. Buffon, Geoffroy Saint-Hilaire, Cuvier and Lamarck in particular, took part, before Darwin, in this extraordinary adventure which changed forever the way in which we look at the world. Another universal is responsible for the synthesis of living beings; not system, not classification, but development. Another irreversible time arrived, progressive and contrary to the reversible time of previous systems; what's more, it's contrary to the irreversible time of thermodynamics. Our thought still moves in this contradiction.

A quarter of a century later, Auguste Comte depicts the particular panorama of Paris in 1800 in the universal panorama of science and history, in which he bases his philosophy on two elements, and two elements alone: the classification of knowledge, system and recapitulation, college and museum, that is to say, gnoseology (see p. 117), and the law of the three states, the history of the human spirit, progressive and transforming, which is gnoseogony. These two words say everything. His own revolution consists in the invention of this new science: sociology. And that indeed is the sociology of the period?

The New Clergy

Around the mid-century, Turgot's teacher, Condorcet, gave a lecture at the Sorbonne, entitled 'Tableau philosophique des progrès scientifiques de l'esprit humain'. Twenty years later, at the Academy of Science, his disciple, friend and successor planned a 'Tableau historique de l'avance de l'esprit humain dans les sciences'. He abandoned it temporarily, but a few months before dying, hiding at Mme Vernet's, condemned to death, he wrote his *Esquisse d'un tableau historique des progrès de l'esprit humain.*

Through irresistible onward movement or progress, in these three texts history and its time flow irreversibly in a straight line – in the last of them in ten periods, for the metric system had just been established. Luckily, the historical descriptions here are no more than sketches, and one doesn't have to look too closely. One often comes across the word *esprit*, meaning mind or spirit, whose good fortune, from Hegel to Bachelard, has been inversely proportional to the clarity of its meaning. Whether it is a question of the human or the scientific spirit, whether in the history of science or in universal history, one can only feel that the word represents a hardly hidden secularized translation of Holy Spirit. When the father has been killed, and one refuses to incarnate oneself as the son, then there remains only the speaking owl, light as a dove, the symbol of Minerva. Its exalted flight allows history to be considered from an altitude great enough to enable one to ignore the details.

Uni-directional time was born with Christianity, which was orientated towards final salvation. After Saint Augustine, who brought together Plato's intuitions and the teachings of the prophet-authors of Israel, history was carried away by its irreversible momentum. In Greece, the demonstration of irrationality (see p. 107) had broken the circle of the Eternal Return, and in Israel a unique and transcendent God had revealed himself in the immanent history of his chosen people. Judaeo-Greek, like its founder's name, Christianity initiates the linear time of theological narrative at the Augustinian confluence of the two sources. Now Christianity is still the only culture within which all the sciences have been developed. In it religion secularizes religion, allowing the sacred to move elsewhere. In the seventeenth century Pascal, a philosopher and a Christian, and surely the greatest scholar of his age, maintained communication between the two, for he thought of the time of salvation and the continual progress of the sciences together. Condorcet published an edition of his work in 1776. It should be noticed in passing that the progressive nature of time is constituted by a historical schema which presupposes it.

In the sciences the accumulation of knowledge is subject to experiment or demonstration, whereas everywhere else it remains a matter of belief. This distinction hasn't changed, except that the meaning of the last word has varied. Instead of believing in history because one believes in God, or rather in Jesus Christ, we believe in history by extrapolating from what goes on in the sciences. For the only proof we have that there is a history is the history of the sciences. As for the rest, everything is doubtful. It was God, infinite reason, who guaranteed progress towards salvation; by virtue of its results, its conquests and its triumphs, science alone, the exercise and store of reason, now guarantees that such a thing as progress actually exists. The structure hasn't changed. God is dead, but is resurrected. Science progresses and, in the general advance of spirit/mind, it is the foundation of belief.

Condorcet's naïvety doesn't go so far as to believe enough to make us believe that everything is always for the best. Progress is braked and delayed by obstacles and setbacks. Just as Gavroche before the firing-squad blamed Voltaire and Rousseau until the bullets put a stop to the words in his throat, so Condorcet with his head almost under the blade, never ceased to insist, in his swansong, on the crimes of the priests and the monks. The street urchin and the philosopher of the French language lived the labour of the negative not in spiritual speculation but in the death-agony of the body. The *Sketch* has a dialectic because in it the religious works as a negative and sometimes reverses itself by working for global progress: from Egypt and India, for example, the priests pass on to Pythagoras the model of the Universe. But Jupiter's fault is general, because trans-historical. I repeat, Jupiter's. Hegel gives the same role to Mars, and Marx gives it to Quirinus. With the help of Dumézil,[6] who gave these three gods a role as totems of our classes or social functions, we can understand these dress rehearsals, these vast verbal manoeuvres extended to the whole of history. Such a gigantic historical schema, extending from the very beginning to the present day, which one refrains from recommending to novice students, lest it lead to the abandonment of all intellectual modesty, seems less naïve when one plunges among the clashing swords of the struggle for mastery or into the manual concrete of productive forces, but more naïve however when one rises towards the empyrean of ideas, when it shifts and finds triple expression in the Indo-European symbols or functions: Jupiter, Mars and Quirinus. Beginners are unable to reproduce a geometrical proof as soon as the letters are changed on the figure: substitute J for M and M or J for Q, and vice versa, in theories of history, and they are no longer similar but exactly the same. One can say anything and prove anything about long-term prospects, when nothing can be falsified. No doubt anything can be proved in history too.

In Condorcet it is Jupiter himself who says he is at fault. The god of gods presents himself as double, king and priest, legislator and potentate. A half-function battles against its symmetrical opposite. Philosophy struggles against the hypocritical superstition maintained throughout history by all the priesthoods of the Earth, and 'this war will continue as long as there are priests and kings.' The scientific function and the scientific class fight against the legal and ritual functions incarnated in the same god or the same image, in the same class. What in fact is the difference between Voltaire and a king or a guru, between Jean-Jacques Rousseau and a pastor? Condorcet the mathematician kills the old pupil of the Jesuits who had taught him mathematics, Condorcet the republican condemns the Marquis within himself, the egalitarian philosopher assassinates the Academician, the Convention arrests this member of the Convention, whose colleagues – the identical ones – will celebrate a triumph for his book only a few months after voting for the death of the author himself. Yes, Condorcet committed suicide and his book resembles his death. Everything happens in such close proximity that it's very difficult to see the differences. The other in the same remains so infinitely close that one might speak of a subtle and continuous dialectic borrowed from the infinitesimal calculus, in which the principle of the excluded middle no longer holds good.

Arrested in the street on the eve of his death, Condorcet, on being questioned, said that he was called Pierre Simon, again twice the same name in a religious context, that is to say on the side of Jupiter; a miserable disciple on the roads of Galilee, with the sinners, the publicans and the prostitutes, flesh and rock, unbaptized and rebaptized by the Word, betrayer, martyr, Pope, first head of the Church and of the whole uni-directional history, almost part of himself, here is Simon-Peter, a double Jupiter, Jew and Roman, like our double Condorcet. History and the Revolution, in Paris in 1800, unfold in the conflict of Jupiter with himself: in the spatio-temporal system of the universe, Pierre Simon Laplace replaces God the creator with the secular god of determinism.

It is still a question of secularizing, but in such a way that one isn't sure who to thank for it: those who converted scientific knowledge into a religion or the religion itself which allows such a translation and such an abandonment of the religious. In the seventeenth century the same symmetry, in Pascal, for example, remained orientated towards the religious, whereas here the orientation is reversed towards science. Just as Pascal, the practitioner of calculus and theorist of probability, humiliates the progress he has just invented, so Condorcet, practitioner of probability and theorist of calculus, humiliates those who have humiliated him. Grandiose identitities are only very subtly different. We ourselves, today definitively secularized, think that this theoretical history, in the literal sense, that is to say, describing a parade or irreversible procession, whether continuous or discontinuous, is only another name for the one Judaeo-Christian God, preceded by his Spirit. So that everything still rests on the function of Jupiter, which we will see that only Auguste Comte was to understand. That's the history: now for the sociology.

Before the French Revolution, the whole field of scholarship and the

443

education of children was more or less the preserve of the Catholic clergy. During the eighteenth century, they even provided the majority of the scientists and philosophers who played their part in the French Enlightenment and collaborated in Diderot and d'Alembert's *Encyclopédie*. The alliance and division between the temporal power, the economy, the army and the institutions, and the spiritual was advantageous to the clergy, as it was later to scientific power, its truly legitimate successor.

When the States General met, scientists were to be found in each of the three classes of the nobility, the clergy and the third estate, but particularly in the second of these. During the Revolution and the Empire, a long series of decisions first of all dissolved the academies and then established the public education system, the responsibility of the State. There now appeared the central schools, the future lycées, and the École Normale and the Polytechnique, but also the Bureau des Longitudes and the Bibliothèque Nationale, the embryos of secondary and higher education and the first research institutions. As for content, these new plans switched attention from the humanities, thought of as having been done to death, towards science and experimental science in particular. The Revolution distrusted the abstract, which, contrary to what is often said, has never had a good reputation in France. The decay of science in the next century followed from the importance given to the Polytechnique and its ideas. The old spiritual power began to lose some of its functions and its dignity, now taken on by a new institution which has never yet been named, while there began an interminable war, pitting teachers against priests, which would last at least 200 years. This was particular to France: disliking their children, they preferred their dogs and civil war.

Classes and Places

The clergy was not a class in the strict sense of the term, but rather an order, divided into two parts, the higher and the lower clergy, one connected to the dominant class and the other to the dominated. It fluctuated between the two in a way that arose from the mixture of the spiritual with the temporal, producing a play of commitment and detachment. It was rich, owning lands, estates and buildings, and it ran important, numerous, well-organized and influential institutions. It made itself credible through its language, whose truth it guaranteed, through time, whose continuity it claimed it ensured and through the immortal beauty of the culture which it inculcated. In return it justified a society closed in on itself by giving it a foundation outside; we shall see how.

Scientists and teachers do not form a class, but rather constitute an order, divided into two parts, from nursery school to the Collège de France. It fluctuates in a relationship which allows a certain independence with regard to the dominant power to which, however, it remains tied. One destroys whatever occupies the place one will take; here it is the territory of the old clergy under the *ancien régime*, enjoying the same alliance or division with the armed forces, the economy and political power, awarding them and itself the same credibility, through the truth of its speech and the continuity of time. Only beauty is absent from the balance. In his *Génie du christianisme*,[7] Chateaubriand derives all the force of his argument from his last point.

The French Revolution initiated this transfer of power. The same men, sometimes without moving an inch, saw the iceberg tip over. The ideocracy[8] didn't change its place or its function, sometimes not even its personnel: only its name was different. But desiring a lack of mysteriousness which could only make it more powerful, it didn't care to find a new one, and so no one now recognizes that clergy in this one, the residue of the longest history and the most ancient religions. The anthropology of science is able to answer the question: what are the scientists, what do they do, considered as a whole? Anthropology has known the answer since the French Revolution, during which they took all the positions previously held by the clergy.

Let us consider, for example, the places occupied by people and their respective distributions. The Church had a hierarchy of cardinals, canons and curates organized around a single pole: Lazare Carnot was like Richelieu or Mazarin, while the senators Laplace, Lagrange, Monge and Fourier, the prefect and the baron, were aristocratic priests, with Arago an ambitious young academician, stirring about like Aramis: between them they centralized France and science as well, dominating from their high rank those who took the place of the lower clergy, the calculator Alexis Bouvard, for example, and a thousand forgotten assistant masters, except for the sombre Auguste Comte, whose isolation and alienation in every possible way certainly enabled him to see science as a whole, as well as the religious *coup d'état* it was carrying out without perhaps even knowing it. The Church excluded heretics: Eugène Cauchy was exiled and travelled in Europe, banished from his own country. A recent plaque, in 1985, encumbered him with the shameful title of 'legitimist mathematician', as if his politics could be seen in his infinitesimal calculus. In the Academy, Condorcet condemned the false science of Marat, who might have taken himself for Goethe and Galileo rolled into one. This religion had its Inquisition: Fourcroy, Berthollet and Guyton de Morveau, chemists close to Lavoisier, who was waiting for the scaffold, went to his house armed with a warrant and took away his apparatus. Condorcet, of course, died in flight or took his own life to escape the guillotine. On the other hand, Lakanal, the reformer of education, died in poverty in the United States, dismissed from his post at the Restoration. The Wars of Religion continued on other terrain, or perhaps the same one. Scientists do not organize as a city, as has been sometimes said, nor as a republic, but as a church. This forgets its saints, victimized during their lives, but canonizes them after their death, so as to survive thanks to them. Évariste Galois,[9] who was often imprisoned for his revolutionary tendencies, was killed in a duel, and Sadi Carnot[10] died mad in the asylum at Charenton: both died very young, before their canonization as heroes and forerunners in algebra or thermodynamics. We have seen, or shall see, the cult devoted to Lavoisier very shortly after his execution. How many founding messiahs became victims in these turbulent times? If we put the big words which cast a shadow, like 'science' and 'religion', into parentheses for a moment, we see that on the whole the scientific and intellectual landscape was hardly changed at all by political change, remaining invariant through subtle variations. Other individuals, or sometimes the same ones, impelled by motivations new or old, occupied places furnished in a millenarian way, from which they spoke of truth, of time and of history, exactly the same things, but in a new language copied from the old.

The same temporal anchorages in institutions and buildings were established: the École normale, the Polytechnique, the Salpêtrière hospital, the Museum of Natural History and the Botanical and Zoological Gardens. Very soon the whole of the eastern side of the hill of Sainte-Geneviève, around the Panthéon, the resting-place of beatified genius, belonged to science – to whom, in reality? – in the same way as it depended on a religious institution before. The regular clergy devote themselves to research while the secular ones occupy themselves with education. There was the same closeness to political power: the scientists threw themselves into the Revolution with enthusiasm. Lazare Carnot organized for victory, the Marquis de Condorcet got himself elected to the Legislative Assembly and the Convention. They lived in palaces, Lagrange at the Louvre, Laplace at the Luxembourg, where he was the guardian of the young Cauchy. They were courtly, aristocratic priests and cardinal ministers.

Distance allows us to see that one order was substituting itself for another because the passage of time renders certain ideas ineffective and others take over like new troops, fulfilling the same social function. The story changes its subject and concerns the discovered rather than the revealed, but it goes on in the same way. From Saint Augustine or Bossuet to Condorcet or Hegel, there is the same prophetic ascension to a spirit, which shame or modesty prevents from being compared to the dove. Images are secularized by disparaging – and reconstructing – them.

Positivism and its End

Auguste Comte's philosophy draws the lessons of the French Revolution in terms of these two components, science and history. Soon there will be only these two sciences and these two philosophies: one can give oneself

PREMIER MOIS. MOÏSE. LA THÉOCRATIE INITIALE.	DEUXIÈME MOIS. HOMÈRE. LA POÉSIE ANCIENNE.	TROISIÈME MOIS. ARISTOTE. LA PHILOSOPHIE ANCIENNE.	QUATRIÈME MOIS. ARCHIMÈDE. LA SCIENCE ANCIENNE.	CINQUIÈME MOIS. CÉSAR. LA CIVILISATION MILITAIRE.	SIXIÈME MOIS. SAINT-PAUL. LA CATHOLICISME.	SEPTIÈME MOIS. CHARLEMAGNE. LA CIVILISATION FÉODALE.
1 Prométhée.	Hésiode.	Anaximandre.	Théophraste.	Miltiade.	Saint-Luc. . . . . . Saint-Jacques.	Théodoric-le-Grand.
2 Hercule. . . . . Thésée.	Tyrtée. . . . . Sapho.	Anaximène.	Hérophile.	Léonidas.	Saint-Cyprien.	Pélage.
3 Orphée.	Anacréon.	Héraclite.	Érasistrate.	Aristide.	Saint-Athanase.	Othon-le-Grand. Henri-l'Oiseleur.
4 Ulysse.	Pindare.	Anaxagore.	Celse.	Cimon.	Saint-Jérôme.	Saint-Henri.
5 Lycurgue.	Sophocle. . . . . . . Euripide.	Démocrite. . . . . . Leucippe.	Galien.	Xénophon.	Saint-Ambroise.	Villiers. . . . . . . La Valette.
6 Romulus.	Théocrite. . . . . . . Longus.	Empédocle.	Avicenne. . . . . . Averrhoès.	Phocion. . . . . . Épaminondas.	Sainte-Monique.	Don Juan de Lépante. Jean Sobieski.
7 NUMA.	ESCHYLE.	THALÈS.	HIPPOCRATE.	THÉMISTOCLE.	SAINT-AUGUSTIN.	ALFRED.
8 Bélus. . . . . Sémiramis.	Scopas.	Solon.	Euclide.	Périclès.	Constantin.	Charles-Martel.
9 Sésostris.	Zeuxis.	Xénophane.	Aristée.	Philippe.	Théodose.	Le Cid. . . . . Tancrède.
10 Menou.	Ictinus.	Empédocle.	Théodose-de-Bithynie.	Démosthènes.	Saint-Chrysostôme. Saint-Basile.	Richard. . . . . . . Saladin.
11 Cyrus.	Praxitèle.	Thucydide.	Héron. . . . . . . Ctésibius.	Ptolémée Lagus.	Sainte-Pulchérie. . . Marcien.	Jeanne-d'Arc.
12 Zoroastre.	Lysippe.	Archytas. . . . . . . Philolaüs.	Pappus.	Philopœmen.	Sainte-Geneviève-de-Paris.	Albuquerque. . . Walter Raleigh.
13 Les Druides. . . . . Ossian.	Apelles.	Apollonius de Tyane.	Diophante.	Polybe.	Saint-Grégoire-le-Grand.	Bayard.
14 BOUDDHA.	PHIDIAS.	PYTHAGORE.	APOLLONIUS.	ALEXANDRE.	HILDEBRAND.	GODEFROI.
15 Fo-Hi.	Ésope. . . . . Pilpaï.	Aristippe.	Eudoxe. . . . . . . Aratus.	Junius-Brutus.	Saint-Benoît. . . Saint-Antoine.	Saint-Léon-le-Grand. . Léon IV.
16 Lao-Tseu.	Plaute.	Antisthènes.	Pythéas. . . . . . Néarque.	Camille. . . . . . Cincinnatus.	Saint-Boniface. . . Saint-Austin.	Gerbert. . . . . . Pierre Damien.
17 Meng-Tseu.	Térence. . . . . . Ménandre.	Empédocle.	Aristarque. . . . . Bérose.	Fabricius. . . . . . Régulus.	St-Isidore-de-Séville. St-Bruno.	Pierre-l'Ermite.
18 Les théocrates du Tibet.	Phèdre.	Zénon.	Érastosthène. . . . Sosigène.	Annibal.	Lanfranc. . . . Saint-Anselme.	Suger. . . . . . . Saint-Éloi.
19 Les théocrates du Japon.	Juvénal.	Cicéron. . . . Pline-le-Jeune.	Ptolémée.	Paul-Émile.	Héloïse. . . . . . . Beatrice.	Alexandre III. . . Thomas Becket.
20 Manco-Capac. . . . Tamehameha.	Lucien.	Épictète. . . . . Arrien.	Albategnius. . . . Nassir-Eddin.	Marius. . . . . . Les Gracques.	Les archiî. du moyen âge.S.-Benezet.	St-François-d'Ass. . St-Dominique.
21 CONFUCIUS.	ARISTOPHANE.	SOCRATE.	HIPPARQUE.	SCIPION.	SAINT-BERNARD.	INNOCENT III.
22 Abraham. . . . . . . Joseph.	Ennius.	Xénocrate.	Varron.	Auguste. . . . . . Mécène.	Si-François Xav. Ignace de Loyola.	Sainte-Clotilde.
23 Samuel.	Lucrèce.	Philon d'Alexandrie.	Columelle.	Vespasien. . . . . Titus.	St-Charles-Borrom. Fréd. Borrom.	Ste-Bathilde. Ste-Math.-de-Toscane.
24 Salomon. . . . . . David	Horace.	Saint-Jean-l'Évangéliste.	Vitruve.	Adrien. . . . . . . Nerva.	Ste-Thérèse.Ste-Cather.-de-Sienne.	Ste-Étienne-de-Hong. Math.Corvin.
25 Isaïe.	Tibulle.	Saint-Justin. . . . Saint-Irénée.	Strabon.	Antonin. . . . . . Marc-Aurèle.	Si-Vinc.-de-Paule. L'abbé de l'Épée.	Sainte-Élisabeth-de-Hongrie.
26 Saint-Jean-Baptiste.	Ovide.	Saint-Clément-d'Alexandrie.	Frontin.	Papinien. . . . . . Ulpien.	Bourdaloue. . . . Claude Fleury.	Blanche de Castille.
27 Haroun-al-Raschid. Abdérame III.	Lucain.	Origène. . . . . . . Tertullien.	Plutarque.	Alexandre-Sévère.	W. Penn. . . . . . . G. Fox.	St-Ferdinand III. Alphonse X.
28 MAHOMET.	VIRGILE.	PLATON.	PLINE-l'ancien.	TRAJAN.	BOSSUET.	SAINT-LOUIS.

HUITIÈME MOIS. DANTE. L'ÉPOPÉE MODERNE.	NEUVIÈME MOIS. GUTEMBERG. L'INDUSTRIE MODERNE.	DIXIÈME MOIS. SHAKESPEARE. LE DRAME MODERNE.	ONZIÈME MOIS. DESCARTES. LA PHILOSOPHIE MODERNE.	DOUZIÈME MOIS. FRÉDÉRIC. LA POLITIQUE MODERNE.	TREIZIÈME MOIS. BICHAT. LA SCIENCE MODERNE.
1 Les Troubadours.	Marco-Polo. . . . . . Chardin.	Lope de Vega. . . . . Montalvan.	Albert-le-Grand. Jean de Salisbury.	Marie de Molina.	Copernic. . . . . Tycho-Brahé.
2 Bocace. . . . . Chaucer.	Jacques Cœur. . . . Gresham.	Moreto. . . . . Guilen de Castro.	Roger Bacon. . . Raimond Lulle.	Côme de Médicis l'ancien.	Kepler. . . . . Halley.
3 Rabelais.	Gama. . . . . Magellan.	Rojas. . . . . Guevara.	Saint-Bonaventure. . . Joachim.	Philippe de Comines. Guicciardini.	Huyghens. . . . . . . Varignon.
4 Cervantes.	Neper. . . . . Briggs.	Otway.	Ramus. . . Le cardinal de Cusa.	Isabelle de Castille.	Jacques Bernouilli, Jean Bernouilli.
5 La Fontaine.	Lacaille. . . . . . . Delambre.	Lessing.	Montaigne. . . . . . . Érasme.	Charles-Quint. . . . Sixte-Quint.	Bradley. . . . . Roëmer.
6 Foé. . . . . Goldsmith.	Cook. . . . . Tasman.	Goëthe.	Campanella. . . . . . Morus.	Henri IV.	Volta. . . . . Sauveur.
7 ARIOSTE.	COLOMB.	CALDERON.	SAINT-THOMAS-d'AQUIN.	LOUIS XI.	GALILÉE.
8 Léonard de Vinci. . . Le Titien.	Benvenuto Cellini.	Tirso.	Hobbes. . . . . . . Spinosa.	Coligny. . . . . . . L'Hôpital.	Viète. . . . . Harriott.
9 Michel-Ange. . Paul Véronèse.	Amontons. . . . . . Wheatstone.	Vondel.	Pascal. . . . . Giordano Bruno.	Barneveldt.	Wallis. . . . . Fermat.
10 Holbein. . . . . . Rembrandt.	Harrison. . . . . Pierre Leroy.	Racine.	Locke. . . . . . . Mallebranche.	Gustave-Adolphe.	Clairaut. . . . . . . Poinsot.
11 Poussin. . . . . Lesueur.	Dollond. . . . . Graham.	Voltaire.	Vauvenargues. . Mme de Lambert.	De Witt.	Euler. . . . . Monge.
12 Velasquez. . . . Alonzo Cano.	Arkwright. . . . . . . Jacquart.	Alfieri. . . . . Métastase.	Diderot. . . . . Tracy.	Ruyter.	D'Alembert . . Daniel Bernouilli.
13 Téniers. . . . . Rubens.	Conté.	Schiller.	Cabanis. . . . . Georges Leroy.	Guillaume III.	Lagrange. . . . . Joseph Fourier.
14 RAPHAEL.	VAUCANSON.	CORNEILLE.	Le Chancelier BACON.	GUILLAUME-le-Taciturne.	NEWTON.
15 Froissart. . . . . . . Joinville.	Stévin. . . . . Torricelli.	Alarcon.	Grotius. . . . . Cujas.	Ximenès.	Bergmann. . . . . . . Scheele.
16 Camoens.	Mariotte. . . . . Boyle.	Mme de Motteville. Mme Roland.	Fontenelle. . . . . Maupertuis.	Sully. . . . . Oxenstiern.	Priestley. . . . . Davy.
17 Les Romancistes espagnols.	Papin. . . . . Worcester.	Mme de Sévigné. Lady Montague.	Vico. . . . . Herder.	Colbert. . . . . . . Louis XIV.	Cavendish.
18 Châteaubriant.	Black.	Lesage. . . . . Sterne.	Fréret. . . . . . Winckelmann.	Walpole. . . . . . . Mazarin.	Guyton—Morveau. . . Geoffroy.
19 Walter-Scott. . . . . Cooper.	Jouffroy. . . . . . . Fulton.	Mme de Staal. . Miss Edgeworth.	Montesquieu. . . d'Aguesseau.	D'Aranda. . . . . . Pombal.	Berthollet.
20 Manzoni.	Dalton. . . . . Thilorier.	Fielding. . . . . . Richardson.	Buffon. . . . . Oken.	Turgot. . . . . . Campomanes.	Berzélius. . . . . . . Ritter.
21 TASSE.	WATT.	MOLIÈRE.	LEIBNITZ.	RICHELIEU.	LAVOISIER.
22 Pétrarque.	Bernard de Palissy.	Pergolèse. . . . . . Palestrina.	Roberson. . . . . . Gibbon.	Sidney. . . . . . . Lambert.	Harvey. . . . . . . Ch. Bell.
23 Thomas A'Kempis. Louis deGrenade.	Guglielmini. . . . . . Riquet.	Sacchini. . . . . . . Gretry.	Adam Smith. . . . . Dunoyer.	Franklin.	Boërhaave. . . . . . . Stahl.
24 Mme de Lafayette. . Mme de Staël.	Duhamel (du Monceau).	Gluck. . . . . Lully.	Kant. . . . . Fichte.	Washington. . . . . Kosciusko.	Linné. . . . . Bernard de Jussieu.
25 Fénelon. Saint-François-de-Sales.	Saussure. . . . . . Bouguer.	Beethoven. . . . . . Handel.	Condorcet. . . . . Ferguson.	Jefferson.	Haller. . . . . . . Vicq-d'Azyr.
26 Klopstock. . . . . . Gessner.	Coulomb. . . . . . . Borda.	Rossini. . . . . . . Weber.	Joseph de Maistre. . . Bonald.	Bolivar. . . Toussaint-Louverture.	Lamarck. . . . . . . Blainville.
27 Byron. . . . . Élisa Mercœur.	Carnot. . . . . . . Vauban.	Bellini. . . . . . . Donizetti.	Hegel. . . . . . Sophie Germain.	Francia.	Broussais. . . . . . Morgagni.
28 MILTON.	MONTGOLFIER.	MOZART.	HUME.	CROMWELL.	GALL.

Jour complémentaire. . . . . **Fête universelle des MORTS.**

Jour additionnel des années bissextiles. . . . . . { **Réprobation solennelle des deux principaux rétrogradateurs** (Julien et Bonaparte), mais seulement pendant la première demi-génération.

Après ces quatre célébrations initiales de la Fête des Réprouvés, ce jour exceptionnel prendra sa destination normale pour le culte abstrait.

The Positivist Calendar: the only tolerant and non-exclusive calendar in history none the less excluded Jesus Christ and any representative of the modern arts to give space to the triumphant modern sciences. (From A. Comte, Catéchisme Positiviste, 1852. Bibliothèque Nationale, Paris.)

447

either to science or to history. The university, which holds the monopoly on the definition of intelligence and its contents will soon recognize no other rational pursuits.

Positivism's successors argued for a long time over the meaning of the word, which they inflected towards experience or towards language, separating themselves into distinct sects. They were all scientists, and all agreed with the encyclopaedia produced by the master, but they joined with his enemies in condemning the delirium which in their eyes spoilt the latest developments in his theory of history and his sociology: science the religion of humanity, with its saints' days and its calendar, its catechism and its encouragement of the emotional. People laugh at the late Comte without ever having read any. If we reverse this superficial reaction, a profound truth will come to light.

How is science to be classified; how will it develop? Here Comte is constantly wrong, and the future of each discipline falsified his predictions. As for the great laws of time, such as the law of the three states, one can always play this game at which one can always win, because it is impossible of falsification. The history of something one neither understands nor controls can always be written without any real difficulty. In other words, positivism turns out to be weak where it is believed to be strong, precisely with regard to science and history, and strong in the areas where everyone condemns it, like an understanding of religion. Who promotes and transmits knowledge; what is its social function? His response to this question, which has been thought weak or crazy, says a very great deal, because it asks a religion to reply to these questions. Auguste Comte the sociologist was describing what was happening in society, and in doing this he founded the anthropology of the sciences. In religion he discovers what transforms ideas into social forces.

The Time of the Calendar

One consequence of the parallelism described here is of interest to the history of science: there is not one model of the latter which has not already been used in religion. Such a repetition makes one laugh, if one thinks of Marx's phrase which compares its appearance in history to farce. All these models were retempered in the crisis or crucible of the French Revolution.

This was the time of messiahs of the new age. Lavoisier is the herald of modern chemistry, obliterating the old; Carnot founds a thermodynamics that has no forerunners, Galois and Philippe Pinel, reforming algebra or psychiatry, tear their new disciplines out of their old frameworks. Nearly all these prophets died tragically, in great suffering. Discontinuous theories of the break here move directly from religious, evangelical or prophetic references to true references, discovery bringing one out from error or from darkness considered as revelation into the light. Neither the Copernican revolution nor, I believe, the 'Galileo affair' had actually been conceptualized until very shortly before this date.

The clergy, I have said, maintains time: the Gregorian calendar was succeeded by the Revolutionary one, evoking flowers, snow and the grape-harvest – nothing but agriculture at the high point of the Industrial Revolution – but the Positivist Calendar was intended to replace it. Days, weeks and months bore the names of great men. Dante, Newton, Archimedes and Lagrange have taken over the places and the niches of the saints. That the calendar didn't succeed doesn't mean that we don't have it in our heads: the history of science canonizes genius just as religion prays to martyrs and to prophets. John Stuart Mill called Positivism 'Catholicism without Christianity'.

The purpose of a calendar is either to wipe out previous history, when hatred gets to such a point that it sets out to kill even the dead, or to celebrate continuity. Tolerant, pluralist and secular, Comte's calendar is a spell for inter-cultural reconciliation. That is why it didn't succeed. It was too universal to inspire a sect or a gang. The first theological month of Moses is connected there with the fourth month of Archimedes, the sixth month of Saint Paul and the last month of Bichat. The religions flow the one into the other, Solomon with Confucius, Saint Bernard with Muhammad, then move towards the sciences, Abraham towards Thales, Sainte Monique towards d'Alembert, without science depreciating religion, equal in memory, and equal too with art or industry, Bernard Palissy and the troubadours. In the academic and political worlds, with their obligatory bloody confrontations, such peace in time and history can only be thought mad, and this breadth of spirit can only be thought illness. Have we ever seen a better table of the best possible times? The only complaint can be the absence of Christ. Question: who first told us our God was dead? Without Christianity. Also deplorable is the total absence of contemporary arts and letters: science has taken up all the room. After Mozart, on the last Sunday of the tenth month, goodbye beauty. A sinister premonition . . .

In the calendar continuous history rejoins the discontinuous and the internal is connected to the external. The lunar time of the months and weeks lines up the stages internal to each field or area, while the time of the Sun associates them externally, without epact. This vignette reproduces in gaudy and naïve colours, agreeable to all, the ignorant, the unbelievers, the learned and the pious all together, the people, that is to say, the whole of the philosophy of science and history constructed by Comte from the tableau of *Paris 1800*, a quarry where many successors would come for their materials, though without admitting it.

On Incompleteness

In his *Critique de la raison politique*, Régis Debray applies or discovers as applicable to social groups the incompleteness theorem valid for formal systems, and shows that societies can only organize themselves on the express condition that they are founded on something other than themselves, outside their own definition or border. They cannot be sufficient in themselves. He calls this foundation religious. With Gödel he completes the work of Bergson,[11] whose *Les Deux Sources de la morale et de la religion* differentiated between closed and open societies. No, he says, internal coherence is guaranteed by the external: the group only closes if it is open. Saints, geniuses, heroes, paragons and all sorts of champions do not break institutions but make them possible.

The history of science uses the words in the opposite way to Gödel, Bergson or Debray, but it discusses the same things, while calling external what they call internal, and vice versa. An externalist history reduces a truth or a scientific proof to the totality of its social conditions or constraints, thus referring them to the interior of the social group, while internal history remains within the limits of the discipline, referring to nothing else. It is certainly the same debate, opposing the open to the closed, but the terms have been reversed.

Since Bergson, the most notable historians have copied from the *Two Sources*, which deals explicitly with the cases of madness and of knowledge. Foucault's *grand renfermement* and Thomas Kuhn's paradigm, to mention only the most famous works, derive from the same sources. Far from copying a model, as they do, Régis Debray solves a problem. Where historians describe the crossing or transgression of social or conceptual limits, without understanding them, because they have borrowed their model ready-made from Bergson, which Bergson constructed on the basis of Carnot and thermodynamics, Régis Debray has constructed his own, and has therefore grasped a new model, based on Gödel and on logical systems.

This decisive contribution from Gödel and Debray frees us from ancient models and from their repetition. Let us consider the most radical possible external history of the sciences, which reduces some discoveries to social factors: the struggle of ideas or of interest, institutional, strategic, financial and economic constraints, the local accumulation of information, the organization of schools and laboratories, briefly, the ensemble of forces in their mutual inter-relations and the concrete conditions mobilized by a certain event as it occurred. A question: why did this mobilization take place? How did these ensembles crystallize? By virtue of the truth. Whose? Carnot's, sympathetic or ambitious, Arago the minister's? The truth of the differentials or of the arc of the meridian? Yes and no, or rather no. In fact, simply the truth, full stop. By virtue of the objectivity, external to all these conditions and constraints and precisely not resulting from them. The whole social process dies off if the objective truth of what is in question does not emerge. The internal (here the external) forms an efficient, dynamic and productive system if and only if it founds itself on this absolutely other thing outside itself, which has to be called the external (here, the internal): the truth or objectivity. Everything is held together around it and through it, is set into motion by its presence and stops in its absence, everything is therefore based on the transcendence of the truth or of this condition of objectivity. Transcendence is what we call this non-belonging to an ensemble and to its constraints.

But if one is afraid of the philosophical terminology or of logical theorems, one can try to understand this through other images. A catalyst, for example, is an element or substance without which a chemical reaction will not take begin, nor *a fortiori* continue, but which none the less is not involved in it: exterior to it, but none the less its condition. In the same way, the veridical elements of science, external to its history, provide its foundation and make possible the social conditions of its emergence.

The argument which opposes the external and internal in our disciplines is therefore evidence of an inadequate analysis of the social relationship, and the history which divides scientific time into moments of openness and ages of closure is doubtless an expression of the same ignorance. In the same way as historians of knowledge and of madness owe their models to Bergson, we owe our solutions to the Gödel–Debray principle.

Open and Closed, Solids and Fluids

This genealogy of ideas and models goes back beyond Bergson, whose thinking is opposed to Positivism. Its two sources come from thermodynamics in the same way as Auguste Comte took the static–dynamic or order–progress pair from classical mechanics. The successor completes its predecessor, while the reference body of knowledge is increased. The question of the open and the closed put in the *Two Sources* has the same origin as the title of the book, and is explained by the passage from Carnot to Clausius and Gibbs. Bergson also criticizes the solid mind and demands a return to flux, to endurance and to consciousness, while Auguste Comte distrusted the cloudy and vague, demanding that only solid and coherent systems be constructed. Here again, the reference science appears, and we move clearly from the mechanics of solids to that of liquids and gases, from the crystal, which geometry was beginning to master in the early nineteenth century, to the molecular disorder which physics was thinking about at the beginning of the twentieth. Almost twenty years ago, I called this development 'the law of the transformation of metaphorical matter' (*Hermès*, III, 1974). The philosophers spoke of solid systems, then of currents or of flux, and finally, of language; crystals first of all, then fluids, all hard things, but becoming softer, material, but changing state, finally speaking of the soft, of programme, speech, discourse, writing. Positivism refers to ordered matter, Bergsonism already refers to disorder or to fluid movement, while after the war all objective reference was abandoned and philosophy devoted itself to language, as much in Europe as elsewhere, where the admirably named logical positivism synthesized all that had gone before.

Science and Religion

In applying Gödel's theorem to questions of the closed and the open, as they relate to sociology then, with one gesture Régis Debray recapitulates and concludes the history and the work of the previous 200 years. After arguing all this, he then appeals to Comte and to the Positivist cult. Why is this?

Because however ridiculous we may think the culminating adventure of his life, the high priest of the religion of humanity had already seen all this, and attempted to express or explain it, however badly, as if in his fond madness, Comte had founded his sociology on a general anthropology without having given it a name. Alone of all the philosophers since Kant, he laboured to acquire all the scientific knowledge of his time: he did not cheat or lie, never invented a concept which might allow him to escape this labour. In the course of this heroic philosophy, he chose from the sciences the very best models and captured the heuristic gesture of local and global science so well that he continued and concluded the itinerary of the *Encyclopédie* with the invention of a new human science ultimately conditioned by all those which had preceded it. Discovery is always the guarantee of the authenticity of the work.

During this authentically heroic journey, he wrote the internal history of each discipline acquired in and through time. No one before or after him made such a connection between the internal and the external, for the history of cultures and societies makes possible an encyclopaedic classification of the sciences whose own history makes possible sociology, which returns to history and culture in an enormous circle which closes the system and in which the external feeds an interior which feeds into the external in its turn.

I believe that here he is codifying an idea or a feeling which one can read or experience in the works and lives, in the social and speculative activities of the scientists who lived through the French Revolution, who were at Paris in the year 1800. Their great treatises, in general, first of all recapitulate the internal history of their respective disciplines, then lay it out according to a systematic plan: progress, then order. Fourier, Laplace, Lavoisier, Haüy, Lamarck, Monge and twenty others knew or felt that they were entering into a new time in which rational science would become the crucial social factor, which would dominate education, the army, industry and agriculture, which in their turn would produce the preconditions of reason: scientific order determines social progress whose order determines scientific progress. Their universal science led them to a unique sociopolitical location through an acute moment of crisis which would give rise at the same time to history, science and society. Comte sanctifies this unique double experience, occupying all the positions at the same time.

He stood back from this perfect gesture and saw that it was a question of religion. Not at all a religion attached to a particular ethnic group or culture, the object of a possible anthropology, but a nascent global relationship, fit to bring about the integration of humanity as a whole. It was properly universal. In other words, the general system of history and the positive science, one and the other producing knowledge both hard and soft, human and exact, ordered and progressive, cannot exist or be understood all by itself. Obscurely, he foresaw Gödel and Debray. There exists something exterior even to the self-sustaining totality, and the second is founded upon the first. He calls this foundation religious. And he is right, even if it is no more than the immanence of humanity or of the Great Being in itself.

What Clerics?

Let us return for a moment to the *ancien régime* and to the societies which Positivism called theological. In those days the clergy occupied a very precise place in society. Dominant and dominated, neither dominated nor dominant, this place, within each dominant or dominated class, belonged to neither one nor the other, to neither the dominated nor the dominant. By virtue of the intermediary position of their order, the priests therefore ensured the closure of the social system with the guarantee they provided of a transcendent truth, absolutely external to any social production. They were internal of course, but they signified the external, and one could call them the civil servants of the truth. On the truth this group is founded. On this rock, the Church was built. What rock? This inert object thrown outside, objectified, then internal to the group when he is called Peter, lying beneath everything.

If one believes in human nature, or in nature simply, one calls this transcendent and foundational truth supernatural. Now it is simply demystified but just as transcendent. The opposition between the internal and external histories of the sciences confirms the continuing transcendence and the act which founds social immanence upon it. This is called religion and Auguste Comte was right, as was Régis Debray.

Now the scientists occupy a very precise place in society. Dominant and dominated, adviser to the prince – or pauper, neither dominated nor dominant – freedom of thought – this place belongs to neither of the two classes. By virtue of the intermediary position of their order, the priests/scientists therefore ensure the closure of the social system with the guarantee they provide of a transcendent truth, absolutely external to any social production.

Each time a court intervenes in this transcendence in the name of another ideocracy, so as to impose its own dogma, it covers itself in shame and opprobrium, from Giordano Bruno to Lavoisier or Lysenko: losing its credibility, it trembles on its foundations. Proof *a contrario* of the capital importance of such transcendence is that one cannot touch it without danger. After having beheaded the chemist, the Revolution was quick to sanctify him.

Defined by a simple or complex intersection of social affiliations, the most politicized of teachers and researchers must designate together with the same gesture this external factor which brings them together, and with which they now bring together contemporary societies, the world as such considered as another world, independent of the social group: one could call them the civil servants of the truth. They have their kingdom thanks to this other world, that is to say the world as such, outside politics. Without this objective truth they are nothing, they can do nothing. They know it, and the whole world knows as well.

There is therefore no functional or structural difference between the world of faith in a transcendent God and the belief that there exists a scientific object independent of us and capable of being expressed in and through a universal truth, binding upon all and objective in itself. Transcendence as such is the same in the two cases and the social consequences are the same. The functionaries, changing or hardly changing at all; the function itself is unvarying and contemporary societies are like primitive ones, which is what I wanted to show. The universal and the particular always meet at the same crossroads, as do the open and the closed. Was there really a revolution?

That said, we are today living through a crisis of the truth. After the death of God, how goes the future of the world? The one and the other are engaged in the same death-agony.

Lavoisier: A Scientific Revolution

Bernadette Bensaude-Vincent

In which we see two careers and two destinies. The conquering hero of the scientific revolution, Lavoisier died a victim of the French Revolution; he was abandoned by his colleagues, yet later venerated as the immortal founder of modern chemistry.

Year One of chemistry. Throughout the nineteenth century chemists, philosophers and historians associated the emergence of scientific chemistry with Antoine Laurent de Lavoisier (1743–94). Contemporary historians in their turn have seen in this event the elements that constitute the structure of a revolution: symptoms of crisis, the emergence of a new paradigm, controversy and division among the community of chemists. The accomplishments of Lavoisier's generation are even talked about, these days, as a 'second scientific revolution'. The word 'revolution' had already been used by Lavoisier, then taken up and popularized by his contemporaries. This gives us a particular opportunity to elucidate the notion of a scientific revolution and to grasp the mechanisms at work in the episodes so described.

The case of Lavoisier is of special interest because it allows a direct confrontation with the political meaning of the word 'revolution'. The chemical revolution culminated in the publication of his *Traité élémentaire de la chimie* in 1789, the year in which the French Revolution began. Lavoisier, who was both a scientist and an administrator under the *ancien régime*, found himself involved in both movements. But while he came out of the chemical revolution victorious, he fell victim to the political revolution. I shall therefore be trying to understand each of Lavoisier's careers in the light of the other.

On 19 Floréal of Year II (8 May 1794), the Tribunal sentenced Lavoisier to death, together with twenty-seven other tax farmers, who were found guilty of 'a conspiracy intended to encourage by all means possible the success of the enemies of France'. That afternoon, they were all guillotined in the Place de la Révolution. Why was this man, who had just brought about an astonishing revolution in science, allowed to die? There was embarrassed silence from some, indignation from others. For more than two centuries, this inglorious episode of the French Revolution has been a matter of historical controversy, always reborn and always passionate. For my part, I shall do no more than underline the contrast between the way he was abandoned and the cult which grew up around him immediately after his death. How was the image of the founding hero forged? By examining the different versions of the events, can one trace the development of representations of the chemical revolution over more than a century?

Two Careers

This intersection of the political and the scientific was not an unprecedented event in eighteenth-century France. Lavoisier appears as a representative of the close connection which had been established between scientists and government towards the end of the *ancien régime*. Understanding the usefulness of the sciences, the monarchy had created great State administrations: the École des Mines and the Ponts et Chaussées. The provincial academies and the Royal Academy of Science encouraged scientists' interest in matters of public importance by means of competitions on the practical problems of national development.

Under the Revolution, interaction between scientists and politics was very intense. Scientists were mobilized *en masse*. Very few appeared among the émigrés, but many took on positions of political responsibility and sat on the Committee of Public Safety. It's true that the academies, the essential organs of scientific life, were dissolved by a decree of the Convention in August 1793, but on the whole, the scientific population profited from this incursion into the field of politics. It brought them fame and legitimacy; it gained them credit, honours and riches (see pp. 422–54).

Few scientists lost from this alliance. If Lavoisier's name stands out among these unfortunate few, it is because he gained much greater glory through science than did his victorious colleagues in the political arena. Lost and won: the contrast is so strong that it calls for a comparison of Lavoisier's attitude to the two fields.

Let us see first how Lavoisier became prominent in both fields. He did not have two distinct or successive careers. Throughout his life he lived both as a royal official and as a scientist. In 1768, at the age of twenty-five, he purchased an office in the Ferme Générale and continued in it until the dissolution of the fiscal institutions of the *ancien régime* in 1791. However important and however intense Lavoisier's scientific researches may appear to be, they were no more than a leisure activity. His main activity, his place in society and the source of his income was the Ferme Générale. Lavoisier seems to have successfully combined his two careers in his daily life. When he arrived in Rouen or Amiens to collect taxes, he would give a paper to the local academy. On one hand he collected the money, and on the other he delivered scientific information and made himself known as a scientist. In 1775 this convenient arrangement soon gave way to another, less tiring and more effective. Lavoisier was appointed a Commissioner of the Régie des Poudres et Salpêtres. He lived at the Arsenal, where he set up, with his own money, a well-equipped laboratory provided with remarkable precision instruments from the best Parisian manufacturers, Fortin and Mégnié. According to his wife, Marie-Anne Paulze – the daughter of a tax-collector, who married him at the age of fourteen, and was entirely devoted to her scientist husband, learning English and engraving in order to translate and illustrate works of chemistry – Lavoisier organized his time according to a rigid time-table: scientific research from 6 till 8 in the morning; administrative and academic activities during the day, then research from 7 till 10 at night; and one full day of science each week. These activities became even more intertwined when Lavoisier, who had been ennobled in 1775, bought an estate at Fréchines, near Blois. He then carried out experiments in agronomy in order to increase its yield. Science became an instrument of agricultural and social experimentation. Lavoisier wasn't then, on one side a progressive scientist and on the other an administrator of the *ancien régime*. He exercised both activities at the same time and made the one more profitable through the other. He managed his time in the same way as he managed his wealth.

In his scientific career, too, Lavoisier was an administrator. He entered the Royal Academy of Science in 1768 at the age of twenty-five, as a 'supernumerary assistant chemist'. He then climbed the rungs of the internal hierarchy, eventually becoming Life Treasurer, managing and administering this scientific institution. In 1785 Lavoisier, as the annual Director, drew up a plan for reform of the Academy. He did this in ten days exactly. Indignant because a royal minister was planning, without consultation, a reform to enlarge the Academy, he drew up a counter-plan. He accepted the creation of two new classes, but he opposed the creation of new posts by reducing the number of Academicians in each class and abolishing the 'supernumerary' posts so as not to diminish the standing attached to the title of Academician.

'It is not that some scientists do not have an Academy, but rather that the Academy cannot find scientists,' Lavoisier added.

The Royal Academy of Sciences

The foundation of the Royal Academy of Sciences in 1666 was part of a royal plan for the development and control of intellectual life, which resulted in the creation of seven academies by Richelieu, Mazarin, and then Colbert. The Academy of Science, richly endowed from the beginning, for the first time provided salaries for doing science full-time, and created the first scientific weekly, the *Journal des Savans*. The Academy of Sciences was also responsible for issuing the 'royal privilege for machines', like a Patent Office in our own day. During the eighteenth century the Academy of Science became even closer to royal power, with the system of competitions introduced in 1720.

When Lavoisier entered the Academy in 1768, there were six classes – geometry, astronomy, mechanics, anatomy, chemistry and botany. In each class the members were arranged hierarchically. As it happened that a member might be appointed by the king against the wishes of the Academicians, custom allowed a 'supernumerary' post to be created to accommodate a real scientist.

The reform of 1785 created two new classes – general physics and mineralogy – but reduced from eight to six the number of Academicians in each class. Under the directorship of Lavoisier, the Academy of Sciences preserved the privileges of a small scientific élite, while demonstrating some slight independence of the royal will.

Later, when the Revolution abolished the Ferme générale, he devoted most of his energy to the Academy. He did everything he could to save it and enable it to escape the decree of 8 August 1793 abolishing all the academies.

Lavoisier was very attached to the life of the Academy, and in fact, it is impossible to understand his life's work without seeing it in the context of that institution. On one hand the Academy stimulated a certain style of research. It encouraged team-work: in each field, Lavoisier collaborated with a specialist in the methods or subjects concerned; Jean Bucquet on air, Pierre Simon Laplace on heat, Jean-Baptiste Meusnier on water, Armand Seguin on the physiology of respiration, Louis Bernard Guyton de Morveau, Claude Berthollet and Antoine François Fourcroy on nomenclature, and finally with the Abbé René Just Haüy to establish the unit of mass for the metric system. Through the competitions, and especially the commissions set up to examine particular problems, the Academy encouraged planned and programmed research. With Lavoisier there was little improvization, but step-by-step work.

He organized his research and planned long series of experiments over several months.

On the other hand the Academy gave direction to Lavoisier's scientific career. If he became a chemist, this was partly because there happened to be a vacancy in the chemistry section. His early work was in geology: under the direction of Jean Étienne Guettard, he carried out field studies in France in 1762–3 and he outlined a theory of stratification. He took up chemistry as an auxiliary science to geology, and on Guettard's advice he followed the courses given by Guillaume-François Rouelle at the Jardin du Roi, as had Diderot, Rousseau and Turgot. But Lavoisier seems to have preferred the experimental sciences to geology. When he wanted a post at the Academy in 1764, he suggested the creation of a chair in experimental physics. It was, then, as a physicist that he first defined himself. This 'physics' side is evident in his early work. His favourite topics were heat and the states of aggregation of matter, two questions which Gabriel-François Venel, in his article on 'Chemistry' in Diderot's *Encyclopédie*, had described precisely as on the frontiers between physics and chemistry. In 1774 Lavoisier entitled his first collection of papers, devoted to the 'aeriform fluids' (gases) which were given off by substances or which combined with them, *Opuscules physiques et chimiques*. The description of the work in the *Histoire de l'Académie* – presumed to have been written by the author himself – presents Lavoisier as the person who introduced the spirit of physics into chemistry.

It is clear already that the science promoted by Lavoisier would not be the one hoped for by Venel. Chemistry's originality did not derive from a full-scale attack on the abstract, speculative and limited spirit of mechanics. On the contrary: Lavoisier was looking for unity and reconciliation. He did not, of course, rehabilitate the mechanistic tradition criticized by Venel, and he left aside the central problem of affinity (see pp. 372–400). In addition, there is hardly any reference to Newton in Lavoisier's published work, though he wanted to import the methods of experimental physics into chemistry.

A Single Method

It has often been said that Lavoisier revolutionized chemistry with the balance. This statement is only partly true. Balances already had their place in chemical laboratories, among the furnaces, bladders, retorts, alembics, flasks and bell-jars. But they only became an essential instrument of experiment around 1770, when people began to study the gases. They gained in importance together with the 'pneumatic chest', a closed piece of apparatus for collecting gases, and the gasometer, which allowed their volumes to be measured. Joseph Black, Carl Scheele, Henry Cavendish and Lavoisier mobilized all the know-how and ingenuity of their countries' craftsmen in the search for increasingly improved and accurate balances. Stimulated by the requirements of Lavoisier, who had the resources to have the most expensive equipment built for him, Paris manufacturers like Mégnié and Fortin gained international renown.

Lavoisier's balances not only brought a gain in the precision of experimental measurement. They were the supreme arbiter in theoretical debate, and took on their full importance within the framework of a methodical programme of research. On each topic dealt with, Lavoisier made a systematic inventory of the work published in France and abroad. He highlighted the uncertainties, the contradictions and the controversies around the issue, and he conceived a series of experiments to be done to decide the matter. Debates were ended in the laboratory and tradition was judged by the balance's beam.

With his balances, Lavoisier turned experimental method into a form of accounting. At each stage he drew up a balance sheet of the reactions carried out. He weighed before the experiment and he weighed afterwards. He weighed everything, every element of the system. This balance-sheet approach presupposed the famous principle, attributed to Lavoisier, that 'Nothing is lost.' It wasn't, in fact, Lavoisier's: a very similar formulation is to be found in the Atomists of Antiquity (see p. 143), Lucretius in particular: 'Nothing is created out of nothing; nothing returns to nothingness.' What is more, Lavoisier never made 'Matter is neither created nor destroyed, it is transformed' into a fundamental principle of chemistry. He mentions this

---

Chronology of Lavoisier's physical and chemical work

1765: Memoir 'On the analysis of gypsum'
1772: Experiments on calcination; sealed letter to the Academy
1773: Calcination of lead and tin in retorts
1774: *Opuscules physiques et chimiques*
1775: Experiments on mercury *per se* (reduction of the mercury calx).
1777: Memoir 'On the respiration of animals'
      Memoir 'On combustion in general'
1780: Experiments on acids
1781: Experiments and work on heat (with Laplace)
1783: Memoir 'On the composition of water'
      'Reflections on phlogiston'
1785: Synthesis and analysis of water
1787: Memoir 'On the necessity of reforming and improving chemical nomenclature'
1789: *Traité élémentaire de chimie*; first volume of the *Annales de chimie*
1792: Work on the metric system

---

formula, in passing, in relation to a study of fermentation, in a chapter of his *Traité élémentaire de chimie* ('Elements of Chemistry'). But it is true that Lavoisier made the conservation of the quantity of matter the fulcrum of his experimental practice. This principle, even if implicit, underlies the whole of the chemistry of balance sheets.

Financier and chemist: the combination of both skills in one person seems to have been highly fruitful. But rather than being a happy accident, does this not rather represent the more or less successful accomplishment, in different fields, of one and the same project?

A Revolutionary Project

Commentators on Lavoisier have been struck by his firm willingness to bring about a revolution in chemistry and his consciousness of doing so. The crystallization of this revolutionary project is usually assigned to 1772, a crucial year, according to the historian Henry Guerlac.

This desire to revolutionize chemistry manifested itself just one year after he wrote an economics essay which was rather disparaging of revolutionary projects. However, looking through Lavoisier's collected works and comparing his writings on economics and politics with the scientific writings of the same period, it is obvious that Lavoisier often thinks these different questions through in the same terms. In his notes for an *Éloge de M. de Colbert*, dated 1771, he describes international trade as a system of exchanges and flows in which the quantity of wealth remains constant. 'Nothing is lost, nothing is created' also operates in economics. But here the system tends spontaneously towards equilibrium. This natural equilibrium, which Lavoisier as Colbert calls 'the physical order', leaves ministers with a very narrow margin of manoeuvre. At the very most they can tip the balance in favour of one country, direct the over flow towards a region or sector. But all possible political effort can only momentarily disturb the equilibrium of the balance of trade. Such a vision of the economic order would hardly incite revolution.

Chemistry seems more favourable. Having investigated various fields, in response to competitions organized by the Academy, Lavoisier seems to have found a promising topic: combustion. On 1 November 1772 Lavoisier sent the Academy a sealed paper on the role of air in calcination, which would only be opened on 5 May 1773, when research was further advanced. Lavoisier carried on with his experiments, and on 20 February 1773 he began a new laboratory notebook, outlining a programme of experiments which, he wrote, would provoke 'a revolution in physics and chemistry'. Let us open the mysterious sealed letter, and try to grasp the germs of revolution it contains:

About eight days ago I discovered that sulphur, in burning, far from losing its weight, on the contrary gained it; that is to say that from a pound of sulphur one could obtain much more than a pound of vitriolic acid, allowance made for humidity from the air; it is the same with phosphorus: this increase in weight arises from the prodigious quantity of air which is fixed during combustion and combines with the vapours. This discovery, which I have made in experiments which I regard as decisive, has made me think that what is observed in the combustion of sulphur and of phosphorus could well take place with regard to all substances which gain weight by combustion and calcination; and I am even persuaded that the increase in weight of metallic calces has the same cause. Experiment has completely confirmed my conjectures; I performed a reduction of litharge in closed vessels with Hales' appara-tus,[1] and I observed that there was given off, at the moment of passage from calx to metal, a considerable quantity of air, and that this air formed a volume a thousand times greater than the quantity of litharge em-ployed. As this discovery appears to me to be one of the most interesting since Stahl, I believed I ought to assure myself of my claim to it, in making this present deposit at the Academy, to remain secret until the time I publish my experiments.

On the basis of two experiments, Lavoisier puts forward a general explana-tion of all combustion and all calcination. He immediately set himself in the history of chemistry, by reference to Stahl. What then was the doctrine that Lavoisier was preparing to overturn?

The Heir to a Tradition

Phlogiston was the principle of fire responsible for combustion: the release of phlogiston explained the phenomena of heat and light produced during combustion. Phlogiston was invisible, hidden and impossible to isolate, be-cause it was always in combination.

The doctrine of Georg-Ernst Stahl (1660–1734) is often reduced to the phlogiston theory, because we are the victims of a sort of screen-effect resulting from Lavoisier's work. Yet many historical studies, initiated in France by Pierre Duhem (1902), Emile Meyerson (1921) and developed particularly by Hélène Metzger (1930, 1932 and 1935), have shown that Stahl's chemistry constituted a powerful system – the first system of chemistry to be adopted through all of Europe – which permitted the interpretation of a great number of experiments: the formation of salts (neutral, acid and alkaline salts) re-sulted from the combination of earth and water, which united with their fellows through affinity; phlogiston, the cause of flammability, allowed the identification of two operations apparently very distant one from the other, the calcination of metals and the combustion of organic materials. Stahl's chemistry included a philosophy of matter which, though corpuscular, was opposed to mechanism. Stahl accepted the existence of indivisible particles but he fought against the idea of a unique and uniform matter. He supposed the existence of different kinds of atoms, highly individualized. These atoms determined the properties of compounds by their quality, by their individuality, not their geometrical characteristics. And it was not by isolating them – as the Cartesians suggested with their shaped, hooked and pointed atoms – but by studying the properties they conferred on mixed substances that the constituent atoms would be identified. These ultimate corpuscles, indeed, were forever unknowable. One could not isolate them, but only sense their presence. It was also pointless to claim to be able to predict the properties of substances from the supposed shapes of the atoms. To all this imaginary to-ing and fro-ing, Stahl preferred the idea of gradation of composition. Starting with 'aggregates', one had to consider each degree of the complexity of matter, going from supercompounds to compounds, to mixtures and then finally to atoms. From this there came a very great attention to subtle degrees, to all the stages of chemical decomposition. The four elements,[2] earth, air, water and fire were essential for the interpretation of chemical properties and reactions in this anti-mechanistic perspective. Thanks to Stahl's success, the Greek conception of elementary principles, universal constituents of matter and bearers of quality, remained current in the eighteenth century. Far from being the merest residue of a worn-out ancient tradition, it formed the basis of an ambitious chemistry anxious to demonstrate its originality. It was more vigorous than ever in France, where it was presented as the result of practical analytic experiment, as witnessed by the article on 'Principles' in Pierre Joseph Macquer's *Dictionnaire de Chymie*, published in 1766: 'It is with astonishment, no doubt, that one will see that we now accept as the principles of all compounds the four elements, fire, air, water and earth, indicated as such by Aristotle very long before one had the chemical knowledge needed to discover such a truth. In fact, however one decomposes substances, one can only ever obtain these substances: they are the final terms of chemical analysis.'

Let us read this carefully. The four elements are not vague principles, the bearers of qualities. They are defined as simple substances accessible to experience. And in the article, 'Element', Macquer adds a notion of relativity: 'It is very possible that these substances, although thought to be simple, may not be so, that they may even be highly compound, resulting from the union of several principles, other simple substances, or that they are transmutable one into another, as is thought by M. le comte de Buffon. But as experiment teaches us absolutely nothing about this, one may without any inconvenience regard, and in chemistry even must regard, fire, air, water and earth as simple substances; because in fact they act as such in chemical operations.'

Thus the doctrine of the four elements triumphed in the eighteenth century, not despite the development of experimental analytic chemistry, but because of it. What is more, with the development of gaseous chemistry, the experimental victories of this young science would reinforce the triumph of the old elementary principles.

Until the middle of the eighteenth century, air was given no role in chemical reactions. With the exception of Robert Boyle, who had hypothesized its role in combustion, it was generally considered as a mechanical agent. Impelled by the study of plant physiology, and thanks to the apparatus built by Stephen Hales – the first pneumatic chest – the 'aeriform fluids' began to attract the attention of chemists. In the 1770s they were subjected to analysis. First of all, 'fixed air' (now carbon dioxide) was studied by the Scottish scientist Joseph Black; then, in 1772, Joseph Priestley published a paper entitled *Observations on Different Kinds of Air*. In 1774, at the same time as Scheele in Sweden, he isolated and described the gas that would be called oxygen. Finally, Cavendish isolated the future hydrogen in 1766. But far from destroying the theory of the four elements, the success of gaseous chemistry first of all helped to strengthen it. Air had certainly been experimentally decomposed, but Priestley interpreted the experiment as the combination of air with phlogiston. He called nitrogen 'phlogisticated air' and oxygen 'dephlogisticated air'. Hydrogen was more or less identified with phlogiston. The phlogiston theory seems to have profited from this, because the hitherto invisible principle could now perhaps be identified with this experimental reality. In the preface to the second edition of *The Critique of Pure Reason* (1787), Immanuel Kant cites Stahl as a hero of the experimental method, alongside Galileo and Evangelista Toricelli. As for the ancient theory of the four elements, it had not been shaken by the decomposition of air and water, two of the Aristotelian elements. But it would be destroyed by fire.

How did Lavoisier come to doubt such a doctrine? Through an experiment, thanks to the balance indeed. But it also has to be said that the idea of this experiment didn't come from a sudden illumination. It was preceded by ten years' thorough study of the chemistry of elementary principles.

The Lavoisierian revolution was so successful that for a long time it hid the connections between Lavoisier and eighteenth-century chemistry, which are, however, the links that give Lavoisier's project its entire meaning. Looking at the succession of Lavoisier's early papers, often inspired by problems submitted to the Academy, one sees that they cover the whole field of the chemistry of principles or elements. These studies, done to order in connection with different practical questions, none the less seem to follow a programme.

His first chemical experiment, at the time of his geological studies, was an analysis of gypsum. After earth, there followed fire. In 1764 Lavoisier submitted to the Academy his first paper, for a competition on 'How to improve the lighting of the streets of Paris'. In it he dealt with all aspects of the problem, the fuel, the form and material of the lamp, choice of wick, means of suspension

... A fine study in optimization, but it won no prize. In 1767 Lavoisier presented a study of the composition of the mineral waters of the Vosges. He then tackled the question of the relationships between earth and water: the problem, connected with the water supply to Paris, was to determine whether the solid residue found in the water was the result, as had been suggested by Boyle and van Helmont, of a transmutation of water into earth. At the end of a series of experiments carried out over 101 days, Lavoisier concluded that the deposit of silica observed in the water was due not to transmutation, but to the entry into solution of a very small amount of glass from the pelican alembic holding the water. At the same time, according to the manuscript notes of 1766–8, Lavoisier was working on the relations between air and fire and, after much reading, he adopted the idea that all substances could exist in three states – solid, liquid or aeriform – depending on the quantity of fire combined. One by one, Lavoisier passed the four elements in review.

At the beginning of 1772 he attacked phlogiston, on the occasion of a paper by Guyton de Morveau which attempted to explain the increase in the weight of calcinated tin and lead. The increase in weight in calcination was indeed difficult to interpret if calcination was a release of phlogiston. But this phenomenon had been known for a long time without damaging the success of the phlogiston theory. It had even been explained in the seventeenth century: in 1630, Jean Rey, a physician from the Périgord, had attributed the increase to the fixation of air, in his *Essays sur la recherche de la cause pour laquelle l'estain et le plomb calcinés augmentent de poids* ('Essays on the investigation of the cause by which calcinated tin and lead increase in weight'); the Englishman John Mayow (1641–79) offered the same interpretation, sketching a general theory of respiration and combustion. But these works were apparently unknown to Lavoisier when he sent his sealed paper to the Academy in 1772.

Although he had already set himself a revolutionary project, Lavoisier did not immediately hold revolutionary opinions. During the ten years that followed, he was extremely cautious in his writings against phlogiston. In 1777, in his paper 'On combustion in general', he stressed at the beginning the need to go beyond the facts in order to form hypotheses, and he presented his own hypothesis at the conclusion of an inductive, generalizing argument, based on a series of experiments methodically carried out, with precise measurements and with repetitions, variations and verifications. Lavoisier had still not broken with the tradition which nourished him.

In fact, his theory of combustion was not really 'a revolution in physics and chemistry'. At least, it did not abolish the elementary principles, the bearers of properties. Lavoisier needed them to explain why heat and light were given off in combustion: he attributed this to the release of the caloric contained in air. Lavoisier's explanation would be seen as a reversal of the phlogiston theory: combustion released the phlogiston contained in the combustible substance, and it is for Lavoisier combination with air; the source of heat is no longer in the combustible (phlogiston), but in the air (caloric). Whereas for Stahl, heat, in fixing itself, provoked combination or condensation, for Lavoisier it provokes an expansion, even a disaggregation. This symmetry did not escape Lavoisier's contemporary, Macquer, who was keen to minimize the impact of the 'revolution'. In addition, the caloric is not a simple residue of the old chemistry, whose system Lavoisier could abandon. It was the key to his conception of the states of matter, because the gaseous state was explained by the proportion of caloric in a substance.

Lavoisier's theory of combustion introduced an inversion of the dominant ideas, rather than a real revolution in chemistry. Nor is Lavoisier's work on acids (1772–6) any more obviously revolutionary. He makes oxygen the acid principle, as its name indicates (acid-generator). In doing so, Lavoisier discarded the Newtonian theory of acids conceived of as extremely reactive substances, to strengthen the Stahlian idea of a universal acid, called *acidum pingue* by Viktor Meyer. This conception of acids demonstrated very clearly Lavoisier's attachment to the chemistry of principles which had triumphed in the eighteenth century.

The Time of Foundations (1783–1789)

Lavoisier's theory was seen as a revolution by most of his contemporaries. It led to a lively controversy among French chemists, and then between the 'French School' and the English and German chemists who remained supporters of phlogiston. The severity and the duration of this confrontation cannot be explained except by the foundational quality that Lavoisier wanted to give his work. Having struck a blow against phlogiston, Lavoisier set out to create a new system of chemistry. This labour of construction, which started in 1783, was practically completed by 1789, with the publication of the *Traité élémentaire de chimie*. In six years, Lavoisier had succeeded in consigning to oblivion the whole of the chemistry that had preceded him, but also in making himself recognized as the sole founder of modern chemistry.

Let us try and reconstruct the major stages of this operation. The campaign began in 1783, when Lavoisier destroyed the last of the ancient elements by demonstrating the composition of water. It has to be said that the 1783 paper 'On the composition of water' is no more than a first step, for it establishes the composition of water by synthesis and not by analysis.

The experiments to analyse the composition of water were conducted within the framework and with the support of the Commission of Study for the Improvement of Balloons, which had been created by order of the King.

Between the hot-air balloon of the Montgolfier brothers and the hydrogen balloon of the physicist, Jacques Alexandre César Charles, the Academicians tended to prefer the second. One of their priorities was: how can hydrogen be manufactured in quantity? At the beginning of 1784, Lavoisier, assisted by Meusnier, a young and dynamic military engineer, developed a first process for the production of hydrogen by the decomposition of water vapour passed over incandescent metal or carbon. Afterwards Lavoisier and Meusnier improved their apparatus, and at the end of February 1785, they called together the luminaries of the world of science and the royal household to witness a solemn experiment, lasting two days: the analysis and synthesis of water. Before this noble audience, proof was given that water was not an element. The fact was now established, and conversions took place one after another. First of all Berthollet, then Fourcroy and Jean Antoine Chaptal and then Guyton de Morveau. Lavoisier felt confident enough to launch an attack against phlogiston before the Academy. The memoir 'Reflections on Phlogiston', read on 28 June and 13 July 1785, clearly stated his theory as an alternative to Stahl's.

No longer is Lavoisier announcing a discovery comparable to Stahl's. He wants to wipe out every trace of his work, to remove it from the inheritance of chemistry. 'I beg my readers, as I begin this paper, to free themselves as much as possible of all prejudice; to see in the facts no more than they show, to banish from them all that reason has supposed, to transport themselves to the time before Stahl, and to forget, for a moment, if it is possible, that his theory ever existed.' He thus asks his readers to be ignorant in order to gain true knowledge. This is the crucial moment, because Lavoisier is completely changing the nature of the event. Up till now, it has been a revolution by inversion of the dominant schemata. Now he puts himself forward as a founder building on virgin territory, in intimate dialogue with the facts.

Two years later, Lavoisier set out to spread his own theory by means of: the reform of the language. Already, for many decades, chemists had been complaining of the imperfection of the nomenclature. The names of chemical substances, developed through the centuries and sanctioned by custom, certainly perpetuated the memory of a tradition, but they sometimes passed on the wrong ideas. In addition, the discovery of new substances during the eighteenth century had made it necessary to coin new names. Moved by a desire to rationalize chemistry, Torbern Bergman and Guyton de Morveau had put forward plans for reform which attempted to forge names modelled on Linnaeus' nomenclature for botany. Lavoisier, convinced by his reading of Condillac of the importance of words in the formation of ideas, seized the opportunity to accomplish his plans for chemistry. To banish the traditional names and to construct an artificial language, based solely on the Lavoisierian theory, would be to finish with the past. Even more: it would be a rebirth through baptism. This was the extraordinary undertaking carried out in a

467

few short months by Guyton de Morveau, Lavoisier, Berthollet and Fourcroy. The result of this collective enterprise, published in 1787 under the title, *Méthode de nomenclature chimique*, reveals how much Lavoisier took over the reform. First of all he persuaded Guyton de Morveau to give up his project of a common nomenclature acceptable to all schools of chemistry for a nomenclature exclusively founded on the antiphlogistic theory. This required a preliminary work of conversion within the team itself. Then, in the internal distribution of the tasks, Lavoisier awarded himself the leading role. It was he who read the first paper to the Academy, which defined the philosophy of the project and set out its main principles. He left their application to Guyton de Morveau, and to Fourcroy he gave the more modest task of drawing up a table of new names. The work was completed by a dictionary establishing concordances between old and new names. It included, as an appendix, a new system of symbols, designed by Pierre-Auguste Adet and Jean-Henri Hassenfratz, to replace the old alchemical symbols. But this was never used. The nomenclature, on the other hand, was adopted everywhere within a very few years.

It is true that it satisfied a very urgent need to escape from the chaos of multiple names. However, its promoters knew how to ensure good distribution. In order to get round the delays in publication at the Academy of Science, Lavoisier used to publish his papers in Rozier's *Observations de Physique*. However, as in 1787 it had passed into the hands of J.-C. de La Métherie, a supporter of the phlogiston theory, Lavoisier, Guyton de Morveau, Gaspard Monge, Berthollet, Fourcroy, Hassenfratz and Adet decided to produce a new journal together, the *Annales de Chimie*, which was immediately distributed in France and in other European countries.

Even if the authors seem to be concerned to maintain continuity in retaining old names which do not carry false ideas with them, this reform was a real revolution because it introduced a new spirit into chemistry. It was more a method of naming than a complete nomenclature. Its basic principle is the logic of composition: to form an alphabet of basic words to designate simple substances and then to designate compound substances by compound words formed by the juxtaposition of simple words. The compound word is always binary, and the proportions of the substances involved are indicated by a suffix. The method stood the test of time: two centuries later, with a few amendments, it is still in force. The nomenclature was the essential element which made the revolution in chemistry a new foundation. It was not simply the manifesto of a school or of a new chemical theory. It gradually abolished tradition with a double rupture. It was an irreversible break with the past: in two generations, chemists had forgotten their natural language forged through centuries of use. Pre-Lavoisierian texts became illegible, and found themselves consigned to an obscure prehistory. It was also a rupture in social space, between the academic chemistry which developed within the framework of the new nomenclature, and the craft chemistry of the druggists and perfumers who continued to speak of spirits of salt or of vitriol. The time of the *Encyclopédie*, when a chemist like Venel could declare proudly that 'chemistry has within its own body a double language, both popular and scientific,' was now over. After Lavoisier, chemistry would boast of another universality, conferred by a rational language.

Extracts from *Méthode de la nomenclature chimique* (1787)

Former names	New Names
Acid of Sulphur	
Vitriolic acid	Sulphuric acid
Oil of vitriol	
Spirit of vitriol	
Caustic vegetable alkali	Potash
Caustic volatile alkali	Ammoniac
Diana	
Moon	Silver
Silver	
Spirit of salt	Muriatic acid
	[the future hydrochloric acid]
Spirit of wine	Alcohol
Mineral kermes	Antimony oxide, red sulphured
Orpiment	Arsenic oxide, yellow sulphured
Oxygen	
Vital base of air	Oxygen
Acidifying principle	
Phlogiston	Stahl's hypothetic principle
Crocus of iron	Iron oxide

The third glorious moment of the Lavoisierian revolution was the publication of the *Traité élémentaire de chimie*, published in 1789. Lavoisier presents it as the logical and necessary sequel to the nomenclature. In fact, after the task of eradicating the past, he turned towards the future and was concerned with the propagation of his theory. Hence the desire to write a really elementary treatise, intended for beginners rather than for fully-fledged scientists. Not only should chemistry be understood rather than learnt, it is more intelligible to those who have not learnt, who haven't suffered from the grip of prejudice. Child's play, so Lavoisier is looking for an entirely new public. A new generation is called up to serve science, and the old sent away. The chemical revolution entails a subversion of the powers conferred by knowledge.

The 'Preliminary Discourse' is truly provocative in this respect. Lavoisier proudly lists everything that is to be found in the traditional textbooks that will not be found in his: nothing on affinity, nothing on the constituent parts of substances, nothing on the history of the discipline. A curious treatise, compared to the great scientific treatises of the period, which began with a historical recapitulation before going on to a logical exposition of the author's own system. Lavoisier rejects any confrontation with the established authorities, or criticism of the authors. He completely denies their competence: 'Thus works in which the sciences are treated with great clarity, with great precision, with great order, will not be within everybody's reach. Those who have studied nothing will understand them better than those who have studied widely, and certainly better than those who have written a great deal about the sciences'.

These ideas are taken from Condillac. Indeed, one might say that Lavoisier claims for his book the patronage of this contemporary philosopher. The 'Preliminary Discourse' opens with a eulogy to Condillac, together with long quotations from his *Logic*, and ends with a page of new quotations. A scientific revolution which lays claim to a philosophy! The case is sufficiently unusual for one to stop and look for a moment. Why, when he is trying to break away from tradition, should Lavoisier subject himself to a philosopher?

But is there real subjection? Lavoisier does not say that he has borrowed from Condillac. He claims that in following his own course he rediscovered the principles put forward in the *Logic*. This is supposed to be a happy encounter between two pathways, one in chemistry, the other in philosophy. Lavoisier's *Elements* appears as an experiment which verifies Condillac's hypothesis; and Condillac legitimizes Lavoisier's boldness with his philosophical support. One can find in Condillac's *Logic* three elements at least which usefully serve Lavoisier's purposes. Lavoisier first of all finds in it an interpretation of the situation, a diagnosis of the difficulties faced by chemistry: the problem is of linguistic origin. False ideas are carried by words; scientific errors are errors of language. So Condillac justifies the nomenclature. Lavoisier also finds in Condillac reasons to distrust tradition, which are necessary to his foundational ambitions: a negative concept of history as a tissue of errors and prejudices which must be swept away in order to rediscover nature. This is made explicit in the quote from Condillac at the end of the 'Discourse':

> Instead of observing the things we wish to know, we have preferred to imagine them. From false presupposition to false presupposition, we have strayed into a multitude of errors; and these errors having become prejudices, we have taken them for this reason as principles; and we have thus strayed further and further ... When things have got to this point, when the errors have thus accumulated, there is only one means

	NOMS NOUVEAUX.	NOMS ANCIENS CORRESPONDANTS.
Substances simples qui appartiennent aux trois règnes, et qu'on peut regarder comme les éléments des corps.	Lumière.	Lumière.
	Calorique.	Chaleur.
		Principe de la chaleur.
		Fluide igné.
		Feu.
		Matière du feu et de la chaleur.
	Oxygène.	Air déphlogistiqué.
		Air empiréal.
		Air vital.
		Base de l'air vital.
	Azote.	Gaz phlogistiqué.
		Mofette.
		Base de la mofette.
	Hydrogène.	Gaz inflammable.
		Base du gaz inflammable.
Substances simples, non métalliques, oxydables et acidifiables.	Soufre	Soufre.
	Phosphore	Phosphore.
	Carbone.	Charbon pur.
	Radical muriatique	Inconnu.
	Radical fluorique	Inconnu.
	Radical boracique.	Inconnu.
Substances simples, métalliques, oxydables et acidifiables.	Antimoine	Antimoine.
	Argent	Argent.
	Arsenic.	Arsénic.
	Bismuth.	Bismuth.
	Cobalt.	Cobalt.
	Cuivre	Cuivre.
	Étain	Étain.
	Fer	Fer.
	Manganèse	Manganèse.
	Mercure.	Mercure.
	Molybdène	Molybdène.
	Nickel	Nickel.
	Or.	Or.
	Platine.	Platine.
	Plomb	Plomb.
	Tungstène	Tungstène.
	Zinc.	Zinc.
Substances simples, salifiables, terreuses.	Chaux	Terre calcaire, chaux.
	Magnésie	Magnésie, base de sel d'Epsom.
	Baryte	Barote, terre pesante.
	Alumine.	Argile, terre de l'alun, base de l'alun.
	Silice	Terre siliceuse, terre vitrifiable.

Thirty-three simple substances, residues of analysis, are classified according to the types of compound they form. Lavoisier presumed that the later ones would soon be decomposed, but he gave a privileged status to the earlier ones, which he called 'elements'. (From the Traité élémentaire de chimie, *in* Oeuvres de Lavoisier, *edited by J.-B. Dumas and E. Grimaux, volume 1, 1862. Bibliothèque Nationale, Paris.)*

of bringing back order into the faculty of thought; that is to forget everything that we have learnt, to take our ideas again from their origin, and,
as Bacon says, to remake the human understanding.

Thanks to Condillac the clean sweep through chemical doctrine can take
on the air of a renaissance, a return to the source.

Finally, and above all, Lavoisier borrows from the *Traité des sensations*
Condillac's theory of the generation of ideas from elementary sensations by
successive associations. The natural course of the formation of ideas, as
described by Condillac, resembles the formation of a compound substance
from simple substances in Lavoisier's chemistry. This relationship is even
clearer when Condillac praises the virtues of analysis, which he calls 'the
lever of the intelligence'.

Lavoisier thus finds a methodology in Condillac. He can take the banal
precept, 'In order to inform ourselves, we can only proceed from the known
to the unknown,' and give it a new meaning. One mustn't start off with
acquired knowledge but with reliable data: the elementary sensations, the
facts. Lavoisier finds in Condillac the certainty that the logic at work in the
nomenclature is 'that of nature', it is 'that of all the sciences'.

We can go further: through Condillac the word 'nature' takes on a new
meaning in chemistry. It is no longer the savage material found in the furious
winds, but an order patiently constructed in a closed space, where measurement reigns. To make compound names or ideas from simple elements
is precisely the inverse of the operation carried out by the chemist in analysing
a body. The elements of chemistry are not taken from nature, they are the
results of an operation in the laboratory. The famous definition Lavoisier
gives in the 'Preliminary Discourse' puts it very clearly:

> Everything one can say about the number and the nature of the elements
> is no more, in my opinion, than metaphysical discussion; these are inde-
> terminate problems susceptible of an infinity of solutions, probably not
> one of which is in accord with nature. I will therefore content myself
> with saying that if by the name of element we intend to designate the
> simple and indivisible molecules which make up the substances, it is
> probable that we do not know them: but if, on the contrary, we attach
> to the name of element, or of principle, the idea of the last term arrived
> at in analysis, all the substances which we have not yet succeeded in
> decomposing by any means are elements for us; not because we can be
> sure that these substances we think of as simple are not composed of
> two or even a greater number of principles, but because these principles
> are never separated, or rather, because we have no means of separating
> them, they act as far as we are concerned like simple substances, and we
> should not suppose them to be compound except when experiment and
> observation have furnished us with proof.

It has sometimes been said that these lines presented the first modern definition of the chemical element. We find here, in fact, the essential requirement of simplicity, conceived in a relative and provisional manner, subordinate to analytical technique. Unfortunately, one could cite a good dozen analogous definitions in chemists who were contemporaries of Lavoisier's, like Guyton de Morveau or Macquer, even without going back to the chemists of the seventeenth century.

And yet, with this definition, Lavoisier founded an entirely new chemistry. The novelty is not in the terms of the definition. It lies in the place it occupies. Lavoisier gives it a central place, for the simple substance is the terminus of the operation of analysis carried out in the laboratory and the starting-point, the alphabet, of the nomenclature. Now this notion takes on its whole value. It is the axis around which chemistry is constructed, the point of articulation between theory and experiment. The name of a substance is indeed, as Lavoisier writes, 'the faithful mirror of its composition', because the name is the inverted image of the analysis carried out in the laboratory. The nomenclature is more than a simple vocabulary, a reflection of the practices of laboratory chemistry. Lavoisier is indeed a founder, in the sense that he repeats the gesture of ancient foundations by tracing out an enclosed space. He defines a new world between the analysis carried out by the experimenter and the catalogue of names drawn up by the nomenclator.

The origin of substances, their distribution in earth or air are of no direct concern to the chemist. What does it matter whether copper comes from Cyprus or elsewhere? The chemist's universe is not in nature. After having broken with the history of chemistry, Lavoisier broke with natural history. 'Chemistry creates its own object,' one might say, anachronistically; it constructs its own universe, transparent to reason.

This is what Lavoisier did. He founded an elementary chemistry, in the double sense of the term; it is constructed on the basis of elements, and it is remarkably simple and accessible to beginners.

Limits and Ambiguities

Such simplicity has to be paid for one way or another. It will entail certain sacrifices. First of all, among all the reactions performed by chemists, Lavoisier accords a privileged status to analysis and the inverse operation of synthesis. For him analysis becomes the unique object of chemistry, its exclusive goal: 'The object of chemistry, in submitting the different substances of nature to experiment, is to decompose them ... Chemistry advances towards its goal and its perfection by dividing, subdividing and subdividing again, and we do not know what will be the limit of its success.' In addition, Lavoisier proposes only one way of reading his analytic experiments: the comparison of the initial and final states. Instead of paying attention to all the degrees of de-composition, marking each stage with fine distinctions as Stahl did, Lavoisier retains only two terms: the compound and the simple. He introduces into chemistry a binary logic with two values, 1 or 0.

And then Lavoisier does not overturn or transcend the whole of eight-eenth-century chemistry. He deliberately leaves the study of affinity out, and invoked the work of Guyton de Morveau to explain his silence on this central topic. This circumstantial reason, however, hides a difficulty more profound: the definition of an element with its negative and wholly provi-sional criterion of the impossibility of decomposing it does not allow one to give an account of the individual behaviour of chemical substances. Only a positive criterion could guarantee the identity of each simple substance.

Finally, let us recall certain ambiguities in the Lavoisierian system. Al-though in the 'Preliminary Discourse' he claims to have finished with the chemistry of principles, he does not eliminate all the elementary principles: do the caloric and oxygen not play the role of true principles, universal mediators in all reactions? Lavoisier condemns the old principles, but gives them a prominent position in his table of simple substances. He proclaims a revolution which is far from complete. In addition, though he claims to have given up the old quest for the elements, he does retain the word; a curious omission for someone so concerned with the errors propagated by language! He bans neither the use of the word 'principle', nor that of the word 'element', which he uses as a synonym for simple substance. But he occasionally introduces a subtle distinction between element and simple substance. So the first group in the 'Table of simple substances' is entitled: 'Simple substances which belong to the three kingdoms and which one can regard as the elements of substances'. Frequency in the natural world re-mains then a pertinent criterion. This is confirmed by a note in an unpublished manuscript entitled *'Cours de chimie expérimentale rangée suivant l'ordre naturel des idées'* (Course in elementary chemistry arranged according to the natural order of ideas): 'It is not enough that a matter be simple, indivisible or at least unable to be decomposed, for it to have the title of element; it is still necessary that it be abundantly distributed in nature and that it enters as an essential and constitutive principle into the composition of a great number of substances.'

One can see then that good old principles still remain within the Lavoisierian system. The break with tradition is neither total nor completely clear, but in the eyes of the majority of chemists, Lavoisier's revolutionary intentions count for more than his acts. His work acted in history as a revolution, a revolution attributed to a single man, though it was the work of an entire generation. It would in fact be only just to mention the names of all those who developed gaseous chemistry – Hales, Black, Scheele, Priestley, Cavendish, etc. – as well as the French chemists who collaborated with Lavoisier. He did, in fact, acknowledge his debts when he sought the recognition of his peers, as witnessed by the dedication in a copy of his *Treatise* addressed to Black in September 1789. But shortly before his death, he jealously claimed the ownership of the new chemistry, and the claim was accepted: 'This theory, then, is not, as I have heard it said, the theory of the French chemists; it is *mine*, and I claim its ownership, before my contemporaries and before posterity.'

The chemical revolution comes to an end with a gesture of appropriation. Lavoisier controls the whole of the territory of chemistry.

A Reformer in the Revolutionary Turmoil

Lavoisier had less success in the management of public affairs. In this field he did, however, have a more flexible strategy. He sought to control and rationalize here as well, but he put forward only reforms. As a liberal economist, he wanted to reduce State intervention in the cattle trade in order to reduce food shortages; to create hemp-spinning workshops to avoid the export of raw materials and to create employment. When he took part in the Agriculture Committee, established in 1785 to deal with the shortage of fodder, Lavoisier waxed indignant at the archaism and low profitability of French agriculture. He insisted that the Minister should bring Pierre Samuel Dupont de Nemours on to the Committee, saying that agriculture required administrative as well as scientific skills. He illustrated this thesis in his analysis of the causes of the slump: the poverty of the farmers, who did not have the capital to invest and to modernize their operations, and the weight of taxes, which acted as a brake on any desire for progress. Lavoisier also denounced the indifference of the government towards the very poorest class in society as hurtful to the rational organization of the French economy. His social policy was motivated above all by economic considerations. Close to the Physiocrats, among whom he had some friends, Dupont de Nemours, the Abbé Emmanuel Joseph Sieyès, Malesherbes, Turgot and Condorcet, Lavoisier thought that agriculture was the main source of wealth, but not the only one.

In 1787, when the States General were re-established, Lavoisier, called upon to represent the Third Estate at the Provincial Assembly of Orléans, drew up a memorandum which tells us of his political opinions on the eve of the Revolution: he put forward a political order based on reason rather than authority. He comes out in favour of a parliamentary monarchy bringing together the will of the people and that of the King. He proposes an equitable representation of the three orders of society in the States General, insists on public debate and the liberty of the Press. Finally, he renews his attachment to Louis XVI, 'restorer of the laws . . . father of the people and benefactor of humanity'.

When the Revolution broke out Lavoisier was very soon involved in the turmoil, for the Arsenal was a strategic point. From 6 August 1789 he had to face riots, as a result of a 'powder-boat'. The people suspected that the gunpowder was destined for traitors, for the émigrés. Lavoisier, taken to the Hôtel de Ville together with another officer of the Arsenal, was threatened with execution on the spot. But he was able to speak, he was listened to, and allowed to leave exonerated.

In January 1791 he was the target of violent attacks made by Marat in the *Ami du Peuple*. Everything conspired to make him seem suspect. But Citizen Lavoisier made himself part of the revolutionary movement; a member of the National Guard, in the Arsenal section, he took part in the demolition of the Bastille. He was elected to the Commune of Paris and then, in September 1789, he joined a group of moderates, the 'Société patriotique de 1789'. In 1791 Lavoisier appeared among the six Commissioners of the new National Treasury, and drew up a report on the state of the French finances: *De la richesse territoriale du royaume de France*.

Finally, and most important, he took part in the great plan to reform weights and measures, called for by numerous *cahiers de doléances* (registers of grievances of the States General), and confided to the Academy in 1791. In this last of his works, Lavoisier brought together both his faces, as a scientist anxious to promote universal standards of measurement and an economist administrator struggling against internal customs barriers and the disparity in systems of weights and measures in order to facilitate commerce.

But all the institutions upon which Lavoisier had built his career were crumbling. After the abolition of the Ferme générale on 20 March 1791, Lavoisier identified himself more and more with the Academy. Elected Treasurer in December 1791, he skilfully negotiated the maintenance of the Academicians' salaries and dealt with the complete disorganization of public finance by advancing money to fill the Academy's coffers.

The decree of dissolution, passed by the Convention on 8 August 1793, was for him a sign of the end. In a desperate effort, he tried to have the Academy transformed into a 'Free and fraternal society for the advancement of science'. He was very close to saving the Commission of Weights and Measures, for on 11 September Fourcroy created a Temporary Commission of Weights and Measures with the same personnel as before – Lavoisier as Treasurer, Jean-Charles de Borda as President and Haüy as Secretary – with a salary for each of the eleven members. But on 24 November 1793 Lavoisier was arrested with all the other tax-farmers, and then on 28 November consigned to the prison of Port-Libre (Port-Royal). On 18 December, in the name of the Commission of Weights and Measures, Haüy protested against Lavoisier's arrest. The Committee of Public Safety replied with a purge of the Commission of Weights and Measures: Lavoisier, Haüy, Borda, Maturin-Jacques Brisson and Delambre were excluded. The preparation for the trial of the tax-farmers dragged on and on, then suddenly the whole affair was dealt with in a few days. Lavoisier was finished.

There is a legend that Lavoisier, asking for a stay of execution to complete his scientific work, was told 'The Republic has no need of scientists.' This 'historic saying' seems to be apocryphal, as Fouquier-Tinville, to whom it is attributed, did not preside over the Tribunal on that day, and the request for a stay of execution is not recorded in the minutes of the trial. But the legend expresses very well the impression of scandal and unease produced by the condemnation.

Were the organs of the Revolution made up of politicians incapable of appreciating Lavoisier's scientific achievements? One would have to forget that Fourcroy and Guyton de Morveau, both of them chemists and collaborators of Lavoisier's, were members of the Committee of Public Education at the National Convention. Lavoisier should have been able to find support and protection from numerous scientists who had taken on political responsibilities: Carnot, Monge, Hassenfratz, Guyton de Morveau, Fourcroy . . . But it was precisely these people who abandoned him. Lavoisier's last meeting with his erstwhile collaborators, in January 1794, resembled the appearance of the accused before representatives of the law. Fourcroy, Berthollet and Guyton de Morveau, authorized by the Committee for Public Safety, came to Lavoisier's house to seize all the apparatus and the files necessary for the Commission of Weights and Measures.

There were, it is true, attempts to rescue Lavoisier from the grip of the Tribunal: by the Consultation Bureau of the Arts et Métiers, which Lavoisier chaired, by various members of the staff of the Poudres et Salpêtres, and by a member of the Convention called Pierre Loysel. Louis Claude de Bessicourt, Cadet and Antoine Baumé – two chemists hostile to Lavoisier's ideas – certified that Lavoisier had never been involved in the fraudulent adulteration of tobacco, a practice current among the tax-farmers. A delegation from the Lycée des Arts went to the Conciergerie and was authorized to speak to Lavoisier on the eve of his execution. But none of his close collaborators made an attempt. Was it fear, cowardice, or revolutionary probity? Passionate explanations abound. Some heap reproaches on Fourcroy, while others make excuses for him.

An explanation begins to emerge, beyond the struggles of partisan historians. The dissolution of the Academy had loosened professional ties, already weakened through the years, says Roger Hahn in his book on the Academy, and political passion overtook them, destroying any trace of solidarity. As Hahn says as well, however, connections between specialists in the same discipline were strengthened. In chemistry, more than anywhere else, where the promoters of the new nomenclature should have formed a bloc against their adversaries, it was precisely these connections which failed.

What kind of relationships did Lavoisier have with his collaborators? Most courteous, if one is to judge from the published parts of his *Correspondence*. But faced with the behaviour of his colleagues, one might well wonder whether he did not inspire ambivalent feelings. It must be recognized that Lavoisier's refoundation of chemistry was carried out with an eye on posterity and with no regard for his predecessors or contemporaries. At the end of the undertaking in which he had appropriated chemistry, Lavoisier must have inspired more esteem and respect, hero worship and veneration, than sympathy or solidarity.

In any case, Lavoisier, abandoned at the fateful moment, was the object of worship immediately after his death. He was eulogized in 1796; on 12 August 1796 there was a great funeral at the Lycée des Arts, and Lavoisier was glorified among the pomp and show of the cult of the Supreme Being. 'To the immortal Lavoisier': this inscription was in front of a pyramid 25 feet high, with a sepulchral portal decorated with white marble caryatids; in the immense hall covered in black hangings scattered with ermine, each column bore a shield evoking one of Lavoisier's discoveries. There was a vibrant eulogy from Fourcroy, a poem set to music sung by a choir a hundred strong gathered in front of the monument. Finally, there appeared a bust, with a crown upon the head. Abandoned, then sent to the scaffold, Lavoisier was immortalized in a statue.

The Emergence of a Myth

The first to give Lavoisier's character a mythological dimension was one of those who had the greatest cause to complain of being dispossessed by him: Guyton de Morveau. He was the editor of volume I of the *Dictionnaire de Chimie* in the *Encyclopédie méthodique*, intended to complement and to improve upon Diderot's *Encyclopédie*. The article on 'Air', written by Guyton de Morveau some years before, had been conceived in terms of the phlogiston theory. When the author was converted to Lavoisier's views he inserted a 'Second Notice' in the middle of the volume to defend Lavoisier's doctrine with the ardour of a neophyte. Lavoisier is portrayed as a saviour, the champion of truth and the enemy of dogmatism, and soon compared to the 'great Descartes'. Guyton describes Lavoisier's achievement as a definitive and unalterable foundation. The order Lavoisier had brought into chemistry was the immutable order of nature: 'Posterity will see the construction of the edifice, of which they [the first chemists] have only been able to lay the foundations; but it will not think to destroy what they have done, except when nature,

478

with the same materials, and in the same circumstances, has ceased to produce the same phenomena.' Already in Lavoisier's lifetime, his work had been assigned to eternity. Hardly had it entered history than it left again.

But the reality of the work done by the French School after Lavoisier's death rather rebuts this idea of foundation. While promoting the new nomenclature, Guyton de Morveau worked on affinity, at the very margins of the system constructed by Lavoisier. Berthollet studied the conditions for chemical reactions, a problem Lavoisier left aside. Of Lavoisier's work, they retained precisely that which was opposed to the myth of the solitary creator: the collective research which he had developed in creating, with Laplace and others from the Arsenal, the Société d'Arcueil. As for Fourcroy, he kept his distance. In the article on 'Chemistry' in the same *Encyclopédie méthodique*, he presented the revolution as the collective achievement of an entire generation.

One should add that shortly after Lavoisier's death, an essential element of his system was challenged, and this in all logic should have led to the abandonment of the word 'oxygen'. In 1810 Humphry Davy, an English chemist, who had remained unconvinced, right up to the end of the century, that Lavoisier had successfully superseded the phlogiston theory, showed that muriatic acid did not contain oxygen and isolated chlorine. It was a major discovery, because it dethroned oxygen as the universal principle of acidity.

But the revisions of Lavoisier's views were not enough to tarnish his heroic image in French memory as the immortal founder of chemistry. On the contrary: towards 1830 Lavoisier was glorified with the title of 'hero of the positive sciences'. Auguste Comte set the tone in the thirty-eighth lesson of his *Course of Positive Philosophy:* Lavoisier was the admirable genius who had brought chemistry from the metaphysical era, full of chimeras and unfounded speculation, to the positive state of a rational, experimental and quantitative science. Comte admits that later research has certainly led to the revision of some too hasty generalizations by the founder, but the 'eminent scientific truths discovered by the genius of Lavoisier have necessarily retained their immediate value'.

One year later, in 1836, Jean-Baptiste Dumas devoted one of his lectures to Lavoisier, on the anniversary of his death (*Leçons sur la philosophie chimique*). Dumas takes up the theme of the founding genius and describes Lavoisier as a being inspired. A primitive intuition had commanded him to reform the science of chemistry, and he carried out this task with method and tenacity, borrowing nothing from others. Dumas thus dramatizes history with a double act of concentration. Not only was the chemical revolution the work of one man; but it is condensed into one moment, one founding intuition. The lecture ends with an emotional account of Lavoisier's death and with a promise of reparation. Dumas committed himself to publishing

Lavoisier's works in these words: 'Yes, I will provide chemists with their Gospel.' Hero with a mission, and sacrificial victim, Lavoisier is ripe for apotheosis: 'A few words on Lavoisier, whom I present to you at the moment when, pronouncing his *fiat lux*, he sweeps away with a bold hand the veils which ancient chemistry has vainly attempted to lift, at the moment when, obedient to his powerful voice, Dawn begins to pierce the darkness which must scatter before the fire of his genius.'

The creative and redeeming word of a god is unassailable: 'You have often been told: Lavoisier's theory has been modified; it has been overturned! A mistake, gentlemen, a mistake! No, it isn't true! Lavoisier is untouched, impenetrable, his armour of steel has not been pierced.'

Dumas thus belabours all those who dare to profane the memory of the creator. Lavoisier embodies the revolution so well that after him no revolution is possible. Hence the paradoxical effect of this cult of the founder: by ejecting pre-Lavoisierian chemistry into the darkness of chaos and pre-history, it is accepted that the history of chemistry begins with Lavoisier. But there is no history after this revolution because everything is definitively embedded in the foundation. And, everything considered, the revolution itself escapes from history, because it is entirely concentrated in a primitive intuition. The positivist mythology abolishes history; past, present and future, all are gathered together and condensed in a miracle formula: revolution.

Considering the whole of French chemistry in the nineteenth century, one sees a second paradox in the Lavoisierian heritage. Far from having been banished by Lavoisier, the quest for the elementary may have been encouraged by the negative and provisional Lavoisierian definition of a simple substance. Because Lavoisier invites chemists to look for substances ever more simple, there is nothing to stop them supposing that the present simple substances could not, with more powerful techniques, be decomposed in turn and derived from a single primordial element. The ambiguity between element and simple substance leaves great latitude for thought. While celebrating the founding hero of positive chemistry, the French chemists, like Dumas, devoted themselves to very un-positive speculations on the primary elements. Prudent, Dumas refused to say anything about atoms, because they go beyond experience: 'If I were master, I would delete the word "atom" from science,' he declared at the Collège de France. But he gave free rein to hypotheses on the primordial unity of matter.

One can see then that the cult of Lavoisier produced in French science a mixture of censorship and freedom, of positivist prudence and unbridled speculation. It favoured equivalentism over atomism. This debate, which divided chemists for half a century, lasted in France until the end of the century. The last bastion of equivalentism was defended by another of

Lavoisier's knights, Marcellin Berthelot, who commemorated the centenary of two revolutions with a book entitled *La Révolution chimique: Lavoisier.*

The rise of nationalist tensions in Europe, together with the greater and greater predominance of the German chemical industry at the end of the nineteenth century, reactivated the myth of the founder and enriched it with an additional theme. Dumas's Lavoisier was the incarnation of a methodology. The scientist's political fate only emphasized the image of the founder by the evocation of his martyrdom. But in 1869, on the eve of the Franco-Prussian war, the positivist scientist became a national hero. 'Chemistry is a French science: it was established by Lavoisier, of immortal memory.'

This is the opening declaration of a *Dictionary of Chemistry.* The author, Adolphe Wurtz, was an Alsatian chemist, a leader of French chemistry, but experienced in German methods after a time spent with Justus von Liebig; he was one of the rare French defenders of atomism. The patriotic fervour of his preface was seen in Germany as a serious provocation. German chemists riposted: Lavoisier was no more than an amateur, a dilettante, claimed Jacob Volhard, a professor at Munich. French Academicians waxed indignant.

The affair took off again in 1914, with the appearance of the French translation of the famous work by Wilhelm Ostwald, *Chemistry: The Evolution of a Science.* The author devoted only one page to Lavoisier, presenting his theory as an inversion of Stahl's. He ended, none the less, by saying that Lavoisier deserved his glory for the 'freedom of spirit' he had shown with regard to the ideas of the time. But this last phrase was omitted from the French translation. In addition, Ostwald was one of the ninety-three signatories of the 'Appeal to the civilized world', launched by German intellectuals in defence of the honour of German soldiers. French patriotism was cut to the quick. Pierre Duhem launched a counter-offensive and chose Lavoisier to defend his country's flag. *La chimie est-elle une science française?* ('Is chemistry a French science?') appeared in 1916. Duhem argued with tremendous skill. He conceded to the enemy that the French exaggerated the importance of Lavoisier when they combined Stahl's doctrine with alchemy in the same obscurantist cloud. But Duhem rehabilitates pre-Lavoisierian chemistry only in order the better to humiliate German pretensions. He claims in fact that if Stahl was indeed the author of the phlogiston theory, it was to a French chemist, Rouelle, that it owed its success: 'To the German idea, the larva of a theory, he gave French wings.' Chemistry had always had one homeland and that was France. Its identity was renewed and confirmed by the cult of Lavoisier.

Three figures thus succeed one another in the French chemical literature of the nineteenth century. The first has no political colouring; Lavoisier as creator of an unchanging order, celebrated as a saviour. Then, in the 1830s, he is the hero of positive science, a victim of politics, his sacrifice calling forth a cult of redemption. Finally, under the Third Republic, Lavoisier is a national hero, the honour of a France too often humiliated.

So, in getting rid of a tax-farmer, the Convention gave birth to an idol. The object of a cult, subjected to varying interpretations, the bearer of many different values, Lavoisier is a memorial, a monument to the glory of science and of France. The figure of Lavoisier derives part of its symbolic power from the interference between his work in chemistry and his tragic destiny. But the essential elements of the myth were composed by Lavoisier himself, in the course of a vast enterprise, masterfully carried out.

Among all the areas in which he was active – administration, finance, agriculture and academic science – Lavoisier spotted first of all a field of operations. Chemistry seemed to him to offer favourable ground, particularly the recent gaseous chemistry, and the chemistry of principles which provided him with his theoretical framework. He took over the territory little by little, thanks to a quantitative, planned and organized experimental method, before declaring his victory over the centuries-old and still lively tradition of the elementary principles. He finalized and completed this revolution by reorganizing chemistry in a way that radically changed the historical significance of the event. This was not the replacement of one body of doctrine or practice with another, but the creation of chemistry as a scientific discipline.

In many respects Lavoisier's undertaking illustrates the favourite themes of the century of the Enlightenment: the overthrowing of tradition, the appeal to the natural, the rationalization of language and the idea of revolution, introduced into scientific literature by Fontenelle, which then become commonplace after Diderot's *Encyclopédie*. But the originality of the revolution carried out by Lavoisier lies, it seems to me, in the importance accorded to a particular task, specifically administrative in its origins: control. Lavoisier's success rests on the multiplication of controls: control of space, with the laboratory substituting for nature; control of objectives, with chemistry having no other purpose than analysis; control of laboratory practice with the balance; control of theory through concepts forged in the light of experiment; control of language through nomenclature; control of the future through the *Traité élémentaire*, which enabled the training, in a very short time, of armies of skilled chemists; and finally control of the past, thanks to a philosophy which justified amnesia. Lavoisier not only changed chemists' working reality, but he transformed their image by remodelling their history. Wiping out the traces, sweeping aside the predecessors, these were the gestures which formed the statue of a founder, and favoured the emergence of a new image of chemistry.

16

In Defence of Geology: The Origins of Lyell's Uniformitarianism

Geof Bowker

In this chapter we will look at the work of Charles Lyell, a geologist whose great work of synthesis, *The Principles of Geology*, has often been seen as founding the scientific discipline of geology in Britain in the 1830s. I will argue that Lyell was indeed a founder, and will try both to uncover the specific strategies that he used to defend the nascent discipline and discuss where those strategies came from.

What does it mean to 'found' a science? One thing it has frequently meant is that a space is created that is not to be touched by the Christian Church. The story of Galileo's fight against the Catholic priesthood is often told as the battle of scientific rationality against religious persecution. When Boyle first tried to define the rules of experimental practice, one of his chief concerns was to make the laboratory a space where religion could not enter. Similarly, Lyell saw his foundation work as taking the history of the Earth out of the hands of religious fundamentalists. These latter used calculations based on the readings of the Bible, in particular the number of generations from Adam until the present day, to give the Earth an age of some 6000 years. The most precise measurement had the earth created at 9.00 a.m. one Monday morning in 4004 BC. No geological evidence had weight against this biblical analysis. Lyell countered this by arguing that the Earth was sufficiently old for there to be no trace of its origin.

The other major foundational work that Lyell saw himself doing was setting down the basic laws that other geologists working empirically could draw on in their own studies. He gave a general rule that the kinds of forces acting in the world at the moment were the same kinds of causes that had always existed, at least as far back as the geological record went. This was a powerful rule. It meant that one could not refer back to a time when there were more earthquakes than at present, or when mountain ranges were thrust up in a single moment and so on. One had to find slow-acting, steady causes in place of the 'catastrophic' causes often referred to by his opponents, religious and other wise. We will look at both these types of foundation.

We will not be arguing that Lyell did in fact found the profession. He certainly tried to offer one possible intellectual foundation: the very title of his work echoes that of Newton's *Principia*,[1] which was the paradigm case of a foundation text at the time Lyell was writing. He was not, however, the first to argue for the extreme age of the Earth; other British and continental geologists of the previous century had done the same. Further, French geology continued at the time and throughout the nineteenth century largely uninfluenced by Lyell's work. It has even been argued that Lyell was not so influential in the later nineteenth century in England, some of his central positions being reversed. Later evidence seemed to indicate that the Earth was only 40,000 years old – far too short a period for Lyell. He would not accept evidence for an interior heat in the Earth (which would have been seen as proof of a molten origin). Lyell was in general seen as something of an extremist by his colleagues and by the generation that followed. His work was foundational simply in the sense that he was writing in England at a time when so-called 'natural theology' – proof of the existence of God's

---

Fluctuations in the age of the Earth since Charles Lyell

One of the remarkable discontinuities in the history of science consists in various estimates of the age of the Earth since those of Lyell. It is possible to think that Lyell won his case and that his conclusions are hardly different from those agreed now. In fact, from about 1880, he seems to have lost some of them. Lord Kelvin (1824–1907) and several other physicists, on the basis of data concerning the present internal heat of the Earth and its speed of cooling, assumed that the age of the Earth could not be more than 40,000 years. Lyell was willing to consider again the hypothesis of an eternal Earth, but nevertheless, not as far as that precise figure. However, with the discovery of the effect of radioactivity on the heat of the world, Lyell's theories were rehabilitated and recent history, which considers him 'the founder of geology', ignores the lost generation of geologists and physicists who thought he was wrong.

---

design in nature through scientific work – was very powerful and he wrote a foundation work so as to counter this trend.

What is of interest to us here is to examine what work he did to try to lay these foundations and to see where his solutions came from. We will see Lyell arguing for a kind of geological time and causality which, if applied by geologists, would effectively exclude the entry of the theologians into the debate. We will then trace some possible origins of this new time and causality in the actual working methods of geologists in the 1830s.

A Time for Geology

There are two types of time at work in Lyell. One we might call time as a passive container: it involves the attempt to give a chronology to the history of the Earth, to trace its origin or to deny that there is any evidence that it has one. The second is time as process: it involves the attempt to pick out certain types of changes that are invariably associated with the history of the Earth at any age and are thus in a sense a feature of time itself. We will now look at how these types of time were articulated within Lyell's work and how they served to create a time that was specific to geology. In particular, we will see how Lyell dealt with religious time (sacred history) and human time (secular history) in order to create a special time for geology that could be dealt with by professional geologists.

Lyell said that to all intents and purposes the Earth could be taken as being eternal. It may once have had an origin, but no sign of this remains. This loss of the origin could be explained by the fact that the Earth was moulded by complementary destructive and creative forces. The latter (flowing water, tides and so on) visited each corner of the Earth, grinding it down, dissolving it. The former (silt deposition, volcanoes and so on) redistributed this formless matter, which thus bore no traces of its state before its dissolution. Each and every part of the Earth only bears traces up to its last dissolution, and since there has been an indefinite number of these, there is no point in trying to discuss the origin of the Earth. Lyell's geology has been taken as the triumph of 'linear' time because it locates the Earth along an indefinitely long line between the past and the future; however, beneath this crust of linearity we find a core of cyclical morphology for the Earth.

In the following collection of citations, we can get some picture of the workings of this calculus of temporal regularity:

> There can be no doubt, that periods of disturbance and repose have followed each other in succession in every region of the globe; but it may be equally true, that the energy of the subterranean movements has been always uniform as regards the *whole earth*. The force of earthquakes

may for a cycle of years have been invariably confined, as it is now, to large but determinate spaces, and may then have gradually shifted its position so that another region, which had for ages been at rest, became in its turn the grand theatre of action . . . In order to confine ourselves within the strict limit of analogy, we shall assume, 1st, That the proportion of dry land to sea continues always the same. 2dly, That the volume of land rising above the level of the sea, is a constant quantity; and not only that its mean, but that its extreme height, are only liable to trifling variations. 3dly. That both the mean and extreme depth of the sea are equal at every epoch; and, 4thly, It will be consistent, with due caution, to assume, that the grouping together of the land in great continents is a necessary part of the economy of nature . . . [On this base, he argued for a climatic 'great year'; the phrase is a reference to the Stoic's Great Year, which marked the period for the repetition of history] . . . We have now traced back the history of the European formations to that period when the seas and lakes were inhabited by a few only of the existing species of testacea, a period which we have designated *Eocene*, as indicating the *dawn* of the present state of the animate creation. But although a small number only of the living species of animals were then in being, there are ample grounds for inferring that all the great classes of the animal kingdom, such as they now exist, were then fully represented . . . Species could, conceivably, survive complete 'revolutions' of the earth's surface.

There is a consistent patterning to the disparate quotes of this text. In each, the part is taken as varying, as liable to be created or destroyed, whereas the whole is immutable and eternal. Mediating between the two is cyclical change: a 'cycle' of years attached to a region, a climatic great year attached to the Earth over time, and 'revolutions' of the Earth's surface attached to species change.

The first volume of the *Principles* gives a series of causes of change and shows how each destructive cause is equally, and in the same degree, constructive. Thus he writes with respect to sea currents that: 'In the Mediterranean, the same current which is rapidly destroying many parts of the African coast; between the Straits of Gibraltar and the Nile, preys also upon the Nilotic delta, and drifts the sediment of that great river to the eastward. To this source the rapid accretions of land on parts of the Syrian shores may be attributed.' Similarly, volcanoes on the surface of the Earth seem to increase the general area of land mass, but submarine volcanoes raise the level of the sea, so the two cancel each other out.

Lyell's assertion of a number of things that never change – land mass, degree of force of volcanic activity and so on – seemed, in the opinion of his contemporaries, directly antithetic to the geological evidence. They also seem a long way from the kind of time we would expect to be associated with the Industrial Revolution, which was reaching its peak as Lyell wrote.

486

The stillness, the ineluctable equilibrium between creation and destruction, contradicted the facts available to Lyell's contemporaries in various ways. One set of contradictions revolved around the whole schema, others around the position of humanity within it. In brief, the overarching problem was this: for all that Lyell might say that 'present causes' explained all past geological occurrences, it was hard to believe that mighty mountain ranges were even now thrusting upwards. Nature had left a series of monuments that looked for all the world like products of cataclysmic change of an order undreamt of today. Whole species disappeared in a flash from the fossil record. It scarcely seemed likely that massive continents had pushed out of the sea at an inch a century. Great truths demanded great causes. More probably, it seemed to most geologists, times had once been different, the world was younger and more lively. This image of an Earth once lively going through a peaceful middle age was commonly used by Lyell's rivals, the catastrophists.

Yet the catastrophic time that these geologists employed was used to reconcile the fossil and geological record with the Bible. The argument here was that it may seem difficult for all the evident changes on the face of the Earth to have happened in 6000 years, yet what really happened is that time went faster then – there were more earthquakes, more volcanoes and so on. Further, this argument was used to give humanity a privileged position within the geological record. For, it was said, God waited until the Earth was in repose before He introduced humanity – for whom it was created – onto its face. Lyell met both these privileged times head on in his work of defending the existence of a separate time for geology.

To do so he developed two sets of metaphors, the first drawing on the statistical societies flourishing in his time and the second drawing on the image, sanctified by long usage, of the Book of Nature. The first dealt with the problem of massive discontinuities in the fossil records which were attached, said most geologists, to catastrophic changes in the past. He asserted that fossils were created only where new strata were being formed, and wrote that:

> these areas, as we have proved, are always shifting their position, so that the fossilizing process, whereby the commemoration of the particular state of the organic world, at any given time, is effected, may be said to move about, visiting and revisiting different tracts in succession. In order more distinctly to elucidate our idea of the working of this machinery, let us compare it to a somewhat analogous case that might easily be said to occur in the history of human affairs. Let the mortality of the population of a large country represent the successive extinction of species, and the births of new individuals the introduction of new species. While these fluctuations are gradually taking place everywhere, suppose commissioners to be appointed to visit each province of the country in succession,

taking an exact account of the number, names, and individual peculiarities of all the inhabitants, and leaving in each district a register containing a record of this information. If, after the completion of one census, another is immediately made after the same plan, and then another, there will, at last, be a series of statistical documents in each province. When these are arranged in chronological order, the contents of those which stand next to each other will differ according to the length of the intervals of time between the taking of each census. If, for example, all the registers are made in a single year, the proportion of deaths and births will be so small during the interval between the compiling of two successive documents, that the individuals described in each will be nearly identical, whereas, if there are sixty provinces, and the survey of each requires a year, there will be an almost entire discordance between the persons enumerated in two consecutive registers ... the comissioners are supposed to visit the different provinces in rotation, whereas the commemorating process by which organic remains become fossilized, although they they are always shifting from one area to another, are yet very irregular in their movements, [so that] ... the want of continuity in the series may become indefinitely great, and ... the monuments which follow next in succession will by no means be equidistant from each other in point of time.

Apparent discontinuity is, then, an effect of want of knowledge and not a sign of real discontinuity. By extension, the heavily accented features of the Earth are a product of the way that the Earth keeps its own records about itself and not a feature of variations over time in the constitution and virulence of its governing forces.

Let us look at a second metaphor that Lyell used to explain the apparent asymmetry between past and present. This metaphor will bring out the peculiar centrality of humanity in Lyell's geology and thus the centrality of human society to his problematic. It revolves around an image sanctioned by long usage in scientific texts: the idea of the Book of Nature. Many writers developed this peculiarly rich theme. In the natural theology that Lyell opposed, the Book of Nature was taken to be fully complementary to the Book (the Bible). Here is his development of the theme:

If, then, there were no spots discoverable which exhibited signs of extraordinary mechanical and chemical changes, the effects at some former period of immense pressure, intense heat, and other conditions far different from those developed on the surface, it might be urged as a triumphant argument against those who are dissatisfied with the proofs hitherto adduced in favour of the mutability of the course of Nature.

In order to set this in a clear light, let the reader suppose himself acquainted with just one-tenth part of the words of some living language, and that he is presented with several books purporting to be written in

the same tongue ten centuries ago. If he should find that he comprehends a tenth part of the terms in the ancient volumes, and that he
cannot divine the meaning of the other nine-tenths, would he not be
strongly disposed to believe that, for a thousand years, the language has
remained *unaltered*? Could he, without great labour and study, interpret
the greater part of what is written in the antique documents, he must feel
at once convinced that, in the interval of ten centuries, a great revolution
in the language had taken place . . . So if a student of Nature, who, when
he first examines the monuments of former change upon our globe, is
acquainted only with one-tenth part of the processes now going on upon
or far below the surface, or in the depths of the sea, should still find that
he comprehends at once the imports of the signs of all, or even half the
changes that went on in the same regions some hundred or thousand
centuries ago, he might declare without hesitation that the ancient laws
of nature had been subverted.

The logic of this passage is not, perhaps, immediately clear – not surpris-
ingly, it was dropped from later editions. What Lyell is saying is that at
present our knowledge of the Book of Nature is highly restricted (to processes occurring on land, and only to a small proportion of these). He argues
that if from our knowledge of these processes we could reconstruct the
history of the Earth, then the past must have been very different – for that
would mean that the small proportion of causes that we know about today
were once all the causes there were. In the first metaphor we looked at,
then, the Earth kept only a limited and random sample of its own records;
here we have access only to a limited and random sample of the words of
the Book of Nature.

Lyell's defence of his geological time against appearances to the contrary
is, first, that these appearances are necessarily deceptive if his system is
right; and second, that there is no way at present that geologists could know
enough to explain past changes. So far as we have gone, we have seen him
arguing against any possible connection between religious time and geological
time by denying an origin to the Earth and buttressing his new geological
time against possible counters by arguments about the nature of the geological record. When he arrived at this point in his argument, he believed that
the bases had been laid for a true science of geology – an argument he
proposed using the contrast between his own true language of geology and
the false language of catastrophists:

> These topics we regard as constituting the alphabet and grammar of
> geology; not that we expect from such studies to obtain a key to the
> interpretation of all geological phenomena, but because they must form
> the groundwork from which we must rise to the contemplation of more
> general questions relating to the complicated results to which, in an
> indefinite lapse of ages; the existing causes of change may give rise.

He made two further moves in order to defend his time. First he tried to legislate for the way that geology would develop as a discipline by trying to attach to it the same time that he attached to the history of the Earth. Secondly he produced arguments to counter the idea that geological time was somehow different since the advent of humanity – more peaceful, or transformed by its presence. We will now look at each in turn.

For Lyell, just as the past history of geology is concerned with catastrophes, so is the past history of the discipline of geology catastrophic; it is 'between new opinions and ancient doctrines, sanctioned by the implicit faith of many generations, and supposed to rest on scriptural authority'. Lyell does not, however, abandon his patterning for geological time when he turns to geologists. The imperceptibly slow operation of simple causes operates for both the Earth and its scientists: 'By the consideration of these topics, the mind was slowly and insensibly withdrawn from imaginery pictures of catastrophes and chaotic confusion, such as haunted the imagination of the early cosmogonists.' To get some idea of just how long a period of time he has in mind, we can turn to his proto-Jungian assertion that: 'The superstitions of a savage tribe are transmitted through all the progressive stages of society, till they exert a powerful influence on the mind of the philosopher.' Thus the catastrophic history of geology is itself underwritten by slow, insensible change. The two rhythms of time (the catastrophic and the uniformitarian) battle it out both within the history of geological ideas and the history of the Earth. Just as our readings of the Book of Nature should become ever more uniformitarian, so should our reading of the history of geology. Lyell signals this change in the nature of the history of geology in a return to the language metaphor. He referred in a lecture to London high society to the former (catastrophic) state of geology:

> While the science was in so fluctuating a state the philosopher who was anxious to discover truth, naturally preferred to enter himself into the field of original investigation, rather than to devote his literary labours; to the comparison and the reduction into order of imperfect observations and a limited collection of facts. One of our poets alluding to the incessant fluctuations of our language after the time of Chaucer complains that:

> > 'We write on sand, the language grows
> > And like the tide our work o'erflows'.

What a contrast with the future, when:

> We shall from year to year approach nearer to the time when the new facts which can be added by one generation of men however important will form but a trifling contribution to the stock of knowledge which had previously been acquired and when that period shall arrive they who

have no opportunity of travelling themselves or of constantly associating with those who are engaged in actual observation will be more on a par.

We can, then, unify Lyell's pictures of the history of geology and the history of the Earth. In the past, knowledge developed catastrophically and analyses were framed in terms of catastrophes; in the present and future, knowledge develops uniformly and analyses are framed in terms of continual steady change. Lyell encourages us in this formulation when he asserts that: 'The connexion between the doctrine of successive catastrophes and repeated deteriorations in the moral character of the human race is more intimate and natural than might at first be imagined.' There is, indeed, a powerful moral force in Lyell's geology which derives from just this quasi-symmetry between the past of geology and of the Earth. It would be better all round, the reader feels, if the time that has eternally framed nature were to frame human society. Thus the same time that is to serve to defend geological inquiry against religious dogma (by operating a separation between religious and geological time, the first concerned with origins and the second not) is also to serve to define the development of the history of the new discipline of geology as opposed to the development of religious theory.

Time and the sciences

From the beginning of the nineteenth century, geology was the first of the sciences to take the nature of time as a central theme. After Lyell, three main phases succeeded each other; in each of them the dominant science of the period concerned itself in one way or another with time. The subject passed first to Charles Darwin (1809–82), whose work provoked debate about the role of historic time in science. Darwin refused to consider its origins, just like Lyell. Then came the Second Law of Thermodynamics, which gave historic time to the history of the Universe. Astronomical bodies could no longer be considered as self-regulating systems, possibly infinitely old, because the quantity of entropy added to time had the effect of making all systems tend towards disorganization. Then, with the theory of relativity and quantum mechanics, physics came back to the centre of the stage. Observers starting from different referents arranged events differently, and present acts of observation could produce events that took place in the past. All this completely upset our intuitive ideas of the past and the present. Although geology and biology had a tendency to deny the specificity of the present and to push humanity to the back of the stage, astronomy and physics gave the present a new specificity and the sentient observer a new centrality. In other words, geology and biology destroyed religious time, while astronomy and physics founded a new time, fit for the religion of science.

There is a final way in which the new time that Lyell is using to found the discipline of geology is applied in his *Principles*. This is in his resolution of the problem of whether the time of the Earth is somehow different since the creation of humanity. Unlike all the other objects in Lyell's geology, humanity irrupts into the picture at a very specific moment. Moreover, this moment is 6000 years ago: precisely the moment that biblical fundamentalists picked for the origin of the whole Earth (including humanity). Not only did humanity make a singular appearance, however, it also set about creating the appearance of singularity. Thus Lyell commented on hybrids that displayed extreme variability in their outward form (and thus changed at a pace too fast for his geology): 'it is easy to show that these extraordinary varieties could seldom arise, and could never be perpetuated in a wild state for many generations, under any imaginable combination of accidents. They may be regarded as extreme cases brought about by human interference, and not as phenomena which indicated a capability of indefinite modification in the natural world.'

Humanity, then, makes time look as if it is irreversible and rapid (even catastrophic), but this serves for Lyell only to highlight the fact that underlying reality is as uniform as can be. In general, humanity has not only a tendency to read the Book of Nature wrongly, it has had a tendency to write it wrongly too, making the same mistake in each instance. Lyell has two strategies for playing down humanity's influence: accreting it to the natural and assigning it to another plane of existence.

In the former, Lyell stresses that changes wrought by humanity are for all that natural changes. Humanity does its work of sowing seeds far afield, but these seeds would have been sown regardless: by the wind or through the agency of a migrating bird. Nature keeps a check on the whole process by organizing flora and fauna into 'nations': nothing can survive long outside its nation. This 'natural' side of humanity is totally divorced from its civilized side, as Lyell stresses in the following passage:

> Were the whole of mankind now cut off, with the exception of one family, inhabiting the old or new continent, or Australia, or even some coral islet of the Pacific, we should expect their descendants, though they should never become more enlightened than the South Sea Islanders or the Esquimaux, to spread in the course of ages over the whole earth, diffused partly by the tendency of populations to increase beyond the means of subsistence, in a limited district, and partly by the accidental drifting of canoes by tides and currents to distant shores . . . Like them [the inferior animals] we unconsciously contribute to extend or limit the geographical range and numbers of certain species, in obedience to general rules in the economy of nature, which are for the most part beyond our control.

Both the spread of humanity and its ability to act as dispersive agent are, then, fully natural and under Nature's control.

There is, however, another aspect to humanity: its ability to transform landscapes and species temporarily. To account for this aspect, Lyell develops his second strategy for playing down humanity's influence: he posits a complete divorce between civilized humanity and Nature. The changes humanity has wrought are:

> not of a *physical* but of a *moral* nature ... It will scarcely be disputed that we have no right to anticipate any modification in the results of existing causes in times to come, which are not conformable to analogy, unless they be produced by the progressive development of human power, or perhaps from some other new relations between the moral and material worlds. In the same manner we must concede, that when we speculate on the vicissitudes of the animate and inaminate creation in former ages, we have no ground for expecting any anomalous results, unless where man has interfered, or unless clear indications appear of some other *moral* form of temporary derangement.

The two arguments about human time can be summarized thus: in so far as humanity interacts with geological time, it is the bestial part of humanity fitting into the economy of Nature (a phrase much used by Lyell), whereas civilized humanity operates in a different dimension to Nature and creates the temporary appearance of an anomaly in Nature's Book.

In general, we have seen that Lyell's foundation work on the creation of geological time operates a series of divorces. The time of the origin is given to religion, the rest of time (effectively all of time) is given to the geologist. Catastrophic change is given to the history of the Earth sciences before the foundation of geology by Lyell; the new discipline of geology will be uniformitarian. Humanity's 'moral' influence is seen as outside geological time and fully reversible; its 'physical' influence is fully within geological time. Thus a single time is created for the history of the Earth, for the development of Earth sciences and for human time – and it is a time whose study is the province of the geologist.

Creating a Kind of Knowledge for Geology

We have seen, then, that Lyell created a separate time for geology, free from religious time. We will trace in this section how he created a causality for geology. Here, we will see, he had two targets in mind. One was the argument that the ultimate causal quest was for God's design in Nature, and the other the argument that the basis of scientific causality was physical causality, with physics being the dominant science. We will see how Lyell defended his foundling discipline against these two threats by creating a distinct kind of causality to operate within geological time.

We will pursue our enquiries by comparing Lyell's work with two from a series of tracts on natural theology published in England during the 1830s, tracts called the Bridgewater Treatises. These formed a major series of books written by leading scientists of all disciplines. They had been commissioned in a bequest made by the dissolute ninth Earl of Bridgewater. The Earl made his fortune in building canals in the industrial north of England, but he squandered most of his money carelessly. Nevertheless, in his will he made provision for the publication of a series of pious works. The President of the Royal Society, with the help of the Bishop of London and the Archbishop of Canterbury, chose eight men of science, who were instructed to shed light on 'the power, wisdom and goodness of God, manifested in his Creation, illustrating the proof with all reasonable arguments'. Charles Babbage, who invented an ancestor of the computer (see p. 638), wrote a ninth renegade treatise, which was not commissioned. The sort of reasoning according to which the existence of God was proved by the fact of the Creation apparently forming a perfectly ordered whole, with even the anomalies appearing as the expression of an intelligent design, is common to all the Treatises in the series. Though they were supposed to consider all the sciences, most of them contain long chapters devoted to geology, which was then expanding fast. We shall look at the Bridgewater Treatise by the Reverend William Buckland, devoted to geology, in order to learn more about the religious causality that Lyell was fighting. Then we shall turn to Babbage to find out more about the subject of physical causality, for which 'the founder of geology' challenged his priority.

Considering the work of Buckland as representing the opionion of High Church religious authorities (as against those of the fundamentalists), we shall see that clergymen were prepared to follow Lyell to a certain point – but without going far enough to raise the question of the infallibility of the Bible or the Book of Nature. Buckland made frequent allusions to the work of Lyell and admitted the principle of the Earth having persisted through an indefinite number of ages. He had two methods of squaring Lyell with the Book of Genesis. The first, which has survived to this day, is to say that: 'there is . . . no sound critical, or theological objection, to the interpreation of the word 'day', as meaning a long period.' Buckland then proposes a more literal reading: 'but there will be no necessity for such extension, in order to reconcile the text of Genesis with physical appearances, if it can be shown that the time indicated by the phenomena of Geology may be found in the undefined interval, following the announcement of the first verse.' The first evening, then, 'may be considered as the termination of the indefinite

The importance of geology in the nineteenth century

Geology was by far the dominant scientific discipline of the period. Aimé Boué has summarized its phenomenal expansion in the following way: 'Comparing the number of books published in 1833 to those in the years 1830, 1831 and 1832, the approximate proportion is established by the numbers 300, 450, 500 and 900' (*Bulletin de la Société Géologique de France*, 1833). In France, according to the *Écho du Monde Savant*, publications on geology and palaeontology in 1833 were far more numerous that those of all other sciences put together: 'Physical and natural sciences (among them astronomy, physics, magnetism, meteorology, chemistry, hydrography and natural history): 144 books, 276 papers; palaeontology and geology: 61 books, 414 papers' (*Écho du Monde Savant*, 20 June 1834). The same tendency was seen in England.

The *Écho du Monde Savant* allows us to follow step by step the peregrinations of Parisian geologists.

> In Paris this year Saturday and Sunday are essentially geological days. Saturday: at 9 in the morning, M. Brongniart begins his lecture on geological mineralogy at the Museum [of Natural History]; at 9 M. Boué delivers his private lecture in the rue Guénégaud; at 2 o'clock M. Élie de Beaumont steps on to the rostrum at the Collège de France; at 7 in the evening M. Boué delivers his public lecture at the Société de Civilisation; and at 8 o'clock M. Rozet starts at the Athénée. Sunday: MM. Constant-Prévost and Boué lead, separately, their troops, armed with hammers, canes and bags for rocks, rousing here and there the fear of the Republic or the edification of a school; while M. Boué explains, from 3 to 4, in the rooms of the Société, the geological relationships of the countries of Europe to those who, unwilling to expose their heads in the villages or their feet on bad roads, prefer to travel on the maps spread out for them by M. Boué. Nevertheless, it must be said that Saturday is going to lose M. Rozet and Sunday M. Boué. These two geologists finished their precious lectures last week, but in compensation M. Cordier's course, which will start soon, will offer a geological meeting and excursions with M. Élie de Beaumont. The organization of these will be announced soon, offering similar advantages on Sundays to a third band of rock-hunters.' (*Écho du Monde Savant*, 10 April 1834)

time which followed the primeval creation announced in the first verse' and the second verse 'may be geologically considered as designating the wreck and ruins of a former world. At this intermediate point of time, the preceding undefined geological periods had terminated, a new series of events commenced.' Genesis, then, is literally right, and so is Lyell . . . about the age of the Earth.

Turning to the Book of Nature, Buckland stressed that it too is literally right. He asserted that:

> The study of these Remains will form our most interesting and instructive subject of inquiry, since it is in them that we shall find the great master key whereby we may unlock the secret history of the earth. They are documents which contain the evidences of revolutions and catastrophes; long antecedent to the creation of the human race; they open the book of nautre and swell the volumes of science . . . [with the help of] recent discoveries in the science of Geology.

He writes of seeing petrified trees in a coal mine in Bohemia, 'little impaired by the lapse of countless Ages; and bearing faithful records of extinct systems of vegetation, which began and terminated in times of which these relics are the infallible Historians.' Even a smooth, rounded pebble is 'fraught with records of physical events'. Whereas for Lyell Nature is profoundly and perhaps irretrievably unknowable, for Buckland it is all in essence already known. Buckland's infallible Book of Nature contains the sure traces of God's design, which provides the true link for geological events; Lyell's fallible book obscures these very traces.

In order to see how Lyell reads this fallible book, we will turn to his reaction to Babbage's use of Laplacean determinism. Laplace had written that:

> An intelligence who at some given moment knew all the forces that animate nature, and the respective situation of the beings that compose it, if it were further sufficeintly vast to submit these data to analysis, could embrace within a single formula, the movements of the largest bodies of the universe and those of the lightest atom: nothing would be uncertain for it, and the future, like the past, would be present to its eyes.

Babbage's ninth Bridgewater Treatise made much of this kind of determinism, which mimicked the operation of his calculating engines – which could even be programmed to contain the numerical equivalent of miracles, if the algorithm were complex enough. He wrote that: 'the air itself is one vast library' because when we speak:

> the waves of air thus raised, perambulate the earth and the ocean's surface, and in less than twenty hours every atom of its atmosphere takes

up the altered movement due to that infinitesimal portion of the primitive motion which has been conveyed to it through countless channels, and which must continue to influence its path throughout its future existence.

He believed that if we knew the original position of every atom in the atmosphere, we could trace its complete future. Every murderer bore a record of his crime, 'some movement derived from that very muscular effort, by which the crime itself was perpetrated'. For a very sensitive organ of hearing: 'all the accumulated words pronounced from the creation of mankind, will fall at once upon that ear.'

Lyell, on receiving a first manuscript of this book from Babbage, fired off a series of criticisms of these passages. Basically these consisted, as in his criticism of design, of stressing over and over the fallibility of the Book of Nature. Thus he wrote to his friend:

> If it be true that all sounds remain in the air, which I cannot help doubting, something should be said for the benefit of the ignorant . . . Can the air be said to be the historian when it is only a mute depositary unread by any one and unheard? Do not the circles on the water cease at last, an ordinary reader (for whom you write) will feel annoyed at not being told how it is that in a resisting medium undulations are not at length destroyed, how it is that they do not combine with others so as to produce new sounds and notes and words.

In general, he considered the book in bad taste, and recommended against publication.

Babbage's is certainly an extreme expression of the theme of complete determinism, but the theme was a common one at the time. Between Buckland and Babbage, then, we have two arguments for the complete knowability of the Book of Nature. One threatened to subsume geology in theology and the second threatened to subsume it in physics. We will now see how Lyell created a picture of the work of geologists that allowed him to know nature without being a theologian or a physicist.

For Lyell, the role of the interpreter of nature is central: God and Nature are both profoundly unknowable, and it is only through an epiphanic moment of profound insight that the scientist can hope to grasp their mysteries. Lyell referred to this moment in a citation from Niebuhr: 'he who calls what was vanished back again into being, enjoys a bliss like that of creation.' Any human attempt, religiously motivated or not, to offer some more direct way of reading or writing the Book of Nature deserved utter scorn. Along with the romantic poets (Keats in his 'Ode on a Grecian Urn' or Byron in 'Childe Harolde', for example), Lyell found reverence and sublimity in the capture of the tension between the fleeting instant and eternal ages; thus anyone who saw 'the summit of Etna often breaking through the clouds for a moment with its dazzling snows, and being then as suddenly withdrawn' must 'form the most exalted conception of the antiquity of the mountain'.

Lyell, by attaching his geological time to a romantic theory of knowledge, frees it finally from the purview of the theologian or the physicist. In so doing he lent the field a grandeur of the sort that is praised in the following analysis of the contribution of the French naturalist, Baron Cuvier, published in 1836:

> The mathematician and the natural philosopher had assumed to themselves the highest locality in the temple of science, and had almost expelled the collector and the classifier from its precincts. Presuming that magnitude and distance ennobled material objects, and invested with sublimity the laws by which they are governed; and taking it for granted that the imponderable and invisible agencies of nature presented finer subjects of research than the grosser objects which we can taste, touch, and accumulate, they have long looked down upon the humble and pious naturalist as but a degree superior to the functionary of a bear garden, or the master of cermonies to a cage of tigers. This intolerable vanity – this insensibility to the unity and grandeur of nature, to the matchless structure of sublunary bodies; and to the beautiful laws of organic life, was perhaps both the effect and the cause of the low state of natural science during the preceding two centuries. Men of acute and exuberant genius were naturally led to invest their intellectual capital in researches that were likely to return them an uxorious interest in reputation; and it must be acknowledged that the richest fields of science were for a long time left to the cultivation of very humble labourers.

In this passage, we see in fine the very movement that we have traced with Lyell in this section: the development of a new kind of knowledge for natural scientists – one based on appreciation of the beauty of nature and its laws – in order to develop the discipline of natural science, which otherwise would languish in the shadow of another dominant form of knowledge – mathematics and natural philosophy in this case.

The Profession of Geologist

So far, we have shown that Lyell created a new time and a new kind of knowledge for geology; now we can start to look at where this new time and knowledge came from. I will argue that both were responses to the information explosion that the foundling discipline was undergoing. They were certainly not the only possible response – indeed, as pointed out above, another model won the day – but they were a response none the less.

To see how they were, we will look at some work where the link is drawn fairly explicitly – in the writings of a French geologist of the time, Léonce Élie de Beaumont – and compare this work with that of Lyell. Here, first of all, are some lecture notes that Beaumont made for an introductory lecture on geology at the Collège de France in 1839:

> today, now that we start to be able to go to St. Petersburg in 5 days, to Constantinople in 8 to 10 and to New York in 14, given that today with the electric telegraph people talk to each other by signs at several hundred leagues distance; we are at the start of a new era when the locality of each person will be much bigger than it has been up to now because the ability to move around will have been much increased and the inconvenience of being away from home will be greatly diminished. We are approaching a time when the locality of each geologist will be the terrestrial globe. It is then that a philosopher will be really able to call himself citizen of the universe ... Buffon ended the heroic age of geology wherein everyone constructed complete systems; it was impossible to go further without making geology the province of a large number of people and as a consequence a profession having its own rules ... it is after him and not though him that geology took its place among the academic sciences, which grow gradually through the successive works of a collection of individuals, it is the application of the principle of the division of labour.

The equivalent lecture in 1834 was entitled 'The Speciality of Geology deduced from the special nature of the geologists' way of, life. Here are some notes from it:

> the nature of geological science deduced from the order which establishes itself in the work of geologists ... the geologist is therefore of all the classes of scientist the most obliged to displace himself ... that fact makes it even more likely to make him part of a distinct class than that this circumstance calls on a particular type of person ... of all the sciences, it is geology that relies most on improvement of the means of transport; means of transport are for the geologist what telescopes are for the astronomer. The new roads that criss-cross Europe make the latter in some way a geological preparation ... Remarks of Cuvier on steam boats; new habits which result from this for the whole population ... geology has in some way become a profession ... where does geology begin and astronomy end? These two sciences are sisters and what above all places a line of demarcation between them is the different way of life that they demand of their cultivators ... one of the things which characterizes and even constitutes the progress of civilization is the division of occupations ... the establishment of railways will have the effect of enlarging geological localities, diminishing the distance between the geologist and the astronomer.

Lyell himself made much of the need for travel as being central to the occupation of the geologist; indeed, in his autobiography he proclaimed that: 'We must preach up travelling as the first, second, and third requisites for a modern geologist.'

One theme of these texts from Beaumont is the idea of the division of labour in modern geology. This ties in in several ways to the idea of specific geological time found in Lyell's *Principles*. Concentrating for a minute on what is common to both authors, we find that they both make exactly the same points about the development of geology. Now is the end of the 'heroic age', of individual systems which are thrown up and hurled down in cataclysmic succession. For both Beaumont and Lyell, the present is the time of slow, piecemeal development by a large group of workers, no one of whom will dominate the field – and both authors in their work tried to lay quasi-mathematical foundations for this field. For Beaumont this reflects the division of labour, for Lyell it connects to uniform geological time.

We can trace a further connection between the principle of the division of labour and the principle of uniformitarianism. Lyell spends large sections of his *Principles* inveighing against what could be called the 'heroic' system of geological change. There was no time when things were different: 'The minute investigations . . . of the relics of the animate creation of former ages, had a powerful effect in dispelling the illusion which had long prevailed concerning the absence of analogy between the ancient and modern state of our planet.' Indeed: 'It was contrary to analogy to suppose, that Nature had been at any fromer epoch parsimonious of time and prodigal of violence.' Here is a direct equivalent of Beaumont's 'heroic system' of geology. What Lyell is doing throughout the *Principles* is to offer a reasonable and rational division of labour between the forces of Nature. Each creative force is also destructive; it takes a myriad of small changes to cause a large change – in other words the political economy of the Industrial Revolution must be written into the Book of Nature, or else Nature is irrational. Thus when humanity appears on the scene, it becomes part of the economy of nature, a worker in Factory Earth:

> we ought always before we decide that any part of the influence of man is novel and anomalous carefully to consider all the powers of other animate agents that may be limited or superseded by man. Many who have reasoned upon these subject seem to have forgotten that the human race often succeeds to the discharge of functions previously fullfilled by other species.

We can, then, use Beaumont's texts to place Lyell with respect to the principle of the division of labour: his geological theory describes the rational organization of time implicit in this princple and ascribes a similar time to Nature.

Why this division of labour, within geology as a discipline, society and Nature? Both Lyell and Beaumont stress that, with current social and economic change, the present is seeing a form of information explosion. If we look at Lyell's geology as a system for the classification of this information, then we can gain another insight into the articulation of his time. Put at its most abstract, Lyell is proposing a change from seeing geology as a litany of an enormous number of singular events (like a huge epic poem) to seeing it as the systematization of a small number of kinds of events. Thus instead of seeing a particular mountain as a sign of a massive upthrusting at some given date in the past, he sees it as a typical example of a kind of change that is occurring today. There are no privileged moments. His geology is a kind of bookkeeping device that allows the storage of vast amounts of information through sorting them into a kind of filing cabinet of different kinds of event. This reading of Lyell brings out why it was easy and natural for Lyell to find the metaphor of the statistical commissioners – after all, his geology is undertaking a version of their task. It brings out, too, how large-scale social change is reflected directly in the writing of geology through the intermediary of the organization of the foundling discipline of geology (the principle of the division of labour) and the handling of the information explosion that all sciences and professions were undergoing (uniformitarian time).

Indeed, Lyell's *Principles* reads like nothing other than a double-entry ledger-book: the sum of creative and destructive forces (credit and debit) is always precisely zero. Lyell carries this principle well beyond the bounds of the available evidence in his four rules of the disposition of land and sea, which we cited above. To recapitulate, these were that the proportion of dry land to sea is always constant, that the volume of land rising above the sea is constant, that the mean and extreme depth of the sea are equal at every epoch, and that 'the grouping together of the land in great continents is a necessary part of the economy of nature.' These rules are frankly absurd unless they are read in the context of Lyell's accounting method. A further justification for the reading lies in Lyell's constant reference to the economy of nature, the plan of nature. Thus it helps us interpret the following enigmatic opinion about the idea some philosophers had that only a few laws produced the 'endless diversity of effects': 'Whether we coincide or not in this doctrine, we must admit that the gradual progress of opinion concerning the succession of phenomena in remote eras, resembles in a singular manner that which accompanies the growing intellingence of every people in regard to the economy of nature in modern times.' The metaphor of the economy of nature is second in his work only to the Book of Nature.

Lyell, then, introduces a principle of the division of labour into the profession of geology and into the economy of nature, and in so doing creates his specific time for both. Let us see finally how this new time was a regular time that mimicked the social time of the industrial revolution. Here we will use another feature of the Beaumont texts to help our reading of time and geological knowledge. In Lyell's *Principles* a close connection is drawn between the nature of geology and the nature of astronomy. Lyell was certainly not averse to connecting the two:

> It was not until Descartes assumed the indefinite extent of the celestial spaces, and removed the supposed boundaries of the universe, that just opinions began to be entertained of the relative distances of the heavenly bodies; and until we habitutate ourselves to contemplate the possibility of an indefinite lapse of ages having been comprised within each of the more modern period of the earth's history, we shall be in danger of forming most erroneous and partial views in Geology.

However, it is a deeper connection than this that I would like to concentrate on now.

To start us on our way, let us look at another text by Beaumont, this time from his sketchy notes for a lecture given on 20 December 1832:

> space without limits . . . time without limits. Astronomical periods . . . periodic oscillations around a mean state. The beauty of this result is a first reason for thinking that it is not a pure abstraction and that there has been an *effective realisation* in nature. *However* the heavenly bodies don't leave in space any trace of their passage . . . solar system a clock . . . It wasn't a clock that the fable had placed in the curia of the time as a symbol of duration. Type of *hourglass.* This *hourglass* is the surface of our globe and the scientists concerned with its functioning instead of calling themselves *astronomers* call themselves *geologists.*

Just as the astronomers had reduced the solar system to a clock, then so could geology reduce the Earth to an hourglass. Looking at Lyell's articulation of time, we see that it is precisely this that he does: every physical operation is made as regular and smooth as clockwork, it was just a case of finding the right periods. Thus he recognized one difficulty with his system:

> It is clear that if the agency of inorganic causes be uniform as we have supposed, they must operate very irregularly on the state of organic beings, so that the rate according to which these will change in particular regions will not be equal in equal periods of time; nor do we doubt that if very considerable periods of equal duration could be taken into our consideration and compared one with another, the rate of change in the living as well as in the inorganic world, would be nearly uniform.

The solar system was hymned as an accurate clock, the clock dominated industry, and between the two the synthesizing geologist Lyell turned the

Earth itself into a clock: ticking away regularly and faithfully when once we understand its workings. Lyell's articulation of the connection between geological and human time can be interpreted in this light. Humanity and geology may have developed raggedly in the past, but with the triumph of industrial society (which was the natural form of association because it ran like clockwork) the two could approximate to the industrial time written into his geology. Thus two basic methods of factory production: the division of labour and the parcelling up of time into regular units, are both written into the time that Lyell created for the new discipline of geology; and we have seen that he used factory production methods precisely because he saw these as best suited to the fruitful exercise of the profession he sought to create.

Time and industry

The clock-making industry was at the head of the Industrial Revolution. Like others, Babbage, a friend of Lyell's, reckoned that the principle of the division of labour was born in the manufacture of clocks, therefore that industry needed machine tools of the greatest precision and was an example of 'the state of the art'. Babbage considered the clock a 'regulator of time' in opposition to 'the negligence and idleness of human agents' and 'the irregular and fluctuating effort of animals or natural forces'. In a wonderfully exaggerated book, Claude-Lucien Bergery remarked: 'The worker then must be mean with his time . . . he can hardly devote 30 years or 262,800 hours to collecting the money he will need in his old age . . . Each minute lost will deprive him of about three-thousandths of a franc . . . every man is capable of at least 5 movements a second, there are 36,000 seconds in a day of ten hours, which will in consequence allow 180,000 movements' (*Économie industrielle*, 1829).

   The working day in a factory was regulated by clocks, from the checking-in clocks that often registered the hours of the workers to the timers that made the machines run regularly (and the workers irregularly). From 1800 to 1820 England produced at least 100,000 clocks a year. When Lyell underlined the importance of the division of the history of the Earth into equal periods of time, he was only reflecting a fundamental obsession of the industrial world then being born.

Conclusion

I have asked the reader to follow some fairly close textual analysis of Lyell's *Principles*. In this concluding section I will offer a few fairly general observations on the results we have achieved.

One thing that stands out is that we have come a long way from the picture of the herioc scientist who rolled back the years of the Earth's origin. Keeping to this one canonic result, we would have been able neither to situate his work within the Industrial Revolution nor to see its link with the Romantic movement sweeping Europe at this time. In particular, we would have missed the thrust of what he says about the nature of time: his stress on the periods of equal duration that govern industry, the solar system, and the world; and his belief in an epiphanic moment wherein all of time is grasped. This result may be generalized: it is only by getting our hands dirty thumbing through the texts that scientific writers produced that we will be able to find traces of the social reality in which their work is embedded. Otherwise we risk getting caught up in the stories that scientists so often tell themselves (and historians so often tell them) of ideas floating in the air and being transmitted intact from generation to generation while all around is change. Thus the obvious connection between Lyell and the Industrial Revolution is the idea of progessive change in society: if we had looked for this we would have failed to find it and might have concluded that his work had no social basis.

In another sense, however, we are not so far removed from our foundation myth, which had Lyell expelling the priests from the temple of science by redefining geological time; in fact, two sorts of religious time are excluded from geology by Lyell. The first is the pagan representation of a time of great heroes bestriding the field of geology like colossi; or of great geological events – earthquakes, floods, storms – dwarfing today's minimal, tranquil variations on the theme of repose. Second, the Christian author of the Book of Nature is denied the right to interpret His works: it is the moment of creation enjoyed by the geologist as the new priest of nature that constitutes the definitive, correct reading of the flawed Book. Where God had been the only being capable of standing outside time and space and able to oversee the whole, now the Geologist could join and effectively supplant Him. The foundation myth is thus shorthand for a much more complex reality which sees modern science, despite its secular assertions; forming itself into the new religion of our times.

This all seems to lend our hero, Charles Lyell, enormous power. Single-handed, armed only with his incisive intellect, he engineered the split of Church and State so as to found the profession of geology. Of course this vision is totally improbable. As I have attempted to demonstrate, the arrow of historical causation in this case is not from towering intellect to society through the mediation of ideas, but from society to intellect mediated by the day-to-day exercise of the profession of geology. The problem of the division of labour and the organization of time in factories and in geology was precisely the same problem. Through the mediation of the creation of the profession of geologist in the image of the middle management of a thriving business, Lyell inscribed the same time scientifically into the history of the Earth as others inscribed socially into industrial society. It is thus scarcely surprising that we find Lyell using the metaphors he does and Beaumont making the connections he does: both were being better historians than an intellectual historian who asserts that all Lyell did was increase the age of the Earth.

17

Mendel in the Garden

Jean-Marc Drouin

The product of the twentieth century, the image of the Czech monk who founded genetics and revolutionized biology while marrying peas in the monastery garden, threatens to obscure what he had in common with the other hybridizers of the nineteenth century.

The story has often been told: in publishing the results of his work on plant heredity, the Dutchman Hugo De Vries stated that the laws he had discovered had already been formulated thirty-five years earlier by Gregor Mendel, a monk from Brünn (now Brno). That same year two other botanists, Carl Correns and Erich Tschermak, one from Tübingen and the other from Vienna, published similar results. All three recognized the priority of Mendel's work, while pointing out that they had arrived independently at the same conclusions. Mendel's paper, read before the Brünn Natural History Society in 1865, thus took its place in the pantheon of the history of science, and its author became the very archetype of the undiscovered genius.

Accepted for a long time, with minor variations of detail, this picture is today being challenged from several directions. This is the result of different kinds of studies. First of all, the cataloguing of references made to Mendel's work between 1865 and 1890 has made it clear that the obscurity in which they were supposed to have been hidden was merely relative. Comparison of his work with that of other authors concerned with hybridization or selection then allowed a clearer distinction to be drawn between the questions Mendel might have been asking and those which are generally implied by modern genetics. Finally, close analysis of the positions and texts of the 'rediscoverers' has provided evidence for a new interpretation of the famous rediscovery. Branningan, in particular, has shown how the reference to Mendel's work was inserted late into Hugo De Vries' paper, probably to avoid a quarrel over priority with Carl Correns and Erich von Tschermak: as none of the three authors could claim the paternity of a discovery they had made simultaneously and independently, it suited them better to attribute priority to a dead and supposedly obscure scientist.

After this, can one still present Mendel's paper as a foundational text, ignored by his colleagues because it was too much out of the ordinary? Shouldn't it rather be seen as one article on hybridization among others, indistinguishable from numerous other similar works before the first geneticists — at the beginning of the twentieth century — reinterpreted it in their own way, and posited it, somewhat speciously, as the origin of their own discipline?

The difficulty in answering these questions will perhaps be lessened by relating the Mendelian text to the technical concerns which gave it its essential significance for his contemporaries. Current developments in biotechnology should not blind us to the fundamental transformations that occurred in agriculture and horticulture during the eighteenth and nineteenth centuries. Mendel may be considered as one of the actors in the history of this transformation, because the laws he stated found their typical exemplification in the selection and hybridization procedures used by livestock breeders and seed-producers. Can it be said, however, that these procedures are an *application* of Mendel's laws? Or should we rather concur with the quip by Bateson himself, who said at the Fourth International Conference on Genetics (Paris 1911), that in this field the scientist 'gets new ideas from the practitioner', which he then proceeds to assimilate? Mendel's work is one of the intersections where the history of agronomy meets the history of biology and this observation may serve us as a guiding thread.

Biographical Sketch

Johann Mendel's childhood is emblematic of this encounter and it belongs to history as much as to legend. He was born in 1822 in a village named Heinzendorf, today Hyncice, in Moravia, one of the regions of the present-day Czech Republic, in those days a province of Austria. One of his great-uncles had been a schoolmaster. His mother came from a family of gardeners

507

from a neighbouring village. His father had been a soldier in the wars against Napoleon. Mendel's parents ran a small farm which they owned. They were still subject to forced labour, which obliged the peasant to work three days a week for the landlord. Mendel's father had an orchard which he tended carefully, encouraged by the parish priest, J. Schreiber. Schreiber, like the schoolmaster, Thomas Makitta, worked to encourage an understanding of natural history and to spread information about techniques for the improvement of fruit-trees. Through paternal example, probably reinforced by the teaching of the schoolmaster and of the priest, Mendel probably acquired during his childhood a substantial body of horticultural knowledge; that is to say, not only practical botanical knowledge, but also the eye and the manual skill the gardener requires to deal with the vegetable kingdom.

The connection between Mendel's childhood and his later preoccupations – hybridization, apiculture and meteorology – has often been stressed. This appears in a new light if, following the example of several recent studies, it is related to the advance of agricultural technique in Moravia in the first years of the nineteenth century.

Several names are associated with this movement, which combined intellectual and economic aspects. Among the most commonly mentioned is that of Ferdinand Geisslern, a livestock farmer who wrote a treatise on the scientific selection of sheep. These methods, which came from England, depended on the measurement and systematic recording of animal characteristics, and the construction of precise and detailed genealogies. Their success illustrated the role of the hereditary transmission of characteristics. The economic impact of this control of reproduction was considerable, and it is reported that, at Brno in 1810, the market value of a ram with a pedigree was a hundred times greater than that of an ordinary beast.

Another name crops up often: that of the naturalist Christian-Carl André (1763–1831), advisor to Count Salm (1776–1861). The Count, promoter of the textile industry in Brno, was the president of a regional Agricultural Society, of which André was the secretary. The Pomological Society,[1] part of the Agricultural Society, was particularly concerned with the artificial pollination of fruit-trees, and it advocated the development of tree nurseries. One of the nurseries established at that time was at the Augustinian priory at Brno, whose superior, Franz Cyril Napp, was a member of the Pomological Society. As President of the Society, Napp summed up the problem of the transmission of characteristics in these two questions: 'What is transmitted and how is it transmitted?' He stressed too the necessity for experimental work to solve the problem, expressing here, most probably, a desire for a certain independence of research in relation to technical practice. It seems, in fact, that landed proprietors provided the Moravian naturalists with effective but somewhat constraining allies. This tension would lead in 1861 to the creation of a Natural Scientific Society independent of the Agricultural Society.

Mendel as student, teacher and monk

Just like their superior, the Augustinians[2] at Brno showed a great interest in agriculture and natural science, and they devoted a substantial part of their time to teaching. Many of them were also interested in philosophy, and one of these, the botanist Matthaeus Klacel (1808–82), was even suspected of pantheism. This intellectual activity was not appreciated by everyone. Taking advantage of the reactionary climate which followed the events of 1848, the Bishop of Brno tried to have the monastery brought back into line or dissolved. Napp defended the proper vocation of his community and, in the end, the Cardinal of Prague did not follow the Bishop's advice. In fact the struggle between the Bishop and the Augustinians would probably have been forgotten had they not included among their members Johann Mendel, who had taken the name of Gregor when he was received as a novice in 1843. Questions have sometimes been raised about Mendel's fundamental attitude towards religion. There is no reason to believe that he was not committed to the Christian faith that he professed, but this never interfered with his scientific work. There is no trace in his writings of the 'natural theology' of the period. In any case, there can be no doubt of his attachment to his order and to his monastery, to which he owed a great deal. Father Schreiber had urged Mendel's parents to enable him to pursue his studies. But despite the efforts of his family – his sister had even given up part of her dowry to him – Mendel, exhausted by hardship, would have given up after his second year at the Institute of Philosophy at the University of Olomuc, if it hadn't been for a professor who got him admitted into the Augustinian monastery at Brno as a novice. There Mendel studied theology and natural science. He taught in a Technical College and as an assistant at a *Gymnasium.*

Between 1851 and 1853 he was sent to the University of Vienna to complete his education. The requirements of secondary teaching – and perhaps his own intellectual tastes – led Mendel to pursue physics and mathematics as well as natural history. At the University of Vienna he attended a course given by Christian Doppler (1803–53), where a small group of students were taught practical experimental physics. For reasons which are not clear Mendel failed certain exams, and did not receive the diploma he had hoped for. Much appreciated for his pedagogical skill, he was none the less able to return to his teaching.

Solitary Research?

It is clear that Mendel started his experiments on the hybridization of peas with a solid scientific education behind him, especially as regards questions of methodology. Here it will be just as well to look at the meaning of the

509

word 'amateur', which one might be tempted to apply to him. If one means by it someone who has no post in research, teaching or administration in a university or similar institution, the term is applicable to him, as it is to Darwin. If, on the other hand, an amateur is defined as someone who has not received a university education in the field concerned, Mendel cannot be considered in this category. In other words, one can regard him a 'voluntary' researcher but not as an 'autodidact'. Finally, he had at his disposal in the monastery an experimental garden and help with cultivation, not to mention the library and exchanges with other naturalist religious; he could indeed be said to have benefited from facilities of which certain academics might well have been jealous.

In addition, the Natural Scientific Society which emerged from the Agricultural Society, and before which he presented the results of his research, is a good example of the local and regional learned societies which flourished in nineteenth-century Europe, whose role was fairly important. On one hand they allowed amateurs and academics to meet, as well as specialists in different fields and, on the other hand, they offered the possibility of publication and represented one of the routes for the diffusion of theories and research programmes. It is not at all surprising then that Mendel's paper on hybridization should have been published in the *Verhandlungen des Naturforschenden Vereines in Brünn* (*Proceedings of the Brünn Natural Scientific Society*). Of course, publication in a journal better known abroad or the publication of a book would have allowed wider diffusion of his results. It remains true that Mendel's work was not completely forgotten. Between 1865 and 1900 it was cited a dozen times, which is evidence of a modest but real audience. The fact that he was then cited as one hybridizer among others presents a different problem.

During the last part of Mendel's life history and legend meet again, as they did in his childhood. Mendel corresponded with the botanist Carl Naegeli (1817–91). The monk and the academic seem to have become friends, but their concerns were too different: Naegeli never recognized the significance of Mendel's work. He encouraged him to produce hybrids of a wild flower (*Hieracium* or hawkweed); the results are difficult to interpret because this plant reproduces parthenogenetically, something which was unknown to Mendel and his contemporaries. Mendel was also active in meteorology and apiculture. Finally, he was elected superior of his monastery in 1868. This important responsibility allowed him to help his family financially, but it imposed constraints on him which left him insufficient time to pursue his research. He found himself involved in a hopeless struggle with the State on the question of the tax status of the religious orders. Disagreeing with his liberal political friends, he exhausted himself in fighting what he believed to be excessive taxation. He died in January 1884, at the age of sixty-one.

In the end, historians have done nothing to alter the picture of the peasant's son obliged to take holy orders in order to continue his studies, a monk who taught the elements of science in a provincial school, at the same devoting years to patient work on hybridization, finally becoming an abbot worn out by his responsibilities, forcing him to neglect even his beloved experiments. They have only added to these naïve pictures – rather like those Mendel had had painted on the ceiling of the chapter-house of his monastery – a few similarly fascinating ones: an agricultural region in full development, agronomists and livestock breeders preoccupied by the problems of heredity, a spell at the University of Vienna, a religious community devoted to intellectual activity, confrontations between liberals and conservatives, learned societies in full swing. At the centre of these elements, which overlap and get tangled up together, the themes of agriculture and gardening are always repeated. Though this is the case, many other authors of the period linked together horticulture and biological research.

Why then does Mendel's work have a special place? Because of his rediscovery? Certainly, but what was it then in his work that made it possible for it to be re-used by his rediscoverers?

Horticulture and botany

Horticulture and botany are today held together by the closest of links; they are so connected one with the other, they are so mutually dependent upon each other, that they ought rather to be considered as two branches of the same body of science than as two distinct sciences. The difference between them is, in reality, only what separates practice from theory. Let a botanist discover a new fact of plant physiology: horticulture grasps it immediately, and soon, in its turn, it gives back to science as much as it received, either confirming by ingenious experiments the truth of what has just been learnt, or by pointing the scientist toward new discoveries. There is the same reciprocity when it is a question of the conquests to be made in the vast field of nature: the success of one benefits the other, for horticulture just like botany has its intrepid and devoted collectors. And finally, this latter frequently also has control of science's decisions in the so complicated matter of species, a question which botany, left to its own devices, is not always able to resolve. (Charles Naudin, *Revue Horticole*, 1852)

Mendel's Papers on Hybridization

According to the historians V. Kruta and Vitezslav Orel, Mendel's scientific writings consist of thirteen articles, his correspondence and about twenty short texts. Of the thirteen articles, there are nine on meteorology, two on

511

insect pests and finally two on hybridization. It is to these last two that reference is usually made. The first, the paper of 1865, 'Researches on Plant Hybrids', published in 1866, rehearses the presentation made in two sessions at Brno, on the 8 February and 8 March 1865. The second, the paper of 1869, 'On certain hybrids of *Hieracium* obtained by artificial pollination', read at the meeting of 9 July 1869, was published at Brno in 1870.

The 'Preliminary Remarks' in the Paper of 1865

Since its introduction into Europe at the end of the eighteenth century, the fuchsia has attracted many amateurs and horticulturalists, who obtained by crossing and selection the numerous forms which may be admired today. Mendel, a member of the Horticultural Section of the Agricultural Society, was also caught up in the craze. The flower he holds in his hand in a photograph of 1861 is a fuchsia. A Brno horticulturalist dedicated a new variety to him in thanks for the work he had done with him. The fuchsia, however, is not the only ornamental plant that interested Mendel. So it is not at all surprising then that he should open his paper with a reference to the cultivation of flowers: 'It was the performance of artificial pollination upon decorative plants which led to the research whose results are to be presented here.'

At the same time Mendel stresses the scope and theoretical importance he understands his work to have: 'The remarkable regularity with which the hybrid[3] forms recur every time pollination takes place between the same types gave rise to the idea of new experiments which would examine the hybrids' descendants.'

Mendel then mentions the 'conscientious observers such as Kölreuter, Gaertner, Herbert, Lecoq and Wichura and others' who had 'devoted part of their lives to the study of these questions'. This homage to his predecessors is accompanied by a critical assessment which is at the same time the announcement of a research programme:

> If one takes a survey of the work done in this field, one will arrive at the conclusion that among all these trials, none have been carried out on a scale and with a method which would allow the determination of the different forms in which the descendants of the hybrids appear, to classify accurately the forms which appear in each generation and to establish the numerical relationships between these forms. It requires, in fact, a certain courage to undertake such a considerable labour. Only this, however, will permit a final solution to a question whose importance for the history of the evolution of living beings should not be mistaken.

Mendel then notes that his paper reports only a first 'experimental trial limited to a small group of plants'. At the end of eight years, this trial is 'essentially complete'.

Hybridization and Evolution

The most surprising thing for us today is the sibylline reference to *evolution*.

It may be thought that *Entwicklung*, the German term employed by Mendel here, signifies only 'the individual development of the organism'. The French term *évolution* was also used in this sense in the last century, at a time when what we now call the theory of evolution was known as transformationism. In this case the phrase 'history of the evolution of living beings' might be understood as 'the natural history of the individual development of plants and animals'. One of the problems concerning the genesis of the individual, which was still causing controversy at that time, was the respective roles of the ovule and of the pollen in the fertilization of plants. Mendel took a position on this question as he presented the results of his experiments: 'If the ovule had on the pollen cell only a superficial effect, if its role was no more than that of a wet-nurse, artificial pollination could have no other result than the production of a hybrid resembling the male plant only, or very like it. This has not been in any way confirmed by our researches so far.'

None the less, and even if the phrase about 'the evolution of living beings' does not refer directly to the debate on evolution as we understand it today, it is still the case that at that time the practice of hybridization was often associated with questions about the stability of species. Because man, in crossing varieties and species, seemed to create new flowers, the fixity of natural forms could be held up to question. To the extent, however, that these new forms proved to be sterile, or that their descendants tended to revert towards the parental type, then the idea of species stability drew new strength from this. In the conclusion to his paper Mendel left the question open, though probably inclining towards Gaertner's creationist theory.

In the 1869 paper on *Hieracium*, a genus of wild flowers in which one finds a disconcertingly large number of similar species, he expounds the theory of those who saw in this multiplication of species the outcome of natural hybridization; he notes too that for other authors, such hybridization was impossible or ephemeral. He goes on to say: 'Recently, the question of the origin of numerous constant intermediate forms has gained not a little in interest since a famous *Hieracium* specialist, adopting the *Darwinian* point of view, has argued that they must be derived from extinct or still-existing species.'

According to Orel, the famous specialist in question was no other than Naegeli, with whom Mendel had been corresponding since 1866.

The Choice of the Pea

In fact, we have very little which allows us to place Mendel in relation to Darwin. Their thought is in agreement, however, on the absence of a

513

fundamental distinction between species and variety. Thus the 1865 paper has no scientific discussion on the classification of the genus *Pisum*. This contrasts with many authors' insistence on distinguishing between hybridization between species and simple crossing between varieties or breeds belonging to the same species. Whether species or varieties, the different forms in which the pea[4] is found offer choice material for experiments in hybridization.

This is how Mendel details the conditions which must be satisfied by the experimental plants: 'They must possess constant alternative characteristics; it is necessary that during flowering, their hybrids are naturally, or may be artificially protected against the intervention of alien pollen; the hybrids and their descendants should not show any notable reduction in fertility through the generations.'

And he adds: 'From the very beginning, attention was drawn to the Leguminosae, because of the particular structure of the flower. Experiments undertaken with many species of this family have led to the conclusion that the genus *Pisum* adequately corresponds to the desiderata outlined.'

All the commentators have noted how this choice of experimental material played a decisive role in the success of the enterprise. In fact, with peas, self-pollination is the rule, while for many flowering plants it is the exception. This means that without the intervention of the experimenter – or the unforeseen arrival of a small insect – the pollen of one pea-flower is deposited on the pistil of the same flower, with the result that Mendel was able to obtain pure strains simply by ensuring that in self-fertilizing through several generations, the peas retained the characteristics of the seeds which he first had. This property, which the pea shares with other Leguminosae, was known to specialists.

Certain biologists have pointed out too that the genes controlling the characteristics he had chosen to observe were all situated on different chromosomes. As Mendel would clearly not have been able to think in these terms, one can either speak of a lucky chance or of the intervention of an intuition based on horticultural empirical knowledge. Many agronomists had already performed hybridizations on peas and some had published their results.

Finally, though this is not mentioned in the paper, the pea is a delicious vegetable, and one of the practical consequences of Mendel's experiments, according to Orel, was that the monastery garden grew particularly tasty varieties of peas.

The Results

Having justified his choice of experimental plant, Mendel presents the seven differential characteristics (*differierende Merkmale* in German) he had chosen, which included the shape of the seed, round or angular, the shape of

514

the pod and the length of the stem. After this he reported the results of his experiments with the numbers in each type, across several generations.

First of all, when he crossed two peas which differed in only one characteristic, and pure-bred in respect of it, for example having a round or angular seed, he obtained hybrids identical to each other, in this case all having round seeds; this characteristic he called 'dominant' (*dominierende*). By then reproducing these hybrids by self-pollination he obtained peas both round and angular, in the ratio of 3 to 1. 'In the second year of the experiment, 253 hybrids gave 7,324 seeds, of which 5,474 were round or rounded and 1,850 angular and wrinkled. From which one obtains a ratio of 2.96/1.'

The angular characteristic therefore remained in the hybrids in a latent state because it could reappear in some of their descendants. Mendel suggested calling it 'recessive' (*recessive*). The forms which exhibited the recessive characteristic were constant for this characteristic in all their descendants. Those which exhibited the dominant characteristic could be divided into two groups: two-thirds behaved like the hybrids (and among their descendants one rediscovered the ratio of 3:1), while the other third gave descendants where the dominant characteristic was constant. Having given the figures for each of the characteristics, Mendel concluded: 'It is now clear that the hybrids of each pair of differential characteristics produce seed of which half

---

Mendel and us

The symbols Mendel used may seem familiar. *A* represents the dominant characteristic, 'round-seededness', for example, and *a* the recessive character, 'wrinkled-seededness'. This is a notation which is still used. However, Mendel's symbols and terminology are different in some respects to our own, and these differences have to be understood in order to be able to read his writing.

We represent peas possessing only the dominant characteristic by *AA* and those possessing only the recessive character by *aa*. Mendel designates them a *A* and *a* respectively, without redoubling the letter.

By crossing these two varieties by artificial pollination, we obtain what is today called the first generation (or F_1) which we designate as *Aa*, because the individuals of which it is composed possess both characteristics, dominant and recessive. Mendel here uses the same symbol as we do, *Aa* in order to designate what he calls the 'hybrids'.

These plants themselves have daughters by self-pollination, which we call the second generation (F_2) but which Mendel called the 'first generation of hybrids' (meaning the first generation bred from the hybrids). In our textbooks these descendants are distributed as follows: a quarter *AA*, one half *Aa* and a quarter *aa*. For Mendel this was expressed: a quarter *A*, a quarter *a* and a half *Aa*.

---

reproduce the hybrid form, while the other half is composed of plants which remain constant, of which an equal number take the dominant and the recessive characteristic.'

Applying this formula, one can predict that the proportion of hybrid forms will tend to diminish constantly among the descendants of the hybrids, as has been shown by experiments conducted across several generations. Mendel notes that these results 'confirm the observation by Gaertner, Kölreuter and other authors, that hybrids have a tendency to return to the original strains'. By calculating the theoretical proportions of the different categories he puts forward what we would call a model to formalize and verify a law which other hybridizers had established in an empirical manner.

Everything up to now has concerned plants which differed in only one characteristic. What happens when one crosses two peas which differ in two characteristics, for example both the shape and the colour of the seeds?

Mendel's response is that the two characteristics come apart and combine entirely independently of each other. For example, by crossing individuals with round yellow seeds with individuals with angular green seeds he obtained a total of 556 seeds which were distributed as follows:

> 315 round and yellow
> 101 angular and yellow
> 108 round and green
> 32 angular and green

Having given the results for peas which differed in three characteristics, Mendel shows how these figures correspond for practical purposes to the theoretical populations calculated by the formulae for the combination of characteristics. He concludes with this general rule:

> If n is the number of characteristic differences between the two original plants, 3^n gives the number of terms in the series of combinations, 4^n gives the number of individuals in the series and 2^n the number of combinations which remain constant. So for example, if the original plants differ by four characteristics, the series contains $3^4 = 81$ terms, $4^4 = 256$ individuals and $2^4 = 16$ constant forms; or, what comes to the same thing, out of 256 descendants of the hybrids there are 81 different combinations, of which 16 are constant.

Mechanism and Generalization

The most original aspect of Mendel's text is the way in which he submits the characteristics to the law of combination and calculates the theoretical numbers in the different groups. Today, his procedure is justified by explaining that each characteristic is borne by one gene and that each gene occurs

---

Were the figures too good?

The independent separation and recombination of characteristics is one of the fundamental principles of Mendelism. From the practical point of view it explains why, by crossing large and tasteless fruit with small and tasty fruit, one might hope to obtain fruit both large and tasty! From the theoretical point of view, the calculation of possible combinations of characteristics depends on it. Paradoxically, however, for present-day geneticists, this principle is only an approximation. Characteristics controlled by genes situated on the same chromosome have in fact a higher probability of remaining associated, and the closer the genes are to each other, the higher this probability will be. In fact, this means that if one performs an experiment combining various characteristics on any particular plant, there is a danger that the figures for the different combinations could be quite different from the classical proportions.

It is in a different sense, however, that various authors have suggested that Mendel's results were too good to be true. R. A. Fisher, the specialist in population genetics, calculated in 1936 that Mendel had – given the size of his sample – only a 5 per cent chance of obtaining a proportion as close as he did to the theoretical figure of 3 to 1. Did Mendel cheat? It would be the more surprising, given that he did not hesitate, on other occasions, to report experiments which had given 'bad results' or provided data which were difficult to interpret.

Indeed, Fisher's calculations are now being challenged by certain authors, who argue that he did not take into account the biological facts, and in particular the behaviour of the pollen grains. R. C. Olby has also shown that the figures given by Erich Tschermak, one of Mendel's 'rediscoverers', were as good as his. According to him, the problem arises from the methods used at that time for counting.

---

twice in the individual. This explanation was inconceivable at a time when the very concept of the chromosome was absent, yet Mendel needed none the less to put forward a mechanical model which would authorize his application of the mathematical model to the products of hybridization. In other words, Mendel had to find a hypothesis which would allow him to treat his characteristics as if they were marbles in a bag. This involved him in a new series of experiments intended to show that there was an even number of pollen cells and of ovular cells, with the result that he felt able to conclude: 'Chance then alone decides which of the two sorts of pollen will combine with each of the two ovular cells. None the less, according to calculations of probability, it should always happen, taking the average of a large number of cases, that each of the forms of pollen A and a should combine with each of the forms A and a of the ovular cell.'

517

Of course, presented in this way, the relationships observed in the genus *Pisum* should be discovered among other flowering plants. Mendel had made a generalization which would pose great problems for his conception. First of all he repeated the experiments he had made, this time on French beans. When one crosses beans whose flowers are not of the same colour, the flowers of the hybrids and those of their descendants present colours which are intermediate between those of the parents.

How to hybridize beans

The Leguminosae represent a privileged experimental material, and so it is not surprising that Henri Lecoq's great treatise *De la fécondation naturelle et artificielle des végétaux et de l'hybridisation* ('On the natural and artificial pollination of plants and hybridization'), published in 1845 and translated into German in 1846, should have devoted one of its finest passages to the genus *Phaseolus*:

> it is always interesting for the amateur or the physiologist to work with plants as susceptible to hybridization as the Beans, and with which one can so promptly assure oneself, by the colour of the seeds, of the success of the operation. Before the flower has blossomed completely, the anthers of the Beans begin to shed their pollen, and at the same time the carina begins to bend away, taking with it the attached filaments which lengthen at the same time, as if they wished to reach the stigma carried by the style and the ovary, which also begin to grow. In any case the pistil lengthens less than the stamens, and pollination will not be very long. To bring it about artificially, it is enough to move the carina a little out of the way and to place the pollen on the stigma with a brush. This retains it easily, and as Bean pollen is not very powdery, but often somewhat pasty, there is no need to remove the stamens, but only to pollinate the pistil as soon as the state of the flower enables this to be done.

To explain this result, which seemed to lend support to the thesis of inheritance by mixture, rather than his own, Mendel had to assume that the colour of flowers was composed of several characters. Lastly, the final pages are devoted to an analysis of the results published by Joseph Kölreuter and Carl Friedrich Gaertner, and a discussion of their conclusions about the concept of species.

The impression one gets from the second part of the text is that Mendel hopes to be able to apply the laws he has discovered with peas to the plant world as a whole; even if, four years later, he seems to doubt that they apply to the wild flowers of the genus *Hieracium* or to the willows, this hope demonstrates the significance he gave to his work on hybridization. Does this mean that we could say that in doing this, he was intending to found a new discipline? In fact, the question of whether Mendel is or is not at the origin of genetics hides another question, which in this case concerns the history of technique as much as the history of science – how is Mendel to be situated in relation to the programme of research, at the edges of botany and of agronomy, which can be summed up in the following way: how can new plant varieties be produced certainly and efficiently? In other words, what was it that Mendel brought to horticulturalists and plant-breeders? How does his work relate to their concerns, and what could it have brought them that they didn't already know themselves?

Mendelism and Plant Improvement

In many textbooks and popular articles, the creation of new breeds and varieties is presented as an *application* of Mendel's laws. If this were the case, the 'rediscovery' would have led to a real revolution in the techniques of plant improvement. This revolution certainly did take place, but well before, at the same time that Mendel was doing his experiments and independent of him. In this connection, the story of the improvement of the sugar-beet is exemplary.

The sugar-beet had been known for a long time as a fodder crop when in the seventeenth century the German André-Sigismond Margraff demonstrated that it contained a sugar similar to that found in cane, which it was possible to extract. At the end of the century, Karl Franz Achard perfected the industrial production of sugar from beet. Napoleonic France, under blockade, was looking for a substitute which could take the place of cane-sugar. In 1812 Benjamin Delessert was able to present the Emperor with samples of beet-sugar which could not be distinguished from cane-sugar. However, the yield of the process was poor, due to the beet's low sugar content (less than 5 per cent). To improve this systematic selection was undertaken: the sugar beets with the highest sugar content were allowed to flower, so that their seed could be replanted. Could this proven technique be further improved?

Louis de Vilmorin and Genealogical Selection

On 3 November 1856 Louis de Vilmorin[5] read to the Academy of Sciences a 'Note on the creation of a new variety of sugar-beet', which he republished in 1859, together with several other articles, in a booklet entitled *Notices sur l'amélioration des plantes par semis* ('Reports on the improvement of plants by seeds'), published by the Librairie Agricole. He described first of all the procedure for precisely measuring a beet's sugar-content without harming it. He notes that this sugar level can be increased by selection, and that 'transmission of the sugary character' is an accepted fact which none the less admits of 'remarkable exceptions'. These exceptions, he says, 'throw a great deal of light on the question of the transmission of characters in plants'.

> Thus, during the first year of the experiment, and at a time then when I was therefore ignorant of the qualities which might have been passed by the ancestors of the plants with which I was working, it happened that I retained for reproduction roots of equal strength in sugar and then observed that the progeny of these roots gave:
>
> – sometimes a batch with high average levels, and little difference;
> – sometimes, with a lower average level, considerable differences producing exceptional maxima;
> – and finally, sometimes a decidedly bad batch, whose progeny must be completely abandoned.

To avoid these variations and to choose the 'breeding stars' in the first category, a plant must be selected by reference to its progeny. Louis de Vilmorin was led, in his own words, to 'have Birth and Marriage certificates and an absolutely correct family tree' for all his plants 'from the beginning of the experiment'. To practise this genealogical selection, one needs only to sow a small sample of seeds from each plant and retain only those batches of seed whose progeny display the characters required. This selection is purely maternal, because the pollen still comes from any old beet, and while it does not exclude chance, it reduces it considerably. Thanks to this technique, sugar-content had risen to 18 per cent by 1870. Applied to plants like wheat in which self-pollination is the rule, and combined with hybridization, genealogical selection would allow the creation of new varieties which would increase yields and contribute to the transformation of the countryside. Seed was no longer that part of the harvest that one saved for planting the next year but became the product of an industry, which, thanks to its control of plant reproduction, was able to position itself upstream of agriculture.

The Transmission of Transmission

It was then by imitating the instruments of the state's control over society – registers of birth and marriage – and the probate archives, covering the inheritance of goods – the genealogy – that the selector controlled plant inheritance and reduced the operation of genetic luck. And this was before there was any biological theory capable of accounting for the mechanisms in operation. Vilmorin indeed felt the need for such a theory. In a 'Note on Inheritance' first intended for the Angers Industrial Society and published in its final form in 1859 in the *Notices sur l'amélioration des plantes par semis*, he attempted an explanation. Inheritance was the resultant of two forces, one of which, 'atavism',[6] connected the individual to its ancestors, and the other, 'immediate inheritance', which expressed the relationship between parents and children. Two individuals endowed with the same qualities could not transmit them to their progeny to the same degree; and what is more, they could endow them '*to very different degrees*, with the capacity to transmit these same characters to the succeeding generation.'

To illustrate this idea, Vilmorin appeals to the experience of stock-breeders: among the qualities a horse may have is that of being a good stallion, that is to say, not only transmitting qualities to progeny, but above all transmitting the capacity to transmit. The explanation depends on the distinction between atavism and immediate inheritance, and on an analogy between animal and plant heredity.

From the Horse to the Melon

The horse had for a long time been the object of attentive selection by breeders, so it is not surprising to find the same argument in a text by another agronomist, Augustin Sageret, devoted to the question of heredity, which appeared in 1826 in the *Annales des Sciences Naturelles* under the title: 'Considerations on the production of hybrids, variants, and varieties in general, and those of the family of Cucurbitaceae in particular.' He notes that, in the human species, facial features or hereditary diseases 'may not appear in the first generation yet reappear in the second and following' and he adds 'It is therefore not without reason that the Arabs are so careful to keep the genealogies of their horses.'

It is not Sageret's conception of atavism that has attracted the attention of historians, but rather his conception of the segregation and recombination of characters. The idea of heredity by mixture, contradicted by Mendel's work, was already being questioned by the French agronomist, who writes: 'It has seemed to me that in general the resemblance between the hybrid and its two parents derived not from an intimate fusion of the various characters proper to each of them individually, but rather in an equal or unequal distribution of the same characters; I say equal or unequal, because it is far from being the same in all the hybrids deriving from the same origin, and between them there is a great diversity.'

To illustrate this he brings together his numerous experiments on the Cucurbitaceae in one 'typical example'. He chooses two varieties of melons, one with yellow flesh and the other with white, and he plots their characters in two parallel lists:

521

Characters of yellow-fleshed melon:	Characters of the white-fleshed melon:
1. Yellow flesh	1. White flesh
2. Yellow seeds	2. White seeds
3. Reticulate skin	3. Smooth skin
4. Strongly marked ribs	4. Slightly marked ribs
5. Sweet flavour	5. Sugary and very acid flavour.

Crossing the two, one might have expected a product with yellow flesh and seeds with only intermediate characters. Instead of which, hybrids were obtained which were as follows:

Characters of the first hybrid	Characters of second hybrid
1. Yellow flesh	1. Yellowish flesh
2. White seeds	2. White seeds
3. Reticulate skin	3. Smooth skin
4. Moderately marked ribs	4. No ribs
5. Acid flavour	5. Sweet flavour.

One should not misunderstand these tables. Sageret is not constructing a mathematical model, he does not show numerical relationships, but this work represents one of the major texts in the literature on hybridization that Mendel knew, at least through Gaertner's books. It would not be wrong to speak of a line of hybridizers, for Sageret begins his own article by noting the agreement between his own results and Kölreuter's. In any event, in his conception of the hybrid Sageret is much closer to Mendel than is the French botanist often suggested as another 'precursor' of genetics, Charles Naudin (1815–99). Naudin, assistant naturalist at the Museum in Paris before becoming the Director of the experimental garden at the 'Villa Thuret'[7] in Antibes, in 1854 began research in hybridization that would last about twenty years and which would win him the praise of the Academy of Science. He had probably never heard of Mendel, who, on his side, was ignorant of the other's work.

The Hybrid Petunias and their Progeny

To present Naudin as the one who failed where Mendel succeeded is certainly neither a new nor a satisfactory manner of approaching his work and risks allowing other aspects of his thought to be forgotten, in particular, his conception of the origin of species. It is none the less true that this is how he was often seen by geneticists at the beginning of the century. In fact, a comparison between the two authors' manner of proceeding is inevitable.

In both, a certain question is raised: what becomes of the progeny of plant hybrids? In order to respond, both have at their disposal horticultural know-how which allows them carefully to choose their experimental materials and to marry their flowers as they wish. Both attempted to classify and count the various forms they obtained. This is how, in 1861, Naudin presented to the readers of the *Revue Horticole* the results of some of his work, under the title 'On hybrid plants'. He describes 'Two clearly distinct species of petunia, the purple (*Petunia violacea*) and the white (*Petunia nyctaginiflora*)' and then explains that they are very easily crossed, giving fertile hybrids, intermediate in colour and form. These hybrids resembled each other. He says that four crossings between these two types of petunia in 1854 gave in the next year 36 hybrids, of which 35 had more or less similar corollas. Sowing the seeds obtained from one of these hybrids, he obtained 47 individuals, of which only one really resembled the hybrid, while 10 had flowers like those of *Petunia violacea* and the others presented intermediate forms, which could be divided into four groups in relation to pollen colour and the form and colour of the corolla. Naudin notes precisely the numbers in each category. He sowed the seeds of the plants which most resembled the hybrid, and thus obtained 116 plants which he analysed as follows:

- twelve individuals which more or less repeated in the shade of their colouring, in the form of the flower and the detail of the pollen the hybrid of 1854;
- twenty-six individuals with white flowers, in which the tube of the corolla is narrow and the pollen yellowish. Many of these cannot be distinguished from *Petunia nyctaginiflora*, and the others are very little different;
- twenty-eight with bright purple corolla, campanulate, with grey-blue or purple-blue pollen, which can hardly or can not be distinguished from *Petunia violacea*;
- finally, fifty other individuals which do not easily fit into any of the previous three categories, and which, in the form and size of the corolla, as well as in their colouring which ranges from pinkish white to purplish lilac, and in the greyish colour of the pollen, seem to be intermediate between the two specific types, some being closer to *Petunia violacea* and others to *Petunia nyctaginiflora*.

One can see that Naudin is short neither of numbers nor of detail. Mendel compared the data he observed to the data he calculated. Naudin, given the purely qualitative character of his hypotheses, can only record the numbers without seeking to predict their ratios. In present-day terms, there is quantification without a quantitative model. Correlatively, the experiment is not performed by isolating one character and ensuring its constancy through several generations before beginning hybridization, but by starting with a handful of examples of a plant commonly to be found in gardens. Nor are the plants utilized (Cucurbitaceae, tobacco, petunia) plants in which self-pollination is the rule and in relation to which one might speak of a pure strain.

In retrospect, all this goes to show why Naudin was not able to observe the constant ratios discovered by Mendel. On the other hand, at that time, this probably gave his work a more concrete character, closer to the immediate concerns of horticulturalists and botanists.

The one who believed in Nature and the one who did not

Apart from the experimentation and the numerical analysis, there are the systems of concepts and hypotheses formulated by both authors, the Parisian naturalist as well as the monk from Brno. For Mendel, the calculation of probabilities was justified by the notion of the dominance or recessiveness of characters and by the principle of the chance pairing of the reproductive cells. Naudin, for his part, reinserts the problem of hybrid constancy within the larger problematic of the species as the fundamental unit of the living world, which he expresses in this way: 'Nature, which produced species because it needed to, and which has organized them for specific functions, does not need to produce hybrid forms which do not correspond to its plan.' This explains the sterility of most hybrids and the return of non-sterile progeny towards the parental types.

Put like this, the hypothesis seems a little wild. More prudent and pragmatic, Henri Lecoq responds to Naudin in his book *De la fécondation naturelle et artificielle des végétaux et de l'hybridation*: 'We do not know why or how nature has made species, and we doubt whether it has great need of all that it has made; and we have much more confidence in the patient and ingenious experiments of this learned naturalist than in his ideas of the needs of Nature.'

Lecoq, however, did not miss this occasion to give a theological significance to the work of the hybridizer: 'Hybrids are a proof of the goodness and might of God who allows man to modify his works, using the divine intelligence which he has granted him during his life. It is impossible to follow the successive mutations of a plant, affected by the varied influences of culture and of hybridization, without being overcome by recognition of him who seems to grant to man part of his rights, and who authorizes him to lift up a small corner of the veil which hides all the secrets of creation.'

The hybridizer is a Prometheus, a Prometheus who may respect the gods but a Prometheus just the same, and Lecoq reports with sadness that certain British horticulturalists were opposed to hybridization, which they thought of as an attempt to change the 'work of the Creator'.

One notices, in contrast, that the strength of the Mendelian model was precisely its poverty. Mendel developed it to solve the same problems faced by Vilmorin, Sageret, Naudin and Lecoq, and he gives an adequate reply to the question about the progeny of plant hybrids. However, his work is not exhausted in this response, and it remains available to be re-used, reinterpreted, extended and modified.

Mendel's laws were not rediscovered, if by this one means formulated in 1865 with their present meaning and then forgotten for thirty-five years and then rediscovered in 1900. They were put forward by Mendel as an experimentally verified mathematical model intended to solve problems faced by all hybridizers, to which they responded in a more empirical manner. Mendel's originality was conscious and incontestable: he himself declares, in a letter to Naegeli of 16 April 1867, that the results he had obtained were 'not easily compatible with the present state of science'. But this originality is not that of a solitary scientist, who asked questions fifty years earlier than others who would ask them later. He was not the only one to ask what became of the progeny of hybrids or to perform artificial pollination, but he was the only one to interpose between the questions and the experiments the filter of the theory of combinations.

On the other hand, for horticulturalists and selectors Mendel's laws represented not a rule of action but a schema which explained what they already did. Horticulture has contributed a great deal to research on heredity, in terms of know-how and empirical knowledge. This is attested by Mendel's own biography and by a reading of the work of his contemporaries. In return, biology explained *a posteriori* the effectiveness of horticultural technique. In one way, nascent genetics was not an applied but an explanatory science.

Pasteur and Pouchet: The Heterogenesis of the History of Science

Bruno Latour

In which it will be seen that it isn't so easy to tell who are the winners and who the losers in the history of science. Félix Pouchet defended the generation of living beings from inanimate matter. Louis Pasteur thought it impossible. For a long time the outcome of the struggle remained uncertain.

From the outside, the sciences often seem cold and inaccessible. Fortunately, the controversies scientists become involved with afford an excellent way of getting inside and rediscovering the heat of history. When one studies the natural history of scientific controversy, one can recognize several different types, which makes a rough classification possible. Some controversies are restricted to what may be called the official forums (the Academy of Science, the specialist press, groups of experts); others overflow extensively into unofficial forums (the mainstream press, the courts, Parliament, public opinion). The quantity of neutrinos emitted by the sun is an example of the first kind, while the question of the mode of transmission of AIDS is one of the second.

Whether official or unofficial, controversies can end in two different ways. Some end up in an implicit rejection, that is, a certain point of view is abandoned, without there being any recognizable point where this happens; the controversy gets lost in the sand, suffocates, or is quietly forgotten while another generation, differently trained, takes over. Others, however, conclude

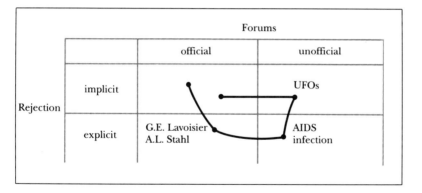

		official	unofficial
	implicit		UFOs
Rejection			
	explicit	G.E. Lavoisier A.L. Stahl	AIDS infection

The movement of a controversy among four quadrants.

in an explicit rejection (crucial experiments, medals, laws, regulations, commissions, reports, trials). Debate about UFOs (Unidentified Flying Objects) is sustained in unofficial forums and is very rarely the subject of an explicit rejection which would put an end to the argument once and for all. The lively controversy between Antoine Laurent Lavoisier and Georg Ernst Stahl (see p. 467), on the other hand, was ended by an explicit judgement, still maintained within the scientific community. But what is interesting, of course, is to follow the sometimes highly complex development of a single controversy. For example, the debate about the danger of nuclear radiation in low doses has gone backwards and forwards, always coming to life again despite various attempts to bring it to an end. The same can be said of the various debates on the inheritance of intelligence, which end and then start again, moving from one forum to another. The diagram above shows the four possibilities.

In this chapter, we shall look at the exemplary mid-nineteenth century controversy between Louis Pasteur and Félix-Archimède Pouchet, on the possibility of demonstrating the spontaneous generation of microscopic living creatures in the laboratory.

When one speaks of a controversy, one already presupposes the existence of clearly identified adversaries; a site where their arguments can engage; a common agenda defining the points for discussion; a series of experiments considered as determinant; judges acceptable to all, to decide who has won and who has lost; an appeal procedure to bring the controversy to a close; and finally, independent historians, not overly committed to either the winner or the loser, to look for explanations for the beginning and the ending of the controversy.

Neither in sport, nor law, nor military strategy, nor science are all these conditions easily satisfied, especially the last. It has been a very infrequent occurrence that an opponent is taken on and indisputably defeated once and for all. Most disputes occur between incommensurable positions. They are combats in a tunnel, in which one is in danger of sometimes ending up fighting one's own shadow. The Pasteur–Pouchet dispute, though, is a simple example and therefore interesting. The controversy was explicit, and the two adversaries met each other and recognized each other as such. It constantly oscillated between official and unofficial forums, but was concluded by a quasi-legal verdict delivered by two successive commissions of the Academy of Science, the first in 1862 and the second in 1864. Pouchet accepted Pasteur's experimental principles, because they were, in his opinion, scientific, but he rejected the decisions because he found the commissions ideologically and politically biased and so skewed in Pasteur's favour that he refused to appear before them.

The most interesting thing about controversies though is that the scientific objects which were their outcome are returned into play. By giving us the discovery before it becomes that, and recreating for us, in the heat of the action, the social groupings committed to or interested and fascinated by the object, they enable historians of science to distinguish many different ways of thinking about relationships between the subjects and the objects of science.

In this chapter, we shall use the controversy between Pasteur and Pouchet as experimental material, in order to compare four different ways of recognizing history *in* the sciences: history as discovery, history as conditioning, history as formation and finally, history as construction.

A Lecture at the Sorbonne

We are in Paris on 7 April 1864, in the main lecture theatre of the Sorbonne:

> Gentlemen, I am going to show you how the mice got in . . .
> *Lights out, please.* Let there be night about us, let everything be dark, and let us illuminate only these tiny bodies, and we shall see them as we see the stars in the evening. *Put on the spotlight.* Ladies and gentlemen, you can see plenty of dust swirling about in the beam of light. *Shine it onto the bench, please . . .*
> If we collect some of this dust on a slide, this is what we see in the microscope. *M. Dubosq, will you project the enlargement . . .*
> Here you see many shapeless objects. But among these shapeless objects, you will see tiny bodies such as these. These, ladies and gentlemen, are the germs of microscopic creatures . . .
> So that you can see the experiment I am going to do on the surface of this mercury-bath,[1] I will light only the tank, and then I sprinkle on a good quantity of this dust. Having done this, I introduce some object into the mercury-tank, a glass rod for example; immediately, you can see the

dust move, making its way towards the point at which I have introduced the glass rod, because mercury does not wet glass . . .

What is the result of this experiment, ladies and gentlemen, so simple, yet so significant for the issue with which we are concerned? It is this: you cannot use the mercury-tank without the dust on its surface entering the flask. It is true that M. Pouchet eliminated dust by using oxygen, artificial air; he eliminated the germs which might be in the water or in the hay; but what he did not eliminate was the dust, and thus the germs which might be on the surface of the mercury. *Lights, please* . . .

But, gentlemen, I am anxious to move on to other experiments, to demonstrations so striking that they are all you will need to consider *Movement. Expressions of approval* . . .

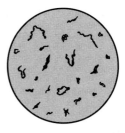

We proved just now that M. Pouchet was mistaken because in his first experiments he used a mercury-tank. Let us eliminate the mercury-tank, because we have seen that it gives rise to inevitable errors. Here, (ladies and) gentlemen, is a perfectly clear infusion of organic matter. *Here is the infusion* . . .

'You can see plenty of dust moving about in the beam of light.'

It was prepared today. Tomorrow, already, it will contain animalcula, tiny infusoria or flecks of mould. *Here is the cloudy solution* . . .

I place some of this infusion of organic matter in a long-necked flask, like this . . .

Suppose that I boil the liquid and then allow it to get cold. After a few days, moulds or infusoria will have developed in the liquid. By boiling it, I have destroyed the germs which might have existed in the liquid and on the surface of the flask-wall. But as this infusion is returned to contact with air, it goes off as all infusions do . . .

Now suppose I repeat this experiment, but before boiling the liquid, I use the burner to draw out the neck of the flask, narrowing it, but leaving the end open . . .

'Here is a perfectly clear infusion of organic matter. Here is the infusion, on the left. It was prepared today. Tomorrow, already, it will contain animalcula . . . Here is the cloudy solution, on the right.'

Having done this, I bring the liquid in the flask to the boil, and then I let it get cold. Now the liquid in this second flask will remain completely unspoilt not for two days, not for three or four days, not for a month or a year, but for three or four years. What is the difference between these two flasks? . . .

They contain the same liquid, they both contain air, they are both open. Why then does this one spoil while this one does not? The only difference between the two flasks, (ladies and) gentlemen, is this: in this one (*on the left*) the dust suspended in the air and the germs it contains

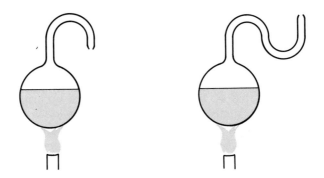

'I place some of this infusion of organic matter in a long-necked flask, like this [on the left] ... Now ... I use the burner to draw out the neck of the flask, so as to narrow it, but leaving the end open.'

can fall into the neck of the vessel and come into contact with the liquid, where they find suitable nourishment, and grow. Here (*on the right*), on the other hand, it is impossible, or at least very difficult, for the dust suspended in the air to enter into the flask ...

The proof that this really is the case is that if I vigorously shake the flask two or three times. *I shake the flask vigorously* ... in two or three days it will contain moulds and animalcula. Why? Because air has come in suddenly, and it has brought the dust in with it. *Murmurs of approval* ...

And consequently, gentlemen, I too can say like Michelet, 'From the vastness of creation I took my drop of water, and I took it full of fecund matter. And I wait, I observe and I question, and I ask it to begin for me the moment of original creation; what a fine sight that would be! *Exclamations.* But it is silent! It has been silent for the many years since these experiments were begun *Murmuring* ... Ah! and this is because I have excluded from it, and still do exclude from it, the one thing it is not given to man to produce; I have excluded the germs which float in the air, I have excluded life, for life is the germ and the germ is life. The doctrine of spontaneous generation will never recover from the mortal blow inflicted on it by this simple experiment. *Sustained applause.*

The End of the Controversy and History as Discovery

With this public lecture, Pasteur did in fact deliver a 'mortal blow' to the theory of spontaneous generation and to Pouchet, its champion. Now what is it that history does? It distinguishes between before and after. It delivers mortal blows, it creates irreversible situations, it makes the past something very different from the present, cutting them apart. Cutting down theories

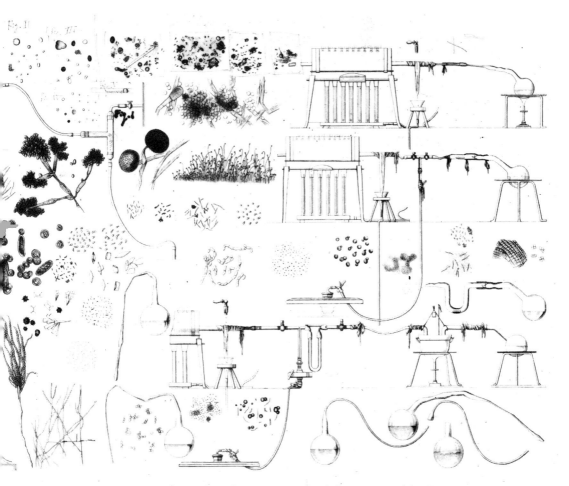

The zoo of micro-organisms created in the experiment and then engraved by Pasteur (with the author's annotations) to illustrate his paper on spontaneous generation. It is possible to recognize the air-filtering apparatus, the swan-necked flasks and various preparations showing air-borne germs. (Musée Pasteur, Paris.)

and cutting off heads prevents others from returning to the situation that existed before; it pushes them beyond the indecision in which years, moments and periods are not differentiated from each other. Pasteur warned: 'You will not leave here without being convinced that the spontaneous generation of microscopic creatures is a chimera.' If all his listeners left the Sorbonne lecture theatre convinced that Pouchet's doctrine could never again raise its head, history had been made, at least in a small way. They could distinguish between two periods, before and after. 'We were there.' If the battle had remained indecisive; if the evidence did not admit of a conclusion; if Pouchet had easily been able to overturn the experiments with the swan-neck flasks, the perplexed audience would have been able to recall

531

neither the day, nor the experiment, nor even Pasteur's position. Nothing would have happened. Nothing would have been superseded. Everything would remain in the same state of confusion.

Pasteur wanted to break with the centuries which had gone before; to single out one day; to mark an epoch; to change his time and change his listeners. It had been believed for centuries that in certain circumstances fairly large animals could develop in a closed environment, without any need of parents like themselves.

In the seventeenth century the Fleming, Jan Baptist van Helmont, had even written that mice would appear in a container filled with wheat whose opening he had sealed with a dirty shirt. Like an involuntary conjurer, van Helmont produced frogs and rats from his container, while no one, not even van Helmont himself, had seen them go in. If their parents hadn't got in hidden in the wheat, these mice had to have come from somewhere. That they should have been generated spontaneously seemed absurd neither to van Helmont, nor to Buffon and Michelet; all attributed force and creative energy to matter, sufficient to create from inanimate or organic matter, if not a mouse, then at least animalcula, primitive creatures like those made visible to astonished eyes by the microscope. Continuous creation. 'Ah! what a fine sight that would be!' What a magnificent discovery it would be, to reproduce in a laboratory flask the conditions prevailing at the origin of life, to send through something like a flash of lightning, and see the primitive elements appear in the primal soup.

But Pasteur was quite the opposite of a conjuror. He unravelled other people's tricks. From his flask he produces neither mice nor life. Rather, he shows us that the others, all the others before him, had been conjurors despite themselves. Through some well-hidden gap they had allowed in various creatures of various sizes, and hey presto! here they are, to the great surprise of public and demonstrators alike. Pasteur is like Rouletabille in *The Mystery of the Yellow Room*. If these little creatures hadn't been able to get in through any crack, then they must have been there from the very beginning. History is, as always, a crossroads: either the little creatures won't quietly get into the flask, and the audience will leave without being convinced that Pasteur has definitely disposed of several centuries of error; or indeed the creatures do in fact enter by some unnoticed ventilator and the audience leaves convinced.

Pasteur mocks the credulity of those who would rather believe in the spontaneous generation of animalcula than feel embarrassed at their own clumsiness. With his mercury bath, Pouchet thought he had taken all possible precautions. His opponent then showed that the mercury itself was covered with dust. Then he showed that this dust contained germs, that is, the parents of the tiny animals that Pouchet saw appearing 'spontaneously' in his flasks.

To further ridicule the theory he was fighting against, Pasteur showed how he himself could control at will the entry and the appearance of the animals. This is the significance of the episode with the swan-neck flask. Keeping the opening unobstructed, Pasteur could obtain flasks which went cloudy and flasks that remained clear. Other things being equal, the only variable was the contact of the particles in the air with the nutritive liquid. When the swan's neck is too long and too curved the air gets through but the dust doesn't, and the liquid remains clear; when the neck is straight or short or has been shaken, the air and the particles it carries come into contact with the nutritive liquid which then goes off. Where Pouchet allowed himself to be manipulated, without understanding anything, by creatures which appeared and disappeared behind his back, Pasteur takes charge of the situation and, by means of a trap, he makes them obey him. The result: he who controls the exit and entry of the animals controls the exit and entry of the audience. 'Pouchet will never get over it; it's over; the case is closed; there is no spontaneous generation.' As always, when one wants to make history, it helps to tell the history of what one has done oneself. Pasteur does not hesitate: 'It has to be said that spontaneous generation has been the belief of all ages of history; universally accepted in Antiquity, it was more disputed in modern times, and more especially in our own day. This is this belief I have come to fight against. That it has lasted more or less throughout the ages worries me very little, for you know, I am sure, that the greatest errors may continue in existence for centuries.' Here he lays down the historical background: on one side, the infinite series of centuries, on the other, this evening now before you; on one side an enduring error and on the other the two champions, Pouchet and your servant, both disciples of the experimental method. M. Pouchet is an estimable man, 'what I admire [in him] is that he proclaims his thought to be *bound* to the results of experiment.' Mine also: 'I approached it [this question] without preconceived ideas, as ready to declare, if experiment had led me to it, that spontaneous generation occurs, as I am today convinced that those who affirm it wear a blindfold on their eyes.' That is the whole problem. Old Pouchet is playing at blind man's buff with microbes whose presence he neither sees nor guesses at. I see, I believe, I understand. End of story.

As always when one makes history, it's a good thing for professional historians to come and verify the chronology, confirm the events, and strengthen, by their independence, the irreversibility one has won. Pasteur, it must be said, is not lacking in historians, not to speak of hagiographers. The story he tells is nearly always embroidered by the professionals. He thought well of Pouchet, recognized that he had done experiments, that he was honest but mistaken. When, in 1875, Rouen wanted to put up a bust of Pouchet, Pasteur subscribed enthusiastically: 'The conscientious scientist deserves the recognition of all for the good and useful things that he has

533

done, and even in his errors, he has a right to every respect.' The historians don't bother with this fair play. Poor Pouchet, what an ass! He couldn't even defend himself. He confounded the brute facts with the hypothetico-deductive method. He mixed everything up. He adhered to the beliefs of another age. He wrote in some sort of gibberish. No, there's no imaginable relationship between Pasteur and Pouchet. Although they were very close in time, the two are separated by an 'epistemological break' which opens like a crevasse between their feet. On one side we have a healthy experimental method and on the other a blind and finicky positivism; on one side someone who discovers microbes, and on the other, someone who insists he has confirmed a discovery when he has discovered nothing at all. To put it very briefly, Pouchet argued with Pasteur and he lost. A good thing too. What did he think he was doing? Exit Pouchet.

With the historians, and even more with the hagiographers, the irreversibility is such that there is no longer any common yardstick between before and after Pasteur; there is no standard which would allow us to compare Pasteur with Pouchet, his adversary. One and the other are in two different spheres, two different states, two incommensurable 'paradigms'. One is in error and the other is in reality. So history has been made and well made. Up till now, we were mistaken; now, thanks to Pasteur, we are no longer mistaken. The discoverer carefully uncovers what had been concealed. Time passes, but its passage has done no more than to dis-cover the true in the false. The history of truth as it emerges from error has certainly helped Pasteur and his supporters, but it doesn't have much history in it or, as the philosophers would say, much historicity. Here the passage of time implies no risk. Before and after are distinguished only in that the second is more true than the first. Time serves to 'correct' positions. Van Helmont was very much mistaken, Buffon a little less, Pouchet less again and Pasteur not at all. As for the microbes, they were never able to breed spontaneously in a sealed flask. What is history for, then? To delay the discovery of what was there in front of them. Some wicked spirit, engaged in a game of hunt the slipper, has hidden scientific truths here and there. Scientists struggle to find them. 'You're getting cold' or 'You're getting hot,' cries the evil spirit. The cleverest wins. The winner takes all; there is no second prize. I will call this history of science, almost entirely bereft of historicity, history as discovery, because its only effect is to bring forward or to put back the date at which the scientist brings an already existing phenomenon to our attention. It is this history which establishes the chronologies one finds at the beginning of certain scientific textbooks, or which still passes for history in certain circles: 'So-and-so became interested in problem x; he published his paper with so-and-so; then he made discovery y.' These are pearls strung into obituary narrative, without there being any history to tell.

The Beginning of the Controversy and History as Conditioning

When in the end nothing happens at all, what is the point of recounting history as discovery? To understand how Pasteur made history, those who want to deserve the honoured name of historian will have to do more than repeat what Pasteur said to himself. To do justice to Pasteur himself, they ought to add a little risk, a little hesitation. That evening at the Sorbonne, the die had not been cast. If he had not lost, Pasteur at least might have failed to convince. Rather than reinforcing Pasteur's position, already very strong, it would be better to strengthen Pouchet's, even if only artificially. At the very minimum, it is a question of applying an elementary justice, which brings the two parties to the controversy together in a sort of court-room, letting each one speak in turn. I will call this application of simple justice to the parties to a scientific controversy the principle of symmetry. Instead of imagining an absolute break between those who were in the wrong and those who were in the right, one will recognize only winners and losers. The winners have no need of the historian's protection, but only the losers, to whom one will, in a way, give a second chance before the tribunal of history. Either they will lose again and the heroes dear to the hearts of the hagiographers will win greater glory; or indeed they will not lose, or not lose as badly, and we will have had the satisfaction of having put right an injustice, of having passed a candidate who was in fact not so much bad as unlucky.

Pouchet was indeed no idiot. An eminent naturalist, professor at Rouen, a correspondent of the Academy of Science, a firm believer, and sixty years old (while Pasteur was no more than thirty-eight), Pouchet was a meticulous experimentalist, as Pasteur politely recognized. The first exchange of letters between the two future protagonists was entirely different from what would come five years later at the Sorbonne. 'You do me great honour, Sir, in seeming to share my own opinion on the question of spontaneous generation. The experiments I have done relating to this are too few, and, I have to admit, too variable in the results obtained, for me allow myself an opinion worthy of being communicated to you.'

After this expression of great prudence, Pasteur goes on in the same letter to offer a confident explanation for Pouchet's 'successful' experiments:

> Do, Sir, adopt the arrangement I have suggested; in less than a quarter of an hour you will be able to set up an experiment, and you will then be convinced that in your recent experiments, you have unknowingly admitted ordinary air, and that the conclusions which you came to are not founded on facts of irreproachable exactitude. I do not think then, Sir, that you are wrong in believing in spontaneous generation, for in

such matters it is difficult not to have some preconceived idea, but wrong in positively affirming it.

Himself applying the principle of symmetry, Pasteur offers a lesson in epistemology both to Pouchet and to adversaries who seem close to Pasteur himself:

> In the experimental sciences, one is always wrong not to remain doubtful when the facts do not require a positive affirmation; but, I hurry to say, when as a result of the experiments I have just mentioned your adversaries claim that the germs of the organized products of the infusions are in the air, they go beyond the results of the experiments, when they should simply have said, that in ordinary air there is something which is a condition of life, that is to say, used a vague word which does not prejudge the question at its most delicate point . . . In my own opinion, the issue is entirely lacking in decisive proof. What is there in the air which provokes organization? Is it germs? Is it a solid substance? Is it a gas? Is it a liquid? Is it a some constituent like ozone? All this is unknown, and it calls for experiment . . .
>
> Despite the invitation you were kind enough to extend to me, I would almost wish to beg you, Sir, to excuse me for having taken the liberty of telling you what I thought on an issue so delicate, which has fallen into my field of study only accidentally, and to a very small degree.

In five years Pasteur moved on from 'preconceptions' to 'prejudices', he made spontaneous generation one of his main topics of research, and replaced the 'vague' words, the whatever-it-is present in the air, with very precise words: no organism is present in a culture medium without being brought there by parents similar to itself. What happened during these five years? Pasteur got involved in the controversy and he developed a theory and an experimental practice which allowed him to redefine what one has the right to expect from micro-organisms.

Becoming involved in a controversy which was 'entirely lacking in decisive proof', and then, after five years of work, changing completely, was Pasteur influenced, conditioned by new factors, not all of which were, as one says 'strictly scientific'?

Extra-scientific Factors

There is no doubt that the controversy was about more than laboratory technique. Politics got into the debate as easily as the rats, mice, flies or microzoans got into the flasks of the supporters of spontaneous generation. Can one imagine a more simply and directly political question than this: 'Are we always exactly the same as our parents? In other words, is creation possible, independent of the millennial conservatism of births and generations?'

The subject was all the more delicate for Pasteur and Pouchet, who approached this difficult question at the height of the controversy over evolution. When, two years later, Clémence Royer translated Charles Darwin's *The Origin of Species* into French, adding a blazing preface in favour of materialism, atheism and the Republic, the controversy over spontaneous generation became linked to that about evolution. For at least half a century, to talk of Darwin or of spontaneous generation was to talk at the same time about biology, the social question, God and forms of government.

Pouchet threw himself into this battle. When in 1859 he published his controversial work *Heterogeny, or a Treatise on Spontaneous Generation*, he made a painstaking critique of materialism and evolutionism. *Homogeny*[2] or homogenesis assumed a chain of exactly similar parents and offspring from the beginning of creation. Geology, however, shows us catastrophic ruptures. How are they to be explained? One must suppose matter to be endowed with a certain plasticity, a certain aptitude which he calls heterogeny, the possibility of giving rise to forms of organized life different from the conditions which gave them birth. His doctrine of spontaneous generation does not demand the chance production of frogs or even of flies from inanimate matter, but only that God should maintain in matter sufficient force to form the eggs of micro-organisms from organic matter. God himself would need to be able to do this to recreate species after each great geological catastrophe. Without this hypothesis, evolutionism becomes unavoidable, because one would not be able to explain how different species occupy different strata without recourse to Darwin's horrible hypothesis on the evolution of species. For Pouchet, to deny spontaneous generation was to take up an atheist position and to embrace Darwinism. Divine creation must continue today. For religious reasons, there must be heterogenesis.

One might ask whether it was necessary to talk of God and of creation, of revolution and conservatism in order to talk about the multiplication of little creatures in a glass flask. Pouchet is clearly mixing up elements which have nothing to do with each other. His knowledge is as heterogeneous as his book; neither the one nor the other displays the fine homogeneity one expects from scientific knowledge. However, if we look at Pasteur, we see that he's not worried at all about mixing things up in this way. The historian of scientific controversy must take great care not to treat this explicit politics in an asymmetrical manner and only analyse the ideologies of the losers. Here, for example, is the beginning of the famous lecture presented earlier on:

> Gentlemen,
> There are very great problems are in the air today, and no-one can ignore them: the unity or multiplicity of the human races; the creation of man some thousands of years ago or thousands of centuries ago; the

fixity of species, or the slow and gradual transformation of one species into another; matter supposedly eternal, and beyond it, nothingness; the uselessness of the idea of God: these are some of the questions which are in these days the subject of dispute between men.

Do not fear that I come here claiming to have resolved any one of these serious matters; but away on one side, neighbouring these mysteries, there is a question which is directly or indirectly linked to them, and concerning which I may allow myself to speak to you, because it is accessible to experiment, and from this point of view I have made it the object of rigorous and conscientious study.

This is the question of so-called spontaneous generation.

Can matter organize itself? In other words, can living beings come into the world without parents, without ancestors? This is the question to be resolved.

In his opinion, spontaneous generation had become the favourite theme of atheists, of those who wished to ascribe to matter sufficient power to bring into existence, without any need of God, the continuous and variable series of living beings. While Pouchet used spontaneous generation to defend God and to attack Darwinism, Pasteur associates his opponent's position with three topics, materialism, atheism and Darwinism, and he places his own research 'in the neighbourhood' of these big questions.

A few minutes later, Pasteur projected his drawings of yeasts onto the big screen, and goes on to say, speaking on behalf of his opponents:

Do you see it here, on the first of these evenings – matter – here in this exhibition of the most beautiful phenomena of nature? Do you see it so powerful and yet so weak, obeying the scientist's every wish. Ah! if we could add to this that other force called life, life which varied in its manifestations according to our experimental conditions, what could be more natural then than to deify this matter? What need would there be for the idea of primordial creation, before whose mystery one must indeed bow? What use is the idea of a creator God?

Ideas are being weighed here not in the balance of the historian of scientific controversy, but in the scales of the Angel of the Last Judgement. To believe in spontaneous generation is to abandon God. But Pasteur is a scientist, not a preacher. Having linked his opponent's position with materialism, having put God on one side of the scales and materialism on the other, he immediately removes them both.

Now you understand the connection between the question of spontaneous generation and the great problems I listed at the beginning. But, gentlemen, on such an issue, that's enough of such poetry, enough of

538

imagination and instinctive solutions; it is time for science, the true method, to regain its rights and to exercise them.

Here neither religion, nor philosophy, neither atheism nor materialism nor spiritualism is relevant. I could even say: as a scientist, they are of no importance to me. Here is a question of fact; I have approached it without any preconceived ideas, as ready to declare, if experiment had led me to it, that spontaneous generation occurs, as I am today convinced that those who affirm it wear a blindfold on their eyes.

Pasteur put forward terrible accusations. Pouchet, a good Catholic and a fierce enemy of Darwinism, is accused of materialism and evolutionism. Then Pasteur suddenly withdraws his accusations. The quivering scales return to balance. Here there are only two poor slaves to experiment, patiently awaiting the result of the races. Which microbes will win? Those carrying the Pouchet colours or those carrying Pasteur's?

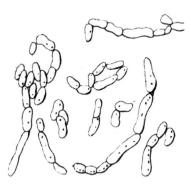

A drawing of yeasts of the genus 'Mycoderma, similar to beer and wine yeasts, etc. with articulations of varied length, and more or less branching.' (In L. Pasteur, Fermentations et générations spontanées, *figure 27E. Musée Pasteur, Paris.)*

Here we have arrived at one of the most sensitive points in the history of science. Apart from the experimental microbes mobilized through projection, demonstration and experiment, did Pasteur bring into the Sorbonne new allies who should not have been there? He did, surely, in tarring Pouchet with the brush of atheism during a third of the lecture. But then he cleared him, recognizing his opponent's qualities and judging that Pouchet, like him, respected only the facts. A clever manoeuvre, one might say. What could be better than to bandy about God's name at the beginning, so as to be able to claim later that we should not be intimidated and that we are absolutely free to end up with spontaneous generation or not? But to speak of a manoeuvre is to impute bad faith. It is to suppose that Pasteur kept his prejudices, his conservatism and his Catholic faith in the balance, and only pretended to withdraw them. If we wanted to take sides, if we were Pouchet's avengers, if indeed we had forgotten the principle of symmetry, we could in fact accuse Pasteur of pretence, of manoeuvring and bad faith. We would have moved from a rationalist version, which attributes to Pasteur only good reasons, to a sociological version which adds some very bad ones to them.

How do we get out of this difficulty and retain the independence of our analysis? By sticking more firmly to the principle outlined above: how do actors explicitly define the invocation and exclusion of the different allies they need in order to win? Pasteur is very clear about the use he makes of God. He is 'in the neighbourhood' of his problem, He is connected to it 'directly and indirectly', but one cannot appeal to Him from experimental failure. That is a position which cannot make victory possible, but one on which his opponents may fall back in the event of his microbes winning. It is an ally which weakens his opponents, 'see how they are influenced by their desire for atheism', without at all contaminating Pasteur's position – 'as a scientist, this is of no importance to me', says this faithful son of the Church, with a fierce independence. On this point, we have no right to say that Pasteur is lying, that he is pretending to be bound to experimentation or, because he is a conservative and a friend of the Emperor's, that he wants to prove that all organisms 'are always born from parents like them'. 'As historians, this is of no importance to us,' for we are not on the side of the true or the false, of bad faith or good faith. We only have to understand what Pasteur put in the balance. How he designed, shaped and refined his invocation of God with the same care that he designed, shaped and refined his mobilization of the microbes in the curves of the swan's-neck flask. In the Sorbonne, in 1864, the invocation of God would have had a negligible effect. But to invoke God, to reinsert the experiment in the context of what was at stake, to award theory its letters of nobility, to recall the moral order and to thus situate the rights and duties of experiment was, before such an audience, in 1864, to ensure a maximum effect for his argument.

The historian's experimental method is gradually becoming clearer. To speak of Pasteur and mention only his experiments, while forgetting his invocation of God and his accusation of atheism against Pouchet, would have been an intolerable act of censorship. But not to give details of the way in which he appealed to theology only in respect of his opponents, and independently of experimental judgement, would also be intolerable censorship. It is too often forgotten that rationalism and sociology are twins. One can only obtain these two versions of the history of science by excising from the texts and the archives the subtle mechanisms by which the actors obtain the assistance of authority, and by segregating these actors into 'extra-scientific factors' on one hand and 'scientific factors' on the other.

To listen to them speak, one would imagine that historians have a special gift for distinguishing, in place of the historical actors, not only the allies they have the right to call upon, but also the sense and the manner in which they must be presented. There is always an etiquette which must not be breached, such a factor always coming before others of such an order, and always having to wear a certain dress and a certain hat! If the authors are rationalizers, they say that Pasteur did not invoke God (forgetting that he did), and if they are sociologizers, they say that Pasteur threw the weight of God into the balance (forgetting that he did not do so). These two families of authors agree on only one thing: that it is possible to draw up two clear compartments, in one of which go the scientific factors and in the other the extra-scientific factors. Then they return to the argument as to which is the most important.

Before seeing how we can get rid of these principles of etiquette, how to avoid operations of excision and segregation, how to put an end to the process of accusation and to imputations of error or bad faith, we must go back to Pouchet. In fact, we have to respect all the actors brought together by the controversy, and see not only how they mobilize their allies, but how they judge their opponent's operations, that is to say, from their point of view, how many allies their opponent has.

Pouchet does not mince his words. The letters he wrote to his collaborators always spoke of official science's 'plots' against him and his microbes. 'My dear friend [he wrote to Joly, professor at the school of medicine in Toulouse], I have reached the apogee of indignation. It is absolutely unheard of, to push impudence to such lengths as has Paracelsus II.[3] What! He says that our experiments on the Maladetta only confirm his own! Really one cannot imagine such boldness and such impudence . . . It goes beyond the permissible.'

And in another letter he wrote:

> I have reorganized my laboratory, my dear and noble friend, in order to defend our holy cause, and there I will hold our standard high.
>   You will not give in, you say! And neither will I! I don't want some scientist, born at Carpentras or Domfront, just like me, to lord it over me just because luck rather than merit has taken him to Paris. He will pay dearly for such an insult. I feel we have the strength of Antaeus, I will not let him go until he is crushed beneath the rocks of heterogenesis.

A simple provincial correspondent of the Academy, he feels he can't make the weight against a Parisian Academician. Twice, in 1861 and in 1864, the Academy set up a commission to decide once and for all the question of spontaneous generation. Since then fallen into desuetude, these commissions had been invented to settle the question of the end of controversies, to avoid the possibility of colleagues being able to reopen debate indefinitely, wasting the time of the scientific community. Although their conclusions had no legal force, they had none the less something of the authority of a legal judgement. Among the mechanisms available to ensure irreversibility, they acted as a fairly effective ratchet. To reopen a debate when two commissions had given definitive judgements was like trying to lift up the heavy slab from a grave. It was to marginalize oneself. But at these commissions were made up of colleagues of Pasteur's, more or less convinced in advance, Pouchet thought he hadn't had a chance.

But what shocked Pouchet very much more was that the commission decided on the experimental agenda without hearing his complaints. In science, as in war, to choose the ground and the arms is already to control the outcome of the battle. Pouchet wanted a discussion of the whole of biology. The commission asked him first of all to repeat an experiment it considered crucial, while following Pasteur's instructions. Pouchet gave up, disgusted, he said, by so much bad faith. In this the commission saw only the proof of his weakness.

Our problem of elementary justice is getting terribly complicated. The Academy commission sat as a court appointed by the history of science. Pouchet rejected its judgements. Twice it congratulated Pasteur and twice it buried the Pasteur–Pouchet controversy. Those of us who want to give the loser a second chance, how should we estimate the weight of the allies in the commission? We haven't the right to be the avengers of a lost cause. We are forbidden to weigh up the 'extra-scientific factors' only in the case of Pouchet and the 'scientific factors' only for Pasteur. But inversely, we can't just take 'extra-scientific factors' into account in Pasteur's case and only the experiments in Pouchet's.

The difficulty of measuring the resources mobilized in a controversy is even greater when one moves from the official to the unofficial forum. In a letter written two years previously to Colonel Favé, ADC to the Emperor, Pasteur tried to bring his research to the attention of His Majesty:

> When I had the honour of seeing you a few weeks ago, you were good enough to say, in passing, that at Vichy you had a chance occasion to speak to the Emperor about my work on so-called spontaneous generation. Since then I have been wondering whether it would not be too indiscreet on my part to offer His Majesty a copy of the paper in which I give an account of all my work on this topic. You know, Sir, that this research has been no more than a necessary digression among those which I have been pursuing for many years into the mysterious phenomena of fermentation, phenomena so close to life, closer perhaps again to those of death and illness, especially the contagious diseases. I am very far from the end of this wonderful research. . . . And I would, too, perhaps be lacking in sincerity if I did not admit that in attempting to bring this work to the attention of the sovereign, I have a secret wish to acquire the means to develop them more freely and more fruitfully.
>
> This little laboratory, Sir, which you one day did me the unexpected honour of visiting to see the results of this work, is no longer enough for the research I am planning.

For, as Pasteur well knew, there is a heterogenesis of science. Research needs laboratories and laboratories need money, support and patronage. To see the little creatures pullulate in the flask or not, one has to bring in not just God but the sovereign, and to involve them both one way or another in the controversy. 'I have great questions, great hopes, about life, death and illness, but only a little laboratory.' We shouldn't say that Pasteur was playing politics because he was trying to interest the sovereign only in order to have a laboratory. But let us not say either that he wasn't involved in politics, because if he hadn't addressed himself to the Emperor's ADC, if he hadn't approached the sovereign, he wouldn't have got a bigger laboratory.

Pasteur and Pouchet agreed on one thing: whatever were the larger questions that could be mobilized around the question of spontaneous generation, it was the laboratory that was the judge. God, the sovereign, the constitution, heritage and morality might all be called upon, but they ought to help to see whether the liquid in the flasks had gone cloudy or not. The common standard Pasteur and Pouchet recognized, without discussion, as the only means of bringing the dispute to an end was the laboratory experiment. It is this agreement which allows historians to distinguish between external factors and the experimental facts. God and the sovereign might condition the debate, but they couldn't directly cloud the liquid in the swan's-neck flasks.

But once the big questions have been taken into the laboratory there are plenty of ways of deciding them. Spontaneous generation is not difficult to observe. Any flask left for a few days fills up like an aquarium. Life pullulated in the laboratories of the 1860s. If Pouchet wanted facts he could have them by the flask. At the beginning, at least, Pasteur willingly admitted the difficulty of his position: how to rarefy the profusion of life and to maintain the sterility of flasks filled with nutritive material. But none the less he did not draw the conclusion that spontaneous generation occurred. He simply said: 'I did not publish these experiments [for] the consequences which had to be drawn from them were too serious for me not to fear some hidden cause of error, despite the care I had taken to put them beyond reproach.' At the beginning, then, it was Pasteur who was short of facts and Pouchet who was accumulating them. But Pasteur knew that this profusion of little creatures in the culture medium was not due to spontaneous generation, but to the contamination of the cultures by foreign bodies. How did he know? Where does this assumption, this prejudice, this *a priori* theory come from?

Most historians recognize that this is a legitimate question. Experiment does not decide completely. Since Pierre Duhem (1861–1916), the idea that experiment always has to be accompanied by something else in order to gain assent has been called 'under-determination'.

Duhem particularly wanted to highlight the role of theory in the reading of experimental results. More sceptical, other contemporary historians of science call what completes and reinforces an experiment, always too weak in itself, a 'paradigm', like Thomas Kuhn, or 'prejudice', like Paul Feyerabend. I will call history as conditioning this history of the sciences which recognizes influences as long as they occur outside the laboratory. This conditioning is not without effect on the product, but in the end, it is not the product itself.

From History as Conditioning to History as Formation

Now that we have re-established the experimental conditions which can make visible the deflection of the balance, now that we can see without prejudice the types of resources which the two camps put into or take away from the scales, the beam is in balance and history hesitates: it can go one way or it can go another. Suspense: thanks to what will Pasteur win?

Let us consider the list of resources and the different schools in the history of science, each of which, like the fairy in the tale, brings to place in the scales the gift thanks to which the hero will win through. Those who offer the hero only scientific or technical resources are called rationalists. The rationalists are themselves divided into two big groups. For the first, who for this reason are called experimentalists, the experiment is always enough to tip the balance; for the second group the experiment is not without importance, but it cannot determine the decision all by itself; you need a theory as well. It is the one who has the most coherent and fruitful theory who will win, even if it means forcing the facts a little. Experiment under-determines. Theory over-determines.

Let us see what lies in our balance if the good fairies' gifts stop here. Pouchet, the conscientious positivist, the great enemy of theory, has the experiments going for him. He goes up into the Pyrenees; he repeats the experiments that Pasteur performed so magnificently on the Montenvers glacier with his swan's-neck flasks. The infusions go cloudy . . . Pasteur has lost. The balance tips towards Pouchet. But no, for theory arrives and falls heavily into the scales. What can one do with Pouchet's theory? Nothing, because the microbiologists' cultures will always be disturbed, corrupted and perverted by spontaneous contamination. What can one do with Pasteur's theory that no organisms are born except to similar parents? Everything. Never mind Pouchet's experiments in the Pyrenees. They have to be wrong. Even if Pasteur can't find the chink in the armour straight away, he is sure that a serious mistake has been made.

What? Against the evidence of indisputable facts proving spontaneous generation in a single matrass,[4] does he prefer a theory which *a priori* affirms its non-existence? But this is the very definition of a prejudice, or if we want to be less rude, of a presupposition. The balance levels out again and finds a new equilibrium. Pasteur's prejudice cannot overcome Pouchet's facts. Here we are faced with a new under-determination: neither the facts or the theories are enough to win the prize. Let us allow in the other fairies, even the wicked one . . .

Those who claim that neither facts nor theories are ever enough to win, and that fairies who bring our heroes only these are sending them to their

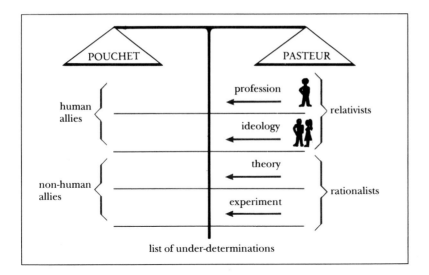

List of under-determinations.

deaths, are called relativists. But there are many tribes of relativists, as there are of rationalists. Those who throw into the balance only large-scale, heavy-weight, extra-scientific factors are called social or macro-sociological relativists. One calls micro-sociological, on the other hand, those who rely on tiny social facts which, although apparently insignificant, are amazingly effective when put in the right place.

Let's see what we get when we apply this to our guinea-pig controversy. Pasteur is a born conservative; there is nothing more conservative than the argument that one is always born resembling one's parents in every point; *ergo*, Pasteur prefers the theory which denies spontaneous generation. Here is a wonderful ideology, of vast size, called 'conservatism', which comes at the right point to tip the balance in Pasteur's favour. The difficulty is that Pouchet rivals Pasteur in his conservatism, while the latter does not hesitate for a moment to upset the society of his time with his ravaging microbes. On the whole, he is a revolutionary, fond of disputes, lacking respect for established organizations – at least when it isn't a question of the Empress but of micro-organisms or of doctors. Ideology, which seemed such an admirable ally, did not succeed. The balance doesn't move at all. The mood of the time is too subtle, too universally distributed, too unpredictable to move a column of mercury one centimetre, or infusions of hay, or swan-neck flasks.

There remain the micro-sociological factors. Pasteur is an Academician and he lives in Paris. Pouchet is a correspondent and he lives in Rouen. Pouchet contradicts Pasteur's experiments. Pasteur rejects Pouchet's facts. Pouchet protests against this prejudice. The Academy appoints a commission, friends of Pasteur's; Pouchet hasn't any. The commission politely asks

Pasteur to repeat an unsurprising experiment and impolitely asks Pouchet to abandon his too general and too vague demands. Pouchet cracks, refuses to debate, returns to the provinces cursing 'official science'. Nothing allows one to say, however, that such professional complicity would be enough to put an end to controversy. Paris and the Academy are substantial resources which discourage poor Pouchet, but they cannot keep him forever confined to the provinces. He could return with other arms. Pasteur himself, for a long time confined to the provinces, did everything to get to Paris and succeeded. At what? Yet another under-determination? Will the issue never be settled? Will the balance not tip one way and for good? How much more must be heaped up in the way of facts, theory, ideological prejudice and professional organizations?

The very form of the question, as it is formulated by history as conditioning, is still not sufficiently historical for us to be able to give a reply. What would one say of a military historian who, having reviewed the cavalry, the engineers, the artillery, the quartermasters' corps, the hussars and the morale of the troops, was still astonished at not being able to understand the outcome of the battle? We would say he had confused the balance of forces with the landscape, with strategy and with movement. Our history is missing something. The principle of generalized under-determination to which we are subject proves it.

What then is missing? What is the ally number '$n + 1$' which we have forgotten to mention? Whether rationalist or relativist, whether they believe experiment or ideology to be in command, all the authors considered above believe that there exists somewhere a repertoire of decisive blows, which, in science, allow one to win the day. Of course, we have left behind the history as discovery with which we started. The microbes aren't waiting, shrouded in darkness, for Pasteur to come and illuminate them with a flash of the spotlight. Now there is suspense, hesitation and the possibility of forks in the road. But this somewhat more lively history, despite everything, will play only a very limited role, if one can say of each controversy 'here it was ideology that carried the day', 'in this other case it was theory', 'in this other one again it was the profession'. The passage of time has no influence on the composition of the list of permissible blows, but only on the use made of them.

History as conditioning turns to this list to explain the delay or otherwise of the discovery of what one ought to have found. It is said, for example, that acceptance of Pasteur's theory was 'facilitated' or 'widened' by virtue of the anti-materialist accent he gave it. The packing doesn't influence the content, but it is not without consequences for sales. I will call history as formation the explanations of controversy which turn to the same list to explain not the acceptance of an argument, but the origin of the argument itself. One can say, for example, that Pouchet developed his concept of heterogenesis to batter atheistic Darwinism. It isn't any longer a question of advancing or delaying the inevitable discovery, but of choosing what might otherwise never have happened. In history as formation there are real forks in the road. The course of things, the course of science, could have been different.

What a Microbe can do: History as Formation

If Pasteur had so many difficulties, if an obstinately maintained prejudice obliged him to reject the hypothesis despite 'the evidence of the facts', and to transform this into an 'experimental error', into 'hidden contamination', it was because Pouchet had the raw materials to construct his hypothesis. Pouchet invented nothing and he nourished himself, as one might say, on the 90 per cent of cases rejected by Pasteur. For him too, there pullulated phenomena on which he might rest his case. And his position was all the stronger for not having to prove that he won at every go. He needed only one positive case to prove his theory, while a single negative case would be enough to ruin Pasteur's. One can see how the symmetry advances. The beam of the balance moves towards the horizontal again. Pouchet was not so mad. It is even because of this that Pasteur fought so hard against him. The quality of Pasteur's victorious experiments to some extent depends on Pouchet's qualities. It is Pouchet's energy, his honesty and his passion which we find anticipated and contradicted in Pasteur's magnificent demonstrations.

In February 1859, in a note to a paper on lactic fermentation, Pasteur had remarked that each fermentation was caused by a specific ferment introduced into the culture medium form outside. 'A great step forward has been taken in the question of spontaneous generation,' he wrote. Pouchet reacted vigorously to this allusion, and after Pasteur's polite but firm response, he incorporated into his own experiments the new precautions prescribed by Pasteur. To understand Pouchet's docility and his younger colleague's self-assurance, one must remember that Pasteur had just come from a very similar debate with Justus Liebig, the greatest chemist of the time. Liebig accused Pasteur of being a vitalist, because he saw in the animalcula which proliferated in wine, vinegar, beer or fermenting milk, the cause of the fermentation, while according to Liebig they could only be the consequence, the initiator or catalyst. Liebig and all the chemists of his day had done their very utmost to counter the vitalist argument by establishing physical causes for the most important transformations of matter, including organic matter; and here was Pasteur, encouraging vitalism, bringing in these tiny creatures which were causing a reaction. The practices that Pasteur imposed upon Pouchet were all the stronger for being the ones that he imposed upon himself in his struggle with Liebig, a struggle in which he occupied a position comparable in some respects to Pouchet's. 'The little creatures are not there,' said Liebig. 'But yes they are,' said Pasteur. 'They are being generated,' said Pouchet. 'But no they aren't,' retorted Pasteur. For him, the reply is the same: 'Whether you wish they weren't there [Liebig], or whether you wish they were [Pouchet], in both cases you are introducing them without your noticing.' You do not control the passages along which they travel: I do. You have not yet reorganized your laboratories, your gestures, your flasks, your beakers, your filters, so as to be able to control their entry and their exit: I have. Your scientific policy has not mastered the frontiers you have established too easily between the organic and the inorganic. Mine, made up of networks and pipes, controls them at will.

For in science, as elsewhere, it is always a question of mastery and of tests, mastery of men testing things, mastery of things testing men. Pasteur challenges Pouchet to submit his cultures to tortures as horrible as he has inflicted on his own. Pouchet takes up the challenge. He heats a little 10g bundle of hay in a drying-oven at 100°C for half an hour. What microbe could survive it? And yet, when he introduces this little of bundle of hay under a mercury bath in a sterile environment, at the end of eight days, the little creatures are pullulating. Pasteur has lost. Pouchet has taken all the precautions demanded by his opponent, but the organisms are there and they can't have come from anywhere else. Pouchet, faithful to the experimental method, is forced to conclude that spontaneous generation occurs. At the Sorbonne lecture, Pasteur recognized this with some slight modesty, before passing vigorously onto the attack;

> In fact, what objection can you make to M. Pouchet? Would you say to him: the oxygen you used contained germs?
>
> 'But no,' he will reply, 'I obtained it from a chemical reaction.'
>
> It's true, it couldn't contain germs. Would you say to him: the water you used contained germs?
>
> But he would reply: this water which was exposed to the contact of the air might well have received some, but I was careful to put it boiling into the flask, and at that temperature, if there were any germs, they would have lost their fertility.
>
> Would you say to him: it's the hay?
>
> But no, the hay came out of a drying oven heated to 100°. One could make one last objection, which is that there are certain examples which do not die when heated to 100°;
>
> But he would reply: its not because of that! And he heated the hay to 200°, 300° . . . He'll even say he went as far as carbonization.
>
> And indeed, *I admit it*, the experiment so conducted is irreproachable, but only on the points the experimenter has drawn attention to.

And Pasteur goes on to show that Pouchet has made a blunder as big as Van Helmont's: dear me, but of course! it was the mercury that was full of germs: 'I am going to show you where the mice got in . . .'

Just the same, it was to reply to Pouchet that Pasteur analysed the contamination of the mercury and stretched the necks of the flasks. Pouchet is a part of the experiment, which has to come to terms with him. To do without opponents, losers, others, to do without polemic, the fire of controversy, to do without people, is to understand nothing of the very content, the form of the experiment. That evening, in front of the cultivated public of the Sorbonne, on the demonstration table, all the objects, the flasks, the matrasses, the enlargements, all designed, in their smallest details, to take account of Pouchet, his colleagues and his microbes. In this sense, the elegant curve of the flask is a historic form, the elongated face of a polemic. As soon as one re-establishes a little symmetry, history is no longer this sort of advanced clockwork which springs the date on which scientists will discover phenomena which owe nothing to history itself. History does something to the phenomena themselves; it conditions them, it forms them, perhaps it constructs them. To re-establish symmetry is to wrench the historians away from the victor's camp, to forbid them to talk in terms of true and false, to give them enough independence to establish a sort of balance whose pans can be loaded little by little. To make Pasteur completely right and Pouchet completely wrong is like throwing Brennus' heavy sword into the balance. '*Vae victis!* Woe to the vanquished!' Who can't see that here we lose not only justice, but understanding? To accept from the very beginning the asymmetry between Pasteur and Pouchet, to claim that the first corrects the concepts of the second, that the first is right and the second wrong, that the first can't see the phenomena which leap to the eyes of the second, it to rub salt into the wound – a great pleasure – but it is not to understand why the balance tipped in Pasteur's favour – perhaps an even greater pleasure. Scientists, especially when they are great scientists, do not ask that their opponents should be thrashed – they know very well how to do that for themselves – but that justice be done to their talents, and thus also to the people against whom they fight. In other words, to do justice to the winner, you have to be symmetrical. It is only if one has first mounted and then loaded an irreproachable balance that the deflection of the beam can mean anything at all. But if one deflects it from the beginning its displacement signifies nothing, if not the enormous prejudice one started with.

Pasteur's microbe is designed to fight against both Liebig and Pouchet. If it attacked only Liebig, and pullulated in cultures without the microbiologist being able to do anything about it, microbiology would not be possible. One would have to content oneself with observing the multiplication of heterogenous beings, and admiring the power of God or of Darwin. If it attacked only Pouchet, and there were in the cultures only chemical phenomena without the activity of organized beings resembling their parents, death and illness would remain out of range. Nobody knew what a microbe could do. For Pasteur it had to be capable of producing fermentation in the real sense wherever necessary – against Liebig – and strictly incapable of making an impromptu appearance in properly prepared cultures – against Pouchet. The history of science now focuses on the microbe itself. It flows from men towards things. Pasteur developed a polemical microbe capable of action on two fronts and continuous resistance to laboratory practice. Real forks in the road are being established. Anyone but Pasteur would have defined another microbe, that is to say an actor capable of other actions, defined by other tests and bringing about other defeats in other disciplines.

The Microbe Network or History as Construction

In history as formation the course of things themselves, and not just the doings of men, begins to gain a little historicity. A Pasteur who wanted to occupy other positions, who dreamed of other goals, would have needed another microbe. However, even in this last form of the history of science, the humans and the non-humans are not treated in exactly the same way. The humans act a lot, the non-humans much less.

The three types of history we have looked at only more or less slow things down and provide more or less clever explanations for the irruption of the inevitable. The things discovered, for their part, do not tell so many stories. They are there. They have always been there already, microbes and phagocytes, antibodies and viruses. There is a history of men, of beliefs and societies, which may discover them or ignore them, but there is no history of the things themselves, impermeable to time. For history to come to the sciences, and for the history of science to fuse with history as a whole, one needs to go a little bit further and provide the things themselves with agitation, uncertainty and passion, that is, with historicity. For that, as always, we must return to the actors, stick closely to Pasteur and Pouchet.

Pasteur didn't know whether Pouchet was right or not, whether micro-organisms were capable or not of resisting temperatures of more than 100°C; he didn't know whether they were sufficiently discrete and taxonomically distinct enough to be really specific; he didn't know either whether the Emperor or the Empress would give enough support, whether one could be both a Darwinian and a believer and whether Pouchet was stronger or weaker than he. Pasteur was in a state of uncertainty, and this is why everything was grist to his mill, seizing on every little argument, looking for support to his friends in the Academy, to God, to the Emperor, and to the creatures which do not get through the neck of the flask. Ally number

'$n + 1$' is uncertainty, which doesn't at all resemble the choice of a factor from a finite repertoire of causes. A hundred years later, historians hesitate over the type of history and the type of explanation which leads to the inevitable; Pasteur himself, as soon as he had defeated Pouchet, would rewrite the story of his experiments as if they were inevitable. But in the heat of the action, just the same, he did not know, and it is uncertainty itself, proper to research as it is to history, which is chased off the field as soon as one claims to be writing the history of science.

This can only be the history of uncertainty about the fate of things themselves. The micro-organism is an actor in the process of being defined, just like the liberal Empire, Pouchet's up-and-down career, the laboratory in the rue d'Ulm in Paris and social Darwinism. How is it defined? Like any actor: by what it does, by what it supports, by what one can get it to do, by what it believes, by what it holds dear. Like Pasteur himself, like Pasteur's audience that evening in the old Sorbonne. We do not know what the micro-organism is. If we knew it we should miss all the importance of the intrigue, all Pasteur's courage. We would be transforming an actor into an essence. We would be retroactively emptying the history of science of its historicity, to replace it with the science of today, displaced by anachronism to a position beneath yesterday's. No one knew in 1864 what a microbe could do.

Pouchet from Rouen, Joly and Musset, his collaborator, from Toulouse, did not allow themselves to be intimidated by the Academy's first commission. Pasteur had been on the Montenvers glacier to prove that the flasks he had taken there would not become troubled. Nor were his opponents troubled. They restarted the experiment in the Pyrenees, on the Maladetta, even higher than Pasteur. They scrupulously repeated the experiments. The indefinite microbe was spontaneously generated in eight of Pouchet's flasks. So it wasn't specific. So all Pasteur's work came to nothing, because specificity was necessary in his argument with Liebig and in his explanation of fermentation: each fermentation has its ferment. So, Pouchet says, one is going to be able to reject Darwinism, crush Pasteur 'beneath the rocks of heterogenesis', maintain the honour of the provinces and alert the press. God sets off catastrophes, but the spontaneous fertility inherent in created matter repeoples the world with new varieties. Novelty can be derived from floods and spontaneous generation rather than from evolution. This is how, at sixty years of age, one can put the good town of Rouen on the scientific map. Here are God, Darwin, Pasteur and Rouen held in place by a network of solid connections. But the microbes? Will they follow? They will, in every respect. Pouchet gets them to pullulate having subjected them to a trial by fire – calcination – that not even his opponent Pasteur had demanded. Pouchet, in alliance with the docile microbes which have passed his opponent's tests, will transform the science of his time, that is to say will give form to his time, create form.

Pouchet's network of alliances brings Pasteur's career to an end by interrupting what Pasteur knows how to do best: stopping and then allowing the continuation of the invisible creatures' journey through the tiny corridors and making them appear under the microscope in well-chosen parades, which he uses as so many spy-holes. If Pouchet is right, all Pasteur's skill is out-dated: spontaneously, and without him being able to do anything about it, the germs appear. Everything which allows him to conquer illnesses, just like so many fermentations, and everything then that he had promised the Emperor as the reward of his gifts, everything he had believed for ten years, everything he can do in his laboratory, finds itself circumvented, betrayed and weakened. To speak of prejudice would be a euphemism. If Rouen shines on the map of the scientific world, Pasteur is no more than a brilliant chemist who threw some light onto the mechanism of fermentation and the asymmetry of crystals. With the form of micro-organisms, their abilities and their performance, go Pasteur's form, abilities and performance. They are connected together, and both have attached their common fate to that of God, of wine, of cheese, of the Empire, to the fate of that little daughter he impotently saw dying before his eyes, to the fate of France, perhaps. The form of the microbe is the point at which the network breaks up or closes up again. No microbe should be able to move or appear without its master knowing, for then, there would be no master.

Pasteur's persistence in at all costs maintaining the integrity of his network and of his skill commands our admiration and forces the microbes to redefine themselves. Pouchet and his friends must have been mistaken. In the 1863 report they sent to the Academy, they forgot to say what had happened to four of their flasks. They had opened them with a file made red hot in a flame, rather than using a pair of pliers. There must have been a mistake somewhere; it was impossible, unthinkable for them to be right. If despite everything they appeared in Pouchet's flask after calcination of the hay, who should give way? Pouchet or Pasteur? Answer: the microbes. Involved in the controversy, they underwent an extreme pressure. Without that, who would have gone and looked for them in the mercury? Who would have spotted their spores in the hay? Mixed up until now with the air that we breathe, here they are distinguished from the air that has borne them up till now, in the experiment with the swan-neck flask. An apparently innocuous experiment, but the more perverse for its using neither fire nor mercury: the air passes through and leaves them stuck in the curves of the neck. Treachery! We shall not pass! They shall not pass! Victory! Pouchet has been defeated, the microbe has been redefined; Pasteur's career goes on; God will not suffer from Darwinism; the audience at the Sorbonne will applaud; the Empress, at Compiègne, astonished, will honour him with a visit. End of controversy.

Pasteur, sure of himself, challenged his opponents to restart their experiment in front of the Academy. They backed out twice, while appealing to the daily press:

> Their retreat in March [Pasteur wrote] should have warned one to expect it in June. None the less, I admit I did not believe it would happen. But what is not less strange is the ease with which these gentlemen have nobbled the daily papers. I do not know if there is a single one which does not denounce the partiality of the Academy's commission and oppression by official science, as they call it.
>
> This June retreat, for me, is something quite beyond words. You know how positive they were when I offered them the challenge of demonstrating before witnesses the experimental proof of their assertions: *We will take up the challenge. If a single one of our flasks remains unaltered, we will loyally accept defeat.*
>
> Is that clear enough? You remember the result to which this sentence refers. This was the experiment with which the commission wanted to begin, because you have to begin somewhere. And how could it have thought about beginning with anything other than this experiment which had given rise to the challenge and led to the nomination of the commission? And who then was it who called the attention of the Academy to this experiment, if it wasn't *them* when they wrote that despite enormous obstacles they had gone and repeated it among the glaciers of the Maladetta, at a height of three thousand metres, while I had only gone to two thousand metres on the *mer de Glace* . . .
>
> Let them do what they like, their experiment is wrong, and once again I challenge them to perform it in front of witnesses with the results they have published.

Here we have arrived at the major difficulty in the history of science, the retrospective reconstruction of the past on the basis of the – always provisional – conclusions of scientific controversy. The microbe, the product of a multiple polemic, becomes an essence which in the end had always been there and which Pasteur discovered, or which he almost saw, or which he could have missed – put a tick next to these useless answers, matching them to the three types of history described above. Spontaneous generation suffered the opposite fate; in the end it was never there, although Pouchet stuck to it so obstinately.

If the microbe isn't an essence, how should it be defined? As the provisional form of networks which, depending on the circumstances, may temporarily or permanently modify the definition of all the actors of which it is made up? Did Pasteur discover the microbes? No, because he made them. What? Did he make them from scratch with his conservatism, his prejudices and his theories? No, because the microbes made him, his career, his conservatism, his liberal Empire and his swan-neck flasks. What? Is it a question of co-production or of compromise? No, it's rather more than that, because the actors who enter into relationships with each other are not the same as the ones that you isolate in advance to bring them into relationship. There isn't a finite list of factors for explaining the history of science any more than there is for history generally. Let us abandon conservatism as an essence, the microbe as an essence. Let's forget about reason. Let's just stick to the networks. From 1860 to 1864 spontaneous generation was derealized and weakened, it lost ground, allies and resources. The micro-organism resistant to 100° and separable from air gained in reality and gained ground in the salons, at the Empress's, at Notre-Dame and with the English surgeon, Joseph Lister.

Change one single point in the network and you change its form. Add a spore which allows resistance to 100° of temperature and you have to change the whole story, granting that Pouchet was right and that Pasteur was blind. From Emile Duclaux to Jerry Geison, the historians have claimed that, if Pouchet had a little more nerve and had redone his experiments in front of the commission, Pasteur would have been put in a very difficult position. The *bacillus subtilis* eliminated from Pasteur's experiments had not been eliminated from Pouchet's infusions of hay. Resistant to heat, it would have reproduced cheerfully as soon as the flasks were opened. If the microbes had been able to resist new tests, they would no longer circulate in the same way as the others, would not bring together the same interests, would not interest the same people and so would not be the same actors.

If one accepts that historicity increases with a decrease in timeless data, then the analysis of controversy has taken us from a history of scientists to a history of science. We no longer have on one side things that can have a history (people, their cultures, their ideas and their tools) and on the other, ahistorical objects. In the eyes of history as construction, it is impossible to understand what a microbe can do without identifying the highly heterogeneous network which constitutes it entirely: a friend of the Emperor's, a tool of microbiology, a response to Liebig, destroyed by heat, transportable by air and by clothing, stopped by the curves in the glass, shatterer of atheism, parent and child of creatures exactly resembling it, anaerobic respirator, the promise of a solution to life, death and illness, absent from glaciers, present in Paris, mastered in the rue d'Ulm, this is how it appears in the heat of the controversy, as far as Pasteur is concerned. It is this list of actions and of tests. The enemy of official science, a friend of the opposition press, present in glaciers, necessary to God, common in Toulouse and Rouen, resistant to calcination, arriving without a similar parent, this is the micro-organism made by Pouchet. This list of actions and of tests is what it is.

The discipline of 'history' too easily allows itself to specialize in studying what has a historicity, long or short: manners, battles, population, the price of grain and ideas. It forgets to question itself about the extra-historical division between what does and what does not have a history. It is the philosophy of science which produces this division, leaving to Clio the vast domain of contingent circumstances, but keeping back, beyond time, the world, number and things. History as construction obliges the historian to go back, to return to this 'Yalta' which carved out his field for him. To the short term, to the long term and to the very long term, he must add innumerable times which have as many different forms as there are sciences and objects. I can now give this way of doing the history of science, which I have called history as construction, its real name: it is history itself, but extended to the things themselves.

19

Mendeleyev: The Story of a Discovery

Bernadette Bensaude-Vincent

In which we will discover a chemistry professor's complex and courageous intellectual adventure. Returned to its context, the discovery of the periodic classification of the elements takes on a very different meaning to the one given to it in our own day.

In 1869 Dmitri Ivanovich Mendeleyev classified all the elements according to a periodic system which followed the order of their increasing atomic weights. Today, his name still appears on the periodic table in all the books and in all chemistry classrooms. Mendeleyev's table is presented as a precursor of present-day classifications based on atomic number. Mendeleyev appears as a diviner, a prophet who in a stroke of genius, a lightning intuition, not only anticipated experimental results with his predictions of unknown elements, but also the quantum theories of the twentieth century.

This interpretation of history from the point of view of present-day science is called 'Whig history' by English historians. 'Whig history' is very common in science courses and conferences. Looking to the past for markers that lead to the theories of today, scientists have a tendency to skew and simplify history. Eliminating the false trails and the dead ends, they discover a royal road leading to the present.

In addition, Mendeleyev's table is an achievement that demands our admiration. Here, gathered together in one table, are the building-blocks of our Universe! How could Mendeleyev have discovered the correct principles of classification when he knew no more than seventy of the elements, knew nothing of the electronic structure of the atom? And what audacity to have provided places for unknown elements, and even more, to predict their properties! The periodic classification has become one of the classical episodes of positivist epistemology. It is always cited, together with the prediction of the existence of Uranus by Urbain le Verrier, to illustrate the difference between an empirical science which collects facts and a rational science which is able to organize experimental data and to predict new facts.

Let us return Mendeleyev to his century, to the issues and debates of the chemistry of the time; let us re-awaken problems now forgotten. What could his project have meant? We will then discover the bold programme of a professor of chemistry at the University of Saint Petersburg, an investigative pathway made up of tension, of patience, of hesitation and of certainty, of decisions – leading to success.

A Truthful Error

Paris, 1902. An old man with a white beard, his eyes alert despite his seventy years, knocks on the door of a laboratory in the rue Lhomond, and in a very strong Russian accent asks if he can speak to Pierre and Marie Curie. This is not some obscure colleague visiting these young research scientists destined for fame, but a well-known and honoured chemist, with doctorates from several universities, and the Director of the Central Bureau of Weights and Measures since 1892. This world-famous scientist is curious, even anxious about radio-activity and wants to have a look for himself. He observes attentively the radiations emitted by uranium, thorium and polonium. He talks, asks questions. Transmutation of elements? It doesn't make sense to him. The individuality of the elements, the immutability of atoms and the indestructibility of matter were the fundamental principles of the chemistry he had studied and then taught, principles on which had been built a positive and realist science. Mendeleyev is inclined to see in these mysterious rays and these 'electrical seeds' or electrons that have been talked about recently an illusion of alchemical fancy and the pretext for an obscurantist offensive.

To Pierre and Marie Curie's still hesitant interpretation Mendeleyev counterposed a more classical explanation. He draws on the blackboard the familiar diagram of a planetary system and he begins: radioactivity is a property exhibited only by heavy atoms. By virtue of their weight, atoms of uranium, thorium and polonium could act like little suns, attracting all the bodies around them. Now the physicists' ether is a substance infinitely light,

557

distributed throughout the Universe. One could then imagine that its behaviour might be changed in the vicinity of these highly attractive heavy atoms. Not that it adopts the harmonious movement of a planet, but it could begin to turn around these heavy atoms a little 'like a comet falling from space into a planetary system turns around the Sun and once more escapes into space'.

Back in St Petersburg Mendeleyev put some flesh onto the bare bones of his explanation and wrote a short treatise, translated into English two years later under the title *Towards a Chemical Conception of Ether*. Here the explanation put forward for radioactivity was the particular consequence of an astonishing intellectual construction which united mechanics, electromagnetic physics and chemistry.

In 1902 the question of the ether was very much alive. Throughout the nineteenth century the ether became more and more important for physics: already needed for the force of gravity to act at a distance, it proved indispensable for explaining the transmission of light waves. And since James Clark Maxwell, it was absolutely necessary for interpreting electromagnetic phenomena. This omnipresent ether none the less defied all the efforts of the physicists. Its nature remained mysterious and contradictory. The most sophisticated of experiments, such as that designed by Michelson and Morley to detect its movement around the Earth, met with failure, despite repeated attempts.

In a few pages Mendeleyev resolved the difficulties: he treated the ether as a chemical element and found a space for it in the periodic table. As ether must not react chemically, it would seem right to place it in the column of the inert gases. The major problem is that it is supposed to be imponderable. Another of the ether's supposed properties, its power of penetrating every substance, prevents ether being isolated and its atomic weight determined, as had already been done for ordinary elements. So, Mendeleyev declares, ether is only weightless in relation to our means of weighing. He has not forgotten Lavoisier's lesson of relativity and he also returns to the arguments that allowed Lavoisier to justify the notion of the caloric (see p. 466). It remained true that ether had to be so light that its atomic weight would be incommensurable with the lightest elements of the periodic table. Mendeleyev resolves this new difficulty by assuming the existence of a new intermediate element, which he hoped to identify with coronium, an element whose existence he thought had been discovered in the spectrum of the solar corona. He then attempted to ascribe an atomic weight value to ether, with a reasonable order of magnitude. He turned to the kinetics of gases and tried to calculate the speed necessary to escape the attraction of the heaviest stars, whose mass is about fifty times greater than that of the Sun. After long calculations, he concluded that this speed had to be somewhere between 2.24 and 3.00 m/s. In this case the atomic weight of ether had to be approximately one millionth of that of hydrogen. The periodic table was thus flanked by two strange newcomers.

This was certainly an audacious operation, but Mendeleyev was no novice at adventurous induction. Risks didn't frighten him: rather they brought him success. And this time he had no choice. Chemistry was in danger. Mendeleyev rebelled and attempted the impossible in order to save the foundations of the science he had engaged in all his life.

We should not, however, imagine that this was some desperate gamble. The stakes were enormous. In this essay Mendeleyev brought together all his struggles and all his hopes. He struck a fatal blow against the spiritualist and the para-scientific, against which he had conducted a relentless struggle. He fulfilled his desire, nurtured so long, for a union of physics and chemistry under the aegis of mechanics. Finally, and above all, he reaffirmed the prestige of the periodic table by extending its empire to the limits of the material Universe.

The rest of the story is well known: it was just a beautiful dream. Mendeleyev was wrong, terribly wrong. In 1905 the principle of relativity, put forward by Albert Einstein, a young physicist from Berne, would eliminate the need for the hypothetical ether, whose brief appearance in the periodic table lasted from 1903 to 1906.

It is tempting to explain this error by invoking Mendeleyev's advancing years. Wasn't he more and more involved in philosophical speculation, devoted to industry and to popularization? He wasn't any longer up to date. The 1902 essay does not, however, reveal any weakness: neither resistance to novelty nor megalomaniac delirium nor absence of critical intelligence. Mendeleyev appears lucid and careful. He knows that even if he is applying the same mode of prediction which had earlier brought him success, he cannot be at all certain.

> I do not regard this imperfect attempt to explain the ether from a chemical point of view as any more than the expression of a series of thoughts which occurred to me, which I have followed through from the simple desire that these thoughts, having been suggested by the facts, should not be altogether lost . . . If they contain any part of the natural truth for which we all search, my effort will not have been in vain; it may then be developed, expanded and corrected, and if my conception proves to be fundamentally mistaken, it will prevent others from repeating it.

By beginning with this late little essay, the least glorious of Mendeleyev's works, I am not trying to tarnish the great man's reputation; I am rather pointing out his courage. But one has to destroy once and for all the too easily accepted idea of the relationship between the periodic classification and modern theories of the atom. This 'error' of Mendeleyev's tells the truth about the periodic table. With a kind of backward movement, it illuminates the principles which organized its construction: his unshakeable faith in the individuality and immutability of the chemical elements; his profound aversion to metaphysics and superstition: and finally, his hope of one day uniting physics and chemistry in a new science, whose foundation would be the periodic classification. No, Mendeleyev was no pioneer of twentieth-century chemistry. Far from anticipating or prefiguring its developments, he fought them with all the strength and all the logic of nineteenth-century science.

The Embarrassment of a Chemistry Professor

It was the development of nineteenth-century chemistry which urgently posed the problem of the classification of the chemical elements. Many were the chemists who tried to provide a solution. Mendeleyev was not a marginal figure, alone in distant Russia.

St Petersburg, October 1867. Mendeleyev had just been appointed professor of chemistry at the University. He was thirty-three, and already had a good record: three important papers presented for his degrees, a period in Germany at the laboratory of Robert Wilhelm Bunsen, studies of Baku petroleum, of fertilizer, experimental agricultural experience on his own property and finally, for three years, a post as professor of chemistry at the Technological Institute in St Petersburg. He had, however, no material prepared for his lectures and he could find no adequate textbook of general chemistry to recommend to his students. Being in the habit of writing – he made a living writing scientific articles for various papers while he was a student – he decided to write his own textbook. From this there emerged two years later the periodic classification. At the origin of the discovery was a pedagogical situation. This fortunate complementarity between teaching and research activity was not unusual at the time, at least in chemistry.

How could one present students with the sum of knowledge accumulated about thousands of substances? Since Lavoisier, the solution had been to relate the properties of compound substances to the simple substances of which they were composed. But this did not solve all the problems. The table of thirty-three simple substances got rather cramped as a consequence of the use of Alessandro Volta's battery. The battery, discovered in 1800, gave birth to a powerful new analytic technique, electrolysis, which led to a series of discoveries: during the two years 1807 and 1808 Humphry Davy isolated sodium, potassium, strontium, boron, calcium and magnesium. He added to this with the discovery of chlorine in 1810, followed by iodine in

The 'population explosion' of elements in the nineteenth century

This table shows the extraordinary 'population explosion' of elements in the nineteenth century. Their numbers more than doubled: 33 simple substances were known at the end of the eighteenth century; 24 were discovered between 1800 and 1849 and a further 24 between 1850 and 1899, hence the pressing need for a classification which would introduce a little order into this multiplicity (after A. Massain, *Chimie et Chimistes*, 1952, modified).

Before 1700	1700–1799	1800–1849	1850–1899
Antimony	Nitrogen	Aluminium	Actinium
Silver	Beryllium	Barium	Argon
Arsenic	Bismuth	Boron	Caesium
Carbon	Chlorine	Bromine	Dysprosium
Copper	Chromium	Cadmium	Gadolinium
Tin	Cobalt	Calcium	Gallium
Iron	Fluorine	Cerium	Germanium
Mercury	Hydrogen	Erbium	Helium
Gold	Manganese	Iodine	Holmium
Phosphorus	Molybdenum	Lanthanum	Indium
Lead	Nickel	Iridium	Krypton
Sulphur	Oxygen	Lithium	Neodymium
	Platinum	Magnesium	Neon
	Strontium	Niobium	Polonium
	Tellurium	Osmium	Praseodymium
	Titanium	Palladium	Radium
	Tungsten	Potassium	Rhodium
	Uranium	Rubidium	Ruthenium
	Yttrium	Selenium	Samarium
	Zinc	Silicon	Scandium
	Zirconium	Sodium	Thallium
		Tantalum	Thulium
		Thorium	Xenon
		Vanadium	Ytterbium

1812 and bromine in 1826. In Sweden, Jöns Jacob Berzelius kept up the competition: after cerium in 1801 he isolated selenium in 1817, silicon and zirconium in 1824, thorium in 1828, and he also played a part in other discoveries. He established the power of electricity over chemistry with an electrochemical theory of chemical combination, which proposed that in each compound, whatever the number of its constituents, there were two parts, one charged with positive electricity and the other with negative. In the earliest editions of his *Treatise on Chemistry* in the 1820s, he listed fifty-four simple substances. In the 1860s a new technique, spectrum analysis, led to new discoveries. In 1869 the first of Mendeleyev's tables classified sixty-three elements and, towards the end of the century, there were nearly a hundred. This proliferation of elements was then an important phenomenon which distinguishes the nineteenth century in the long history of chemistry. But it hardly made the chemistry teacher's job any easier. Would he be condemned to present his subject in the form of an endless collection of detailed descriptive monographs on every known element?

Most chemists organized their textbooks according to a distinction between metals and metalloids, but the order within these two groups was not fixed. The balance between theoretical and pedagogical requirements achieved by Lavoisier in his *Elements of Chemistry* was broken. The preface to Berzelius' *Treatise* reveals a divorce between systematics and didactics. Berzelius looked for a compromise but did not manage to escape from the endless string of monographs:

> The plan I have adopted is not perfectly conformable to the systematic spirit. I decided to abandon this order on all occasions when it seemed to me that by sacrificing it I could make the science more easily accessible. There are two ways of expounding chemistry in a book written for beginners . . .
>
> Sometimes one seeks, so far as this does not cause inconvenience, to put together a collection of monographs on simple substances, and as for the combinations into which each of these substances is capable of entering, they are arranged in some sort of order, but one which has been fixed in advance, so that one will not be at all obliged to describe a compound twice, or even more times. In my opinion, it is in this form that science is reduced to its most simple expression, and is most easily inculcated in the memory . . .
>
> Sometimes one deals first of all with the simple substances, and then one examines, in a given order, the combinations of all these substances with all the others, and then one looks at the combinations of these various combinations with each other, in such a way as to proceed from the simple to the compound. At first glance this method seems to be the one which best fulfils the requirements of a book written for beginners.

The more chemistry progressed, the more the ideal of logical and rational presentation seemed to recede. The rapid development of organic chemistry from the 1840s onward also added to the problem of the population of elements a host of new compounds which needed to be urgently classified, at the risk of finding very soon that the transmission of chemical knowledge had become impossible. So chemistry found itself faced by the difficulty of managing the runaway proliferation of substances. The problem already encountered and successfully dealt with at the end of the eighteenth century by the reform of nomenclature was back again a half-century later.

Of course, the amount of knowledge accumulated about each of these known substances did help to control the situation. But nineteenth-century chemistry was not, as is often suggested, a triumphant pure science, solidly established on the unshakeable foundations of Lavoisier's system, progressing steadily and giving rise, like a cornucopia, to innumerable applications in industry and agriculture.

The international community of chemists was deeply divided in the first half of the nineteenth century, shaken by controversy on matters as fundamental as the existence of atoms or the nature and number of the chemical elements. On each of them Mendeleyev had taken very clear positions which determined his approach to classification.

The Shock of a Conference

The atomic hypothesis, formulated in 1805 by John Dalton, could be seen as the simplest explanation of the conjunction of two laws which governed chemical combination: the law of definite proportions, stated by Joseph Proust in 1802, and the law of multiple proportions put forward by Dalton himself. Chemical combinations, suggested Dalton, were made atom by atom. Dalton's atom was first and foremost the minimal unit of combination, and no longer, like its homonym in ancient Greek physics, an ultimate constituent of matter. Its real existence remained just as hypothetical as that of the classical atom, but the Daltonian atom made possible a quantitative interpretation of chemical combination: in his *New System of Chemical Philosophy* (1808–10) Dalton in fact constructed a system of relative atomic weights[1] based on a conventional standard unit (H = 1 in modern terms). For oxygen, sulphur, phosphorus, carbon and nitrogen, Dalton determined their atomic weights from their hydrogen compounds, always assuming that one atom of hydrogen was joined to one atom of the other substance. Water, for example, had an atomic weight of 8, which is to say one atom of hydrogen which weighed 1 and an atom of oxygen which weighed 7 (in modern notation, its formula would be HO). In the case of an element which had several hydrogen compounds, Dalton determined the atomic weight from the least hydrogenated. The weight of carbon, for example, was fixed on the basis of olethylene gas, made up of one atom of hydrogen weighing 1 and an atom of carbon weighing 5, giving a total atomic weight of 6. Dalton's system thus allowed each element to be identified by a positive quantitative character.

Given the simplicity and numerous advantages of this hypothesis, one would predict its universal acceptance, the more so that it was rapidly confirmed by laws on gases. The fixed proportions between gaseous volumes defined by Gay-Lussac's Law lent support to the atomic hypothesis. Another law, independently put forward by Avogadro in 1811 and by Ampère in 1814, brought together Dalton's and Gay-Lussac's views,[2] determining atomic weight on the basis of the gas or vapour density. But rather than favouring the adoption of the atomic hypothesis, this evidence from gas provoked disputes and blockages. Dalton himself did not accept Gay-Lussac's Law and criticized it mercilessly. His attitude did not prevent other chemists, particularly the powerful and influential Berzelius, from using the two theories as support for each other and putting forward in 1818 a system of atomic weights which related ratios of volume to ratios of weight.

The atomic hypothesis seems to have become the victim of its heuristic power: atomic weight was a notion indispensable to the chemist for the characterization of newly discovered elements. But why bother with the idea of the atom, which goes beyond experience, bringing with it its heavy metaphysical baggage? The dominant positivist injunctions to distrust hypotheses invited chemists to prefer a more neutral expression. 'Equivalent weight', the alternative, was put forward in 1814 by William Hyde Wollaston, another English chemist. Having zealously defended Dalton's hypothesis, and having illustrated it in his own work, Wollaston became his principal adversary. While exploiting all the resources of the conception of atomic weight, Wollaston criticized the uncertainties of the hypothesis and criticised Dalton for the arbitrary nature of his system of weights. He proposed a system of 'equivalent weights' founded on another conventional standard: oxygen, symbolized by O and equal to 100. Up to here the argument between atomists and equivalentists seems purely formal. The disagreement is about words, and conversion from one system of units to the other is easy.

The affair became more serious in the years between 1840 and 1850, when supporters of Avogadro succeeded in gaining acceptance for his hypothesis. The chemists had rejected it for a long time, believing it to be too complicated and too hypothetical. On one hand, it did not rest directly upon experimental data but upon the coexistence of Gay-Lussac's law and Dalton's hypothesis, which had led to another hypothesis: 'The number of whole molecules in any gas is always the same for the same at equal volumes or is always proportional to the volume.' On the other hand, gaseous densities gave atomic weights different to Dalton's (oxygen 15 as opposed to 7 for Dalton). This led Avogadro to introduce an additional hypothesis: to reconcile gas density as measured by experiment with the sum of the weights calculated according to the hypothesis, Avogadro assumed that the molecule which entered into combination split into two molecules of the same kind. He therefore made a distinction between 'integral molecules', of which there are equal numbers in equal volumes of gas, and 'elementary molecules'. Ampère, who formulated the same hypothesis in 1814, proposed another terminology which distinguished between 'particles' and 'molecules'. Avogadro and Ampère's law thus introduced two key notions which would later be called molecules and atoms. It was hard to accept such a distinction in the early nineteenth century because the prevailing duolestic electrochemical theory precluded the combination of two identical atoms in one 'integral molecule'.

Moreover, in some cases like sulphur, phosphorus, mercury, the atomic weight values determined with the help of Avogadro's hypothesis were in conflict with the values determined with the specific heat method. To escape such contradictions, Jean-Baptiste Dumas formally condemned Avogadro's hypothesis is 1836.

But a young and unruly chemist, Charles Gerhardt, refused to bury Avogadro's hypothesis. In order to reconcile the results obtained from various methods,[3] Gerhardt doubled the atomic weights of a certain number of metals. His *Traité de chimie organique* (1853–4) revived Avogadro's distinction, using the terms 'atom' and 'molecule'; it put forward a new system of atomic weights in which C = 12 rather than 6 and O = 16 rather than 8. No reconciliation was possible between this system and the system of equivalent weights. One had to opt for one or the other.

By 1856 Mendeleyev had chosen. In his master's thesis, submitted to the University of St Petersburg, he used Gerhardt's system. Mendeleyev declared himself a supporter of his unitary theory of chemical combination and expressed his aversion for Berzelius' dualist electrochemical theory. Gerhardt's system was also supported and perfected by Stanislao Cannizzaro, but it was far from receiving the unanimous agreement of the chemists. At the end of the 1850s, the multiplicity of systems in use had become such a problem for international communication that several chemists put forward the idea of an international congress to define a single system of atomic weights. The first International Chemical Congress, organized by August Kekulé, was held at Karlsruhe (Germany) in 1860. It was a decisive event. It inaugurated a new type of co-operation by the scientific community, which spread to other disciplines during the course of the century. International meetings were repeated regularly, and developed into institutions for the arbitration of conflict and the standardization of terms and units. In chemistry this first International Congress put an end to forty years of disagreement about atomic weight values. In Karlsruhe 140 chemists from a dozen countries discussed the advantages of different notations, and in doing this attempted to reach agreement about the basic concepts: atom, molecule and equivalent. By the end of the Congress no official decision had been taken. But thanks to Cannizzaro's energetic campaign, fought to present Avogadro's hypothesis as a solid induction from a host of experimental data and calling on chemists not to agree on a clear distinction between atoms and molecules, the Avogadro–Gerhardt system emerged triumphant. Mendeleyev, in any case, was one of those who left convinced and completely converted. While he had reservations about the Daltonian atom, which seemed to him conventional, he thought of Avogadro's hypothesis 'as the most important basis for the study of natural phenomena'. He was not interested in the reality of the atom and the molecule, being more concerned with the distinction between the two:

With the application of the Avogadro–Gerhardt law, the idea of the molecule is completely defined, and by virtue of this, atomic weight. One calls a particle, or chemical particle, or molecule the quantity of substance that enters into chemical reaction with other molecules and which occupies, in the gaseous state, the same volume as two parts by weight of hydrogen ... The atoms are the smaller quantities, or indivisible chemical masses of the elements which form the molecules of simple or compound substances.

After 1869 Mendeleyev would write in many articles that it was the Karlsruhe Congress which put him on the right path, giving him the intuition of a possible periodicity in the properties of elements arranged in increasing order of atomic weight. He thus suggests that from then on he had only to develop this intuition and overcome the obstacles presented by certain still incorrect atomic weights. Mendeleyev was probably happy enough to reconstruct the history of his discovery with the illusion of continuity and clear evidence, but it is none the less the case that the Karlsruhe Congress was a decisive factor for the elaboration of the periodic table.

Regaining a Lost Unity

Mendeleyev took an even more clearly defined position in debate about the plurality of the elements. There were two ways of looking at the multiplication of the elements. Some suggested that this multiplicity should be reduced to a single primordial element: others attempted to subordinate the multiple to a single law.

The idea that the diversity of the elements should be derived from hydrogen had been put forward at the beginning of the century by an Englishman, Dr William Prout. It rested, apparently, on 'a sense of the harmony of things', but it received unexpected help from Daltonian atomism. Although the choice of hydrogen as the basic unit was purely conventional, it was very soon interpreted as confirmation of the hypothesis. It resisted the most vigorous attacks and experimental refutation. When the atomic weights of too many elements ceased to be expressed in whole numbers, Prout proposed in 1831 a modified version of the hypothesis: atomic weights had to be whole multiples of a fraction of the hydrogen. Put in this way, the hypothesis escaped any possibility of experimental falsification and met with considerable success.

At first this hypothesis found followers essentially in England, its country of origin. Continental chemists, influenced by Berzelius, an energetic defender of the atomist and pluralist orthodoxy, had formed a united front against Prout's hypothesis. But towards the end of the 1830s Berzelius' authority declined a little, equivalentism was stronger than ever and the hypothesis gained ground in France, Switzerland and Germany.

During the 1850s, when the precision of experimental determinations of atomic weights led to doubts about integral multiples, the hypothesis found new support in organic chemistry: correspondences were established between the series of organic radicals and the series of elements in inorganic chemistry. Around 1860 two other factors intervened in favour of Prout's hypothesis: the Darwinian theory of evolution (see p. 420) and the spectroscopic study of the elements combined to encourage the idea of the evolution of inert matter from a single primary element.

Prout's hypothesis plays a double role in the history of the classification of the elements. First, it encouraged research on the determination of atomic weight. If it is true that supporters of Prout's idea, like Thomas Thomson in his *System of Chemistry*, tended to round the numbers obtained by experimentation so as not to be bothered by inconvenient decimals, their adversaries, such as Berzelius or the Belgian chemist, Jean Servais Stas, struggled to obtain more and more precise results and conscientiously countered arithmetical speculation with experiment. But this classical strategy failed, because Prout's partisans themselves began cultivating precision and relying upon experiment. Thus a Swiss chemist, Jean Charles Galissard de Marignac, as well known as Berzelius for the exactitude of his results, did not hesitate to fix the atomic weight of chlorine as 35.5. In his eyes, what was important was not 'the size of the unit which may serve as the common divisor for the weights of simple substances . . . Whether this weight is that of an atom of hydrogen, of half or a quarter of an atom, or whether it might be a smaller fraction, a 1/100th or a 1/1000th for example, all these considerations retain the same degree of probability. The only difference would be less simple constitution ratios between the elements.'

Whatever the ratios established between the elements, the hypothesis is saved because it draws its strength not from the numerical value of the ratio established but from its arithmetical form. Many chemists thus considered Prout's hypothesis an advanced stage in the development of quantitative chemistry. It opened the fascinating perspective of an arithmetical science of matter which would free the chemist from the multiple and the diverse, which could save the phenomena with a simple calculation. Thus the rigid rules of experimental method and the draconian conditions of positivist epistemology gave way before a Pythagorean dream.

In addition, Prout's hypothesis inspired attempts to systematize knowledge about the elements. On one hand, it imposed the prevalence of atomic weight as a criterion of classification and so discredited a classification of Ampère's which had rested on more or less artificial chemical properties. On the other hand, it directed attention towards the search for structures of inter-relationship between elements. Chemical classification could of course be distinguished from the classifications of the naturalists, because the atomic weight provided a single and quantitative principle of classification; but groupings by family were also made in the search for evidence of filiation and an attempt was made to build up a kind of family tree of inert matter.

Such was the objective, with one or two exceptions, of all attempts at classification until Mendeleyev and even after him. The exhaustive list of systems drawn up by the historian J. W. van Spronsen (1969) shows that the waves of classification clearly follow the fashion for Prout's hypothesis. The first attempts were made in Germany. In 1817 a professor from Jena, Johann Döbereiner, discovered a remarkable relationship between the equivalent weights of certain substances: that of strontium oxide (50) was equal to the arithmetical mean of those of calcium oxide (27.5) and barium oxide (72.5), when H = 1 and O = 7.5. Twelve years later, armed with Berzelius' more accurate atomic weights, Döbereiner generalized this three-substance relationship and proposed a series of triads which depended on a correlation between chemical similarity and arithmetical ratios:[4]

$$Br = Cl + 1/2, \; Na = Li + K/2, \; Se = S + Te/2.$$

Shortly afterwards Leopold Gmelin, a professor at Heidelberg, moved on from triads to families of elements. As his purpose was to demonstrate the truth of Prout's hypothesis, he concentrated all his attention on the numbers, to the detriment of chemical similarity. He thus grouped the elements under three heads: those which clearly had the same atomic weight; those whose atomic weights were multiples of each other; and finally those whose atomic weights formed arithmetical means, in the manner of Döbereiner.

After these two attempts there was a quiet period in the history of classification. Then suddenly, in the 1850s, dozens of chemists from different countries suggested classifications. They were mostly based on the comparison of the series of organic radicals and the series of elements, which they did not hesitate to rename as 'inorganic radicals', the better to suggest their complexity. This idea came from Dumas who, despite his prudence in matters of theory, none the less supported Prout's hypothesis in its most unfalsifiable form. He assumed, in fact, that all the diversity of simple substances was derived from a single element, still unknown, having an atomic weight equal to a half or a quarter of that of hydrogen. He based his conjecture on a comparative classification of the organic radicals and the families of simple substances which displayed a regularity of arithmetical progression.

It has to be admitted that this attempt hardly represented any advance in classification. Dumas constructed no new family and he was far from classifying all the known elements. The same could be said of many attempts undertaken between 1850 and 1860 and listed by van Spronsen: Gladstone,

Cooke, Lenssen, Carey Lea, who allowed negative atomic weights, and then Hinrichs, who established relationships between elements' spectra and the planets, all managed to construct families of elements but no one could connect them together into an overall system.

It was, however, an advocate of Prout's hypothesis, not a chemist but a mineralogist, who succeeded in identifying a function governing the whole. In 1862 Alex Béguyer de Chancourtois put forward to the Academy of Sciences the basic outline of a periodic system of the elements, and declared: 'It is only by taking Prout's law into account that I arrived at a perfectly demonstrable system.'

This involved a helix along whose axis lay the series of whole numbers, corresponding to the atomic weights of the elements. Béguyer de Chancourtois called his system the 'telluric screw' for two reasons: because 'tellurium plays an essential part in the system, and the adjective "telluric" very happily recalls its geognostic origin, because *tellus* signifies earth in the most positive, most familiar sense, in the sense of the nourishing earth.'

Was it perhaps the cult of the telluric that put the chemists off? This classification remained unnoticed, totally ignored by the chemists of the 1860s. It had two weaknesses: it mixed up simple and compound substances, and its graphical representation was very complicated. It seemed so little illuminating to the gentlemen of the Academy that they didn't even think it right to print it together with the paper published in the *Comptes rendus*.

We can see then that, if Prout's hypothesis did encourage attempts at classification, it did not help them to succeed. It seems to me that it created two major obstacles to the successful conclusion of these projects: it focused interest on numerical relationships to the detriment of the analogies of chemical properties; the quest for a primordial unity led to the privileging of local family relations rather than a treatment of the problem as a whole. This is at least the interpretation suggested by comparison with the few rare attempts at classification which were not inspired by the Prout hypothesis.

The Search for a General Law

There were three who swam against the stream: John Alexander Newlands, William Odling, both English, and Mendeleyev. They worked independently, but their projects share certain common features which allow us to discern a style of classification very different from that of Prout's supporters. All three adopted the Gerhardt–Cannizzaro system of atomic weights, hallowed by the Karlsruhe Congress. They subordinated the whole taxonomic enterprise to the quest for a general law. As a result of this, they alone made predictions of unknown elements.

In 1865 Newlands stated a 'law of octaves', according to which chemical properties were repeated every seven elements, in the same way as the notes of the musical scale. In order to respect analogies of chemical properties, he reversed the order of tellurium and iodine, as did Mendeleyev. He was the first to predict new elements: one of his predictions, of an element of atomic weight 73 between silicon and tin, very strongly resembles Mendeleyev's prediction of eka-silicon, with an atomic weight of 72. Why then was such an audacious system forgotten? It did, of course, contain many incorrect atomic weights, and it classified only about fifty of the elements. But the principal reason for its eclipse does not lie in Newlands' system but in the reception it received from his colleagues. When in March 1866 Newlands submitted his discovery to the respected Chemical Society of London, Professor George Carey Foster jokingly asked whether arranging the elements in alphabetical order would not have revealed coincidences just as interesting.

In 1865 Odling, then a professor at Oxford, independently constructed a more comprehensive system – fifty-seven elements out of the sixty known at the time – and strictly ordered according to increasing atomic weights. He drew attention both to the regularities and to the differences in atomic weight, and firmly concluded that 'among the members of each well-defined group, the sequence of properties and the sequence of atomic weights are strictly parallel to each other.' Odling left a great number of vacant places, particularly between the atomic weights 40 and 50 and between 65 and 75. But curiously, instead of improving his system in later publications, Odling seems to have retreated, and in 1868 he proposed a new, less comprehensive table.

Of the three it was Mendeleyev who went furthest. Hardly had he glimpsed the idea of periodicity in March 1869 than he worked out all its possible consequences, and sought to test it in relation to the known and even the unknown: predictions of elements, corrections of atomic weights, reversals . . . To allow himself such liberties, Mendeleyev must have been extremely confident. In his eyes, the periodic law rather than the table itself was the real discovery. He drew his confidence from a very demanding conception of scientific law. A law had to be general or it wasn't a law: 'Natural laws do not allow exceptions, and it is in this that they are different from rules such as those of grammar, for example. A law can be confirmed only when all the consequences that may be drawn from it have been sanctioned by experiment.'

This epistemology is linked to a rejection of Prout's hypothesis. Of all the chemical classifiers, Mendeleyev is the most hostile to the idea. It was not enough for him to reject it or to criticize it. He fought against it with all his strength. But how could one fight against a hypothesis which benefited from experimental results while evading all refutation? Just to oppose it with pluralism as a philosophical conviction would get one nowhere. Then

Mendeleyev thought he had a response: the only weapon against the seductive Prout hypothesis was a general law which governed all of the elements. Not only because the unity of a law governing the multiplicity of the elements could compensate for the loss of material unity and put an end to this frenetic quest, but also, and above all, because the periodic law was, for Mendeleyev, a rival to Prout's hypothesis. 'The periodic law can explain all the facts and suggests developments to the philosophical principle which governs the mysterious nature of the elements. This tendency falls into the same category as Prout's law, with this essential difference, that Prout's law is arithmetical while the periodic law draws its inspiration from that combination of mechanical and philosophical laws which constitutes the character and force of the present progress of the exact sciences.'

Mendeleyev presented himself as a rival of and a victor over Prout. But the irony of the story is that Prout's supporters saw in his table a spectacular confirmation of their hypothesis; a very bitter irony for Mendeleyev. For ten years his discovery was scorned or criticized. Then, after the confirmation of his predictions of elements, it generated enthusiasm but served a cause he had always fought against. William Crookes, for example, used the periodic law as support for his own hypothesis of the protyle.

Faced with these attempts at recovery, Mendeleyev never ceased to protest and emphatically proclaim his faith in the individuality of the elements:

> Kant thought that there existed in the Universe two objects which provoked the admiration and the veneration of men: 'The moral law within and the starry sky above us.' With the discovery of the nature of the elements and the periodic law a third object must be added: 'the nature of the elementary individuals which is expressed everywhere around us', given that without these individuals, we could have no idea of the starry sky, and that the idea of atoms reveals both the singularity of individuality, the infinite repetition of individuals and their subjection to the harmonious order of nature.

Simple Substances or Elements

Mendeleyev's pluralism was also involved in another hidden motivation behind his discovery. Let us read carefully his statement of the periodic law: 'The properties of simple and compound substances depend on a periodic function of the atomic weights of the element for the simple reason that these properties are themselves the properties of the elements from which these substances are derived.'

It is usual to note a periodic function between chemical properties and atomic weight, without specifying to what these properties and these atomic weights belong. Mendeleyev's formulation, though, is explicit on this point. The periodic function establishes a relationship between simple and compound substances on one hand and elements on the other. Mendeleyev was careful to define distinctly the basic terms at the head of the article in which he published his discovery:

> In the same way as before Laurent and Gerhardt the words 'molecule', 'atom' and 'equivalent' were used one for the other, indistinctly, today in the same way the expressions 'simple substance' and 'element' are often confused. Both however have a distinct meaning, which it will be best to state, in order to avoid confusion in the terms of chemical philosophy. A simple substance is something material, a metal or metalloid, endowed with physical properties and capable of chemical reactions. To the expression 'simple substance' corresponds the idea of the 'molecule' . . . The name of element must on the other hand be kept to characterize the material particles which form the simple and compound substances and which determine the way in which they behave from the physical and chemical point of view. The word 'element' corresponds to the idea of the atom.

This restatement of vocabulary shows that the Karlsruhe Congress had not only brought about the adoption of a more correct system of numerical values for atomic weight, but also the clarification of fundamental concepts. Starting with the distinction between atom and molecule, Mendeleyev establishes a parallel distinction between element and simple substance. Apparently less important, because more intuitive and even obvious in certain cases, such as that of carbon which can be found in the form of three simple substances – graphite, diamond and amorphous carbon – or nitrogen, inactive in the free state but reactive in combination, or to explain the reactions of substances in the nascent state, or finally in the case of isomerism.

In this almost trivial clarification of vocabulary, however, there takes place a complete change in chemistry's theoretical landscape, for the difference between element and simple substance is the organizing principle of Mendeleyev's work, and it offers chemistry its programme: 'To understand more thoroughly the relationships between the composition, reactions and the qualities of simple and compound substances, on one hand, and the intrinsic qualities of the elements contained in them, on the other, in order to be able to deduce from the already known character of an element the properties of all its combinations.'

The simple substance, which since Lavoisier had been the key concept of chemistry, and its goal, to be attained by more and more refined analysis, was dethroned by the element. The distinction between simple and compound which had been the axis of Lavoisier's system now became secondary. It was

no longer a question, as it had been with Lavoisier, of explaining the properties of compounds by those of simple substances. The simple substance
explained nothing; and with the compound substance it belonged to the
phenomenological realm. The element was the sole explanatory principle.

How did the distinction between an element and a simple substance work
within the periodic classification? A first common-sense remark is that, if one
wants to classify, it's better to know what is to be classified. Since Lavoisier's
famous definition, element and simple substance had been more or less
synonymous. Classifications were made of elements or simple substances.
The confusion was nurtured by Prout's disciples, whose quest for a unique
primordial element benefited from the lack of individuality of the simple
substances. Mendeleyev, on the contrary, classified the elements in a search
for an explanation of the behaviour of simple and compound substances.
The distinction established by Mendeleyev was thus a chief weapon in the
struggle against Prout because it imposed the plural on the word element.

It was also needed to attain the level of abstraction required by the
operations of classification and prediction. The simple substance was a
concrete thing with physical and chemical properties determined by experiment. The element, on the other hand, has no phenomenal existence; it is
always hidden in the simple or compound substance and it circulates, is
exchanged and conserved in chemical reactions. It is an abstract entity,
constructed by the mind to account for the conservation and permanence
of individual properties. About elements, predictions were possible. The
simple substance was too concrete for this: purely phenomenological by
definition, it cannot exist before being isolated as a result of analysis. Only
the element can be predicted, because it is defined by its place in a network
of relations.

A rapid comparison between Mendeleyev's table and Julius Lothar Meyer's
emphasizes the role of this conceptual distinction in Mendeleyev's famous
predictions. Meyer, like Mendeleyev, was a university professor. Like him he
adopted the atomic weights supported at the 1860 Congress. Like him, he
decided to write a textbook for his students, to present chemistry in a rational
and ordered manner. But he had a good lead on Mendeleyev, for the first
edition of his textbook appeared in 1864, with a table of elements based on
valency. In 1868 Meyer, preparing the second edition, drew up a table which
gives a real periodic classification of all the known elements, including the
transitional metals between nickel and iron, with vacant places for elements
yet to be discovered. Unluckily for Meyer, publishing delays meant that the
table conceived of one year before Mendeleyev's was published one year
after, in 1870. In that year Meyer verified the periodicity of properties in a
very remarkable case, the atomic volume, and illustrated it with a curve.

There was an inevitable priority dispute, which tormented Mendeleyev for many years. Without attempting any arbitration of the quarrel, one can none the less point out the essential differences between the two rivals. Like Mendeleyev, Meyer was confident in his system, and in his 1872 textbook he reorganized the whole of the section on inorganic chemistry on the basis of his periodic classification. Mendeleyev himself did not do this. Meyer, on the other hand, did not make corrections to atomic weights, and above all, did not predict the properties of the elements for which he had reserved the empty spaces. Not having distinguished between element and simple substance, he had no means of making predictions. This attitude went along with doubts on the individuality of the elements, and the hope, which he maintained for the rest of his life, that a primordial element might be discovered which would account for the analogies recorded in the table.

One can see then that Mendeleyev's project, profoundly rooted in the debates and problems of the chemistry of the time, is marked off from those of his precursors and rivals by the firmness of the philosophical convictions by which it was inspired. His commitment to a material pluralism was expressed in the distinction between element and simple substance, as it was in his demand for an absolutely general law.

The Paths of Discovery

Between the first 'intuition of periodicity' at the Karlsruhe Congress in 1860 and the periodic table, there lies a great gulf. To get to the system presented before the Russian Chemical Society in 1869 under the title 'Relations between properties and atomic weights of the elements' he had needed long research, a patient filling-in marked by trial and error, compromises and occasional successes.

Mendeleyev loved to tell his story of the way he went about things in a host of articles between 1869 and 1889, all with more or less the same title and all reproducing, as if it were a sacred text, the conclusions of the article of 1871. More than those later reconstructed narratives, the textbook which led to the discovery, his *Principles of Chemistry*. composed between 1868 and 1871, affords us a precious guide to understanding the process of discovery. The periodic table is presented at the end of the first section, and it provides a more systematic organization for the second half.

To begin with Mendeleyev gives a definition of chemistry centred on the notion of the simple substance, accompanied by a vibrant homage to Lavoisier its founder. We may note in passing that Mendeleyev seems to be entirely ignorant of the work of his compatriot Mikhail Vassilievich Lomonossov, who is however presented by Soviet historians as a sort of national Lavoisier

of their own. Mendeleyev ends the introduction with a list of simple substances 'intended to reflect the current state of our knowledge'. It is a sort
of provisional classification, a review of established knowledge and a basis
for further work. The distance which separates this initial list and the classification given at the end of the second volume gives some idea of the work
that Mendeleyev had to do. The first table already contained seventy simple
substances and, in the later editions of the 1880s, Mendeleyev included in it
the elements whose discovery had been made possible by the classification.
That is to say the difference was not of a quantitative kind. On one hand it
lies in the criteria of arrangement: the introductory table makes no pretence
at a rational order. Mendeleyev did not up-date Lavoisier's table of simple
substances or any other more recent one. He deliberately adopted multiple
and more or less pragmatic criteria: the occurrence of simple substances in
nature or their degree of importance in human activity. On the other hand,
the table in the introduction lists substances and not elements. Working with
simple substances, all one can do is collect and multiply collections in an
attempt to match all their observable properties. Only a displacement of the
problem onto the element, defined by its atomic weight, allows one to arrive
at a general law and a systematic classification. One can see then that the
construction of the periodic table and the element/simple substance distinction represent the same approach which is carried through chapter by
chapter.

Before presenting the periodic law Mendeleyev looks at water, then air,
then some carbon compounds and finally kitchen salt. He proceeded from
the most familiar substances. But this apparently pedagogical choice offered
the opportunity of reviewing the properties of hydrogen, oxygen, nitrogen,
carbon, sodium and chlorine, all elements which stand at the head (except
for sodium, which is second) of the groups of similar elements in the future
classification. Mendeleyev begins then by studying elements with very well-
defined properties which can provide patterns, models to order the others.
Mendeleyev later called them 'typical elements'.

Presumably Mendeleyev was influenced by Gerhardt, whom he admired
and to whom he pays homage throughout the *Principles* in the choice of the
term 'type' as well as in the selection of typical elements, since they all
belonged to Gerhardt's four types.

In this collection of monographs, an order gradually emerged in chapter
10, devoted to sodium chloride. Chlorine and sodium, in fact, present a
specific problem: they are 'typical' of two families already known, the 'halogens' and 'alkaline metals', which have the same valency of 1. But they form
very different chemical combinations. Despite the very clear contrasts in
their chemical behaviour, Mendeleyev decides to compare them, and so
discovers an interesting regularity in the differences of atomic weights:[5]

	Na = 23	K = 39	Rb = 85.4	Cs = 133
Li = 7				
	F = 19	Cl = 35.5	Br = 80	Te = 127

Here is the skeleton, the framework of the periodic table: the series of typical elements and the two groups of opposite properties. The style of the chapters suddenly changes. Instead of the random, there is now a systematic progression. Beginning with the extremes and working towards the middle, Mendeleyev little by little constructed a network: in chapter 11 he presents the halogens; in chapters 12 and 13 the alkaline metals. In order to decide which metals should appear in chapter 14, and take their place beside the alkaline metals, Mendeleyev now relies on the regularity of the atomic weights, and completes his initial little table by adding three elements:

		Ca = 40	Sr = 87.6	Ba = 137
	Na = 23			
Li = 7		K = 39	Rb = 85.4	Cs = 133
	F = 19			
		Cl = 35.5	Br = 80	Te = 127

The metals called 'alkaline earths' are thus ranked after the alkaline metals. After these three ranks there arrives, as an immediate consequence, the periodic law, formulated in chapter 15.

This quick survey of the first part of the *Principles* calls for some remarks. The sequence of the chapters strictly follows the course of Mendeleyev's own thoughts, as reconstructed by B. M. Kedrov from the papers deposited at the Mendeleyev Museum in St Petersburg. Kedrov tells how Mendeleyev, who on 1 March was getting ready to leave St Petersburg in order to visit a cheese factory, was trying to think what chapter he would put after the alkaline metals; he suddenly had 'the crucial idea' of arranging the groups of elements in the order of their atomic weights. He would have seen the periodic regularity and glimpsed the possibility of placing the elements of intermediate atomic weight. What happened next could be described, in a phrase Mendeleyev was very fond of, as 'chemical solitaire', a sort of game of patience, with the cards each bearing the name and the properties of an element, which must be put in order and moved about until the whole is solved. The typical elements and the two extreme families gave the overall structure of the table; local analogies filled in the gaps.

This narrative of the process of discovery also finds confirmation in Mendeleyev's writing: he thought that the secret of his success was the decision to compare the halogens and the alkaline metals: 'Around 1860 the ground had already been prepared for this law, and if it was only stated later, the cause, in my opinion, was that similar elements were being compared to each other, and dissimilar elements were ignored.'

Here we see the importance of Mendeleyev's warnings about Prout's hypothesis. The chemists who classified elements to draw up their genealogy were more attentive to resemblances than to differences. They managed, at best, to list families of similar elements, isolated groups. Mendeleyev, who was looking for an absolutely general law, first of all concentrated his attention on the contrasts and dissimilarities, so as to obtain the most inclusive scheme possible.

The second remark concerns the successive editions of the *Principles*. Not only does the text of the first edition reproduce Mendeleyev's itinerary, but the eight editions which appeared during his lifetime faithfully reproduced the original text. Given his initial, pedagogical concerns, one might have thought Mendeleyev would revise the whole work in order to give a systematic account of chemistry on the basis of the periodic classification, as Meyer did in 1872. Mendeleyev chose an alternative path: he added numerous footnotes to bring results up to date, to respond to objections, to refute his critics and correct misinterpretations, but also to give additional information and develop thoughts on agriculture, industry, astronomy, biology ... As the years passed, the notes grew so much that they ended by overpowering the text and threatening its educational effectiveness. Stranger still, the second section of the textbook – which should have given an account of chemistry systematized by the periodic law – appears in neither the English nor the French edition of the *Principles*. And it seems to have been abandoned in the Russian, from the fifth edition onward. Paradoxically, the most modern part of the work fell out of use. What lay behind this editorial decision, we don't know. No doubt Mendeleyev thought that a systematic exposition was less educational than a narrative which told of the adventure of discovery, the experiences of a chemistry professor, and his ideas on Russian development.

From the Law to the Periodic Table

Mendeleyev privileged the periodic law over the table. Does this mean that the table is no more than a secondary elaboration, simply an illustration of the law?

In one sense the law dictates the table. It is the law which gives the general framework, imposes the criterion of increasing atomic weights, and demands several hardly negligible corrections for indium (an atomic weight of 114 instead of 75), uranium (240 instead of 120), cerium and others. Even more, it commands local breaks in this general order, putting tellurium before iodine, for instance. Mendeleyev never questioned this inversion, though he never managed to explain it. And it is the periodic law which commands that empty places be left for elements yet to be discovered.

But the periodic law does not draw up the table, and it leaves uncertain the way it should be filled in at a local level. The famous prediction of the three elements eka-aluminium, eka-boron and eka-silicon do not derive from a blind, mechanical application of the law. To determine the properties of each element, Mendeleyev had to undertake some delicate work of approximation, taking into consideration the four elements by which it was surrounded. All this step-by-step reasoning was forgotten when, some years later, the predicted elements were discovered and found to have properties which coincided almost exactly with those Mendeleyev had predicted. The precision of correspondence should not make us forget that the periodic law, in its generality, cannot perfectly govern each individual case. Mendeleyev stressed this flexibility when he called the discoverers of the elements predicted 'reinforcers of the periodic law' and he sometimes recalled it explicitly: 'As in the function between atomic weights and properties only the character is known, one cannot at the moment account for individual divergences. One can only determine the narrow interval in which the value of the atomic weight of an element should fall.'

He also recognized that the periodic law failed to account for certain obvious analogies: in particular, between lithium and magnesium, beryllium and aluminium, boron and silicon. These analogies, today called 'diagonal' on account of the elements' positions in the table, show the limited power of the periodic law.

As much as Mendeleyev was determined and confident in his statement of the periodic law, so uncertain and hesitant was he in the construction of the periodic table. First of all he hesitated between a table and a spiral. On this point it was the periodic law that won: the periodic law, he said, implied the discontinuity in the atomic weights, so one couldn't then adopt a presentation like a curve which suggested continuity. He hesitated longer about the shape of the table. In 1871 he put forward two solutions, and seems not to have decided between them (p. 581, top and centre): in one he places the series horizontally, thus stressing the periodic variation in properties, particularly the inverse variation in the forms of oxides and acid hydrates which is mentioned above the table. The alternative presentation, putting the series into vertical columns, has the advantage of highlighting the typical elements. In 1879 Mendeleyev recommended a third, exploded version (p. 581, bottom). The compact table is divided into three blocks, the typical elements, the even series and the uneven series. This brings out even stronger analogies between elements belonging to alternate series. Above all, this presentation allowed a more satisfactory place to be found for the elements of group VIII: Fe, Co, Ni . . . On the other hand, it altered the unity of the function and interrupted the increasing series of atomic weights. Compact or exploded, both had advantages and disadvantages. Mendeleyev did not find an ideal table. He chose, it seems, a compromise solution, which, given the minimum number of assumptions, could give the most information with maximum clarity.

Predictions about elements by Dmitri Ivanovich Mendeleyev and later determinations of their properties

Predictions		Determinations
Eka*-aluminium		Gallium (discovered in 1875 by Lecoq de Boisbaudran)
Atomic weight:	68	69.9
Specific gravity:	6.0	5.96
Atomic volume:	11.5	11.7
Eka-boron		Scandium (discovered in 1879 by Nilsen)
Atomic weight:	44	43.79
Oxide:	Eb_2O_3	Sc_2O_3
Specific gravily of oxide	3.5	3.864
Sulphate:	$Eb_2(SO_4)_3$	$Sc_2(SO_4)_3$
Eka-silicon		Germanium: (discovered in 1866 by Winkler)
Atomic weight:	72	72.3
Specific gravity:	5.5	5.469
Atomic volume:	13	13.2
Oxide:	EsO_2	GeO_2
Specific gravily of oxide	4.7	4.703
Chloride:	$EsCl_4$	$GeCl_4$
Boiling point of oxide	< 100°C	86°C
Density of chloride	1.9	1.887
Non-gaseous fluoride	EsF_4	$GeF_4.3H_2O$ (white solid)
Ethyl compound	$EsAe_4$	$Ge(C_2H_5O)_4$
Boiling point of ethyl compound	160°C	160°C
Specific gravity of ethyl compound	little less than 1	0.96

* the prefix designating the number 1 in Sanskrit

Critical Points

Before the periodic law, Mendeleyev made an important discovery about gases: he identified the phenomenon of 'critical temperature' (the temperature at which a gas or a vapour can be liquified by a simple application of pressure). This concept can serve as a guide in specifying the historical context of his most famous discovery, the periodic system.

Let us sum up. The problem of classification arose following the increase in the number of elements in the first half of the century and the increasing amount of knowledge about them. At the end of a period of agitation during which new ideas were debated, a critical point seems to have been reached in 1860, the date of the Karlsruhe Congress. After the outcome of the Congress, if Mendeleyev's account is to be believed, it was enough to apply sufficient intellectual pressure and the periodic classification arrived to gather, organize and systematize the whole of this scattered body of information. It inaugurated what Mendeleyev called the 'systematics of elements'. This expression can be understood in two senses: on one hand, the periodic classification organized chemistry into a system; on the other, it made the element the organizing concept of the system, differentiating it from Lavoisier's concept of the simple substance.

But this critical point was no more than a narrow interval of a few years. In the 1870s came many discoveries that weakened the assumptions behind Mendeleyev's work, which might have made success if not impossible then at least improbable.

First came the isolation of many 'rare earths' in the 1870s: a series of elements very close in atomic weights and properties, nowadays called the 'lanthanides', from their position in the table. In 1869 Mendeleyev knew of only five and already found himself in serious difficulty. He was continuously modifying their positions and atomic weights; he introduced and then removed terbium, left empty places between cerium and erbium. He couldn't do better for, according to Mendeleyev's criteria, the rare earths were practically invisible. First of all they lacked individuality, while showing very marked group characteristics. They presented another difficulty: they breached the regular succession of periods and would later find their place in a discontinuity in the table. At a time when nothing was known of the electron structure of the atom, these elements were so difficult to integrate that it is reasonable to think that the periodic table could be so rapidly constructed after Karlsruhe only because not all the rare earths were known.

There was also a second series of problematic discoveries: the rare gases. In 1895 William Ramsay isolated argon and helium, two inert gases, which challenged the periodic table. Mendeleyev had not predicted them and the absence of chemical properties hardly enabled one to find analogous elements. What's more, with an atomic weight of 40, argon ought to have been placed between potassium and calcium, which was inconceivable because there were no spare places between groups I and II.

The supporters of the periodic system were perplexed. Mendeleyev tried to rescue the situation by suggesting, as others did, that argon was not an

Groupe I	Groupe II	Groupe III	Groupe IV	Groupe V	Groupe VI	Groupe VII	Groupe VIII
—	—	—	RH^4 RO^2	RH^3 R^2O^5	RH^2 RO^3	RH R^2O^7	
R^2O	RO	R^2O^3					RO^4
—	—	—	—	—	—	—	
H = 1	»	»	»	»	»	»	
Li = 7	Be = 9.4	B = 11	C = 12	N = 14	O = 16	F = 19	
Na = 23	Mg = 24	Al = 27.3	Si = 28	P = 31	S = 32	Cl = 35.5	
K = 39	Ca = 40	— = 44	Ti = 48	V = 51	Cr = 52	Mn = 55	Fe = 56; Co = 59; Ni = 59; Cu = 63.
(Cu = 63)	Zn = 65	— = 68	— = 72	As = 75	Se = 78	Br = 80	
Rb = 85	Sr = 87	?Yt = 88	Zr = 90	Nb = 94	Mo = 96	— = 100	Ru = 104; Rh = 104; Pd = 106; Ag = 108.
(Ag = 108)	Cd = 112	In = 113	Sn = 118	Sb = 122	Fe = 125	I = 127	
Cs = 133	Ba = 137	?Di = 138	?Ce = 140	»	»	"	
"	"	"	"				
"	"	?Er = 178	?La = 180	Ta = 182	W = 184	»	Os = 195; Ir = 197; Pt = 198; Au = 199.
(Au = 199)	Hg = 200	Tl = 204	Pb = 207	Bi = 208	"	"	
"	"	"	Th = 231	"	Ur = 240	"	" " "

LES GRANDES PÉRIODES

K = 39	Rb = 85	Cs = 133	»	»	
Ca = 40	Sr = 87	Ba = 137	»	»	
»	?Yt = 88?	?Di = 138?	Er = 178?	»	
Ti = 48?	Zr = 90	Ce = 140?	La = 180?	Th = 231	
V = 51	Nb = 94	»	Fa = 182	»	
Cr = 52	Mo = 96	»	W = 184	Ur = 240	
Mn = 55	»	»	»	»	
Fe = 56	Ru = 104	»	Os = 195?	»	
Co = 59	Rh = 104	»	Ir = 197	»	
Ni = 59	Pd = 106	»	Pt = 198?	»	
Cu = 63	Ag = 108	»	Au = 199?	»	
»	Cd = 112	»	Hg = 200	»	
»	In = 113	»	Tl = 204	»	
»	Sn = 118	»	Pb = 207	»	
As = 75	Sb = 122	»	Bi = 208	»	
Se = 78	Fe = 125?	»	»	»	
Br = 80	I = 127	»	»	»	

ENTS TYPIQUES

1	Li = 7	Na = 23
	Be = 9.4	Mg = 24
	B = 11	Al = 27.3
	C = 12	Si = 28
	N = 14	P = 31
	O = 16	S = 32
	F = 19	Cl = 35.5

ÉLÉMENTS TYPIQUES.

I	II	III	IV	V	VI	VII
H						
Li;	Be;	B;	C;	N;	O;	F;
Na						

ÉLÉMENTS PAIRS — **ÉLÉMENTS IMPAIRS**

II	III	IV	V	VI	VII	VIII				I	II	III	IV	V	VI	VII
											Mg	Al	Si	P	S	Cl
Ca	..	Ti	V	Cr	Mn	Fe	Co	Ni	Cu		Zn	Ga	..	As	Se	Br
Sr	Yt	Zr	Nb	Mo	...	Ru	Rh	Pd	Ag		Cd	In	Sn	Sb	Fe	I
Ba	La	Ce	..	...	...	...	...	..	..		..	..	..	..	..	..
..	Er	Di?	Ta	W	...	Os	Ir	Pt	Au		Hg	Tl	Pb	Bi	..	..
..	..	Th	..	Ur	...	...	...	..	..		..	..	..	..	..	..

Mendeleyev's firmness and confidence in stating the periodic law contrast with his hesitations about the layout of the table. Each form has advantages and disadvantages. His first table (above) highlights the typical elements which played a leading part in the process of discovery. The compact form (centre) stresses the regular variation of forms of combination. In 1879 Mendeleyev recommended an exploded presentation (below; the Roman numbers designate the groups or the forms of combination); it makes available additional information on the similarity of the even periods, but upsets the order of increasing atomic weights. (Tables in G. Mendeleyev, 'La loi périodique des éléments chimiques' in Le Moniteur Scientifique, *March 1879, pp. 692–3, 700–1. Bibliothèque Nationale, Paris (above and centre). Photograph, Jean-Loup Charmet (below).)*

element, but perhaps a tri-atomic molecule of nitrogen, for nitrogen was known for its lack of chemical activity.

Finally, it was the periodic law which enabled the difficulty to be resolved. Trusting to the regularity of the periodic function, Ramsay and John William Rayleigh forecast an element intermediate between helium and argon, and added a group 0, which was quickly filled by the successive discoveries of neon, krypton and xenon. For Mendeleyev these strange elements remained strangers. This additional group upset the symmetry of the table, whose two ends had been composed of two groups, both highly reactive with contrasting properties. The rare gases were easy to insert, but Mendeleyev himself was very upset. The inertness of these gases challenged the idea of an individuality expressed in exchanges and relationships. The relationship between atomic weights and properties suddenly lost something of its relevance.

With the rare gases we are reaching the extreme limits of the 'systematics of elements'. Was it not the strangeness of these gases which led Mendeleyev, in 1902, to consider the ether as a chemical element and insert it into the periodic table?

The periodic system thus marks the apogee of a chemistry centred on the elements: it summed up facts and laws, systematized knowledge and oriented the development of the chemistry of elements until the point where its own limits became clear.

Far from being a prophetic discovery by an isolated individual whose genius had carried him ahead of the science of his time, the periodic system appears, on the contrary, as a response to a specific problem in nineteenth-century chemistry and the conclusion of a long story punctuated by repeated attempts at a systematic classification.

If Mendeleyev was on the margin of the dominant currents in the chemistry of his period, it was not because he looked forward to the twentieth century and anticipated the future of chemistry; on the contrary, it was because he denied the transmutation of elements. Far from being a precursor, Mendeleev is rather the heir of eighteenth-century chemistry, reviving faith in the individuality of elements and interest in the study of exchanges and combinations. But discoveries escape their authors. Welcomed by Mendeleyev's contemporaries as evidence for the complexity of chemical elements, the periodic table would be rapidly reinterpreted on the basis of modern atomic theory.

20

Manufacturing Truth: The Development of Industrial Research

Geof Bowker

In which it is shown why the history of science in industry and its heroes are so little known.

History is a success story. We historians talk about great warriors and victors, the rich and the famous. We do not in general talk about the poor, about women, about slaves. In the history of science we have anecdotes about crazy inventors, but our bread and butter is the towering genius: for every book about the dead-ends, the failed experiments, the frauds there are a thousand about Kepler, Newton or Einstein. Yet within the success story of science there are successes that are not recounted. This chapter is about one such: the rise of industrial research. We are going to track through some dense historical undergrowth, so let us equip ourselves with a compass. This will be the fact that the overwhelming majority of science that is done today, and the overwhelming majority of science that has ever been done, has been industrial and/or military in origin. We may change what we mean by science or by industry, but this will remain a stable fact. We will look two ways at once. We will look toward the changes in social organization that accompanied, as cause or effect, the development of industrial research and we will try to see simultaneously why this story has been little told.

To put our questions more graphically, consider the problem of military science. The development of the atomic bomb provides a nice example. The bomb replaced reliances on human resources in the form of the foot soldier by reliance on atomic theory as developed by leading physicists. During World War II, faith in science was so marked that 120,000 people worked on the Manhattan Project[1] at its peak – and this before anybody 'could be absolutely sure that atomic theory could be embodied in the engineering of an explodable warhead'. Yet this massive deployment of resources used to devestating effect gets no mention in physics textbooks studied in school and university and little attention from historians of science.

There is a kaleidoscope of possible reasons for this: the story is a shameful one; it goes too much against the self-image of science; its full retelling involves access to documents that remain confidential; the science is not so interesting innately; narrative histories of this sort do not belong in science and so on. Let us start to pick our way through this maze of factors so that we can block out the development of industrial and military science and audit its accounts. Our story will fall into two parts. In the first we will look at the discourse of industrial science: how do its forms of analysis differ from those of 'pure' science and what does this have to do with its non-appearance in the annals of history? In the second we will look at what industrial science does, and how its history is tied in to that of the Industrial Revolution.

The Discourse of Industrial Science

Our uniting thread through both parts will be a consideration of the industrial science produced by Schlumberger, a company that could be said to have scientific research at its base. In particular we will look at its early years, the period from 1920 until World War II, when the geophysical techniques and inventions that made its success were developed.

The central techniques developed by the company were in the field of well-logging, a way of using electrical readings to gain a picture of the contents of a hole that was being drilled. Anyone who has ever dug a hole in the ground knows that what comes out on the other end of the spade is messy, heterogeneous and pretty difficult to describe. As you dig, bits of the hole start caving in and as your spade comes up it scrapes the side of the hole, so that what you get is not what you dug but that plus an indeterminate amount of other stuff. Oil wells have the same problem magnified, for you also need to circulate 'drillers' mud', so as in part to bring up and filter out the latest cuttings the bit has drilled through and in part to stop the well from blowing out when high-pressure oil strata are reached. And this turbulent fluid often leaches the oil you were interested in from the samples you tried to take. This is why the taking of actual samples could often be superseded by the act of electrical logging. In fact, it led to a curious reversal

584

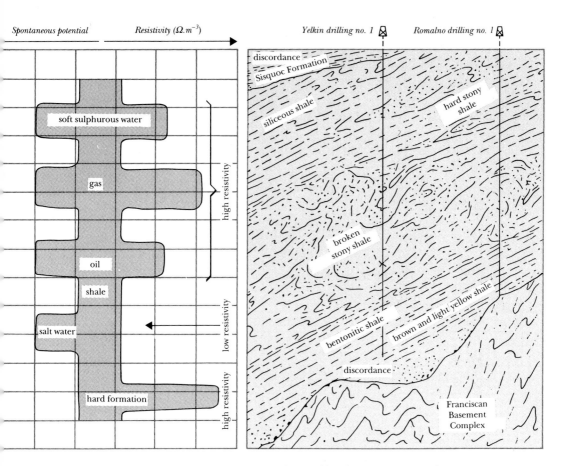

Ideal and real representations of the subsoil. Left, a Schlumberger diagram of resistivity and spontaneous potential; right, a detailed section to illustrate the structural nature of the productive cherty shale zone. (From Schlumberger papers.)

whereby getting your hands dirty by pulling up the oil sand was relegated to the psychological plane and getting an electrical log was the real material evidence. A consulting petroleum engineer questioned in 1939 brought this out:

> It is a very peculiar situation. Many of these people have a feeling that they want to look at it, and to the so-called practical oil man that maybe was a driller and now is an independent operator, a lot of wiggly lines don't mean so much to him as seeing something taken out of the ground. So it is quite frequent that they take cores that I don't particularly care about their taking. Q. Why do they want to look at the mechanical cores that they take? A. To make them feel better, to see the oil sand, I guess. Q. And that is the only reason that they take them, so that they can feel better when they see the oil sand? A. I try to talk them out of it quite frequently, and often they say 'Well I will feel better if I look at it.'

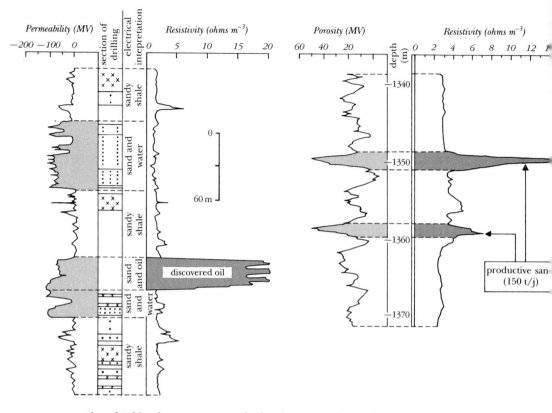

Examples of Schlumberger curves. Left, the discovery of a productive oil-bearing sand (United States, 1933); right, identification of a productive sand missed by the drillers (Romania, 1934). (From Schlumberger papers.)

The electrical logs Schlumberger produced were the new material baseline – they provided peaks on paper that were easier to interpret than shovelfuls of mud.

The wiggly lines that Schlumberger marketed were based on two principles. One set was a 'resistivity' reading: one electrode was grounded at the surface and a second was lowered into the well. A current was sent between the two through the ground. Imagine that down to 500 metres the subsoil was a uniform highly conductive block and then suddenly it became highly resistant (granite, for example). At this point there would be a change in the current between the two electrodes, just as there would at each subsequent variation of resistivity of the surrounding strata. A pair of subsidiary electrodes would be used to pick up the resulting variations in the electrical field as the first electrode descended the well hole. Now, oil can be found in porous layers that are relatively highly resistant. If the layer is non-porous,

586

The main Schlumberger measurements carried out during drilling in the 1930s

The first curve, known as the curve of resistivity, according to the principle in which measurements of the resistance of a given volume of soil crossed by an electric current serve to distinguish layers conducting very well (for example, strata containing salt water or various metals) from resistant layers (for example, strata containing petroleum, which is strongly resistant). It is obtained by placing an electrode B in the mud of a borehole, beside the well at ground level, and lowering a second electrode A into the well being drilled, then passing a current between the two. Two secondary electrodes, M and N, are lowered with A into the well – M and N close to each other, above A. A potentiometer, placed on the surface, measures the difference in potential between M and N all down the hole. The resulting diagram consists of the marks of the fluctuations of the needle in the potentiometer. A break in the curve is interpreted as a sign of a change in the resistivity of the strata traversed by M and N.

For several years only this curve could be used. In certain areas it allowed the identification of petroleum-bearing sands, which are much more resistant than sands containing, for example, salt water. In other ground the curve was no use – for example, when there were too many other very resistant strata close to the petroleum-bearing ones. In this case it could still be used to construct correlations between wells in a given oil field. Curves of resistivity provide electric 'horizons' which form so many signatures of a particular stratum, able to be adjusted from well to well. It is then possible to make a three-dimensional map of a field being considered. On the other hand, the curves are useless in some fields – those that are too complex (with too many faults and too few horizons) or too simple (so that the geologists have no need of external help).

The second curve – called the curve of porosity, the curve of permeability, the curve of spontaneous potential, the curve of discharge potential or simply the left curve – consisted in the beginning of a way of stabilizing the reading of the first curve. It was inspired by the general principle according to which, for reasons unknown, electrical activity is more intense on the level of permeable layers than on the level of impermeable ones. This curve looks different from the first one. It produces strong breaks when it encounters petroleum-bearing sands, just like the curve of resistivity, but it does not react when it crosses very resistant impermeable strata. Even if the two curves combined do not always allow the discovery of the oil that is there, they form a method far superior to any other.

587

there is no room for the oil. If it is porous but with a very low resistance, then it probably contains salt water, which is highly conductive. If it is porous and resistant, then it might well contain oil.

This first curve was good as far as it went, but it could not tell the difference between a layer that was porous and resistant (possibly oil-bearing) and one that was non-porous and resistant (say granite). It had its uses in determining the point at which oil stopped and salt water started – underground, as in the laboratory, oil floats on water – and this helped in planning the optimal depth to drill to. The second curve, however, made the difference. With this one, you could tell the difference between porous and non-porous layers. This worked by lowering the electrode without passing any current. What was measured by it was the local variation in electrical activity. The two major causes of jumps in this curve were electrofiltration – electrical reactions caused by the passing of conductive fluid like the drillers' mud into the surrounding strata – and electrochemical reactions – the difference in salinity between the drillers' mud and the surrounding fluid in the rocks, for example, causing a current to flow. Clearly these two causes were attached to porous layers. On a bad day they might cancel each other out entirely (perhaps being opposite signs), but often they permitted drillers to map oil-bearing layers with a precision that was unthinkable when all they could do was try to interpret the mess of (messy) data available at the surface of the well hole.

In sum, a group of scientists and engineers working in the oil industry produced radically new solutions to old problems based on the application of their scientific knowledge. What they did was archetypically industrial science. In this chapter, as stated above, what we are interested in is how people talked about and described industrial science and how this reflected on the lack of history we have of the field. Our starting-point here will be a problem that is faced by much industrial science: defending your patents against competitors. In particular, we will look at a trial for patent infringement brought by Schlumberger against their competitors Halliburton in 1939, the former claiming that the latter copied the methods they used in the generation of the two curves just described.

The Trial

We just want the truth. Let us try to work out who was right in the court case Schlumberger vs Halliburton. It is quite simple really. Take a curve, any curve, from a Schlumberger log, and compare it with a curve from a Halliburton log of the same well. If the two logs are equivalent then we have established a *prima facie* case that the two methods are the same. We will then look at the Schlumberger patent. If it describes the way to draw the curves, and if it has priority over any other methods used, then Schlumberger was in the right. This is a game for high stakes though, so we will have to go slowly. Whatever Schlumberger were doing, they were doing a lot of it. From June 1932 to October 1940 they logged 108 million feet, the equivalent of some 2.5 times the diameter of the Earth, or some 400 times the round trip between the judge's home town and the court-room. They had over 95 per cent of the business going in the field.

So let us start the trial. First we take a curve, any curve. That's easily said. The first problem was that the curves to be used in the trial were in general the property of the oil companies, who did not necessarily want to broadcast valuable information about their wells. Thus Ennis of the Oil Well Water Locating Company, used by Halliburton to prove prior art (the common existence of a method before its patenting), said that he could only produce in evidence published logs, and

> anything that we have published was records that we were allowed to publish, and any changing in the formations in the records were suggestions made by whoever was in authority to allow us to publish those records, so when I make the statement as to the accuracy as to the depth of the well, I couldn't actually say that was the depth of the well . . . Q. And you couldn't say that the so-called open hole was the correct footage on that factor, could you? A. Well, within possibly 25, 30, 40 or 50 feet.

So we can start with a curve, but we can't be sure it's a real one. And nobody necessarily knew what the real one was – to the extent that Dr Rust had to admit that it turned out that the curves published in the body of one rival patent were in fact Schlumberger logs.

So there is a bit of confusion about reality, but let us not exaggerate the problem. Both parties could and did go to the oil companies that were supporting them and get permission to use real logs. However, we have not yet exhausted the problem. Take a curve, any curve. But what do we mean by 'any curve'? All curves are not the same. The defence asked the prosecution expert witness, Dr Aiken to read an imaginary log:

> Q. Suppose I told you that the resistivity of a formation was four ohm meters. Would you deduce the nature of that formation without any other information? A. In Texas here? Q. Without any other information. A. You always have other information, Mr Martin. You cannot use any physical method utterly divorced from everything else. That just doesn't happen. The user of anything always knows something.

The curve only exists within a context.

We cannot just take any curve: if we did we would be prejudging the outcome by lining up for Halliburton against Schlumberger. For Halliburton, if the patents described a valid method then they should be universally applicable, without differentiation by area. For Schlumberger, context was prime – for the interpretation of both curves and patents. Thus Aiken, having made his point about the specificity of curves, does the same for words. He is asked to define the word 'parameter': 'what we mean by parameter would depend on what we are talking about. Here we are talking about geophysical exploration . . . So when we consider a word we must consider it in connection with its context, and not arbitrarily make it mean something the patent obviously did not intend it to mean.' The same problem of the words and drawings depending on context appeared in Halliburton's attempt to prove that the Schlumberger method had in fact first been used by Fox in the tin mines of Cornwall in 1830.

These hermeneutic stances by Schlumberger and Halliburton also need to be put into context. For both plaintiff and defendant, reality and its interpretation attract different modalities in the court-room, in scientific literature and in advertising literature. And this introduces us to another problem of taking a curve, any curve. The court-room itself is no isolated arena. There are competitors in the back rows taking notes, picking up ideas. Worthington Campbell, who represented Schlumberger in a case against Geoanalyzer (which was settled out of court) made the point when summarizing the strategy used in this earlier case:

> It was also decided to include the porosity patent in the belief that we might later obtain more convincing evidence of clear infringement. In the absence of such clear and convincing evidence we always feared that a judge might be confused by a composite curve in which self potentials and resistivity values were mixed, since the same curve would be the basis for the charge of infringement of both porosity and resistivity patents. Geoanalyzer in its answers to interrogatories admitted the resistivity or resistance curve but denied the influence of self potentials . . . the Court might have said some things that would have told others how to obtain a mixed curve which would not infringe, although to a skilled man it might well serve for porosity indications.

Schlumberger might indeed want to talk about and defend their curves, but there were things best left unsaid. Thus Leonardon, managing director of Schlumberger in the United States, refused to go too far in describing optimal electrode spacings for an ideal cubic meter: 'Now it is by a long and costly experiment that we determined which are the best spacings in different areas so I am not inclined to give you all the details.' A later interview suggests that an unpatented new idea did slip out at the trial and was taken up by the Halliburton, so that in the end Schlumberger had to buy back the use of their own invention. The counsel for the Oil Well Water Locating Company was quite explicit about the need for a secret space:

> I have cautioned the witness against furnishing you anything; in fact, when you first requested the drawings or anything like that I told you 'No', and it was only after you took depositions here the other day that I finally consented to permit Mr Ennis to see if he had any advertising literature or the like which might show what he was doing at a certain period of time so far as his public operations were concerned.

In the context of the trial we cannot just 'take any curve'. Neither the oil companies nor Schlumberger nor the defendants will allow us to and, if we believe Schlumberger, no one curve is just any curve. Tracing the difference between a 'typical' curve and an actual curve, we will now see to what extent each curve was in turn constitutively local, decorative and open to radically different interpretations from the actors who dealt in it – the well surveyors and the oil companies. The typical curve in question was produced by the Houston Geological Society Study Group on Electrical Logging, and was entitled 'Typical S.P. and Resistivity Diagram'. This diagram is indeed typical of a whole range of representations of electrical logging, giving as it does a clear picture of the ideal effect. It is a typical textbook presentation.

There are concessions to realism, in that the gas is on top of the oil is on top of the water and there is a hard bedrock. But just looking at the curves themselves, we know immediately that something is wrong. Any curve, as opposed to the typical curve, is saw-toothed. This is due to several factors, an important one being the phenomenon known as 'measurement hiccups'. The three electrodes in question – A, M and N, say – encounter the different strata in turn. Imagine that A and M are in a very resistant layer and that N encounters a conductive layer. There is a kick on the curve. Then either M and N come together in the conductive layer or N has already passed through by the time that M arrives. In the meantime A itself may or may not have passed through a conductive layer higher up. At each boundary (and the boundaries are not necessarily well defined) there is a kick in the curve, so that instead of having an effect opposite a given layer, there is a series of effects. The curve is a composite reading of these variations and Schlumberger engineers had a subset of ideal curves to show them which possible sequence they might be dealing with. In principle, the result was unreadable if you had: 'configurations of electrodes whose length was of the order of the thickness of the layers traversed', but 'this clearly supposes that we know what kind of terrain we are dealing with and what magnitude of effect we will get'. This result is known as the hermeneutic circle in human science and has been called 'experimental regress' in hard science. A Texas sonde was not the same as an Oklohoma sonde; you needed different configurations to get the same curve. The left-hand curve was also a composite curve. As we have seen, there were two effects at least determining its shape:

591

electrochemical and electrofiltration; and they might have opposite signs and cancel each other out. Further, spontaneous potential increased relatively uniformly with depth as the temperature of the drilling fluid went up – another local factor that varied from field to field, from well to well. To make matters worse, salt layers could dissolve locally in the drilling mud, so that the effective range of both curves was limited to the mud itself. Thus in both cases, we are dealing with composite curves that had to be interpreted by the engineer or the geologist on the spot. The actual curve – the tailored, local curve – was sufficiently fine-tuned for Halliburton to charge that in fact: 'the plaintiff in this action, the Schlumberger Well Surveying Corporation, has concealed from the American public and has preserved in secrecy the methods and apparatuses it has actually successfully used in the field.' Far from there being an obvious kick in the right places in accordance with mutually agreed theory, there was actually suspicion in the early years that the curves were a decorative blind covering the actual garnering of information elsewhere: 'When the Schlumberger was first introduced, part of our duties was to go out on the truck, regardless of the clemency of the weather, and sit there while they ran their log, to see whether the Frenchmen were trying to put something over on us.' One accusation was that Schlumberger added dummy electrodes to their equipment to conceal the actual configurations that they were using.

In the context of the oil field, however, it was not appropriate to argue the infinite interpretability of the curve and the skill of the individual interpreter. The rhetoric of the time and the impulsion of the oil companies was to exclude the human – thus J. Boyd Best claimed of the Schlumberger: 'It is an accurate log, without any personal element involved, and as such we use it to find sands not located by the driller's log, and use it to map faults and structures, and for correlation.' And this impersonal effect was enhanced by the layout of the logs given to the companies. This comes out particularly clearly in the case of another element lacking from the typical curve – the scales appended to the top of every log. The S.P. curve was given in millivolts and the resistivity curve in ohms per cubic metre by Schlumberger, in impedance ohms by Halliburton. The scales at the top gave the impression that specific parameters were being measured, so that the judge got quite confused by Blau's testimony that.

> those things which are measured are not specific resistivities or the contact potentials. The Court: You mean by that then that the plaintiff measures something, and the defendant measures something, but you are not willing to name them? [Thus Turner, a Halliburton engineer, was asked about his relationship with the curves:] Q. And you help the customer interpret them? A. In interpreting what the different curves mean. Q. And what purposes do these numbers at the top of the two scales serve,

under potential, on this Exhibit 124, graduated in millivolts, minus 50, minus 150, minus 200, what is the purpose of that, so far as you are concerned? A. So far as I am concerned, that on the potential side is of no value. Q. On the impedance side on this Exhibit 124, under 'Impedance Ohms', appear the figures 75, 150, 225, 300. A. To me that represents the scale on which the log was run. Q. I understand. [Turner was pushed on the question as to why there were different numbers for dry and for wet holes:] Q. Well now, I don't believe I still quite understand the difference between those two scales there. What is your explanation for that? A. I think that plainly shows you that the scale really doesn't have any value as far as the scale is concerned. It is just run on that scale so as to bring out the magnitude of the kick.

The numbers on the top of the logs were decorative, then, in the sense that they were not essential to the interpretation of the logs. What they did serve to do was to send a message that the curves were a scientific production and not some French person's artistic impressions. They also served to distinguish a particular opus, to mark it as original, to sign the *oeuvre*. One of Schlumberger's best coups during the trial was to scale down one of their logs to match a Halliburton log taken of the same well on the same day – the tracings were sufficiently identical for Halliburton not to challenge the similarity. This raised a question of the scales used, put by the defence attorney to Schlumberger's expert witness:

Q. In view of your last statement then, if the scale the defendant puts on suggests he is measuring something else than resistivity then the scale does not indicate the value of the resistivity, does it, the specific resistivity? A. We have been over that before. This is a resistivity curve. The only reason I could see for putting such an arbitrary scale on it might be the hope that it would suggest that the defendant was measuring something else.

A second decorative dimension of the curves was attached to the process of habituation. The loggers' strategy was to get the oil companies used to seeing the curves, whether or not they got useful information from them. Thus the director of Schlumberger operations in Venezuela wrote in 1934 to the central office in Paris that:

I propose to leave the prices for this style of operation [logs of holes being repaired] as they are. As it turns out, the results furnished by such runs are sometimes influenced by the presence of pieces of 'liners' or other 'fish'. But it is in my opinion good politics to help and encourage companies such as Gulf, who are having us log systematically all the holes that they repair, despite the risk of obtaining distorted diagrams which therefore yield incomplete information.

This habituation factor – get them used to it and give them what they are familiar with – went back to the original form of the curves: 'A set of Schlumberger logs (Resistivity and S.P.) is similar to a normal lithological log "translated into an electrical code" which, with proper experience, can easily be read.' The logging companies did not want to continually change the curves, even if this would bring improvements. For example, there was a certain configuration of electrodes for which, when fresh-water mud was used by the drillers in the well-hole, there was 'an increase in the size of the kicks opposite water-layers and a diminution opposite oil-bearing layers. You can easily see that this singularly complicates the fundamental problem of the distinguishing of oil-bearing layers.' And yet the configuration in question was good for correlation purposes, used little current and: 'In the end, we need to keep using it for reasons of continuity, because a lot of diagrams have been made with it in the past and the geologists are used to it.' Given the often precarious relationship between geologist and geophysicist – the former feeling threatened by the latter – it was worth some sacrifices to keep the geologists háppy, to give them a feeling of control. The solution was to add another curve which did not have the same problems and to leave the vestigial curve in place.

The Secrets of Science

Where are we, then, with respect to the discourse of industrial science? An outstanding feature of the above account is that industrial scientific knowledge is not 'transparent' to the user. It is in the interest of the company, Schlumberger in this case, to keep parts of its knowledge secret, to vary the accounts it gives of its own processes according to the reader expected. This stands in direct contrast to the discourse of pure science. When the first research laboratory was created in England by Robert Boyle, he proposed a series of rules about the presentation of data. His laboratory would be the converse of the alchemists' 'elaboratories'. The latter were secret places that none could gain access to: no description of their work that alchemists gave could be used to reproduce it elsewhere. The research laboratory was to be open to the public, and published accounts of experiments made should be made sufficiently clear and explicit so that anyone could replicate any of the experiments at will. In fact such replication is notoriously difficult and pure science laboratories are generally as closed to the public as those of industrial science, yet the discourse of transparence remains a key feature of pure science. To this extent, industrial scientists are more within the alchemical tradition, producing a result whose genesis none can know but the initiate. Thus, for example, in the account above Leonardon refused to give optimal electrode spacings; Halliburton could be charged with giving a false name to the thing that they were measuring; Schlumberger could be charged with adding knobs to their measuring equipment whose only purpose was to confuse.

The first industrial research laboratories

The historian David Kevles says that the first industries to really get going in industrial research, in the 1890s, were the electrical, iron and steel, fertilizer, sugar, drugs, dyes and petroleum products industries. Steven Rae gives the following chronology: 1875 Pennsylvania Railroad (his choice for the first industrial research laboratory), 1876 Menlo Park (Thomas Edison's laboratory), 1889 Standard Oil, 1886 Eastmann Kodak, 1890 DuPont (the chemical manufacturers), 1900 General Electric at Shenectady and 1903 Westinghouse. This quasi-simultaneity of the introduction of research laboratories is suggestive. Alternatively we could follow Partington and see the first industrial research being carried out by Egyptian priests.

But it is not enough to point to this difference in discourse. We want to characterize and explain it. Our approach to this will be to consider for the nonce what is the chief product of industrial science – the patent – and then perhaps to compare it with what is the chief product of pure science – the scientific paper. Patents lay at the origin of the dispute in the Schlumberger story above. What, then, are patents? How do they get used in industrial science? And how can these answers contribute to a solution to the problem posed in this section: why is such a success story as industrial science so little sung by historians?

In a lovely turn of phrase, David Noble asserts that: 'Patents petrified the process of science, and the frozen fragments of genius became weapons in the armories of science-based industry.' The early history of research at Bell (later A. T. & T.) will bring out this strategic role of patents. Alexander Graham Bell had originally, in 1877, tried to sell his two major patents for the telephone to the Western Union telegraph company, but because of doubts about the validity of the patents (there was a rival claim) and about the commercial value of the telephone, the latter refused. So Bell, with his partners, incorporated a company. Western Union set up a subsidiary, American Speaking Telephone Company, to exploit the rival patent. A legal battle started. Soon Western Union moved out of the field for the period of Bell's patent (twenty years), in return for 20 per cent of their revenue. Bell had a virtual monopoly of the field, one which it maintained through two strategies: repeated patent infringement suits (600 over the period of the validity of the two main patents) and trying to acquire all possible patent rights to improvements in telephone design (collecting some 900 telephone-related patents up to 1904). Thus the strategy was plain: to maintain control

Thomas Alva Edison (sitting in the middle, wearing a beret) in his laboratory in Menlo Park in 1880. The laboratory is like a factory. The lamps were among the first electric ones, invented in this laboratory. (PPP-IPS-NASA.)

over the development of the telephone field through the use of patents as arms in a continuing commercial war. Over this early period, Bell spent little money on its own industrial research. Reich cites a 1906 report which points out that Bell was still committed to 'the often risky policy of acquiring important patent rights from outside inventors', a policy which had generally worked to date 'because its already strong patent position assured inventors would have difficulty exploiting their patents outside the Bell system'.

Given the amount of time and energy that companies like Bell were having to spend on this defensive policy of acquiring and protecting patents, we can see the impulsion to set up industrial research laboratories and guarantee a continual string of these weapons. And indeed, once in 1907 A. T. & T. – Bell's new name – started serious industrial research, they were quickly able to move onto the offensive. The case of the telephone repeater clearly demonstrates this. A major problem of long-scale telephone communication was the need to regularly boost the signal passing along the lines. At the same time, as research director Carty pointed out, successful development of the repeater would permit the domination of a rival, nascent industry: 'A successful telephone repeater, therefore, would not only react most favorably upon our service where wires are used, but might put us in a position of control with respect to the area of wireless telephone should it turn out to be factor of importance.' This involved calling on physicists who were aware of 'recent advances in molecular physics, and who are capable of appreciating such further advances as are continually being made'.

This centrality of the theme of control through patents is not, of course, peculiar to Bell – it can be found throughout the history of industrial science. Thomas Hughes brings out the significance of patent considerations for the establishment of General Electric's laboratory. Indeed the GE patent attorney was one of its key advocates. He had argued that this would give GE the options of producing its own patents, and of 'inventing round' external patents that blocked the way and were too expensive to acquire. Patents figured largely in day-to-day laboratory work, laboratory heads spending a large part of their time preparing briefs for patent infringement suits. Thus in general it was the search for control and security that led Bell and GE to set up industrial research laboratories; soon after their inception the patent as manufactured article became the laboratories' central product.

The Development of Industrial Science

One more factor that goes hand in hand with the manufacture of patents is the role of the research laboratories in enabling diversification in their parent companies, as we will exemplify by looking at GE's research laboratory work. GE's laboratory was set up in 1900, with the avowed aim of maintaining the company's control of the American electrical lighting market. At the time their bulbs used the inefficient carbon filament, producing too much heat and not enough light. Two types of lamp were invented in Germany in the 1890s that threatened GE's position: Nernst lamps (the progenitors of our fluorescent lamps) and lights using an osmium filament. After much negotiation within GE, a laboratory was set up on the recommendation of Charles Steinmetz and under the direction of Willis Whitney, both of whom had been educated in Germany. Progress on lighting was slow in the early years: Whitney protected the laboratory by doing problem-solving for the production line and by manufacturing articles for the production process that needed the laboratory's special resources. This strategy worked so well that, by 1903, its personnel consisted of nineteen researchers and twenty-six support staff.

Whitney's laboratory soon managed to improve the efficiency of the carbon filament lamp, but results from Europe with osmium then tantalum lamps suggested that a new filament was needed. Whitney, who had trained as an electrochemist, chose tungsten, which fitted the bill but was too brittle to be conjured into the right form. For the period 1907 to 1912, the laboratory concentrated on ways of overcoming this problem. The work often involved cut and try methods: at one stage a skilled blacksmith was taken on to try to beat the tungsten into shape. Ultimate success allowed GE to retain control of the lighting industry through its patents – this not only in the face of European competition but against the antitrust legislation that had already forced the splitting up of huge concerns like Standard Oil.

The extent of the success is shown by the fact that in 1928 GE had 96 per cent of the market for incandescent lamps. From this secure base the laboratory widened its range of interests. To cite Hughes: 'Whitney engendered an atmosphere that was found congenial by PhD scientists: weekly colloquia, publication of scientific papers, and encouragement to acquire a greater fundamental understanding of technological phenomena.' Whitney maintained an up-to-date library, encouraged membership of professional organizations of physicists and chemists, and – to the extent that patents permitted – encouraged publication of results. Everything was done to create an atmosphere of pure research. The patent lawyers looking over their shoulders could take care of commercialization.

Reich gives the picture of GE's patent lawyers going through laboratory reports and frequently picking out makeshift solutions to subsidiary problems and making successful patents out of them. Thus a circuit rigged up to help in X-ray crystal analysis turned into a patent that became an obligatory passage point for radio manufacturers. X-ray and radio work was undertaken by the company under the impulse of laboratory developments. The early GE laboratory had been problem-orientated: Whitney, as laboratory head, had managed successfully to create a space within the large corporation where pure research could be done. Industrial science had come into its own.

Industrial Science and its History

Patents then did two things. They played a defensive role in protecting industry already in place and served to buttress the creation of new industries. We can see both these factors with Schlumberger. Before the development of the two patents defended in the infringement suit, there was no such thing as a well-logging industry. Once the industry had been created, they served to protect it from competitors such as Halliburton. Let us put ourselves into the position of the head of an industrial research laboratory and see what these two factors do to our view of the history of our research effort.

Clearly, the history we want written becomes part of the battle to protect ourselves. Schlumberger did not want to give too detailed an account of their processes of industrial research. This might give ideas to competitors (as it did in the case of the trial discussed above). Worse, it might actually invalidate the patents – thus the Halliburton claim that close historical analysis showed that techniques used in the field were not the same as those represented in the patent. The case of Rudolf Diesel provides a good example here. Diesel's work is classically based on his attempt to render the Carnot cycle,[2] which was a mathematical treatment of the workings of an ideal heat engine, manifest. He had learnt of the cycle as a student at the Berliner Technische Hochschule; and on his own account became obsessed with the problem of giving it physical form. When he applied for his first patent in 1892, he (and many other experts) thought that he had achieved this – and yet in its Carnot form the engine did not work. During the 1890s at the M.A.N. factory a whole series of variations were made to the engine; these variations altered both the form of the engine and the theory of its operation. This led Diesel into difficulties, for in his patent, Diesel had committed himself to a non-workable theory of the operation of his engine; and when he applied for a second patent based on a functional engine, he had to gloss over theoretical changes in order to protect the first, more basic, patent. The two patents formed the cornerstone of his fortune, and the gaps between them and the functioning engines were their Achilles heel.

Thus we can see that there is in fact an inbuilt interest for the producers of industrial science in keeping their own products' history as open and as flexible as possible, while trying to brake the trajectory of their competitors'. The former is the best possible position for meeting attacks, which could come from any direction. Compare this with one of the classic roles of the historian: tracing the genesis of the objects they discuss, be these social classes, religions or ideas. The best patent, the best product of industrial science, is one without a history. It was therefore natural for Schlumberger to deny any filiation between their work and that of Fox, and Halliburton denied that their measuring devices were in the tradition of Schlumberger's. These constraints work within the general organization of the research laboratory. Stories are rife of frustrated industrial scientists having to watch others make 'their' discoveries because their corporation would not let them publish results that might give an edge to the competition.

We can see, then, that industrial science seeks to destroy its own past and, failing that, to control it. There is a difference here with respect to academic science, but in order to discern it we will have to see the extent to which the two are structurally and discursively virtually equivalent.

'Pure' and 'Applied' Science

Take first of all the discourse of 'pure' versus 'applied' science: this is in itself a historical invention associated with the creation of academic disciplines in the nineteenth century. It has by no means always been maintained by scientists and is arguably less and less so today. Further, we should consult our compass again; it reminds us that the overwhelming majority of scientific work done is industrial in origin, for even that done at universities is often tied to external contracts from the military or from large corporations. To complicate matters more, industrial scientists themselves maintain a discourse of pure science. Thus we saw Whitney at the GE laboratories encouraging an atmosphere of pure research. When patent lawyers permitted, corporation scientists were permitted to publish. If we look at the boom in industrial research in the first half of this century, we can discern some of the reasons for this. Students had been filled at university with stories of great science and the pursuit of knowledge for its own sake. Attached was a myth of the community of science as a rational community that openly shared its results for the greater good of the whole. Industry was seen as a second-class choice. What it could offer was much more money, but at the price of glory. Corporations tried to sweeten the deal by making the work background as similar to that of academic research as possible. Bright young graduates were encouraged to come and work on subjects that interested them. They were offered the prize of autonomy if their work was successful in corporation terms. So not only did academic science often have an industrial origin but also industrial science was often dressed up to look academic.

Another process, attached to the discourse of pure science, deepens the connection between the two. For in pure science, too, there is a process of the destruction of one's own past. If one looks at a typical scientific paper we see two specific historical processes going on: the presentation of the paper within the history of the scientist's own discipline and the destruction of historical context. The former provides the key to the distinction between academic and industrial science, and we will discuss it below. The latter is the site of a powerful link between the two. To spell out what I mean by the destruction of historical context, consider the difference between the following two statements:

A: On a cold winter's morning in 1911, Conrad Schlumberger performed some experiments at the École des Mines in a converted bathtub that had been used by his daughter – these experiments indicated that one could locate hidden metal in copper bathtubs.
B: Conrad Schlumberger proved that ore deposits could be located by charting lines of equal resistivity at the surface of the ground.

Clearly statement B is the one that we are more likely to find in a scientific paper: recent analysis in the sociology of science indicates that it is precisely the progressive destruction of historical context that is the mark of scientific writing. The rival scientist, like the rival company Halliburton, will try to show that the historical contingencies mattered – that the fact that it was cold mattered in the first case, that the fact that Schlumberger included a

600

circuit-breaker not mentioned by the patent mattered in the latter. Thus the discourse of industrial and pure science both in a sense involve the destruction of historical context.

However, the difference between them lies in the second function of the scientific paper: to insert the work done within the context of other work. This can be done locally by referring to recent papers in the field that seem relevant, or more globally by claiming to contribute to ever larger research programmes. This process of historical insertion is one that creates the giants of science, who furnish relevant traditions for a huge range of work. And the result of this process is that anyone can name a dozen 'pure' scientists, but few can name a single industrial scientist. Industrial science destroys its own past in the act of creating capital, pure science destroys its own past in the act of creating intellectual capital. The two orientations overlap with respect to the process of abstraction; they differ in that one creates the canonical, highly individualized object given a brand name and the other the canonical, highly indivualized piece of knowledge given the name of its creator. The same process, but different results. On one side lies history, on the other technology. Industrial science does not have a past; its empire is the present.

The Nature of Industrial Science

We now have some idea of why there is so little history of industrial science, despite its being far the dominant mode of scientific practice. Let us now look at what such a history might look like. What is this success story that does not get told?

We will again take as our basis a discussion of the two curves Schlumberger developed in the period 1920 to 1940. This was a period when industrial science reached its maturity. Although it was only in World War II that scientific research became central to the war industries, already in World War I it had made great strides forward. The chemical industries in the Allied countries, for example, were invigorated during the war by munitions work and by desire to break the German monopoly of dyes. The first ten years after the war was a period of boom in industry and in industrial science; nowhere was this boom more marked than in the oil industry. Control and development of petroleum resources had been a key factor during the Great War – in a much-quoted phrase, Lord Curzon said that the allies floated to victory on a wave of petrol. Bérenger, the French Oil Commissioner for War, wrote that: 'He who owns the oil will own the world, for he will rule the seas by means of the heavy oils, the air by means of the ultra-refined oils, and the land by means of petroleum and illuminating oils.' The industry's increase was phenomenal: from the period 1918 to 1936, petrol and gas went from 21 per cent to 43 per cent of all energy produced. Scientists made significant contributions, from the process of the discovery of new deposits to the refining of the crude oils pumped out of existing ones.

Schlumberger's early years need not concern us in much detail here. A few strands, however, can be briefly brought out. The nature of the method is important. At its origin, it was used for trying to determine the nature of the subsoil from the surface. An electric current was run between two electrodes, thus setting up an electrical field. Measurement of this field gave some indication of the kinds of deposits to be found beneath – just as they had worked in the bathtub in the basement of the École des Mines for Conrad Schlumberger before the war. There was no initial idea of prospecting for oil: indeed, metallic seams (of low resistivity) seemed the more natural choice. However, as I have pointed out, oil was where the money was, so the two brothers Schlumberger – with a handful of staff – sought ways of prospecting for it. In the 1920s there was a battle royal between three different methods of prospecting: seismic, gravitational and electrical. Companies developing seismic methods won for two reasons: they were highly accurate in finding the salt domes often associated with oil reserves, and they furnished the possibility of continual improvements and thus the development of new patents and the possibility for a given company of retaining control of the technique. Gravitational methods were efficient but static. They thus came to be colonized by the other companies, becoming methods used by electrical and seismic prospectors. Electrical methods lost: despite numerous variations in the measuring equipment, the signals received remained too ambiguous.

The battle lost, Conrad Schlumberger thought of abandoning the project. There followed a period of development not unlike the one I have described for the Diesel engine: a series of measurements were made in the hope that they would measure something. There was not any governing theory. Finally, the two devices described above were developed. Both were transformations of surface techniques. Instead of the two electrodes being placed on the surface they were placed inside a well that had already been drilled. Thus what started as a tentative method of prospecting for metal then became a failed method of prospecting for oil and ended up as an enormously successful technique for helping the oil drillers. Along the way, Schlumberger took out a series of patents that gave them a virtual monopoly of electrical methods in a range of activities in the oil industry. Laboratory and field work consisted precisely in trying to tie together a demand from the prospectors, a signal that could be separated from the enormously complex electrical noise surrounding a drilling operation, and a measurement device.

There was thus a whole series of mutual adjustments that went on before textbooks about the science of petroleum could be written. What I want to do now is to bring out, necessarily sketchily, some general features of these adjustments. These features are, I will argue, able to be generalized to science and industry as a whole and give us a way of understanding the development of industrial science.

Adjustments of Industrial Science

For the moment, though, let us return to Schlumberger. In order to get some idea of the kind of adjustments Schlumberger made, we will compare how it acted in this new space that was the oil well and how oil companies acted in the new territories they opened up for their industry. The oil companies set up new kinds of social space, time and energy configurations that allowed them to operate within the traditional state. Thus roads networked the country – constituting very thin filaments that connected energy centres (the cities and the oil wells). The advantage of the network configuration of roads, railways and pipelines for the oil companies was that it permitted them to do their work while coming into only minimal contact with the states they operated in and their inhabitants. A nice example of this minimal contact comes from a Schlumberger engineer's reminiscences of his period in Burma, where the oil companies had literally created an infrastructure that undercut local industry:

> The Burmese had a very particular method for exploiting this petrol. They dug a square well (about 1 m across). It was a manual operation with all sorts of short-handled tools being used – and even hands when the formations lent themselves to it. There was a tower over the well, with cranks which served to lift up the debris using a bucket. This bucket was also used to lower and bring up the Burmese well-digger. The time that each Burmese spent digging was variable, but fairly short for there was gas everywhere and the problem consisted above all in bringing back a person rather than a corpse. . . . The English, respectful of the local ways, left these rudimentary installations completely untouched and did not impinge on the rights of the prior owners. They decided that all hydrocarbons to be found above the maximum depth of these wells were not B.O.C. property. They didn't even interfere with the market for this artisanal petrol: to such an extent that in 1936 you could see a certain number of these wells being exploited or in the process of being dug in the middle of a forest of B.O.C. wooden towers.

I have given this citation at some length, because it brings out the major features of the oil industry's operations that I want to stress. It shows the minimal contact with the traditional state in the form of the isolated Burmese well-diggers being left alone, and in the fact that the British Oxygen Company dug below their domain. It indicates the vastly greater amount of energy that was available within the oil network (the forest of towers against the individual digger). It shows the new configurations of social space associated with the network – the international market for BOC's petrol is tacitly

opposed to the previous local market. Finally it hints at the new configurations of social time: the Burmese worked erratically, BOC twenty-four hours
a day. If we want, then, to describe the way in which the oil companies
operated in new territories, we can look at the way that they channelled
energy in highly concentrated form along vast networks that operated according to a different social time from the traditional one.

This same process of the reconfiguration of space, time and energy can
be discerned when we analyse the content of the science developed by the
oil prospectors. Here, for example, is a typical passage from a 1930s work
of industrial geology about a field in Venezuela:

Rio Tarra Field

Location and accessibility – State of Zulia, District of Colon. Accessible
by shallow draft boats via Lake Maracaibo; Rio Catatumbo; and Rio Tarra
to La Poloma. Narrow gauge railway to camp.
Date of discovery – Toldo No. 1 of the Colon Development Company
entered production August 27, 1916.
Producing horizons – Fifty feet of sandy shale in middle of First Coal
horizon carry light oil . . .
Structure – Rio Tarra anticline is asymmetrical throughout its entire length;
has locally vertical and slightly overturned east flank; and more gently
inclined west flank . . .
Character of oil – Asphaltic oil, varying from 23° Bé to 32 Bé. Higher
gravity oil obtained from Tabla sands at base of Third Coal horizon of
Eocene age.
Production – Production has been consumed in drilling operations.
Tankage being erected. No facilities for exportation of oil. About 25,000
barrels production to December 1925.

There are references here to three different kinds of map of the oil domain:
a transport map, a map of geological structure and a map of subsurface
structure. There are three kinds of date: the era of the formation of the oil,
the date of its discovery and the statistical year 1925. With respect to energy,
there is the ironic happenstance that all of the oil produced was consumed
in the act of drilling. The industrial geologist is an integral part of the process
that the oil companies are involved in – locally reconfiguring space, time
and energy. What the oil companies are doing is, very abstractly, to create
ways of redistributing energy about the surface of the globe. How they do
it is to change the way that people work, the way that they relate to their
environment. What this involves in part is a new accounting of the nature
of the subsoil, the industrial geologist doing for inanimate nature what the
oil company does for its workers – providing a time and a space within
which the resources can be exploited rationally.

So how does this relate to the series of adjustments made by Schlumberger to their invention as it progressed from a failed surface technique to a successful subsurface one? As they began working with the oil companies, they had frequently to change their methods to accommodate the time and space that the oil companies were operating in. Their surface methods failed in part against seismic methods because you could do a seismic test without entering a given piece of property and thus without having to buy prospecting rights first. Seismic methods in turn succeeded against geological surveys, which were at first more exhaustive, because they could turn out the quick results needed in the highly competitive atmosphere – seismologists could give a result in a week that might take a geologist a year. So Schlumberger had to juggle its methods till they could find a space and a time within the working procedures of the oil companies. What they did was to change the scope of their measuring devices – from long lines on the surface to short lines descending down wells. They learnt to measure variations in electrical activity that had been caused by the act of drilling itself – the phenomena of electrofiltration and electrochemical reactions being caused by the act of drilling and the circulation of the drillers' mud. In successfully accommodating themselves to these spatial and temporal constraints, Schlumberger became in turn part of the process of the larger work done by the oil companies: figuring out rational ways of exploiting energy resources. They guaranteed control of the subsoil. What had been heterogeneous shovelfuls of mud and assorted debris became layered strata with oil well-marked.

This account of the development of Schlumberger's methods suggests that in looking at the development of industrial science we should look first at what industry does and is, and to what extent industrial science is an integral part of the industrial process. This will provide us with a way of bypassing traditional pictures of science coming to industry like the fairy godmother came to Cinderella. It is not so surprising that industry and science can be seen as aspects of the same process. After all, the two in their modern form grew up together – what is surprising, as our compass reminds us, is that they have become separated in historical discourse. On the face of it, what happens in laboratories is closely akin to what happens in factories. Both turn out uniform products (mathematical laws in one case, manufactured goods in the other). Both do so by operating a series of reproducible processes on raw materials designed to produce these uniform products. Finally, these processes involve the use of extremes (of temperature or pressure and so on) and a precision in the regulation of time and often a speed of action not found in nature. The site where these extremes are produced is the factory or the laboratory. Of course the analogy cannot go on for ever, though it does resonate nicely with the discursive similarity we uncovered between the manufacture of patents and technical objects and the writing of scientific papers.

Science and Industry

In order to give a bit more form to the fusion between science and industry, we will look at the story of the development of the dye industry. It has frequently been argued that the first industrial research laboratories grew out of this industry in the last twenty years of the nineteenth century and thus mark the moment when the individual inventor began to give way before the industrial scientist.

Appropriately, the story starts in England with a lone English inventor: W. H. Perkin, who discovered the new dye 'aniline purple' in 1857, while on holiday from the Royal College of Science – he had been looking for a way to synthesize quinine. Perkin was sure that this new colour would make his fortune, and despite the doubts expressed by Hofmann, the head of his college, he left in order to set up his own firm. After the inevitable teething troubles he made his fortune. Perkin's story is thus a latter-day version of the self-help stories so beloved in Victorian Britain, the difference being that he started from a college rather than a cottage. Though this was an accidental discovery, it tied in with new developments in organic chemistry – and indeed once Kekulé and others had developed the theory of the benzene ring, the way was open for a systematic search for new dyes.

There was already, at this early stage of the story, a strong German connection. Hofmann, the head of Perkin's college, was a student and protégé of Justus von Liebig, the German analytic chemist whose work achieved great renown during his lifetime. L. F. Haber quotes Liebig as saying that: 'Every student . . . must devote himself from the morning until the evening to analytical investigations of every description,' and goes on to comment: 'Liebig rightly attached great importance to mastery of qualitative and quantitative analysis and he devised equipment which speeded up and simplified the procedure. The time taken for an analysis was cut from months or weeks to days, and he estimated that some 400 analyses were made annually at Giessen.' There was, then, an emphasis on routine; and a number of Liebig's students were of an eminently practical bent. Many, like Hofmann, spent periods overseas, during which time they taught or trained foreign chemists and learnt industrial procedures later introduced into the nascent German state.

At home, Heinrich Caro, influential in setting up BASF (Badische Anilin und Soda Fabrik), Carl A. Martinus at Agfa and Wilhelm Meister at Hoechst had all learnt their chemistry either from Liebig or one of his students. And it is on German soil that we will find the continuation of the story of the dye industry. The historian Beer marks the mid-1870s as the period when companies such as Hoechst, Agfa and so on started hiring academically trained chemists. Bayer, a little late on the scene, had 15 research chemists in 1881, 58 in 1890 and 104 in 1896. A spacious, three-storey laboratory was built to house them in 1891. The early chemists were expected to improve production methods, the discovery of new dyes being a secondary concern. During the 1880s, however, their value in this regard was recognized. The period 1890 to 1914 was one of massive development and diversification: inorganic chemists were taken on to work on the inorganic chemicals that furnished the raw materials; the companies diversified into pharmaceuticals, insecticides. Bayer and Agfa went into the photographic film industry and the Haber process of nitrogen fixation permitted success in the fertilizer and munitions markets.

There are two aspects to the routine work of the industrial chemists that I would like to stress. The first is that of the generation of patents. In Beer's words: 'In the two decades that preceded the First World War, patents became more than mere guarantees that provided their owners with a legal monopoly over newly discovered manufacturing processes, they became a manufactured article, turned out by the research laboratories of the big companies.' Thus although it was by no means the prime goal of the setting up of research laboratories, routine innovation soon became a central part of their work. We will see the same development below for laboratories in the electronics industry. The patents that were manufactured by the German chemists came under the 1876 Patent Law – the differences between patent laws providing for Beer and others a key difference between England and Germany. The second aspect of the chemists' routine work is that they were, just like Liebig's students, carrying out a huge number of tests. In around 1900 Hoechst would test 3500 of its own new colours, of which 18 reached the market. Each colour had to be tested against every fabric and under every condition that that fabric might experience in normal operation. Further, the laboratory tested each of the major competitors' dyes in the same way. The sheer number of tests involved was massive. So, however, were the stakes. In the late 1860s, Landes notes, the dye industry was small and dispersed; scarcely a decade later the burgeoning German industry held about half of the world market; by the turn of the century this was up to 90 per cent.

The work of the industrial chemists had grown out of a rationalization and standardization of factory procedures. In turn, it provided a rationalized and standardized search for new chemical products, with the systematic search through variations to the benzene ring. Thus industrial chemists had taken for themselves a method that strongly resembled that of factory work – the splitting-up of a task into its component parts (for example, the split between organic and inorganic chemistry effected in the factories in the 1890s) and the routine repetition of tasks being key features. And their work underwrote the success of the companies they worked for. This routinized, rational search for new dyes had no competition from the traditional indigo gatherers of India; who operated socially and naturally within a different time and space. As with Schlumberger, when we look behind the science that is produced in the industrial context we discover new ways of working.

New Methods

How general is this assertion, and how does it help us sing the unsung glories of industrial science? A look at the context of arguably the first industrial laboratory in the United States, at the Pennsylvania Railroad Company in 1875, makes the point. I will draw on the work of the business historian, Alfred Chandler, here. What we have first of all in the States is a railroad boom during the 1840s and 1850s, such that: 'the consolidated railroad systems remained the largest business enterprise in the world.' Management of this huge system required new methods. Chandler singles out two, standardization and control of information flow. For the former, he summarizes some of the major changes:

> On the night of May 31–June 1, 1886, the remaining railroads using broad-guage tracks, all in the south, shifted simultaneously to the standard 4' 8.5" guage. On Sunday; November 18, 1883, the railroad men (and most of their fellow countrymen) set their watches to the new uniform standard time. The passage of the Railroad Safety Apliance Act of 1893 made it illegal for trains to operate without standardized automatic couplers and air breaks. In 1887 the Interstate Commerce Act provided for uniform railroad accounting procedures that had been developing for a quarter of a century. All four of these events resulted from two decades of constant consultation and cooperation between railroad managers.

This standardization facilitated the control of information flow and he notes that, for the new professional managers, 'control through statistics quickly became both a science and an art. This need for accurate information led to the devising of improved methods for collecting, collating, and analyzing a wide variety of data generated by the day-to-day operations of the enterprise.' The industrial research laboratory created at Penn was part of this move to standardization (through the testing of the standard materials used) and to the control of information flow.

Just like the chemical industry, the move was from standardization within the industry to control of that standardization through the industrial research laboratory to the recognition of the independent value of industrial research as standardizing the natural world in the image of the new social world. The same steps have been traced in military science, stretching back from the pioneering work of Jacquette de Gribeauval in standardization to the creation of global, industrialized armaments industries in the 1860s to the concept of 'command technology' in the 1880s. We have seen in fine the same process occurring with the development of Schlumberger: the oil industry in its expansion outwards to new countries and down to greater depths beneath the surface of the earth sought ways of standardizing its practice, rationalizing the time used by the workforce and the drill-bits. A new standard time was imposed on workers and, on the subsoil, the capricious log kept by the individual driller was replaced by the reliable electrical log produced by Schlumberger – a log, moreover, that could be understood by the middle managers of the oil industry. Statistical methods of comparison of logs became possible. Schlumberger's science was an integral part of the introduction of new working methods to the distribution of energy effected by the oil industry. In general, industrial science was consequent on and reflective of a new way of working; its true filiation is not with the mythology of great scientists but with the might of the Industrial Revolution.

The rise of the industrial research laboratory in America

David Noble remarks that: 'Before 1900 there was very little organized research in American industry, but by 1930 industrial research had become a major economic activity. In a 1928 survey of nearly six hundred manufacturing companies, 52 per cent of the firms reported that research was a company activity, 7 per cent stated that they had established testing laboratories, 29 per cent were supporting cooperative research activities of trade associations, engineering societies, universities, or endowed fellowships, and 11 percent of those doing little or no research indicated they they intended to introduce research work.' Discounting the last figure, he demonstrates a rise from very little organized research to 88 per cent of the 600 surveyed companies being actively engaged in research. Leonard Reich cites a 1931 survey showing that 1600 American companies said that they supported research laboratories and that these employed a total of almost 33,000 people. The research effort was highly concentrated – Noble reckons that in 1938 thirteen companies accounted for one-third of research workers. This effect of concentration can be seen by looking at a few companies. To cite Noble's figures again, the General Electrical laboratory had 8 people on its staff in 1901, 102 in 1906, 301 by 1920 and 555 in 1929. Bell Laborotories was incorporated in 1912, had expenditure totalling $2.2 million in 1916 and $22 million in 1930; its staff totalled over 3600 in 1925. What we can retain from this string of figures is that over the first three decades of the twentieth century there was an exponential increase in industrial scientific work in the United States and this work was concentrated in the large companies.

And this gives us a second reason why industrial science is not spoken about in the annals of history. We saw in the first part of this chapter that it is in the interest of industrial corporations to keep the science that they do close to their chests, to leave its history as open and flexible as possible. We have seen that the closer we look at the sweep of that history, the more we see it departing from the myth of the disinterested pursuit of knowledge by great thinkers. In so doing it also departs from the myth that sustains its own practitioners, one fostered by the industrial corporation and respected by the scientists they employ.

We have seen that industrial science does not have and probably does not want a history. In conclusion then, we can ask ourselves what is the value of giving it one? The principal advantage we can gain from providing such a thing is that we can start to re-integrate science within our culture as a social artefact. This will give us a richer understanding of both our society and the natural world.

21

Joliot: History and Physics mixed together

Bruno Latour

In which we see how what we call society and what we call science were indissolubly combined in the work of numerous scientists, politicians and soldiers. Joliot the physicist contributed to both the history of science and the history of France.

The Example of Joliot

In May 1939 Frédéric Joliot, advised by his friends in the Ministry of War and by André Laugier, the Director of the recently established CNRS (Centre National de la Recherche Scientifique, the National Centre for Scientific Research), entered into an extremely subtle legal agreement with a Belgian company, the Union Minière du Haut-Katanga. Thanks to the discovery of radium by Pierre and Marie Curie and the discovery of uranium deposits in the Congo, this company had become the most important supplier to all the laboratories in the world that were feeling their way towards the production of the first artificial nuclear chain reaction. Joliot, like Marie Curie before him, had found a way of getting the company involved. In fact, the Union Minière only used its radioactive ores to extract the radium, which was then sold to doctors; there were immense heaps of uranium oxide lying about

611

more or less all over its waste sites. For his planned atomic reactor, Joliot needed an enormous quantity of uranium, which made something useful from what until now, for the Union Minière, had been the waste product from radium production. The company promised Joliot five tons of uranium oxide, technical assistance and a million francs. In return, all the French scientists' discoveries would be patented by a syndicate which would distribute the profits fifty-fifty between the Union Minière and the CNRS.

In his laboratory at the Collège de France, Joliot and his two main collaborators, Hans Halban and Lew Kowarski, were looking for an arrangement just as subtle as the one which had brought together the interests of the Ministry of War, the CNRS and the Union Minière. But this time it was a matter of co-ordinating the apparently irreconcilable behaviours of atomic particles. The principle of fission had just been discovered. Bombarded by neutrons, each atom of uranium broke in two, liberating energy. This artificial radioactivity had a consequence which was immediately grasped by several physicists: if under bombardment each atom of uranium gave off two or three other neutrons which in their turn could bombard other atoms of uranium, they would in this way initiate a chain reaction. At the time, this was merely an intellectual possibility, but Joliot's team immediately set to work to prove that such a reaction could be produced and that this would open the way to new scientific discoveries and to a new technique for producing energy in unlimited quantities. The first team able to prove that each generation of neutrons did indeed give birth to an even greater number would gain considerable kudos in the extremely competitive scientific community, in which the French occupied a position of the first rank.

Certain above all that what was at stake was an important scientific discovery, Joliot and his colleagues continued to publish, despite the telegrams Leo Szilard was sending them from America. In 1934 Szilard, an émigré from Hungary and a visionary physicist, had taken out a secret patent on the principles of construction of an atomic bomb. Worried by the idea that the Germans too would develop an atomic bomb as soon as they could be certain that the neutrons emitted were in fact greater than the number at the beginning, Szilard fought to encourage self-censorship by all anti-Nazi researchers. He could not, however, prevent Joliot from publishing a final article in the English journal *Nature* in April 1939, which showed that it might be possible to generate 3.5 neutrons per fission. On reading this article, physicists in Germany, England and the Soviet Union all thought the same thing at the same time: they reorientated their research towards bringing about a chain reaction in practice and immediately wrote to their governments to alert them to the extreme importance of this research, to inform them of its dangers and of the need to supply immediately the enormous resources needed for the first feasibility studies.

Across the world about ten teams became passionately engaged in the attempt to produce the first artificial nuclear chain reaction, but no one yet, except Joliot and his team, was already set up to turn this into industrial or military reality. Joliot's first problem was to slow down the neutrons emitted by the first fissions, for if these were too fast they would not set off the reaction. Joliot and his friends looked for a moderator which could slow down the neutrons without, however, absorbing them or bouncing them back. The ideal moderator would have a set of properties very difficult to reconcile. In the workshop fitted out for the project at Ivry, they tried different moderators and different configurations, for example, paraffin and graphite. It was Halban who drew their attention to the decisive advantages of deuterium, an isotope of hydrogen, twice as heavy but with the same chemical behaviour. It could take the place of hydrogen in the water molecule, which then became 'heavy'. From the work he had done before on heavy water at Copenhagen, Halban knew that it absorbed very few neutrons. Unfortunately, this ideal moderator had one major drawback: there was only one atom of deuterium in every 6000 atoms of hydrogen. It cost a fortune to obtain heavy water, and it was produced on an industrial scale at only one plant in the world, belonging to the Norwegian company, Norsk Hydro Elektrisk, immortalized for cinephiles in the film *La Bataille de l'eau lourde* ('The Battle for Heavy Water').

Raoul Dautry, a *polytechnicien* and senior civil servant who became Minister for Armaments only too shortly before French defeat in World War II, was also kept informed of Joliot's work from the very beginning. He had favoured Joliot's agreement with the Union Minière and did everything he could to support the team at the Collège de France and the early days of the CNRS, as much as the French tradition allowed, integrating military and advanced scientific research. Although he shared none of Joliot's political opinions, he had the same confidence in the progress of knowledge and the same passion for national independence. Joliot promised an experimental reactor for civilian use which might eventually lead to the construction of a new type of armament. Dautry and other technocrats offered him enormous support while asking him to change his priorities: if the bomb was practicable, it was this that must be attended to and very quickly. Halban's calculations on the slowing down of neutrons, Joliot's on the feasibility of the chain reaction, and Dautry's on the necessity of developing new armaments became even more closely entwined when it came to obtaining the heavy water from Norway. Right in the middle of the 'phoney war', spies, bankers, diplomats and German, English, French and Norwegian physicists fought over the twenty-six containers the Norwegians had given the French to prevent the Germans getting hold of it. After an eventful few weeks, the containers were in Joliot's possession. Halban and Kowarski, both foreigners and therefore suspect, had been put out to grass by the French secret service for the duration of the operation. They were authorized to return to the laboratory at the Collège de France and, under the protection of Dautry and of the military, they set to work to combine the uranium from the Union Minière and the heavy water from the Norwegians with the calculations which Halban re-did every day with the confusing data from the Geiger counter.

The History of France and the History of Science

How should one understand this story, so magisterially narrated by the American historian Spencer Weart, of which I have given no more than a summary of a single episode? There is a temptation to divide it into two parts. One could put on one side, in one column, the legal problems with the Union Minière, the 'phoney war', Dautry's nationalism, the German spies . . . In another column one would then put neutrons, deuterium, the absorption coefficient of paraffin . . . One would then have two lists of characters corresponding to two stories and two histories: the first the history of France from 1939 to 1940 and the second the history of science in the same period. One would deal with politics, law, economics, institutions and passions, the other with ideas, principles, knowledge and procedures.

A professional historian would have no problem in dealing with the first list, but would leave the second to the scientists themselves, or to the philosophers of science.

Human actors	Non-human actors
Raoul Dautry	Deuterium
CNRS	Cross-section
Union Minière	Chain reaction

Of course, once this division between human and non-human actors has been produced, there remains a slightly muddled area of hybrids, which might be found perhaps in one column, perhaps in the other, or perhaps in neither: Joliot, Halban, Szilard, their articles, patents, letters and discourse. To deal with this grey and uncertain area, one would have to call sometimes on one column and sometimes on the other. One can say, for example, that Joliot mixed up political concerns with purely scientific interests. Or one might say that the plan to slow down neutrons with deuterium was, of course, a scientific project, but that it happened to be influenced by extra-scientific factors. Szilard's project of self-censorship is not strictly scientific, we say, because it introduces military considerations into the free communication of pure science. What is mixed up is explained by reference to one or other of two equally pure constituents: to put it briefly, politics or science.

We might even imagine two professions of historians, one preferring explanations by pure politics, the other by pure science. The first kind of explanation is usually called externalist and the second internalist. In considering this single period of 1939–40, these two histories can have no intersection. One speaks of Adolf Hitler, Raoul Dautry, Édouard Daladier and the CNRS, but not of neutrons, deuterium or paraffin; the other talks about the principle of the

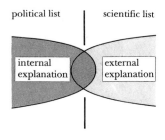

chain reaction, but not about the Union Minière or the banks which owned Norsk Hydro Elektrisk. Like two teams of civil engineers working in two parallel valleys in the Alps, they both do an enormous amount of work without ever meeting or even knowing of each other's existence.

I call the intellectual project whose aim is to resist this division the social history of science. The story of Joliot as told by Spencer Weart is a seamless fabric which cannot be torn in two without making both the politics of the time and the atomic physics equally incomprehensible. Instead of following the parallel valleys, the purpose of the social history of science is to dig a tunnel between them by putting together two teams, each of which attacks the problem from opposite ends, but which hope to meet in the middle. By following Halban's arguments on cross-sections, which conclude on the decisive advantages of deuterium, without prejudice and without a great divide, the historian is led through an imperceptible transition and into Dautry's office and from there into the plane of Jacques Allier, a banker and flying officer who was the secret agent sent by France to outwit the fighters of the Luftwaffe. Starting on the science side of the tunnel, the historian ends up on the other side, with the war and the politics. But in the course of this journey he or she will meet a colleague who has started with the industrial strategy of the Union Minière and ended up, through another imperceptible transition, very interested indeed in the method of extraction of uranium 235 and then in Halban's calculations. Starting on the politics side this historian, willingly or not, becomes involved in mathematics. Instead of recounting two histories which do not intersect at any point, we now have people who tell two symmetrical stories which include the same elements and the same actors, but in the opposite order. The first thought to follow Halban's calculations without having to deal with the Luftwaffe, and the second expected to look at the Union Minière without having to do atomic physics.

They were both mistaken, but the path they traced, thanks to the opening of the tunnel, is much more interesting than they believed at the beginning. In fact, by following without prejudice the interconnected threads of their reasoning, the historians will reveal *a posteriori* the work the scientists and the politicians had to do to become so inextricably bound up together. It wasn't laid down in advance that all the elements of Weart's account should be mixed up together. The Union Minière could have carried on producing

Frédéric Joliot acting out his own story in the laboratory re-created in 1947 for the film La Bataille de l'eau lourde *('The Battle for Heavy Water'): a mixture of physics and politics complicated by a fictional re-creation. (Musée Curie, Paris.)*

and selling copper without bothering about radium or uranium. If Marie Curie and then Frédéric Joliot had not worked at getting the company interested in the work done in their laboratories, an analyst from the Union Minière would never have had to do nuclear physics. When speaking of Joliot, Weart would never have had to speak of the Upper Katanga. Conversely, once he had envisaged the possibility of a chain reaction, Joliot could have directed his research at some other topic, without having to mobilize, in order to produce a reactor, everyone that France had in the way of industrialists and enlightened technocrats. Writing about pre-War France, Weart would not have had to mention Joliot.

In other words, the project of this social history is not to state *a priori* that there exists some connection between science and society, because the existence or otherwise of this connection depends on what the actors have done to establish it or not. Social history merely provides itself with the means of tracing this connection when it exists. Instead of cutting the Gordian knot – on one hand pure science and on the other pure politics – it struggles to unravel it. The social history of the sciences does not say: 'Look for society hidden in, behind or underneath the sciences.' It simply puts forward this principle for unravelling: in a given period, how long can you follow a policy before having to deal with the detailed content of a science? How long can you follow the reasoning of a scientist before having to get involved with the details of a policy? A minute? A century? An eternity? One second? Do not cut the thread of history. All the answers are interesting and count as a major datum for anyone who wishes to understand this imbroglio of things and of people – our own history.

The Translation of Science in History

It is not enough to say that the connections between science and politics form a very tangled web. To refuse any *a priori* division between the list of human or political actors and that of ideas and procedures is no more than a first stage, and entirely negative. We must now be able to understand the series of operations and transformations by which an industrialist who wants to do no more than develop his business finds himself forced to do calculations on the rate of absorption of neutrons by paraffin; or how someone who wanted nothing but a Nobel Prize set about organizing a commando operation in Norway. In both cases the initial vocabulary is different from the final vocabulary. There is a translation of political terms into scientific terms and vice versa. For the managing director of the Union Minière, 'making money' now means, to some extent, 'investing in Joliot's physics'; while for Joliot, 'demonstrating the possibility of a chain reaction' now means in part 'look out for Nazi spies.' It is the analysis of these translation operations which makes up the essence of the social history of science. The idea of translation provides the two teams of historians, one coming from the side of politics and going towards the sciences and the other coming from the side of the sciences and going to meet them, with the guidance and alignment system which gives their project some chance of meeting in the middle. Nothing would be more ridiculous, indeed, than to set two teams to digging a tunnel without providing them with the means of meeting!

Let us follow an elementary operation of translation, so as the understand how in practice one passes from one register to another. Dautry wants to ensure France's military strength and energy self-sufficiency. This is his goal. Joliot wants to be the first in the world to produce controlled artificial nuclear fission in the laboratory. This is his goal. To say of the first ambition

political side   science side

system of alignment

team A   team B

that it is purely political and of the second that it is purely scientific is completely pointless, because it is the 'impurity' alone which will allow both to be attained. Indeed, when Joliot met Dautry he didn't try particularly to change Dautry's goal, but to put his own project in such a way that the nuclear chain reaction became for Dautry the shortest and most certain way of achieving national independence. 'If you use my laboratory,' Joliot says, 'it will be possible to gain a significant lead over other countries, and perhaps to envisage the production of an explosive which goes beyond everything that we know.' This discussion is not of a commercial nature. For Joliot it is not a question of selling nuclear fission. This doesn't yet exist. The only way of making it exist is to obtain from the Minister of Armaments the personnel, the premises and the connections which will enable him, in the middle of war, to obtain the tons of graphite, the uranium and the litres of heavy water that are needed. Both of them believe that, because it is impossible for either to achieve his goal directly, because political and scientific purity is vain, it will be best to negotiate an arrangement.

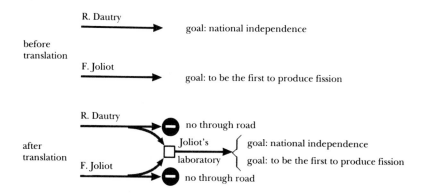

The operation of translation consists of combining two hitherto different interests (making war, slowing down neutrons) so as to make one. Of course, there is no guarantee that one or other party isn't cheating. Dautry may be running down precious resources to enable Joliot to play about with his neutrons while the Germans are massing their tanks in the Ardennes. Conversely, Joliot might have seen himself forced to build the bomb before the civilian reactor. Even if the balance is equal, neither of the parties, as is shown in the diagram, is able to arrive exactly at the goal they had aimed for. There is a drift, a slippage and displacement, which, depending on the case, may be tiny or infinitely large. In the case we are using as an example, Joliot and Dautry only achieved their goal fifteen years later, after a terrible defeat, when General de Gaulle created the CEA, the Commissariat à l'Énergie Atomique (Atomic Energy Commission). What is important in such an operation of translation is not only the fusion of interests which it allows, but the composition of a new mixture, the laboratory. In fact, the shed at Ivry became what allowed the joint realization of the national independence so close to Dautry's heart and Joliot's scientific project. The laboratory's walls, its equipment, its staff and its resources were brought into existence by both Dautry and Joliot. Once the first loop has been knitted, it is no longer possible truly to tell, among the complex of forces mobilized around the copper sphere filled with uranium and paraffin, what belongs to Joliot and what to Dautry.

By itself, to study a single stitch, one negotiation, one meeting, would be useless. In fact, Joliot's labours could not be confined to ministerial offices. Now he had to go and negotiate with the neutrons themselves, and hard. Was it one thing to persuade a minister who wanted to save France to provide a stock of graphite and quite another to persuade a neutron to slow down enough to hit a uranium atom so as to provide three others? Yes and no. For Joliot it wasn't very different. In the morning he dealt with the neutrons and in the afternoon he dealt with the minister. The more time passed, the more these two problems became one: if too many neutrons escaped from the copper vessel and lowered the reaction output, the minister might lose patience. For Joliot, to contain the minister and the neutrons in the same project, to keep them in activity and to keep them under discipline were not really distinct tasks. He needed them both. He crossed and recrossed Paris, moving from mathematics to law and to politics, preventing the others from letting up, sending telegrams to Szilard so that the flow of publications needed for the project would continue, telephoning his legal adviser so that the Union Minière would carry on sending uranium, and recalculating for the nth time the absorption curve obtained with his Geiger counter. This was his *scientific* work: holding together all the threads and getting favours from everybody, neutrons, Norwegians, deuterium, colleagues, anti-Nazis, Americans, paraffin ... To be intelligent, as the name indicates, is to hold all these connections. To understand science is, with Joliot's help (and Weart's), to understand this complex network of connections.

It is now easier to see the difference between the social history of science and the two parallel histories which it replaces. In order to explain all the political and scientific imbroglios, the two teams of historians have always to see in them a regrettable confusion between two equally pure registers. All their explanations were therefore produced in terms of distortion, of impurity, or at best of juxtaposition: to properly scientific factors there were 'also' added purely political or economic factors. Where these historians see only confusion, the social historian sees a continuous and entirely explicable substitution of a certain kind of concern and a certain kind of practice by another. There are in fact moments when, if one holds firmly on to the calculation of the cross-section of deuterium, one also holds, through substitutions and transfers, the fate of France, the future of industry, the destinies of physics, a patent, a good paper, etc.

With the help of another diagram, it is possible to draw out further the contrast between these two poles of research. Here the separation between science and politics that I talked about earlier is visualized in its most common form (above): there is a nucleus of scientific content surrounded by a social, political and cultural environment, the context. On the basis of such a separation, it is possible to offer social explanations or scientific ones. The first use the vocabulary of context and attempt (sometimes) to penetrate as far as they can into the scientific content; the second use the vocabulary of content and remain within the central nucleus. In the first, what explains science is society – although usually only the surface of science is in question; in the second the sciences explain themselves, without anything being left out and without need of external assistance. They are their own commentary upon themselves. They develop from their own inner forces. The social environment can only hinder or encourage their development. It never forms or constitutes the sciences.

In the other, the translation model (below), it is impossible to define a context or content with any precision. The only thing one can say is that the successive chains of translation involve, at one end, exoteric resources (which are more like what we read about in the daily papers), and at the other end, esoteric resources (which are more like what we read about in university textbooks). But the two ends are hardly important. Everything important happens between the two and the same explanations serve to follow the translation in both directions. In this second diagram, the same methods are used to understand science and society – which probably implies that we still understand little enough about science and even less about society.

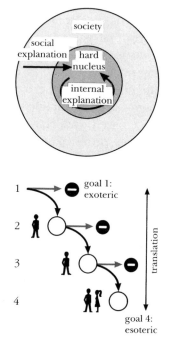

How to Convince

Joliot found himself mixed up at the same time in stories of spies, patents, publications, atoms and calculations. Why this mixture? Deep down, historians of science, like those of politics, would really prefer to do without it. Things would be clearer and narrative more easily managed if, when speaking of the development of societies, one lot could do without science and technology, or if, when they spoke of the development of the sciences, the others could do without all the rest. In order to understand why these two symmetrical dreams are impossible (even though a great deal of the historical literature is written in accordance with these fairy-tales), we must look a little closer at the operation of convincing.

Take the sentence: 'Every neutron in its turn liberates 2.5 neutrons.' This is what one reads today in encyclopaedias. This is what is called a 'scientific' fact. Now let us take another sentence: 'Joliot claims that each neutron liberates 3 to 4 neutrons, but it's impossible; he has no proof; he's far too optimistic; that's the French all over, counting their chickens before they're hatched; and in any case, its incredibly dangerous; if the Germans read that, they'll believe it's all possible and get down to it seriously.'

Unlike the previous sentence, this does not conform to the stylistic rules governing the appearance of scientific facts: it cannot be read in any encyclopaedia. Its dated character is easily visible (somewhere between 1939 and 1940) and it might be ascribed to a fellow physicist (probably Szilard, who had found a haven at that time in Enrico Fermi's laboratory). We may note that both these sentences have a section in common: 'each neutron liberates x neutrons,' the statement; and a very different part, made up of an ensemble of situations, people and judgements, called the modifier. It's enough to clear away this second part for a scientific fact to appear. The elimination of these modifiers is the result and sometimes the goal of scientific controversy, which is thus also the elimination of its own operation. For example, if Joliot and his group have done their work successfully, his colleagues will move on imperceptibly from the second sentence to a third, more respectful one: 'The Joliot team seems to have proved that every neutron liberates three neutrons; that's very interesting.' A few years later we will read sentences like this: 'Numerous experiments have proved that each neutron liberates between 2 and 3 neutrons.' One more effort, and here we are with the phrase we started off with: 'Each neutron liberates 2.5 neutrons.' A little later this sentence, without a trace of qualification, without author, without judgement, without polemic or controversy, without even any allusion to the experimental mechanism which founds it, will enter into a state of even greater certainty. Atomic physicists will not even speak of it, will even stop writing it – except for an introductory course or a popular article – so obvious will it have become. From the most lively controversy to tacit knowledge, the transition is progressive and continuous – at least when everything goes well, which is, of course, very rarely.

How can Joliot get rid of the qualifications with which the scientific fact he wishes to establish is hedged about? It is the answer to this question that explains why there can be no other history of science than the social history (in the sense defined above). Joliot may be convinced in his own mind that the nuclear chain reaction is feasible and that it will lead in a few years to the construction of an atomic reactor. If, however, each time he states this possibility, his colleagues add to what he says qualifications such as: 'It is ridiculous to believe that . . . (statement),' 'It is impossible to think that . . . ,' 'It is dangerous to imagine that . . . ,' 'It is contrary to theory, to claim that . . .' Joliot would have found himself impotent. He cannot by himself transform the statement he is proposing into a scientific fact accepted by the others; by definition, he needs them to bring about this transformation. It was Szilard, himself, who had to admit: 'I am now convinced that Joliot can make his reactor work' even if he immediately added: 'As long as it isn't before the Germans occupy Paris.' In other words, the statement's fate is in the hands of others, dear colleagues, who are for this reason both loved and hated (all the more loved or hated the fewer they are, and the more esoteric or important the statement in question). Here there is no question of a rather regrettable 'social dimension' which proves only that scientists too are only human. Controversy is not something one could do without if researchers were really scientific. One might as well imagine Joliot immediately writing an encyclopaedia article on the operation of a nuclear power station. It is always necessary to convince the others. The others are always there, sceptical, undisciplined, inattentive, uninterested; they form the social group that Joliot cannot do without. If you take a historian, trained in the analysis of hunger riots in the eighteenth century or in the study of the meaning of degeneration in the nineteenth, and put him in front of the article 'Neutron' in the *Larousse Encyclopaedia*, he will feel completely at a loss. On the other hand, if you confront such a person with the controversy about neutrons, then he will have the feeling of knowing what's going on and be able, without stepping much out of character, to go on and retrace this history, full of sound and fury, which is his daily bread.

One can imagine the collective situation as a chain of people (speakers), who pass on a message, a bit like a game of Chinese whispers. Joliot starts the game, and says: 'Each neutron should be able to liberate 4 neutrons; pass on the message.' What will the next colleague say? He is not perhaps a faithful carrier of the message. In fact, he can say a lot of things: first of all, criticizing the statement: 'It's another of the Curies' crazy ideas,' 'That's altogether too optimistic,' 'At the most one might expect 1 neutron, not enough to start off a chain reaction.' He might also, and this is frequent and more serious, not understand the message, or worse, not be interested, and substitute another entirely different: 'I have developed a new standard to define the international unit of radium, pass it on.' He might also pass on the message but attribute it to himself: 'I'm wondering whether each neutron might not liberate 3 or 4 neutrons,' which would pass on the statement as such, but it's no good to Joliot, who will no longer be thought of as the author. If one imagines a fairly long chain, in which each speaker behaves like the second, one will have some idea of the scientific field and of the difficulty of convincing anyone in it. The ideal case, in which each colleague passes the message on to the next, without altering it and while expressing agreement with it, using it and upholding Joliot as its author, is extremely rare.

Joliot, like all researchers, needed the others, needed to discipline them and to convince them; he wasn't able to do without them and lock himself up in the Collège de France, convinced by himself that he was right. But he is not, however, completely without arms of his own. He can introduce other resources into his discussions with colleagues. This is even the reason he was in such a hurry to slow down the neutrons with deuterium. Alone, he could not force his colleagues to believe him. If his reactor could get going for only a few seconds, and if he could get sufficiently clear evidence of this event for no one to accuse him of seeing what he wanted to see, then Joliot would no longer be alone. With him, behind him, disciplined and supervised by his collaborators and properly lined up, the neutrons of the reactor would be contained in the form of a diagram. The experiment, in the shed at Ivry, was very expensive, but it was this expense precisely that would force his esteemed colleagues to take his article in *Nature* seriously. For six months he was the only one in the world to dispose of the material resources allowing him to mobilize both colleagues around and neutrons inside a real reactor. Joliot's opinion might be swept aside with a wave of the hand; Joliot's opinion, supported by Halban's and Kowalski's diagrams, themselves obtained from the copper sphere suspended in the shed at Ivry, could not so easily be cast aside – the proof being that three countries at war, which had hardly done a thing so far, immediately set to work. Disciplining men and mobilizing things, mobilizing things by disciplining men; this is a new way of convincing, sometimes called scientific research.

The imbroglio we started with at the beginning of this chapter is not a regrettable aspect of scientific production, but a result of that very production itself. At every point one finds people and things mixed up, opening a controversy or putting an end to one. If, after Joliot had outlined his project, Dautry had not obtained a favourable reaction from his advisers, Joliot would not have had the resources to mobilize the tons of graphite necessary for his experiment – and, not having been able to convince Dautry's advisers, he would not have been able to convince his own colleagues. It is the same scientific work which leads him to go down to the shed at Ivry, to go up to Dautry's office, to approach his colleagues, to go back over his calculations. The same labour of discipline which leads him to concern himself with the development of the CNRS – without which he would not have physicist colleagues modern enough to be interested in his arguments; to give lectures for the workers in the Communist suburbs – without which there would not have been widespread support for scientific research as a whole; to frequent ministerial offices; to get the directors of the Union Minière to visit his laboratory – without which he could not hope to receive the tons of radioactive waste needed for his reactor; to write articles for *Nature* – without which the very goal of his research would have been in vain; but above all else, this damned reactor must start up. The energy with which Joliot pushed Szilard, Kowarski, Dautry and all the others is proportional to the number of resources and interests he had already mobilized. If the reactor dies off, if each neutron liberates no more than one other neutron, then all these accumulated resources will scatter and disperse. It will no longer be worth going to all this bother. This line of research will be costly, useless or premature. Can one say that such an assessment is scientific, pure, applied, political or military? It doesn't matter; such a division has no interest. On the other hand, the work by which the problem of national independence becomes a problem of slowing down neutrons, that is important.

History of Science or History of Scientists?

Translation operations transform questions of politics into questions of technique and vice versa; operations of conviction mobilize a mixture of human and non-human agents in a controversy. The result of these two operations obliges us to define a sort of 'right of pursuit': there can be no comprehensible general history if the historian does not follow all the scientific and technical contents which have become indispensable to the unfolding of this history; there can be no history of science if the historian does not rediscover the multiplicity of agents, resources and goals with which it is involved. Instead of defining *a priori* the distance between the nucleus and the context, a distance which will render incomprehensible the numerous short-circuits between ministers and neutrons, the social history of the sciences provides itself with leads, nodes and paths. The historian does not have to establish in advance the distance which will allow us to travel, as if through successive circles, from the hell of social relations to the empyrean of mathematical theory. Neither does he or she have to define, in advance, a continuous and repetitive *rapprochement* which sees society always underlying science. In this history full of sound and fury, it's less a question of distance than of often unpredictable and heterogeneous connections. Sometimes one may be able to follow reasoning for several minutes by going from one equation to another, then jump suddenly to a problem of national defence, then turn rapidly towards the oil and grease of an item of equipment before slipping off, just as suddenly, into a long series of technical speculations, to return after a little while to questions of offended professional honour or perhaps to the big money.

If it is impossible, by definition, to give a general description once and for all of the unpredictable and heterogeneous links which explain the formation of a given technical content, it is possible to outline rapidly the different preoccupations that all researchers will simultaneously have in mind. For this overview, it will be enough to return to the episode in Joliot's career that we have already been using as an example. All at the same time, Joliot must get the reactor to work; convince his colleagues; interest the military, politicians and industrialists; give the public a positive image of his activities; and, last but not least, he must understand what is happening to these neutrons that have become so important. These are five perspectives which provide a pretty good framework for the historian's work: the instruments, colleagues, allies, public, and finally, what I will call the links or bonds, so as to avoid the historical baggage associated with the words content and concept. Each of these five activities is as important as the others and each reacts on itself and on the four others: without allies no graphite, and thus no reactor; without colleagues, no favourable opinion from Dautry and thus, no graphite; without a way of calculating the neutrons' rate of reproduction, no assessment of the reactor, so no proof, and thus no colleagues convinced. Joliot's work can be schematized in the diagram below, distinguishing it once again from the model made up of a nucleus and a context.

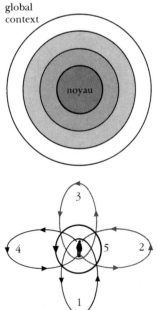

global
context

noyau

What Joliot (in the centre below) has to do is to find a way of keeping together, all at the same time, his instruments, his colleagues, the interests he has got involved and the public; he can only succeed by understanding the chain reaction and quickly, before Szilard does it himself, before the Germans arrive in Paris, before the 200 litres of heavy water from the Norwegians run out and before Halban and Kowarski are obliged to flee, denounced by their neighbours as foreigners. There is indeed a nucleus, as in the diagram, but this is not defined by the preoccupations furthest away from the others; on the contrary, it is what keeps them all together, to strengthen their cohesion, to accelerate their circulation. There's something about this famous nucleus which makes one think of a motorway interchange; this fifth circle is what allows one to move rapidly from one preoccupation to another. All the care one devotes to it, all the energy one spends in discovering it and attending to it, is not a function, as in the other model, of its distance, but of the number of heterogeneous elements which it is capable of holding together. If etymology has any meaning, the content, paradoxically, should be called the container, which 'holds together'.

This paradox appears to be one only because of our habit of thinking of the history of science in terms of the model above, that is to say, in terms of nucleus and context. In this model it seems absurd to consider the most technical aspects of Joliot's reasoning as those which are most directly connected to social, political and instrumental preoccupations. However, the choice between the two models can be made quickly enough if one is a historian and if one is therefore interested in the growth and development of the socio-technical imbroglio. Why is there a nucleus or a fifth circle? The supporters of the first model – some faithful to the context, others to the content – have real difficulty in answering this question because they deal with two histories, one of the context and the other of the content, each of which has its own logics, dynamics and periods.

At best, the two groups could give two replies, one from the side of science, the other from the side of society; at the worst, they might give none, except that there exists a nucleus, because, when everything is considered, the sciences are outside time. For a social history, on the other hand, there is in principle only one answer to this question: the existence of a nucleus, of an esoteric technical content, is a direct function of the amplitude of the other circles. If one considers, for example, the growth of Joliot's research programme from the discovery of artificial radioactivity to the fifties and the creation of the CEA and the French nuclear arms programme, one might define very roughly the several states of technique, but it would also be necessary to define several ensembles of equipment, alliances and policies. These grow with the others. Rather, they grow because the others grow. At the very beginning the discovery of artificial radioactivity mobilized a cloud-chamber,[1] a few collaborators, the Nobel committee, the French physicist Jean Perrin and a few others. At the end, the French nuclear programme mobilized the CEA, a new profession of atomic physicists, General de Gaulle and all the international relations of the Cold War. Was there a tremendous increase in technical content? There was indeed, but now it is the whole of France that must be contained. It is the same with the central circle as it is with junctions on roads: simple crossroads are enough for a couple of minor roads, but if vast eight-lane motorways are to have a junction, then a sophisticated interchange is necessary. Those who want to have two histories, one for society and the other for science, are like those who claim to understand the history of road junctions without having to deal with roads, or even more bizarre, the history of roads without the junctions! Technical contents are not astounding mysteries put in the way of historians by the gods so as to humble them by reminding them of the existence of another world, a world which escapes from history; nor are they provided just for the epistemologists to enable them to look down on all those who are ignorant of science. They are part of this world. They only grow there because they are partly what makes it up.

An Abridged Social History of Science

Now we can see what society is doing in science: it is there because science and society are two versions of the same thing in two different translations. We also have a clearer idea of the project of the social history of the sciences, of the main ideas it uses and of the types of connection that it attempts to describe – and in doing this, we have seen more clearly the parallel histories from which it must be distinguished. By generalizing the schema that we have developed in studying the example of Joliot, we can now give a rough sketch of the field this history covers and the type of objects it uses – objects, which for the most part are already familiar to historians. To organize and simplify this overview, I will content myself with a very brief description of the five circles whose ensemble defines the state of the operations of translation and persuasion proper to a scientific community. Of course, such a description has no meaning outside a particular empirical example, but it is none the less not entirely useless to consider, if only rapidly, the whole field of the social history of science, simply to remind us of the immensity of which we remain ignorant. It goes without saying that to describe this sort of rose pattern, all starting-points are equally valid, as long as it is covered in its entirety.

Mobilization of the World

The first field of the social history of science concerns the mobilization of the world and bringing it into contact with controversy. I will call this circle that of mobilization. It is a matter of moving towards the world, making it mobile, bringing it to the site of controversy, keeping it engaged and making it available for rhetorical use. In certain disciplines, such as Joliot's nuclear physics, it's a question of doing the history of the instruments and the great installations, which, since World War II, have made up the history of megascience. For many others it is a question of the history of expeditions sent across the world over the past three or sometimes four centuries to bring back plants, animals, trophies and cartographical observations. For other sciences, it will not be a matter of instruments or expeditions, but the history of surveys which have allowed the accumulation of knowledge about the state of a society or an economy. In all these cases, it is a matter of operating what Immanuel Kant, as a philosopher, called a Copernican Revolution. Instead of moving around the world's objects, the scientist makes them move around him. The geologist is lost in the indecipherable landscape which he surveys, hammer in hand. Once all the geological formations are mapped, once the geologist has surveyed them and, field-guide in hand, has assembled an ordered and labelled collection of specimens, which are then brought together in a single place, he or she will have become the master of the Earth and its history.

The historian of this first circle has to deal with expeditions and surveys, with instruments and with large-scale equipment, but also with the sites where all the objects of the world thus mobilized have been assembled and contained. The galleries of the Muséum d'Histoire Naturelle, the collections of the Musée de l'Homme, the maps of the Service Géographique, the databases of the CNRS, the files of the police and the equipment of the physiology laboratories of the Collège de France: all these are so many necessary objects of study for those who wish to understand through what mediation people speaking to each other as people begin little by little to speak of things. Thanks to a new survey and new data, an economist hitherto without resources will start spitting out reliable statistics at thousands of columns a minute. An ecologist nobody really used to take seriously is now able to intervene in a debate with beautiful satellite photographs in artificial colours, which allow him, without budging from the Jussieu laboratory, to observe the advance of the desert in Burkina Faso. A doctor, used until now to treating clients case by case at the surgery, now finds herself confronted with tables of symptoms based on hundreds of cases, obligingly provided by the hospital service. If we want to understand why these people begin to talk more loudly and with more assurance, we have to follow the story of this mobilization of the world, thanks to which things now present themselves in a form which allows them to be used in their arguments.

Apart from the instruments – in the widest sense – and the places in which they are brought together, the historian of the first circle also has to deal with the metrological systems which make possible a regular supply of data. By this I mean not only metrology in the narrow sense, that is to say the maintenance of constants of measurement, but also in a slightly wider sense: the establishment and maintenance of chains of equivalence. In the narrow sense there is the history of the calculation of weights and measures and, more widely, the history of the calculation of costs and prices. One can study the development of the occupational categories of the INSEE (Institut National de la Statistique et des Études Économiques, National Institute for Statistics and Economic Research), and also the professional training of naturalists sent on expeditions, to enable them to collect their specimens without damaging them, as well as the way that signals from atomic clocks maintain through the years the rhythm of universal time. Places like the Greenwich Observatory or the one in Paris will be the object of monographs as detailed as those on the Muséum National d'Histoire Naturelle or the famous Botanic Gardens at Kew in the western suburbs of London. To sum it up in one sentence, the history of the first circle is the history of the transformation of the world into mobile, stable and combinable elements. If you like, it is the history of the writing of the 'great book of nature' in characters legible to scientists. It is the history of logistics. Logistics, de Gaulle said, will always follow, but the world?

Autonomization

To persuade one needs data, but also someone to persuade. The object of the historians of the second circle is to show how a researcher finds colleagues. I call this circle the circle of autonomy, because it concerns the way in which a discipline, a profession, a clique or an invisible college becomes independent and forms its own criteria of evaluation and relevance. We always forget that specialists are produced from amateurs in the same way as soldiers are made out of civilians. There have not always been scientists and researchers. It was necessary, with great difficulty, to extract chemists from alchemists, economists from jurists, sociologists from philosophers; or to obtain the subtle mixtures which are neurobiologists from biologists and neurologists, as well as social psychologists from psychologists and sociologists. The conflict of disciplines is not a damaging element of science, but one of its motors. The only way of increasing the cost of an experiment presupposes a colleague capable both of criticizing and using it. What would be the use of obtaining ten million coloured images from a satellite, if there were only two specialists in the world who could interpret them? An isolated specialist is a contradiction in terms. No one can specialize without the concurrent autonomization of a small group of peers.

The history of scientific professions (preferred in the English-speaking countries) and the history of scientific disciplines (preferred in France) is certainly the most highly developed part of the social history of the sciences. It deals with both the history of the corps and that of the learned societies, with great associations like the societies for the advancement of science and with little cliques and other clusters which form the grain of relationships between researchers. More generally, it deals with what allows one to distinguish, in the course of history, between a scientist and a virtuoso, an intellectual or an amateur. How does one establish the values of a new profession, the meticulous control of titles and of barriers to entry? How does one impose a monopoly of competence, how regulate the internal demography and place the students and disciples? How does one resolve the innumerable conflicts of competence between the profession and its neighbouring disciplines?

In addition to the history of professions and disciplines, the second circle also includes the history of scientific institutions. There must be organizations, resources, statutes and regulations in order to keep the crowds of colleagues together. It isn't possible to think of French science without the Académie, the Institut, the *grandes écoles*, the CNRS, the Bureau de Recherches Géologiques et Minières and the Ponts et Chaussées. The institutions are as

629

necessary for the resolution of controversies as is the regular flow of data obtained in the first circle.

Alliances

No instruments can be developed, no discipline can become autonomous, no new institution can be founded without the third circle, which I call the circle of alliances. Groups which previously couldn't care tuppence must be made to be interested in controversy. The military must be made interested in physics, industrialists have to get interested in chemistry, kings in cartography, teachers in educational theory, parliamentarians in political science . . . Without this labour of making people interested, the other circles would be no more than armchair travelling; without colleagues and without a world, the researcher won't cost a lot but won't be worth a lot either. Immense groups, rich and well-endowed, must be mobilized for scientific work to develop on any scale, for expeditions to multiply and go further abroad, for institutions to grow, for professions to develop, for professorial chairs and other positions to open up. We should remember that here we have the operation of translation, and it is impossible to say in advance who will win and who will lose in these alliances, who is taking whom for a ride, who reaches their goals and who gets side-tracked. It's not a matter of studying the impact of the economic base on the development of the scientific superstructure, but of how an industrialist transforms his products by investing in a solid-state physics laboratory, how a state geological service expands by attaching itself to a department of transport. It is also a matter, at the conclusion of highly complex translation operations of every kind (involving connivance, treason, contracts or runaway enthusiasm) of studying the question of global responsibility. Who led whom? Who was the unmoved mover in all the agitation? Science or politics? Administration or knowledge? Law or fact? This new controversy, which refers not to the composition of alliances but to their final outcome, is added to all the others, and its effects are to speed up or slow down overall scientific activity.

Depending on the circumstances, these alliances can take innumerable forms, but there are four which cover most of the history of science: alliances with the State, with the Army, with industry and with the educational system. The first is dealt with by the study of the 'technocracy' and the 'bureaucracy', that is to say, the creation of a governmental power which is at the same time scientifically competent. Innumerable disciplines, some prestigious, others less so, have found a home in the administration, adding their own weight to the State apparatus. From the point of view of numbers and of scale, it is the next two, almost inseparable one from the other, which are the most important; no science, or at least hardly any, without the Army and without industry, hardly more than a handful of scientists. This enormous labour of persuasion and liaison isn't self-evident; there is no natural connection between a military man and a chemical molecule, between an industrialist and an electron. They didn't meet each other according to some natural inclination. This inclination had to be created, the social and material world had to be worked on to make these alliances inevitable. This presents an immense and passionately interesting history, probably the most important for understanding our own societies, and which nearly all still remains to be written. The fourth alliance, finally, less spectacular, is just as necessary to the reproduction and extension of the other circles. Nothing prepares the town or country child to receive and absorb either mathematics, chemistry, physics, natural sciences or literary criticism. Without the enormous resonating chamber of the educational system, even if all the rest were in place, the sciences would remain incomprehensible, isolated, perhaps even suspect. The history of the relationship between a discipline and the educational system is one of the decisive elements fortunately more closely studied than many other fields of social history by the historians of education.

Representation

Even if the instruments were in place, if peers had been trained and disciplined, if well-endowed institutions offered a home to this wonderful world of colleagues and collections, and if State, industry, Army and education offered the sciences wide support, there would still be a great deal of work to be done. This whole mobilization of novel objects – atoms, fossils, bombs, radar and new mathematics – all this agitation and all these controversies would overturn the normal system of belief and opinion. It would be astonishing if it were otherwise, for what else is science for? The same scientists who had to travel the world to make it mobile, to convince colleagues and lay siege to ministers and to boards of directors, now have to deal with their relations with the public. I call the history of this fourth circle the history of representation. Here we find the history of the ways that societies have represented scientific certainties to themselves, the history of their spontaneous epistemology. What trust was accorded to science? How can this confidence be measured in different periods and for different disciplines? Here also is the troubled and revealing history of society's reception of a new theory or discipline. How was Isaac Newton's theory received in France? How was Charles Darwin's theory taken on board by English clerics? How was Taylorism accepted by French trade-unionists during the Great War? How did the economy little by little become one of the stock topics of journalism? How was psychoanalysis gradually absorbed into day-to-day teaching practice?

But the most important question in this circle, a question still untouched, is that of the active resistance offered by millions of people to the growth, privileges and pretensions of innumerable scientific disciplines. How is it that people don't believe, do not understand, do not want the results of scientific controversy that scientists want so much to see us pass on as if we were so many faithful and reliable conductors. I say that this history is untouched, because the learned always think it a scandal that their education is not universally shared. Unaware of the enormous labour needed for the extension of their knowledge, they never see the resistance of the multitude as another kind of work, as interesting to study as their own, even if it is aimed at undermining their morale. They speak of ignorance and popularization instead of extension and active resistance. The history of science, from the point of view of those who actively reject it, is still to be made although, by definition, it is part of a system of research.

Ties and Bonds

To reach the fifth circle is not to reach science at last. From the first circle on, not for a moment have we left the course of scientific intelligence at work. Even so, reaching the circle that I call, for want of a better label, the one of ties and bonds, is teaching something much harder. We know the reason for this extra hardness. Taking hold at the same time of all the resources mobilized in the other four circles is no picnic. Now it's a matter of taking up all these threads that are still scattered and tying them firmly together, before they give way to centrifugal force. This whole heterogeneous heap asks only to be revealed; the world wants it to go back to being distant and indecipherable; colleagues do only what they like; allies lose patience or interest; the public longs not to understand. The more aspects collected, the more necessary is an idea, an argument, a theory to bring them all together. The strength of the links is what will allow a long-lasting connection.

The Enucleation of History

This essential connection between difficulty and duration is why the history of science is so hard to practise. Indeed, it is by the work belonging to the fifth circle that, for the first time, looser networks can be distinguished from more tightly knit ones. The first will give us what philosophers and historians call circumstances or historical contingencies, the second necessity. Not only is social history stronger than those it claims to replace, sticking closer to the scientific intelligence at work, not only is it more reasonable, but in addition, it is capable of understanding as a result of what events have occurred in the other models of science and for what reasons.

In fact, if one does not pay very close attention to the entirety of the scientific endeavour – symbolized by the rose illustrated on p. 625 – one might have the impression that there exists on one hand a history of contingencies

– the corona – and on the other hand, at the centre, a necessity which itself is not historical. Here it needs only the slightest lapse of attention, the slightest carelessness, and that's it! The closest links will be cut and distanced from those things which they connect and resemble. A little more inattention, and the nucleus of scientific content is separated from what becomes, by contrast, a contingent historical 'context'.

The model of content and context that I have so extensively criticized is obtained from the heterogeneous and multiple labour of the scientists by inattention and cutting up. The whole of this labour then becomes opaque, because one no longer sees the essential point, which is what the theories and concepts theorize and bring together. Instead of a continuous and tortuous path, the historian sees only an iron curtain separating the sciences and extra-scientific factors. Just as, in the centre of Berlin, a wall of shame used to divide the fine fabric of lanes and neighbourhoods. Not understanding either the theories or what they are theories of, the historian, discouraged in the face of these objects so hard and so durable that they seem to come from somewhere else, can only send them to a Platonic heaven and connect them together in an entirely fantastical history, sometimes called 'history of science', despite the fact that there is no longer anything historical and thus nothing scientific about it either. The damage has been done; long trajectories of ideas and principles traverse contingent history like so many foreign bodies. Historians, however, used to studying all the collective factors I have just listed, are then discouraged in the face of all this strangeness and leave the sciences to the scientists and philosophers, modestly contenting themselves with the study of battles, of daily life, of popular belief and the price of wheat.

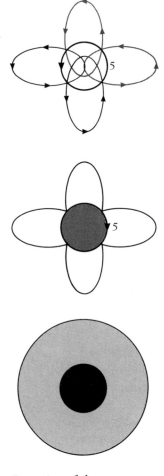

Extraction of the content/context model.

This modesty would do them honour if, in abandoning scientific and technical contents, they had not also rendered equally incomprehensible this small-scale history which they claim to study and to which they claim to restrict themselves. Indeed, what is most serious in this separation between the nucleus and its corona, of theories from what they theorize, is not that it enables this endless unfolding of an intellectual history of scientific ideas. It lies in the exactly corresponding belief among historians that by lining up previously 'enucleated' contexts it is possible to account for the social history of our societies without having to deal with science and technology. The first situation, which results in the dreams of epistemology, is simply annoying and puerile; the second, which results in the illusion of a social world, far too social, is far more damaging. It is the whole of modern history which is made impossible to understand.

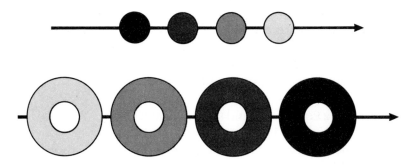

The creation of an 'intellectual history of science' by the alignment of nuclei ... and also of a social history of science by the alignment of contents.

Let us suppose, for example, that translation operations have made Joliot's laboratory indispensable to the conduct of French military affairs, and that Joliot himself could not get his reactor to start except by discovering a new radioactive element, plutonium, which starts off this reaction far more easily. The historian of military affairs, following the series of translations, must inevitably become interested in the history of plutonium; more precisely, this inevitability is a function of Joliot's work and his success. Given scientists' activity over the last three or four hundred years, how long can one study a military man before finding oneself in the laboratory? This would have to be settled by investigation, but to be going on with, let us say a quarter of an hour. Consequently, to write military history and not look at the laboratories that make up this history is an absurdity. It is not a matter of knowing whether or not one has the right to approach history without paying attention to science and technology; it is a question of fact, whether or not the players followed by historians mixed their lives and their feelings with non-human players mobilized by laboratories and scientific professions. If the answer is yes, it is unthinkable not to put back in the game the plutonium that Joliot and the military used in their fashion to make war and peace.

The drama of a previous enucleation of history is to make science appear incomprehensible, but above all to make social science imbecile, in its etymological sense of lacking support. Why did the French military concern themselves with the controversy between Joliot and his colleagues in order to resolve the differences between themselves and their German or English adversaries? The same question might be asked of Joliot: why did he have to go through the calculation of the cross-sections of deuterium before resolving the problem of the possibility of a chain reaction? The obvious response is too schematic all through: because, once in possession of deuterium, Joliot made the position irreversible. It was possible to doubt the reaction before 1940 but it could no longer be doubted afterwards. The cost of the test had grown enormously. Going on with physics without an atomic pile was becoming impossible for all the colleagues. But the reply to the first question was in the same general form: once sure of the feasibility of an atomic bomb, the French military also wanted to make their position irreversible. People could make fun of them with their carrier pigeons and their Maginot Line full of holes like a sieve; they could be turned back and routed. This became much more difficult once they had atomic weapons, but the cost of the war would be vastly increased. It became impossible to make war without atomic physicists. You wanted to make war with the Maginot Line, you have been overturned. You will have to end it with atomic physics. Reliable, disciplined reserves would not be found among generals ready to betray, but among neutrons. A general and a Maginot Line and two or three traitors made a reversible position. A general plus the CEA might make a strong position, in any case more difficult to reverse (whatever the number of traitors and the accusations brought later against the loyalty of Joliot during the Cold War).

A historian who is deprived of the non-human factors mobilized by science and technology in human battles is forbidden to understand irreversibility, that is to say the very passage of time, or what is historical in history. If history was social in the ordinary sense of the word, that is to say the product simply of human beings, it would be reversible, entirely reversible, and nowhere would it manifest the acute passage of time.

One can see that the history of science is not a particular branch of general history, like a history of costume, of climate, of tears, of fear, or of electoral law. Too easily we define the terrain of history in opposition to those rocks that nothing can erode, the scientific facts. The history of science, here redefined, forces one to consider the preliminary separation between what is or what has a history (short or long, it matters not) and what cannot have one. In other words, it has made us look again at the division between what is contingent and what is necessary, between what belongs to people and what to things.

22

The Invention of the Computer

Pierre Lévy

In which we see that the computer makes its appearance only at the end of a whole series of twists and turns, of reinterpretations of heterogeneous materials and of different technical arrangements, at the conclusion of a chance succession of opportunities and local circumstances, more or less successfully exploited by a multitude of actors.

Histories of Computing

The immediate postwar period: in everyone's memory the monstrous atomic mushroom clouds rise into the sky above Hiroshima and Nagasaki. From the concentration camps liberated by the Allies come unbearable images. They say that American scientists have created an electronic brain, perhaps precisely to build the bomb. Perhaps the computer was born here, within the barbed wire at Los Alamos,[1] just before the Germans built one themselves, with the debris of a giant machine lying in the ruins of a bombed Berlin. One can imagine no more emotional a birth for this symbol of technological modernity; as if the absolute and horrifying evil revealed to humanity in the course of that war might be redeemed by the spin-off from military research; as if the bomb and the computer, baptized in the same river of blood, should

636

now watch over our well-being, one indefinitely delaying the next world conflict, the other by multiplying our intelligence and our powers.

The Founding Event

A history of the beginning of computing could put World War II among the main 'causes' of the invention of the computer. It is a condition, an enormous determining factor on the same scale as the economic and cultural transformations brought about by the computer. Wouldn't it have been necessary to mobilize all the intellectual and financial resources of nations at war to produce a tool of such power? This is the thesis of the foundational event; but it suffers from two weaknesses. At the level of its underlying intention, it isn't always true that causes and effects are proportional to each other. Small events can have important consequences. But, above all, the thesis of the founding war does not stand up very well to a detailed analysis of the forerunners of automatic calculation in the thirties and forties. The war did not favour the invention of computers in a simple and unequivocal manner. It did not produce them in Japan. In Germany the career of Konrad Zuse, the inventor of the first programmable binary electromechanical machines, was almost interrupted by his conscription. But thanks to the support of his friend Helmut Schreyer, an engineer and a Nazi, he returned from the front and reversed the situation by getting hold of military funds to use them for his own ends. Not without set-backs: neither Zuse nor Schreyer was able to convince the authorities of the need to begin construction of an ultra-rapid electronic machine. The General Staff refused to finance a project of no immediate benefit to the war effort. But Zuse and Schreyer did at least manage to convert the war into good conditions of work ... until Allied bombing destroyed the machines and suddenly gave the world conflict another meaning than that of an opportunity for funding.

In France, Couffignal's project for a programmable electro-mechanical calculating machine, in association with the Logabax company, was interrupted by the outbreak of hostilities.

In the United States, John Atanasoff's conscription into a US Navy research centre put a sudden end to the construction of his electronic calculating machine, already far advanced. In this case, the mathematician's ingenuity had been tapped by the military, rather than the inventor tapping military funds. Still in the United States, the International Business Machines (IBM) company interrupted research on an electronic multiplier being done by its engineers, Ralph Palmer and Byron Phelps. Priority had to be given to government orders orientated towards military applications.

The British case also demonstrates the ambiguous role played by World War II. Brought together by the government code-breaking service, a large team of scientists and technicians succeeded in building the first large-scale programmable electronic calculators. Great Britain was thus provided with an invaluable capital of experience and know-how in the field. But the ultra-secret and purely military character of the project proved to be double-edged. It prevented the English from pressing home their advantage. Unlike the Americans, they had difficulty in moving into the scientific and industrial registers, which implied a certain openness about what had been achieved.

Two factors connected with the conflict did, however, play an undoubtedly positive role in technological discovery generally and in the appearance of electronic calculators in particular. First of all, of course, there was considerable growth in the sums allocated to research and development. Military funds allowed certain teams to make remarkable break-throughs in the field of electronic calculation. Second, large concentrations of researchers were trained and worked together on big projects, as at Los Alamos, or in the construction of improved radar systems. The first large electronic calculators were built within such concentrations: at Bletchley Park, the home of the British code-breaking service, at the Ballistics Research Laboratory (BRL) in the United States. These concentrations reached a 'critical mass' particularly favourable to invention. In fact, they created the conditions in which a mass of ideas, theories and technological items could be hijacked, reinterpreted and captured for different purposes by a multitude of actors. These chain reactions within the scientific-technical order make one think of those occurring in reactors or atomic bombs. Atoms of uranium only begin to emit and receive enough neutrons to give off usable energy when they are assembled in a sufficiently dense mass. But the relationship between the first electronic calculators and the uranium bomb which exploded above Hiroshima stops with this metaphor. The atomic bomb was built without the help of any computer.

Forerunners and Founding Geniuses

A second, altogether classical, temptation presents itself to the historian of science and technology who wishes to deal with the invention of the computer: that of the founding individual (or genius indeed). Here many candidates jostle together, among them Charles Babbage, Alan Turing and John von Neumann, each one with different reasons for being considered for the title.

A scientist of great ability, a mathematician, astronomer and economist involved in the whole of British scientific life in the first half of the nineteenth century, Babbage (1792–1871) would make an excellent forerunner. He owes the privilege of appearing in all the histories of computing to his having designed two calculating machines, the Difference Engine and the Analytical Engine. The first, a calculator connected to a printer, specialized in certain operations particularly frequent in astronomical calculation and in drawing up mathematical tables (logarithms, sines, square roots, etc.). Babbage thought of it around 1830. Its power and complexity far exceeded that of the adding and multiplying machines available at the time. He did not succeed in building it, despite the substantial sums of money he received for it, among them grants from the British government. It was built only thirteen years later, by the Swede P. G. Scheutz; a few more examples were copied up to the beginning of the twentieth century. The plans for the Analytical Engine, which was never completed, describe a non-specialized calculator whose layout seems to prefigure that of the computer. In fact, Babbage separated the functions of memory and calculation for the first time. In particular, he provided for operation under the control of programmes, coded onto punched cards.

It is only one step to go on from here to claim that the English astronomer invented the principle of the computer but was unable successfully to bring it about in practice, because he was too far ahead of the ideas of his time and of the technological possibilities of his period . . . and it is a step we shall not take.

First of all, Babbage's project belongs to an intellectual universe very different from that of the 1950s, when the first computers appeared. For him there was no question of constructing a mechanical brain, nor even a universal machine capable of capable of processing information in any way at all. In his mind, the Analytical Engine was not a support for a programme, in the way that computers are for us.

Babbage's objective, above all, was the construction of accurate nautical, astronomical and mathematical tables, those of his own time being riddled with errors. To construct precise numerical tables, he had at the same time to mechanize calculation and printing. This wasn't a matter of a thinking machine, but a proposal for an industrial method to shorten production times, eliminate errors in calculation and typesetting, and reduce the publishing costs of tables.

Babbage did not envisage the complete suppression of human intervention in operations of calculation. The Analytical Engine was an enormous instrument for calculating and printing in the hands of specialized personnel. In his writing there was never any question of automation as we understand it today. The Analytical Engine, which was to have been powered by steam, certainly belongs to the industrial atmosphere of the nineteenth century.

It is usual to define the computer as a programmable electronic calculator with stored programmes. The idea of the stored programme is fundamental for, as we shall see, this is what accounts for the flexibility and truly universal character of the computer. Now the stored programme is only a technical advantage when very high speeds of calculation are available, and thus only after the appearance of electronic machines. The computer *per se* was therefore unimaginable for Babbage. Certain marginal notes in the writings

639

of Lady Augusta Lovelace (Babbage's main collaborator) can be read as allusions to the possibility of storing the programmes for the Analytical Engine. But the meaning this could have when the machine had only an internal mechanical memory, and therefore a very slow one, cannot be identified with the meaning it has for us.

Inventor, founder or forerunner, a great man's action is supposed to look towards the future. It heralds, precedes, inaugurates. But looking backwards shows that the founder is also an inheritor, that he uses, converts and re-employs a host of ideas and materials already available at the time.

In Babbage's case, the idea of the programmable machine and even of the punched card as support for the programme were obviously borrowed from Jacquard, whose invention triumphed in the textile industry at the beginning of the nineteenth century. Babbage had studied such machinery in every detail. (One might mention that he himself possessed a portrait of Jacquard woven by means of a programme made up of 24,000 punched cards.) We should note too that, quite independently, punched cards were converted from their initial use a second time, around 1883, by Hermann Hollerith, the inventor of mechanical data-processing and founder of the company that would become IBM. Punched cards were still in use in computing during the 1970s, when they had already all but disappeared from industry a long time before, though they are still used in carpet manufacture.

Another borrowing was from Marie Riche, Baron de Prony's idea for a 'logarithm factory'. De Prony, the Director of the Cadastre at Paris at the turn of the eighteenth and nineteenth centuries, had organized a real factory for calculating mathematical tables by the strictest application of the principle of the division of labour. Following the method of calculation by finite differences, a general plan of work was developed by mathematicians, worked out in more detail by mathematicians of a lower grade, and executed by operatives who could do no more than perform additions. Babbage, who had met de Prony, drew on this directly for his Difference Engine and indirectly for the Analytical Engine. A series of adders could perform the most complex calculations, provided they followed rigid and well-designed plans. Babbage substituted mechanical adders, already commonly used, for human ones.

One could add to the list of re-uses, borrowings and conversions by Babbage, who was an indefatigable observer of all the machines and working methods of his time, in both the scientific and industrial fields. But despite all the debts he owed to his period and to his forerunners, surely the creator (and Babbage cannot be denied this title) produced something original himself? Certainly he did, and precisely by giving a new meaning to the heterogeneous materials and ideas which he re-used and adapted to his own purposes. That is why the idea of a founder or forerunner is useless. For the actors who come afterwards, the work of the so-called founder or forerunner is in its turn material to be re-used. The point where he had proved creative, the new meaning imposed on a multitude of different materials, is exactly what will be the least respected, for the innovative arrangement will itself be material for reinterpretation, an entity whose original meaning will be forced, converted and distorted to incorporate it in a new achievement and new projects. Babbage might well be presented as the venerable ancestor of computer scientists, but he could just as well be considered as a ruin from which stones were stolen to build a temple to a new religion.

And what is more, to continue with the metaphor, the stones in question will be used far more to decorate the façade afterwards than to hold the roof up. In fact, Babbage would directly inspire the construction of ten or so machines in the nineteenth century, but the effective influence of his work on the inventors of the great digital calculators[2] of the 1930s and the computers of the forties was almost non-existent.

The business of Alan Turing is more difficult. Turing (1912–54) was an English mathematician and logician whose strange and tragic death corresponds well enough to the stereotype of the sacrificed scientist, illustrated in this book by Archimedes and Lavoisier (see pp. 124 and 455). In this respect he is a particularly well-qualified candidate for the founder's role. Found guilty of sex with another man by a British court, he was given a choice between prison or hormone injections, supposed to improve his condition. He opted for the hormones. He killed himself two years later by eating an apple poisoned with cyanide. His biographer, Andrew Hodges, suggests that the British and American services, whom he had served as an expert during the war, had something to do with this brutal death, because they feared that the Russians might use his 'weak spot' to obtain information from him or make him change sides.

Turing comes into the history of computing first of all as a pure theorist. In 1936 he published an article on a complex problem of mathematical logic. This article contained the description of an abstract automaton – the universal machine – capable of performing all imaginable calculations. Thanks to the precision of his description, Turing succeeded in demonstrating the existence of problems insoluble by the universal machine, and therefore insoluble by any method of calculation whatever.

Some years later Turing contributed in various ways to the design of English electronic calculators and then, after the war, to the construction of some of the first computers.

From a purely formal point of view, and restricting oneself to the question of calculability, the universal machine described in the 1936 article is the exact logical equivalent of the computer, although it precedes it by a decade. One finds in it, in particular, the idea of the stored programme, of calculations based on the programmes and of maximum generality which one sought in vain in the work of Babbage . . . It is therefore tempting to see the

641

computer as the realization of the logical automaton of 1936, the more so that the same individual was successively involved in producing one and then the other. According to this version, the computer existed first in ideal form, before being embodied in a real machine. The temptation, however, must be resisted. Turing's universal machine in the 1936 article is essentially no more than an ingenious ideal model in a proof relating to a fundamental problem of mathematics. It was only after having seen ultra-rapid electronic calculators in operation during the war (the speed factor, as ever!) that Turing came to accord an entirely different significance to the universal machine. In the event, he made it not only a possible, but also a concrete and easily constructed support for artificial intelligence.

Nor did Turing's universal machine ever inspire the builders of any concrete computer. The field in which it made its appearance was too distant from the problems encountered by the designers of calculators of glass and metal for them to be able to turn it to advantage. After the fact, but only after the fact, was a history of logicist inclination able to name Turing as one of the founders of computer science, as if the abstract functional isomorphy between computers and the Turing machine allowed one to infer a relation of condition to possibility, of cause and effect or of derivation. All other things being equal, though the universal machine had never been imagined, computers would have been built all the same. Theoretical fundamentals have to be distinguished from genetic foundations.

While Turing embodies the type of the unhappy scientific hero, often misunderstood and persecuted during his lifetime, the mathematician John von Neumann presents, on the other hand, the image of the recognized and prestigious scientist, granted every kind of honour and distinction, integrated into the ruling circles of the American army and administration. Von Neumann is generally considered to be the founder of computer science because he drew up the plans for EDVAC (the Electronic Discrete Variable Automatic Calculator), the first document to describe the internal arrangement and principles of operation of the modern computer. Today the phrase 'von Neumann architecture' is still used to designate the classical organization of the computer, as it has existed since World War II.

Von Neumann's case is clear, because it is clearly a story of predation and capture. His plans for EDVAC synthesized and gave an attractive form to all the ideas which had come from the work of the Moore School of Engineering at the University of Pennsylvania. This group, which had just developed the first American electronic calculator, was looking for new principles of organization and operation for a second machine when von Neumann joined the School. We know that the concept of the recorded programme, as well as many other original ideas which enter into the EDVAC plans, were formulated for the first time by John Eckert and John Presper Mauchly, the two main leaders of the Moore School team. By signing the plans for EDVAC, von Neumann, the distinguished mathematician, brought them an unexpected audience and legitimacy, very useful in obtaining military credits, but at the same time he attracted to himself all the glory of having invented the computer. At this time, indeed, people were already aware of the importance that might accrue to the computer in the postwar world.

Unlike Turing, von Neumann succeeded in his own lifetime in an operation of capture which redefined the nature and function of electronic calculation machines and put him, together with his discipline, at the 'true' origin of computer science. From now on, computers would belong to the field of the theory of automata, which deals not only with logico-mathematical objects and electronic machines, but also with the nervous system of living beings. This is the significance of the cybernetic enterprise with which the Princeton mathematician associated himself. Eckert and Mauchly would now appear only as the 'engineers', concerned only with the equipment. In fact, von Neumann had converted the work done by the Moore School team to his own ends.

Here the supposed founder had brought about a brilliant synthesis of ideas, most of which were the work of others, but which he organized within the framework of a general theory of automata. It's not a question of denying the reality or the originality of von Neumann's contribution to the history of computer science, but of challenging the very idea of foundation or points of origin, which each time one looks at a particular case shows itself to be an operation of reinterpretation or capture of past or contemporary work. It is an operation of which the supposed founder may in turn become the victim, chased from the dawn glory of the beginning into the indistinct half-light of prehistory.

Logicism: Engineers and Mathematicians

For the history of computer science, the general problem of foundation is paralleled by the particular question of logicism. We have noticed, in fact, that the three main candidates for the title of founder were all mathematicians. They are famed for having conceived the abstract structures characteristic of computers today and which make their astonishing performances possible.

By doing this one consigns to the shadows the mechanics, the engineers and skilled workers whose contribution was not any less essential, although it lends itself less easily to the scenario of 'invention'. For the gallery of individual portraits traditional in the history of science, the history of computer science must substitute a series of couples, whose relationships were often stormy.

At Babbage's side is Joseph Clement. Babbage's failure was not the result of the limitations of the technology of the time, but of the bad relationship

he had with one of England's best mechanical engineers. Among other reasons
for their dispute, Babbage wanted to keep
a monopoly of the Analytical Engine,
while Clement had made its parts, designed some of them, and had built all
the machinery necessary to produce them.
The disagreements mounted up to such
an extent that the engineer pulled out of
the project. Babbage had already sunk too
much of his money with Clement to start
again with someone else, and the English
government, having burnt its fingers once,
refused to finance the construction of the

Reconstruction of a part of Babbage's Analytic Engine. (IBM Collection.)

Analytical Engine when Babbage had not succeeded in building the Difference
Engine.

One sees this polarity between mathematician and engineers in the war-
time England of the 1940s, although no open conflict had been declared. It
must be said that the achievements of the Bletchley Park team remained
secret for thirty years, so there could be no question of quarrels over priority
or attribution. The general design of the decoding machines was confided to
Turing and Max Newman. Newman, a specialist in topology and the foundations of mathematics, had been Turing's professor at Cambridge. But the
first electronic calculators were in fact constructed by the physicist C. E.
Wynn-Williams, a specialist in electronic meters, and by the engineers T.
H. Flowers, S. W. Broadhurst and W. W. Chandler. Before the war, all these
had worked in the field of telecommunications. The decisions taken by the
engineers, to adopt binary numeration or to synchronize all the machine's
operations by means of an internal clock, cannot be considered as unimportant technical details. Without the experience of electronics and the inventive resources deployed by Flowers and his technicians, the famous English
Colossus used to decode German communications in the period before the
landings would never have seen the light of day.

As for the United States, we have already mentioned the open conflict
between von Neumann on one side and the physicist Mauchly and Eckert,
the inventive technical genius, on the other. The former felt that the computer was a scientific discovery (his own), which belonged in the public
domain. The others maintained firmly that they were faced with a series of
technological inventions (their own), which had to be protected by patents.

The historian's hesitations about the identity of the inventors (engineers or
mathematicians) expresses an ambiguity in the status of computing today –
is it a science or a technology? To choose Babbage as a forerunner and
Turing or von Neumann as founders is to opt for science. But this choice

cannot account for the effectiveness of the computer, for logical structures and internal architecture would have remained ineffective in practice if certain materials and physical arrangements had not made it possible to reach a speed where it needed only a millisecond to perform a large calculation (see p. 655). As we will show, the question of the speed and duration of operations, although not relevant from the logical point of view, becomes crucial when it is a question of actually building a programmable calculating machine suitable for every purpose.

Logical positivism thinks of time and matter as accessory or accidental features when, on the contrary, speed and materials can play a constitutive, foundational role, as is the case, indeed, in the history of computer science. In the positivist version of history, the essential principles of computer science were already contained in the brilliant intuitions of forerunners such as Gottfried Wilhelm Leibniz (late seventeenth century), Babbage, or Turing in 1936, who could not, however, have possibly imagined the use that would be made of these intuitions with vacuum tubes and transistors.

History through the Generations of Materials

But should we do what was done for a long time and adopt a history of computer science aligned on the series of material techniques employed for automatic calculation? In this case there would be a succession of 'generations': mechanical machines first of all, then electro-mechanical and finally electronic, which would open the era of the true computer. Such a succession could be discovered within electronics itself: the vacuum tube at the beginning, the transistor and then the printed circuit would measure the linear progression of the computer towards ever greater lightness, speed and reliability. But this simple vision, widely propagated by the big manufacturers, ignores other dimensions of computer science, such as programming languages, programmes and methods of communication between machines and their users, not to mention their social uses.

The microcomputer has helped to challenge this generational schema. Was the development of the microprocessor the essential 'cause' of the success of the personal computer? No, this was no more than one event among others, interpreted and mobilized in the service of a struggle against the giants of the computer business. Let us quote from the heterogeneous list of actors drafted in by the founders of the first microcomputer firms: the Basic programming language, communication interfaces designed for users who would not be professional computer people, the counter-cultural movement which was at its height in 1970s America, venture-capital companies in search of fast profits, etc. The innovative Silicon Valley companies brought onto the stage of computer history social actors other than the State, science and big business. In 1976 IBM didn't attribute the same meaning to the microprocessor as Apple did, and did not engage in the same network of alliances. This example suggests that explanation in terms of 'causes' (here, a new stage in the increasing density of printed circuits) sometimes lacks pertinence in the history of technology. It seems more interesting to bring out the way in which actors interpret, convert or capture situations, technological set-ups and social forces for their own advantage. Though they are clearly important determinants of major turning-points in the history of the computer, material possibilities do not provide a less unambiguous or reassuring determination than do great events (like World War II) or the genius of mathematicians.

What is more, representation in terms of successive material generations does not give a good account of the real succession of events. This is the case for the relationship between electromechanical machines using relays and electronic machines. The first are supposed to have preceded the second, but in 1935, even before the first trials with telephone relays by the German Konrad Zuse and the American Georg Robert Stibitz, Atanasoff had imagined the construction of diode-based circuits for electronic calculation. Atanasoff knew the work of the Englishmen, W. H. Eccles and F. W. Jordan, who had already described a very simple version of such a circuit in 1919. He took their ideas and developed them on a larger scale. By 1942 the electronic part of his calculator was operational, but his mobilization interrupted the project. One can see clearly here that the ideas of 'generations' or 'stages' do not describe the normal unfolding of technical progress, but indeed hide its multiple overlaps or even its chaotic nature.

The Future Cause

In our look at the different ways of writing the history of the computer, we have eventually arrived at the teleological account. According to this version, the computer as we know it today was already the obscure goal of all those who were more or less involved in the improvement of methods of calculation, from the first builders of abacuses to the inventor of logarithms. The whole of history, from the very beginning, converges on the current state of our technology and science. There never were any bifurcations, any choices or reinterpretations of the past to the advantage of new projects. Our present and, even more, our way of understanding this present, are a sort of Omega point which, from the future, has organized the whole unfolding of history. The multiple hazards and contingencies from which history is moulded are no more than means in the service of this project. Circumstances are thus denied all power of determination or orientation.

Robert Ligonnière's book offers an excellent illustration of this tendency, and opens with the following sentence: 'The creation of the computer is the fulfilment of an ancient dream, unconscious at first, then matured through twenty centuries, in which intellectual labour could be executed by machine. The whole of humanity has been engaged upon this venture, from the most distant centuries to the contemporary period, with its collection of powerful technological solutions.'

How could it be that the same dream inspired, unbeknown to them, a Chinese builder of abacuses in the tenth century, Nicolas Chuquet (see p. 262), Leibniz and Steve Jobs (the founder of Apple)? Objects and concepts change meaning as a function of the use made of them by different actors. Each of the creations in the history of calculation is caught up in different social, technological and imaginative world. The very ideas of calculation and automatism have changed completely since theatrical wonders were constructed by the mechanics of Alexandria or the first arithmetical checker-board was sketched in the sands of Egypt. The critic of future causes in history has to use almost the same arguments as the critic of teleology in the theory of evolution. There is no need to spend too long on this point.

With the notion of foundation one is in search of a determination from the past, one is looking for a point from which the future can be fixed once and for all. With the idea of a prescience of the future that obscurely orientates the activity of our forerunners, we believe we have found the fixed point, the magnetic pole of history, in a present on which everything converges. At bottom, this is the myth of progress. In both cases the indeterminacy proper to duration, the unexpected bifurcations where the future is constantly re-decided and the contingency which presides over these decisions are all eliminated.

From 1936 to 1946: A Possible Account

The ASCC (Automatic Sequence Controlled Calculator)

In 1937 Howard Aiken (1900–73), an assistant professor at Harvard, was engaged in writing his doctoral thesis in physics. Spending long and laborious hours in solving his equations, he had the idea of a calculating machine specially designed for the solution of scientific problems. At this time, in fact, there were available only mechanical machines to carry out the four operations, electric multipliers, tabulators intended for accounting purposes and analog computers.[3] None of these could process negative numbers, and even less were they able automatically to call on mathematical tables while calculating, or automatically bring up to date all the variables of a function after each stage in the calculation. In the paper he wrote that year, Aiken outlined the general plan of the machine. Although he invoked the precedent of Babbage's Analytic Engine, the means of control provided were only very distantly related. For example, conditional branching,[4] possible with Babbage, was absent from Aiken's paper. In fact, the plans for the ASCC combined a multitude of different arrangements, such as mechanical registers, electrical counters and readers for cards or for the punched tape already in use in 1937, but put them at the service of an original ambition. Aiken insisted too on the immediate availability of all the parts for his machine, so that he could more easily convince a potential manufacturer of the feasibility of his project. Taking advantage of his connections at Harvard, Aiken succeeded in meeting Thomas Watson, president of IBM, and in convincing him to produce the ASCC. At this time IBM was the biggest producer of tabulators, ahead of Remington-Rand and Bull. Although Watson did not believe there was a large market in scientific calculation, he hoped that the construction of the ASCC would bring prestige to his company and new know-how for his engineers. In addition, not all the costs would have to be carried by IBM, because there was the possibility of a contract with the Navy, which had a very large requirement for scientific calculation. It was agreed that the machine should be built at IBM's factory at Endicot (in the state of New York), under the direction of experienced in-house engineers. As the representative of Harvard, Aiken played the role of client but also acted as adviser, since he had drawn up the initial plans for the machine.

The ASCC was officially inaugurated in August 1944, although it had already been in use for a few months by the Navy. In his introductory speech, Aiken made no mention of IBM's role in the funding and especially the construction of the machine. This omission was the cause of a violent conflict between Aiken and Watson, each one accusing the other of stealing his ideas and his know-how. The argument over paternity between Aiken and IBM would drag on for years.

As IBM had made Harvard a gift of the machine, it was rebaptized the Harvard Mark 1. Technologically mixed, both electronic and electro-mechanical, the Mark 1 was 16m long, 2.60m high, and consumed several hundredweight of ice every day for cooling. It could multiply two decimal numbers 23 figures long in 3 seconds, and it gave the answer in coded form on a punched card.

After the United States entered the war, Aiken's Computation Laboratory at Harvard received several orders for new machines from the US Air Force and Navy, first electro-mechanical and then electronic. Aiken, however, never ventured to use a technology which had not first of all been proven by somebody else. None of the machines of the Harvard series served as a model for later commercial production, because after the war all attention turned to ENIAC, EDVAC and the machine at the Princeton Institute of Advanced Studies, which were almost a thousand times faster than the Harvard Mark 1.

The Bell Relay Calculators

Unlike Aiken, George Robert Stibitz had not made an exhaustive study of automatic calculation when he began the development of his first relay adding-machine, an extremely modest machine which he put together in his kitchen over a few weekends. This activity was, however, the point of departure for one of the biggest producers of large calculating machines during the forties.

Stibitz, born in 1904, is associated with the first massive and systematic use of telephone relays in digital calculation. The telephone relay was a switching[5] mechanism in use since the end of the nineteenth century. This technology is called electro-mechanical because an electrical signal produces the movement of a mechanical part. As with punched cards and tapes, this electro-mechanical technology is another example of conversion. To claim that a very long time (more than forty years) was taken to 'understand' the possibility of using telephone relays for digital calculation would be absurd. Before capture of the relay mechanism had been attempted by the calculation specialist, this equipment belonged to an altogether different field. In fact, various scattered plans and schemes for using it in calculating machines had circulated around Europe at the end of the twenties, but they weren't followed up. The only large-scale use of relays anywhere other than in telecommunications was on English and American race-courses. On the basis of information gathered from hundreds of machines dispensing betting slips, the Totalizator was able immediately to give the number of bets per horse. It should be noted that the operation of the Tote reminds one of the telephone because it includes instantaneous communication at a distance. The American Bell company was perhaps too involved in the normal use of relays to play the role of an avant-garde in using it for another purpose; it was through the unexpected conversion of one of its engineers' hobbies into an official project that it moved into the vanguard.

Stibitz's mother, a mathematics teacher, had taught him binary mathematics at a young age. Thanks to his familiarity with this, the Bell engineer discovered what could be done with the binary mechanism of the telephone relay. Stibitz had first of all to deal with scepticism on the part of the company hierarchy, as they couldn't see what you could do with machines that did calculations in binary. Having tried without success to design a machine which would work in decimal, he had the idea of constructing a little mechanism to translate decimal into binary for input into machine and binary into decimal for the output. His special keyboard re-employed (for a new purpose) a code already used in telecommunications since the beginning of the thirties: the binary coding of decimal. This translation ensured the success of Stibitz's projects. The Complex Computer, or Model 1 was operational by January 1940. It was used within the Bell company to design networks to

649

minimize noise. When the war came, there came orders and thus money from outside. Stibitz's team developed four new models of calculator for the National Defense Research Council, each more powerful than the last and with wider and wider fields of application. Model 5, the last of the series, was a true programmable 'universal' calculator with the possibility of conditional branching. Inaugurated in 1946 by the US Air Force, the Model 5, weighing 10 tons, occupied an area of 300 square metres and contained 9000 relays.

Zuse's Machines

Konrad Zuse, born in 1910, while still a civil engineering student at the Berlin Technische Hochschule, was overwhelmed by the amount of calculation he had to do and dreamt of inventing some way of mechanizing the process. He worked on this topic from 1934 and two years later took out a patent for a new calculating machine. This manipulated numbers expressed in base two, possessed a register of sixteen words and was controlled by a programme coded on perforated paper tape. Each programme instruction was made up of an operating code, two operand addresses[6] and an address for the result. This standard for instructions is still used today in programming computers. In his 1936 plans Zuse also included what is today called floating-point calculation. This is a procedure allowing the registration of very large or very small numbers without, however, increasing the size of the standard word.[7]

Having approached several people, Zuse found Kurt Panke, a manufacturer of calculating machines, who agreed to produce his machine. The Z1 was finished in 1938. Almost entirely mechanical and not very reliable, it hardly ever worked. Not discouraged, Zuse, with the engineer Helmut Schreyer, began to build another calculator, the Z2, based this time on telephone relay technology. For this they bought a supply of second-hand relays. The relay was not the only equipment converted from its normal use. A cinema projectionist at night, to help make ends meet, Schreyer had plenty of opportunity to notice the resemblance between the sequential unfolding of images in the films he projected and that of the instructions on the perforated tape in his friend's machine. He gave Zuse the idea of punching his programmes onto waste 35mm film, which was less expensive and much more durable than paper tape.

The two men's collaboration was brusquely interrupted by Zuse's mobilization. But Schreyer sent the military authorities a memorandum in which he described in glittering terms all the possible applications of Zuse's machine, which he presented as a 'universal' machine: for laboratory and engineering calculations and even for real-time[8] calculation if the switches were purely electronic. As the machine promised to be useful for the production and operation of military equipment, the authorities let Zuse go and provided him with a small grant to put his plans into action. The Z3 was intended, in particular, to deal with problems of aircraft wing vibration, but it was in fact a programmable universal calculator, the first in Europe. Zuse then constructed for the Luftwaffe two specialized calculators for the development of flying bombs. All these machines were destroyed in the bombing of 1944 and 1945.

The German government refused Zuse and Schreyer's request to build an entirely electronic machine. The project, which required 1500 vacuum tubes (ENIAC, completed in the United States in 1946, had 19,000), was considered to be too adventurous. And even if it was successful, it was thought that the war would be finished before it could be built! This was before the battle of Stalingrad.

The Z4 was still a relay-based machine, financed by the Luftwaffe, and the only one of Zuse's machines to survive the war; it was also the most advanced and the most powerful. It was rented by the Zurich Technische Hochschule and contributed to the training of the first generation of Swiss computer scientists.

The First Electronic Calculators; the Colossus Series

The first major English achievements in the field of automatic calculation were marked by the exclusively military context in which they were planned and in which they remained confined.

In 1940 the continent of Europe was almost entirely controlled by the Axis powers. Great Britain survived thanks only to the sea connection with the United States. The Germans were trying to interrupt traffic between America and England. The American convoys were particularly vulnerable to attacks by U-boats, the German submarines. At that time there was still no means to detect them except by the interception and decoding of their communications.

From the moment that war was declared, the British decoding service had tried to take maximum advantage of the information it had on the coding system used for German military communications. This system relied in part on a machine, the Enigma, which could produce millions of different keys and which made decoding almost impossible by the usual methods. For the English it wasn't just a question of decoding a few messages thought to be important, but rather the entirety of enemy communication and the complete German coding system. A team under Turing's leadership, made up of mathematicians, linguists, engineers and programmers (more than a hundred), was therefore faced with the new problem of mechanizing the process of decoding. On average more than 6000 messages a day were decoded. On the basis of calculations of probability and of certain linguistic considerations, Turing designed a machine which could reconstruct the initial position of the Enigma machine's rotors. Electro-mechanical in its technology, the first Bomb was completed towards the end of 1940. The machine's programming was constantly improved and its different versions succeeded in decoding more and more quickly the always increasing message traffic.

Turing's team occupied only 'Hut No. 8' on the decoding services' site at Bletchley Park in Buckinghamshire. More than 10,000 people worked there, day and night, among them many linguists, mathematicians, physicists and engineers. A group led by the mathematician Max Newman worked in the greatest secrecy on the design of new calculators intended to break the coding system of the German Army and to speed up the decoding of the Navy communications. These calculators, called Robinson, used some 80 electronic valves. Construction was under the direction of Wynn-Williams, a specialist in electronic counters and equipment for high-speed communications: also working on the design of radar, he brought to the calculators improvements borrowed from radio-electrical detection systems.

While the Robinsons were being developed, Flowers, a Post-Office engineer, had the idea of getting rid of the mechanism for reading punched paper tape, too slow, too fragile and difficult to synchronize, and replacing it by internal registration in an electronic memory. This was the beginning of a new series of machines, the Colossus, which can be considered as the first large-scale electronic calculators ever to have operated.

Before the war, Flowers had designed one of the first telephone switching circuits to use vacuum tubes. His role, as well as that of Wynn-Williams, shows the importance of lateral borrowings from other fields than calculation properly speaking. Geiger counters, telephone circuits and radar do not figure among the 'roots' of the 'family tree' usually drawn up for the computer. This dendrographical metaphor has to give way to an image of horizontal, rhizome-like development which follows only the logic of encounter and proximity. Turing's earlier work wasn't any more 'at the origin' or 'at the root' of the computer than the work of the specialist in telephone communications or the inventors of radar we have just mentioned.

The group of communications technicians under Flowers' leadership built a series of machines with remarkable characteristics. In addition to certain instructions contained in the memory, they were programmable by means of a patch-board similar to the one found in the telephone switchboards familiar to Post-Office engineers. Programmes could contain conditional branching. The Colossus series worked in binary. All operations were synchronized by an internal clock. To avoid the valves blowing too often, the pragmatic engineers who built the first electronic calculators decided to leave the machines permanently switched on.

The first Colossus became operational in 1943. With more than 1500 electronic switching elements, it was the most powerful calculating machine ever constructed. It worked more than a thousand times more quickly than the electro-mechanical Bombs. More than a dozen of the Colossus series were completed before the end of the war, each one more rapid and more highly developed than its predecessor.

A good part of the first Allied victories on the Western Front should be attributed to the speed of decoding allowed by the different versions of the English ultra-rapid calculator. But in order to keep its electronic weapon secret the British military sometimes agreed to enormous sacrifices, such as, for example, allowing cities to be bombed without any response, so as to pretend to be surprised. The achievements at Bletchley Park were kept absolutely secret for thirty years. England and its Commonwealth allies had adopted a coding system analogous to Enigma . . .

The ENIAC (Electronic Numerical Integrator and Computer) was the first American calculator. Initially designed for the calculation of tables for shooting and bombing, it was used early on to solve fluid dynamics problems for the Los Alamos centre. It was then used for calculations in various problems of nuclear physics and meteorology. Its maintenance and programming manual consisted of no less than five 300-page volumes! (PPP-IPS-NASA.)

So as to be able to continue to capture communications, the English intelligence services prevented English industry from taking hold in its turn on the electronic and logical weapons developed during the war. This is why the first British civilian computers were built on the model of the American machines, despite the efforts of scientists such as Turing, who were unable to publish the work they had done.

The ENIAC (Electronic Numerical Integrator And Computer)

The Ballistic Research Laboratory (BRL), the main artillery research laboratory in the United States, commissioned the construction of the first American electronic calculator. Among this laboratory's numerous functions, the calculation of ballistic tables (for firing guns and dropping bombs) represented an extremely heavy burden of calculation. Firing tables were essentially used to reply to questions of the type: 'At what angle should I elevate the barrel of the gun, given that the target is at such and such a distance?' But the table did more than convert a distance into an angle; it also had to take account of air temperature and pressure, wind direction, projectile weight, propulsive force, which depended on the explosive charge, etc. Bomb tables met similar needs. Each time a new model of bomb, gun, shell or explosive emerged from the American Army's arsenal, it remained unusable until tables had been calculated.

Now an average table required the calculation of 3,000 trajectories, and the calculation of one trajectory required 750 multiplications of two ten-figure numbers. The most rapid (analog) calculator of the time, the differential analyser, could calculate a trajectory in twenty minutes, which meant thirty full days to draw up one table, without taking account of breakdowns and the sometimes very long time required to reprogramme the machine between each calculation.

In 1935 the BRL had acquired the only existing example of the differential analyser and had built another at the Moore School of Engineering, on condition that the School should lend it back in case of emergency. Before the beginning of the war, Paul Gillon, assistant director of the BRL, had got in touch with IBM in order to obtain digital punched-card machines designed especially for ballistic calculation, machines which in the event operated perfectly. When war was declared, Gillon set up at the Moore School a special course to train all the women volunteers who possessed science degrees in ballistic calculations. This course took place within the framework of Engineering Science Management War Training (ESMWT), an enormous training programme which illustrates the constitution of an immense

Comparative table of speeds of calculation		
	Speed of multiplication of 2 10-figure numbers	Calculation of trajectory for a range table
By hand or by Babbage's machine	5 minutes	2.6 days
Man with office calculator	10–15 seconds (20 to 30 times faster than by hand)	12 hours
Harvard Mk 1 (mechanical)	3 seconds (3 to 5 times faster than a man with a calculator)	2 hours
Model 5 (electro-mechanical)	2 seconds (10 to 15 times faster than a man with a calculator)	40 minutes
Differential analyser (analog)	1 second (20 to 30 times faster than a man with a calculator)	20 minutes
Harvard Mk 2 (electro-mechanical)	0.4 of a second (25 to 40 times faster than a man with a calculator)	15 minutes
ENIAC (electronic)	0.001 of a second, 1 millisecond (one million times faster than a man with a calculator, one thousand times faster than the Model 5)	3 seconds

system for the integrated management of all the country's resources, in which the objectives of defence, science and industry ('management') were all combined together. Despite its anticipation of the burden of calculation which would fall upon it, the first months of the war found the BRL overwhelmed. In 1942 Gillon appointed Hermann Goldstine to head BRL activities at the Moore School. With his wife Adèle, this young liaison officer, a specialist in ballistic mathematics, reorganized the training of the BRL's (human) calculators. At the Moore School, Goldstine met the physicist Mauchly and the engineer Eckert, who told him about their plans to construct an electronic calculator with vacuum tubes for switching elements. The new technology would allow calculations to be done a thousand times faster than by electro-mechanical or analog machines. Goldstine saw the benefits of using this electronic machine to calculate the firing and bombing tables for which he was responsible. He was not slow in persuading the leaders of the BRL, who exerted themselves to obtain the enormous credits needed for the new project within just a few months.

Why was electronic technology a thousand times faster than electro-mechanical? The basic components of the big digital calculators of the forties were two-state automata, flip-flops, relays, etc. whose ingenious arrangement in complex circuits allowed the machine to perform various mathematical or logical operations. The time the little two-state automaton took to change from one to the other, to open or shut a circuit, is called the switching-time.

The mechanical part which performed the switching in telephone relays took about a millisecond (10^{-3}s) to change position. This part weighed about one gram and therefore resisted the movement with a certain inertia. In addition, the millimetre distance to be covered was not negligible. When switches were circuits based on vacuum tubes, on the other hand, all the movements took place on a microscopic scale. The mass of an electron being 9.10^{-28}g, this represents practically no inertia and responds almost instantaneously. The electronic circuit switched in 1 microsecond (10^{-6}s). But the normal field of use for vacuum valves in the first half of the twentieth century was radio broadcasting and reception, a field apparently without any connection with automatic calculation. The engineers and firms most familiar with the use of electronics dealt with sound, not multiplication.

What is more, vacuum tubes had never been used more than a few at a time except in certain complex electronic apparatus, such as Wynn-Williams's nuclear counters. They had a well-established reputation for fragility: they usually had to be replaced almost every day. The construction of a machine made up of thousands of vacuum tubes seemed *a priori* destined to fail, for obvious reasons of maintenance and reliability. The telephone relay, on the other hand, had already been used on a very large scale, and the numerous engineering problems connected with its massive utilization had been provided with a number of satisfactory solutions.

John Atanasoff, a mathematician at Iowa State College, is generally considered to have been the first to design an electronic calculator. Having made a long study of the different means of automatic calculation which existed in the early thirties, he set out in 1935 to construct an electronic machine. In 1939 he received a small grant from the state of Iowa; but in 1941 his machine still didn't work, on account of a fault in its card-reader. Mobilized in 1942 and sent to the Naval Ordnance Laboratory, Atanasoff abandoned his project. Things would perhaps have remained like that for a long time if, in 1941, he had not had a long discussion with the physicist Mauchly about electronic calculation in general and the plans for his machine in particular. The year after, Mauchly was appointed to the Moore School of Electrical Engineering at the University of Pennsylvania, where he met Eckert, a twenty-year-old technician and master of electronics, who spent his time dismantling early television receivers and repairing radios, and who owned an electronic organ, at that time one of the items of electrical equipment which contained the greatest number of valves (170).

In 1942 Mauchly sent Goldstine and Gillon a report in which he brought together his own ideas and those of Atanasoff and Eckert. At the beginning of 1943, despite the numerous reservations of the different scientific committees consulted on the matter, the American government signed a contract with the University of Pennsylvania for the construction of an electronic calculator. A particular objection was that the war would already be finished before the machine could be constructed (if it ever was), and that it would be better to use the funds for more certain and more immediately operational projects. Events would prove the ENIAC's opponents right: the machine would only be completed in 1946.

Under the leadership of Eckert and Mauchly, the project was transformed several times in the course of the three years it took to build the machine. The Bell Company was called on to produce the different parts (card-readers, trans-coders, printers, etc.), and the Radio Corporation of America (RCA) produced the valves designed by Eckert. A large number of scientific consultants were used, among them engineers from IBM and RCA. As in the English case, the great experience in electronics possessed by one of the team leaders was decisive. Eckert knew all the radio engineers' tricks of the trade. He knew in particular that to ensure a long life for the valves of his electronic organ he had to keep it always switched on and never use the valves at more than 80 per cent of their maximum voltage. The operation of all the organ's components was synchronized by the beat of an electric clock. It was these same principles which ensured the reliability of the first American electronic calculator. Once more, technical principles developed in a particular field (here the radio and the electronic organ) were converted or captured for the benefit of an entirely different project. But the change of purpose, and especially of scale, changed the significance of the recycled ideas. One moved on from a rule of thumb, a trick of the trade, to systematic rules for the construction of enormous highly developed pieces of equipment.

The production of each of the components for the ENIAC was supervised with the greatest care, and each of them was used at only a half or a quarter of its capacity, so as to reduce the frequency of breakdown to a minimum. The ENIAC contained 19,000 valves connected according to a labyrinthine plan, which made it one of the most complex machines ever built. The numbers were coded in the calculator in 'unitary decimal' (each figure being represented by the corresponding number of pulses on a 'ring' of ten places, and there was a units ring, a tens ring, etc.). The ENIAC carried out several lines of calculation in parallel, and it contained various tables which avoided the need to programme certain repetitive calculations, as well as several wired sub-programmes. Programming it was therefore extremely complicated and each programme required different particular procedures. The programming was the machine's weak point, since for each new problem it had to be rewired on a sort of enormous switchboard by connecting and disconnecting dozens of plugs. In consequence, the altogether more manageable electro-mechanical Bell Model 5 was not 'made redundant by the new electronic technology', because it could solve certain problems more rapidly, when programming time was taken into account.

Initially designed to perform only ballistic calculations, that is to say essentially to solve differential equations, in the course of its construction the ENIAC developed towards being a kind of universal calculator without, however, finally becoming one. Convinced by their short experience of the viability of electronic calculation, Eckert, Mauchly and Goldstine dreamed in 1944 (even before the ENIAC was completed) of another machine, which would be truly universal and easy to programme. This was the EDVAC.

The EDVAC (Electronic Discrete Variable Automatic Computer)

When he met Goldstine by accident on a railway platform, von Neumann had already been a consultant at Los Alamos for more than a year, working on problems of fluid dynamics, more particularly on detonations and shock-waves. The solution of such problems, necessary for the construction of the atom bomb, required enormous quantities of long and painstaking calculations. In addition, von Neumann helped the Los Alamos physicists to construct mathematical models of their physical problems and played the role of chief calculator. He had helped to organize the immense Los Alamos calculation laboratory, which was equipped with hundreds of calculating machines operated by means of punched cards. This explains why the Hungarian mathematician was so interested in Goldstine's description of the ENIAC, which was meant to calculate a thousand times faster than the fastest machines available at the time. This was the first time he had heard of the top-secret plans for the electronic calculator. A few days after the two mathematicians had met, von Neumann went to visit the team at the Moore School. Two weeks later he was appointed as a consultant to the BRL for the construction of the ENIAC.

Even before von Neumann's arrival, the team at the Moore School was already thinking about building a new machine with less tubes, which would be both cheaper and easier to programme. They had examined with the greatest interest the operation of Stibitz's Model 5, whose programme was coded in digital form on punched tape. As for the reduction in the number of tubes, Eckert had invented, for a new model of radar, an electronic device based on the piezo-electric effect:[9] the mercury delay line. He hoped to re-use this technique for the memory of the new machine. Properly employed, the mercury delay line allowed a hundred-fold reduction in the number of tubes required to store information in the machine's registers.

In the technologies employed for the memories of the first computers one discovers these re-employments, conversions and reinterpretations characteristic of technological inventiveness. For the memory of his machine at the Institute of Advanced Studies, von Neumann got the idea from RCA of using an 'iconoscope' or selectron, a sort of cathode-ray tube of the kind used in early television receivers. At the Macy conferences,[10] he also put forward a model of human memory inspired by the iconoscope. The selectron never worked properly.

The Moore School team's thinking progressed rapidly. Until then the task of designing a new calculating machine appeared more or less as follows: given such and such a type of calculation to be performed, what material devices could carry it out most effectively? Eckert, Mauchly and von Neumann asked a new question: what machine would be capable of reading, interpreting and executing any programme whatsoever? Here, technological know-how was to serve the objective of maximum generality and optimal coding or decoding. Von Neumann summed up the new approach in a sentence of the *First Draft of a Report on the EDVAC*:

> if the apparatus has to be elastic, that is to say as general as possible, then one must distinguish between the specific instructions for the definition and solution of a particular problem and the general organs of control which read and execute these instructions whatever they may be. The first must be recorded in one way or another, the second are represented by an organ in the apparatus. This function, and this function alone, we call the 'central control'.

To tell the truth, inventors like Zuse, Aiken and Stibitz had certainly been orientated towards maximum generality, but for reasons of speed and effectiveness they hadn't been able to pursue this tendency to the very end. The more universal the machine, the longer it takes to programme a particular problem and the more steps of calculation it takes to solve it, because no wiring and no technical tricks enable the machine to take short cuts. A universal machine is infinitely slower in the execution of a calculation than specialized equipment. Only a thousand-fold increase in speed could make it seem reasonable to construct a universal machine in practice, and this is why Turing's theoretical machine did not inspire anyone to produce a concrete example. The new orientation of automatic calculation from 1945 is illustrated by two fundamental principles: sequentiality and the stored programme.

The plans for EDVAC provided a remarkably simple logical organization explicitly inspired by a simplified model of the human nervous system. The arithmetic unit carried out the calculations, the memory contained the programmes and the data, while the input and output organs managed the computer's communication with its environment and the control unit ensured the execution of programmes and communication between different organs of the machine. This organization differed from that of the great new digital calculators of the early forties. They had been made up of a multitude of specialized organs of calculation working simultaneously, to which distinct registers (or memories) were attributed. Having only a single organ of calculation, the EDVAC was obliged to carry out its operations one after another in a sequential manner. It no longer worked in parallel. The very high speed made possible by electronics had made multiplication by simultaneous processing unnecessary. The material devices for calculation could be as simple as possible and be composed of only a minimum of switching elements. All the complexity was put into the programmes, that is to say into memory. On the technological and financial levels, the savings were considerable and the machine became more reliable. The plans for the EDVAC testify to a reversal of the tendency followed until then, which had always been in the direction of an increasing complication of the material part of the machines. The memory, the location of the programme, was the only material part to be enlarged. The EDVAC possessed a hundred times more internal memory than the ENIAC but 10 times fewer switching elements.

The storage of instructions in the internal memory of the machine made it theoretically possible for the programme to modify itself during its own execution and in particular, automatically to change the addresses of the numbers on which the instructions were operating. In this way, the same portions of the programme could be used to calculate different numbers located in different parts of the memory. In 1964 C. Ergot and A. Robinson demonstrated that machines with non-stored programmes (without any possibility of programme self-modification) could not calculate all calculable functions. But the stored programme didn't only transform the digital electronic calculator into a universal machine; it made it a computer, in the current sense of the term, that is to say a device for processing information in which the programme part would become preponderant. In fact, once the storage of the programme became possible, computer systems would essentially be made up of a complex hierarchy of programmes translating and commanding each other. This is why the usual programmers and users of a computer can be ignorant of nearly everything about the material infrastructure and wiring of their machines. In this way there opened up a space for the indefinite development of the 'languages', which were crude programming codes at the end of the forties, so-called 'developed' languages in the fifties, followed by the flourishing of multiple formal languages in the sixties.

It certainly seems that it was Eckert and Mauchly who invented the stored programme, in the sense that they were the first to put it to work successfully in a real machine. But at that time it was only a matter of taking advantage of the speed of the electronics and not of opening the way for software development. In fact, the reading of instructions from an independent support, such as punched tape, had slowed down calculation considerably, and the manual rewiring of the apparatus between each calculation, as was necessary with the ENIAC, had been recognized as too long and impractical.

The idea of treating instructions as data was so bizarre and counter-intuitive that not even von Neumann thought immediately of any other practical advantage of registering the programme, other than the speeding-up of calculation. He had, however, assiduously practised a formal logic of a Gödelian type – in which digital coding allows the integration of elements from different logical levels – and had been aware of Turing's work, in which the idea of calculation on the programme explicitly plays a part. The coding of instructions and numbers respectively in the internal memory of the EDVAC was initially done in such a way that arithmetic operations were not applicable to instructions, which denied any possibility of programme self-modification! It was only much later that the Hungarian mathematician grasped all the advantages which could be derived from storing the programme, according to Goldstine many years after the plans for the EDVAC, during the construction of the Institute of Advance Studies machine. Of course, none of the actors at this time could see that the storage of instructions in the same format as the data would be interpreted in the mid-fifties as a way of getting the computer to calculate its own programme on the basis of indications provided in a 'developed language' and that this, together with other improvements, would enormously increase the number and variety of programmes available.

So the computer, as a support for the programme or a concrete universal machine, was never a goal as such. It appeared only at the end of a succession of conversions and reinterpretations of heterogeneous materials and various arrangements, of a chance succession of opportunities and local circumstances exploited with more or less success by a multiplicity of actors.

After the War

The violent conflict at the end of 1945, between Goldstine and von Neumann on one hand and Eckert and Mauchly on the other, broke up the Moore School team. The quarrel over priority (who was the real author of the plans for the EDVAC?) was paralleled by a divergence of strategy. As far as Eckert and Mauchly were concerned, it was necessary to take out patents and exploit the computer commercially. For von Neumann and Goldstine, however, the computer was a scientific discovery to be developed in the universities and research centres (which is what indeed they did at Princeton), requiring the widest possible diffusion without any legal obstacles. And indeed the Institute of Advanced Studies machine, the university prototype designed by von Neumann and Goldstine, was freely used as a model for a number of machines sold on the market.

On their side, Eckert and Mauchly established their own company, which was responsible in particular for marketing the Univac (Universal Automatic Computer). Faithful to their initial orientation, they sued other computer companies for the wrongful use of patented inventions. At the end of a ten-year trial they lost their case, the court holding that the technical devices at the centre of the litigation now belonged in the public domain. The trial began to go against them as soon as Atanasoff began to give evidence. With documents to hand, he revealed to the astonished jury that he was the author of a number of inventions of which Eckert and Mauchly claimed to be the owners, and which they had learned of before the ENIAC was built . . .

The ENIAC was finished only in 1946. It helped the Allied victory neither by calculating ballistic tables nor by speeding up the calculations necessary for the construction of the atomic bombs exploded on 6 and 9 August 1945 over Hiroshima and Nagasaki. As for the EDVAC, it was finished only in 1951.

Even before the end of the war, an uninterrupted stream of visitors from Europe and from every corner of the world came to see the ENIAC at the Moore School. Unlike the British, who kept the existence of the Colossus secret, the American military authorities wished to see the widest possible diffusion of the scientific and technological innovations of the war. Copies of the EDVAC plans were in free circulation and numerous conferences and lecture series were given at American universities to expound the fundamental principles of the construction and programming of computers. Thus it was that the Englishman Maurice Wilkes, one of the inventors of radar, very quickly exploited the American experience by building the EDSAC (Electronic Delay Storage Automatic Computer) at Cambridge in 1947. This was the first electronic machine with a stored programme. Built according to the EDVAC plans, it used mercury delay lines for its memory and included the beginnings of an operating system.[11]

The history of computer science (like perhaps all history) can be seen as an indefinite distribution of creative times and places, a sort of torn and irregular meta-network, full of holes, in which each node, each actor, defines the topology of its own network as a function of its own goals and interprets in its own manner everything which comes to it from neighbouring nodes. Each of the living links in this fabric reinterprets the past which it has received from the others, as if it had to lead to its own choices, and projects a future in which these options will be continued. But the future, just as much as the image of the past, is in the hands of the following links and so on indefinitely.

In this vision of things the notion of a forerunner or founder, in the absolute sense, has very little relevance. On the other hand, one can detect certain operations carried out by actors who wish to impose themselves as founders or which designate in the near or recent past the prestigious ancestors they appropriate and whose descendants they proclaim themselves to be.

Not univocal 'causes' or 'factors', but circumstances and occasions to which persons or particular groups ascribe various meanings. No peaceful 'lines of succession', but rather scoops of the net from every direction, attempts at appropriation, and endless court cases around the inheritance (Aiken–IBM, Eckert–von Neumann, Mauchly–Atanasoff).

Technological discovery is revealed as a disorderly swarm of pottering about, re-utilization and a precarious stabilization of functional arrangements. Among all these agglomerations of irregular devices and disparate ideas, some, often for contingent reasons, will be used by the greatest number and will establish themselves through time. They will then appear as homogeneous and coherent technological objects and display their functional self-evidence as if it were a natural attribute. The computer, for example, seems today to be the earthly embodiment of an eternal Platonic idea. A certain skill at theatrical presentation is not entirely uninvolved in this result. It seems that von Neumann, in particular, had a particular gift for convincing presentations. He imposed his own idea of the computer (and the idea of the computer as his) before the slightest electronic machine had ever been built.

Captures, conversions and reinterpretations are in strong contrast to the ideas of algorithm or predetermined mechanism which are, justifiably, associated with computer science. But the history of computers can not be at all identified with the realization of a plan, a programme, or even a dream, were this the dream of Leibniz, Babbage or Turing. This is precisely because it is history.

Chronology

Michel Authier

This chronology finds its meaning in and through the book it accompanies. It is a tool intended to provide a context for scientific discoveries and the forces that produced them. The operation by means of which one cuts out from culture, society or history the part called science cannot be performed without violence: a skeleton is not a body. None the less, as science is neither the product of the best-known scientists alone nor simply the accumulation of the discoveries our own time considers as fundamental, the greatest possible number of actors and discoveries have been included (within the limits of the field covered by the book, that is to say, of Western science).

These individual and collective, human and institutional actors are listed in the central column. On the left appear the 'scientific events' which correspond to them . . . or not. To provide each period with meaningful background, the third column lists selected artistic, social, political and religious landmarks.

The problem of dates is central to a chronology and, unfortunately, it is one of those which poses most problems. In the field of science it is often difficult and sometimes impossible to discover exact dates. At what point should a discovery be located? At the moment of intuition, of theoretical or experimental confirmation, or of oral or written communication? And then there are the extreme cases! Here is just one of them: in different dictionaries and chronologies, the mechanist Hero of Alexandria is listed in every century from the third BC to the third AD!!!

Sources and Thanks

A new chronology is developed within the matrix of works – each bringing a new perspective, correcting mistakes, stressing particular aspects and offering a new presentation – which makes it possible. I therefore have to acknowledge my debt to the chronologies of science or technology by F. Russo, M. Daumas, G. Canguilhem, R. Caratini, J. Rosmorduc and B. Gilles, and to the more general chronology by J. Boudet. It goes without saying that in the labyrinth of scientific discovery, with its overlap in culture in the widest sense and the uncertainty of the data, it is a comfort to be able to turn to a book which can serve as a reference point. So I cannot go without mentioning two major works: R. M. Gascoigne's *Historical Catalogue of Scientists and Scientific Books* (Garland, New York, 1984), with its supplement on periodicals, and W. Stein's *Kulturfahrplan* (F. A. Herbig Verlagsbuchhandlung, Munich, Berlin and Vienna, 1977).

CHRONOLOGY

This chronology also owes a great deal to those who have collaborated in the book, who each in their own way have contributed something of their knowledge, and also to X. Polanco and P. Doray.

Instructions for Use

* As no scientists are known for the first millennia, the first page has only two columns: 'science and crafts' and 'history, culture and agriculture'.

* From about 750 BC to 1400 AD, the 'scientific events', on the left, appear at their date, opposite the 'individual and collective actors' responsible for them; the 'background information' is arranged in parallel for each century.

* From 1401 to 1947, the date at which the chronology ends, as does the book, actors are listed at their date of birth, and in the period between the birth of the scientist and his or her production are shown the events, discoveries and contemporaries of the period.

For ease of reading, titles of works and the names of discoveries are given in the shortest or most common forms.

Names of scientists are printed in bold, and the titles of works (books, journals, paintings, sculptures or films . . .) in italics.

Given that dates, as we have seen, are often uncertain, they are, when needed, preceded by the abbreviations *c.* (*circa*) and fl. (*floruit*: indicating the period of influence).

Before ~3500. Settlement, plant selection (wheat, barley, millet, 7th millennium, Iraq, Palestine). Grain store (7th millennium), domestication of animals: sheep (~9000), goats (~7500, northern Iraq), pigs, cattle (Proto-neolithic, Thessaly), donkey, horse, buffalo, zebra, elephant (before ~3500). Beginning of culture of the vine (5th millennium).

Pottery (~6200 in Thessaly, ~5200 in Cyprus), ceramics (6th millennium, Anatolia, Iran, Syria, Thrace). Mirrors of obsidian, then lead and copper (Anatolia). Spread of pottery in the Near East (6th millennium). Weaving of linen, glass beads in Egypt.
~5000 about 5 MILLION HUMAN BEINGS.

Science and crafts	History, culture and agriculture
~3500. Copper tools, stone vaulting, first [hydraulic works]. Copper in China. Bronze at Ur. Potter's wheel in Mesopotamia.	About ~3500, 20 MILLION HUMAN BEINGS. First dolmens. Foundation of Uruk, urbanization, pre-dynastic period in Egypt. Use of hoe and plough. Tillage and stock-raising in the Nile valley.

Science and crafts	History, culture and agriculture
c. ~3300. Beginning of writing in Mesopotamia and metrological systems.	
	after ~3200. Unification of Egypt, followed by Archaic period. Foundation of Troy, Tyre, Carthage, beginning of Cretan civilization.
c. ~3100. Beginning of writing in Egypt and metrological systems. *c.* ~2900. First known map (of Egypt). Beginning of systematic observation of the heavens (Mesopotamia, Egypt, India, China).	after ~2900. Archaic dynasty in Mesopotamia (Classical Sumerian civilization). First money of copper and silver. Stepped pyramid at Saqqara.
c. ~2850. First instance of solar-lunar calendar (Troy). *c.* ~2770. Egyptian calendar of 365 days. *c.* ~2750. Extraction of gold replaces washing. *c.* ~2700. Evidence of reduction of fractures.	*c.* ~2700. *Gilgamesh* epic. 900-towered wall at Urul (9.5 km). after ~2700. Old Empire, Classical civilization in Egypt. *c.* ~2650. Sphinx of Giza: first Egyptian mummies. *c.* ~2600. Cheops, construction of pyramids.
c. ~2500. First Mesopotamian mathematical texts. Development of astronomy in Babylon.	*c.* ~2400. The spoked wheel replaces the solid wheel (Near East). *c.* ~2300. Akkadian Empire under Sargon. *c.* ~2050. Third Empire of Ur. Oldest instance of the caduceus. First codes of laws and of divination (Ur).
c. ~2000. Medical texts (Ur III).	*c.* ~2000. Middle Empire in Egypt. Records of Mâri period in Mesopotamia. Palaeo-Babylonian, invention of loan against harvest at 33% (Babylon). The Sesostris. Influence of Egypt in Nubia, Crete, Palestine and the Red Sea.
c. ~1850. Medical texts and first mathematical texts (Egypt). *c.* ~1750. Stonehenge, megalithic [ensemble]. Writing in China.	*c.* ~1730. Ḥammurapi. Two-wheeled Assyrian chariot.
c. ~1700. Mathematical problems and tables (Babylon). *c.* ~1650. 5.4 km dam in India.	*c.* ~1600. Hyksos invasion (Egypt). First Intermediary period.

Science and crafts	History, culture and agriculture
c. ~1500. Medical texts of the New Empire (Egypt). *c.* ~1450. Solar obelisks. Portable sundial of **Thutmosis III.**	*c.* ~1500. Kassite invasion of Babylon. New Empire (Egypt).
	c. ~1375. Akhnaton (first monotheistic religion). *c.* ~1350. Assassination of Tutankhamen. *c.* ~1350. Ramses II.
c. ~1300. Hittite and Middle Assyrian medical texts.	
	c. ~1250. Flight of the Jews from Egypt.
c. ~1200. Magic square (China). Development of iron in Greece.	*c.* ~1200. Egyptian expedition as far as the iron-ore mines of Sumatra. Phoenician alphabet. *c.* ~1100. Ziggurat of Ur.
c. ~1090. Use of the gnomon and measurement of the angle of the ecliptic. *c.* ~1000. Abacus and book of arithmetic in China.	*c.* ~1000. The horse appears in Greece. Solomon (*c.* ~970, ~931). *c.* ~900. First Greek colony in Asia Minor. *c.* ~860. Zoo and royal library of Assurnasirpal II. Homer (*fl. c.* ~850).
c. ~800. Neo-Assyrian medical texts. ~8th c. Recorded birth of Chinese calendar and astronomy.	*c.* ~800. Horseshoe and cart with four spoked wheels (Celts).
	c. ~790. Separation of doctors from the caste of priests (India).

Scientific events		Individual and collective actors	Background information
Up to 1400, events are listed at their date; those responsible are listed opposite. The background information can be read independently within each century.	750	100 MILLION HUMAN BEINGS	~776. Olympie games. ~753. Foundation of Rome. ~733. Foundation of Syracuse. First Greek coins.
	~700		
According to myth, Daedalus invents numerous tools.			Hesiod (*fl. c.* ~700). Greek *Theogony.* Assurbanipal (~668–~626)
	~600		
Nature becomes the object of science. Thales predicts the eclipse of the Sun of . . . Water is		School of Miletus: **Thales** (*c.* ~625–*c.* ~547), **Anaximander** (*c.* ~610–~545).	~570. Temple of Artemis at Ephesus. Attempt to make a canal through the Isthmus of

Scientific events		Individual and collective actors	Background information
the primordial element in his cosmogony. Voyage from Egypt. Birth of Greek geometry.	~600	**Anaximenes** (*fl. c.*~546).	Corinth. The legendary Eupalinos builds the tunnel of Samos (1.5 km). ~546. Ionia submits to Cyrus.
'All is number'. Duplication of the square. Discovery of incommensurables.		School of Crotona: around **Pythagoras** (*c.* ~560–480), [**Alkmenes**], **Hippasus of Metapontus.**	The Greeks introduce the vine into Gaul. ~522. Darius becomes King of Persia. Aesop writes his *Fables.* ~508. The Roman Republic.
	~500		
Paradoxes of movement and of the unity of being		**Heraclitus** (*fl. c.* ~500) Eleatic School: **Parmenides** (*c.* ~544– *c.* ~450), **Zeno** (*c.* ~490– ~425), **Xenophanes, Melissos of Samos. Anaxagoras** (*c.* ~500– ~428).	~490. First Persian War: Aristeides (~550–~469) and Miltiades (~540– ~489) win at Marathon. ~480. Second Persian War. Leonidas dies at Thermopylae, Themistocles (*c.* ~525– *c.* ~460) wins at Salamis and Pausanias (in ~479) at Plataea. Births of Phidias (~490– ~430), Protagoras (~485– ~411), Herodotus
Theory of the four elements. Problems of the quadrature of the circle. Duplication of the cube. Trisection of the angle. Attributed to him is the discovery of the obliquity of the ecliptic and the period of the Earth–Moon cycle (59 years).		**Empedocles** (*c.* ~490– ~435). School of Chios: **Hippocrates** (*fl. c.* ~460). **Oenopidus.**	(~484–~425). Plays by Aeschylus (~525–~456). Thucydides (~460–395). Peloponnesian Wars (~431–~404). Construction of the great Greek temples. Laws of the Twelve Tables at Rome. Plays by Sophocles (~496–~406).
Atomism. Infinitesimal algorithm (volume of a cone).		School of Abdera: **Leucippus** (~460– *c.* ~370), **Democritus** (~460– ~370).	~443. Pericles (~495– ~429) in power. Parthenon. Plays by Euripides (~480–~406). ~430. Plague of Athens. Alcibiades (~450–~404).
Medicine: theory of the four humours. Discovery of the quadratrix by trisection of the angle.		**Hippocrates of Cos** (~460–*c.* ~370). Sophists and Megarics: **Euclid of Megara** (*c.* ~450–*c.* ~380), **Hippias of Elis** (*fl. c.* ~400).	~408. Plato follows the teahing of Socrates (~470–~399) until ~399, then leaves for Sicily. Plays by Aristophanes (~450–*c.* ~387).

Scientific events	Individual and collective actors	Background information
	~500	~404. Fall of the Athenian democracy, Council of the Thirty Tyrants.
	~400	
Problem of proporortional means and duplication of the cube (mesolabe).	**Archytas of Tarentum** (*c.* ~430–*c.* ~348).	~399. Condemnation of Socrates. Dionysus tyrant of Syracuse. First Dialogues of Plato.
Work on incommensurables. Research on polyhedra.	School of Athens: **Thodore of Cyrene, Plato** (~428–~438), **Theaetetus** (~417–~369), **Speusippus** (~408– ~339) **Meno.**	*c.* ~390. Birth of three sculptors; Praxiteles, Scopas, Lysippus. ~381. Brennus in Rome: 'Vae victis!' ~375. Temple of
Earth's rotation about its own axis. Arithmetic, conic sections. Planetary epicycle. Year of $365\frac{1}{4}$ days.	**Heraclitus of Pontus** (~388–~312). ~387. Foundation of the Academy at Athens. School of Cyzicus: **Eudoxus of Cnidus** (~400–~347), **Callippus.**	Epidaurus. ~359. Accession of Philip in Macedonia. ~351. Demosthenes (~384–~322), first philippic. Menander (~340–~292).
Encyclopaedia of scientific knowledge.	~335. Peripatetic school (foundation of the Lyceum): **Aristotle** (~384–~322), **Eudemus of Rhodes** (*fl.* ~320), **Autolycus of Pitane** (*fl. c.* ~300). Beginning of the garden of **Epicurus** (~378– ~287).	~338. Battle of Chaeronea. ~331. Alexander (~356– ~323) becomes King of Egypt, beats the Persians, founds Alexandria. The navigator Pytheas explores the seas of northern Europe.
On conic sections	**Menaechmus** (~374– ~325)	
	~300	
The *Elements* (collective), proportion, plane geometry, optics, music . . .	School of Alexandria: **Euclid** (~322–~285).	Opening of the Porch, Stoic School of Zeno of Citium (~355–~264).
Heliocentric theory.	~290. Museum of Alexandria (and library). **Aristarchus** (~310– ~230).	Theatre of Epidaurus (15,000 seats) ~280. Pharos of Alexandria.
Study of compressed air and hydraulics Mathematics, astronomy, mechanics.	**Ctesibius** (~296–~228) **Archimedes** (~287– ~212).	~264. First Punic War (–~241). The *Argonauts* of Apollonius of Rhodes (~269–~186).
Geometry and astronomy. Work in mechanics.	**Conon of Samos** (*fl.* ~245) **Philo of Byzantium** (*fl. c.* ~250).	Comedies of Plautus (~254–~184). ~219. Hannibal (~247– ~183) wages the Second Punic War.

Scientific events		Individual and collective actors	Background information
Evaluated circumference of Earth as 252,000 stadia and wrote a *Great Geography*.	~300	**Eratosthenes** (~276–~195)	~212. Marcellus takes Syracuse.
Essential work on conics		**Apollonius of Perga** (~262–~180)	
Work in mathematics.	~200	**Diocles** (*fl.* ~190)	~186. 7,000 involved in the scandal of the Bacchanalia in Rome. Comedies of Terence (~194–~159).
Explanation of the tides.		**Seleucus of Babylon.**	
Application of geometry to astronomy (Earth-Sun distance, prediction of eclipses).		**Hipparchus** (*fl.* ~147–~127).	~167. Polybius (~205–~125) arrives in Rome. ~161. Rome: expulsion of the philosophers.
Medical works.		**Aesculapius of Bithynia** (*c.* ~130–~40)	~149. Third Punic War (–~146)
On nature. *Geography* (physical and human) in 17 books.	~100	**Lucretius** (*c.* ~95–*c.* ~55). **Strabo** (~63–+5).	~73. Slave revolt. ~58. Caesar (~100–~44) in Gaul, [Battle of Alesia (~52). ~48. Battle of Pharsala (Caesar beats Pompey).
~31. Publication of *Architecture* in 10 books; mechanics, hydraulics, gnomonics.		**Vitruvius** († *c.* ~25).	~46. Julian Calendar. ~31. Battle of Actium (Octavius beats Antony).
Work in botany.		**Nicholas of Damascus.**	~29. *Aeneid* of Vergil (~70–~19), friend of Maecenas (~69–+8) and of Ovid (~43–+17). Livy (~59–+17) and Diodorus of Sicily (~30–+30) write their histories.
Natural History.	0	**Pliny the Elder** (23–79).	1. Jesus is seven years old.
Botany and medicine. Development of the sciences of nature.		**Dioscorides** (*fl.* 50–70).	Pax Romana. Seneca (~2–+65). 57. Saint Paul (5–67) preaches at Ephesus.
Spherical geometry.		**Menelaus** (*fl.* 100).	79. Destruction of Pompeii, 80. Inauguration of the Coliseum. Plutarch (50–125) returns to Chaeronea.
Astronomy and optics (the *Mathematical*	100	**Ptolemy** (90–170).	122. Hadrian's (76–138) Wall.

670

Scientific events	Individual and collective actors	Background information
Composition or *Almagest).* Applied mathematics (catoptrics, automata, pneumatics, measuring instruments, machines of war). *Mathematics useful in reading Plato.* Anatomy, medicine, psychopathology. First Hebrew book of geometry.	100 **Hero of Alexandria** (*c.* 150). **Theo of Smyrna** (*fl.* 150). **Galen** (130–200). **Nehemiah.**	Beginning of Christian theology, Tertullian (160–145), Origen (185–154). Jewish rebellion against Rome. 129. Perpetual Edict drawn up by Salvius Julianus. 160. Montanist heresy (end of the world and moral intransigence). 175. *Meditations* of Marcus Aurelius.
On Mixture Theory of numbers. Multiple equations, principle of the unknown represented by a sign. Neo-Pythagorean theory of numbers.	200 **Alexander of Aphrodisia** (*fl.* 200). **Diophantus** (*fl.* 270). **Iamblichus** (250–333)	Development of Christian theology. The Emperor Probus abolishes restrictions on the culture of the vine in Gaul. Neo-Platonism: Porphyry (233–282) writes the life of his teacher, Plotinus (205–270).
Mathematical Collection. Alchemical research [Great work]	**Pappus** (280–340). **Zosimus of Panopolis.**	Beginning of barbarian invasions.
Commentaries on Euclid, Archimedes, Ptolemy. *Encyclopaedia of medical knowledge.* Work in optics. First (?) principle of the minimum. *Commentaries* on Plato, Aristotle, Diophantus, Apollonius, Ptolemy.	300 **Theo of Alexandria** (*fl.* 364–377). **Oribasius** (*c.* 325– *c.* 403). **Heliodorus of Larissa** and his son **Damianus.** **Hypatia** (370–415).	312. Conversion of Constantine. 325. Council of Nicaea (condemnation of Arianism). 354. St Monica gives birth to St Augustine († 430), author of *(The City of God)* (427). 393. Last Olympic Games. Vulgate Bible of St Jerome (347–420).
Commentaries on Plato. India: use of negative numbers, of 0, circle, trigonometry (sine). Positions of Moon and Sun. Knowledge of rotation of the Earth. Encyclopaedia of Greek science.	400 200 MILLION HUMAN BEINGS **Proclus** (412–485). **Aryabhata** (476–535). **Boethius** (480–525).	410. Rome falls to Alaric. 415. Hypatia killed by Christians. Latin translations of Plato and Aristotle. 449. The Council of Ephesus condemns Ibas for Nestorianism (double personhood of Christ).

Scientific events		Individual and collective actors	Background information
Mathematical *Commentaries* on Archimedes, Apollonius, Eudemus, Eratosthenes. The *Wedding of Mercury and the Sun* (encyclopaedia).	400	**Eutocius** (480–560).	451. Defeat of Attila (*c.* 395–453). 486. St Geneviève at Paris during the siege laid by Clovis, who was baptized by St Rémi in 496.
		Capella.	
First attack on Aristotle's physics and cosmology.	500	**John Philoponus.**	Theodoric (455–526) imposes Roman law on the barbarians.
Mathematics, and architecture of the first Santa Sophia in Constantinople, inaugurated 537.		**Isidore of Miletus** (*fl.* 532).	529. The emperor Justinian (483–552) closes the School of Athens, where the neo-Platonist Simplicius (*fl.* 525–545) teaches.
Religious classification of knowledge. Mathematics and astronomy (India).		**Isidore of Seville** (560–636). **Brahmagupta** (598–after 665).	531. Chosroes I, Persian emperor (?–578). 532. The Church institutes the Christian calendar conceived by Dionysus Exiguus. 540. Monte Cassino founded by Saint Benedict (480–547).
Astronomy. A geography in Syriac. Medicine and surgery.	600	**Severus Sebokht.** **Paul of Aegina** (640).	622. Flight of Mohammed (575–632). Muslim conquests: Syria (633–640), Mesopotamia (633–637), Egypt (639–646).
Universal chronology based on the Christian era. Treatise on metrics. Astronomical scholarship and natural history.		**The Venerable Bede** (672–735).	661. Ummayyad dynasty (?–750). 670. Founding of Kairouan.
Work in alchemy.	700	**Heliodorus** and **Theophrastus.**	Conquests: North Africa (687–702), Spain (711–716), Eastern Iran (714).
Indian science reaches the Arabs.		**Al-Fazari** (father † *c.* 777, son *c.* 800).	750. Abbassid power until 1258. 762. Foundation of Baghdad.
Alchemy, esoteric knowledge, numerology.		**Jabir ibn Hayyan.**	768. Charlemagne king (?–814). 786. Haroun al-Rashid caliph (?–809).
First translation of Euclid into Arabic.	800	**Al-Hajjâj** (*fl.* 786–833).	813. Beginning of reign of al-Mamoun (786–833),

672

Scientific events		Individual and collective actors	Background information
Plan for a star-catalogue. An institute for the translation of scientific texts.	800	832. Foundation of the 'House of Wisdom', under the direction of **Hunayan ibn Ishâk** (808–873).	who establishes a scientific policy: laboratories, collections, library. High period of Carolingian art.
First Arab philosopher (encyclopaedic knowledge). Principle of positional numeration. *Treatise on Algebra.* Astronomical tables (sines). Stars and celestial motions. Idea of the tangent in trigonometry. Mathematics, mechanics, medicine. Translates science and philosophy. Alchemy, medicine, pharmacy. Observes the eccentricity of the solar orbit.		**Al-Kindi** (*c.* 801–*c.* 866). **Al-Khwarizmi** (*fl.* 800– 847). Foundation of the School of Salerno. **Al-Farghani** († after 861). **Thâbit ibn Qurra** (836– 901). **Ishâq ibn Hunayn** († 910). **Al-Razi** (Rhazes, *c.* 860– *c.* 923). **Al-Battani** (856–929).	845. The Vikings lay siege to Paris. The theologian John Scotus Erigena arrives in Paris. 867. Photian schism between Rome and Byzantium. 868. First known woodcut book (sûtra). *Cantilena of St Eulalia* (written in French). Many cathedrals, abbeys, archives and centres of study established.
Book of fixed stars.	900	**Al-Sufi** (903–986).	Foundation of the Abbey of Cluny.
Calculated table of tangents and cotangents. First influence of Arab science in the West. Optics, vision, astronomy, meteorology. Measurements of the specific weight of many substances.		**Abu-al Qafâ** (940–998). **Gerbert of Aurillac** (950–1003). **Ibn al-Haytham** (Alhazen, 965–1039). **Al-Birûni** (973–after 1050).	*Thousand and One Nights.* Gerbert became Pope Sylvester II in 999. 962. Restoration of the Western Empire. 975. In Egypt, the Caliph al-Aziz found a library which grew to 1,400,000 books.
Anatomy and physiology of the eye. *Canon of Medicine, Book of Healing.* The 'Brothers of Purity' were to produce a compendium of all knowledge.		**Alî ibn 'Isâ** (*c.* 940– 1010). **Avicenna** (Ibn Sîna, 980–1037). Foundation at Basra of the 'Ikhwân al-Safá'.	987. Capetian dynasty in France.
	1000	Bishop Fulbert founds the school at Chartres.	Leif Ericsson discovers Vinland.
Philosophical and Scientific Encyclopaedia. The first great agent of		**Michael Psellus** (1018– 78).	1022. Rise of Catharism. 1055. Baghdad captured by the Turks.

Scientific event		Individual and collective actors	Background information
transmission of Arab science to the West,. Translation, medicine. Astronomy: author of the *Toledan Tables*. Mathematics and astronomy. Encyclopedia of theology, logic, mathematics, philosophy: Greek and Arab learning pass into Latin.	1000	**Constantine Africanus** (*fl.* 1065–85). Ibn-Sa'id (1029–70). **Al-Khayyami** (1048–1131). **Adelard of Bath.**	1061. The Normans conquer Sicily (–1091). 1066. Battle of Hastings. 1077. Canossa. 1079. Each cathedral must have a school. 1085. Reconquest of Toledo. 1099. First Crusade in Jerusalem.
Equations on the planetary periods. Systematic policy of translation (Al-Khwarizmi, Ptolemy, Euclid . . .).	1100	**Bhâshara** (1114–85). **Hermann the Dalmatian** (*fl.* 1138–43). **John of Seville** (*fl.* 1135–53). **Gerard of Cremona** (1114–87). From 1150 on, first steps towards the establishment of the Universities of Paris, Oxford, Bologna . . .	Abelard (1079–1142), scholar and philosopher. 1148. The Council of Rheims condemns Gilbert de la Porrée. 1163. Construction of Notre-Dame de Paris begins. *Guide of the Perplexed*, by Maimonides (1135–1204). 1179. Introduction of paper into Europe.
On the equations of 2nd degree. Fractions.		**Savasorda** (Abraham Bar-Hiyya, *fl.* 1133–6).	1180. Reign of Philippe Auguste (–1223).
Translation of the *Almagest*.		**Averroes** (Ibn-Rusd, 1126–98).	
Book on the abacus (1202). Introduces Arabic numbers, algebraic computation. Book of geometry (1220). Univeral and experimental optics.	1200	**Fibonacci** (Leonard of Pisa, 1170–1240).	1208. Crusade against the Albigensians. Extensive use of the magnetic compass. 1212. The Christians crush the Muslims at las Navas de Tolosa.
Mathematics, art, mineralogy (1230). Mathematical works.		**Robert Grosseteste** (1175–1253) founds the University of Oxford. **Al-Tusi** (1201–74). **Sacrobosco** (John of Halifax, or Hollywood, 1190–1250). **Albert the Great** (1206–80).	*c.* 1220. Palermo the 'trilingual' capital of Frederick II (1294–1350). 1231. Lay constitution of Melfi. 1244. Montségur. 1258. Mongol invasion, sack of baghdad by the grandson of Genghis Khan (1160–1227).
Geology, botany, biology, base-acid 'chemistry'. Optics, acoustics . . . Rainbow. Many translations.		**Roger Bacon** (1219–92). **John of Palermo** (*fl.* 1221–40).	1271. Voyage of Marco Polo (1254–1324). 1277. Condemnation of

Scientific events		Individual and collective actors	Background information
	1200	1229. University of Toulouse.	the works of St Thomas Aquinas (1226–74).
Trans. of Al-Haytham and study of optics (refraction).		**Witelo** (1230–75). (Foundation of the Sorbonne.)	1289. Dante (1265–1321) sides with the Guelphs in the war against the
Book of arithmetic.		**Alexandre de Villedieu** († 1280).	Ghibellines. 1290. Cimabue (1240–
De magnete (1269).		**Pierre de Maricourt.**	1302) takes Giotto
Opposition to Aristotle's physics.		**Campanus of Novara** († 1296).	(1276–1336) as his pupil.
Very many translations (i.e. Archimedes).		**William of Moerbeke** (1230–86).	
Alchemical compilations.		**Ramon Lull** (1235–1315).	
Medical work at Montpellier.		**Arnald of Villanova** (1240–1311).	
Theory of the rainbow.		**Theodoric of Freiberg** (1250–1310).	
Astronomy, optics, medicine.		**Al-Shîrâzi** (Qutd al-Din, 1236–1311).	
c. 1270. King Alfonso has astronomical tables drawn up. (Judas ben Moses, Isaac ibn Sid).		1289. University of Montpellier. Universities of Padua, Naples, Toulouse, Rome.	
	1300		
Astronomical Tables.		**Levi ben Gerson** (1288–1344).	1307. Dante writes the *Divine Comedy* (–1321).
		1308. University of Cracow.	*c.* 1320. First weight-driven clocks.
Work in optics and mathematics.		**Al-Farizi** (Kamal al-Dîn, †1320).	1327. Beginning of the Hundred Years' War.
Important thinking in physics.		**Bradwardine** (1290–1349).	1346. Battle of Crécy. 1348. Black Death
		Buridan (1295–1358).	arrives in the West; a
Theory of impetus.		**Nicole Oresme** (1302–82).	third of the population will die of it.
Mathematics and movement (circular).			*c.* 1355. Boccaccio (1313–73): the
Ptolemaean astronomy.		**John of Lignières** (*fl.* 1320).	*Decameron*.
Theory of music and calculation of eclipses.		**John of Murs** (*fl.* 1340).	1389. Birth of Cosimo de Medici († 1464).
Natural philosophy and mathematics.		**Dominicus of Clavasio** (*fl.* 1346–1357).	
		1386 University of Heidelberg.	
From 1401, events and discoveries are listed	1401 1405		Windmills in Holland. Portable firearms.
next to their own dates, while actors are listed at their dates of birth.	1409	Universities of Aix-en-Provence and Leipzig.	

Scientific events		Individual and collective actors	Background information
Principle of the rod and crank.	1410		Battle of Tannenberg.
	1415		Agincourt. Execution of Jan Hus (1369–1415).
	1418		Cupola of Duomo at Florence by Brunelleschi (1377–1416).
	1420		Portuguese caravel.
Bourges astronomical clock.	1423	**Puerbach** († 1476).	
	1426	University of Louvain.	
Anonymous treatise on mechanics: hydraulic mills, boring and polishing machines.	1430		
	1432		Malatesta (1417–68), condottiere at Rimini.
Nicholas of Cusa hypothesizes the movement of the Earth.	1435		The Medici in power in Florence. Zuiderzee polders.
	1436	**Regiomontanus** († 1476). University of Caen.	Invention of movable type by Gutenberg at Strasbourg.
	1439		Development of the Venetian coast.
	1445	(c.) **Pacioli** († 1517), **Chuquet** († 1500)	
	1446	Columbus († 1506).	
	1447		Nicholas V (Pope); vast policy of patronage.
M. Taccola's treatise on machines.	1449		
	1450	(c.) **Brunschwig** († 1512).	Gutenberg at Mainz. Vatican Library.
	1451		End of the Hundred Years' War.
	1452	University of Valence. **Da Vinci** († 1519).	
	1453		Constantinople taken by the Turks.
	1454	**Vespucci** († 1512).	RENAISSANCE
	1455	Gutenberg's printing-press increases the diffusion of scientific works.	Italy – *painting*: Fra Angelico (1387–1455), Masaccio (1401–64), Botticelli (1445–1510),
	1460	Destruction of the observatory of Samarkand (end of mediaeval Arab science).	Piero della Francesca (1416–92), Mantegna (1439–94), da Vinci (1452–1519), Raphael
	1462	(c.) **Widman** († after 1498).	(1483–1520), Michelangelo (1475–

676

Scientific events		Individual and collective actors	Background information
	1463	University of Bourges.	1564); *architecture, sculpture*: Bramante
Regiomontanus perfects plane and spherical trigonometry.	1464		(1444–1514); *letters*: Nicholas of Cusa (1401–
	1465	**Scipione del Ferro** († 1526).	64), Pico della Mirandola (1463–95), Machiavelli
Production of improved vegetables in Italian gardens (artichoke, carrot, French bean, cauliflower).	1466		(1469–1527), Netherlands – *painting*: Van der Weyden (1400– 64), Memling (1433–94), Bosch (1450–1516);
First scientific work printed at Venice: **Pliny**.	1469	Da Gama († 1525).	*letters*: Erasmus (1469– 1536).
	1470	First printing press at the Sorbonne. **Vergilius** († 1555), Magellan	Germany – *painting*: Dürer (1471–1528). France–*painting*:
	1471	(† 1521). (*c.*) Bartolomeo Diaz	Fouquet (1415–80), Clouet (1475–1540);
Edition of **Aristotle's** works (Venice).	1472	(† 1500), **Dürer** († 1528).	*sculpture*: Collombes (1430–1514); *letters*:
First edition of **Galen** (700 editions by 1600).	1473	**Copernicus** († 1543).	Villon (1431–90), Ph. de Commynes (1447–1511),
Treatise on mechanics: ball governor, water turbines, lifting tackle, automobile carriage.	1475		Budé (1467–1540), Rabelais (1494–1553), Marot (1496–1544).
	1476	University of Uppsala.	
	1478		Viso tunnel (alt. 2,000m).
	1479	**Calcagnini** († 1543).	
	1480	**Biringuccio** († 1539).	
Edition of **Euclid's** *Elements* (Venice).	1482		
Chuquet: *Triparty sur la science des nombres*.	1484	**Scaliger** († 1558).	Improvement of mining engineering.
Edition of **Lucretius** *De natura rerum* (Venice).	1486	**Agrippa** († 1535).	Diaz reaches the far south of Africa.
Widman uses the signs + and –.	1489		
(Between 1469 and 1575 nearly all known scientific works in Greek, Arabic, Hebrew and Latin are printed.)	1491	Jacques Cartier († 1557).	
	1492	**Riese** († 1559).	Columbus 'discovers' the island of Guanahani. Fall of Granada.
	1493	**Paracelsus** († 1541).	
	1494	**Agricola** († 1555), **Maurolico** († 1575).	Savonarola (1452–98) head of the Florentine republic.

Scientific events		Individual and collective actors	Background information
Dürer, work on proportions and perspective.	1495	(*c.*) **Vicary** († 1561).	
	1497	**Holbein** († 1543).	
	1498		Vasco da Gama opens the route to the Indies.
Vergilius publishes a work of scientific popularization.	1499		Vespucci arrives in 'Guiana'. First edition of *Celestina*, by Rojas (*c.* 1473–after 1538).
Leonardo da Vinci's flying machines.	1500	**Tartaglia** († 1557), **Besson** († 1576).	Appearance of the first wooden screw-presses.
	1501	**Cardan** († 1576).	
	1503		Invention of the Venetian mirror.
	1507	**Rondelet** († 1566).	
	1508	**Piccolomini** († 1578).	
First publication of textbooks of science and technology.	1509	**Commandino** († 1575), **Paré** († 1590).	First watches.
	1510	**Palissy** († 1590).	
	1511	**Servet** († 1553).	
	1514	**Vesalius** († 1564), **Rheticus** († 1576).	
	1515	**Ramus** († 1572).	François I (1494–1547), beginning of reign.
	1516	**Gesner** († 1565).	Machiavelli (1469–1547): *The Prince*. First African slaves in America.
	1517	**Belon** († 1564).	Luther (1483–1546): publication of the theses.
	1518		Importation of cochineal (Mexico).
	1519	**Cesalpino** († 1603).	Charles V (1500–1558) emperor. Cortés (1485–1547) in Mexico.
	1520		Fall of Aztec empire and of Mexico.
	1521		Excommunication of Luther.
	1522	**Ferrari** († 1565), **Aldrovandi** († 1605).	S. de Cano (*c.* 1460–1526) ends the voyage begun by Magellan (1470–1521) in 1519.
Dürer: *Treatise on Perspective.*	1525		Introduction of the potato. Birth of Brueghel the Elder († 1569); followed

Scientific events		Individual and collective actors	Background information
	1525		by his two sons called Hell Brueghel (1564–1638) and Velvet Brueghel (1568–1625).
	1526	**Bombelli** († 1572).	
Paracelsus (chemistry, alchemy, magic, medicine, surgery) settles in Basle.	1527	**Dee** († 1608).	The sculptor-goldsmith B. Cellini (1500–71) organizes the siege of Rome. Use of powder in mines.
	1529	Collège royal (the future Collège de France).	Pizarro (1475–1541) in Peru.
Agrippa: occultism and relections on the sciences.	1530		
	1531		Fall of the Inca empire.
Botanical garden at Padua.	1532	**Xylander** († 1576).	
	1533	**Fabricio** († 1619).	Holbein (1497–1543): *The Ambassadors*.
Tartaglia: equation of 3rd degree.	1534	**Della Porta** († 1615).	Ignatius de Loyola (1491–1556) starts the Society of Jesus. J. Cartier (1509–64) explores the Saint Lawrence. Anglican schism.
	1535		Calvin (1509–64) publishes *The Institutions of the Christian Religion*.
	1537	**Clavius** († 1612).	
	1538	**Olivier de Serres** († 1619).	
Botanical garden at Touvois (near Le Mans). *De la pirotechnia* (a metallurgy by **Biringuccio**).	1539		
Rheticus publishes a summary of the work of **Copernicus.**	1540	**Viète** († 1603), **van Ceulen** († 1610). First foundation of Academy of Science at Padua.	The Jesuits recognized by Rome.
Piccolomini: *De la sfera del mundo.* Posthumous edition of **Copernicus:** the sun is the centre of the celestial spheres. **Vesalius:** *The Epitome,* (human anatomy). Botanical garden at Pisa.	1543		Arrival of the Portuguese in Japan.
Cardan: *Ars magna,*	1544	**Gilbert** († 1603).	Royal ordinance on the

679

Scientific events		Individual and collective actors	Background information
equation of 3rd degree. Theory of terrestrial impetus, **Calcagnini** (1479–1541).	1544		cutting of forests (France).
Paré: *Method for treating wounds.* J. **Bock:** *Kreuterbuch.*	1545	**Gerard** († 1612).	Council of Trent (–1563).
Tartaglia: investigation of the trajectory of projectiles.	1546	**Tycho Brahe** († 1601).	
	1548	**Stevin** († 1620), **G. Bruno** († 1600).	
	1549		Francis Xavier (1506–1552) arrives in japan. Du Bellay (1522–1560), *Défense et illustration de la langue française.*
Ferrari: equation of 4th degree. **A. Reise** recommends written calculation rather than counters.	1550	**Beguin** († ca. 1620), **Napier** († 1617).	Goujon sculpts the *Fontaine des Innocents.*
Cardan: investigation of falling bodies. **Gesner:** *Historia animalium.* **Rheticus:** idea of the cosine. **P. Belon:** book on 'Strange Sea-fish'; then on 'Birds' (1555) and 'The Culture of Plants' (1558).	1551		Standardization of printed letters. First import licence in France (glassware).
	1552	**Ricci** († 1610), **Sarpi** († 1623).	Ronsard (1524–85) writes his *Amours.*
M. Servet: first idea of the circulation of the blood.	1553		Execution of M. Servet (1511–53).
Rondelet: *Libri de piscibus marinibus.*	1554		*Lazarillo de Tormes* (first picaresque novel).
Palissy burns his furniture to fire ceramics.	1555		Peace of Augsburg. First edition of 7 *Centuries* of Nostradamus (1603–66). Birth of Malherbe († 1628).
Agricola: *De re metallica* (treatise on metallurgy). **Tartaglia:** *Traité des nombres et des mesures.*	1556		Formation of the *Pléiade.*

Scientific events		Individual and collective actors	Background information
	1557		The Portuguese reach Macao.
Della Porta: *Magia naturalis.*	1558	University of Jena.	Reformer John Knox (1505–72), the disturbances at Perth.
	1559		Amyot (1513–93) translates Plutarch. Nicot introduces tobacco.
First revolving observatory	1561	**Bacon** († 1626), **Roomen** or **Romanus** († 1615).	Theresa of Avila (1515–82). Reform of the Carmelites.
	1562	Foundation of Collège de Clermont.	Beginning of the slave trade. Beginning of the Wars of Religion.
	1563		Beginning of the construction of the Escorial.
	1564	**Galileo** († 1642).	Invention of the pencil.
Commandino: centre of gravity of solids.	1565		First commercial exchange in London.
Scaliger: Commentary on Aristotle's *On Plants* (posthumous). **Rondelet** at Montpellier, first anatomical lecture theatre.	1566		Revolt of the United Provinces (–1574).
	1568	**Campanella** († 1639).	
Ramus: *Geometriae libri XXVII*	1569		
	1571	**Kepler** († 1630), **Branca** († 1640).	Conquest of the Philippines. Cervantes (1547–1616). Battle of Lepanto.
Bombelli: Algebra, study of Diophantus.	1572		St Bartholomew's Day Massacre.
	1573	**Scheiner** († 1650).	
Clavius: annotated edition of *Euclid.* **Fabricio:** bases of physiology of circulation.	1574		
	1576	**S. de Caus** († 1626).	Bodin (*c.* 1530–96) publishes his *Republic.*
	1577	**Van Helmont** († 1644), **Guldin** († 1643).	Voyage of Sir Francis Drake (1545–92), until 1580.
	1578	**Harvey** († 1657).	
Viète: *Canon mathematicus,* trigonometry.	1579		
B. Palissy: *Etude sur les fossiles.*	1580	**Snellius** († 1626), **Fabri de Peiresc** († 1637).	Montaigne (1533–92), *Essays* (–1588).

Scientific events		Individual and collective actors	Background information
Galileo: Isochronism of pendulum's oscillations.	1581	**Gunter** († 1626), **Bachet** († 1638).	
	1582	**J. Rey** († 1645).	Gregorian calendar (according to calculations by Clavius).
Cesalpino: first coherent classification of the plants.	1584	**P. Vernier** († 1638).	Voyage of Walter Raleigh (*c.* 1554–1618), who brought back the tobacco plant and in 1614 wrote a *History of the World*.
T. Brahe: abandonment of solid spheres. **Benedetti:** intuition of the principle of inertia.	1585		*Atlas of the World* (–1590) by Mercator (1512–94).
Stevin: *Decimal Arithmetic*. **Stevin:** investigation of the inclined plane.	1586	**Zucchi** († 1670).	El Greco (1541–1614) paints *The Burial of Count Orgaz*.
Stevin: principles of mechanics.	1587	**Froidmont** († 1653), **Fabricius** († 1615).	
	1588	**M. Mersenne** († 1648), **Beeckman** († 1637), **Hobbes** († 1672).	The Invincible (Spanish) Armada.
Galileo: *De motu* (investigation of falling bodies).	1590		
Jansen inaugurates the development of the compound microscope.	1591	**Désargues** († 1661).	
Viète: use of letters, algebraic formulae in geometry, work on the circle.	1592	**Gassendi** († 1655), **Schickard** († 1635).	
Roomen gives π to 15 places.	1593		'Paris is worth a mass!'
Posthumous edition of *Théâtre des instrumens mathématiques et mécaniques* by the instrument-maker **Besson** (1500–76).	1594		
	1595	**A. Girard** († 1632), **Beaugrand** († 1640), **Linus** († 1675).	
Gerard: *Catalogue of trees* (London).	1596	**Descartes** († 1650).	
Gerard: *Herball* (London).	1597		

Scientific events		Individual and collective actors	Background information
Aldrovandi: Zoological encyclopaedia (vocabulary, classifications, monsters).	1598	**Cavalieri** († 1649).	Edict of Nantes. End of the Wars of Religion.
	1599		Gathering of silk (O. de Serres). East India Company (Britain).
Gilbert: *De magnete.* **O. de Serres:** *Théâtre de l'agriculture.* **Fabrici:** first study of embryology.	1600	**Carcavi** († 1684), **De Lalouvère** († 1664), **Montmor** († 1679).	Execution of Giordano Bruno (for atomist heresy).
	1601	**Fermat** († 1665).	Ricci meets thre Emperor of China.
	1602	**Roberval** († 1673), **De Guericke** († 1686), **Kircher** († 1650), **De Billy** († 1679), **Bosse** († 1676).	The Gobelins manufactory.
Van Helmont attacks the theory of the four elements, invents the word 'gas'.	1603	**Digby** († 1665). Academia dei Lincei.	Shakespeare (1564–1616): *Hamlet.* Tokugawa Shogunate (–1868), transfer of capital from Kyoto to Tokyo.
Galileo: first (incorrect) statement of the law of falling bodies. **Kepler:** *Paralipomena to Vitellion* (optics). **Joït Brughi:** first attempt at logarithms. **Bacon** begins publication of his enormous encyclopaedia.	1604	**Glauber** († 1668).	
	1605	**Frénicle** († 1675).	Bellarmine (1542–1621), Librarian at the Vatican. Gunpowder plot. Cervantes: *Don Quixote.* Shakespeare: *Macbeth, King Lear.*
Galileo: operations with the compass.	1606		
	1607		Monteverdi (1567–1643): *Orfeo.* Navarre returned to France.
Stevin: hydrostatics and the principle of virtual work.	1608	**Torricelli** († 1647).	Foundation of Quebec. Mathurin Régnier (1573–1613): *Satires.*
Metius: astronomical telescope. **Kepler:** *Astronomia nova*, the circle is broken: 1. The planetary orbits	1609		Independence of the United Provinces. Port-Royal reform. Creation of the Bank of Amsterdam.

Scientific events		Individual and collective actors	Background information
are ellipses. 2. The area swept by the planet-Sun radius is proportional to time.	1609		Beginning of the reduction of Paraguay.
Bequin: *Eléments de chymie.* **Galileo** discovers satellites of Jupiter, and sunspots with **Scheiner.**	1610	**Bourdelot** († 1685).	Assassination of Henri IV. François de Sales (1567–1622) founds the Order of the Visitation.
Maurolico: *Photismi de lumine* (posthumous edition), optics. **Fabricius:** *De maculis in sole observata.*	1611	**Hevelius** († 1687), **Pell** († 1685), **P. Perrault** († 1680).	
	1612	**Tacquet** († 1660), **Gascoigne** († 1644).	Construction of the Mosque of Isfahan.
Beeckman: the conservation of motion.	1613	**Perrault** († 1688).	
Napier: invention of logarithms.	1614	**Wilkins** († 1672), a promoter of English science.	States-General. Majority of Louis XIV.
Kepler: *Stereometria*, calculation of area and volume. **S. de Caus:** conceives the steam engine. Exhortation of **Galileo.**	1615	**Glaser** († 1672).	Rubens (1577–1640) paints *The Battle of the Amazons.*
	1616	**Willis** († 1703), **Bartholin** († 1680).	Deaths of Cervantes and Shakespeare. Van Dyck (1599–1641) paints the *Crucifixion.*
Napier: principle of the calculating-machine. First microscope.	1617	**Ashmole** († 1692).	
	1618	**Grimaldi** († 1663), **Blondel** († 1686).	Beginning of the Thirty Years' War (–1648).
Kepler: *Harmonices mundi* (3rd law). **Galileo:** *Discorso sulle comete.*	1619	**Wing** († 1668).	
Gunter: table of sines and tangents, principle of the slide-rule.	1620	**Brouncker** († 1684), *c.* 1620 **Mariotte** († 1684).	The *Mayflower.* First weekly paper (Amsterdam). Establishment of the libertinism. Philip IV (1616–65) king of Spain.
Snellius establishes the Law of Refraction. **Zonca:** *Théâtre des machines.*	1621		
Campanella: *Apologia pro Galileo.*	1622	**Varenius** († 1650), **Viviani** († 1703).	

Scientific events		Individual and collective actors	Background information
Schickard builds a calculating machine.	1623	**Pascal** († 1662).	Urban VII (1568–1644) Pope.
	1624		Beginning of the law of patents in Britain.
Girard: statement of the fundamental theorem of algebra (on the roots of equations).	1625	**Cassini** († 1712).	Van den Vondel (1587–1679) has *Palamède* performed.
	1626	**La Quintinie** († 1688), **Redi** († 1697). Creation of the Jardin du Roi (the future Jardin des Plantes).	Duelling forbidden (France). Urban VIII pardons Campanella, condemned since 1599.
Froidmont: *Meteorologicorum libri sex.*	1627	**Boyle** († 1691), **J. Ray** († 1705).	Richelieu (1585–1642) lays siege to La Rochelle.
Harvey: circulation of the blood. **Descartes:** idea of the conservation of motion.	1628	**Malpighi** († 1694).	Velasquez paints the *Crucifixion.*
Branca: *Le machine.*	1629	**Huygens** († 1695).	Peace of Alès with the Protestants.
Rey: *Essays* (on the calcination of tin and lead).	1630	**Barrow** († 1677), **Richer** († 1696).	Public postal service throughout France.
Invention of the vernier	1631	**Wilson** († 1711).	Beginning of the construction of Versailles. Renaudot's (1586–1653) gazette.
Galileo: *Dialogue on the two chief world systems,* invention of the water-thermometer.	1632	**Leeuwenhoek** († 1723).	Rembrandt (1606–69), *The Anatomy Lesson.* Christina queen of Sweden.
Trial of **Galileo.**	1633	**Boccone** († 1704), **Vauban** († 1707).	Bernini (1598–1680): tabernacle of St Peter of Rome.
Mersenne translates Galileo's *Mechanics.*	1634	**Amman** († 1691).	Vincent de Paul (1581–1660): the Daughters of Charity. Poussin (1594–1665): *Helios and Phaeton.*
Cavalieri: *Geometria indivisibilibus.* **Mersenne:** first measurement of the speed of sound.	1635	**Hooke** († 1702), **Becher** († 1685).	Esatblishment of the Académie Française. Closure of Japan. Lope de Vega (1562–1635) dies having written 1800 plays.
Roberval: treatise on mechanics and method for the drawing of tangents.	1636	**Glanvill** († 1680). Harvard founded.	Corneille (1606–84), *le Cid.*

Scientific events		Individual and collective actors	Background information
Descartes: The Three Essays of the *Discourse*. **Fermat:** method for the discovery of tangents.	1637	**Swammerdam** († 1680).	Descartes: *Discourse on Method*.
Galileo: *Discourse on two new sciences*.	1638	**Malebranche** († 1715).	The Dutch sweep the Portuguese from Ceylon.
Desargues: *Draft Project* (projective geometry).	1639	**Grundel** (*fl.* 1670–80).	The English found Madras.
Pascal: *Essay on Conics*.	1640	**La Hire** († 1718). Académie de Toulouse, private academy. **Bourdelot,** Paris.	Jansenius (1585–1638) publishes the *Augustinus*. G. de La Tour (1593–1652): *Saint Sebastian*.
Guldin: on centres of gravity.	1641	**Grew** († 1712).	Corneille: *Polyeucte*. Descartes: *Meditations*.
Pascal: calculating machine.	1642	**Newton** († 1727).	Rembrandt: *The Night Watch*.
Torricelli: the barometer.	1643	**Gabrieli** († 1705).	Foundation of New Amsterdam (New York). Reign of Louis XIV (–1715). Molière founds the famous Théâtre.
Torricelli: studies on the weight of air, centres of gravity and motion. **Digby:** *Two Treatises*, alchemy.	1644	**Römer** († 1710).	The Manchus established in China.
Bartholin: work on anatomy.	1645	**Lemery** († 1715). Foundation of the Philosophical College (the future Royal Society) in London.	Bosse: treatise on engraving. Condemnation of the 'Chinese rites'.
Kircher: *Ars magna lucis*.	1646	**Leibniz** († 1716), **Flamsteed** († 1719).	Bernini: *The Ecstasy of St Teresa*.
Pascal: new experiments on the vacuum.	1647	**D. Papin** († 1714).	C. Lorrain: *The Flight into Egypt*.
Pascal: the Puy de Dôme experiment.	1648	**Duverney** († 1730).	Treaty of Westphalia. Beginning of the Fronde (–1653).
Redi: observations on snakes.	1649	**Bidloo** († 1713).	Execution of Charles I (England). Mexico, population 1.5 million: in 1521 it had been 11 million.
De Guericke: invention of the air-pump. First observation of a double star.	1650	500 MILLION HUMAN BEINGS The Montmor Academy active.	
Mersenne: optics and catoptrics.	1651	**Bion** († 1733).	Hobbes (1598–1679): *Leviathan*.

Scientific events		Individual and collective actors	Background information
Ashmole: *Theatrum chemicum Britannicum.*	1652	**Rolle** († 1749). Academia Naturae Curiosorum (Leipzig, then Breslau, Nuremberg, Bonn . . . see 1682).	Scarron marries Françoise d'Aubigné (1635–1719), grand-daughter of the poet Agrippa d'Aubigné and the future Mme de Maintenon, wife of Louis XIV (1638–1715).
Pascal: studies in hydraulics.	1653	**Hoffmann** († 1727), one of its members.	Cromwell (1599–1658) takes power.
Pascal, Fermat: calculus of probabilities.	1654	**J. Bernouilli** († 1705).	De Scudéry: *Clélie.*
Magdeburg hemispheres experiment.	1655	**Varignon** († 1722).	Murillo (1617–1682): *The Birth of the Virgin.*
Huygens: improvement of optical glasses. **Wallis:** arithmetic of the infinitely small. Discoveries: rings of Saturn, Orion nebula.	1656	**Halley** († 1742), **de Maillet** († 1738), **Tournefort** († 1708).	Pascal: *Provinciales.* Fouquet (1615–80): construction of the Château of Vaux-le-vicomte by Le Vau (1612–70), Le Nôtre (1613–1700), Lebrun (1619–90). Velasquez (1599–1659): *Las Meniñas.*
Huygens: pendulum clock.	1657	**Fontenelle** († 1757), Academia del Cimento.	Scarron: *Roman comique.*
Huygens: geometrical studies (evolutes, catenary).	1658		Hobbes: *De Homine.* Spinoza (1632–77) expelled from the synagogue.
Boyle: vacuum-pump. **Vivianni:** *De maxima et minima.*	1659	**Gregory** († 1708).	Molière: *les Précieuses ridicules.*
Hevelius: very accurate catalogue of 1500 stars.	1660	**Stahl** († 1734).	Vermeer (1632–75): *Young Girl Reading a Letter.*
Hobbes: *Dialogus physicus* (on the nature of air; on the duplication of the cube). **Boyle:** *The Sceptical Chymist.*	1661	**de L'Hôpital** († 1704).	Death of Mazarin (1601–61). Famine in France.
Fermat: optical principle of least time. **Boyle** establishes his Law: PV is constant.	1662	Royal Society (Oxford).	Cavalli (1602–76): *Ercole Amante* (opera).
Glaser: *Traité de la chimie.* **Pascal:** *On the Equilibrium of Liquids.*	1663	**Newcomen** († 1729).	Nouvelle-France in Canada.
Newton discovers the calculus of fluxions.	1664		'Fêtes de l'île enchantée' at Versailles.

Scientific events		Individual and collective actors	Background information
Varenius: *Geographia generalis.* **Glanvill:** *Scepsis scientifica*, on the new experimental philosophy.	1664		Colbert creates the Compagnie des Indes. Crisis at Port-Royal.
Grimaldi: *Physico-mathesis de lumine.* **Kircher:** *Mundus subterraneus.*	1665	**Camerarius** († 1721). Creation of the *Journal des Savants* (Paris), and of the *Philosophical Transactions* (London).	Great Plague in London (100,000 dead). La Rochefoucauld (1613–80), *Maxims*.
Malpighi: beginnings of microscopic biology. The invention of the micrometer (**Gascoigne**) improves the accuracy of astronomical telescopes.	1666	Foundation of the Académie des Sciences by Colbert.	Leibniz: *De arte combinatoria.* Great Fire of London.
Swammerdam: microscopic biology.	1667	**Jean Bernouilli** I († 1748), **Moivre** († 1754), **Saccheri** († 1733).	Lighting of the Paris streets. Racine (1638–99), *Andromaque.*
Wallis: theory of the impact of bodies. **Redi:** experiment on the generation of insects.	1668	**Boerhaaave** († 1738).	La Fontaine (1621–95), *Fables.*
Wing: *Astronomia Britannica* (posthumous). **Cassini:** Ephemerides of Jupiter's satellites. Recognition of double refraction.	1669	Foundation of the Paris Observatory (meridian).	French royal decree on rivers and forests. Academy of Music and Opera, Paris. Stradivarius (1644–1737) violin-maker at Cremona.
Newton: corpuscular theory of light. **Leibniz:** theory of motion. Project to measure the solar system.	1670	*Misellanea curiosa medico-physica* (Leipzig).	Pascal: *Pensées* (posthumous.) Spinoza: *Tractatus theologico-politicus.*
Leibniz: calculating machine (+, −, ×, :). **C. Perrault:** *Histoire naturelle des animaux.*	1671	First memoir of the Académie des Sciences.	Milton (1608–74), *Paradise Lost.* Marquise de Sévigné (1626–96), *Letters.*
Newton: first telescope with parabolic mirror. **De Guericke:** machine producing static electricity.	1672	**E. F. Geoffrey** († 1731).	Franco-Dutch War (–1678).
Huygens: isochronism of the pendulum, cycloid, live force, centrifugal force.	1673		
P. Perrault: *On the Origin of Springs.*	1674	**Pontchartrain** († 1747).	Foundation of Pondicherry.

688

Scientific events		Individual and collective actors	Background information
Leibniz: infinitesimal calculus. **Huygens:** spring regulator for clocks. **Lemery:** *Cours de chimie.* **Brandt:** phosphorus.	1675	**Clarke** († 1729). Foundation of the Greenwich Observatory (meridian).	Population of Paris 540,000. Racine: *Iphigénie in Aulide.*
Römer measures the speed of light. **Mariotte:** *Sur la nutrition des plantes.*	1676	**Threlkeld** († 1728).	The poison affair.
Leeuwenhoeck: observation of spermatozoa.	1677	**S. Hales** († 1761), **L. Lemery** († 1743).	Spinoza's *Ethics.*
Cassini: *Découverte de deux nouveaux planètes autour de Saturne.*	1678	**Henckel** († 1744).	Mme de La Fayette (1634–93), *La Princesse de Clèves.*
Halley: catalogue of southern stars. **Mariotte:** law of gases. **Fermat:** *Works* (posthumous).	1679	**Wolff** († 1754).	By 1780 more than two million Africans had been sent to the Americas.
Borelli: study on muscles. **Groundel:** microscopy.	1680	**Chambers** († 1740).	French forbid Protestant synods. Establishment of the Comédie Française. Appearance of the terms 'Whigs' and 'Tories'.
Grew: *The Anatomy of Plants* [1682]. **Leeuwenhoeck** discovers bacteria.	1681	Moscow Academy of Science.	Bossuet (1627–1704): *Sur l'histoire universelle.* W. Penn (1644–1788) founds Pennsylvania.
Halley: observation of the comet. **J. Ray:** first systematic botanical method.	1682	Leopoldina Academy (Vienna). *Acta Eruditorum* (Leipzig).	Water-raising machine at Marly. Halley's comet appears.
Blondel: *L'art de jetter les bombes.*	1683	**Réaumur** († 1757), **Rameau** († 1764).	Fontenelle: *Dialogue avec les morts.*
Grégory: *Exercitatio geometrica de dimensione figurorum.* **Amman:** *Character plantarum naturalis.*	1684		Puget (1620–94) sculpts *Andromeda.* Completion of the Canal du Midi, begun in 1666.
Bidloo: *Anatomia humani corporis.*	1685	**Taylor** († 1731), **Clifford** († 1760).	Food shortage in France. Louis XIV secretly marries Mme de Maintenon. Revocation of the Edict of Nantes (economic and commercial consequences).

Scientific events		Individual and collective actors	Background information
Beginning of the argument of live force (–1740).	1686	**Fahrenheit** († 1736).	Establishment of the Gobelins factory.
Mariotte: movement of water, study of colour. **Newton:** *Principia mathematica* (principle of universal attraction, action-reaction, composition of forces in mechanics).	1687	**Simson** († 1768).	Fontenelle: *Sur la pluralité des mondes.* Famine in France. Fénelon (1651–1715) the leader of quietism. Death of Lully (1632–87), Surintendant de la musique.
Varignon: composition of forces in statics. **Leeuwenhoek** observes the red corpuscles.	1688	**Bragelone** († 1744).	La Bruyère (1645–96): *Caractères.*
Tournefort: *Schola botanica; sive catalogus plantarum.*	1689		Purcell (1659–95): *Dido and Aeneas* (opera).
Jacques Bernouilli: differential calculus (differential equation). **Huygens:** wave theory of light. **D. Papin:** the pressure cooker. **La Quintinie:** work on fruit and vegetables.	1690	**Goldbach** († 1764). Bologna Academy.	Pianoforte. Locke (1632–1704): *Essay on Human Understanding.* Creation of the Génie by Vauban (1633–1707).
Rolle: theorem of continuous functions.	1691		Gabrielli founds the Physiocratic Academy (Siena).
	1692	**Stirling** († 1770), **Musschenbroek** († 1761).	Salem witch trials (19 executed). Foundation of Saint-Gobain glass company.
Leibniz: theory of determinants.	1693	**Bradley** († 1762), **Harrison** († 1776).	Food shortage in France. Scarlatti (1659–1725): *Theodora.*
Camerarius: *Epistola . . . de sexu plantarum.* **Tournefort:** *Élémens de botanique.* **Hoffman:** foundations of medicine.	1694	**Quesnay** († 1774).	Creation of the Bank of England. *Dictionnaire de l'Académie française.*
Leibniz: dynamics. **Papin:** *De novis quibusdam machinis.* **Leeuwenhoek** publishes the results of his microscopical observations.	1695	**N. Bernouilli** († 1726).	Dom Perignon invents champagne process.
De L'Hospital: infinitesimal calculus.	1696	**Christine Kirch** († 1782).	

Scientific events		Individual and collective actors	Background information
Stahl: theory of phlogiston. **Jean and Jacques Bernouilli:** Brachistochrone and calculus of variations. **Boccone:** museums of physics and of plants.	1697	**Bélidor** († 1761).	French decree on public lighting. Bayle (1647–1706): *Dictionnaire historique et critique.*
	1698	**MacLaurin** († 1746), **Maupertuis** († 1759), **Bouguer** († 1758).	Savary invents a steam-pump for the mines.
Jacques Bernouilli: calculus of probabilities.	1699	*Mémoires de mathématiques et physique* (Paris). **B. de Jussieu** († 1777).	Peace of Karlowitz (Austro-Turkish).
Huygens: study on the impact of bodies. **Tournefort,** voyage in the East with **Aubriet** (1665–1742).	1700	**Daniel Bernouilli I** († 1782), **Trembley** († 1784). Lyons Academy of Science. Foundation of the Observatory and Academy in Berlin.	
Cassini undertakes the measurement of the meridian (–1718). **Ralphson:** *Mathematical Dictionary.*	1701	**La Condamine** († 1774), **Celsius** († 1744). First issue of *Mémoires de Trévoux.*	War of the Spanish Succession (–1714).
	1702	University of Breslau.	Camisard Rebellion (–1714).
	1703	**Hall** († 1771), **F. Rouelle** († 1770).	Foundation of St Petersburg.
Newton: optics, analytic geometry.	1704	**Cramer** († 1752), **J. de Jussieu** († 1779).	Galland (1646–1715) translates the *Thousand and One Nights.*
Vauban: *Traité des fortifications.*	1705		Newcomen: first steam engine.
	1706	**Franklin** († 1790). Montpellier Academy of Science.	Hardouin-Mansart (1646–1708): the Invalides.
Stahl: theory of medicine.	1707	**Euler** († 1783), **Buffon** († 1788), **Linnaeus** († 1778).	J. S. Bach (1685–1770), musician at Mühlhausen.
Boerhaave: *Institutiones medicae.*	1708	**Von Haller** († 1777).	Handel (1685–1759) meets Scarlatti (1685–1757).
	1709	**Vaucanson** († 1782).	First use of coke in a blast furnace.
Halley: *Synopsis of the Astronomy of Comets.*	1710	**Simpson** († 1761). Royal Society of Sciences (Uppsala).	Hard winter, famine in France.
Moivre: calculus of probabilities.	1711	**Wright** († 1786), **Lomonossov** († 1765),	The end of Port-Royal.

Scientific events		Individual and collective actors	Background information
	1711	**Boscovitch** († 1787).	
Réaumur: study of steel.	1712	**Venel** († 1778). Bordeaux Academy.	Wolff (1679–1754). Beginning of the Aufklärung.
Flamsteed: catalogue of stars.	1713	**Clairault** († 1765). Spanish Royal Academy of Sciences.	Couperin (1666–1733), music for harpsichord.
Fahrenheit: mercury thermometer.	1714	Institute of Arts and Sciences, Bologna.	Vivaldi (1678–1741), 24 concerti for violin.
Taylor: differential calculus.	1715	**Guettard** († 1786).	Idea of a typewriter. Treaty of Utrecht.
	1716	**Daubenton** († 1800).	Bank-notes issued by Law (1671–1729).
Jean Bernouilli: generalization of the principle of virtual work. **Wolff:** first mathematical dictionary. Recognition of the proper movement of the stars. **Geoffroy:** first tables of affinity.	1717	**D'Alembert** († 1783).	Growth of Paris (Faubourgs of St Germain and St Honoré).
	1718	**Macquer** († 1784), **H. Rouelle** († 1799).	London: first Masonic lodge. Foundation of New Orleans.
	1719		Defoe (1630–1731): *Robinson Crusoe.*
	1720	**Bonnet** († 1793).	Famine in France. Plague in Marseille, 85,000 dead. Law bankrupt.
Halley's diving-bell.	1721	**Bichat** († 1802).	Freemasonry arrives in France.
Réaumur studies the conversion of iron into steel.	1722		Construction of the Palais-Bourbon.
Fahrenheit: the boiling-point of water varies with pressure.	1723	**Mayer** († 1762), **A. Smith** († 1790), **Brisson** († 1806).	Bach: *St John Passion.*
	1724	St Petersburg Academy of Sciences. **Kant** († 1804).	
Bradley: studies on the aberration of light.	1725	**Montucla** († 1799), **d'Arcy** († 1779).	First machine operated by punched cards. Vico (1688–1744): *La Scienza nuova.*
	1726	**Hutton** († [1797]).	Swift (1667–1745): *Gulliver's Travels.*
Hales: on gaseous exchange between plants and the surrounding air. **Threlkeld:** *Synopsis Stirpium Hibernicarum.*	1727	**Adanson** († 1806), **Commerson** († 1773). Refoundations of the Toulouse Academy.	The 'possessed' people at Saint-Médard.

Scientific events		Individual and collective actors	Background information
Clairault: studies on gauche curves.	1728	**Lambert** († 1777), **Black** († 1799), **Baumé** († 1804).	E. Chambers: first modern encyclopaedia.
Gray: on the distribution of electrical charge. **Moivre:** trigonometrical formula.	1729	**Bezout** († 1783), **Spallanzani** († 1799), **Bougainville** († 1811), **Foster** († 1798).	Opium banned in China.
Réaumur: rules for the construction of a thermometer. **Clairault:** study of celestial mechanics.	1730	*Acts of the Royal Scientific Society* (Uppsala).	Salon of Marie du Deffand (1679–1780). Boucher (1703–70) returns from Rome.
Bélidor: *Traité de ballistique.* Linnaeus' travels in Lapland. Soon after, began research on a classification of living things.	1731	**Cavendish** († 1810).	Abbé Prévost (1679–1763): *Manon Lescaut.*
Boerhaave: *Elementia chemiae.*	1732	**Lalande** († 1807), **Köhlreuter** († 1806).	Birth of Washington († 1799).
Saccheri: studies on Euclid's postulate. **Réaumur:** study on insects.	1733	**Borda** († 1799).	War of Succession in Poland (1738).
Expeditions to Peru (1770, **La Condamine, Bouquer ...**), to Lapland (**Maupertuis, Clairault**), to investigate the flattening of the Earth.	1734	**Priestley** († 1804), **Mesmer** († 1815), **Rozier** († 1793).	Picardy Canal.
	1735	**Waring** († 1798), **T. O. Bergmann** († 1784), **Vandermonde** († 1796). University of Göttingen.	J. Kay's flying shuttle (industrial weaving). First exploitation of crude oil. First coke blast furnace.
Euler: *Mechanica sive motus scientia analytica.*	1736	**Lagrange** († 1813), **Coulomb** († 1806), **Watt** († 1819).	Pergolesi (1710–36): *Stabat Mater.* Apogee of Manchu Empire.
D. Bernouilli: kinetic theory of gases, studies in fluid dynamics.	1737	**Galvani** († 1798), **Parmentier** († 1813), **Guyton de Morveau** († 1816).	Rameau (1683–1768), *Castor and Pollux.*
Rouelle: *Cours de chimie.* **Bélidor:** *Architecture hydraulique.*	1738	**W. Herschel** († 1822), **Wolf** († 1794).	Vaucanson builds his duck (automaton). Hume (1711–76), *Treatise on Human Nature.*
Buffon director of the Jardin du Roi.	1739	**Samuel** († 1817). Swedish Academy of Sciences, Stockholm	Frederick II (1712–86), King of Prussia writes his *Anti-Machiavelli.*

Scientific events		Individual and collective actors	Background information
Trembley: study of the hydra.	1739		
	1740	**Saussure** († 1799).	War of the Austrian Succession (–1748). Sign language for deaf-mutes.
MacLaurin: *The Treatise of Fluxions.*	1741	**Pallas** († 1811), **La Pérouse** († 1788).	First bascule bridge.
Celsius: thermometric scale. **Clairault:** shape of the Earth, study on fluids. **Maupertuis:** principle of least action. **D'Alembert:** fluid mechanics. **Euler:** calculus of variations, movement of the planets.	1742	**Schelle** († 1786), **Le Blanc** († 1806).	Invention of rifled barrel.
	1743	**Lavoisier** († 1794), **Condorcet** († 1794), **Haüy** († 1822), **Lamétherie** († 1817). Copenhagen Academy.	Birth of J. Balsamo (alias Cagliostro).
Musschenbroek, von Kleist: Leyden jar. **Euler:** work on astronomy.	1744	**Lamarck** († 1829), **L. Crell** († 1816), **Marat** († 1793).	Workers' revolt in Lyons. Berkeley (1684–1755): *Siris.*
Bornet discovers parthenogenesis. Irregularities in the motion of Saturn.	1745	**Volta** († 1827). Memoirs of the Berlin Academy.	J. A. Poisson (1721–64), the future Mme Pompadour, becomes the mistress of Louis XV (1710–74).
Bradley: nutation of the Earth's axis.	1746	**Monge** († 1818), **Charles** († 1823), **Bucquet** († 1780).	Works by Diderot (1713–1804), Condillac (1715–80), La Mettrie (1709–51) and Vauvenargues (1715–47) condemned to be burnt. Industrial production of sulphuric acid.
Euler: *Introductio ad analysin infinitorum.* **D'Alembert:** precession of the equinox.	1747	Von Haller publishes the *Göttingen Zeitungen.*	Foundation of the École des Ponts et Chaussées. Montesquieu (1689–1755): *l'Esprit des Lois.* Machine for carding cotton.
	1748	**C. Berthollet** († 1822), **A. L. de Jussieu** († 1836), **D. Cassini** († 1845).	
Buffon: *Histoire naturelle* (. . . 1789). **Needham:** on spontaneous generation. **Cramer:** systems of linear equations.	1749	**Laplace** († 1827), **Jenner** († 1823), **Goethe** († 1832). **Delambre** († 1822), **Sonnerat** († 1814). **Werner** († 1817), **Mascheroni** († 1800), **Fortin** († 1831), **Baudin** († 1803).	Treaty of Aix-la-Chapelle. *Discours sur les arts et sciences* (Rousseau).

Scientific events		Individual and collective actors	Background information
Adanson writes a natural history of Senegal. Between 1750 and 1770 many works on machines by Euler (Archimedes' screw, gear-wheels, windmills . . .)	1750		Invention of the mechanical seed-drill. Goldoni (1707–93): *Le Café*. Beginning of coal-mining at Le Creusot. Fashion for chinoiserie in France and Europe.
D'Alembert and others: work on the *Encyclopédie*. **Maupertuis** conceives of transformism. Construction of new telescopes.	1751	**Daniel Bernouilli III** († 1834), **Loysel** († 1813), **Geisslern** († 1824). *Acta Helvetica.* Göttingen Royal Society.	*Encyclopédie* (Diderot). Famine in the South of France.
Calculation of distance and irregularities of the Moon. **Réaumur** investigates the digestion of birds as a chemical reaction.	1752	**Blumenbach** († 1840). *Commentaires sur les choses faites en sciences naturelles et médecine*, Leipzig (the largest journal of the time).	Franklin invents the lightning-conductor. Vaucanson: flute-player (automaton).
Linnaeus' nomenclature of living things.	1753	**L. Carnot** († 1823), **Jacquart** († 1834), **Adet** († 1834), **Achard** († 1821). Foundation of the British Museum.	Rousseau: *Dictionnaire du Musique.*
Euler: principles of fluid mechanics. **Black:** quantitative method in gaseous chemistry and discovery of carbonic anhydride.	1754	**Proust** († 1826), **Meusnier** († 1793), **Bonald** († 1840). Erfurt Academy.	Condillac: *Traité des sensations*. Gabriel (1698–1782) lays out the Place Louis XV, the future Place de La Concorde.
Euler: *Institutiones calculi differentialis*. **Kant**'s *Theory of the Heavens*.	1755	**Fourcroy** († 1809), **La Billardière** († 1834), **Hassenpratz** († 1827).	First appearances: the physiocrats, the masonic Grand Loge de France, the sewing machine.
Black identifies 'fixed air', the future carbon dioxide.	1756	**Chaptal** († 1832), **Lacépède** († 1825).	Invention of cement. Diderot: *Le Fils naturel*.
Von Haller: birth of modern physiology.	1757	*Acts of the Erfurt Academy.*	
Production of achromatic lenses.	1758	Known species: 1,222 vertebrates, 677 molluscs, 2,119 arthropods.	*De l'esprit* by Helvetius (1715–71) is condemned to be burnt.
Montucla: *Histoire des mathématiques*. Return of the comet predicted by Halley. **Wolf** founds embryology.	1759	Academy of Science at Munich. *Mélanges philosophiques. et mathématiques*, Turin.	Voltaire (1694–1778): *Candide*.

Scientific events		Individual and collective actors	Background information
Lambert: geometry, trigonometry, series. **Euler:** study of rotating bodies. **Black:** work on calorimetry.	1760		Blackwell breeds improved sheep (–1795). Intensive construction of toll-roads. Appearance of cast-iron rail (England).
Lambert: irrationality of π.	1761	*Acts of the Siena Academy.*	Rousseau: *La Nouvelle Héloise.*
	1762	**Richter** († 1807).	Execution of Calas (1698–1762).
Euler, Lagrange: calculus of variations. Catalogue of 10,000 stars. Cook's voyage round the world, many species studied.	1763	**Chappe** († 1805).	Kant: *Dreams of a Visionary.* The Treaty of Paris decides the fate of the French colonies. Free trade in corn in France.
Adanson publishes *Les Familles des plantes.* **Harrison:** chronometer.	1764		First steam automobile. The Jesuits expelled from France. Salon of Mme de Lespinasse (1732–1776).
	1765	**Niepce** († 1833).	Lyons: first veterinary college. Voltaire has Calas rehabilitated. Rousseau (1712–78) writes the *Confessions* (–1770). Watt's first steam engine.
Cavendish isolates 'inflammable air', the future hydrogen. Beginning of **Bougainville**'s voyage. **Euler:** *Algebra.*	1766	**Dalton** († 1814), **Wollaston** († 1828). *Acta physica* of the Mannheim Academy.	Execution of the Chevalier de la Barre (1747).
	1767	**W. von Humboldt** († 1835), **Seguin** († 1835), **Bouvard** († 1843).	Gluck (1714–87), *Alceste* (opera).
Exploration of Siberia (discovery of mammoths).	1768	**J. Fourier** († 1830), **Schreiber** († 1850).	*Encyclopaedia Britannica.* Corsica becomes French.
Monge: beginning of descriptive geometry.	1769	**G. Cuvier** († 1832), **A. von Humboldt** († 1859), **Bonaparte** († 1821). *Transactions* of the American Philosophical Society.	Watt takes out patent on the steam engine. Arkwright (1732–92): water-frame to spin cotton.
Euler: *Institutiones calculi integralis.*	1770	**A. Brongniart** († 1847).	Holbach (1723–89): *Système de la nature.*

Scientific events		Individual and collective actors	Background information
Vandermonde studies the equation of 5th degree.	1771	F. Rozier publishes the future *Journal de physique, chimie, histoire naturelle.* **Gergone** († 1839), **Bichat** († 1802).	Haydn (1723–1809): six string quartets. Gainsborough (1727–88) paints *The Market Cart.*
Lagrange: work in algebra (notion of invariant). **Cavendish:** study on electricity.	1772	**Geoffroy de Saint-Hilaire** († 1844). Brussels Academy.	First boring and turning mill. First partition of Poland.
Lavoisier: experiment on combustion. **Priestley** isolates 'dephlogisticated air', the future oxygen, at the same time as **Scheele.**	1773	**Young** († 1829), **Bonpland** († 1858), **Delessert** († 1847). Philadelphia Museum.	Clement XIV dissolves the Society of Jesus.
Werner: mineralogical studies (–1791).	1774	**Biot** († 1862). First specialist journal (natural history).	Louis XVI (1754–93) King of France.
Bergmann: Table of elective affinities.	1775	**Ampère** († 1836), **Malus** († 1812), **T. Thomson** († 1852).	Beaumarchais (1732–99): *The Barber of Seville.*
Jenner: first experiment in vaccination.	1776	**Dalton** († 1844), **Avogadro** († 1856), **Sophie Germain** († 1831).	American Declaration of Independence. First railway (in a mine). First daily newspaper in Paris. A. Smith (1723–90): *The Wealth of Nations.*
Spallanzani: artificial insemination of batrachians. **Lavoisier:** composition of air.	1777	**Gauss** († 1855), **Poinsot** († 1859), **Oersted** († 1851), **Thénard** († 1857).	J. Priestley: *On Matter and Spirit.*
Rumford: on the friction/heat relation.	1778	**Gay-Lussac** († 1850), **Davy** († 1829), **Herbert** († 1847), **A. P. de Candolle** († 1841), **Raffeneau** († 1850). Lorenz Crell founds the *Chemisches Journal.*	First threading lathe. France joins the insurgent Americans.
Scheele: glycerine. **Bezout:** general theorem of algebra (attempt at a proof, see 1625).	1779	**Berzelius** († 1848). Royal Academy, Naples. Second foundation of the Padua Academy of Sciences.	Goethe: *Iphigenia in Tauris.* Gluck: *Iphigenia in Tauris.*
Laplace, Lavoisier: *Mémoire sur la chaleur.* **Haüy:** study on crystals. **Herschell:** Uranus, and proper motion of the sum.	1780	American Academy of Science (Boston). **Crelle** († 1855), **Döberei** († 1849).	Lessing (1729–81): *The Education of the Human Race.* Fortin and Mégnié, first precision instruments.

Scientific events		Individual and collective actors	Background information
Coulomb: friction and electricity.	1781	**Bolzano** († 1848), **Poisson** († 1840), **Laennec** († 1826).	Abolition of the 'question préparatoire'. Kant: *Critique of Pure Reason.*
Scheele: hydrocyanic acid. **L. Carnot:** *Essai sur la puissance des machines.*	1782	*Memorie di Matematica e Fisica* (Verona).	C. de Laclos (1741–1803): *Les Liaisons dangereuses.*
Cavendish: synthesis of water.	1783	Royal Society of Edinburgh. **Magendie** († 1855). Creation of the École des Mines.	Watt: double-action rotary engine. Treaty of Versailles.
Charles: work on the expansion of gases. **Marat:** *Sur le feu, Sur la lumière, Sur l'électricité.* **Herschel:** catalogue of 711 double stars. Start of **La Pérouse's** expedition, which came to a dramatic end in 1788.	1784	**Bessel** († 1846). Bengal Asiatic Society. Royal Irish Academy [1785], Dublin. **Buckland** († 1771).	Beaumarchais: *The Marriage of Figaro.* David (1748–1827): *The Oath of the Horatii.* Gas-lighting. First railway in France. First mechanical weaving machine.
	1785	**Dulong** († 1838), **Brianchon** († 1864), **W. Prout** († 1850), **Grothus** († 1822), **Audubon** († 1851). *Memoirs of the American Academy of Arts & Sciences. Memoirs of the Manchester Literary and Philosophical Society.*	
Berthollet: bleaching by chlorine. **Scheele:** *Chemical Essays.*	1786	**Arago** († 1853), **Fresnel** († 1827), **Chevreul** († 1889).	First iron frame for building. First ascent of Mont Blanc. Metal ships.
Lagrange: *Mécanique analytique.* Reform of chemical nomenclature by **Lavoisier, Guyton de Morveau, Fourcroy, Berthollet.**	1787	**Bergerie** († 1863).	Constitution of the United States. Mozart (1756–91): *Don Giovanni.* Schiller (1759–1805): *Don Carlos.*
Fourcroy: *Elémens d'histoire naturelle et de chimie.*	1788	**A. C. Becquerel** († 1878), **Boucher de Perthes** († 1868), **Poncelet** († 1867). Société philomatique, Paris.	Invention of the parachute between 1785 and 1797. Abolition of the 'question préalable'.
Jussieu proposes a classification of plants by 'natural families'. **Lavoisier:** *Traité élémentaire de la chimie.*	1789	**Gmelin** († 1853), **Ohm** († 1854), **Cauchy** († 1857). *Botanisch Magazin* (Zurich). *Botanical magazine* (London).	Parmentier: *Traité sur la Pomme de Terre.* Beginning of the French Revolution (–1799). Bentham (1748–1832): *Panopticon.*

Scientific events		Individual and collective actors	Background information
	1789	*Annales de chimie* (Paris). *Annali di chimica.*	
Leblanc: procedures for the artificial production of sodium.	1790	**Champollion** († 1832), **Möbius** († 1868), **Daniell** († 1845). *Journal der Physik* (Halle, Leipzig).	Chappe's telegraph. American law on patents. Slave trade: 70,000 each year.
Galvani: studies on muscular electricity in the frog. **Goethe:** articles on optics.	1791	**Faraday** († 1867), **Morse** († 1872), **Petit** († 1820). *Bulletin des sciences* of the Société philomatique. *Transactions* of the Linnean Society (London). Société d'histoire naturelle de Paris.	Creation of the Commission des Poids et Mesures. Scientific expedition sent to look for La Pérouse breaks up in political disagreement. French law on patents.
J. B. Richter: *Stoichiometry.*	1792	**Napp** († 1867), **Coriolis** († 1843), **Baer** († 1876), **Lobachevsky** († 1856), **Babbage** († 1871).	French Republican Calendar. Commission temporaire des arts (France).
Herschel plans a giant telescope.	1793	**Chasles** († 1880). The Jardin du Roi becomes the Muséum d'Histoire Naturelle.	L. Carnot the Revolution's 'organizer of victory'. David paints *The Death of Marat.*
Legendre: *Eléments de géométrie.* **Blumenbach:** *Bibliothèque médicale*, studies on comparative anatomy.	1794	**Lesson** († 1848), **Boué** († 1881), **K. M. Marx** († 1864), **Mitscherlich** († 1863). Creation of the Écoles Supérieures in France: Travaux Publics (becoming the Polytechnique in 1795), Arts et Métiers, Centrale, Normale. *Journal polytechnique* (of the Polytechnique). *Journal de l'Agence des mines de la République.*	The Revolution organizes courses in the production of saltpetre and gunpowder. Lavoisier is guillotined. Condorcet dies in prison. Fichte (1762–1814): *On the Doctrine of Science.*
Lagrange: *Géométrie analytique.* **Mascheroni:** *Geometrio del Compasso.* **J. Hutton:** *Geological Theory of the Earth.*	1795	**Lamé** († 1870). Establishment of the Institut de France (the five Academies). Bureau des Longitudes.	Metric system of weights and measures. Appert (1750–1841) invents tinned food. Condorcet: *Esquisse des progrès de l'esprit humain.*
Laplace: *Exposition du système du monde.*	1796	**S. Carnot** († 1832).	First public vaccination in England.
L. Carnot: *Métaphysique du calcul infinitésimal.* **Wessel:** geometrical	1797	**Lyell** († 1875).	German encyclopaedia. First steam locomotive (on the road).

Scientific events		Individual and collective actors	Background information
representation of complex numbers. **Lagrange:** lectures on the functional calculus.	1797		
The expedition to Egypt **(Monge, Berthollet, Geoffroy Saint-Hilaire, Raffeneau-Delile)** **Legendre:** *Théorie des nombres.*	1798	*Allgemeines Journal der Chemie* (Leipzig). First issue of the *Philosophical Magazine.*	Paper-making machine. Malthus: *Essay on the Principle of Population.*
Voyage to South America by **Humboldt and Bonpland** (study of plant geography).	1799	**Clapeyron** († 1864), **Argelander** († 1875). Establishment of the metric system. *Annalen der Physik.* 334 authors have written in the *Philosophical Transactions* (London). A third of the journals are specialized.	Bonaparte First Consul, end of the Revolution. Watt sells more than 300 machines in England and Europe. Schelling (1774–1854): *Philosophy of Nature.*
Cuvier: *Leçon d'anatomie comparée.* **Bichat:** physiological research on life and death. **Volta's** pile, first electrolysis.	1800	**J. B. Dumas** († 1884), **Goodyear** († 1860), **Pouchet** († 1872), **Milne-Edwards** († 1885), **Wöhler** († 1882). There are about a hundred scientific periodicals.	Creation of the Banque de France. First metal printing-press. Act of Union between Great Britain and Ireland.
Gauss: *Disquisitiones arithmeticae.* **Lalande:** catalogue of 50,000 stars.	1801	**Cournot** († 1877), **Plücker** († 1868).	Concordat between Napoleon and Pius VII. Schiller: *The Maid of Orleans.*
Young: *On the theory of light and colours.* **Lamarck** rejects the immutability of species.	1802	**Bolyai** († 1860), **Wheatstone** († 1875), **Abel** († 1829), **Balard** († 1876), **Lecoq** (1871). *Annales* of the Muséum d'Histoire Naturelle.	France: creation of *lycées* for boys. First labour legislation: in England children of less than 9 years may not work more than 12 hours per day.
Poinsot: elements of statics. **Berthollet:** *Statique chimique.* **L. Carnot:** positional geometry.	1803	**Liebig** († 1873), **Strum** († 1855), **Doppler** († 1853).	De Sade (1740–1814) moved from Saint Pélagie to Bicêtre. On the Seine, the first screw-driven boat. Louisiana is sold to the Americans (80 million francs).
Richter isolates nickel.	1804	**Jacobi** († 1851).	Napoleon Emperor of France.
Grothus: theory of ions.	1805	**Hamilton** († 1865),	Trafalgar, Austerlitz.

Scientific events		Individual and collective actors	Background information
A. von Humboldt: *Essai sur la géographie des plantes.*	1805	**Lejeune-Dirichlet** († 1859).	
Proust: the chemical law of definite proportions. **Argand:** on imaginary numbers.	1806	**A. P. de Candolle** († 1893). Society of Naturalists, Moscow.	France: Université impériale (public educational service). Blockade of the Continent.
Chaptal: *Chimie appliquée aux arts.* **Davy** obtains sodium and potassium by electrolysis. **Brongniart:** *Traité de minéralogie.* **Monge:** infinitesimal geometry.	1807	**Laurent** († 1853), **I. Holden** († 1897), **D. Alter** († 1881). Société d'Arcueil and first publication of its *Mémoires de physique et de chimie.* Geological Society of London.	Slave trade forbidden by the English. Fichte: *Addresses to the German Nation.* Hegel (1770–1831): *The Phenomenology of Mind.*
Malus: polarization of light. **Dalton:** law of multiple proportions and atomic hypothesis. **Gay-Lussac:** law on volumes of gases. **Berzelius:** *Treatise on Chemistry* (–1818).	1808	**C. Pritchard** († 1893), **Klacel** († 1882). Physikalisch-Medicinische Societät (Erlangen).	Metternich (1773–1859) Austrian Minister of Foreign Affairs (–1848). Murat (1767–1815) King of the Two Sicilies. Peninsular War (–1812).
Lamarck: *Philosophie zoologique.* **Gauss:** theory of heavenly bodies. **Goethe:** *Theory of colours.* **Davy** isolates chlorine, theory of acids.	1809	**Darwin** († 1882), **Grassmann** († 1877), **J. D. Forbes** († 1868), **Liouville** († 1882).	Wagram. Goethe: *Elective Affinities.*
	1810	Foundation of Berlin University. *Annales de mathematiques pures et appliquées* (Gergonne). **Regnault** († 1878), **Kummer** († 1893), **Walter** († 1847).	A. von Arnim (1781–1831): *Countess Dolores.*
Biot, Arago: chromatic and rotatory polarization. **Avogadro:** law on the density of gases. **Fourier:** series and partial differential equations. **Berzelius** isolates silicon.	1811	**Galois** († 1832), **Le Verrier** († 1877), **Bunsen** († 1888), **Dzierzon** († 1906).	Foundation of the Krupp steelworks. Industrialization of the weaving machine. Blockade of the Continent leads to research on beet sugar.
Laplace: *Théorie analytique des probabilités.*	1812	**Joly** († 1895), **Galle** († 1910), **Davoine** († 1882). Academy of Natural Sciences (Philadelphia).	Napoleon's Russian campaign. Grimm (1785–1863): *Snow White and the*

701

Scientific events		Individual and collective actors	Background information
Cuvier: *Recherche sur les ossements fossiles.*	1812		*Seven Dwarves.* Hegel: *The Science of Logic.*
Fresnel: first work on light. First spectroscopic analysis of the Sun and a star.	1813	**C. Bernard** († 1878), **Stas** († 1891).	Expedition to explore Australia.
Laplace: *Essai philosophique sur les probabilités.* **Cauchy:** *Etude sur les integrales définies.*	1814	**Mayer** († 1878). **Lermontov** († 1841).	Steam locomotive on metal rails. Opening of the Congress of Vienna (–1815).
Gay-Lussac isolates cyanogen. **Poinsot:** *Sur la rotation des corps autour d'un point fixe.*	1815	**Weierstrass** († 1897), **Boole** († 1864), **Naudin** († 1899). **C. F. Gerhardt** († 1856),	German Confederation. The Hundred Days . . . Waterloo.
Fresnel studies diffraction. **Magendie:** experiments in animal physiology.	1816	**Vilmorin** († 1860). *Annales de chimie et de physique* (see 1789). Universities of Liège and of Ghent. Vienna Polytechnic Institute.	Schubert: (1797–1828): *Symphonies 4 and 5.* Gold Standard Act. Drais bicycle. Argentine independence.
Cuvier: *Le Règne animal.* **Freycinet**'s expedition around the world.	1817	**Wurtz** († 1884), **Wichura** († 1866), **Naegeli** († 1891), **Galissard** († 1894), **Borchardt** († 1880).	Byron (1788–1824): *Manfred.* Rothschild Bank (Paris). Catholic reaction in France.
Geoffroy de Saint-Hilaire becomes a supporter of transformism. **Thénard** discovers hydrogen peroxide.	1818	**Joule** († 1889).	M. Shelley (1797–1851): *Frankenstein.* Creation of the Caisse d'Epargne. MacAdam (1756–1836) develops a new road surface.
Dulong and **Petit:** law on the specific heat of simple substances. **Laennec:** on auscultation (the stethoscope). **Mitscherlich:** law of isomorphism. **Faraday, Oersted, Ampère:** numerous studies on electricity-magnetism relationship. **De Candolle** defines a research programme for plant geography.	1819	**Fizeau** († 1896), **Foucault** († 1868), **Stokes** († 1903). University of St Petersburg.	Epic of Bolivar (1783–1830). Schopenhauer (1788–1860): *The World as Will and as Representation.* Population of the United States: 10,000,000. Lamartine (1790–1869): *Méditations poétiques.* Scott (1771–1832): *Ivanhoe.* P. B. Shelley (1792–1822): *Prometheus Unbound.*
	1820	**Béguyer** († 1886), **A. Ed. Becquerel** († 1891), **Tyndall** († 1893). (Royal) Astronomical Society, London.	
Cauchy begins his lectures on analysis.	1821	**Helmholtz** († 1894), Boncompagni († 1894), **Cayley** († 1895).	Mexican independence. Birth of Saint-Simonianism.

Scientific events		Individual and collective actors	Background information
	1821	University of Buenos Aires.	Weber (1786–1826): *Der Freischütz* (opera).
Fourier: *Théorie analytique de la chaleur.* **Fresnel:** *Théorie ondulatoire de la lumière.* **Poncelet:** *Propriétés projectives des figures.*	1822	**Pasteur** († 1895), **Mendel** († 1884), **Clausius** († 1888), **Hermite** († 1901).	Champollion (1790–1832) deciphers hieroglyphs. Brazilian independence. First photo (Niepce). Comte (1798–1857): *Plan of Scientific Work for the Reorganization of Society.*
Faraday: liquefaction of certain gases. **Chevreul:** research on fats.	1823	**Kronecker** († 1891), **Eisenstein** († 1852), **Fabre** († 1915), **Wallace** († 1913), **Carrylea** († 1897).	Industrial production of soap. Monroe Doctrine. Settling of Australia. Beethoven (1770–1827): *9th Symphony.*
Dirichlet: first work on number theory. **S. Carnot:** *Sur la puissance motrice du feu.* **Abel:** on equations. Appearance of clockwork-driven astronomical telecopes.	1824	**Kirchoff** († 1887), **Kelvin** († 1907).	First candles (Chevreul). Construction of first suspension bridge. Leopardi (1798–1837): *Canzoni.*
Faraday discovers benzene. **Oersted** isolates aluminium. **Bessel's** function. **K. M. Marx:** *Geschichte der Krystallkunde.*	1825	**Charcot** († 1893), **Bates** († 1892).	France: reactionary laws 'of sacrilege' and 'of the émigrés' billion'. First public railway (England).
Ampère: mathematical theory of electrodynamics. **Lobachevsky:** Hyperbolic geometry. **Gauss** studies the probability of errors.	1826	**Riemann** († 1866), **Z. Gramme** († 1901), **J. Thomson** († 1909), **Lannizzaro** († 1910). Linnaean Societies of Bordeaux and Calvados. *Journal für die reine und angewandte Mathematik* (Crelle).	Defeat of the Greek revolt. Schubert: *Death and the Maiden.* Mendelssohn (1809–1847): *Midsummer Night's Dream.*
Ohm's Law. **Möbius** strip. **Legendre:** *Traité des fonctions elliptiques.* **Gauss:** *On Curved Surfaces.* **Baer** recognises the role of the egg in mammalian reproduction.	1827	**Berthelot** († 1907), **Villemin** († 1892), **Gladstone** († 1902), **Cook** († 1894), **Lister** († 1912). University of Helsinki.	Naval battle of Navarino for the liberation of Greece. Ingres (1780–1867): *The Apotheosis of Homer.* Heine (1797–1856): *Buch der Lieder.*
Wöhler: first synthesis of an organic substance (urea).	1828	**Cohn** († 1898). University of London.	Independence of Uruguay.

Scientific events		Individual and collective actors	Background information
	1828		Delacroix (1798–1863): *The Death of Sardanapalus.*
A. von Humboldt: expedition to Siberia. **Jacobi:** study of elliptical functions.	1829	**Kékulé** († 1896), **Odling** († 1921), **Moutier** († after 1894). 300 scientific periodicals created since 1665.	Greek independence. C. Fourier (1772–1837): utopian socialism. Inauguration of the Liverpool–Manchester railway.
Cuvier and Geoffroy Saint-Hilaire controversy on mammalian organization. **Lyell:** *Principles of Geology.* **Galois:** thesis on equations.	1830	**Marey** († 1904), **C. W. Thomson** († 1882), **Royer** († 1902), **Meyer** († 1895). Arago opens the sessions of the Académie des Sciences to journalists. Société géologique de France, and the first number of its *Bulletin.*	July Revolution. Poland crushed. *Hernani*, by Victor Hugo (1802–85). Comte: *Cours de philosophie positive* (–1842). Stendhal (1783–1842): *le Rouge et le Noir.*
Gauss: on complex numbers. **Darwin** on the *Beagle* for a five-year voyage. **Brown** identifies the nucleus of a cell. **Faraday:** law of induction.	1831	**Maxwell** († 1879), **L. Meyer** († 1895), **Dedkind** († 1916), **Suess** († 1914).	Bellini (1801–35): *Norma.* Schumann (1810–56): *Papillons.* First combine harvester (US). Rebellion of the Lyons silk-weavers.
Galois: letter-testament on a new algebra. **Coriolis** forces.	1832	**Koenig** († 1901), **Crookes** († 1919). *Abstract of Philosophical Transactions*, first of its kind. *Entomological Magazine*, London. Société entomologique de France. Zurich University. **Liebig** founds the *Annalen der Pharmacie.*	Cholera kills 18,500. Pixii (1808–35): electric generator. Berlioz (1803–69): *Symphonie fantastique.*
Gauss, Weber: electromagnetic telegraph. **Babbage:** Analytical Engine. **Bolyai:** non-Euclidean geometry. **Faraday:** law of electrolysis.	1833	**Nobel** († 1896), **P. Bert** († 1886), **Waage** († 1990). First French scientific congress.	Pushkin (1799–1837): *Eugene Onegin.* Owen (1771–1858) creates his 'Trade Union'. Abolition of slavery in the English colonies. France: law on primary education.
Milne-Edwards: *Éléments de zoologie.* **A. C. Becquerel:** *Traité d'électricité et de*	1834	**Mendeleyev** († 1907), **Haeckel** († 1919), **Weismann** († 1914), **Volhard** († 1910).	Balzac (1799–1850): *La Recherche de l'absolu.* Exodus of the Boers. Abolition of the

Scientific events		Individual and collective actors	Background information
magnetisme (–1840). **Clapeyron:** *Sur le rendement de machines mécaniques.* **Hamilton:** mechanical equations.	1834		Inquisition in Spain. First European steam railway network.
Berzelius: Theory of chemical proportions, introduction of the word 'catalysis'. Return of Halley's comet.	1835	**Baeyer** († 1917), **Foster** († 1919), **Schiaparelli** († 1910).	Invention of the Colt (1814–62) revolver. First issue of the *New York Herald.* Quetelet (1769–1874): social statistics.
Appearance of the concept of a vector. **Dumkas:** chemical theory of substitution. **Daniell:** non-polarizing battery with two different solutions.	1836	**Guldberg** († 1902), **Hinrichs** († 1923). Liouville founds the *Annales de mathématiques pures et appliquées.* *Bulletin* of the St Petersburg Academy of Science.	Chopin (1810–49): Waltzes and Nocturnes. First bladed propeller.
Lobachevsky: *Imaginary geometry.* **Jacobi, Spenecer:** electro-deposition. **Chasles:** studies on geometrical methods.	1837	**Kuhne** († 1900), **Lenssen** († after 1870). More than 70% of chemistry periodicals are primarily devoted to pharmacy.	Dickens (1812–70): *Oliver Twist.* Victoria accedes to the throne. Morse (1791–1872): telegraphic system. Paris–St Germain-en-Laye railway.
Schleiden-Schwan: cellular theory of life. **Poisson:** theory of probability. **Boucher de Perthes:** work on prehistory.	1838	**Jordan** († 1922), **Solvay** († 1922), **Newland** († 1898), **Perkin** († 1907), **E. Mach** († 1916), **Morley** († 1923). *Transactions of the Bombay Geographical Society.*	Daguerre (1787–1851): photography. Beginning of organic, agricultural chemistry (Liebig). Crossing of the Atlantic without sails. Chartism.
Boole: analytical transformations. **Liebig:** theory of fermentation.	1839	**Gibbs** († 1903). *Cambridge Mathematical Journal.*	German Law: children of less than 16 years may not work more than 10 hours a day. Goodyear: vulcanization of rubber.
Bessel: first measurement of the distance of a star. **Regnault:** specific heat of compound substances.	1840	**Dunlop** († 1921), **Kohlrausch** († 1910), **Duclaux** († 1904). Hungarian Congress of Physicists and Natural Scientists.	Opium Wars in China. Braille (1809–52) invents a script for the blind. First power-hammer. First postage stamp.
Boole: theory of invariance and covariance.	1841	**Graebe** († 1927), **J. Murray** († 1914).	Industrial manufacture of chemical matches.

705

Scientific events		Individual and collective actors	Background information
Jacobi: work in mechanics. **Joule**'s law.	1841	Chemical Society of London.	Invention of the saxophone.
Mayer-Joule: principle of the conservation of energy. **Doppler** effect.	1842	**Rayleigh** († 1919), **Darboux** († 1917), **J. Deward** († 1923). **S. Lie** († 1899), **Horstman** († 1929).	State plan for seven railway lines from Paris. Gogol (1809–52): *Dead Souls*.
Hamilton: discovery of quaternions. **Wheatstone** bridge.	1843	**Bonnier** († 1922), **W. Flemming** († 1906), **Koch** († 1910).	Viollet-le-Duc (1814–79): restoration of Carcassonne begins. Eugène Sue (1804–57): *Les Mystères de Paris*.
Grassmann: 'theory of extension' (intuition of mathematical structures).	1844	**M. Noether** († 1921), **Branly** († 1940), **Bolzmann** († 1906), **Golgi** († 1926).	A. Comte: *Discourse on the Positive Spirit*. First telegraph line (Washington–Baltimore).
Cayley: matrices. Giant telescope allows the discovery of spiral nebulae. **Adams** takes the first steps in the hypothesis of an 8th planet (Neptune).	1845	**Cantor** († 1918), **Metschnikov** († 1916), **Röntgen** († 1923). *Scientific American.*	Agricultural crisis brings about the emigration of 2 million Irish to the US.
Le Verrier calculates the position of Neptune, observed by **Galle** one month later. **Faraday** suspects the electromagnetic character of light.	1846	**Picard** († 1941), **Van Beneden** († 1910), **Tesla** († 1943).	The US reaches from the Atlantic to the Pacific. First general anaesthesia. First sewing machine.
Kirchoff's Law. Notion of the ideal in mathematics (**Kummer**). Mathematical analysis of logic (**Boole**).	1847	**Le Bel** († 1930), **Edison** († 1931), **Bell** († 1922). *Die Fortschritte der Physik* (abstracts).	English labour law: women and children not to work more than 10 hours a day. Surrender of Abd el-Kader (1807–83).
Bates spends nearly 11 years in Amazonia, studies insects, and deduces a theory of mimicry.	1848	**Eötvös** († 1919), **de Vries** († 1935), **Henry** († 1905), **Meyer** († 1897). Société de Biologie (Paris), Geological Society (Berlin), American Association for the Advancement of Science.	Marx (1818–83) and Engels (1820–95): *The Communist Manifesto*. Revolution in Europe. Swiss Federal Constitution. France abolishes slavery.
Evaluation of the difference in speed of light in water and in air (**Foucault and Fizeau**). **Weierstrass:** continuous totally undifferentiable function.	1849	**Pavlov** († 1936), **J. A. Fleming** († 1945), **Klein** († 1925), **Hertwig** († 1922). *Naumannia*, first ornithological journal.	Proclamation of the Roman Republic (Year I). End of Britain's 200-year-old monopoly on maritime trade between its islands.

Scientific events		Individual and collective actors	Background information
Redefinition of the yield defined by Carnot. Second Principle of Thermodynamics (**Kelvin and Clausius**). **Gerhardt:** theory of chemical types.	1850	1.2 BILLION HUMAN BEINGS. **Richet** († 1935), **Goldstein** († 1930), **J. Milne** († 1913), **Le Chatellier** († 1936), **Kowalevskaya** († 1891). About 1000 scientific periodicals and 200 universities in existence.	Gold rush in Australia. 6,000 km of railway in Britain. Invention of prismatic binoculars. Dumas (1802–70): *Le Vicomte de Bragelonne*.
Riemann: function of a complex variable. Experiment with **Foucault**'s pendulum.	1851	**W. Reed** († 1902), **Balfour** († 1882). *Annali di scienze matematiche e fisiche* (Rome).	Australian population increases by 150% a year. Victor Hugo's exile in Jersey. First hydraulic lift.
Bunsen discovers magnesium.	1852	**H. Becquerel** († 1908). **Lindemann** († 1939), **Ramsay** († 1916), **Michelson** († 1931), **Van't Hoff** († 1911), **Flahault** († 1935), **Kitasato** († 1931). *Cosmos*, review for scientific progress, Paris.	Second Empire. Beecher-Stowe (1811–96): *Uncle Tom's Cabin.*
J. Thomsen: birth of thermochemistry. **Gerhardt:** *Traité de chimie organique.*	1853	**H. A. Lorenz** († 1928), **W. Ostwald** († 1932), **Brillouin** († 1948), **G. Ricci** († 1925), **Roux** († 1933). *Report and Transactions* of the Adelaide Philosophical Society (Australia).	Haussman, Prefect of the Seine, begins his reconstruction of Paris. Aspirin (Gerhardt). Crimean War (–1856).
Boole: *Investigation of the Laws of Thought.* **Helmholtz:** on the thermal radiation of the Sun. **Riemann:** integrals, non-Euclidean geometry.	1854	**H. Poincaré** († 1912), **Behring** († 1917). *Deliberations* of the California Academy of Sciences.	Industrial production of aluminium. Mommsen (1817–1903): *History of Rome.*
Fabre: first publication on insects. **C. Bernard:** *Leçons de physiologie expérimentale.* **Berthelot** seeks to synthesize alcohol.	1855	**Michurin** († 1935), **Appel** († 1930).	Verdi (1813–1901): *Sicilian Vespers.* Universal Exhibition at Paris. First artificial dye (Perkin).
Discovery of Neanderthal Man. **Wallace** spends from 1854 to 1862 in the Malay Archipelago, later proposing a theory of evolution.	1856	**J. J. Thomson** († 1940), **Markoff** († 1922). *Review of Maths and Physics*, Leipzig.	Liszt (1811–86): *Hungarian Rhapsodies.* Ibsen (1828–1906): *The Feast at Solhoug.*

Scientific events		Individual and collective actors	Background information
Vilmorin: develops the 'genealogical method' for seed selection.	1856		
Pasteur: on lactic fermentation. **Riemann** works on what will become topology. **Kirchoff:** spectroscopic analysis of the stars. **Boncompagni** publishes the works of **Fibonacci.**	1857	**H. Hertz** († 1894), **Larmor** († 1942). Museum of London (1857).	Sepoy Rebellion. Baudelaire (1821–67): *Les Fleurs du Mal.* Flaubert (1821–80): *Madame Bovary.*
Kékulé: theory of valency. **Plücker:** cathode rays.	1858	**Planck** († 1947), **Peano** († 1932), **Diesel** († 1915), **E. Dubois** († 1940). *The Geologist,* popular magazine, London. *Bulletin* of the Société de Chimie, Paris.	B. Juares (1806–76) President of Mexico (–1863). Lourdes: 'Apparition of the Virgin'.
Darwin: *The Origin of Species.* **Cayley:** synthesis of non-Euclidean geometry.	1859	**P. Curie** († 1906), **Arrhenius** († 1927). Known species: 18,660 Vertebrates, 11,600 Mollusca, 5,770 Arthropoda.	E. Drake (1819–80) drills his first oil-well. V. Hugo: *La Légende des siècles.*
Berthelot: an organic chemistry based on synthesis. Synthesis of acetylene.	1860	**Volterra** († 1940), **Lummer** († 1925). *Revue de géologie,* Paris.	Garibaldi (1807–61), Cavour (1810–61), Italian unification. First ammonia refrigerator.
Weierstrass: relationship of continuous and derived functions. **Pasteur:** study of organisms living in the atmosphere. Dispute with **Pouchet** about spontaneous generation. **Argelander:** catalogue of 324,000 stars of the Northern Hemisphere.	1861	**G. M. Hopkins** († 1947), **Whitehead** († 1947), **Zsigmondy** († 1885). Entomological Society, St Petersburg. Creation of Massachussets Institute of Technology.	Dostoyevsky (1821–81): *Injury and Insult.* Abolition of serfdom in Russia. Lincoln (1809–65) President. American Civil War.
	1862	**Hilbert** († 1943).	Bismarck (1815–98) Prime Minister. Internal combustion engine. Hugo (1802–85): *Les Misérables.*
Solvay: process for obtaining sodium. **Lyell:** *Geological Evidences of the Antiquity of Man.*	1863	**Painlevé** († 1933), **Yersin** († 1943), **Correns** († 1933), **A. Lacroix** († 1948). *Revue des cours*	*Dictionnaire de la langue française* (–1872), Littré (1801–81). Renan (1823–92): *Life of Jesus.*

CHRONOLOGY

Scientific events		Individual and collective actors	Background information
	1863	*scientifiques* in France and abroad.	
Spencer: *Priciples of Biology.* **Maxwell:** dynamic theory of electromagnetic fields. **Weierstrass:** function of a complex variable.	1864	**Minkowski** († 1909), **Wien** († 1928), **Hernst** († 1941). *Annales* of the École Normale Supérieure (founded by **Pasteur**). Moscow Mathematical Society. First International Botanical Congress.	First International Beginning of the Indian Wars. London Underground.
Mendel: research on plant hybridization. **Pasteur:** patent on the preservation of wine. **C. Bernard:** *Introduction à l'étude de la médecine expérimentale.*	1865	**Hadamard** († 1963), **Weiss** († 1940), **Zeeman** († 1943), **Steinmetz** († 1923). London Mathematical Society. *Repertorium für physicalische Technik,* Munich.	Wagner (1813–83): *Tristan.* Manet (1832–1883): *Olympia.* The velocipede. War of Paraguay (ended in 1870): there remained only 1 man to 28 women. Lewis Carroll (1832–98): *Alice in Wonderland.*
Nobel discovers dynamite. **Haeckel** invents the term 'ecology'. **Solvay** process for the production of sodium.	1866	**Morgan** († 1945), **E. W. Brown** († 1938), **La Vallée-Poussin** († 1962).	Larousse (1817–75): *Grande dictionnaire universel du XIXe siècle.* First transatlantic cable.
Livingstone discovers the source of the Congo. Discovery of alizarin allows the replacement of madder. **Kronecker:** number theory.	1867 1868	**M. Curie (Sklodowska)** († 1934). **Landsteiner** († 1943), **Sommerfeld** († 1951), **Hausdorff** († 1942), **Wood** († 1955), **Millikan** († 1953). *Geological Survey of India. Transactions and Proceedings* of the New Zealand Institute.	Marx: *Capital.* Zola (1840–1902): *Thérèse Raquin.* Production of celluloid, the first plastic. Disraeli (1804–81), fall of his government. End of the Shogunate in Japan.
Mendeleyev: periodic table of chemical elements.	1869	**C. T. R. Wilson** († 1959), **E. Cartan** († 1951), **C. Thomson** († 1959). *Nature,* celebrated scientific weekly, London.	Suez Canal. Tolstoy (1828–1910): *War and Peace.* Zola makes plans for the *Rougon-Macquart* series.
Jordan: substitution groups for algebraic equations.	1870	**J. Perrin** († 1942), **Harrison** († 1959). First modern journal of bacteriology (Breslau).	First Minister for Industry in Japan. Dogma of papal infallibility.

709

Scientific events		Individual and collective actors	Background information
Maxwell: theory of heat. **Pasteur:** patent on the preservation of beer.	1871	**Borel** († 1956), **Zermelo** († 1953), **Tschermak** († 1962), **E. Rutherford** († 1937).	Paris Commune. Schliemann (1823–90) rediscovers Troy. Proclamation of the German Empire.
Klein: Erlangen Programme for geometry. **Dedekind:** on irrational numbers. **Haeckel:** hypothesis of the pithecanthropus.	1872	**B. Russell** († 1970), **P. Langevin** († 1946). *Popular Science Monthly*, New York. Société mathématique de France. Association pour l'avancement des sciences (France). Marine zoology research station, Naples.	First automobile carriage (5 tons, 12 seats, 40km/hr). Savoy attached to France. Tchaikovsky (1840–93): *Symphony No. 2.* Economic crisis in England, Germany and the US.
Hermite: 'e' is a transcendental number. Nerve-fibres studied by **Golgi**. **Kelvin** and **W. Thomson:** analogue calculator for differential equations.	1873	**W. D. Coolidge** († 1975), **A. Carrel** († 1944), **Levi-Civita** († 1941), **K. Schwarzschild** († 1916). International Bureau of Weights and Measures.	Charcot (1825–93) at the Salpêtrière. Rimbaud (1854–91): *Une saison en enfer.* Nietzsche (1844–1900): *Untimely Meditations.*
Kirchoff: spectral analysis of elements. Stereochemistry and molecular chemistry (**Le Bel, Van't Hoff**).	1874	**Marconi** († 1937), **A. Debierne** († 1949). Société française de physique.	J. Verne (1828–1905): *The Mysterious Island.* Monet (1840–1926): *Impression, soleil levant.*
W. Flemming discovers the chromosomes. **O. Hertwig** establishes a link between the cell-nucleus and fertilization. **Suess:** *Die Entstehung der Alpen.*	1875	**M. de Broglie** († 1960), **Lebesgue** († 1941). Physical Society of London.	Bizet (1838–75): *Carmen.* Constitution of the Third Republic.
Wallace: *The Geographical Distribution of Animals.* **Ramsay** studies the Brownian motion of molecules.	1876	**Noguchi** († 1928), **O. Diels** († 1954). American Chemical Society.	Bell invents the telephone. First Ring Cycle at Bayreuth. Sitting Bull victorious over Custer at Little Big Horn.
Gibbs: *On the [thermodynamic] Equilibrium of Heterogeneous Substances.* **Boltzmann:** kinetic theory of gases.	1877	**H. N. Russell** († 1957), **Soddy** († 1956). *Transactions* of the South African Philosophical Society.	Edison's cylinder phonograph. Theoretical conception of a television.
Schiaparelli: observes the 'canals' on Mars.	1878	**J. Becquerel** († 1953), **Fréchet, G. Bertrand** († 1953), **K.**	Kodak creates gelatino-bromide film.

710

Scientific events		Individual and collective actors	Background information
Pasteur: *Microbes:* theory of germs and applications to medicine and surgery. **Kuhne** proposes the term 'enzyme'.	1878	**Schlumberger** († 1936). *Journal of Pure and Applied Mathematics*, US. Chemical Society of Japan. *Brain*, neurological journal, London.	16 million visitors to the Paris Universal Exhibition. Nihilist revolt in Russia.
Baeyer: synthesis of indigo. **Crookes:** cathode-ray tube (electrical discharge in a rarefied gas). **Berthelot:** *Essai de mécanique chimique.*	1879	**Einstein** († 1955), **O. Hahn** († 1968). *Memoirs of the Science Dept.*, University of Tokyo. *Circulars* of the Johns Hopkins University.	First international convention on patents. Edison's electric bulb. Electric locomotive (Siemens).
Balfour: *Comparative Embryology.* **P. and J. Curie:** piezo-electrical properties of quartz. **Hermite:** elliptical function, number theory. **Charcot:** on the diseases of the nervous system.	1880	**Wegener** († 1936), **Dautry** († 1951), **Freundlich** († 1941). *Ciel et terre*, popular astronomical magazine.	Start of the Panama Canal. Dostoyevsky: *The Brothers Karamasov.* Electric lift. C. Flammarion (1842–1915): *Astronomie populaire.*
Pasteur: vaccine against anthrax. International definitions of electrical units. **Poincaré:** Fuchsian (automorphic) functions. **Michelson:** negative results of the ether-drift experiment (repeated with **Morley** in 1887).	1881	**A. Fleming** († 1955), **Teilhard de Chardin** († 1955), **H. Standinger** († 1965), **Langmuir** († 1957). Società Geological Italiana.	France, ministry of J. Ferry (1832–93), then Gambetta (1838–82): laws on compulsory schooling, freedom of the Press, freedom of association.
Lindemann demonstrates the transcendence of π. **Koch** identifies the tuberculosis bacillus.	1882	**Eddington** († 1994), **Geiger** († 1995). **Born** († 1970), **E. Noether** († 1935).	Electric lighting of the streets of New York. End of the Indian Wars. Bertillon (1853–1914): anthropometry.
Tesla studies alternating current. **Koch:** cholera bacillus. **Cantor:** foundations of set-theory. **Van Beneden:** constancy of the number of chromosomes.	1883	**V. F. Hess** († 1964), **Haworth** († 1949). *Science*, New York. *Archiv für Hygiene*, Munich.	Brahms (1833–97): *Symphony No. 3.* The shock of the Krakatoa explosion is felt around the world. Mach: *Mechanics.* Convention on industrial protection.
Van't Hoff: on chemical equilibrium. Distinguishing between soma and germ,	1884	**C. Funk** († 1967), **A. Piccard** († 1962), **D. Birkoff** († 1944). Tokyo Mathematico-	Automatic machine gun. Artificial silk. First photographic film in rolls.

Scientific events		Individual and collective actors	Background information
Weismann is persuaded of the non-transmission of acquired characteristics.	1884	Physical Society. First International Ornithological Congress.	French law on professional associations.
Appel: on Abelian functions. **Pasteur** treats rabies in **J. Meister** (1876–1941).	1885	**N. Bohr** († 1962), **H. Weyl** († 1955).	Zsigmondy: *La Traversée des Alpes.* Railways in Africa and Asia. Statue of Liberty. Charcot: functional centres of the brain.
Hertz: first work on electromagnetic waves (radio). **Goldstein** discovers positive rays.	1886	**E. D. Kendall** († 1972), **von Frisch** († 1982), **Trumpler** († 1956).	Rodin (1840–1917): *The Kiss.* Fauré (1845–1924): *Requiem.* World tonnage of steamships greater than that of sailing-ships.
Volterra: functional analysis. **Kronecker:** foundations of arithmetic. **Forbes:** ecological study of a lake. **Arrhenius:** ionic theory of electrolytes. **Weismann:** studies on chromosomes.	1887	**Moseley** († 1915), **Schrödinger** († 1961). *Société française d'astronomie. Acts* of the Pontifical Academy dei Nuovi Lincei. *General Guide* of the London Museum. Journal of Bacteriology and Parasitology, Jena.	Debussy (1862–1918): *le Printemps.* Start of the Eiffel Tower (–1889). First car with four-stroke petrol engine. Krafft-Ebbing (1840–1902): *Psychopathia Sexualis.*
Dedekind; arithmetization of analysis. **S. Lie:** theory of continuous transformation groups. **Berthelot** publishes the Greek alchemists.	1888	**Waksman** († 1973), **Tupolev** († 1988), **Baird** († 1946), **Zernike** († 1966). Institut Pasteur. New York Mathematical Society. *National Geographic Magazine,* New York.	Van Gogh (1853–90): *Self Portrait with Ear Cut Off.* Cézanne (1839–1906): *La Montagne Sainte-Victoire.* Marey discovers the principle of the cinema. Dunlop invents the inner-tube.
Behring: antitoxins. **Peano:** *Axiomatization of Arithmetic.* **Branly** invents the coherer necessary for radio reception.	1889	**Hubble** († 1953), **Ramanujan** († 1920), **Brillouin** († 1969). First International Physiological Congress. Geological Society of America.	Bergson (1859–1941): *Les Données immédiates de la conscience.* Foundation of the Second International. Maupassant (1850–1893): *Fort comme la mort.*
Behring, Kitasato: Anti-tetanus serum. **Peano** curve passing through all the points of a square.	1890	**R. A. Fisher** († 1962), **Holmes** († 1965), **Nishina** († 1951). *Revue de mathématiques spéciales.* US Department of Agriculture.	Christophe: *Le Sapeur Camember.* C. Ader (1841–1925), first flight of the *Eole.* Koch's tuberculin produces mass deaths. First submarine.
Dubois discovers a pithecanthropus in Java.	1891	**Chadwick** († 1974), **G. Banting** († 1941).	Construction starts on Trans-Siberian Railway.

Scientific events		Individual and collective actors	Background information
H. Poincaré: *Nouvelles méthodes de la mécanique céleste* (–1895). **Frege** (1848–1925): mathematical logic.	1892	**Banach** († 1945), **L. de Broglie** († 1987), **C. P. Thomson** († 1975).	Panama Canal scandal. First petrol tractor.
Behring: anti-diphtheria serum. Invention of the photo-electric cell in Germany. **Poincaré:** lectures on probability.	1893	**W. Baade** († 1960), **Urey.**	First Diesel motor. First film by Edison and the Lumière brothers (Auguste 1862–1954 and Louis 1864–1948). Durkheim (1858–1919): *The Division of Labour in Society.*
E. Cartan: thesis on **Lie**'s algebras. **Roux, Yersin:** anti-plague serum. **Lorentz:** electron theory of matter. Liquefaction of air. **Röntgen** discovers X-rays. **H. Becquerel** discovers radioactivity. **Hadamard** and **La Vallée-Poussin**: work on prime numbers. **Zeeman:** effect of magnetic field on light.	1894 1895 1896	**J. Rostand** († 1977), **W. Weiner** († 1964). First International Chemical Congress. **Domagk** († 1964), **Dam** († 1976), **P. P. Grassé.** *Comptes rendus* of the Congrès des sociétés savantes. **Carothers** († 1937), **G. T. Cori** († 1957), **C. F. Cori** († 1984).	Beginning of the Dreyfus (1859–1935) Affair. Ncholas II (1868–1918) becomes Czar. Creation of French West Africa. Foundation of the Confédération Générale des Travailleurs. Return of the Olympic Games (Athens). First film by Méliès (1861–1938). Lumière brothers: first public cinema performance. Herzl (1860–1904) publishes *The Jewish State.* Hollerith (1860–1929) creates TCM, the future IBM.
Hilbert: on algebraic numbers. **J. J. Thomson, Wein Weichert:** measurement of charge/mass ratio of the electron. **Larmor:** calculation of the radius of the electron. **P. and M. Curie** discover polonium and radium. **Dewar** liquefies hydrogen. **Hilbert:** *Foundations of geometry.* Chemical fertilization (sea-urchin's egg). **Rutherford:** alpha and beta rays.	1897 1898 1899	**I. Joliot-Curie** († 1956), **P. M. Blackett.** First International Mathematical Congress. **Lysenko** († 1976), **I. I. Rabi.** *Archives de Parasitologie*, Paris. **C. H. Best** († 1978). International Geodesic Association. *Enseignement Mathématique*, published by the International	E. Rostand (1868–1918): *Cyrano de Bergerac.* First wireless: above the Channel (15km). Chekhov (1860–1904): *Uncle Vanya.* Sudanese War (Fashoda). Zola: *J'accuse.* Gold-rush in Alaska. First modern submarine. Industrialization of aspirin production (Bayer).

Scientific events		Individual and collective actors	Background information
	1899	Commission on the Teaching of Mathematics. Nearly 10,000 scientific periodicals have been created.	Ravel (1875–1937): *Pavane pour une infante défunte.*
M. Planck: quantum of action. **De Vries, Tschermak, Correns** formulate the laws of hybridization. **Hilbert** puts 20 problems to the Paris Mathematical Congress.	1900	**W. Pauli** († 1958), **F. Joliot** († 1958), **H. Aiken** († 1973). The *Comptes rendus* of the French Académie des Sciences are cited more than 2000 times a year.	Japan annexes Manchuria. Paris Universal Exhibition and first Metro line. Zeppelin (1838–1917), first dirigible. Freud (1856–1939): *The Interpretation of Dreams.* Boxer rebellion in China.
Levi-Civita, G. Ricci: research on the tensor calculus. Discovery of the first hormone (adrenaline) and of blood-groups (**Lansteiner**).	1901	First award of the Nobel Prizes. **Heisenberg** († 1976), **Fermi** († 1954), **Schreier** († 1929), **G. G. Pincus** († 1967), **E. O. Lawrence** († 1958), **Huggins, Pauling.** Caisse des recherches scientifiques (France).	First Concours Lépine (1846–1933). First Mercedes. French law on associations. Wireless message across the Atlantic (Marconi). M. Weber (1864–1920): *The Protestant Ethic and the Spirit of Capitalism.*
Poincaré: *la Science et l'Hypothèse.* **Lebesgue**'s integral. **De Vries:** hereditary mutation. **Mendel**'s law extended to the animal kingdom. **Rutherford:** work on radioactivity.	1902	**P. Dirac** († 1984), **A. Lwoff, Kastler** († 1984), **Barbara McLintock, L. Neel.** *Annalen der Physik*, the first journal of physics, 200 to 300 references a year.	Ostwald: *Leçons sur la nature de la phlosophie.* Méliès: *le Voyage dans la Lune.* Debussy: *Péleéas et Mélisande.* France, ministry of Combes (1835–1921).
J. A. Fleming invents the diode. The chromosomes are recognized as the bearers of heredity. Electrocardiogram.	1903	**Delsartes** († 1968), **K. Lorenz, G. E. Hutchison, von Neumann** († 1957), **Kolmogorov.** *Berichte* of the German Chemical Society. International Seismological Association.	Meeting of Schönberg (1874–1951), Mahler (1860–1911), Klimt (1862–1918). Philosophy, medicine, jurisprudence. J. Conrad (1857–1924): *Typhoon.* First Tour de France. Wright brothers (W. 1867–1912 and O. 1871–1948), first powered aeroplane.
Principle of radar. **Zermelo:** axiom of choice. **Lorentz:** transformation groups.	1904	International Solar Union. **H. Cartan, J. R. Oppenheimer** († 1967), **Cherenkov, Gamow** († 1965).	Hesse (1877–1962): *Peter Camezind.* Work restarts on the Panama Canal. Russo-Japanese War.

Scientific events		Individual and collective actors	Background information
	1904	*Revue de mathématiques et physique*, 50 to 60 references a year.	
Three articles by **Einstein** on: *Probability and Brownian motion; light and the photon; special relativity, mass-energy relationship* ($E = mc^2$).	1905	**C. D. Anderson, Kuiper** († 1973), **E. Segré, M. S. Livingstone, Frohlich.**	First Russian Revolution. Picasso (1881–1973), pink period until 1907. Fauvism in painting. France: separation of Church and State.
Hopkins discovers what will be called vitamins. **Fréchet:** abstract space, general topology.	1906	**Majorana** († 1938), **H. A. Berthe, Parenago, J. Dieudonné, K. Gödel** († 1978), **A. Weil, Tomonaga** († 1979), **Maria Gopper Mayer.**	Lagerlöf (1858–1940): *Nils Holgerson.* Invention of the triode. First gas turbine. San Francisco earthquake.
Markov: concept of a chain in probability. **Harrison** produces the first tissue-culture.	1907	**Yukawa** († 1981), **Tinbergen, Jensen** († 1973), **Kowarski, J. Bernard.**	Hague Convention on international conflicts. Cubist exhibition in Paris. Maeterlinck (1862–1949): *l'Intelligence des Fleurs.*
Liquefaction of helium. Gigantic explosion of probable comet [nucleus] in Siberia.	1908	**Landau** († 1968), **Alfen, Ambartsumian.**	Pu-Yi becomes Emperor of China at two years old. Action Française.
Geiger designs a particle-detector. **H. Baekeland** invents Bakelite. **Suess:** completion of *Der Antlitz der Erde,* begun in 1883.	1909	**C. Chevalley** († 1984), **J. Herbrand** († 1931). Indian Scientific Institute.	Peary (1856–1920) arrives at the North Pole. Blériot (1872–1936) crosses the Channel in an aeroplane. Ballets Russes in Paris.
B. Russell and **Whitehead:** *Principia mathematica* (foundations of mathematics). **M. Curie** and **A. Debierne** isolate radium.	1910	**Chandrasekhar, J. Monod** († 1976), **Shockley.** Marine Biological Laboratory (Massachussets).	Growth of abstract painting. TNT. Creation of the Union of South Africa. Péguy (1873–1914): *Jeanne d'Arc.*
C. T. R. Wilson: ionization chamber to detect particle trajectories. **Rutherford** proves the existence of the atomic nucleus. **C. Funk** discovers Vitamin B.	1911	**Leroi-Gourhan** († 1986), **Alvarez, P. Kutz.** Species known: 34,400 Vertebrates, 62,300 Mollusca, 394,000 Arthropoda. First Solvay Conference, on physics.	M. Curie receives a second Nobel Prize. Amundsen (1872–1928) and Scott (1868–1912) at the South Pole. Sun Yat-Sen (1866–1925), first President of the Republic of China. Kandinsky (1866–1944): *Composition.*

Scientific events		Individual and collective actors	Background information
V. F. Hess observes cosmic rays. **Brouwer** develops algebraic topology.	1912	**A. Turing** († 1952), **W. von Braun** († 1977), **Weizsäcker.** Geological Service, China. Foundation of the Institut Curie.	F. W. Taylor: *Principles of Scientific Management.* First and last voyage of the *Titanic.* C. G. Jung (1875–1961): *Psychology of the Unconscious.* First passage through the Panama Canal. M. Proust (1871–1922)
J. Perrin: *Les Atomes.* **Bohr** and **Rutherford** propose a planetary model of atomic structure. **Pavlov:** on conditioned reflexes. Formulation of the genotype-phenotype distinction. **Moseley:** X-ray spectra of the elements.	1913	**M. Duchesme, R. W. Sperry, W. E. Lamb, G. Haro.**	publishes at his own expense the first part of *À la Recherche du temps perdu.* Stravinsky (1882–1971): *Rite of Spring.* P. Duhem (1861–1916): *Le Système du monde* (–1959). End of the Balkan Wars. France: re-establishment of three years' military service. The British occupy
Soddy: the idea of an isotope. **Hausdorff:** principles of general topology. **Kendall** isolates the thyroid hormone.	1914	**Abragam, Salk, A. L. Hodgkin.** Commission Supérieure des Inventions Intéressant la Défense Nationale (France, August).	Basrah, where Al-Haytham was born. The dollar doubles in value against the franc.
Wegener: *Die Entstehung der Kontinente und Ozeane* (theory of continental drift). **Sommerfeld:** theory of the atom.	1915	**S. F. Hoyle, L. Schwartz, C. H. Townes, Hofstadter.** National Research Council, US. Consultative Council on Research, GB. Direction des Inventions Interessant la Défense, France.	France and England take over the German colonies. M. Curie: radiological delivery 1915. D. W. Griffith (1875–1948): *Birth of a Nation.* First use of poison gas (Germany).
Von Frisch shows that bees perceive certain colours. **Einstein:** *General relativity.* **Borel:** on the calculus of probabilities. **Hardy** and **Ramanujan:** number-theory. **P. Langevin:** ultrasonic sensor.	1916	**Crick, Wilkins, R. Lindeman, A. M. Prokhorov, Shannon.** [Department of Industrial and Scientific Research], G.B.	The pacifist R. Rolland (1866–1944) receives the Nobel Prize for Literature. First tank (British). Tzara (1896–1963), beginning of Dada.
	1917	**L. J. Rainwater, R. R. Porter, C. de Duve, A. Selberg, I. Prigogine.**	February and October Revolutions in Russia. Mutinies on all fronts. The United States enters

Scientific events		Individual and collective actors	Background information
	1917		the War. Balfour Declaration.
Discovery of genes in the chromosomes. **Rutherford:** first artificial disintegration of an element and hypothesis of the neutron.	1918	**Ryle** († 1984), **R. Feynman, J. S. Schwinger.** Institutes of physics, aeronautics, Moscow; physico-technology, optical, Leningrad. Research Information Service, US.	Peace. O. Spengler (1880–1936): *The Decline of the West.* Epidemic of Spanish flu (1,000,000 dead). Universal suffrage in Britain, with women's right to vote.
Observation of bending of the rays of the sun. **Eccles** and **Jordan** design the first electronic circuit. **E. Noether:** arithmetical theory of algebraic functions.	1919	International Research Council. International Astronomical Union. International Time Bureau. Direction des recherches scientifiques et industrielles des inventions (France).	Nature reserves in the Urals and in Astrakhan. Spartacist Revolt in Germany. First crossing of the Atlantic by plane. Gropius (1883–1969): Bauhaus architecture.
H. Standinger founds macro-molecular chemistry. Catalogue of 250,000 star spectra.	1920	**Shatzmann, F. Jacob.** The Throop Technical Institute (est. 1891) becomes the California Institute of Technology. University of Rio de Janeiro.	Beginning of radio broadcasting. Creation of the SDN. Beginning of campaign by Gandhi (1869–1948). P. Valéry (1871–1945): *Le Cimetière marin.*
Langmuir: model of the structure of helium. Nerve impulses conceived of as chemical exchange. **F. Dahl:** *Grundlagen einer ökologischen Tiergeographie.*	1921	**Herbig, G. Wilkinson, R. Yalow.** Physico-mathematical Institute, Leningrad. Institut National de la Recherche Agronomique (France).	Husserl (1859–1938): on the philosophy of inter-subjectivity. Rorschach (1884–1922): *Psychodiagnosis.* NEP in the USSR. US immigration quotas. Partition of Ireland. First tuberculosis vaccine.
E. Cartan: generalization of Riemannian geometry (relative spaces), theory of generalized spaces. **Banting** and **Best** discover insulin.	1922	**A. Sakharov, C. Barnard, A. Bohr, N. G. Bassov.** Office nationale des recherches scientifiques et industrielles et des inventions. International Geographical and Geological Union.	Egyptian independence. Mussolini (1883–1945) in power in Italy. Pirandello (1867–1936): *Henry IV.* Joyce (1882–1941) publishes *Ulysses.* Wittgenstein (1889–1951): *Tractatus Logico-philosophicus.*
S. Banach: new theory of measurement. **Zworykin** (1889–1982)	1923	**M. Schwarzschild, H. Kahn, Ph. Anderson, R. Thom.**	Ataturk (Mustafa Kemal) comes to power in Turkey.

Scientific events		Individual and collective actors	Background information
designs a cathode-ray tube to produce images (television/iconoscope). **P. M. Blacket:** first transmutation (nitrogen-oxygen).	1923	National Research Council, Italy.	First record-player. First Domestic Arts exhibition in Paris. 1 dollar is worth 18 billion marks.
Methodological development in mathematical physics (i.e. **Heisenberg** in quantum mechanics (**Cayley**'s matrices)). **Levene** and **Mori** identify DNA. **L. de Broglie:** wave mechanics.	1924	**A. Hewish, R. Guillemin, A. Cormack.**	A. Breton (1896–1966): *Surrealist Manifesto.* M. Mauss (1873–1950): *The Gift.* Puccini (1850–1924): *Turandot.* Berg (1885–1935): *Chamber Concerto.* Proclamation of the Greek Republic. The TMC (see 1896) becomes IBM.
Morgan: *Genetics of Drosophila.* Appearance of the concept of electron-spin. **O. Schreier** works on the theory of general topological groups.	1925	**L. Esaki, J. Lederberg, S. van der Eer.** International Centre for Radiological Protection. Arctic Institute, USSR.	Eisenstein (1898–1948): *The Battleship Potemkin.* Kafka (1883–1924): *The Trial.* Hitler(1889–1945) writes *Mein Kampf* in prison. Chaplin (1889–1977): *The Gold Rush.*
First liquid-fuel rocket. **Schrödinger** proposes a synthesis of wave and quantum mechanics. Construction of the first electron lens.	1926	**Glaser, Tsung Dao Lee, M. Walker, J. P. Serre, Abdus Salam, P. Berg.** First Congress of the Pan-Pacific Science Association.	Salazar (1899–1970), Portuguese Finance Minister. Abel Gance (1889–1981) finishes his *Napoléon.* Baird, first demonstration of television. F. Lang (1890–1976): *Metropolis.* J. Renoir (1894–1979): *Nana.*
Artin: abstract theory of algebra. **Dirac** generalizes the notion of spin and introduces special relativity into quantum mechanics. 5th **Solvay** Conference (electron and photon).	1927	**C. Milstein, J. R. Vane, M. Nisenberg, M. Eigen.**	Heidegger (1889–1976): *Being and Time.* S. Zweig (1881–1942): *Volpone.* B. Traven: *The Treasure of the Sierra Madre. The Jazz-Singer:* first talking film. Execution of Sacco and Vanzetti. Invention of synthetic rubber.
Hubble observes the Doppler effect in galactic	1928	**A. Grothendieck, D. Nathans.**	Alain (1868–1951): *Propos sur le bonheur.*

Scientific events		Individual and collective actors	Background information
radiation and deduces his Law of Proportionality. **Einstein** proposes a unified-field theory.	1928	National Academy, Beijing. Academia Sinica (science).	Carnap (1891–1971): *The Logical Structure of the World.* D. H. Lawrence (1885–1837): *Lady Chatterley's Lover.* Lovecraft (1890–1937): *The Call of the Cthulhu.* French law on health insurance.
A. Fleming discovers penicillin. **Lyot** invents the coronograph to observe the solar corona. **Herbrand:** work in mathematical logic.	1929	**Watson, M. Gell-Mann, I. Giaever, Mössbauer.** Academy of Agricultural Sciences, USSR.	Hergé (1907–1983): *Tintin in the Land of the Soviets.* Liquidation of the kulaks in the USSR. Wall St Crash. Malinowski (1884–1942): *The Sexual Life of Savages.*
Van der Waerden: synthesizes the algebra of structures. Discovery of Pluto. **Trumpler:** *Statistical Astronomy,* (on the mass/clustering of the stars).	1930	**L. N. Cooper, R. H. MacArthur** († 1972), **S. Smale.** Service nationale de la recherche, Caisse nationale de la recherche (France).	Haile Selassie (1892–1975) becomes Emperor of Ethiopia. Musil (1880–1942): *The Man without Qualities.* Mermoz (1901–1936): Dakar–Recife air postal service. Von Sternberg (1894–1969): *The Blue Angel.*
First radio-telescope. **Urey:** deuterium. **Gödel's** incompleteness theorem. **Pauli** hypothesizes the existence of the neutrino. **A. Piccard** reaches 16,000m in a stratospheric balloon. **Carothers:** neoprene. **W. Bush:** differential analyser.	1931	**R. Schieffer, J. W. Cronin, Hamilton Smith.** International Council of Scientific Unions.	H. Broch (1886–1951): *The Sleepwalkers.* M. Planck: *Positivism and the Real World.* The Japanese occupy Manchuria. Formation of the Commonwealth. Proclamation of the Spanish Republic.
E. O. Lawrence and **M. S. Livingstone** build the first cyclotron. **Anderson** hypothesizes the existence of the positron (anti-electron with positive charge). **Chadwick** discovers the predicted neutrons.	1932	**S. L. Glashow, W. Gilbert.**	Céline (1894–1961): *Voyage au bout de la nuit.* A.L. Huxley (1894–1963): *Brave New World.* Creation of Saudi Arabia. End of the first phase in the reclamation of the Zuiderzee.

Scientific events		Individual and collective actors	Background information
Morgan: first experimental mutation.	1932		Colonial exhibition at Vincennes. De Valera (1882–1975) Prime Minister of Ireland.
Yukawa: hypothesis of the meson. **Kolmogorov:** abstract theory of probability. First electron microscope.	1933	**S. Weinberg, A. A. Penzias.** Between 1933 and 1938, more than 1800 German scientists were excluded from the universities. Science Advisor Board, US. Conseil supérieur de la recherche (France).	T. Mann (1875–1955): *Joseph and his Brethren* (–1943). A. N. Whitehead: *Adventures of Ideas.* First quartz astronomical clock. F. D. Roosevelt (1882–1945) President of the US. Hitler Chancellor of Germany.
Dam: Vitamin K. **I. and F. Joliot Curie:** first artificial radioactivity. **C. F. and G. Cori** synthesize glycogen.	1934	**P. J. Cohen, R. Rubbia, N. Bourbaki.**	Einstein: *The World as I See It.* Citroën (1878–1935) produces the first front-wheel drive. The Long March in China (–1935). Nicaragua: Sandino (1895–34) [overthrown] by Somoza (assassinated 1975).
V. Volterra and **U. d'Ancona** propose a mathematical model of prey-predator population ratios. **G. Domagk** dicovers the sulphonamides.	1935	Caisse nationale de la recherche scientifique.	Hartmann (1882–1950): *Zur Grundlegung der Ontologie.* Nuremberg Laws against the Jews. Stakhanov digs 14 times as much coal as the norm lays down. Italian invasion of Ethiopia.
Gentzen: logical research on the coherence of arithmetic. **G. Reber:** first radio-telescope. Experimental confirmation of the existence of **Yukawa's** mesons. **Carothers** creates polyamine 6–6 (nylon).	1936	**R. W. Wilson, S. Chao Chung Ting, K. G. Wilson.** Service national de la recherche scientifique (France). Creation of the Fields (1863–1932) Medal (mathematics).	Hitchcock (1899–1980): *Sabotage.* London: first television transmitter. Stalin's purges. Popular Front in France. Rome–Berlin Axis. Beginning of Spanish Civil War.
Shannon: relationship between	1937	**D. Munford, R. Hoffmann.**	Cassirer (1874–1945): *Determinism and*

Scientific events		Individual and collective actors	Background information
binary logic and electrical contacts. **Turing** develops the concept of the Universal Machine. **H. Aiken:** plan for an electronic calculator.	1937	Creation of an Under-Secretariat for scientific research (France).	*Indeterminism in Modern Physics.* *Hindenburg* airship disaster, after one year in use. Total independence of Southern Ireland. Sino-Japanese War: Communist-Kuomintang alliance.
E. D. Kendall: produces cortisone. **O. Hahn, L. Meitner, Stressmann:** fission of uranium. **H. A. Bethe:** stellar energy is thermonuclear. Radioactive iodine allows the investigation of glands (thyroid). **K. Zuse:** Z1 'computer'.	1938	**S. Novikov.** Centre nationale de la recherche scientifique appliquée (France).	J.-P. Sartre (1905–1980): *La Nausée.* Disappearance of Majorana. Invention of ball-point pen. First commercial air crossing of the North Atlantic. *Anschluss* (German annexation of Austria).
Rabi: method of magnetic resonance. **N. Bourbaki:** first installment of the *Elements of Mathematics.* **L. Brillouin:** diffraction of light by ultrasound. **G.G. Pincus:** brings about the first mammalian parthenogenesis. **G. Stibitz:** Bell Laboratories calculator.	1939	**J. M. Lehn.** Centre nationale de la recherche scientifique – CNRS (France).	E. Jünger (1895): *On the Marble Cliffs.* B. Bartok (1881–1945): *String Quartets.* Nazi-Soviet pact. Invasion of Poland. The old Apostolic Nuncio in Germany becomes Pope (Pius VI). Industrial production of nylon.
Lansteiner discovers the Rhesus factor. The **Joliot-Halban-Kowarski** team demonstrates the possibility of chain-reaction and the production of neutrons. Radar becomes usable. **Landau:** quantum analysis of liquid helium.	1940	**B. D. Josephson, D. Quillen.** From 1940 to 1942, Nobel prizes for science were not awarded.	Greece resists Italian aggression. Helicopter and experimental colour television in the US. Discovery of the Lascaux caves. Vichy decrees the 'Aryanization' of Jewish property.
	1941	Office de la recherche et du développement scientifique, Institut national d'hygiène, France.	World war and its consequences. Penicillin begins to be used.

Scientific events		Individual and collective actors	Background information
Grassé: study on the swarming of termites. **Fermi** starts the first atomic pile, in Chicago (uranium/graphite). **Aiken** puts the Mark 1 electronic calculator into service. **G. D. Birkoff:** lattice theory. Quantitative interpretation of energy exchanges within an ecosystem.	1942		A. Camus (1913–60): *The Stranger.* Vercors (1902): *Le Silence de la mer.* B. Brecht (1898–1956): *Galileo Galilei.* M. Ernst (1891–1976): *Anti-Papa.* Intensive use of DDT. The war continues . . .
Britain builds the *Colossus* computer for the Decoding Service (**A. Turing**).	1943	Academy of pedagogical Sciences, USSR.	J.-P. Sartre: *Being and Nothingness.* M. Carné (1909): *Les Enfants du paradis.* Ferhat Abbas (1899–1985) claims Algerian autonomy.
Waksman and **Schatz** discover streptomycin. Inauguration of the ASCO automatic calculator. A team at the Rockefeller Institute recognizes the role of DNA in the genotype. **L. Schwartz:** *Theory of Distributions.*	1944	**P. Deligne.**	Liberation of some European countries. Anouilh (1910–1987): *Antigone.* Death of Kandinski (1866), Maillol (1861), Mondrian (1872), E. Munch (1863), Giraudoux (1882), Saint-Exupéry (1900),
The first atomic bomb explodes at Alamogordo, having cost 2 billion dollars. **McMillan** and **Veksler:** principle of the synchrotron.	1945	Commissariat à l'énergie atomique.	R. Rolland (1866). R. Bresson (1907): *Les Dames du Bois de Boulogne.* C. Levi (1902–75): *Christ Stopped at Eboli.*
The ENIAC computer is operational. Bikini: experimental detonation of A-bombs. **Frenkel:** *Kinetic Theory of Liquids.* **Von Frisch** discovers the meaning of the dance of the bees.	1946	**W. T. Thurston, G. A. Margoulis, G. Köhler.**	Bathyscaphe of A. Piccard (1884–1962), which will reach 10,600m. First session of the UN and creation of UNESCO. Proclamation of the Italian republic. In France: closure of brothels. Truman (1884–1972): 'An iron curtain has fallen

Scientific events		Individual and collective actors	Background information
Von Neumann and **Morgenstern:** *Theory of Games and Economic Behaviour.* Invention of the vacuum-valve (kenotron). Construction of the EDVAC. Construction of the EDSAC, first electronic machine with a stored program. Discovery of the transistor. **Shannon:** *Mathematical Theory of Communication.* **Gamow** thinks that the Universe began with a primordial explosion – the BIG BANG!	1947	2.5 BILLION HUMAN BEINGS. **A. Connes.** Nearly 10,000 scientific periodicals, including 200 abstracts. UN inquiry into international scientific research organizations.	across Europe.' A. Toynbee (1889–1975): *A study of history,* vol. 6. De Chirico (1888–1978): *Perseus and Andromeda.* H. Moore (1898–1986): *Family Group.* B. Britten (1913–76): *Albert Herring.* Marshall (1880–1959) Plan for revival of the European economy.

Notes

Introduction

1 Auguste Comte (1798–1857), positivist philosopher.
2 *Devisement*: a word invented by Jacques Cartier to describe the exploration of the land and the sea.
3 Percolation: a statistical description of systems made up of a great number of objects which can be linked together. In such a system long-distance communication is either established or it is not.

Chapter 1   Babylon −1800

1 Diacritical signs and their pronunciation in the transcription of Akkadian names and words:
   j, J – 'y' as in '*y*es'
   ḫ, Ḫ – 'ch' as in German 'a*ch*tung'
   ṣ, Ṣ – 'ts' as in 'bi*ts*'
   š, Š – 'sh' as in '*sh*ip'
   ṭ, Ṭ – 'emphatic t'
   g, G – always hard as in '*g*one'

   Punctuation used in text transcriptions:
   ( ) – words added to help the comprehension of the text
   [ ] – lacunæ in the text
   < > – word forgotten by the scribe
   . . . – word of unknown meaning
   . . . . . . – clause of unknown meaning
2 *wāṣitum* – a word whose meaning is currently being discussed among Assyriologists.
3 *nindan* – The definition of this unit of length is given on p. 60.

Chapter 2   Measure for Measure

1 Diacritical signs and their pronunciation in the transcription of Akkadian and Egyptian words:
   ḥ, Ḥ – 'emphatic h'
   ḫ, Ḫ – 'ch' as in German 'i*ch*'

š, Š – 'sh' as in '*ship*'
g, G – always 'hard' as in '*gone*'

Punctuation used in text transcriptions:
( ) – words added to help the comprehension of the text
[ ] – lacuna in the text
. . . – word of unknown meaning
. . . . . . – clause of unknown meaning

2   1 *ḫar* ≈ 48 litres
    1 hecta-quadruple-*ḫeqat* ≈ 960 litres
3   1 *pānum* ≈ 60 litres
    1 *sūtum* ≈ 10 litres
    1 *qûm* ≈ 1 litre
4   For an example of a Babylonian mathematical table text, see the illustration on p. 42.

Chapter 3   Gnomon

1   *Diogenes Laertius*: Hieronymus says that Thales determined the height of a pyramid by means of the length of its shadow: the shadow was measured at the hour of the day when a man's shadow is the same length as himself.
    Plutarch: The height of a pyramid is related to the length of its shadow in precisely the same way as the height of any measurable vertical object is related to the length of its shadow at the same time of day.

2   πολοσ or *polos*: a concave portion of a sphere in the hollow of which the gnomon's shadow is projected.

3   Alphonsine or Toledan tables: made on the orders of Alfonso X the Learned (1221–84), King of Castile and Leon, these were constructed by a group of astronomers under the supervision of Isaac ben Said, completed in 1252 and kept in print until the sixteenth century.

4   Apex: the point in the sky towards which the Sun appears to be moving.

5   Compare the Greek word επιστημη, science, and the word of the same family, επιστημα, which means the funeral cippus, a vertical column over a tomb.

6   Cathode: in Greek and literally, a path going from top to bottom or a descent.

7   Anabasis: in Greek and literally, movement from bottom to top or an ascent. The march of Cyrus the younger from Sardis to Cunaxha described by Xenophon.

8   On page 81 b-c of *Meno*, Socrates cites a fragment of Pindar on the rebirth of the soul. Pindar (518–438 BC), author of odes, the *Olympia* and *Pythia*, etc.

9   Duplicate: not to replicate to obtain a copy, but to construct the same form with double the surface.

10 Algorithm: contrary to appearances, the word does not come from Greek but from Arabic and means a finite set of elementary operations for a computational procedure or the resolution of a problem.

11 φαρμακον: a medicine or a harmful drug; φαρμακοσ: a poisoner, a sorcerer, the victim of a sacrifice.

12 The paradoxes of Zeno: leaving one point to go to another, the traveller or moving body must first pass through the middle of the segment that separates them, then through the middle of the remaining segment, and so on to infinity, so he will never arrive at his goal.

13 ριζα: root; ριζωμα: something rooted, therefore roots, foundations, elements.

14 Prosody: a set of rules relating to the quantities of vowels that govern the composition of verse, especially in Greek and Latin poetry.

15 Epistemology: the theory of scientific knowledge; gnosiology: the theory of knowledge in general.
16 Transcendent: beyond the world and bearing no relation to it.
17 Transcendental: bearing a relation to the *a priori* conditions of knowledge.
18 The quotation is taken from Plato's *Gorgias*, edited by G. P. Goold and translated by W. R. M. Lamb (Loeb Classical Library, Heinemann, 1977).

Chapter 4   Archimedes

1 The works of Archimedes: *On the Sphere and the Cylinder*, Books I and II, *The Measurement of the Circle, On Conoids and Spheroids, On Spirals, On the Equilibrium of Planes*, Books I and II, *The Sand-Reckoner, Quadrature of the Parabola, On Floating Bodies*, Books I and II, *The Stomachion, The Method, The Lemmas* and *The Problem of the Cattle*.
2 *Spolia opima*: in Rome, the spoils given to a general who had killed the opposing general with his own hand,
3 The heliocentric system of Aristarchus (around 290 BC): in this system the Earth revolves around its own axis while describing a circular orbit around the Sun. The first great astronomer of the school of Alexandria, Aristarchus of Samos was the precursor of Copernicus, some seventeen centuries later.
4 Thucydides (*c.* 460–*c.* 395 BC), the Greek historian, author of the *History of the Peloponnesian War*.
5 Atomism: a cosmological theory originating in the fifth century BC (Leucippus, Democritus) and developed by Epicurus (fourth century BC), which holds that the Universe is made up of discontinuous atoms, infinitely hard, immutable and eternal.

   Elementarism: a cosmological theory originating in the sixth century BC, which holds that the essence of things is an element: water (Thales, Anaximander), air (Anaximenos), fire (Heraclitus) or the combination of several of them, that is, earth, water, air and fire (Empedocles, Aristotle).

Chapter 5   Stories of the Circle

1 Mantra: a verse or phrase credited with magical or religious powers; a Vedic hymn.
2 Trigonometry allows the determination of the relations between the sides of a triangle as functions of its angles and vice versa. These functions (sine, cosine, tangent ...) are called trigonometric or circular.
3 Differential calculus studies the infinitely small local variations of functions. Integral calculus, on the other hand, is used to discover functions on the basis of their local variations. An infinite series is a sequence of numbers or mathematical expressions considered as a its sum. A function can be developed as a series (or is analytic) if one can write it as the (infinite) sum of powers of the variable, like the sine function:

$$\sin x = x - \frac{x^3}{2 \times 3} + \frac{x^5}{2 \times 3 \times 4 \times 5} - \frac{x^7}{2 \times 3 \times 4 \times 5 \times 6 \times 7} + \dots$$

4 Transformation: mapping from one figure or mathematical expression to another. Projection, rotation and permutation are examples of transformations.
5 The author would particularly like to thank Guy Mazars, Edwige Regenwetter and Bernard Vitrac for their valuable assistance.

Chapter 6   The Arab Intermediary

1   *Canon of Medicine*: an encyclopaedic compilation offering a broad, complete and methodical synthesis of Graeco-Arab medical knowledge. It owes its fame to its rigorous
construction, to the scale of its theoretical contribution and to its demand for rationality.
2   For a description of Aristotle's *Organon*, the reader should refer to note 2 of Chapter 7
below.

Chapter 7   Theology in the Thirteenth Century

1   *Scolares*: those who attended the schools, that is, the University.
2   *Organon*: a Greek word meaning an instrument, an organ; the name under which Aristotle's thirteenth-century editors grouped together his treatises on logic, that is, the *Categories, On Interpretation*, the *Analytics*, the *Topics* and the *Refutation of the Sophists*.
3   Subalternation: the relation between two sciences when one borrows its principles from
another. Aquinas' example was that of optics, which took its principles from geometry.
4   Angelic Doctor or Common Doctor: it was the custom in the Middle Ages to give such
titles to university teachers of great fame. Albertus Magnus was the Universal Doctor,
Bonaventure the Seraphic Doctor, Roger Bacon the Admirable Doctor and Thomas Aquinas
first of all the Common Doctor, in recognition of the importance of his thought, though
later he was given the more beatific name of Angelic Doctor.
5   The author would particularly like to thank Jean-Marc Drouin, who helped considerably
in the preparation and writing of this text.

Chapter 8   Algebra, Commerce and Calculation

1   Alum: used as a mordant in the textile industry. The dye would only take or bite (French
mordre, to bite) on the cloth or thread after it had been treated with alum. A product
indispensable to the textile industry, alum was a very important article of trade in the
Middle Ages.
2   Double-entry book-keeping: a complex system of book-keeping, utilizing numerous
accounts, which requires two entries to be made for each transaction, one credit and one
debit, so that the balance is always zero.
3   False position enabled the solution of a number of problems without the use of algebraic
symbols. The method finds an unknown on the basis of one or two solutions, one or two
'positions' chosen arbitrarily. These calculations, difficult to perform, which might lead to
mistakes, allowed a correct solution to be found. Replaced by algebraic equations, the
method of false positions has today been abandoned.
4   Opposition and remotion: this technique of calculation, badly understood by many
authors, is a means of finding complete solutions to indeterminate problems of two
equations with three unknowns.
5   Positional notation is the system we use ourselves. It is linked to the use of Aabic numerals. It takes its name from the fact that in the number each figure has a value related to
its position. When placed on the right of a number 1 is a unit, when second from the right
a ten and so on.
6   Mediation, that is to say division by two, and duplication, multiplication by two, were still
much used in this period. They are the legacy of a time when paper calculation was
unknown. The considerable difficulty of multiplication and division led to the use of
simplified operations, multiplication or division by two being repeated several times over.

7 Barter in the *Kadran aux marchans* (1485): 'Two merchants want to exchange or barter their merchandise, the one with the other. One has cloth, the other has pepper. The one with the cloth wants to sell it at 12 shillings a yard in exchange, which is worth 10 shillings in cash. How much pepper must the other sell him, if it is worth 9 shillings a pound in cash?'

8 The utilization of mathematical results in other fields of activity, such as business, thus encouraged the development of a scientific profession. But other scenarios are possible. It was a different kind of mathematician, connected this time with the University, who made his appearance in the nineteenth century (see pp. 344–71).

Chapter 9   The Galileo Affair

1 Galileo's main works: *De Motu* ('On Motion') was written but not published while Galileo was a professor at Pisa (1589–91); *Sidereus Nuncius* ('The Starry Messenger'), 1610; *Il Saggiatore* ('The Assayer'), 1623; *Dialogo sopre i due sistemi del mondo, Ptolemaico e Copernico* ('Dialogue on the Two Chief World Systems, Ptolemaic and Copernican'), 1632; *Discorsi e dimostrazioni mathematiche intorno a due nuove scienze attenente alla mecanica e i movimenti locali* ('Discourses and Mathematical Demonstrations . . .'), 1638.

2 The modern formula for uniformly accelerated rectilinear motion is written as follows: $v = \gamma t$, $x = 1/2\gamma t^2$ where γ, t and x are velocity, acceleration, time and the distance travelled by the moving object.

3 The Italian text allows one to think that the velocities here are the subject: 'The velocities therefore increase the degree of velocity.'

4 'Potential' function: introduced in Joseph Lagrange's *Analytical Mechanics* (1788), it gives an overall description of a mechanical system as a function of the point masses of which it is composed and of the distances between the masses. The interactive forces acting on each mass at every moment are defined as derivatives of the potential function. This function has the dimensions of an energy (it is also called the potential energy). It allows the statement at the highest level of generality of the mechanical conservation of the cause in its effect: every increase in potential energy is paid for by a diminution in the kinetic energy (related to the velocities of the masses making up the system) and inversely.

5 Fourier's Law (1822) describes the velocity with which heat diffuses between two points of a body as a function of the difference in temperature between the two points. The diffusion of heat is the very type of a process which cancels its own cause; when it has led to uniformity of temperature, it stops.

Chapter 10   Refraction and Cartesian 'Forgetfulness' has no notes.

Chapter 11   Working with Numbers

1 One of the problems Fermat set his correspondents in 1657 was to find integers x and y such that $x^2 - Ny^2 = 1$ for a given integer N. To Frenicle he suggested, in particular, the study of N = 61 and N = 109 'so as not to give [him] too much difficulty'. These were precisely cases for which the smallest solutions were numbers of more than nine digits.

2 *Privatdozent*: a lecturer in German universities, paid by the students who attend his courses.

3 Viète's *Zetetics*: Viète saw 'analysis' as the real source of mathematical discovery (as opposed to the synthetic exposition of Euclidean geometry). He distinguished three kinds of analysis: Zetetics corresponded to the expression of the problem as an equation, Poristics to the verification and Exegetics to the determination properly speaking of the solutions to a problem. He applied this in particular to an algebraic rereading of Greek works, among them Diophantus' *Arithmetic*.

Chapter 12   Ambiguous Affinity

1 In the formula $1/r^2$, r represents the distance between any two bodies.
2 In modern terms the reaction would be written

$$2NaCl + CaCO_3 \rightarrow Na_2CO_3 + CaCl_2$$

Chapter 13   From Linnaeus to Darwin

1 Example of a diagnosis: 'Vinca caulibus procumbentibus, foliis lancolata ovatis, floribus pedunculatis' which means 'Periwinkle with rampant stem, oval lanceolate leaves and pedunculate flowers' (referring to the lesser periwinkle in C. Linnaeus, *Species Plantarum*, 3rd edn, 1764).
2 Cotyledon: a simple embryonic leaf in seed-bearing plants, which, in some species, forms the first green leaf after germination.
3 Phanerogamous species: species which bear flowers at some point in their development and which reproduce themselves by seeds. Etymologically, phanerogamous species are those which display their sexual organs, as opposed to the cryptogams, whose mode of reproduction was obscure for a long time.

Chapter 14   Paris 1800

1 Hubert Robert (1733–1808), a French painter who liked gardens and ruins.
2 Ephraim Chambers (1680–1740): in 1728, by means of a subscription, he launched an *Encyclopaedia, or Universal Dictionary of the Arts and Sciences*, which met with great success in England and served as one of the principal sources for Diderot and d'Alembert's own *Encyclopédie*.
3 Cosmology: the science of the general laws that govern the Universe.
4 Cosmogony: a theory explaining the formation of the Universe.
5 Oxymoron: a figure of speech which ingeniously combines contradictory terms (from two Greek adjectives meaning sharp and dull, therefore acuteness behind a foolish appearance).
6 Georges Dumézil (1898–1986) includes under Jupiter the class or function of priests and judges, under Mars the class and function of the military and under Quirinus those of the producers. See, for example, his *Mythes et Épopées* (1968–73).
7 In *Le Génie du christianisme* (1802) Chateaubriand demonstrates that, more than any other religion, Christianity favours intellectual and artistic creation.
8 Ideocracy: power or government founded on ideas or dogma.
9 Évariste Galois (1811–32) discovered the role of groups in the solution of algebraic equations.
10 Sadi Carnot (1796–1832) wrote *Reflections of the Motive Power of Fire* in 1824.

Chapter 15   Lavoisier

1   Stephen Hales's 'pneumatic chest' was constructed to collect the gases extracted from different substances, such as plant materials and saltpetre. When these substances were heated, the air given off was conveyed by a long tube to a container filled with water. The bubbles thus produced collected under the lid. It was the receiving part of the apparatus which enabled Cavendish, Priestley and Lavoisier to undertake the chemical study of gases.

2   For a definition of the cosmological theory of elements, the reader should refer to p. 143.

Chapter 16   In Defence of Geology

1   The full title of the first edition of his book (1830–3) is *Principles of Geology: being an attempt to explain the former changes of the Earth's surface, by reference to causes now in action.*

Chapter 17   Mendel in the Garden

1   Pomology: that part of arboriculture than concerns fruit trees.

2   What is an Augustinian? The religious order of the Augustinians was founded in 1256 by Pope Alexander IV. Its members are not, properly speaking, monks, because they are not subject to vows of enclosure. It is a religious order which may be compared to the Franciscans and the Dominicans. Inspired by the writings of St Augustine (354–430), it counts among its theologians Giles of Rome (d. 1316), a commentator on Aristotle and a disciple of St Thomas Aquinas. The intellectual vocation of the order was particularly marked among the Augustinians of Brno because an imperial decree had imposed upon them the obligation of providing some of the teaching in the educational establishments of the region.

3   Hybrid: in Latin *ibrida* or *hybrida*. Levrault's *Dictionnaire des sciences naturelles*, 1821, gives as its origin the Greek υβρισ (*hubris*), genitive υβριδοσ (*hubridos*), which often means insult, affront, adultery, as if interbreeding could only arise from a breach of the law or from violence done to the order of things. Ernout and Meillet's *Dictionnaire de la langue latine* does not confirm this etymology, but none the less supposes that the spelling *hybrida* was 'influenced by a false literary identification with υβρισ'.

4   The peas of which Mendel spoke, the seeds of which are eaten as 'garden peas', belong to the genus *Pisum*.

5   Louis de Vilmorin (1816–60) was the heir to the nursery gardeners and seedsmen Vilmorin-Andrieux and Co. His wife and many of his descendants were also well-known botanists. In 1929 H. F. Roberts counted 360 articles published by seven successive generations of the Vilmorin family.

6   Atavism: a didactic term. In botany, the tendency of hybrid plants to return to their original type. In physiology, resemblance to ancestors, 'more particularly, the reappearance of a primitive character after an indeterminate number of generations. Etymology: *Atavus*, from *ad* and *avus*, ancestor, according to the Latin etymologists' (E. Littré, *Dictionnaire de la langue française*). Augustin Sageret has simply: from the Latin *atavus*, ancestor. He attributes the formation of the word to Duchesne, a naturalist and horticulturalist of the eighteenth century.

7   Villa Thuret: in 1856 the botanist Gustave Adolphe Thuret (1817–75) had a magnificent botanical garden laid out around his villa, into which he introduced a great number of exotic ornamental plants. In our own day these plants are common in all the gardens of the French Riviera.

Chapter 18   Pasteur and Pouchet

1   Mercury is used to prevent the entry of the ambient air into the bath.
2   Homogeny, heterogeny: Pouchet uses the term heterogeny ('other-birth') to designate the birth of an organism from a parent which does not resemble it (*heteros*) and to distinguish this from the usual process of the birth of offspring which resemble their parents, which he calls homogeny. Pouchet's terms are used metaphorically here to speak of the birth of scientific ideas from ideas which resemble them (homogenesis) or from highly dissimilar practices (heterogenesis).
3   Paracelsus II: this was the name that Pouchet and his collaborators gave to Pasteur.
4   Matrass: a long-necked flask, spherical or egg-shaped in form.

Chapter 19   Mendeleyev

1   Dalton called atomic weight the weight of an element which combined with another element in fixed or multiple proportions to form a compound. As it was impossible to weigh an atom, the atomic weight values were referred to a conventional unit, the atomic weight of hydrogen. Here are some examples:

Hydrogen: 1	Zinc: 56
Nitrogen: 5	Copper: 56
Carbon: 5	Lead: 95
Oxygen: 7	Silver: 100
Phosphorus: 9	Platinum: 100
Sulphur: 13	Gold: 140
Iron: 38	Mercury: 167

2   Gay-Lussac's Law (1808): when two gases combine together, their volumes are in a simple ratio to each other.
     Avogadro's Hypothesis (1811): equal volumes of gas, at the same temperature and pressure, contain the same number of molecules.
3   Dulong and Petit's Law of Specific Heat (1819): the atoms of all simple substances all have the same capacity for heat.
     Mitscherlich's Law of Isomorphism (1821): isomorphic substances have an analogous chemical composition and generally contain the same number of atoms or 'equivalents'. Consideration of isomorphism is therefore an additional tool for the determination of atomic weight.
4   Br = bromine; Cl = chlorine; I = iodine; Na = sodium; Li = lithium; K = potassium; Se = selenium; S = sulphur; Te = tellurium.
5   Li = lithium; Na = sodium; F = fluorine; K = potassium; Cl = chlorine; Rb = rubidium; Br = bromine; Cs = caesium; Te = tellurium; Ca = calcium; Sr = strontium; Ba = barium.

Chapter 20   Manufacturing Truth

1   Project Manhattan: the project for the manufacture of atom bombs. Started in 1942, it led to an experimental explosion on 16 July 1945 at Alamogordo, in the desert of New Mexico.
2   Carnot cycle: a reversible, dithermic, thermodynamic cycle, made up of two isothermic transformations (at a constant temperature) and two adiabatic transformations (without an exchange of heat with the surrounding environment).

Chapter 21   Joliot

1   Cloud-chamber: as the particles are invisible, physicists recognize them by the traces they leave in something visible. The English physicist Charles Wilson, who was a meteorologist, had the idea of tracking the particles thanks to the condensation of tiny raindrops in a cloud of water vapour created inside a closed chamber. By photographing the course of the droplets, the particles may be identified. Joliot perfected the cloud-chamber.

Chapter 22   The Invention of the Computer

1   Los Alamos: an American nuclear physics research station in the state of New Mexico, where the first atom bomb was developed.
2   Digital calculator: a distinction is usually made between analog and digital calculating machines. Digital machines proceed by separate steps and code the quantities to be treated in a discontinuous manner (cogs, sprockets, holes in punched card, closed or open circuits, the presence or absence of electrical pulses).
3   Analog computer: analog machines code the quantities to be processed continuously (rules sliding one against the other, discs turning on mobile tables, voltage . . .); an analog arrangement works by simulating a particular function, for example, 'an integration, and the result is obtained by measuring one of the physical magnitudes put into play by the machine (length, angle, voltage); analog computers are almost always specialized; before the arrival of the computer, analog machines were quicker than digital ones.'
4   Conditional branching: the possibility of subordinating the execution of an instruction to a test condition (example: if $x < 0$, perform $x + 1$; if not, pass on to the next instruction).
5   Switching: modification of the configuration of an electrical circuit which makes or breaks certain contacts; in a computer the equipment used for switching (for calculation) can be different from that used for storage (for memory).
Switching element: an element capable of being in two different states, making or breaking a circuit.
6   Addresses: codes which indicate the location of words in the machine's memory or registers; the instructions which make up programs include in particular the addresses of the data to be processed.
7   Word: a group of bits (0 or 1) of standard length; for a given type of computer a word generally corresponds to a number or to a letter of the alphabet.
8   Real time: the almost instantaneous production of results as soon as the order to execute the program is given.
9   Piezo-electric effect: deformation of a crystal under the influence of an electrical charge.

10   Macy Conferences: interdisciplinary conferences organized in particular by W. McCulloch, Norbert Weiner and John von Neumann between 1944 and 1954; most of the ideas of cybernetics were discussed there.
11   Operating system: programs which manage the internal organization and resource allocation in computers.

Bibliography

General Bibliography

Bibliographies of the History of Science and Technology, R. Multhauf and E. Wells (eds), New York, Garland Publishing, 1984. (Annotated bibliography by discipline.)

Dictionary of Scientific Biography, C. C. Gillispie (ed.), New York, C. Scribner's Sons, 1970–80, 16 vols. (This dictionary contains biographical and bibliographical notices on most of the key figures in the present work.)

Élément de bibliographie de l'histoire des sciences et des techniques, F. Russo (ed.), Paris, Hermann, 1954; 2nd edn 1969.

Histoire générale des sciences, ed. R. Taton, Paris, Presses Universitaires de France, 1966– .

Chapter 1   Babylon –1800

Citations

Niniveh: First text: PARPOLA, S., *Letters from Assyrian Scholars . . .* , Neukirchen-Vluyn, Butzon and Bercker Kevaler, 1970, vol. 1, no. 2; second text: LEHMANN-HAUPT, C., *Samassumukin, König von Babylonien*, Leipzig, Hinrichs, 1892 pl. 13: 10–18.

The Professionals: First text: OPPENHEIM, A. L., *Letters from Mesopotamia*, Chicago and London, University of Chicago Press, 1967, no. 84; second text: FINET, A., 'Les Médecins au royaume de Mâri', *Annuaire de l'Institut de philologie et d'histoire orientales et slaves*, 1954–7, vol. 14, p. 133; third text, ibid., pp. 135–6; fourth text: FINET, A., in DOSSIN, G. et al., *Archives royales de Mâri*, Paris, Geuthner, 1950, vol. 2, no. 15; fifth text: JEAN, C. F., *Archives royales de Mâri*, Paris, Geuthner, vol. 2, no. 15; sixth text: OPPENHEIM, A. L., *op. cit.* no. 23; seventh text: DOSSIN, G., *Archives royales de Mâri*, Paris, Geuthner, 1978, vol. 10, no. 20; eighth text, ibid., no. 94.

Education: First text: SJÖBERG, A., 'The Old Babylonian Eduba', *Sumerological Studies in Honor of Thorkild Jacobsen*, Chicago and London, University of Chicago Press, 1974 'Assyriological Studies' series, vol. 20, p. 159; second text: CIVIL, M., 'Sur les "livres d'écolier" à l'époque paléobabylonienne', *Miscellanea Babylonica* (Mélanges Birot), J.-M. Durant and J.-R. Kupper (eds), Paris, Recherches sur les Civilisations, 1985, pp. 71–2.

734

Divination: First text: PETTINATO, G., *Die Ölwahrsagung bei den Babyloniern*, Rome, Instituto di Studi del Vicino Oriente, 1966, vol. 2, pp. 24–7; second text: KÖCHER, F. and OPPENHEIM A. L., 'The Old Babylonian Omen Text VAT 7525', *Archiv für Orientforschung*, 1958, vol. 18, pp. 65–6.

Medicine: First text: VAN DIJK, J., *Tabulae cuneiformes a F. M. Th. de Liagre Bohl collectae*, Leyden, Nederlands Instituut voor het Nabije Oosten, 1957, vol. 2, no. 21: 15–21; second text: LABAT, R., *Traité akkadien de diagnostics et pronostics médicaux*, Leyden, Brill, 1951, pp. 130–3; third text: KÖCHER, F., *Die babylonische-assyrische Medizin in Texten und Untersuchungen*, Berlin, Walter de Gruyter, 1971, vol. 4, no. 393: 1–26; fourth text: THOMPSON, R. C., 'Assyrian Prescriptions for Treating Bruises or Swellings', *American Journal of Semitic Languages and Literatures*, 1930, vol. 47, pp. 5–6.

Mathematics: Procedural text BM 13 901: THUREAU-DANGIN, F., 'L'équation de deuxième degré dans la mathématique babylonienne d'après une tablette inédite du British Museum', *Revue d'Assyriologie*, 1936, vol. 33, pp. 29–30; multiplication table AO 10 762: GENOUILLAC, H. de, *Premières recherches archéologiques à Kich*, Paris, Champion, 1925, vol. 2, no. D3.

General works and articles

BOTTÉRO, J., *Mesopotamia: Writing, Reasoning and the Gods*. Chicago, University of Chicago Press, 1992. NEUGEBAUER, O., *The Exact Sciences in Antiquity*, Providence, Brown University, reprinted, New York, Dover. OPPENHEIM, A. L., *Ancient Mesopotamia*, rev. edn, Chicago and London, University of Chicago Press, 1977. PIRES, C., 'Le cheminement rationnel en Babylonie: Divination, jurisprudence, mathématiques', master's thesis, Université de Paris-VIII. RITTER, J., 'Les pratiques de la raison en Mésopotamie', in MATTEI, J.-F., *La Naissance de la raison en Grèce*, Paris, PUF, [1984]. ROUX, G., *Ancient Iraq*, Harmondsworth, Penguin, 1993.

Specialized works and articles

Divination: BOTTÉRO, J., 'Symptômes, signes, écritures', in VERNANT, J.-P. (ed.), *Divination et rationnalité*, Paris, Le Seuil, 1974. C.E.S.S., Université de Strasbourg, *la Divination en Mésopotamie ancienne*, Paris, PUF, 1966.

Medicine: CONTENAU, G., *la Médecine en Assyrie et Babylonie*, Paris, Maloine, 1938. GOLTZ, D., *Studien zur althistorischen und griechischen heilkunde*, Wiesbaden, Franz Steiner, 1974. HERRERO, P., *Thérapeutique mésopotamienne*, (ed. M. Sigrist, Paris), Recherches sur les Civilisations, 1984.

Mathematics: BRUINS, E., and RUTTEN, M., *Texts mathématiques de Suse*, Paris, Geuthner, 1961 (Mathematical texts found at Susa). NEUGEBAUER, O., *Mathematische Keilschrift-texte*, Berlin, Springer, 1935–7, repr. 1973, 3 vols. NEUGEBAUER, O., and SACHS, A., *Mathematical Cuneiform Texts*, New Haven, American Oriental Society, 1945 (Texts from American collections). THUREAU-DANGIN, F., *Textes mathématiques babyloniens*, Leyden, Brill, 1938. VAIMAN, A., *Shumero-vavilonskaya Matematika*, Moscow, IVL, 1961 (Texts from the Hermitage Museum, St Petersburg).

Chapter 2   Measure for Measure

Citations

'Hymn to Sulgi' from SJÖBERG, A., 'The Old Babylonian Eduba', *Sumerological Studies in Honor of Thorkild Jacobsen*, Chicago and London, University of Chicago Press, 'Assyriological

Studies' series, vol. 20, pp. 172–3. Text of the 'satirical letter' transcribed in hieroglyphics in Fischer-Elfert, H.-W., *Der satirische Streitschrift des Papyrus Anastasi I*, Wiesbaden, Otto Harrassowitz, pp. 106–11. All the Egyptian mathematical problems are from Rhind Papyri BM 10 057 and 10 058 (see below); the two problems relating to a granary: nos 41 and 42 respectively; the 'rule' for calculating $\frac{2}{3}$ of a number: no. 61b; the table of $2/N$ in Griffith, F. L., *Hieratic Papyri from Kahun and Gurob*, London, Bernard Quaritch, 1898, pl. 8, no. IV-2; the table of $\frac{2}{3}$: Rhind Papyrus, no. 61a.

For the Babylonian problems: the two first problems on Haddad tablet 104 in Al-rawi, F. and Roaf, M., 'Ten Babylonian Problems from Tell Haddad, Hamrin', *Sumer*, 1987, vol. 43, pp. 175–218, I, 1–9 and III, 2–9; the 'division' of 1.10 by 7, tablet BM 85 200 and VAT 6 599 in Neugebauer, O., *Mathematischer Keilschrift-texte*, Berlin, Springer, 1937, vol. 2, pl. 39–40; the last problem, tablet YBC 6 967 in Neugebauer, O., and Sacgs, A., *Mathematical Cuneiform Texts*, New Haven, American Oriental Society, 1945, pl. 17; the copy of the table of squares HS 244 in Hilprecht, H., *Mathematical, Metrological and Chronological Tables from the Temple Library of Nippur*, Philadelphia, University of Pennsylvania Press, 1906, no. 26 and that of the table of *igigubbû* IM 53 961 in Edzard, D. O., *Texts in the Iraq Museum*, Wiesbaden, Harrassowitz, 1971, vol. 7, no. 236, the table of reciprocals MLC 1 670 in Clay, A. T., *Babylonian Records in the Library of J. Pierpont Morgan*, New Haven, 1923, vol. 4, no. 37.

General works and articles

Trigger, B. G., et al., *Ancient Egypt: A Social History, Cambridge*, Cambridge University Press, 1983. Gardiner, A. H., *Egypt of the Pharaohs*, Oxford, Clarendon Press, 1961.

Mesopotamia: (see bibliography for chapter 1, general works and articles.)

Specialized works and articles

Egypt: Chace, A. et al., *The Rhind Mathematical Papyrus*, Oberlin, Mathematical Association of America, 1927–9, 2 vols; an abridged edition, Reston (USA), National Council of Teachers of Mathematics, 1979. Couchoud, S., 'Recherches sur les connaissances mathématiques de l'Égypte pharaonique', doctoral thesis, Université de Lyon-II, 1983. Gillings, R., *Mathematics in the Time of the Pharaohs*, Boston, MIT Press, 1972; paperback edn 1987. Parker, R., *Demotic Mathematical Papyri*, London, Lund Humphries, 1982 (for demotic fragments). Peet, T., *The Rhind Mathematical Papyrus*, London, Hodder and Stoughton, 1923; reprinted Liechtenstein, Kraus, 1970. Struve, W., *Mathematischer Papyrus des Staatlichen Museums der schönen Kunst in Moskau*, Berlin, Springer, 1930; reprinted Würzburg, JAL, 1973.

Mesopotamia: (see bibliography for chapter 1, mathematics.)

Chapter 3   Gnomon

Sources and citations

M. R. Cohen and I. E. Drabkin, *A Source Book in Greek Science*, Harvard University Press, 1958. PLATO, *Dialogues*, trans. and ed. B. Jowett, 4th edn, Oxford, Clarendon Press, 1964. *Les*

Présocratiques, Paris, Gallimard, 1988. VEREECKE, P. ed. and trans. *Apollonius*, Paris, Blanchard, 1963; *Archimède, Diophante et Euclide*, 1959; *Pappus, Serenus*, 1969; *Théodose*, 1959.

Historians and commentators:

CAVEING, M., *La Constitution du type mathématique de l'idéalité dans la pensée grecque*, Université de Lille-III, 1982, 3 vols. FOWLER, D. H.. *The Mathematics of Plato's Academy: A New Reconstruction*, Oxford, Clarendon Press, 1987. KNORR, W. R., *The Evolution of the Euclidean Elements*, Dordrecht, Reidel, 1975. MUGLER, C., *Platon et la Recherche mathématique de son époque*, Strasbourg, 1948; *Dictionnaire de la terminologie géometrique des Grecs*, Paris, Klincksieck, 1964. SZABO, A., *Les Débuts des mathématiques grecques*, trans. M. Federspiel, Paris, Vrin, 1977; *Les Débuts de l'astronomie, de la géographie et de la trigonométrie chez les Grecs*, trans. M. Federspiel, Paris, Vrin, 1986. SERRES, M., *Hermès*, Paris, Minuit, 1969–80, 5 vols: t. 1, *La Communication*, Genèse intersubjective de l'abstraction, pp. 39–46; t. 2, *L'Interférence*, Ce que Thalès a vu au pied des Pyramides, pp. 161–81; t. 5, *Le Passage du Nord-Ouest*, Origine de la géométrie, 3, 4, 5, pp. 165–95; *Le Parasite*, Paris, Grasset et Fasquelle, 1980, pp. 235–43; *Rome, le Livre des fondations*, Paris, Grasset et Fasquelle, 1983, pp. 254–63; *Détachement*, Paris, Flammarion, rev. ed. 1986, pp. 97–116; *Statues*, Paris, F. Bourin, 1987.

Chapter 4   Archimedes

Works cited:

ARCHIMEDES, *The Works*, ed. and trans. T. L. HEATH, Cambridge, 1897–1912, two vols reprinted in one, New York, 1953. CICERO, *Tusculan Disputations; Republic.* LIVY, *Roman History.* PLATO, *Collected Dialogues of Plato, including the Letters*, ed. E. Hamilton and H. Cairns, Princeton, Princeton University Press, 1973. PLUTARCH, *Lives.* POLYBIUS, *History.* LIVY, *History.*

General works and articles:

DIJKSTERHUIS, E. J., *Archimedes*, Princeton, Princeton University Press, 1987. TATON, R., (ed.), *Histoire générale des sciences*, Paris, PUF, 1966, vol. 1. DAHAN-DALMEDICO, A. and PEIFFER, J., *Une histoire des mathématiques*, Paris, Le Seuil, 1986. DEDRON, P. and ITARD, J., *Mathematics and Mathematicians*, trans. J. V. Field, Milton Keynes, Open University Press, 1978.

Specialized works and articles:

CARAMATIE, M. C. and DELBREIL, B., 'Sur le pas d'Archimède', *Rigueur et calcul*, ed. Cedic, 1982. CLAGETT, M., 'Archimedes', *Dictionary of Scientific Biography*, C. C. GILLESPIE (ed.), New York, C. Scribner's Sons, 1973. YOUSCHKEVITCH, A. P., 'Remarques sur la méthode antique d'exhaustion', *Mélanges A. Koyré, I: l'aventure de la science*, Paris, Hermann, 1964. SIMMS, D. L., 'Archimedes and the invention of artillery and gunpowder', *Technology and Culture*, vol. 28, no. 1, 1987. THUILIER, J., 'Une énigme: Archimede et les miroirs ardents', in *La Recherche en Histoire des sciences*, Paris, Le Seuil, 1983.

Chapter 5   Stories of the Circle

Sources and citations

ARISTOPHANES, *The Birds*, in *Sulba-Sutra*, in SARASVATI, S. S. P. and JYOTISHMATI, U. (eds), *The Sulba-Sutras: texts on Vedic geometry*, Allahabad, Dr. R. K. S. Sansthana, 1975. CHUQUET, N., *La*

Géometrie, ed. H. L'Huillier, Paris, Vrin, 1979, pp. 101, 102, 104, 105, 108, 328 and 329. *Nicolas Chuquet, Renaissance Mathematician*, trans. and ed. H. G. FLEGG, C. M. HAY, B. MOSS DESCARTES, *La Géométrie*, trans. D. E. SMITH and M. L. LATHAM, Dover, 1954, EUCLID, *Elements*, in HEATH, T. L., *The Thirteen Books of Euclid's Elements*, Cambridge, Cambridge University Press, 1925. EUTOCIUS, *Commentary on Archimedes' Sphere and Cylinder*, tr. I. E. DRABKIN, in M. R. COHEN and I. E. DRABKIN (eds), *A Source Book in Greek Mathematics*, Cambridge (Mass.), Harvard University Press, 1948. GAUSS, *Disquisitiones arithmeticae*, trans. A. A. CLARKE, New Haven, Yale University Press, 1966. KLEIN, F. *Erlanger Programm*, trans. Padé, Paris, Gauthier-Villars, 0000. MASCHERONI, L., *Geometrio del Compasso*, Paris, Daprat, 1804. MONGE, G., *Géométrie descriptive*, Paris, Gauthier-Villars, 1799. MONTUCLA, J. E., *Histoire des recherches sur la quadrature du cercle*, Paris, 1754. SEN, S. N., and BAG, A. K., *The Sulvasutras of Baudhyana, Apastamba, Katyayana and Manava with Text, English Translation and Commentary*, Indian National Science Academy.

General works

BOYER, C. B., *A History of Mathematics*, Princeton, Princeton University Press, 1985. DAHAN-DALMEDICO, A. and PEIFFER, J., *Histoire des mathématiques – Routes et dédales*, Paris, Le Seuil, coll. 'Points sciences', 0000 (one of the best general histories of mathematics). FAUVEL, J. and GRAY, J. (eds), *The History of Mathematics: A Reader*, London, Macmillan/Open University, 1987 (gives extracts from historical texts concerning mathematics, with commentary setting them in their social and cultural context, etc.). *La Rigueur et le Calcul, documents historiques et epistemologiques*, Paris, Editions du Cedic-Nathan, 1982 (includes several interesting studies of geometry, equations, the historicity of mathematics.)

Specialized works

BAG, A. K., *Mathematics in Ancient and Medieval India*, Chaukhhambra orientalia. LINDBERG, D. C. (ed.), *Sciences in the Middle Ages*, Chicago, Chicago University Press, 1978. *Le Matin des mathématiciens*, Paris, Belin-Radio France, 1985. *Petit Archimède*, special number on 'Pi', May 1980 (gives various approaches to this number, enjoyable to read, but not always reliable from the historical point of view.)

Chapter 5   The Arab Intermediary

Citations

IBN HAYYAN, J., *le Livre des balances*, in M. Berthelot, *la Chimie au Moyen-Age*, Paris, 1893, t. 3, pp. 141 and 147. ARISTOTLE, *De generatione*, in *The Complete Works of Aristotle, the revised Oxford translation*, ed. J. Barnes, Princeton and Guildford, Princeton University Press, 1984. AL-SOUYOUTI, *Livre de la miséricorde dans l'art de guérir les maladies*, French trans. By Pharaon, Paris, 1856, p. 11.

General works

BRILL, L.-J. (ed.), *Encyclopédie de l'Islam*, Leyden and Paris, Maisonneuve et Larose, 1955– (articles on the great scholars and main disciplines). CROMBIE, A. C., *Augustine to Galileo*, 2nd edn, London, Heinemann Educational, 1979. HASSAN, A. Y. AL-, 'L'Islam et la science', *La Recherche*, no. 134, June 1982. MICHEAU, F., 'L'âge d'or de la médecine arabe', *Les maladies ont une histoire*, special no. of *L'Histoire*, 1984. NASR, S. H., *Science et Savoir en Islam*, Paris, Sindbad, 1979 (with beautiful photos by R. Michaud). *Islamic Science, World* of Islam Festival, Paris, Sindbad, 1976. RASHED, R., *Entre arithmétique et algèbre. Recherches sur l'histoire des mathématiques arabes*, Paris, Belles-lettres, coll. 'Science et philosophie arabes', 1984. ROSENTHAL, F., *The Classical Heritage in Islam*, London, Routledge and Kegan Paul, 1980. SAYILI, A., *The Observatory in Islam and its Place in the General History of the Observatory*, Ankara, 1960. TATON, R. (ed.), *Histoire générale des sciences*, Paris, PUF, 1966, t. 1, *la Science antique et médiévale*, articles 'La science arabe' by R. Arnaldez, A. P. Youschkevitch, and L. Massignon, 'La science dans l'Occident médiéval chrétien' by G. Beaujouan, 'La science indienne médiévale' by J. Filliozat. ULLMANN, M., *Islamic Medicine*, Edinburgh, 'Islamic Surveys' series, 1978. VERNET, J., *Ce que la culture doit aux Arabes d'Espagne*, Paris, Sindbad, 1985. WATT, M. M., *L'Influence de l'Islam sur l'Europe médiévale*, Paris, Geuthner, 1974.

Chapter 6   Theology in the Thirteenth Century

Citations

AQUINAS, *Summa theologiae*, (English trans.), London, Methuen, 1991, Prologue and Questions. DENIFLE, H. and CHATELAIN, E., *Chartularium Universitatis Parisiensis*, Paris, Delalain, 1889, t. 1, p. 136. DROUIN, J.-M., 'La théologie comme science chez saint Thomas d'Aquin et Guillaume d'Occan', master's thesis, Université Paris-I, 1970. RADELET-DE GRAVE, P. and SPEISER, D., *Le De magnete de Pierre Maricourt*, French trans. with commentary, *Revue d'histoire des sciences*, 1975, pp. 193–234.

General works

BEAUJOUAN, G., articles on the medieval West in *Histoire générale des sciences*, ed. R. Taton, t. 1, *La Science antique et médiévale*, Paris, 1966. CROMBIE A. C., *Augustine to Galileo*, 2nd edn, London, Heinemann Educational, 1979. LE GOFF, *Les Intellectuels au Moyen Age*, Paris, Le Seuil, 1985. LINDBERG, D. C., *Science in the Middle Ages*, Chicago, University of Chicago Press, 1978. PAUL, J., *Histoire intellectuelle de l'Occident médiéval*, Paris, A. Colin, 1973. VERGER, J., *Les Universités au Moyen Age*, Paris, PUF, 1973.

Specialized works and articles

BENJAMIN F. S., and TOOMER, G. J., *Campanus of Novara and Medieval Planetary Theory. Theorica planetarum*, Madison, University of Wisconsin Press, 1971. BONNASSIE, P. and PRADALIE, G., *la Capitulation de Raymond VII et la fondation de l'université de Toulouse, 12229–1279*, Toulouse, Service des publications de l'université de Toulouse, 1979. CHENU, M. D., *Towards Understanding St Thomas*, trans. A.-M. Landry and D. Hughes, Chicago, Henry Regnery Co., 1964; *Saint Thomas d'Aquin*, Paris, le Seuil, 1959. CLAGETT, M., *The Science of Mechanics in the Middle Ages*, Madison, University of Wisconsin Press, 1959. DUHEM, P., *Etudes sur Léonard da Vinci*, Paris, A. Hermann et Cie, 1913, reprinted Paris, Edition des Archives contemporaine, 1984. POULLÉ, E., 'Astronomie théorique et astronomie pratique', in *Conférence du Palais de la Découverte*, no. 119, 1967. VAN STEENBERGHEN, F., *La Philosophie au Moyen Age*, Louvain, Publications universitaires, and Paris, Béatrice-Nauwelaerts, 1966.
[This paper is much indebted to two as yet unpublished works by Guy Beaujouan: 'L'émergence médiévale de l'idée de progrès', Report to the Panel on the History of Science and Philosophy in the Middle Ages, at the 8th International Congress of Medieval Philosophy, August 1987, and a paper given at the Colloquium on 'Sciences et techniques au Moyen Age', Orléans-la Source, May 1988].

Chapter 7   Algebra, Commerce and Calculation

Citations

BORGHI, P., *Nobel opera de arithmetica*, Venice, 1484. CHUQUET, N., *La Géométrie*, ed. H. L'Huillier, Paris, Vrin, 1979; *Manuscrit français 17346*, Paris, Bibliothèque Nationale, 1484, fol. 285 v. LIBRI, B., *Histoire des sciences mathématiques en Italie*, Paris, 1838–1841. *Manuscrits français 1339 and 2050*, Paris, Bibliothèque Nationale *c.* 1460. *Manuscrit latin 7287*, ibid., 1449. MARRE, A., 'Notice sur Nicolas Chuquet et son Triparty dans la Science des nombres', *Bolletino di Bibiographia e di Storia delle Scienze Matematiche e Fisiche*, XIII, 1880, pp. 555–804. PACIOLI, L., *Summa de Aritmetica, geometria, proportioni et proportionalita*, Venice, 1494. PELLOS F., *Compendion de l'Abaco*, ed. and intr. by R. Lafont and G. Tournerie, Montpellier, Edition de la Revue des langues romanes, 1967. SALOMONE, L., 'Me. Benedetto da Firenze, *La Reghola de Algebra Almuchabale'*, *Quaderni del Centro Studi della Matematica Medioevale*, Siena, 2, 1982.

Other treatises have also been published, particularly in Italy, but also in Germany. For over ten years, the University of Siena's *Quaderni del Centro Studi della Matematica Medioevale* have been providing scholars with algebraic texts hitherto to be found only in manuscript.

General and specialized works and articles

ARRIGHI, G., 'La matematica en Italia durante il Medioevo', [*Proceedings of the International Congress on the History of Science*], Warsaw, 1965, pp. 165–8. BAXANDALL, M., *Giotto and the Orators: Humanist Observers of Painting in Italy and the Discovery of Pictorial Composition 1350–1450*, Oxford, Clarendon Press, 1971; *Painting and Experience in Fifteenth-century Italy*, Oxford, Clarendon Press, 1972. BEAUJOUAN, G., 'Les arithmétiques en langue française de la fin du Moyen Age', *Proceedings of the 8th International Congress of the History of Science*, Florence, 1959, pp. 84–7. BEC, C., *Les Marchands-Ecrivains, affaires et humanisme à Florence 1375–1434*, Paris, Mouton, 1967. BENOÎT, P., 'La formation mathématique des marchands français à la fin du Moyen Age, l'exemple du *Kadran au marchans*, les entrées dans la vie, *Actes du XIIe Congr des médiévistes de l'endeignement supérieur public*, Nancy, 1982, pp. 209–24. DJEBBAR, A., 'Enseignement et recherches mathématiques dans le Maghreb des XIIIe et XIVe siècles (étude partielle)', Orsay, *Publications mathématiques d'Orsay*, no. 81–02. FANFANI, A., 'La préparation intellectuelle et professionnelle à l'activité économique en Italie aux XIVe et XVe siècles', *Le Moyen Age*, 1951, pp. 327–46. FLEGG, G., 'Nicolas Chuquet – an introduction', *Mathematics from Manuscript to Print, 1300–1600*, ed. C. Hay, Oxford, Clarendon Press, 1988, pp. 59–72. FRANCI, R. and TOTI RIGATELLI, L., *Introduzione all'Arithmetica mercantilo del Medioevo e del Rinascimento*, Siena, Quatro Venti, 1982. HAY, C. (ed.), *Mathematics from Manuscript to Print, 1300–1600*, Oxford, Clarendon Press, pp. 11–29, 59–88, 96–116, 127–144. HEERS, J., 'L'enseignement à Gênes et la formation culturelles des hommes d'affaires em Méditerranée à la fin du Moyen Age', *Revue des études islamiques*, XLIV, 1976, pp. 229–44. GOLDTHWAITE, A., 'Schools and Teachers of Commercial Arithmetic in Renaissance Florence', *Journal of European Economic History*, 1, 1972–3, pp. 418–31. LE GOFF, J., *Marchands et Banquiers au Moyen Age*, Paris, PUF, Coll. 'Que sais-je', 1986, p. 31. L'HUILLIER, H., 'Elements nouveaux pour la biographie de Nicolas Chuquet', *Revue d'histoire des sciences*, 1976, pp. 347–50. RENOUARD, Y., *Les Hommes d'affaires italiens au Moyen Age*, Paris, A. Colin, 1968. SESIANO, J., 'Une arithmétique médiévale en langue provençale', *Centaurus*, vol. 27, 1984, pp. 26–75. SMITH, D. E., *Rara arithmetica*, Boston, 1908, p. 459; *History of Mathematics*, New York, 1925, reprinted New York, Dover, 1958. VAN EGMOND, W., 'Practical Mathematics in the

Italian Renaissance: A Catalogue of Italian Abbacus Manuscripts and Printed Books to 1600', [*Annali dell Istituto e Museo di Storia della Scienza di Firenze*], Florence, 1980; 'New Light on Paolo dell'Abbaco', *Annali del Istituto e Museo di Storia della Scienza di Firenze*, Florence, 1977, pp. 3–21.

Chapter 9   The Galileo Affair

Sources and citations

The Leibniz-Clarke Correspondence, ed. H. G. Alexander, Manchester, Manchester University Press, 1956. DRAKE, S., *Galileo at Work – His Scientific Biography*, Chicago, University of Chicago Press, Phoenix Ed., 1981, pp. 88–90, 97–104, 116. DUHEM, P., *Etudes sur Léonard da Vinci – les précurseurs parisiens de Galilée*, Paris, A. Hermann, 1913, reprinted Paris, Édition des Archives contemporaines, 1984, pp. xii–xiv, 398. FEYERABEND, P., *Against Method*, rev. edn, London, Verso, 1988. GALILEO, *Dialogues, Concerning Two New Sciences*, trans. S. Drake, Madison, University of Wisconsin Press, 1974. KOESTLER, A., *The Sleepwalkers*, London, Hutchinson, 1952. KOYRÉ, A., *Galileo Studies*, trans. J. Mepham, Atlantic Highlands, Humanities Press and Hassocks, Harvester Press, 1978. PERRIN, J., *Les Atomes*, 1912, reprinted Paris, Gallimard, Coll. 'Idées', 1970, pp. 16–17. WHITEHEAD, A. N., *Science and the Modern World*, 1925, reprinted New York, Free Press, Macmillan, 1967, p. 8.

General works and articles

ARBOUR, A., *Galileo Studies*, University of Michigan Press, 1970. CLAGETT, M., *The Science of Mechanics in the Middle Ages*, Madison, University of Wisconsin Press, 1959, pp. 344–5. DRAKE, S. and MACLACLAN, J., 'Galileo's Discovery of the Parabolic Trajectory', *Scientific American*, 232, 1975, pp. 102–10. GEYMONAT, L., *Galileo Galilei*, tr. with add. notes and appendix by Stillman Drake, New York, McGraw-Hill, 1965. GUEROULT, M., *Dynamique et Métaphysique leibniziennes*, 2nd edn, Paris, Aubier, 1967, pp. 63–6, 111–13. MAURY, J.-P., *Galilée, le messager des étoiles*, Paris, Gallimard, Coll. 'Découvertes/sciences', 1986. SHARRATT, M. *Galileo: Decisive Innovator*, Oxford, Blackwell, 1994.

Specialized works and articles

COHEN, I. B., *The Newtonian Revolution*, Cambridge, Cambridge University Press, 1980. DRAKE, S., 'Galileo's 1640 Fragment on Falling Bodies', *British Journal for the History of Science*, 4, 1969, pp. 340–58. ELKANA, Y., *The Discovery of the Conservation of Energy*, London, Hutchinson Educational, 1974. LERNER, L. S. and GOSSELIN, E. A., 'Galileo and the Specter of Bruno', *Scientific American*, 255, 1986, pp. 126–33. FERRONE, V. and FIRPO, M., 'From Inquisitors to Microhistorians: a critique of Pietro Redondi's *Galileo Eretico*', *Journal of Modern History*, 1986, pp. 485–524. HANKINS, T., 'The reception of Newton's Second Law of Motion in the Eighteenth Century', *Archives international d'histoire des sciences*, XX, 1965, pp. 42–65. REDONDI, P., *Galileo Heretic*, trans. R. Rosenthal, Princeton, Princeton University Press, 1987; London, Allen Lane, 1988.

Chapter 10   Refraction and Cartesian 'Forgetfulness'

Citations

ARISTOTLE, *The Complete Works of Aristotle, the revised Oxford translation*, ed. Jonathan Barnes, Princeton and Guildford, Princeton University Press, 1984. DESCARTES, *A Discourse on Method*, in *Essential Writings*, tr. with introduction by J. J. Bloom, New York and London, Harper and Row, 1977. EUCLID, *Optics and Catoptrics*. IBN AL-HAYTHAM, 'Le discours de la lumière', tr. with intro. by R. Rashed, *Revue d'histoire des sciences*, XXI, 1968. KEPLER, [Paralipomena to Vitellion]. LUCRETIUS, *De rerum natura*, tr. and ed. by C. Bailey, rev. edn, Oxford, Clarendon Press, New York and London, Garland, 1977. PLATO, *Dialogues*, trans. with analysis and introduction by B. Jowett, 4th rev. edn, Oxford, Clarendon Press, 1964.

General works and articles

DUGAS, R., *A History of Mechanics*, New York, Dover, 1988. P. DUHEM, *Système du monde*, Paris, Hermann, 1913–55; English translation (selections) in *Mediaeval Cosmology*, trans. and ed. by R. Ariew, Chicago and London, Chicago University Press, 1985. MAITTE, B., *La lumière*, Paris, Le Seuil, 1981. RONCHI, V., *The Nature of Light: An Historical Survey*, tr. V. Barocas, London, Heinemann, 1970. TATON, R., ed., *Histoire générale des sciences*, Paris, PUF, vol. 1, 1966.

Specialized works and articles

BRUHAT, G., *Optique*, Paris, Masson, 1954. FEYNMAN, R., *QED: The Strange Theory of Light and Matter*, Princeton and Guildford, Princeton University Press, 1985. GILLISPIE, C. C., ed., *Dictionary of Scientific Biography*, New York, C. Scribner's Sons, 1973, articles by A. I. Sabra on 'Ibn al-Haytham', J. Murdoch on 'Euclid', O. Gingerich on 'Kepler'.

Chapter 11   Working with Numbers

Citations

DESCARTES, *Oeuvres*, ed. C. Adam and P. Tannery, Paris, Vrin & Cerf, 1897–1913, rev. and corr. edn, vols. 1 to 5 for the *Correspondance*, 1969–74. FERMAT, P. de, *Oeuvres complètes*, ed. P. Tannery and C. Henry, Paris, Gauthier-Villars, 1891–1912; [new edition planned by Blanchard]. FUSS, N., ed., *Correspondance mathématique et physique de quelques célèbres géomètres de XVIIème siècle*, Saint Petersburg, 1843; New York and London, Johnson Reprint, 1968. GAUSS, K. F., *Disquisitiones arithmeticae*, tr. A. A. Clarke, New York, Springer Verlag, 1966. KLEIN, F., *Vorlesungen über die Entwicklung der Mathematik in 19ten Jahrhundert*, Berlin, 1926–7. KUMMER, E. E., *Collected papers*, ed. André Weil, vol. 1, *Contributions to Number Theory*, Berlin, Heidelberg, New York, Springer Verlag, 1975. MONTUCLA, E., *Histoire des mathématiques*, 1799–1802, reprint, Paris, Blanchard, 1960.

General works

BOS, H., MEHRTENS, H. and SCHNEIDER, I. (eds), *Social History of Nineteenth Century Mathematics*, Boston, Basel, Stuttgart, Birkhauer, 1981. ITARD, J., *Arithmétique et théorie des nombres*, Paris, PUF, Coll. 'Que sais-je?', no. 1093, 1963. MANDROU, R., *From Humanism to Science, 1480–1700*, tr. B. Pearce, Harmondsworth, Penguin, 1978, reprinted 1985.

Specialized works and articles

BELHOSTE, B. and LUTZEN, J., 'Joseph Liouville et le Collège de France', *Revue d'Histoire des sciences*, XXXVII, no. 3/4, 1984, pp. 253–304. BEN-DAVID, J., *The Scientist's Role in Society: A Comparative Study*, Englewood Cliffs NJ, Prentice-Hall, 1970. DICKSON, L. E., *History of the Theory of Numbers*, Carnegie Institute of Washington, 1919–23, reprinted New York, Chelsea, 1971. EDWARDS, H., 'The Background of Kummer's Proof of Fermat's Last Theorem', *Archive for History of the Exact Sciences*, 14, 1975, pp. 219–36. EDWARDS, H., *Fermat's*

Last Theorem, New York, Heidelberg, Berlin, Springer Verlag, 1977. Fox R. and Weisz, G. (eds), *The Organisation of Science and Technology in France in 1808–1914*, Cambridge and Paris, Cambridge University Press and Maison des Sciences de l'Homme, 1980. Merton, R. K., *Science, Technology and Society in XVIIth Century England*, New York, Harper Torch Books, 1970. Paul, R., 'German Academic Science and the Mandarin Ethos', *British Journal of History of Science*, 17, 1984, pp. 1–29. Opolka, H. and Scharlau, W., *From Fermat to Minkowski*, Berlin, Heidelberg, New York, Springer Verlag, 1984. Schubring, G., 'Mathematics and teacher-training; plans for a Polytechnic in Berlin', *Historical Studies in the Physical Sciences*, 12, pt 1, pp. 61–94. Taylor, E. G. R., *The Mathematical Practitioners of Tudor and Stuart England 1484–1714*, Cambridge, Cambridge University Press, 1954. Weil, A., *Number Theory: An Approach through History*, Boston, Basel, Stuttgart, Birkhauser, 1983.

Chapter 12   Ambiguous Affinity

Citations

Bachelard, G., *le Matérialisme rationnel*, Paris, PUF, 1953, p. 35. Daumas, M., *L'Acte chimique. Essai sur l'histoire de la philosophie chimique*, Bruxelles, Édition du Sablon, 1946, p. 61. Diderot, D., *Diderot, Interpreter of Nature*, selected writing trans. J. Stewart and J. Kemp, ed. J. Kemp, London, Lawrence and Wishart, 1937. Dobbs, B. J., *The Foundations of Newton's Alchemy*, Cambridge, Cambridge University Press, 1975, p. 209. Goethe, J. W. von, *Elective Affinities*, tr. E. Mayer and L. Bogler, Chicago, Henry Regnery Co, 1963. Metzger, H., *Newton, Stahl, Boerhaave et la doctrine chimique*, Paris, Blanchard, 1930, p. 50. Thackray, A., *Atoms and Power. An Essay on Newtonian Matter Theory and the Development of Chemistry*, Cambridge (Mass.), Harvard University Press, p. 204.

General works

Bachelard, G., *l'Activité rationaliste de la physique contemporaine*, Paris, PUF, 1951. Ben-David, J., *The Scientist's Role in Society: A Comparative Study*, Englewood Cliffs NJ, Prentice-Hall, 1971. McMullin, E., *Newton on Matter and Activity*, Indiana, University of Notre-Dame Press, 1978. Ostwald, F. W., *L'Évolution d'une science. La chimie*, Paris, Flammarion, 1910.

Specialized works and articles

Haber, L. F., *The Chemical Industry during the Nineteenth Century*, Oxford, Clarendon Press, 1958. Holmes, F. L., 'From Elective Affinities to Chemical Equilibrium: Berthollet's Law of Mass Action', *Chymia*, 8, 1962, pp. 105–45. Kapoor, S. C., 'Berthollet, Prout and Proportions', *Chymia*, 10, 1965, pp. 53–110. Morrel, J.-B., 'The Chemist Breeders: the Research School of Liebig and Thomas Thomson', *Ambix*, 19, 1972, pp. 1–46. Westfall, R. S., 'Newton and the Hermetic Tradition', in *Science, Medicine and Society*, ed. A. G. Debus, London, Heinemann, 1972; 'The Role of Alchemy in Newton's Career', in *Reason, Experiment and Mysticism*, ed. M.-L. Righini Bonelli and W. R. Shea, London, Macmillan, 1975.

Chapter 13   From Linnaeus to Darwin

Citations

Bougainville, L.-A., *Voyage autour du monde*, 1st edn 1771, republished Paris, Gallimard, Coll. 'Folio', 1982, with introduction by Jacques Proust. Buffon, G. L., *De la manière d'étudier et de traiter l'histoire naturelle*, 1749, republished Paris, Bibliothèque Nationale, 1986,

p. 45. DARWIN, C., *The Voyage of the Beagle*, 1st edn London, 1845; *The Origin of Species*, 1st edn London, 1859 (published in Penguin Books, 1968), *The Life of A Naturalist in the Victorian Age*, London, 1876. *Dictionnaire des sciences naturelles*, Paris, Levrault, 1816–30, 60 vols., vol. XVIII, pp. 392, 401, 404, 410, 417, 418; vol. XXX, pp. 426–468; vol. CII, pp. 353–436. HUMBOLDT, A. von, *Essai sur la géographie des plantes*, Paris, Levraut, 1807. LAMARCK, J.-B., 'Discours préliminaire', *Flore française*, Paris, 1778, p. lii (both a pedagogical exposition and a critical discussion of problems of systematics in botany). LINNAEUS, *Species plantarum*, 3rd edn, 1764, p. 304; *L'Équilibre de la nature*, Paris, Vrin, 1972, p. 40 (brings together numerous papers by Linnaeus and his disciples, tr. B. Jasmin and introduced by C. Limoges); *Travels*, ed. D. Black, London, Elek, 1979. ROUSSEAU, J.-J., 'Fragments pour un dictionnaire des termes d'usage en botanique. Introduction', *Oeuvres complètes*, vol. 4, Paris, Gallimard, Coll. 'Bibl. de la Pléiade', 1969, pp. 1206, 1207, 1209. TOURNEFORT, J.-P. de, *Voyage d'un botaniste*, Paris, Maspéro, La Découverte, 1982 (abridged edition of the posthumous *Relation d'un voyage au Levant*, Lyons, 1717).

General works

BROSSE, P., *les Tours du monde des explorateurs*, Paris, Bordas, 1983. BROWNE, J., *The Secular Ark: Studies in the History of Biogeography*, New Haven and London, Yale University Press, 1983. DAGOGNET, F., *Une épistémologie de l'espace concret: néogéographie*, Paris, Vrin, 1977. DAUDIN, H., *De Linné à Lamarck. Méthodes de classification et idée de série en botanique et en zoologie (1740–1790)*, Paris, Alcan, 1926, 2 vols., reprinted EAC, 1983 (a monumental work, of great rigour; very useful, and often cited). DUVAL, M., *La Planète des fleurs*, Paris, Robert Laffont, 1977, reprinted Seghers, 1980; *The King's Garden*, trans. A. Tomarken and C. Cowen, Charlottesville, University of Virginia Press, 1982. (An overview, a good read, and fairly well-stocked with information, but without references or bibliography). LIMOGES, C., *La Sélection naturelle*, Paris, PUF, 1970. TAILLEMITE, E., *Sur des mers inconnues*, Paris, Gallimard, Coll. 'Découvertes', 1987.

Specialized works and articles

BLUNT, W., *The Compleat Naturalist: A Life of Linnaeus*, London, Collins, 1971, reprinted 1984. BROC, N., 'Voyages et géographie au XVIIIème siècle', *Revue d'histoire des sciences*, XXII, 1969, pp. 137–54. DAGOGNET, F., *Le Catalogue de la vie*, Paris, PUF, 1970. DEAN, J., 'Controversy over classification: a case study from the history of botany', in *Natural order: Historical Studies of Scientific Culture*, ed. B. Barnes and S. Shapin, Beverly Hills, Sage, 1979, pp. 211–30. FOURNIER, R., *Voyages et Découvertes scientifiques des missionaires naturalistes français*, Paris, Lechevalier, 1932. GAZIELLO, C., *L'Expédition de La Pérouse, réplique au voyage de Cook*, Paris, CTHS, 1984. LAISSUS, Y., 'Les voyageurs naturalistes du Jardin du Roi et du Muséum d'histoire naturelle: essai de portrait robot', *Revue d'histoire des sciences*, XXXIV, 3–4, 1981, pp. 259–317. RICHARD, H., *Une grande expédition scientifique au temps de la Révolution française: le voyage de d'Entrecasteaux à la recherche de La Pérouse*, Paris CTHS, 1986.

Chapter 14   Paris 1800

CARNOT, L., *Réflexions sur la métaphysique du calcul infinitésimal*, 1797, Paris, Gauthier-Villars, 1921, 2 vols. CARNOT, S., *Réflexions sur la piuissance motrice du feu et les machines propres à développer cette puissance*, 1824, Paris, Blanchard, 1953, Vrin, 1978. COULOMB, C. A., *Recherches sur la meilleure manière de fabriquer des aiguilles aimantées*, 1777–88; *Construction et Usage d'une balance électrique*, 1785. DEBRAY, R., *Critique of Political Reason*, tr. D. Macey, London, Verso/NLB, 1983. FOURIER, J., *Théorie analytique de la chaleur*,

1822, Paris, Gauthier-Villars, 1888. LACROIX, S. F., *Rapport historique sur le progrès des sciences mathématiques depuis 1789 et sur leur état actuel*, Paris, 1810. LAGRANGE, J. L. de, *Mécanique analytique*, 1788; *Théorie des fonctions analytiques*, 1797; *Traité de la résolution des équations numériques*, 1798; *Leçons sur le calcul des fonctions*, 1799, in *Oeuvres*, Paris, Gauthier-Villars, 1867–92, 14 vols. LAMARCK, J.-B. de, *Recherches sur l'organisation des corps vivants*, 1802, in *Corpus des oeuvres de philosophie en langue française*, Paris, Fayard, 1986. LAPLACE, P. S., *Exposition du système du monde*, 1796; *Mécanique céleste*, 1798–1825; *Essai philosophique sur les probabilités*, 1814, in *Oeuvres complètes*, Paris, Gauthier-Villars, 1864–1893 and *Corpus des oeuvres de philosophie en langue française*, Paris, Fayard, 1986. LAPLACE, P. S. and LAVOISIER, A. L. de, *Mémoire sur la chaleur*, 1780 in *Oeuvres complètes* above. LAVOISIER, A. L. de, *Traité élémentaire de chimie présenté dans un ordre nouveau*, Paris, 1789, 2 vols, in *Oeuvres*, Paris, 1864–93. MONGE, G., *Géometrie descriptive. Leçons donnés aux Écoles normales*, Paris, Gauthier-Villars, 1799; *Application de l'analyse à la géométrie*, Paris, 1807. POINSOT, L., *Éléments de statique*, Paris, Bachelier, 7th edn, 1837; *Théorie et détermination de l'équateur du Système solaire*, pp. 385–425. SERRES, M., *Hermès*, vol. 3, *La Traduction*, Paris, Minuit, 1974, pp. 190–282.

Historians and commentators

COMTE, A., *Cours de philosophie positive*, 1830–42, 6 vols, republished Paris, Hermann, 1975, 2 vols; *Catéchisme positiviste*, 1852, republished Paris, Flammarion, 1966. CONDORCET, J. A. N. CARITAT de, *Esquisse d'un tableau historique des progrès de l'esprit humain*, in *Oeuvres*, 1847–9, 12 vols, republished Paris, Flammarion, 1988. HEGEL, G. W. F., *Phenomenology of Spirit*, trans. A. V. Miller, with analysis and foreword by J. N. Findlay, Oxford, Clarendon Press, 1977; *Science of Logic*, tr. A. V. Miller, London, Allen and Unwin, New York, Humanities Press, 1969: [Encyclopaedia].

Chapter 15   Lavoisier

Sources and citations

BERTHELOT, M., *La révolution chimique, Lavoisier*, Paris, Alcan, 1890. CONDILLAC, É. de, *La Logique ou l'art de penser*, Paris, 1780. DUHEM, P., *La chimie est-elle une science française?* Paris, 1916. DUMAS, J.-B., *Leçons sur la philosophie chimique professées au Collège de France en 1836*, Paris, 1837. DUMAS, J.-B. and GRIMAUX, E. (eds), *Oeuvres de Lavoisier*, 6 vols, Paris, Imprimerie nationale, 1862–93, DUVEEN, D. I. and KLICKSTEIN, H. S., *A bibliography of the works of Antoine Laurent Lavoisier 1743–1794*, London, 1954; *Supplement*, D. I. Duveen, London, 1965. FRIC, R. (ed), *Oeuvres de Lavoisier. Correspondances*, 5 fasc., Paris, 1955–94, GUYTON de MORVEAU, L. B., LAVOISIER, A. L. de, FOURCROY, A. F. and BERTHOLLET, C., *Méthode de nomenclature chimique, nouveau système de caractères chimiques, adaptés a cette nomenclature, par Hassenfratz et Adet*, Paris, 1787, republished Paris, Seuil 1994. LAVOISIER, A. L. de, *Opuscules physiques et chimiques*, 1774; tr. *Essays Physical and Chemical*, London, 1776, reprinted 1970. *Traité élémentaire de chimie*, Paris, 1789; *Elements of Chemistry*, tr. R. Kerr, Edinburgh, 1790, reprinted New York, 1965; *Mémoires de chimie* (a posthumous collection edited by Mme Lavoisier) 2 vols, 1803; laboratory registers, various papers, archives of the Académie des Sciences (originals and microfilm); letters, Chazelles Collection, Clermont-Ferrand Municipal Library. MACQUER, J., *Dictionnaire de chymie*, 2 vols, Paris, 1766. WURTZ, A., *Dictionnaire de chimie pure et appliquée*, Paris, 5 vols, 1869–78.

General works and articles

BENSAUDE-VINCENT, B., *Lavoisier, mémoires d'une révolution*, Paris, Flammarion, 1984; English trans. forthcoming from Harvard University Press. BERETTA, M., *The Enlightenment of Matter*,

Science History Publications. CROSLAND, M. P., 'Lavoisier, le mal-aimé', *La Recherche*, 14, 1983, pp. 785–91. DAGOGNET, F., *Tableaux et Langages de la chimie*, Paris, Le Seuil, 1969. DAUMAS, M., *Lavoisier, théoricien et expérimentateur*, Paris, PUF, 1955. DONOVAN, A., *Antoine Lavoisier: Science, Administration, and Revolution*, Oxford, Blackwell, 1993. DONOVAN, A. (ed.), 'The Chemical Revolution: Essays in Reinterpretation', *Osiris*, ser. 2, 4, 1988 (see the article 'Lavoisier and the origins of modern chemistry'). FAYET, J., *la Révolution française et la Science, 1789–1795*, Paris, Rivière, 1960. FOX, R., *The Caloric Theory of Gases: From Lavoisier to Regnault*, Oxford, 1971. GILLISPIE, C. C., *Science and Polity in France at the End of the Old Régime*, Princeton University Press, 1980. GRIMAUX, E., *Lavoisier, 1743–1795, d'après sa correspondance, ses manuscrits, ses papiers de famille et d'autres documents inédits*, Paris, 1888. GUERLAC, H., 'Lavoisier', *Dictionary of Scientific Biography*, ed. C. C. Gillispie, New York, C. Scribner's Sons, 1973, vol. 8, pp. 66–91. HAHN, R., *The Anatomy of a Scientific Institution: The Paris Academy of Science 1663–1803*, Berkeley, University of California Press, 1971. KAHANE, E., *Lavoisier, pages choisies*, Paris, Édition Sociales, 1974. KUHN, T., *The Structure of Scientific Revolutions*, Chicago and London, University of Chicago Press, 1962, 2nd edn 1970. McKIE, D., *Antoine Lavoisier, Scientist, Economist and Social Reformer*, London, 1952. MACQUER, J., *Dictionnaire de chymie*, 2 vols, Paris, 1766. METZGER, H., *Les Doctrines chimiques en France du début du XVIIème à la fin du XVIIIème siècle*, Paris, Blanchard, 1923; *Newton, Stahl, Boerhaave et la doctrine chimique*, Paris, Blanchard, 1930. MEYERSON, E., *De l'explication dans les sciences*, Paris, 1921, republished Fayard, 1995. POIRIER, J. P., *Lavoisier*, Paris, Pygmalion, 1993. POUCHET, G., *Les Sciences pendant la Terreur*, Paris 1896.

Specialized works and articles

ANDERSON, W. C., *Between the Library and the Laboratory: The Language of Chemistry in the Eighteenth Century*, Baltimore, Johns Hopkins University, 1984. BERTHELOT, M., *La Révolution chimique, Lavoisier*, Paris, Alcan, 1890. CROSLAND, M. P., *Historical Studies in The Language of Chemistry*, Dover, New York, 1962, 2nd edn 1978; 'Lavoisier's theory of acidity', *Isis*, 64, 1973, p. 223. DAUMAS, M., 'Les appareils d'expérimentation de Lavoisier', *Chymia*, 3, 1950, pp. 45–62; 'Justification de l'attitude de Fourcroy pendant la Terreur', *Revue d'histoire des sciences et de leurs application*, 11, 1958, pp. 273–4. DUHEM, R., *Le Mixte et la Combinaison chimique*, Paris, 1902, republished Fayard, 1985. DUVEEN, D. I. and KLICKSTEIN, H., 'Some new facts relating to the arrest of Antoine Laurent Lavoisier', *Isis*, 49, 1958, pp. 337–48. DUVEEN, D. I. and VERGNAUD, M., 'L'explication de la mort de Lavoisier', *Archives internationales d'histoire des sciences*, 9, 1956, pp. 43–50. GUERLAC, H., *Lavoisier, the Crucial Year. Background and Origin of his First Experiment on Combustion in [1722]*, New York, Cornell University Press, 1961; 'Chemistry as a Branch of Physics: Laplace's collaboration with Lavoisier', *Historical Studies in Physical Sciences*, 7, 1976, pp. 193–276. GUILLAUME, J., 'Un mot légendaire: la République n'a pas besoin de savants', in *Révolution française*, 38, 1900, pp. 385–99, and *Études révolutionnaires*, first series, Paris, 1908, pp. 136–55. KERSAINT, G., 'Fourcroy a-t-il fait des démarches pour sauver Lavoisier?' *Revue générale des sciences pures et appliquées*, 65, 1958, pp. 151–2. MELHADO, E. N., 'Chemistry, Physics and the Chemical Revolution', *Isis*, 76, 1985, pp. 195–211. METZGER, H., 'Introduction à l'étude du rôle de Lavoisier dans l'histoire de la chimie', *Archeion*, 14, 1932, pp. 21–50; *La Philosophie de la matière de Lavoisier*, Paris, Hermann, 1935.

Chapter 16   In Defence of Geology

Sources and citations

BABBAGE, C., *The Ninth Bridgewater Treatise; a fragment*, London, 1837. BUCKLAND, W., *Geology and Mineralogy considered with regard to Natural Theology*, 2 vols, London, 1836.

Bulletin de la société géologique de France, Paris, 1833, t. 5, p. 501. Dufrénoy, P. A. and Élie de Beaumont, L., 'Mémoires pour servir à une description géologique de la France; Pt. 1: Élie de Beaumont, Observations géologiques sur les différentes formations qui, dans le système des Vosges, séparent la formation houillère de celle du Lias; Pt. 2: Dufrénoy, mémoire sur l'existence dugypse et de différents minerais métallifères dans la partie supérieur du Lias du Sud-Ouest de la France', Paris, 1830, t. 1. *Écho du monde savant*, Paris 1er avr. 1834, t. 1, p. 2; ibid., 20 juin 1834, t. 1, p. 54. Élie De Beaumont, L., *Papiers boîte 1: Mécanique-chimie; boîte 7: Manuscrits divers, Analyse de Géologie; boîte 9: Collège de France, 1832–1843*, Paris, Archives de l'Institut de France. Laplace, F. de, *Exposition du système du monde*, 6th edn, Paris, 1836, t. 1. Lyell, C., *Lectures on Geology, King's College, London, Royal Institution, 1833 etc.*, Lyell Manuscripts 7/8, University of Edinburgh; *Principles of Geology*, 3 vols, London, 1830–2. *Life, Letters and Journals of Sir Charles Lyell*, London, 1881, vol. 2.

General works and articles

On the idea of time: Attali, J., *Histoires du temps*, Paris, Fayard, 1982. Sohn-Rethel, A., 'Science as Alienated Consciousness', *Radical Science Journal*, 1975. Thompson, E. P., 'Time, Work, Discipline and Industrial Capitalism', *Past and Present*, 36, 1967. Young, R., 'The Historiographical and Ideological Contexts of the Nineteenth Century Debate on Man's Place in Nature', in *Changing Perspectives in the History of Science: Essays in Honour of Joseph Needham*, ed. M Teich and R. Young, London, 1973.

On geology: Burchfield, J. D., *Lord Kelvin and the Age of the Earth*, London, 1975. Oldroyd, D. R., 'Historicism and the Rise of Historical Geology', *History of Science*, 17, 1979, pp. 191–257. Porter, R., *The Making of Geology: Earth Science in Britain, 1660–1815,* Cambridge, Cambridge University Press, 1977. Rossi, P., *The Dark Abyss of Time: The History of the Earth and the History of Nations from Hooke to Vico*, Chicago, University of Chicago Press, 1984. Rudwick, M. J., 'The Strategy of Lyell's *Principles of Geology*', *Isis*, 61, 1970. Wilson, L. G., *Charles Lyell: The Years to 1841: The Revolution in Geology*, New Haven, 1972.

Chapter 17   Mendel in the Garden

Sources and citations

Lecoq, H., *De la fécondation naturelle et artificielle des végétaux et de l'hybridation*, 1845, trans. into German 1846, 2nd edn, Paris Librairie agricole de la maison rustique, 1862, pp. XII, 66. Mendel, G. Original texts of the papers in *Verhandlungen des naturforschen-den Vereines in Brünn*, IV, 1865 (1865 paper), vol. VIII, 1869 (1869 paper); English trans. in Stern, C. and Sherwood, E., *The Origin of Genetics: A Mendel Source Book*, San Francisco and London, W. H. Freeman, 1966. Naudin, C., 'Sur les plantes hybrides', *Revue horticole*, 1861, pp. 397, 398; 'Nouvelles recherches sur l'hybridité dans les végétaux', *Annales des sciences naturelles. Botanique*, XIX, 1863, pp. 180–203. *Revue horticole*, 1852. Sageret, A., 'Considérations sur la production des hybrides, des variantes et des variétés en général, et sur celle de la famille des Cucurbitacées en particulier', ibid., VIII, 1826, pp. 302, 303, 306. Vilmorin, L. de, *Notices sur l'amélioration des plantes par semis*, Paris, Librairie agricole, pp. 27, 28, 44. Vilmorin, P. de (ed.), *Comptes rendus et rapports*, IVe Conférence internationale de génétique, Paris, 1911, Paris, Masson, 1913.

General works

CIEEIST (Centre interdisciplinaire d'étude de l'évolution des idées, des sciences et techniques), *Le Cas Mendel*, Orsay, Université Paris-XI, 1984, (contains a French translation of Mendel's papers on hybridization). ILTIS, H., *Life of Mendel*, tr. E. and C. Paul, London, Allen and Unwin, 1932 (the classical biography). OREL, V. and ARMOGATHE, J. R., *Mendel. Un inconnu célèbre*, Paris, Belin, 1985 (French translation of Mendel's letters to Carl Naegeli appended).

Specialized works and articles

BLANC, M., 'Gregor Mendel: la légende du génie méconnu', *La Recherche*, 1984, no. 151, pp. 46–59. BRANNIGAN, A., 'L'obscurcissement de Mendel', in *Les Scientifiques et leurs alliés*, ed. M. Callon and B. Latour, Paris, Pandore, 1985, pp. 53–87; first published in *Social Studies of Science*, 1979, no. 9, pp. 423–54 (very interesting for its analysis of geneticists' strategies at the beginning of the twentieth century, but disappointing from the point of view of a comparison between Mendel and the hybridizers). CALLENDER, L. A., 'Gregor Mendel: an opponent of descent with modification', *History of Science*, 1988, vol. 26, 1, pp. 41–75. DAGOGNET, F., *Des révolutions vertes. Histoire et principes de l'agronomie*, Paris, Hermann, 1973. FISCHER, J.-L., 'Contribution à l'histoire de la génétique en France (1900–1915): le monde des praticiens et de l'abbé Germain Vieules (1846–1944)', in *Histoire de la génétique, pratiques, techniques et théories*, Actes du colloque international, Paris, 1987. LENAY, C., 'Le hasard chez Mendel', in *Le Hasard dans les théories biologiques dans la seconde moitié du XIXème siècle*, thesis supervised by J. Roger, Paris, Université de Paris-I, 1988. L'HÉRITIER, P. et al., *La Grande Aventure de la génétique*, Paris, Flammarion, 1984 (indispensable in making the connection between Mendel and contemporary genetics). OLBY, R. C., *The Origins of Mendelism*, London, Constable, 1966. PIQUEMAL, J., *Aspects de la pensée de Mendel*, Paris, Palais de la Découverte, 1965. ROBERTS, H. F., *Plant Hybridization before Mendel*, New York, Hafner, 1926; reprinted 1965 (without doubt the fundamental work in the field; it is extensively used by all the other authors). SERRE, J. L., 'La génèse de l'oeuvre de Mendel', *La Recherche*, 1984, no. 158, pp. 1072–80.

Chapter 18   Pasteur and Pouchet

Sources and citations

PASTEUR, L., *Oeuvres complètes*, Paris, Masson, 1922–39; *Correspondance*, Paris, Grasset, 1940, republished Flammarion, 1951, 4 vols.

General works

BULLOCK, W., *The History of Bacteriology*, New York, Dover Books, 1960. FEYERABEND, P., *Against Method*, rev. edn, London, Verso, 1988. GEISON, G., 'Pasteur', *Dictionary of Scientific Biography*, ed. C. C. Gillispie, New York, C. Scribner's Sons, 1974. KUHN, T., *The Structure of Scientific Revolutions*, Chicago and London, University of Chicago Press, 1962, 2nd edn 1970. BLOOR, D., *Knowledge and Social Imagery*, 2nd edn, Chicago, University of Chicago Press, 199.

Specialized works and articles

LATOUR, B., *The Pasteurization of France*, trans. A. Sheridan and J. Law, Cambridge and London, Harvard University Press, 1988. ROLL-HANSEN, N., 'Experimental Method and Spontaneous Generation: the Controversy between Pasteur and Pouchet', *Journal of the History of Medicine*, XXXIV, pp. 273–92. SALOMON-BAYET, C. (ed.), *La Révolution pastorienne*, Paris, Payot, 1986. FARLEY, J. and GEISON, G., 'Science, Politics and Spontaneous Generation in 19th-century France: The Pasteur-Pouehet debate: *Bulletin of the History of Medicine*, 48, 1974, pp. 161–98. GEISON, G., *The Private Science of Louis Pasteur*, Princeton, Princeton University Press, 1995.

Chapter 19   Mendeleyev

Citations

AMPÈRE, L. M., *Annales de chimie*, 90, 1814, pp. 43–86; *Annāles de chimie*, 1, 1816, pp. 195–373; 2, 16, pp. 5–105. AVOGADRO, A., *Journal de physique*, 73, 1811, pp. 58–76. DALTON, J., *A New System of Chemical Philosophy*, 2 vols, London, 1808–10. DUMAS, J. B., *Leçons sur la philosophie chimique*, Paris, 1837, republished Brussels, 1972; 'Sur les équivalents des corps simples', *Annales de chimie et de physique*, 55, 1859, pp. 129–210. GALISSARD DE MARIGNAC, 'Sur les rapports réciproques des poids atomiques', *Bulletin de l'Académie de Belgique*, 10, 1860, no. 8, in *Les Poids atomiques*, Paris, 1941, p. 100. GERHARDT, C., *Traité de chimie organique*, 4 vols, Paris, 1853–4.

Works by Mendeleyev

The complete works, published in Russian under the title *Sochinenia* (Works), Leningrad, 1934–52, 25 vols with index, include the *Principles of Chemistry*, articles on chemistry, studies of petroleum in North America and in Russia, on the mineral resources of the Donetz basin, etc. as well as numerous encyclopaedia articles; *Principles of Chemistry*, English trans. of the Russian 5th, 6th and 7th eds., London, Longmans, 1891–1905, reprinted New York, 1969; *Towards a Chemical Conception of Ether*, tr. G. Kamensky, London 1904.

General works and articles

BACHELARD, G., *La Pluralisme cohérent de la chimie moderne*, Paris, Vrin, 1930, reprinted 1973. KEDROV, B. M., 'Mendeleyev', *Dictionary of Scientific Biography*, ed. C. C. Gillispie, New York, C. Scribner's Sons, 1973. KOLODKINE, R., *D. I. Mendeleev et la loi périodique*, Paris, Seghers, 1963. PARTINGTON, J. R., *A History of Chemistry*, vol. IV, London, Macmillan, 1964 (a reference work, with a full account). VAN SPRONSEN, J. W., *The Periodic System of Elements. A History of the First Hundred Years*, London and Amsterdam, Elsevier, 1969 (a reference work, with a full account).

Specialized works and articles

BROCK, W. H., *From the Protyle to the Proton, William Prout and the Nature of Matter, 1785–1985*, Boston, Adam Hilger, 1985. BRUSH, S. G., 'Should the History of Science be Rated X?' *Science*, 183, 1974, pp. 1164–72. BUTTERFIELD, H., *The Whig Interpretation of History*, London, Bell, 1968. CARDWELL, D. S. L., *John Dalton and the Progress of Science*, Manchester, Manchester University Press, 1968. FARRAR, W. V., 'Nineteenth century speculations on the complexity of elements', *British Journal for the History of Science*, 2, 1965, pp. 297–323. HARRISON, E., 'Whigs, Prigs and Historians of Science', *Nature*, 329, 1987, pp. 213–14. KUHN, T., 'The History of Science', *International Encyclopaedia of Social Science*, vol. 14, New York, Macmillan, 1968, pp. 74–83. LEICESTER, H. M., 'Factors which led Mendeleyev to the Periodic Law', *Chymia*, 1, 1948, pp. 67–74. MENSCHUTKIN, B. N., 'Early History of Mendeleyev's periodic law', *Nature*, 133, 1934, p. 946, RUSSEL, C., 'Whigs and Professionals', *Nature*, 308, 1984, pp. 777–8. KNIGHT, D. A. (ed.), *Classical Scientific Papers, Chemistry II*, London, Mills and Boon, 1970. KNIGHT, D. A., *The Transcendental Past of Chemistry*, London, Dawson, 1978. ROCKE, A. J., *Chemical Atomism in the Nineteenth Century: From Dalton to Cannizaro*, Columbus, Ohio State University Press, 1984. NYE, A. J., *The Question of the Atom: From the Karlsruhe Congress to the Solvay Conference, 1860–1911*, Los Angeles, Tomash, 1983. BENSAUDE-VINCENT, B., 'Mendeleev's Periodic System of Chemical Elements', *British Journal for the History of Science*, 19, 1986, pp. 3–17. BENSAUDE-VINCENT, B., 'L'ether, élément chimique: an essai malheureuse de Mendeleev', *British Journal for the History of Science*, 15, 1982, pp. 183–8.

Chapter 20 Manufacturing Truth

Sources and citations

The sources in this chapter come from the archives of the Schlumberger company; they may be consulted at the École des Mines in Paris.

General works and articles

BEER, J. J., 'The Emergence of the German Dye Industry', *Illinois Studies in the Social Sciences*, vol. 44, 1959. BIJKER, W. E., HUGHES, T. P. and PINCH, T. (eds), *The Social Construction of Technological Systems: New Directions in the Sociology and History of Technology*, Cambridge, MIT Press, 1985. CHANDLER, A. D., *The Visible Hand: The Managerial Revolution in American Business*, Cambridge, Mass., Belknap, 1977. HUGHES, T. P., *Networks of Power: Electrification in Western Society 1880–1930*, Baltimore, Johns Hopkins Press, 1983. LANDES, D., *The Unbound Prometheus*, Cambridge, Cambridge University Press, 1970. LOCKE, R., *The End of Practical Man: Entrepreneurship and Higher Education in Germany, France and Great Britain, 1880–1940*, London, JQI Press, 1984. McNEILL, W. H., *The Pursuit of Power*, Chicago, University of \Chicago Press, 1982. NOBLE, D. F., *America by Design: Science, Technology and the Rise of Corporate Capitalism*, New York, Knopf, 1977. PARTINGTON, J. R., *A History of Chemistry*, London, Macmillan, 1969–70. REICH, L., *The Making of American Industrial Research: Science and Business at G. E. and Bell, 1876–1926*, Cambridge, Cambridge University Press, 1985. SMITH, M. (ed.), *Military Enterprise and Technological Change: Perspectives in the American Experience*, Cambridge, Mass., MIT Press, 1985. THOMAS, Jr, D. E., *Diesel: Technology and Society in Industrial Germany*, Tuscaloosa, University of Alabama P., 1987.

Chapter 21 Joliot

Sources and citations

WEART, S., *Scientists in Power*, Cambridge, Harvard University Press, 1979.

General works

KEVLES, D., *les Physiciens*, Paris, Économica, Anthropos, 1988. ROUBAN, L., *L'État et la Science*, Paris, Éditions du CNRS, 1988. SALOMON, J., *Science and Politics*, tr. N. Lindsay, London, Macmillan, 1973. DELASSUS, M. (producer), *La Course à la Bombe*, series broadcast on TF1, 1987. KEVLES, D. *The Physicists*, New York, Knopf, 1978.

Specialized works

CALLON, M., LAW, J. and RIP, A., *Mapping the Dynamics of Science and technology*, London, Macmillan, 1986. LATOUR, B., *Science in Action. How to Follow Scientists and Engineers through Society*, Cambridge, Mass., Harvard University Press, 1987. LATOUR, B., and POLANCO, X., *Le Régime français des sciences et des techniques. Bibliographie raisonnée de la littérature secondaire de langues anglaise et française sur l'histoire sociale des sciences et des techniques françaises, de 1666 à nos jours*, Paris, Ministère de la Recherche, CPE, 1987, and data-bank, École des Mines, Paris, CSI, 1988. SZILARD, L., *His Version of the facts. Selected Recollections and Correspondence*, Cambridge, Mass., MIT Press, 1978.

Chapter 22 The Invention of the Computer

Sources and citations

AIKEN, H., 'Proposed automatic calculating machine', 1937, unpublished paper, reproduced in B. Randell, *The Origins of the Digital Computer*, Berlin, Heidelberg and New York, Springer Verlag, 1973, pp. 191–8. LIGONNIÈRE, R., *Préhistoire et Histoire des ordinateurs*, Paris, Robert Laffont, 1987, p. 7. NEUMANN, J. von, *First Draft of a Report on EDVAC*, contract no. W 670 ORD 4926, Philadelphia, Moore School of Engineering, University of Pennsylvania, 30th June 1945, reproduced in B. Randell, above. ELGOT, C., and ROBINSON A., 'Random Access Stored Program Machines, an Approach to Programming Languages', *Journal of the Association for Computing Machinery*, 11, 1964, pp. 365–99. TURING, A. M., 'On Computable Numbers, with an application to the entscheidung problem', *Proceedings of the London Mathematical Society*, 42, 1936, pp. 230–67.

General works

BRETON, P., *Histoire de l'informatique*, Paris, La Découverte, 1987. GOLDSTINE, H., *The Computer from Pascal to Von Neumann*, Princeton, Princeton University Press, 1972. HODGES, A., *Alan Turing, the Enigma*, London, Burnett Books and New York, Simon & Schuster, 1984. LÉVY, P., *La Machine univers. Création, cognition et culture informatique*, Paris, La Douverte, 1987. MOREAU, R., *Ainsi naquit l'informatique*, Paris, Dunod, 1982.

Index

Page numbers in *italics* refer to illustrations or boxed sections

abacus 177, 250, *253–4*, 264, 647
Abbasids 198
Abd el-Malik 198
Abelard, Pierre 225
Abu al-Wafa 175
Academy of Sciences (French) 365, 382, *458*, 569
  and Lavoisier 458–9
acceleration 307, 308
Acradina 132, 139
Adanson, Michel *406*, 692, 695, 696
Adelard of Bath *210*, 216, *217*, 674
administration, development 227
aeriform fluids 459, 464
affinities and circumstances 384–5
affinity 372–400, 474
  elective or functional 390–3
  and force of attraction 380
  outmoded? 374–6
Agfa 606, 607
aggregate union 388, 389, 390
Aiken, Howard 647, 659, 721, 722
air, dephlogisticated 464
Akkadian language 19
Al-Battani 203, 673
Al-Biruni 198, 673
Al-Farabi 233, 235
Al-Farizi 328, 672, 675
Al-Hajjaj, translator of *Elements* *216*, *217*, 672
al-Haytham *see* Alhazen
Al-Karaji 213
Al-Khayyam 212, 218
Al-Khwarizmi, Mohammed ibn Musa 218, 264, 265, 273, 274, 673
  Al-Jabr wa l-muqabala 210–12, 213
  Arithmetic 210, 216
  mathematician and astronomer 202
  Zif al-Sindhind 203
Al-Kyndi, philosopher geometer 202, 673
Al-Maghribi, Muhyi *217*
Al-Ma'mun, Caliph 197, 198, 199, 202, *206*, 216
Al-Mansur, Caliph 198, *216*
Al-Rashid, Harun *216*
Al-Razi
  physician 202, 673
  The Book which Contains All 208
Al-Shirazi 328, 675
Al-Suyuti *209*
Alberti, Leon Battista 183
Albertus Magnus *229*, *230*, 241, 244
alchemy 206–8
Alexander of Hales *228*, 236, 245
Alexandre de Villedieu, *Carmen de algorismo* 216

Alexandrian School *76*
algebra 181, *211*, 246–79, 361, 439
  and commerce 263–70
  in Europe 181, 272–6
  geometric 103
  Italian 264
algebraic calculation, development 265–70
algebraic language (Chuquet) *271–2*
algorism 9
algorithm 37, 39, 52
  bifurcation 71
  first infinitesimal *76*
algorithmic presentation 50
algorithmic thought 85, 104–5, 108
Alhazen 323, 324, 327, 328
aliquot parts *352*
alkaline earths 576
alkaline metals 575, 576
Allier, Jacques 615
Almagest (Ptolemy) 174, 199, 203, 216, 218
Alphonsine (Toledan) tables 87
America, and industrial research *609*
American Speaking Telephone Company 595
Ampère, André Marie 563, 565, 697, 702
analysis 181
  real 427
Analytical Engine 638, 639, 640, *644*, 648
anapaests 113
Anaximander *76, 82*
Anaximenes *76*
André, Christian-Carl 508
angle, trisection *76*, 133
angle of deviation 333
angles in refraction 324
angles of refraction 244
animals 433
  classification 411–15
  dispersal *419–20*
  distribution and genealogy 415–21
anthypheresy *84–5*
antipheresis, imitation 109
Apastamba
  construction of square 162–3
  transform square to circle 163, 164
Apollonius of Perga *76*, *216*
  Conic Sections 183
  Conics 218
Arab conquest, and scientific development 197–202
Arabs
  and ocular lenses 329
  originality and inheritance 202–13
  and science 191–221
Arabs *see also* Islam, Muslims

Arago, Pierre-François *426*
Arbogast, Louis *425*
Archimedes *76*, 124–59, 669
  area of a parabola *141–3*
  bath 150
  catoptrics 144, 321
  death, three versions 154–6
  drawing a vessel towards him 136
  Measurement of the Circle 173
  mechanics of levers 137
  military inventions 132
  On Conoids and Spheroids 157
  On the Equilibrium of Planes 137, *142*
  On the Face that Appears on the Surface of the Moon 129
  On Floating Bodies 151
  On Method 127, 157
  On the Method relating to Mechanical Propositions 140
  On the Quadrature of the Parabola 140
  On the Sphere and the Cylinder 155–6
  Problem of the Cattle 130, 137
  ratio of weights *143*
  and refraction 316
  as scientist 147–51
  screw *149*
  solution of practical problems 132
  spiral 158
  The Measurement of the Circle 126
  The Sand-Reckoner 135, 137, 143
  three propositions 126–8
  tomb commissioned by Marcellus 155, 159
Archytas of Tarentum 132, 133, 174, 319
Argand, Jean-Robert *425*
Aristarchus of Samos 135, *283*, 669
Aristophanes, *The Birds* 172
Aristotelianism 232–5, 239
Aristotle *75*, *76*, 110, 111, 113, 669
  and Aristotelianism 232–5
  Categories and On Interpretation 193
  elements *207–8, 207*
  four causes 233
  and light 319
  Logic *208*, 225
  Meteors 223
  Organon 220
  second law of mechanics 136
arithmetic 45–9, *76*, 181, 196
  and algebra *76*
  and commercial practice 261–3
artifices 90–1
artificial intelligence 91, 92, 93, 94
artificial memory 105
Aryabhata, and decimal places 196

INDEX

ASCC 647–8
asipu 21
associations 261–2
Assurbanipal 17, 18, 47
    library, tablets 33–4
astrolabe 175, 189, 205, 215, 243
    by Ahmad ibn Khalaf *204*
astrology 25, 27, 43
astronomy 203–6, 274, *426*, 439
    and geology 502
    Indian 196, 202
    Islamic 203, 205–6
    and Newton 377–8
    observational 205, 316
    without eyes 85–7
asû 20, 21, 22, 23
asymmetry, between past and present 487–8
A.T. & T. 595, 596
Atanasoff, John 646, 662
    and electronic calculator 637, 656, 657
atheism 537
Athenian School 76
atomic bomb 583–4
atomic volume 573
atomic weights 563–5, 569, 575
atomism 143, 287, *288*
atomists 317, 319
Atomists of Abdera 76
attraction, universal 435
Aubriet, Claude 404
auctores 244
auctoritates 231, 243
Audubon, Jean-Jacques, painter 406, 698
Augustine, Saint 193
    City of God 224
    De Doctrina Christiana 224
Augustinians, at Brno 508, 509
Autolycos of Pitane 76
automata, theory 643
Automatic Sequence Controlled
    Calculator 647–8
automaton, and perpendicular 92–5
Averroes (Ibn Roch) 218, 233, 239, 674
Avicenna (Ibn Sina) 199, 218, 233, 673
    Kitab al-Shifa 217
Avogadro, A. *397*, 563, 564, 565

Babbage, Charles 494, 496, 497, *503*, 638,
    699
    and construction of accurate tables 639
Babylon 19–20
Babylonia, Old 17, 18, 19
    school exercise 25
    texts from 27–32, 36–7, 50
Babylonian clay tablets 164
Bachelard, Gaston 375, 376
Bachet de Méziriac, Claude-Gaspard 347,
    348
bacillus subtilis 554
Bacon, Francis, father of inductive
    method *381*
Bacon, Roger 234–5, *241–2*, *326*, 328, 674
    and Augustinian Platonism 229
Badische Anilin und Sada Fabrik 606
Baghdad, Muslim capital 197
Bailly, Jean Sylvain *426*
ballistic calculation 655
Ballistic Research Laboratory (BRL) 638, 654,
    655
ballistic tables 654
balloons 467
bankruptcy, and Italian capitalists 248
bârûm (seer) 23, 27, 31–2
BASF 606
Bates, H. W. *406*
Baudin, Nicolas 405, *406*
Bayer, pharmaceutical company 606, 607
beams, equilibrium 140
beans, hybridization *518*
Beau Dieu, at Amiens Cathedral 228
Beaugrand, Jean 346
Beaujouan, Guy 213
Bell, Alexander Graham 148, 595
Bell relay calculators 649–50
Bellarmine, Robert, Cardinal 283, 286
Ben-David, Joseph 381

Benedetto of Florence 247, 272–3
    geometrical proof of equation *268–9*
    Trattato di praticha d'arismeticha 264
    typical equations *266–8*
benzene ring 606, 607
Bergman, Torbern 382–3, 390–1
Bérguyer de Chancourtois, Alex 569
Bernoulli family *348*
Bernoulli, Jacques, tomb 155, 692
Berthollet, Claude-Louis
    expedition to Egypt 392
    and French Revolution 391
    and saltpetre for gunpowder 381
    Statique chimique 393
Berzelius, Jons Jacob 561, 562, 565, 566
Bible, The 231
Bichat, Marie-François-Xavier *426*
Billy, Jacques de 346, *348*, 353
binary numeration 644
binocular vision 332
biogony 439
biology *426*
biquadratics 354
bird observers *see dâgil-işşûrë*
Blainville, Henri Decrotay de *426*
Bletchley Park 638, 644, 652, 653
bodies, falling 290, 292, 301
Boethius 193, 273, 671
    and Aristotle's writings 218–19
    De trinitate 236
    Geometry 215
Bombelli, Raffaele 272, 277, 346, *348*
    Algebra 218
Bonaventure, Saint 229
Bonpland, Aimé 406, 417
Book of Nature 487, 489, 494
Book of Squares 218
book-keeping, double-entry 248, 263, 264,
    273
Borchardt, Carl Wilhelm *348*
Borda, Jean Charles *426*
Boscovitch, Roger 378, 380
Bosse, Abraham 184
botany and horticulture *511*
Bougainville, Louis Antoine de 406, *412*
Boyle, Robert 464, 594
Brahe, Tycho 183, 205, *286*, 331, 680
Brahmagupta 196–7, 672
Bravais, Louis *426*
Bréhier, Emile 136
Bridgewater Treatises 494, 496
British decoding service 651
British Oxygen Company (BOC) 603, 604
BRL 638, 654, 655, 658
Broadhurst, S. W. 644
Brongniart, Alexandre *426*
Brouillon Project 183
Broussais, François Joseph Victor *426*
Brunelleschi, Filippo 183
Buckland, William 494, 496
Buffon, Georges Louis Leclerc, comte de
    380, 412–13, 693, 694
Bunsen, Robert Wilhelm 560
bureaucracy 630
business: learning 250–1
Byzantine Empire 197

Cabanis, Pierre Georges *426*
calculation 246–79
    in Christian Europe 278
    Indian 196
    new art 253–4
    speeds 655
calculators
    first electronic 651–4
    programmable electronic 638
calculus 185, 342, 435
calendar
    Gregorian 449
    Revolutionary 449
caloric 558
camera obscura 328, 330, 331
Campanella, Tommaso *288*
Campanus of Novara, *Theoria planetarum*
    217, 244

Candolle, Augustin Pyramus de 416, 417,
    418, *426*
canonical lists 87–8
Capella, Martianus 193, *217*
carbon filament lamp 597
Carcavy, Pierre de *348*
Cardan, Jerome 272
Carnot, Lazare 187, *425*
Carnot, Nicolas Leonard Sadi *426*
Caro, Heinrich, and BASF 606
Cartesian 'Inversion' 335–43
Cartesians 311, 373
    and Newtonian chemistry 376–9
Cassiodorus 194, *217*
catastrophists 484, 487, 489, 490
catoptrics 321, 322
Cauchy, Augustin-Louis 359, *425*, 702
CEA 618, 626, 635
Certain, Jehan, *Kadran aux marchans 252,
    254, 257, 258,* 259
Ceruello, Pedro Sanchez 277
Chambers, Ephraim 429
Chandler, W. W. 644
chemical affinity 375
    tables 379
chemical bond 374, 395
chemical combination, unitary theory 565
Chemical Congress, International (Karlsruhe)
    565, 566, 572, 574, 580
chemical language, reform 467–70
chemical nomenclature 473
chemical proportions, simple *397*
chemical solitaire 576
chemistry
    analytic 398
    Cartesian 373
    eighteenth century 372–400
    learning 385–8
    limits and ambiguities 473–5
    nineteenth century, development 560
    practical benefits 381
    research programme 379–83
    science of experimentation 399
    science of the heterogeneous 388–90
chemistry laboratory *386*
Chenu, Marie-Dominique 232
Cheops 78, 122
Chios School *75*, 76
Christianity and Islam, contacts *214*
chronology 664–723
Chuquet, Nicolas 128, 177, 252, 270–2, 647
    inscription of pentagon in circle *179*
    Measurement of Areas 177
    practical geometry 177, 178
    problem of Association *262*
    short and simple rules 261
    Triparty en la science des nombres 179
Cicero 155, 158
circle 127, 157
    areas 171, 181
    calculation of area from
        circumference 178
    calculation for circumference *127*
    and conic sections 184
    definition 187
    divided into odd prime number of sides
        187
    eighteenth century 185–6
    history 160–90
    as line 183
    measurement *167–9*
    in Medieval Europe 174–9
    modern images *189*
    nineteenth century 186–90
    as plane figure 183
    ratios 171
    sixteenth-seventeenth centuries 179–83,
        183–4
    as square *190*
    squaring *76*, 133
    within rectilinear figure 173
circle of alliances 630–1
circle of autonomy 629–30
circle of movement 627–9
circle of representation 631–2
circle of ties and bonds 632

INDEX

circles, sixth century BC to second century AD
166–74
Clagett, *Questions on Euclid's Elements* 296
classification 411–15
common descent 421
Clement, Joseph 643, 644
clergy, new 441–54
classes and places 444–6
on incompleteness 449–51
positivism and its end 446–8
solids and fluids 451–2
time of the Calendar 448–9
clocks 96
coins 75
Colossus 644, 651–4, 662
combustion 461, 465, 466
Commandino, Federico 217
commerce 246–79
and algebra 263–70, 272–3
and arithmetic 261–3
Commerson, Philibert 406, 407, 412
Commissariat à l'Énergie Atomique (CEA)
618, 626
communications, and oil companies 603
community, equality 116
Complex Computer (Model 1) 649–50
compound union 388, 389, 390
compounds, types 467
computer
Complex (Model 1) 649–50
creation, and Ligonnière 646–7
definition 639
invention 636–63
computer languages 660
computer science 645
history 662–3
computers (1936–46) 647–63
computing, histories 636–47
Comte, Auguste 125, 426
Condillac, Etienne de 472
Condorcet, Marie-Jean Nicolas Caritat de
425, 441–3
conductive layer 591
cone, volume 143
congruences 361
conic section 76
definition 184
theory 140
conics 183
Constantine Africanus 215
continuity 296
Copernican Revolution 627
Copernicus, Nicolaus
orbits of planets 434
*Six Books on the Revolutions of the
Celestial Orbs* 283
Correns, Carl 506, 507
Corvisart, Jean Nicolas 426
Cosists 347
cottabus 146
Couffignal, programmable electro-mechanical
calculating machine 637
Coulomb, Charles 426
Council of Trent 288
couplings 207
Crelle, August Leopold 348, 358, 367–8
critical temperature, of gases 579–80
Crookes, William, hypothesis of protyle 571
crystals, asymmetry 552
cube 344, 354
duplication 76
cube roots 178
cubit 50, 60, 62
cuneiform writing 45–6, 47
Curie, Pierre and Marie 557, 611, 711
currency 262–3
curve
actual 592
composite 592
left 592
resistivity 592
saw-toothed 591
curve of discharge potential 587
curve of permeability 587
curve of porosity 587
curve of resistivity 586

curve of spontaneous potential 587
Cuvier, Frédéric 412, 427
and Lamarck, controversy 402
Cuvier, Georges 427
cyclic points 188
cylinder 67, 157
Cyzicus School 76

dactyls 113
dāgil-iṣṣūrē (bird observers) 20, 25
d'Alembert, Jean 385, 692, 694
Dalton, John 397, 563
Damianus 324
Dante Alighieri, *Convivio* 317
Dardi of Pisa, *Aliabra argibra* 264, 265
Dāriš-lībūr 22
Darwin, Charles Robert 491, 708
Galapagos finches 418–19
Origin of Species 402, 418, 420
voyage of *Beagle* 403, 436
Daubenton, Louis 427
d'Aurillac, Gerbert *see* Sylvester II, Pope
Dautry, Raoul 613, 614–15, 617–19, 623
Davy, Sir Humphry 479, 561
de Vries, Hugo 506, 507
decimals 180, 196
decupling technique 54, 55, 68
dedecupling technique 54, 55
Dedekind, Richard 357, 710, 712
Dee, John 347, 348
degrees of velocity 299, 300
Delambre, Jean-Baptiste Joseph 426
dell'Abbaco, Paolo 251, 274
Della Porta, Battista 330, 679
Demetrius and siege of Rhodes 138
Democritus 76, 143
denominator 58
Desargues, Girard 183, 184, 188, 682
Descartes, René 12, 346, 348, 682
Dioptrics 338, 341
Discourses 182, 336, 337
as modern scientist 337
and movement of light 339–40
principle of inertia 292
determinism 435
detripling 65
deuterium 613
diagonal 101
diagonal relationships 578
dialectic 224
Diderot, Denis 385, 386–7
Diesel, Rudolf, and Carnot cycle 599
Difference Engine 640, 644
differential equations, system 435
differentials 427
Digby, Sir Kenelm 346
digital coding 661
digital punched-card machines 655
Diodorus of Sicily 138, 144
Diophantus of Alexandria 76, 246, 279,
348
Arithmetic 213
in Bombelli's *Algebra* 218
translation 347
and trigonometry 75
Dirichlet, Gustav 348, 357, 457, 703
dispersal of animals and plants 420
displacement reactions 377
distances 331
distich 13
divination 23, 27–32
division
modern 257
old style 256–7
in Rhind Papyrus 53
Döbereiner, Johann 568
dominant characteristic 515
Dominicans 228, 229
Dominicus of Clavasio 177
Doppler, Christian 509
double-entry book-keeping 248, 263, 264,
273
doubling technique 55, 68
Drake, Stillman 300, 303, 304
Duhem, Pierre 239, 240, 543
Études sur Léonard de Vinci 293, 295

Dulong, Pierre Louis 426
Dumas, Jean-Baptiste 395
Dupuytren, Guillaume 427
Dürer, Albrecht 183
Dutrochet, Henri 427
dye industry 606
dynamics 440

Earth
age 483–5, 496
as eternal 485
as hourglass 502
origin 492
and Sun 79–82
earths, rare 580
Eccles, W. H. 646
Eckert, John 644, 656, 659
as engineer 643, 655
and recorded programme 642
eclipse and camera obscura 331
Edict of Nantes 288
Edison, Thomas Alva 596
EDSAC 662, 723
see also computer(s)
eduba texts 26
EDVAC 642, 648, 658–61, 662, 723
see also computer(s)
Egypt 46, 161–6
Einstein, Albert 559, 715, 716, 719
Eisenstein, Ferdinand Gotthold Max 348
eka-aluminium 578
eka-boron 578
eka-silicon 578
Eleatic School 76
electric lighting 597–8
electrochemical reactions 588
electrofiltration 588
electrolysis 561
electromagnetic phenomena 558
electromechanical machines 637
see also computer(s)
electronic calculation, diode-based circuits
646
electronic calculators 651–4
electronic circuit 656
Electronic Delay Storage Automatic Computer
see (EDSAC)
Electronic Discrete Variable Automatic
Computer *see* (EDVAC)
Electronic Numerical Integrator and Computer
see (ENIAC)
electrons 557
elements 76, 571–4
atomic weights 556
families 568, 569
first 76
four 207, 463
multiplication 566
periodic system 569
plurality 566
population explosion 561
predictions 579
spectroscopic study 567
systematics 580
transmutation 557
triads 568
Élie de Beaumont, Léonce 499–500, 502
Empedocles 112, 318
encyclopaedia 76
Engineering Science Management War
Training *see* ESMWT
engines, calculating 496
ENIAC 648, 653, 654–8, 662
see also computer(s)
Enigma 651, 652
Enlightenment 9, 10, 429
Entrecasteaux, Antoine d' 405, 406
environment, weight 276–9
epicycles 174
Epipolae 132, 139
episteme 94, 114
epistemology 94, 238, 536
equality, community 116
equations 182, 439
of Benedetto of Florence 266–8
eighteen basic types 269

equations (cont'd)
  geometrical proof *268–9*
  indeterminate 196
  with integer coefficients 186
  lists 264–5
  negative solution 275
equilibrium 307
equivalent weight 564
equivalentism 567
  and atomism 480
Eratosthenes of Cyrene 76, *82*, 130, 140
Erigeron canadensis 416
Esarhaddon, King of Assyria 17, 18
escalation, logic 139
ESMWT 655
Esquirol, Jean *427*
ether 558, 582
Euclid of Alexandria 74, *75, 76*, 669
  algorithm 85
  Axioms 114
  circle in square 172
  circle with square 172
  definition of gnomon 90
  Definitions 114
  Elements 112, 113, 114, 120, 166
    arrival in Europe *216–17*, 218
    definition of circle 160, 169
  Postulates 114
  square in circle 171–2
  square with a circle 172
  Treatise on Optics 183
Euclid of Megara 76, 668
Eudemus *76*
Eudoxus of Cnidus *76*, 132, 133
Eugene of Palermo 216
Euler, Leonhard *348, 352*, 691, 694, 696
  definition of force 309
Euphrates 18, 19
Eureka! eureka! 150
Europe
  algebra in 272–6
  Christian, and science 213–21
  Christian calculation 278
  emergence of science 192
  technological changes in 219
  and trade with Muslims 247
Eutocius of Ascalon 133, 173
evolution
  Darwinian theory 401, 567
  and hybridization 513
exchange, bill *248*
expeditions, by travellers *406*
experimental regress 591
experimentalists 544
experts, in Mesopotamia 17, 18, 19, 21, 24
externalism 615
extispicy 23, 25

false positions, simple and double 254
fauna, island 420
Ferguson, Adam 137
Fermat, Pierre de 342, 346, 347, *348, 352*, 353,
  683
  Challenge to Mathematicians 365
  theorem 360
fermentation 547, 552
Fermi, Enrico 621, 722
Fertile Crescent 73, 104, 109, 122
Feyerabend, Paul, *Against Method* 290,
  291
Fibonacci, Leonardo *see* Leonard of Pisa
finite sexagesimals 67, 68
fission 612
flask, long-necked 529, *530*, 531
flora
  island 417, 420
  key *414–15*
Flowers, T. H. 644, 652
Fontenelle, Bernard le Bovier de 373
form and matter 95–7
form of procedure 52, 54, 55–6, 63
fossil records, discontinuities 487–8
foundations, scientific 120–1
Fourcroy, Antoine François de *426*
Fourier, Joseph *425, 426*
  law for diffusion of heat 311, 313

fractions 258–9
  division *258*
  Egyptian *57–8*, 57–60
  sexagesimal *61*
  unit 57, 58, *61*, 68
Français, Jacques-Frédéric *425*
Franciscans *228*, 229
Francon, schoolmaster, squaring the circle
  175
Frederick II of Sicily 218, 219
Freiberg, Thierry de 326, 328
Frenicle de Bessy, Bernard 346, *348*, 352
fuchsia (Mendel) 512
fundamentalists 492

galaxies 439
Galen *201*
Galileo Galilei 280–314, 681
  Affair (1604) 11, 297–304
  Dialogue on the Two Chief World Systems
    283
  Dialogue on Two New Sciences 180, 297, 305
  Discourses and Mathematical
    Demonstrations 297
  falling bodies 183
  Il Saggiatore 287, *288*
  Letter to Castelli 282, 283, 286
Gall, Franz Joseph *427*
ganita (calculation) 196
gases
  critical temperature 579–80
  rare 580, 582
Gauss, Karl Friedrich *348*, 703
  Disquisitiones Arithmeticae 186, 357
  tomb 156
Gaussian integers 362
Gay-Lussac, Louis-Joseph 395, *397, 426*, 701
  law 563, 564
Geber *see* Jabir ibn Hayyan
Geiger counters 652
Geisslern, Ferdinand 508
Gelon, son of King Hiero II 130, 135
General Electric 597
generalization 111–14
generations of materials 645–6
Genghis Khan *206*
geodesy 76, *189*
Geoffroy, Claude-Joseph 376, 377
geography, botanical 416, 417
geologist, profession 498, 503
geology 483–505
  and astronomy 502
  new knowledge 493–8
  in nineteenth century *495*
  time for 485–93
geometry 76, 88–90, 209–10, 440
  analytic 433
  computational 264
  and Fermat 346
  in Greece 73–123
  separated from mechanics 132
Gerard of Cremona 215, 216, *217*
Gergonne, Joseph-Diaz *425*
Gerhardt, Charles 565, 707
Germain, Sophie 367, *426*
Gherardi, Paolo 265
Gillon, Paul 655, 656
Girard, Philippe de 346
Gmelin, Leopold, families of elements 568
gnomon 73–123, *81*, *206*
  ruler and compass 115–16
  vertical 95
Goethe, Johann Wolfgang von, *Elective*
  Affinities 375, 384
Goldbach, Christian *348*
Goldstine, Hermann 655, 656, 658, 661
granaries, two 49–67
gravity 308
Greek Empire 73–6
Greeks, and light 317
Gregory IX, Pope, *Parens scientiarum* 226,
  229
Gregory VI, Pope 226
Gregory VII, Pope, and literary arts 220
Grosseteste, Robert, Bishop of Lincoln *229*,
  241, 325, 327, 674

guild 225, 226
Gundelsheimer, André de 404
Gundissalinus, Dominicus 235
Gupta Empire 196
Guyton de Morveau, Louis Bernard 376,
  383, 693, 698
gypsum, analysis 464

Haber, L. F., and nitrogen fixation 607
habitats 417
habituation 593, 594
Halban, Hans 612, 613, 614–15
halogens 575, 576
halving technique 54, 55, 71
Ḫammurāpī 19, 22, 27
  Code of 43
  Mesopotamia in time of *21*
Harvard Mark 1 ASCC 648
Hassenfratz, Jean-Henri *426*
Ḫattušili 23
Haüy, Rene-Just *426*
heat 332, 565
Heath, Thomas L. 83
heavy water 613
heliocentric theory *285*, 286
Heraclides 135
Heraclitus *75*
heredity 420
Hermann of Carinthia 216
hermeneutic circle 591
hermeticism *288*
Hermite, Charles 186, 710
Hero of Alexandria 102, *216*, 321, 324, 671
  Definitions, Pneumatics, Metrics 174
Herodotus 79, *82*, 137
Hesiod, *Theogony* 317
heterogenesis 537, 551
  history of science 526–55
Hieracium sp. (Mendel) 510, 513, 519
Hiero, King 135
hieroglyphic writing 47
Hieron II of Syracuse 130
Hieronymus, grandson of Hieron II 130
Hilbert, David, *Foundations of Geometry* 190
Hipparchus of Alexandria 75, *76*, *82*, 670
Hippias of Elis 76
Hippocrates of Chios *76*, 112, 208, 668
  squaring the lune 172
Hiroshima 636, 662
history: enucleation 632–5
history of science 624–6
  brings together different people 2–3
  as cornerstone of contemporary culture 4
  intellectual *634*
  social *634*, 635
history of sciences, spontaneous 5
history of scientists 624–6, 629
Hittites, king of 20–1
Hiyya Ha-Nasi, Abraham bar 275
Hoechst, pharmaceutical company 606
Hollerith, Hermann 640
Holzmann, Wilhelm *see* Xylander
Homo faber 356–64
Homo ludens 346–56
homogenesis 537
homothesis 88, 89
horse, selected breeding 521
horticulture and botany *511*
House of Tablets 25
Hughes, Thomas 597
Hugo, Victor 148
human sciences 223
Humboldt, Alexander von *348, 397, 406*, 416
humours 208, *209*
Huygens, Christiaan 307, 308, 340, *348*, 685,
  688
  wave model *343*
hybridization 507
  and evolution 513
  Mendel's papers on 511–19
hydraulics 219, 440
hydrogen 464, 467
Hypatia, female mathematician 194

IBM 637, 645, 655
Ibn al-Banna 276

INDEX

Ibn al-Nafis 209
Ibn Badr 276
Ibn Roch see Averroes
Ibn Sīna see Avicenna
iconoscope 659
ideal figures 108
ideal numbers 360, *363*
ideas
  battle 146–7
  generation 472
igibûm 71
igigubbû 41, 49, 62, 64
India (2000–1000 BC) 161–6
industrial chemists, two aspects 607
industrial research, development 583–610
Industrial Revolution, and new chemist 381
industrial science 584–8
  adjustments 603–5
  development 597–8
  and its history 598–9
  nature of 601–10
  new methods 608–10
  and World Wars 601
industry and science 606–8
infinite descent 354–6, 359, 364
infinite divisibility 141
infinity of primes 361
information flow, control 608
Innocent III, Pope 228
inorganic radicals 568
INSEE 628
Institut National de la Statistique et des
  Études Économiques see INSEE
Institute of Advanced Studies, computer 661
integers 53, 186, 361
  series 102
internalism 615
International Business Machines see IBM
International Chemical Congress (Karlsruhe)
  565, 566, 572, 574, 580
interpolation, in Babylon 39
Interstate Commerce Act 608
intervals *84*
invariants 424
inverses 65, 66, 68, 70
inversion 68
  in Rhind Papyrus 53
iron metallurgy 75
Ishaq ibn Hunayn 200, *201, 216*, 673
Isidore, Bishop of Seville, *Etymologies* 194
Islam
  and Christianity *214*
  and science 191
Išme-Dagan, king of Assyria 22, 23
isomorphism 565
Ištar-šumu-ēreš 17, 18
Italian companies (*corpo*) 248, *249*
Itard, Jean Gaspard *427*

Jabir ibn Hayyan (Geber) 218, 672
  The Book of Balances 207, *208*
Jacobi, Carl 348, 368–9, 704
Jacquemont, Victor *406*
Jain religion: cosmography 164, *165*
Jasmaḥ-Addu 22, 23
Jean de Saint-Gilles *228*
Jesus Christ, redrawing sphere of world *176*
Job of Edessa 197
Jobs, Steve 647
John of Halifax 231
John of Palermo 218, 674
John of Salisbury 224
John of Seville 216, 674
Joliot-Curie, Frédéric 124, 611–35
  and atomic reactor 612, 617–19
Jordan, F. W. 646
Jordanus de Nemore 244
journals, mathematical 358
jurisprudence 27
Jussieu, Antoine Laurent de 413, *427*
Jussieu, Joseph de *406*
Justinian, Emperor 193

Kadašman-Enlil, brother Išme-Dagan 23
Kadran aux marchans, Jacques Certain
  252, 254, 257, 258

Kairouan 197
Kant, Immanuel 10, 692, 695
Kekulé, August 565
Kelvin, William Thomson, Lord *484*
Kepler, Johannes *286*, 681
  Astronomia Nova 286, 331, 333
  definition of circle 180
  Dioptrics 334
  and ellipse of planets 183
  founder of modern optics 330, 331–5
  laws 87
  Paralipomena to Vitellion 315, 331, 332,
    334
Khosroes Anushirman 196
Kilwardby, Robert 235
kinematics, study of movement 279
Klacel, Matthaeus 509
Klein, Felix 188
Koestler, Arthur 286
Kowarski, Lew 612, 613, 614–15
Koyré, Alexandre 292, 296
Kronecker, Leopold *348*
Kummer, Ernst Eduard *348*, 356–7, 368, 369

La Hire, Philippe de 184
La Pérouse, Jean-François 405, *406, 407*
La Ramée, Pierre see Ramus
Lacepède, Étienne *427*
Lacroix, Sylvestre-François *425*
lactic fermentation, and Pasteur 547
lacunae 263
Laënnec, René Théophile Hyacinthe *427*
Lagrange, Joseph-Louis 310, *348*, 426, 693,
  696, 697
Lalande, Joseph Jérôme François de *426*,
  693
Lalouvére, Antoine de 346
Lamarck, Jean-Baptiste de Monet de *412*,
  427, 438–9, 694, 701
  Flore française 414, *415*
Lambert, Jean Henri 185
Lamé, Gabriel 359
Laplace, Pierre-Simon de *426*, 436, 701, 702
Laplacean determinism 496
latitudes, calculation 83
Latreille, Pierre-André *427*
Lavoisier, Antoine Laurent de 11, 391, *426*,
  455–82, 694, 697
  arrest and death 476–8
  and balances 459–60
  chronology of work *460*
  emergence of myth 478–82
  hero of positive sciences 479
  national hero 482
  reform of chemical language 467–70
  reform of weights and measures 476
  reformer in revolutionary turmoil 475–8
  Traité elementaire de chimie 383
  two careers 456–9
law of distances 331
learning
  development 227
  as recollection 97
Lecoq, Henri 524–5
lectio (reading) 232
Legendre, Adrien-Marie *348*, *426*
lenses
  convergent 329, 333
  divergent 330
  spherical 326
Leonard of Pisa 265
  Flos Leonardi 218
  Liber abbaci 218, 274–5
  Practica geometriae 176, 218
Leonardo da Vinci 183, 330, 676
Lesson, René 409
Leucippus 76, 668
levers 140
  bent 244
  concept 157
  mechanics 137
  technique 139
Liebig, Justus von 547, 548, 606
Liebniz, Gottfried Wilhelm 308, *348*, 647, 688
  momentum of descent 307
  principle of sufficient reason 309

light
  and Mediterranean peoples 321
  physics 319–22
line
  solid 182
  straight 91, *189*, 190
Linnaeus, Carl
  journey in Lapland 404–5, *406*
  and nomenclature 410, 695
  Species plantarum 411
  Systema Naturae 401, 411, 412
Lions, Jâcques-Louis 15
Liouville, Joseph 359, 701
littera (letter) *232*
logarithm factory 640
logicism: engineers and mathematicians
  643–5
logos 75, 107, 122
Lomonossov, Mikhail Vassilievich 574
Los Alamos calculation laboratory 658
Louis XIII, King of France *288*
Lovelace, Lady Augusta 640
Lucretius 143, 318
Lyell, Charles
  age of Earth *484–5*
  four rules 501
  The Principles of Geology 483, 485, 492,
    500, 502
  and uniformitarianism 483–505

machine and memory 83–5
Macquer, Pierre Joseph 382, 463
Macy conferences 659
madhyaloka, and Jain cosmology *165*
magic square *351*, 353
Mahasiddhanta, Indian astronomical treatise
  202
Malus, Étienne-Louis *426*
management techniques 608
Manhattan Project 584
manubalista 145
Marcellus 132
  and Archimedes' tomb 155, 159
  blockade of Syracuse 144–6, 151–4
Mari, kingdom of 21
Maricourt, Pierre de 243
Marignac, Jean Charles Galissard de 567
Mascheroni, Lorenzo 186
materialism 537
mathematical recreation 140–4
mathematicians 25
  exchange of information 130, 349
  famous *347–8*
mathematics 36–42, 44–5, *425*
  in ancient Greek 77
  Babylonian 38–9, 49, 60–4, 70–2
  Egyptian *41*, 48, 50–5, 57–60, 67–70
  as foundations of physics 120
  French School 439
  networks 348–50
  problems 133
  publication 348–9
  pure 369
  in 17th and 19th centuries 344–71
matter, metaphorical 457
matter and form 95–7
Mauchly, John Prosper 642, 659
  and Atanasoff 656
  physicist 643, 644, 655
  quarrel 661
Maupertuis, Pierre Louis Moreau de 310, 367
Maurolico, Francesco 334
Maxwell, James Clark 558
measurement, and position 105
Mechain, Pierre *426*
mechanics
  rational 308–11
    relevance 311–14
  separated from geometry 132
Medici, Cosimo de' 248
medicine 32–6, 208–9, *426*
Mediterranean, eastern 74
Meister, Wilhelm 606
Melissos 76
melon 521–2
memory 83–5, 103

men
  and mathematics 346–8
  theory of numbers in everyday life 356–8
Mendel, Gregor 506–25, 709
  biography 507–11
  and plant improvement 519–25
  and us *515–16*
  were the figures too good? *516–17*
Mendel, Johann *see* Mendel, Gregor
Mendeleyev, Dmitri Ivanovich 556–82, 709
  and gases 579–80
  layout of periodic table *581*
  periodic law 570, 571
  predictions about elements 579
  Towards a Chemical Conception of Ether 558
mendicant orders *228–9*
Meno, from Socrates 97–103, 149
mercury delay line 658
Mersenne, Marin de 337, 346, *348*
mesolabe 133, *134*
Mesopotamia 18, *21*, 66
Messier, Charles *426*
metrology 60
Meyer, Julius Lothar 573
microbe 547–50
microbe network 550–5
microcomputer 645
  see also computer(s)
micro-organisms *531*
minerals, classification 207
Minos, king of Crete 125
mirror
  Archimedes 143
  parabolic 244
  position of images 332
  reflection 333
mnemotechnics 84
model, content/context *633*
molecule 564
momentum of descent 307
money *75*, 254
Monge, Gaspard 187, *426*
monomials 265
Montucla, Jean-Etienne 129, 137, 185, 367, 695
Moore School of Engineering 642, 643, 655
motion
  accelerated, measurement 304–8
  laws of naturally accelerated 294, 295, 296
  uniform 303
Mugler, Charles 136
multiplication 58, *256*
  in Rhind Papyrus 53
  sequential 255
multiplication table 38
mūsarum 62
music *84–5*
Muslims
  defeat at Las Navas de Tolosa 220
  expansion 197
  and trade with Europe 247
Muslims *see also* Islam
mutual attraction 331

Naegeli, Carl 510
Nagasaki 636, 662
narcosis 105–6
narke 106
Nasir al-Tusi 203, *206, 217*
  Exposition of Euclid 210
  School of Maragha 206
naturalists and travellers 401–21
nature: economy 325
Near East, on eve of Arab conquest *195*
neccessitarianism, Greek *240*
Nemore, Jordanus de 176–7
Nernst lamps 597
Nestorians 196
neutrons: chain reaction 612
Newlands, John Alexander 569, 570
Newman, Max 644, 652
Newton, Sir Isaac
  and astronomy 377–8
  force 378
  laws 87, 434, 439

Principia 484
  research 377
  second law 309
  and universal attraction 12
Nicomachus of Gerasa 273
Niepce, Nicéphore *426*
Nile Valley 46–7
nindan 60, 62
Nineveh 17–19
Norman conquest, in Mediterranean 219
Norsk Hydro Elektrisk 613
notation, positional 255
nuclear chain reaction, artificial 612–13
number theory 342, 344, 369, 370
  in eighteenth century 367
  as queen of mathematics 363
numbers 76, 181
  decomposition into sums of squares 361
  ideal *363*
  irrational 76
  large 76
  laws 360–4
  negative 273
  odd 102
  prime 76
  and their powers 351–6
numerals, Arabic 196, 214
numeric presentation 50

oaks 410
odd and even 106–8
Odling, William 569, 570
Oenipides of Chios 76, *82*
oil 584–8
  and water, Babylonian text of 28–30
oil companies, and communications 603
oil reserves, and salt domes 602
Oklahoma sonde 591
Oppenheimer, Jacob Robert 124
optics
  Arabic 323–5
  & Bacon *242*
  & painting 329–30
  divine 325–9
  market 329–30
Oresme, Nicolas 278, 294, 295, 675
Ortygia 129, 159
oscillation 435
over-determination 544
Oxford school 325, 326
oxygen, and acidity 479

Pacioli, Luca 250, *271*
Pallas, Peter Simon 405, *406*
Palmer, Ralph 637
Panke, Kurt 650
pantheism 509
pānum 49, 62
paper-making 197–8
Pappus of Alexandria 76, *216*
  Mathematical Collection 194
papyrus 46, 50, 59
parabola, definition 140
paradox 111–12
parallelogram *83*
parimandala (round figure) 162
Paris
  (1800) 422–54
  university (13th-century) 228
Parmenides 76
Parmentier, Antoine Augustin *426*
Pascal, Blaise 126, 184, 342, *348*, 686
  triangle *91*
Pasteur, Louis
  and God 540
  and micro-organisms *531*, 708
  and Pouchet 526–55
  Sorbonne lecture 528–30
patents 595, 598
pattum 22, 23
Paul of Aegina 194
peas
  characteristics 516
  hybridization 509–10
  and Mendel 513–17
Pellos, Francès 259, 261

Peloponnesian Wars 138
pendulum 306–7
pentadecagon *82*
pentagon: line of diagonals *110*
percussion 304, 305
periodic law to periodic table 577–9
periodic table
  of elements 373, 374, 556, 566, 574
  layout *581*
peripatetics *76*
Perkin, W. H. 606
perpendicular and automaton 92–5
Perrin, Jean 280
Persian School 195
perspective 322
Peter of Lombardy 224, *230*
  Book of Maxims 231
  Sententiae 236
Petunia nyctaginiflora 523
Petunia violacea 523
petunias, hybrid 522–4
pharmacopoeia 22
pharmakos 106
Phaseolus sp. *518*
Phelps, Byron 637
Philoponus, John 194
phlogiston 464, 465, 466
  and Lavoisier 462
physicists of Miletus 76
physics 121–3, 202–3, *426*, 440
  Galilean 308
  Iorian 122
  of light 319–22
  mathematical 292–7
phytogeography 418
pi (π) 185, 209
Piero della Francesca 277
piezo-electric effect 658
Pinel, Philippe *427*
Pisum sp. *514–17*
  see also peas
plane 114, 182, 244
plant improvement 519–25
plants 433
  classification 411–15
  dispersal *419–20*
  distribution and genealogy 415–21
Plato *75*, 76, 110, 111, 130
  Gorgias 123
  Letter VII 137
  and light 318
  Republic 153, 223
  Timaeus 220
Platonic knowledge 79–80
plough
  with asymmetric share 219
  swing 219
plumb-line 92, 93, 94, 95
Plutarch 131, 144, 153
  Life of Timoleon 128
  On the Sphere and the Cylinder 134
plutonium 634
Poinssot, Louis *426*
point at infinity 188
Poisson, Siméon-Denis *426*
Polybius 139, 144
polygon *126*, 187
polyhedrons 76
polynomials 265
polytheism 193
Poncelet, Jean-Victor *187*, 188, *426*
Porphyry, *Isagoges* 233
positivism 451, 452, 453, 563
  and its end 446–8
Positivist Calendar *447*, 449
Pouchet, Félix-Archimède
  and anti-Darwinism 539
  as conscientious positivist 544
  credentials 535
  and Pasteur 526–55
  and spontaneous generation 530
Prévost, Abbé 407
principles, first 116–20
Priscian 224
probability 425
procedure texts 36, 41, 42, 43

Proclus 113, 194, *216*
  Athenian school closed by 76
  and plane and solid lines 182
professionals 20–7
Prony, Gaspard Marie Riche de *426*, 640
proportions 395, 563
prosody 113
prospecting methods 602
protyle, hypothesis 571
Proust, Joseph Louis 394, *426*, 563, 695
Prout, William 566, 567, 568, 569
Prussia 368
Ptolemy of Alexandria 75, 76, *82*, 321, 670
  Almagest 174, 199, 203, 216, 218
  Great Composition 203
  Mathematical Syntax 174
punched cards 640
Punic Wars 130, 131
Pythagoras of Samos 76
  theorem 8, 106, 351
Pythagorean arithmetic 90, 107
Pythagorean table 102
Pythagoreans of Croton 76
Pytheas of Marseilles *82*

quadratrics 76
quadrivium 193, 225, 273
quaestia (question) *232*
quality
  uniform 294
  uniformly deformed 294
  what is it? 294–5
quantum mechanics *491*
qûm 49, 62

radar 652
Radio Corporation of America (RCA) 657
radioactivity 557
Raffenau-Delile, Alire 408
railroads 608
Raimond, Bishop of Toledo 215
rainbow *322*, 326, 328
Ramsay, William 580, 582
Ramus (Pierre La Ramée) *348*
Rashed, Roshdi 212
rathacakracit (chariot wheel) 162, 163
rationalist pseudo-realism *326–7*
rationalists 544, 545
ratios 140
Rayleigh, John William 582
RCA 657
reactions, complete 392
reading 45–9
reason in power 430–2
recapitulation 432–3, 437, 440
recessive characteristic 515
rectangle *83*
refraction 315–43
rejection
  explicit 527
  implicit 526
relativists 545
relativity *491*
religion
  and defeat of Syracuse 152
  and science 452–3
reminiscences 103–5
Renan, Ernest 191
research laboratories, industrial *595*
Rhind Papyrus *51*, 53, 55
  circles 164
  reference tables 59
  text 69–70
Richter, Benjamin 394
right angle 91
Rio Tarra oil field 604
Robert of Chester *210*
Robert de Courcon 226, 234
Roberval, Gilles Personne de *348*
Robinson calculators 652
Roland of Cremona *228*
Rome de L'Isle, Jean-Baptiste *426*
Ronchi, Vasco 334
Rouelle, Hilaire 385

Rousseau, Jean-Jacques 409–10
Royal Academy of Science *see* Academy of Sciences (French)
Royal Hymns 26
Royal Society 365, 377, 382
rule of apposition and remotion *270*, 275
rule of firsts *270*, 272
rule of mean numbers *270*
rule of opposition and remotion 254
rule of position *270*
rule of three 255, 259–61, 276
  compound 254, 260
rule of two positions *270*
rupture 296
Rutebeuf of Joinville 231

Sacrobosco, *Algorismus* 216
Sa'id al-Andalusi 202
Saint-Victor, Hugues de 235
Sakharov, Andrei Dimitrievitch 124, 717
salt domes, and oil reserves 602
sambuca 144, 145, 146
Sassanid Empire 195, 196, 197
Saussure, Théodore de *427*
Savart, Félix *426*
Scheutz, P. G. *639*
Schlumberger 584–610
  patents and electrical prospecting 602
  Schlumberger and Halliburton trial, for patent infringement 588–94
Schlumberger curves *586*, 601
Schlumberger diagram of resistivity and spontaneous potential *585*
Schlumberger measurements *587*
Scholastics 231–2, *232*, 325
School of Maragha 203, *206*
School of Salerno 215
Schools
  Florentine 250–1
  Greek *76*
  in medieval Italy 250
  twelfth-century 224–5
science
  and religion 452–3
  and time *491*
  applied 132
  and Arabs 191–221
  basic *76*
  Christian 193
  in Christian Europe 213–221
  classical 192
  Greek 192–7
  history of, and history of battles 132–5
  industrial 597–9, 601–10
  and industry 606–8
  modern 289–92
  nature as subject-matter *76*
  and politics 619
  and power 428–9
  pure 132
  pure and applied 599–601
  rational 382
  and religion 452–3
  secrets 594–7
  and time *491*
  where does it come from? 7
  without frontiers 429–30
sciences of calculation 210–13
scientific development, and Arab conquest 197–202
scientific exchange 349–50
screw *149*
  telluric 569
scribal dispute 48
scribe *see* tupsarrum
seeds, transport 417
seer *see* barum
seismology 605
selection *see* hybridization
selectron 659
Semitic dynasty 19
sensus (sense) *232*
Sergius of Ras el-Ain 194, 196
Serres, Michel 143, 341
set square 83, 90, 91, *95*
  and ruler and compass 115–16

sexagesimal fractions *61*
sexagesimal inverse 67
sexagesimal semicolon *61*
sexagesimal system *61*
sexagesimals 68
Siberia: scientific expeditions 405
Siger de Brabant 239
Silicon Valley 645
Silius Italicus 136
silo, calculation of capacity 49–50
silo problem, second 55–60
similars 318
simple substances *471*, 571–4, 575
simplex 81
simulacra 317, 318
slave, and Socrates 97–101, 103–4
Snell, Willebrord (Snellius) 335, 375, 681
Snellius *see* Snell, Willebrord
Snell's Law 335, 341
Société d'Arcueil 392
sociology 440
Socrates
  and the slave 97–101, 103–4
  torpedo 111
sodium chloride 575
solar system 76, 436
  as clock 502–3
solids: ratios of volumes and areas 157
Sophists and Megarians 76
space 75
  Galilean 308
  homogeneous and isotropic 308
species
  naming and classifying 409–15
  origin 401
Speusippus 76
spontaneous generation 530–8, 543–8, 554
square *83*
  duplication 76, 101, 102
  extraction 83
  magic 351, 353
  reducing circular figures to 178
  side and diagonal 102, *111–12*
square numbers, complement of successive 90
square roots 65, 71, 178
  table 38, *42*
squares 83
  perfect 102
squaring the circle 172, 185
  Francon 175
Stahl, Georg Ernst 379
standardization 608
stars 87
Stas, Jean Servais 567
stathme *see* plumb-line
statics 114–15, 307
steicho 113
Steinmetz, Charles 597
Stevin, Simon 140, 346
Stibitz, Georg Robert 646, 649–50, 658, 659, 721
stichomythia 113
Stoicheia 112
stomachion 143
Strabo, Greek geographer 151
students: exercise tablets 26
subalternation 238
subsoil *585*
sugar-beet 519–20
Sulgi, literary hymn 47
Sulvasutras 162, 165, 166, 170, 171
substitution 395
Sumeria 19
Sumerian metrological systems *61*
Sun, and Earth 79–82
sundial 79, *81*, 83
sûtum 49, 62
Sylvester II, Pope 214, 215
Syracuse 129–31
  battle 138–40
  causes of defeat 152
  siege of 131
Syriac 200

INDEX

system 436
 definition 435
 and evolution 437-8
Szilard, Leo 612, 614-5

table texts 36, 41
tables 87-8
 of affinity 382
 of chords 84, 878
 of refraction 335
tākiltum 71
Tartaglia, Niccolo Fontana 277
taxidermy *408*
teacher, conditions of employment (1517)
 251
teaching and research 357
techniques and operations 53
technocracy 630
technological change, in Europe 219
telephone 595-7
telephone circuits 652
telephone relay technology 649, 650, 656
telescopes, astronomical 333
Tempier, Étienne, Bishop of Paris 239,
 241
Texas sonde 591
text
 of *bārum* 31-2
 Neo-Assyrian 36
 Old Babylonian 27-32, 35
 procedure 36, 41, 42, 43
 table 36, 41
textbooks, of commercial arithmetic 251-2
Thabit ibn Qurra 195, 205, *216*
Thales of Miletus 74, *75*, 76, 667
 and pyramid 77, 85-6
 theorem 78
Theaetetus of Athens 76
Thénard, Louis Jacques 395, *426*
Theodorus of Cyrene 76
theology
 of the four causes *233*
 natural 236, 484
 as science 222-45
 the word and the thing *223*
Theon of Alexandria *216*
Theon of Smyrna, gnomonic numbers 91
Theophilus of Edessa, translating Aristotle
 198
theorem
 incompleteness 449
 Pythagoras' 8, 106, 351
thermodynamic affinity 375
thermodynamics 374, 427, 440
 chemical 374
 second law *313*, *491*
Third Lateran Council 220, 226
Thomas Aquinas, Saint 236-8, 245
 Angelic Doctor *230*
 as Dominican 229
 Eternity of the World 239
 sacra doctrina 224
 Summa Theologica 222, *230*, 236
 Treatise on Animals 241
 Treatise on Plants 241
 Unity of the Intellect 239
 Utrum sacra doctrina sit scientia 237
Thomson, Thomas 567
three body problem 425
Tigris 18, 19

time *75*
 and defeat of Syracuse 152
 human 485
 and industry 504
 rational organization 500
 religious 485, 493, 504
 and sciences *491*
Toledan tables 87
Toledo 215, 220
topology *189*
torpor and narcosis 105-6
Tournefort, Joseph Pitton de 403, *404*, *406*,
 408
trade, Mediterranean 247-50
tradition 77
transformationist hypothesis 438
transistor 645
translation 198
 of Arab science 221
 in twelfth century 216
transubstantiation *288*
trapeze 114
travellers and travels 402-9
travels, equipment *408-9*
treatises 251-2
 great 433-4
trial, for patent infringement 588-94
triangle 127
 equilateral 169
trigonometry *76*, 84, 175, 196
trivium 193, 225
trochees 113
Tschermak, Erich von 506, 507
tungsten filament lamp 597, 598
tupšarru-astrologer 25
tupšarrum (scribe) 24, 25
Turing, Alan 638, 651, 721
 and computer science 645
 and decoding machines 644
 and Enigma 651, 652
 and universal machine 641, 642

Umayyid caliphs 197
ummânū (expert) 17, 18, 19, 21, 24
under-determination 543, 544
ungula 113
uniformitarianism 483-505
uniformity of matter 379
uniformly accelerated movement 295
uniformly deformed movement 300-1
uniformly deformed quality 294
Univac 662
Universal, The 429-34
Universal Automatic Computer (Univac)
 662
universal machine 659
universality of force 379
universe
 profile *81-2*
 three models *284*, *285*
Universitas magistrorum et scolarium
 225
universities
 faculties in thirteenth century 229
 and mendicant orders *228*
 new *227*
University of Paris 225, 226
uranium 557, 611
uranium oxide 611, 612
Urban VIII, Pope 283, *288*

vacuum tables 656
valency, and table of elements 573
Van Helmont, Jan Baptist 532, 534
vanishing quantities 427
velocity 298-303
Venel, Gabriel-François 385
 defines affinity 388-9
 and force of attraction 393
vibrations 435
Viète, François *348*, 679
 and algebra 246, *272*
 counsellor 346
 expression for area of circle *180-1*
Vilmorin, Louis de 519-20
vitalism 547
Vitellion *see* Vitello
Vitello of Silesia 244, 328, 675
 and optics 328
Vitruvius *82*, 320
Volta, Alessandro 561
volume 150
von Lindemann, Ferdinand 186
von Neumann, John 643, 658, 663, 723
 founder of computer science 642
 iconoscope or selectron 659
 and invention of computer 638
 and Mauchly and Eckert 644
von Neumann architecture 642
voyages 433

Wallace, Alfred Russel 402, *406*
Wallis, John *348*
wāšipū 20, 21
wāṣitum 37, 39
water
 analysis and synthesis 467
 composition 466
 heavy 613
 production 297
water-mills 219
Watson, Thomas 648
wave model, of Huygens *343*
Weierstrass, Karl Theodor Wilhelm 363
weights and measures 26
well-logging 584-8
Western Union telegraph company 595
Whitehead, Alfred North 290
Wilkes, Maurice 290
William of Moerbacke 219
Witelo *see* Vitello
Witney, Willis 597, 598
Wittgenstein, Ludwig 161
Wollaston, William Hyde 564
World War II 627
 computers after 661-3
World Wars, and industrial science 601
writing 45-9, *75*
Wynn-Williams, C. E. 644, 652, 656

X-ray crystal analysis 598
Xenophanes of Colophon 76
Xylander (Wilhelm Holzmann) *348*

Yahya ibn al-Batriq 198
yeasts *539*

Zeno of Elea 109-11
Zimri-Lim 22, 24
zodiac, signs of 80-1
Zuse, Konrad 637, 646, 650-1, 659, 721